Advances in Nonlinear Geosciences

Anastasios A. Tsonis
Editor

Advances in Nonlinear Geosciences

With a Foreword by Michael Ghil

Springer

Editor
Anastasios A. Tsonis
Department Mathematical Sciences
Atmospheric Sciences Group
University of Wisconsin - Milwaukee
Milwaukee, WI, USA

Hydrologic Research Center
San Diego, CA, USA

Additional material to this book can be downloaded from http://extras.springer.com.

ISBN 978-3-319-58894-0 ISBN 978-3-319-58895-7 (eBook)
DOI 10.1007/978-3-319-58895-7

Library of Congress Control Number: 2017947896

Printed on acid-free paper

This Springer imprint is published by Springer Nature
The registered company is Springer International Publishing AG
The registered company address is: Gewerbestrasse 11, 6330 Cham, Switzerland

To the memory of my brother, Panagiotis
"Takis." A scientist and a poet.
1953–2016

Foreword

In early July 2016, the meeting "30 Years of Nonlinear Dynamics in Geosciences" was hosted by Aegean Conferences in Rhodes, Greece, and organized by Anastasios A. ("Tasos" to most of us) Tsonis. This was ten years after the first "20 Years of Nonlinear Dynamics in Geosciences" conference that also took place in Rhodes in June 2006.

Standing on the shoulders of giants, I will go a step further than Ed Lorenz, ten years ago, and ask, "Why 30 and not 3000"? One of the oldest nonlinear problems in the geosciences is certainly drawing a right angle on the face of the Earth, e.g., between a meridian and a parallel: it is equivalent to solving the Diophantine equation $a^2 + b^2 = c^2$. And the ancient Egyptians, who had to solve it to build the pyramids, from the basis up, knew the particular solution $a = 3$, $b = 4$, $c = 5$ very well and used it in order to build the great pyramids of Gizeh (also spelled in the Roman alphabet as Giza or Jizah), and many other temples, palaces, and tombs.

But that's, of course, not what we all have in mind, since we know well the saying, oft attributed to Stanislaw Ulam, that "nonlinear dynamics is akin to non-elephant zoology," or words to that effect. What we mean by tracing back the rapid rise of nonlinear dynamics, nonlinear sciences, or what not to some time after World War II is the following fact: according to the well-known story of the lamppost, and of attempts to find the forlorn keys in its circle of light, a superb development of methods for solving linear algebraic and differential equations in the nineteenth century led to great emphasis on solving such problems in the first half of the twentieth century.

Basically, linear problems are easily separable, and hence solvable, due to the superposition principle, projection onto orthonormal bases, and so on. Thus, many such problems were solved over 200 years, and quite important ones, at that. And these methods are still of great use to us, in deriving and determining the properties of tangent linear equations, adjoint operators, and many other mathematical approximations of real-world problems.

It is, as Ed Lorenz already pointed out in his letter to the participants of the previous anniversary conference, the rise of more-and-more powerful computational devices after World War II that changed our way of thinking about what a solution really is, i.e., not necessarily an analytical expression but an algorithm for obtaining information about a solution with prescribed accuracy. I would add that the improvement in observational methods—in the geosciences and elsewhere, whether in vitro, i.e., in the lab, or in vivo, i.e., outdoors—has also contributed greatly to our appetite for going beyond linear approximation to model, simulate, understand, and predict the complexity of the phenomena under study.

The nonlinear way of thinking about problems, in the geosciences and many other sciences—physical sciences in general, biosciences, socio-economic sciences—still needs to operate within the circles of light projected into the night of our ignorance by a certain number of lampposts. These lampposts include the theory of dynamical systems, statistical mechanics, scale invariances, the theory of localized coherent structures, and several others. Some lampposts that have been added or whose light circle has expanded in the last decade or so are network theory and the theory of non-autonomous and random dynamical systems. The program of the conference—and the table of contents of this volume—indicates that all of these lampposts were well represented by those who, luckily for them, were able to attend the meeting in person.

May all the light circles of nonlinear dynamics in the geosciences expand, overlap, and generally increase our delight in what we are all trying to accomplish, for our own enjoyment and for the benefit of humanity. And I fondly hope and trust that those who were present in the flesh, and not just in spirit, did not forget to revel in the blue waters of the Aegean and in the delights of the local architecture, customs, food, and wine. Nonlinear dynamics in the geosciences and the Aegean—immortalized in Greek poetry of all ages—are both very dear to my heart, as they should be to the heart of every geoscientist and civilized person, respectively.

Ecole Normale Supérieure, Paris, France
and University of California, Los Angeles, CA, USA

Michael Ghil

Preface

From July 3 to 8, 2016, a group of scientists from around the world met in Rhodes, Greece, 10 years after the meeting "20 Years of Nonlinear Dynamics in Geosciences" held at the same place in June 2006. The purpose of the meeting was to discuss the new advances in Nonlinear Geosciences since then and to propose future research directions.

A lot has happened since 2006. Most notably, the introduction of networks in geosciences studies, advances in chaos synchronization, topological data analysis, new insights on fractals, multifractals and stochasticity, climate dynamics, extreme events, complexity, and causality, among other topics.

This volume is the result of this meeting. I would like to thank all the contributors for their effort to produce this book. I am honored to host all of you and I hope that there will be a "40 Years in Nonlinear Dynamics in Geosciences" in 10 years from now.

By the way, the sequence of the papers is based on the alphabetic order of the first author.

Milwaukee, WI, USA　　　　Anastasios A. Tsonis

Contents

Contributors

Orhan Akyılmaz Faculty of Civil Engineering, Department of Geomatics Engineering, Istanbul Technical University, Maslak, Istanbul, Turkey

Bjarne Andresen Niels Bohr Institute, University of Copenhagen, Copenhagen, Denmark

M. Barreiro Instituto de Física, Facultad de Ciencias, Universidad de la República, Igua, Barcelona, Spain

Pavel Berloff Imperial College London, London, UK

H. Chang School of Marine Science and Policy, University of Delaware, Newark, DE, USA

Mickaël D. Chekroun Department of Atmospheric and Oceanic Sciences, University of California, Los Angeles, CA, USA

Stuart Crampin British Geological Survey, The Lyell Centre, Edinburgh, Scotland, UK

L. Del Rio Amador Department of Physics, McGill University, Montreal, QC, Canada

Jonathan Demaeyer Institut Royal Météorologique de Belgique, Brussels, Belgium

Ethan R. Deyle Scripps Institution of Oceanography, University of California San Diego, La Jolla, CA, USA

J.I. Deza Instituto de Física, Facultad de Ciencias, Universidad de la República, Igua, Barcelona, Spain

Henk A. Dijkstra Department of Physics and Astronomy, Institute for Marine and Atmospheric Research Utrecht, Utrecht University, Utrecht, The Netherlands

Panayiotis Dimitriadis Department of Water Resources and Environmental Engineering, School of Civil Engineering, National Technical University of Athens, Zographou, Greece

Gregory S. Duane Geophysical Institute, University of Bergen, Bergen, Norway

Department of Atmospheric and Oceanic Sciences, University of Colorado, Boulder, CO, USA

Deniz Eroglu Potsdam Institute for Climate Impact Research (PIK), Potsdam, Germany

Institute of Physics, Humboldt-Universität zu Berlin, Berlin, Germany

Christopher Essex Department of Applied Mathematics, The University of Western Ontario, London, ON, Canada

Qingyi Feng Department of Physics and Astronomy, Institute for Marine and Atmospheric Research Utrecht, Utrecht University, Utrecht, The Netherlands

Yuan Gao British Geological Survey, The Lyell Centre, Edinburgh, Scotland, UK

Michael Ghil Geosciences Department, Ecole Normale Supérieure and PSL Research University, Paris, France

Department of Atmospheric and Oceanic Sciences, University of California, Los Angeles, CA, USA

R. Hébert Department of Physics, McGill University, Montreal, QC, Canada

H.S. Huntley School of Marine Science and Policy, University of Delaware, Newark, DE, USA

A.C. Iliopoulos Department of Electrical and Computer Engineering, Research Team of Chaos and Complexity, Democritus University of Thrace, Xanthi, Greece

Igor Kamenkovich RSMAS, University of Miami, Miami, FL, USA

L.P. Karakatsanis Department of Electrical and Computer Engineering, Research Team of Chaos and Complexity, Democritus University of Thrace, Xanthi, Greece

Noel Keenlyside Geophysical Institute, University of Bergen, Bergen, Norway

Sung Yong Kim Environmental Fluid Mechanics Laboratory, Department of Mechanical Engineering, Korea Advanced Institute of Science and Technology, Daejeon, Republic of Korea

A.D. Kirwan School of Marine Science and Policy, University of Delaware, Newark, DE, USA

Dmitri Kondrashov Department of Atmospheric and Oceanic Sciences, University of California, Los Angeles, CA, USA

Institute of Applied Physics of the Russian Academy of Sciences, Nizhny Novgorod, Russia

Demetris Koutsoyiannis Department of Water Resources and Environmental Engineering, School of Civil Engineering, National Technical University of Athens, Zographou, Greece

Witold F. Krajewski IIHR-Hydroscience and Engineering, Department of Civil and Environmental Engineering, The University of Iowa, Iowa City, IA, USA

Sergey Kravtsov Department of Mathematical Sciences, Atmospheric Sciences Group, University of Wisconsin-Milwaukee, Milwaukee, WI, USA

Ramazan Alper Kuçak Faculty of Civil Engineering, Department of Geomatics Engineering, Istanbul Technical University, Maslak, Istanbul, Turkey

Jürgen Kurths Potsdam Institute for Climate Impact Research (PIK), Potsdam, Germany

Institute of Applied Physics of the Russian Academy of Sciences, Novgorod, Russia

Ivan L'Heureux Department of Physics, University of Ottawa, Ottawa, ON, Canada

Federico Lombardo Dipartimento di Ingegneria Civile, Edile e Ambientale, Sapienza Università di Roma, Rome, Italy

S. Lovejoy Department of Physics, McGill University, Montreal, QC, Canada

Michael D. Madsen Department of Mathematical Sciences, Atmospheric Sciences Group, University of Wisconsin - Milwaukee, Milwaukee, WI, USA

Ricardo Mantilla IIHR-Hydroscience and Engineering, Department of Civil and Environmental Engineering, The University of Iowa, Iowa City, IA, USA

Norbert Marwan Potsdam Institute for Climate Impact Research (PIK), Potsdam, Germany

Mahesh L. Maskey Department of Land, Air and Water Resources, University of California, Davis, CA, USA

C. Masoller Departament de Fisica, Universitat Politecnica de Catalunya, Barcelona, Spain

Francis C. Motta Mathematics Department, Duke University, Durham, NC, USA

J. David Neelin Department of Atmospheric and Oceanic Sciences, University of California, Los Angeles, CA, USA

C. Nicolis Institut Royal Météorologique de Belgique, Brussels, Belgium

Nurcan Meral Ozel British Geological Survey, The Lyell Centre, Edinburgh, Scotland, UK

Ibrahim Ozken Department of Physics, Ege University, Izmir, Turkey

Milan Paluš Department of Nonlinear Dynamics and Complex Systems, Institute of Computer Science, Czech Academy of Sciences, Prague, Czech Republic

G.P. Pavlos Department of Electrical and Computer Engineering, Research Team of Chaos and Complexity, Democritus University of Thrace, Xanthi, Greece

E.G. Pavlos Department of Electrical and Computer Engineering, Research Team of Chaos and Complexity, Democritus University of Thrace, Xanthi, Greece

Gabriel Perez IIHR-Hydroscience and Engineering, Department of Civil and Environmental Engineering, The University of Iowa, Iowa City, IA, USA

Gulten Polat British Geological Survey, The Lyell Centre, Edinburgh, Scotland, UK

Carlos E. Puente Department of Land, Air and Water Resources, University of California, Davis, CA, USA

Paul Roebber Department of Mathematical Sciences, Atmospheric Sciences Group, University of Wisconsin-Milwaukee, Milwaukee, WI, USA

Daniel Schertzer Hydrology Meteorology and Complexity (HM&Co), Champs-sur-Marne, France

William Seitz Department of Marine Sciences, Galveston Campus of Texas A&M University, Galveston, TX, USA

Frank Selten Royal Netherlands Meteorological Institute (KNMI), De Bilt, The Netherlands

Mao-Lin Shen Geophysical Institute, University of Bergen, Bergen, Norway

Bellie Sivakumar School of Civil and Environmental Engineering, The University of New South Wales, Sydney, NSW, Australia

Department of Land, Air and Water Resources, University of California, Davis, CA, USA

Thomas Stemler School of Mathematics and Statistics, The University of Western Australia, Crawley, WA, Australia

Spencer Stevens Independent Researcher, London, UK

George Sugihara Scripps Institution of Oceanography, University of California San Diego, La Jolla, CA, USA

Noriyuki Sugiyama Department of Mathematical Sciences, Atmospheric Sciences Group, University of Wisconsin-Milwaukee, Milwaukee, WI, USA

David B. Taylor British Geological Survey, The Lyell Centre, Edinburgh, Scotland, UK

Ioulia Tchiguirinskaia Hydrology Meteorology and Complexity (HM&Co), Ecole des Ponts ParisTech, Champs-sur-Marne, France

G. Tirabassi Instituto de Física, Facultad de Ciencias, Universidad de la República, Igua, Barcelona, Spain

Anastasios A. Tsonis Department of Mathematical Sciences, Atmospheric Sciences Group, University of Wisconsin - Milwaukee, Milwaukee, WI, USA

Hydrologic Research Center, San Diego, CA, USA

Raşit Uluğ Faculty of Civil Engineering, Department of Geomatics Engineering, Istanbul Technical University, Maslak, Istanbul, Turkey

Stéphane Vannitsem Institut Royal Météorologique de Belgique, Brussels, Belgium

Geli Wang Key Laboratory of Middle Atmosphere and Global Environment Observations, Institute of Atmospheric Physics, Chinese Academy of Sciences, Beijing, China

Wim Wiegerinck SNN Adaptive Intelligence, Nijmegen, The Netherlands

Department of Biophysics, Donders Institute for Brain, Cognition and Behaviour, Radboud University, Nijmegen, The Netherlands

Karl-Heinz Wyrwoll School of Earth and Environment, The University of Western Australia, Crawley, WA, Australia

Peicai Yang Key Laboratory of Middle Atmosphere and Global Environment Observations, Institute of Atmospheric Physics, Chinese Academy of Sciences, Beijing, China

Hao Ye Scripps Institution of Oceanography, University of California San Diego, La Jolla, CA, USA

Xiaojun Yuan Lamont-Doherty Earth Observatory of Columbia University, Palisades, CA, USA

He Zhang International Center for Climate and Environment Sciences (ICCES), Institute of Atmospheric Physics, Chinese Academy of Sciences, Beijing, China

Feifan Zhou Laboratory of Cloud-Precipitation Physics and Severe Storms (LACS), Institute of Atmospheric Physics, Chinese Academy of Sciences, Beijing, China [illegible] Facultad de Ciencias, Universidad de la [illegible], Spain

Anastasios A. Tsonis Department of Mathematical Sciences, Atmospheric Sciences Group, University of Wisconsin-Milwaukee, Milwaukee, WI, USA

[illegible]

Pullback Attractor Crisis in a Delay Differential ENSO Model

Mickaël D. Chekroun, Michael Ghil, and J. David Neelin

Abstract We study the pullback attractor (PBA) of a seasonally forced delay differential model for the El Niño–Southern Oscillation (ENSO); the model has two delays, associated with a positive and a negative feedback. The control parameter is the intensity of the positive feedback and the PBA undergoes a crisis that consists of a chaos-to-chaos transition. Since the PBA is dominated by chaotic behavior, we refer to it as a strange PBA. Both chaotic regimes correspond to an overlapping of resonances but the two differ by the properties of this overlapping. The crisis manifests itself by a brutal change not only in the size but also in the shape of the PBA. The change is associated with the sudden disappearance of the most extreme warm (El Niño) and cold (La Niña) events, as one crosses the critical parameter value from below. The analysis reveals that regions of the strange PBA that survive the crisis are those populated by the most probable states of the system. These regions are those that exhibit robust foldings with respect to perturbations. The effect of noise on this phase-and-parameter space behavior is then discussed. It is shown that the chaos-to-chaos crisis may or may not survive the addition of small noise to the evolution equation, depending on how the noise enters the latter.

Keywords Chaos-to-Chaos crisis • El nino-southern oscillation • Pullback attractors • Satistical equilibrium

---

M.D. Chekroun (✉) • J.D. Neelin
Department of Atmospheric and Oceanic Sciences, University of California, Los Angeles, CA, USA
e-mail: mchekroun@atmos.ucla.edu; neelin@atmos.ucla.edu

M. Ghil
Geosciences Department, Ecole Normale Supérieure and PSL Research University, Paris, France

Department of Atmospheric and Oceanic Sciences, University of California, Los Angeles, CA, USA
e-mail: ghil@atmos.ucla.edu

A.A. Tsonis (ed.), *Advances in Nonlinear Geosciences*,
DOI 10.1007/978-3-319-58895-7_1

1 Introduction and Motivation

The El Niño–Southern Oscillation (ENSO) is a dominant mode of climate variability on seasonal-to-interannual time scales and affects the climate over a great portion of the globe on interdecadal and longer time scales. A major aspect of ENSO is the strong coupling between the Tropical Pacific ocean and the atmosphere above, and the physical mechanisms that give rise to ENSO are fairly well understood (Neelin et al. 1998; Philander 1992).

A key mechanism, originally proposed in Bjerknes (1969), is the positive atmospheric feedback on the equatorial sea surface temperature (SST) field via the surface wind stress. Still, ENSO's unstable, recurrent but irregular behavior implies challenges for prediction (Cane 1986), even at subannual lead times. Conceptual numerical modeling plays a prominent role in understanding ENSO variability and developing prediction methods for it (Chekroun et al. 2011a; Ghil and Jiang 1998; Mechoso et al. 2003; Neelin et al. 1998). The delayed oscillator description of ENSO has led to a hierarchy of models of increasing complexity that include delay effects taking various forms (Battisti and Hirst 1989; Galanti and Tziperman 2000; Neelin et al. 1998; Suarez and Schopf 1988), via negative (Ghil and Zaliapin 2015; Ghil et al. 2008a) and positive (Tziperman et al. 1994) feedbacks.

Seasonal forcing has been suggested as a crucial ingredient in explaining ENSO's irregularity (Jiang et al. 1995; Jin et al. 1994, 1996; Tziperman et al. 1994, 1995). In this approach, the intrinsic ENSO oscillator may enter into nonlinear resonance with the seasonal forcing. In the case of exact frequency locking with the seasonal cycle, such resonant behavior is characterized by perfect periodicity. ENSO's irregularity occurs when the nonlinear effects are stronger, and several resonances may coexist. In this case, the ENSO oscillator is not able to lock to a single resonance, and it jumps irregularly between several resonances, while the resulting irregular behavior still bears the fingerprint of the underlying frequency-locked regimes that now coexist. Dynamically, this phenomenon corresponds to the overlapping of nonlinear resonances also known as Arnold tongues in parameter space (Arnold 1988; Ghil et al. 2008b; Jensen et al. 1984). Noise due to atmospheric internal variability has also been shown to be an important factor in ENSO irregularity (Blanke et al. 1997; Eckert and Latif 1997; Kleeman and Moore 1997).

To study the effects of the seasonal cycle in the aforementioned ENSO models, direct numerical integrations and examination of return maps in a low-dimensional, reconstructed phase space[1] are often preferred to a rigorous mathematical analysis, which is typically challenging to carry out. Recently, continuation methods for bifurcations in delay differential equations (DDEs) have also been used to analyze the interactions of the seasonal cycle with the ENSO oscillator (Keane et al. 2015, 2016; Krauskopf and Sieber 2014). Rigorous approximation techniques of DDEs by systems of ordinary differential equations (ODEs) Chekroun et al. (2016b) offer another path to the analysis of such interactions.

[1]Relying, for instance, on the Takens embedding theorem (Takens 1981).

In this study, we propose yet another approach, which relies on the theory of pullback attractors (PBAs) (Carvalho et al. 2013; Chekroun et al. 2011b) and the statistical equilibria they support Chekroun and Glatt-Holtz (2012), Lukaszewicz and Robinson (2014). The application to DDEs herein uses careful numerical approximations of the PBAs (Chekroun et al. 2011b), along with visualization in low-dimensional, embedded phase space; see Ghil and Zaliapin (2015), Ghil (2017) for preliminary DDE results. Here, PBAs are used to analyze a complicated chaos-to-chaos transition.

We focus in this chapter on the seasonally forced ENSO model of Tziperman et al. (1994) that includes delayed positive and negative feedback mechanisms. For the sake of the nonspecialist reader, this model is outlined in Sect. 2.1 below. We compute for this model approximations of the PBA and of the statistical equilibrium it supports, both of which are represented in a natural two-dimensional (2-D) embedded phase space. Recall that, loosely speaking, a global PBA $\mathscr{A}(t)$ describes the states in the system's phase space X that are reached at a time t, when the system is initiated from an asymptotic past, $s \to -\infty$, and the initial states are varied within a collection of bounded sets of X (Carvalho et al. 2013). The statistical equilibrium μ_t supported by the PBA, as defined in Sect. 2.4, is crucial for the description of the distribution of current states at time t (Chekroun et al. 2011b; Ruelle 1999).

After recalling in Sect. 2.2 some fundamentals about PBAs, in particular in the context of DDEs, we first numerically show the "strangeness" of an embedded version of $\mathscr{A}(t)$ in Sect. 2.3. In particular, the folding and stretching that is typical of nonlinear, chaotic dynamics in the autonomous setting are observed in various regions of this PBA. After proving the periodicity of $\mathscr{A}(t)$ with the same period as that of the seasonal forcing, the time evolution of $\mathscr{A}(t)$ within a calendar year is then analyzed in Sect. 2.3. There, we show that the PBA provides a natural global geometric view of the dynamics, consistent with variations in ENSO phase-locking that occur within a given frequency-locked regime, as previously documented in the literature (Galanti and Tziperman 2000; Neelin et al. 2000). In Sect. 2.4, we provide a brief but still rigorous description of the aforementioned statistical equilibrium μ_t.

Section 3 contains a parameter-dependence study of the PBA $\mathscr{A}(t)$ and of the statistical equilibrium μ_t it supports. Numerical experiments allow us to conclude that a chaos-to-chaos crisis takes place as the intensity of the positive feedback crosses a critical value; see Sect. 3.1. The crisis separates two different types of overlapping of nonlinear resonances. In Sect. 3.2, we analyze the changes in the PBA and in the statistical equilibrium across the crisis. Both these mathematical objects change relatively smoothly, until reaching eventually an abrupt, discontinuous change as the critical parameter value is crossed. The crisis manifests itself by a brutal change not only in the size but also in the shape of the PBA, which keeps its strange character across the transition.

Dynamically, this abrupt change in the PBA is associated with the sudden disappearance of extreme warm (El Niño) and cold (La Niña) events, as one crosses the critical parameter value from below. The analysis of the statistical equilibrium μ_t supported by the PBA $\mathscr{A}(t)$ reveals that the regions of the strange PBA that survive the crisis are those populated by the most probable states of the system. These regions are those that exhibit robust foldings with respect to perturbations.

Two dynamical mechanisms are proposed in Sect. 3.3 to explain the origin of the chaos-to-chaos crisis identified herein. One consists of the crossing of a crisis line within an overlapping region of two Arnold tongues (Mori and Kuramoto 2013) that separate two coexisting PBAs. The other consists of a PBA-widening scenario suggested in Grebogi et al. (1987) for low-dimensional autonomous maps. In our case, an unstable pullback periodic orbit would collide with $\mathscr{A}(t)$ as one crosses a critical value of the control parameter, causing the PBA widening reported hereafter.

Finally, the effect of noise on this phase-and-parameter space behavior is discussed in Sect. 3.4. It is shown that the chaos-to-chaos crisis may or may not survive the addition of small noise to the evolution equation, depending on how the noise enters the latter. These noise effects find a natural interpretation within each of the aforementioned possible crisis mechanisms.

2 PBAs and Statistical Equilibria in a Periodically Forced ENSO Model with Delays

2.1 *The Model*

We focus hereafter on the nonlinear delay oscillator mechanism, and analyze a statistical crisis occurring in this model as a certain control parameter varies. The model takes its root in the following conceptual description.

A positive SST perturbation along the eastern equatorial Pacific weakens the easterly trade winds above the equator. The change in the winds excites a downwelling wave in the thermocline that travels eastward to the South American coast as equatorial Kelvin waves and an upwelling signal that travels westward as equatorial Rossby waves. The downwelling Kelvin waves enhance the warming off the coast of South America, starting an El Niño event. Subsequently, the westward-traveling upwelling Rossby waves are reflected from the western boundary of the Pacific Ocean as upwelling Kelvin waves, which travel eastward to counter the downwelling Kelvin waves. This negative feedback ultimately terminates the El Niño event.

A simple model of such a delay mechanism, including one Kelvin wave, one Rossby wave mode, and a dynamic link from mid-Pacific wind stress anomalies to these equatorial wave modes has been proposed in Tziperman et al. (1994). The model includes an idealized seasonal forcing term that represents the effects of the numerous seasonally varying features of the equatorial Pacific ocean and atmosphere, such as wind amplitude and SST variations. The single dependent variable in the equation is $h(t)$—the thermocline depth deviation from seasonal depth values at the eastern boundary—and the model reads as follows:

$$\frac{dh}{dt} = aR\Big[h\Big(t - \frac{L}{2C_K}\Big)\Big] - bR\Big[h\Big(t - \frac{L}{C_K} - \frac{L}{2C_R}\Big)\Big] + c\cos(\omega_a t + \varphi). \quad (1)$$

A version of this model with only the negative feedback included was studied in Ghil and Zaliapin (2015), including its PBA.

In Eq. (1) L is the basin width, ω_a denotes the annual frequency of the seasonal forcing, and φ denotes its phase. The wind-forced Kelvin mode that travels eastward at a speed C_K is represented by the first term in the right-hand side of Eq. (1). It takes this wave a time $L/(2C_K)$ to reach the eastern boundary from the middle of the basin. The second term is due to the Rossby wave that travels westward at a speed C_R; this wave is excited by the wind at a delayed time, namely $t - (L/C_K + L/(2C_R))$, and it is reflected as a Kelvin wave off the western basin boundary.

The function $R[h]$ relates wind stress to SST, and SST to thermocline depth. We follow here (Münnich et al., 1991), where the nonlinear form of $R[h]$ is given by

$$R[h] = \begin{cases} b_+ + \frac{b_+}{a_+}\Big(\tanh\big(\frac{\kappa a_+}{b_+}(h - h_+)\big) - 1\Big), & \text{if } h_+ < h, \\ \kappa h, & \text{if } h_- \leq h \leq h_+, \\ -b_- - \frac{b_-}{a_-}\Big(\tanh\big(\frac{\kappa a_-}{b_-}(h - h_-)\big) - 1\Big), & \text{if } h < h_-. \end{cases} \tag{2}$$

The specific form of $R[h]$ reflects the non-uniform stratification of the ocean; it is fashioned after the shape of the tropical thermocline. The slope of $R(h)$ at $h = 0$, set by the parameter κ, provides a measure of the strength of the ocean-atmosphere coupling. Based on Münnich et al. (1991), we consider here $a_\pm > 1$, and

$$h_+ = \frac{b_+}{\kappa a_+}(a_+ - 1), \quad h_- = -\frac{b_-}{\kappa a_-}(a_- - 1). \tag{3}$$

These values ensure that $R[h]$ is continuous at h_+ and h_-. As $h \to \pm\infty$, we get $R[h] \to b_\pm$. The parameters $a_\pm$ control the curvature of $R[h]$, and the greater $a_\pm$, the faster the limits $b_\pm$ are reached as $h \to \pm\infty$. The values used in our numerical simulations are reported in Table 1.

The parameter κ, cf. (Tziperman et al. 1994, Fig. 1), is a key parameter in the control of the model's dynamical behavior. For small values of κ, the time series $h(t)$ is, for instance, perfectly periodic with the annual period of the forcing. Besides

Table 1 Glossary of model's parameter

Parameter	Interpretation	Numerical value
L	Basin width	1
ω_a	Frequency of the annual cycle	$2\pi/360$
φ	Phase of the forcing	$\pi/2$
C_K	Kelvin wave speed	1/69 days^{-1}
C_R	Rossby wave speed	1/207 days^{-1}
$a_{+/-}$	Control parameters of the curvature of $A[h]$	1
b_-	Limit of $A[h]$ as $h \to -\infty$	-0.44
b_+	Limit of $A[h]$ as $h \to +\infty$	2.2
a, b	Magnitude of the feedbacks	$a = (1.12 + \delta)/180$, $b = 1/120$
c	Magnitude of the periodic forcing	$c = 2.2/180$
κ	Slope at the origin in Eq. (2)	2.6

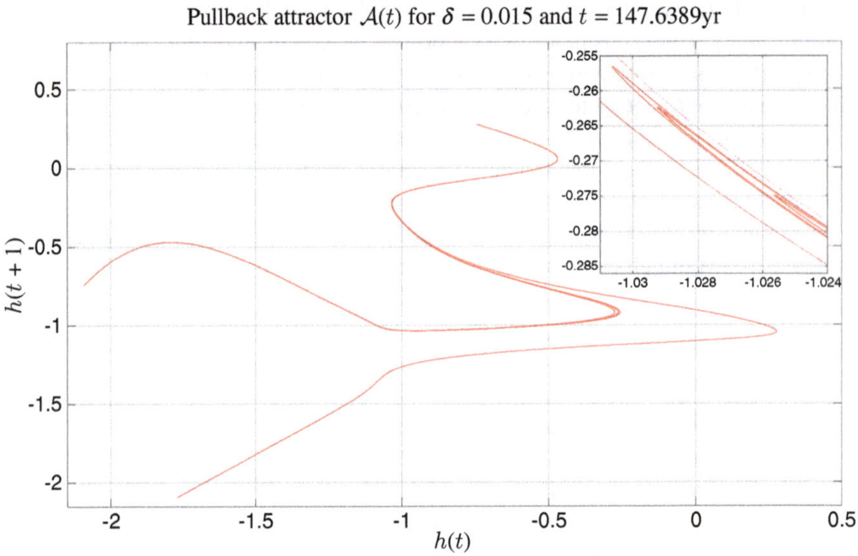

Fig. 1 A strange pullback attractor (PBA) $\mathscr{A}(t)$ associated with the periodically forced delay differential equation (DDE) (1). The PBA is projected onto the delay coordinates $(h(t), h(t+1))$

this simple periodic behavior, three dynamical regimes are typically exhibited by the model. For the parameter values used in Tziperman et al. (1994), these regimes are classified as follows:

(I) **Irregular quasi-periodic dynamics.** As κ increases, an internal frequency ω_i appears; it characterizes the natural oscillator of the Tropical Pacific's ocean-atmosphere system (Jin and Neelin 1993; Neelin et al. 1998). This second frequency is, in general, incommensurable with the annual frequency; the superposition of two incommensurable frequencies creates a quasi-periodic time series. The resulting oscillations are irregular but not chaotic; the power spectrum shows two dominant frequencies with several subharmonics; see again (Tziperman et al., 1994, Fig. 1).

(II) **Frequency-locked dynamics.** For a steeper slope of $A[h]$ at $h = 0$, the system becomes frequency-locked: The frequency of the nonlinear delay oscillator changes slightly to a simple rational multiple of the driving annual frequency: $\omega_i = \omega_a p/q$, with p and q integers. This regime corresponds to a nonlinear resonance between the driving annual frequency ω_a and the internal oscillatory frequency ω_i. The time series is periodic, and the phase-space diagram (in a Poincaré section) is a set of points whose number depends on the values of p and q. The parameter regimes corresponding to the frequency-locked solutions are also known as Arnold tongues; see, for instance, Ghil et al. (2008b, Fig. 7).

(III) **Chaos by overlapping of resonances.** For certain values of κ, the time series $h(t)$ becomes irregular, and it is associated with a strange PBA and a power spectrum that is broad and no longer contains sharp peaks, as in Regimes (I) or (II). Two or more frequency-locked solutions—that is, solutions with different ratios p/q—may coexist; the nonlinear resonances are said to overlap in this case. The chaotic behavior is caused by the irregular "trapping" of the system among the different possible resonances. This characterization of Regime (III) in terms of a strange PBA is provided in Sect. 2.2 below.

In what follows, we denote by τ_1 and τ_2, the basin-crossing times $L/(2C_K)$ and $L/C_K + L/(2C_R)$, respectively.

2.2 *PBAs of Delay Models with Time-Dependent Forcing*

Recall that the standard theory of global attractors in the autonomous case (Temam 1997) requires one first to define a phase space in which the solutions of a given evolution equation are well-defined. It is necessary to proceed in the same way for non-autonomous dynamical systems (NDSs) and their PBAs.

In the case of nonlinear DDEs, such as Eq. (1), several function spaces can be used as a state space. Among the most standard ones, those that start with the space of continuous functions on the interval $[-\tau, 0]$ play an important role; see, for instance, Diekmann et al. (1995), Hale and Verduyn-Lunel (1993). Hilbert spaces, though, are better adapted to the approximation of DDEs by systems of ordinary differential equations (ODEs) (Chekroun et al. 2016b).

The reformulation of Eq. (1) as a retarded functional differential equation (RFDE) is classical and proceeds as follows. Let us denote by h_t the time evolution of the history segments of a solution h to Eq. (1). In other words, for each t, h_t is a function from $[-\tau, 0]$ into $\mathbb{R}$ defined as

$$h_t(\theta) := h(t+\theta), \qquad t \geq 0, \qquad \theta \in [-\tau, 0]. \tag{4}$$

Introducing the phase space $X := \mathscr{C}([-\tau, 0], \mathbb{R})$ of continuous functions from $[-\tau, 0]$ into $\mathbb{R}$, with $\tau = \tau_2 > \tau_1$, and the nonlinearity $\mathscr{F}$ defined for all ψ in X by

$$\mathscr{F}(\psi) = aR\left[\psi(-\tau_1)\right] - bR\left[\psi(-\tau_2)\right], \text{ with } R \text{ given in (2)}, \tag{5}$$

Eq. (1) can be recast into the following RFDE:

$$\frac{\mathrm{d}h}{\mathrm{d}t} = \mathscr{F}(h_t) + g(t), \tag{6}$$

in which the time-dependent forcing $g(t)$ is given by

$$g(t) = c\cos(\omega t). \tag{7}$$

Note that the nonlinearity $\mathscr{F}$ in (5) is bounded as a mapping from X into $\mathbb{R}$,

$$|\mathscr{F}(\psi)| \leq \max(b_+, |b_-|), \quad \text{for all } \psi \text{ in } X. \tag{8}$$

Furthermore, $\mathscr{F}$ is globally Lipschitz on X endowed with the uniform-norm topology, i.e., the topology induced by the supremum norm

$$\|\phi\|_\infty := \sup_{\theta \in [-\tau, 0]} |\phi(\theta)|. \tag{9}$$

Since $\mathscr{F}$ is continuous, due to Eq. (3), as well as bounded and Lipschitz continuous, the general theory of RFDEs (Hale and Verduyn-Lunel 1993) applied to Eq. (6) ensures that, for any (s, ϕ) in $\mathbb{R} \times X$, there exists a unique solution to Eq. (6), defined on a maximal interval $[s, T_{\max}(\phi))$, $T_{\max}(\phi) > s$, such that

$$h_s(\theta) = \phi(\theta), \quad \theta \in [-\tau, 0]. \tag{10}$$

Moreover if $T_{\max}(\phi) < \infty$, then the solution blows up at time $T_{\max}(\phi)$, i.e.,

$$\lim_{t \to T_{\max}(\phi)^-} \|h_t\|_\infty = \infty, \tag{11}$$

On the other hand, an integration of Eq. (6) between s and t for $s \leq t < T_{\max}(\phi)$ and the bounds (8) with $g(t) \leq c$ lead to the estimate

$$\|h_t\|_\infty \leq \|\phi\|_\infty + (T_{\max}(\phi) - s)(c + \max(b_+, |b_-|)). \tag{12}$$

This latter inequality is incompatible with (11) and therefore $T_{\max}(\phi) = \infty$ for all ϕ in X. As a consequence, solutions to Eq. (6) are guaranteed to exist in X for all positive times t, and to be uniquely determined by an initial history ϕ in X, taken over any anterior time interval $[s - \tau, s]$, with $s \leq t$.

In other words, one can define a nonlinear process (Hale and Verduyn-Lunel, 1993, Chap. 4), i.e., a solution map U defined by

$$(t, s, \phi) \mapsto U(t, s)\phi := h_t \in X, \; t \geq s, \; \phi \in X, \tag{13}$$

where h_t denotes the unique solution to Eq. (6) that emanates from ϕ at a time $s \leq t$, i.e., such that $h_s = \phi$. The existence and uniqueness property translates here into the process composition property, which replaces the more traditional (semi-)group property (Temam 1997, Chap. I, Sect. 1.1) and is formulated here as

$$U(t, s) \circ U(s, r) = U(t, r), \quad t \geq s \geq r. \tag{14}$$

The solution map U can be thus referred to as a two-parameter semigroup of transformations of X. It provides a two-time description of the dynamics associated with Eq. (1): the time s describes when the system was initialized, while the other time t is associated with the current state of the system. In the autonomous case, only the amount of time separating s and t, i.e., $t - s$, matters and a one-parameter (semi-)group suffices to entirely determine the dynamics; e.g. Chekroun et al. (2011b, 2016b). In the non-autonomous case, the history of the forcing between the time s and the time t—which we call a forcing snippet—is an important ingredient of the dynamics and may drive the system differently between a time s' and a time t', even though $t' - s' = t - s$.[2]

Note also that the phase space X on which the process U is acting is infinite-dimensional as a function space. Even in this setting, a PBA can be rigorously defined (Caraballo et al., 2005; Carvalho et al., 2013). A family of compact[3] sets $\{\mathscr{A}(t)\}$ of X is then said to be a (global) PBA for U, if it satisfies

(i) (Invariance property) $U(t,s)\mathscr{A}(s) = \mathscr{A}(t)$ for all $t \geq s$; and
(ii) (Pullback attraction property) $\lim_{s\to-\infty} \text{dist}_X(U(t,s)B, \mathscr{A}(t)) = 0$, for all bounded subsets B of X.

The pullback attraction property (ii) considers a collection of states of the system at time t when the system is initiated in a distant past s, as s goes to $-\infty$ and for initial states lying in B. As B is varied in the collection of bounded subsets of X, a useful explicit PBA characterization in terms of the omega limit set is available (Carvalho et al. 2013, Theorem 2.12); see also (18) below.

Note that $\text{dist}_X(E, F)$ denotes here the Hausdorff semi-distance between the subsets E and F of X,

$$d_X(E,F) := \sup_{x\in E} d_X(x,F) \text{ with } d_X(x,F) := \inf_{y\in F} \|x - y\|. \tag{15}$$

One calls $d_X(E,F)$ a semi-distance since, in general, $d_X(E,F) \neq d_X(F,E)$ and $d_X(E,F) = 0$ merely imply $E \subset F$. From (ii) above, one can thus say, loosely speaking, that, for any set B of initial data, $U(t,s)B$ is "almost included" in the pullback attractor $\mathscr{A}(t)$, whenever $t - s$ is sufficiently large. Intuitively, for B spanning a sufficiently large ensemble of possible initial data, one can reasonably say that $U(t,s)B$ constitutes a good approximation of a significant portion of the pullback attractor $\mathscr{A}(t)$.

In the nonlinear physics literature, $U(t,s)B$ is often called a "snapshot attractor" (Bódai and Tél 2012; Bódai et al. 2011, 2013; Drótos et al. 2015; Romeiras et al. 1990). In practice one lets, roughly speaking, a cloud of points—each driven by the same segment of the forcing—flow forward in time. However, to justify this procedure, one needs to ensure the existence of a global PBA, which allows for a rigorous characterization of dissipation in the presence of time-dependent forcing, either deterministic (Carvalho et al. 2013) or random (Crauel and Flandoli 1994; Crauel et al. 1997). This rigorous treatment has to be valid also in the infinite-dimensional context of partial differential equations (PDEs), such as the

[2]Still, a segment $[s', t']$ of the forcing may drive the system in a way that is similar to that over the segment $[s, t]$, even when $g(t)$ is a white noise, provided the system's solutions exhibit recurrent patterns as time evolves; see Chekroun et al. (2011a), Kondrashov et al. (2013).

[3]Here compact set is understood in the sense of point set topology (Kelley 1975).

2-D Navier–Stokes equations, subject to time-dependent disturbances, or to that of the delay differential ENSO model considered here. Remarkably, global PBAs support meaningful invariant measures that characterize the statistics of the nonlinear, non-autonomous dynamics, as explained in Sect. 2.4 below. Global PBAs are thus natural objects to describe both the statistics and the geometry of non-autonomous dynamics, and identifying conditions for their existence is theoretically crucial.

Useful conditions—expressing often a form of balance between the forcing and the intrinsic dissipative effects of the underlying autonomous dynamics—may be identified within the PBA framework to ensure their existence; see, for instance, Pierini et al. (2016, Appendix) and the proof of Kondrashov et al. (2015, Theorem 3.1). In the context of the DDE model (6), the nonlinearity $\mathscr{F}$ defined in (5) is responsible for autonomous dissipative effects in X and the periodic forcing $g(t)$ permits their translation into a pullback dissipation.

We do not address the rigorous existence of such a pullback dissipation here, nor of a global PBA, and refer to Caraballo et al. (2001, 2005) for techniques to prove the existence of PBAs for DDEs. Instead, we illustrate next, by means of numerical simulations, geometric features of the global PBA associated with Eq. (6). These features are studied in the chaotic regime that corresponds to the value $\delta = 15\times 10^{-3}$ of the parameter δ that affects the magnitude of the feedback a in the model (1), while the other parameters take the values listed in Table 1.

2.3 A Strange PBA and Its Time Evolution

Characterizing Strangeness of a PBA

To analyze the structure of the global PBA for the parameters considered in this chapter, we first computed approximations of the PBAs $\mathscr{A}(t)$ for different values of t and for $\delta = 15\times 10^{-3}$. To do so, we have integrated numerically Eq. (1), using a set B of $N = 5 \times 10^5$ initial histories ϕ that have been sampled over $[-\tau, 0]$, according to a distribution described in Sect. 3.2, below. The results are shown in Fig. 1 for $t - s$ sufficiently large: we found that $t \approx 147.64$ years and $s = 0$ are sufficient to ensure convergence, that is to have $U(t, t - s)B$, with $\phi \in B$, "quasi-contained" in $\mathscr{A}(t)$, i.e., $\text{dist}_X(U(t, t - s)B, \mathscr{A}(t)) \approx 0$. Thus we do not distinguish $U(t, t - s)B$ from $\mathscr{A}(t)$ in the discussion below.

The PBA $\mathscr{A}(t)$ is plotted in Fig. 1 in the embedded phase space of the delay coordinates $(h(t), h(t + 1))$, where the unit delay corresponds to 1 year. The PBA's global structure is indicative of nonlinear effects, with characteristic folds occurring in several locations. To simplify the discussion, we often make hereafter no distinction between $\mathscr{A}(t)$ and its embeddings, such as that shown in this figure. A zoom at a specific location of $\mathscr{A}(t)$, depicted in the inset of Fig. 1, shows finer structure with several interleaved stretchings and foldings that occur over a very narrow region of the embedded phase space. Several other regions of the PBA (not shown) reveal the same fine filamentation when put under this kind of magnifying glass.

It is, in fact, not surprising to find a complex structure associated with stretching and folding in the global PBA of a system exhibiting chaos when subject to a time-dependent forcing: such PBA geometry was illustrated in Chekroun et al. (2011b), Pierini et al. (2016) for dynamical systems of lower dimension than considered here; see also Ghil (2017). The emergence of strange attractors in periodically forced dynamical systems has even been addressed rigorously recently for a broad class of evolution equations, including some parabolic PDEs (Lu et al., 2013). At the core of this approach is a geometric mechanism for the production of chaos that has been first identified in Wang and Young (2008) and generalized in subsequent works of the same authors (Wang and Young, 2001, 2003). In particular Wang and Young (2003, Theorem 1) shows that when suitably kicked by an external periodic forcing, a limit cycle can be turned into a strange attractor. Further details of this theory are discussed below in Sect. 2.4. We turn next to the PBA's time evolution.

Time Evolution of the PBA

First, note that, due to the periodicity of the forcing, the process U is T-periodic, with $T = 1$ years. Indeed, given $s < t$, it follows that integrating Eq. (6) from $s + T$ to $t + T$ is equivalent to integrating it from s to t, since the vector field $\mathscr{F}$ (on X) is time independent and g is T-periodic. In other words, for all ϕ in X and any $s \leq t$,

$$U(t + T, s + T)\phi = U(t, s)\phi. \tag{16}$$

From this property we conclude that the pullback omega limit set of any bounded subset B of X (Carvalho et al. 2013, Definition 2.2) satisfies[4]

$$\omega_B(t) := \bigcap_{\tau \geq 0} \overline{\bigcup_{s \geq \tau} U(t, t - s)B} = \bigcap_{\tau \geq 0} \overline{\bigcup_{s \geq \tau} U(t + T, t + T - s)B} = \omega_B(t + T), \tag{17}$$

where $\overline{E}$ ($E \subset X$) denotes the set of points of X that can be obtained as limit of elements in E. Thus, recalling the characterization of the global PBA in terms of omega limit sets (Carvalho et al. 2013, Theorem 2.12), we have

$$\mathscr{A}(t) := \overline{\bigcup_{B \in \mathscr{B}(X)} \omega_B(t)} = \overline{\bigcup_{B \in \mathscr{B}(X)} \omega_B(t + T)} = \mathscr{A}(t + T), \text{ for all } t \in \mathbb{R}, \tag{18}$$

where $\mathscr{B}(X)$ denotes the collection of bounded subsets of X.

[4]This set is equivalently defined as the set of elements ψ in X obtained as the pullback limit $\psi = \lim_{k \to \infty} U(t, s_k)\phi_k$, with $s_k \to -\infty$ and $\phi_k \in B$.

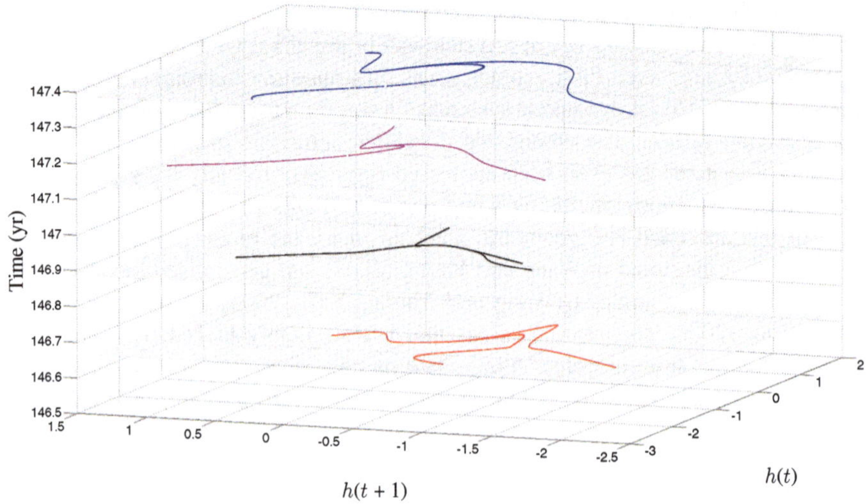

Fig. 2 Time evolution of the PBA $\mathscr{A}(t)$ in Fig. 1 throughout a calendar year. Each snapshot is represented—for $\delta = 15 \times 10^{-2}$—by a heavy curve, hiding the fine-scale details and foldings shown in Fig. 1 for $t \approx 147.64$. Note that the snapshot of $\mathscr{A}(t)$ that is represented in *red* at the *bottom* of the figure for $t \approx 146.64$ years is actually exactly the same as that shown in Fig. 1, due to the periodicity of the seasonal forcing; see (18)

The time evolution of $\mathscr{A}(t)$ within a year is illustrated in Fig. 2; the four snapshots are shown at 3-month intervals. The periodicity of the global PBA can be seen by comparing the bottom snapshot in Fig. 2 (red curve) for $t \approx 146.64$ years with the PBA shown in Fig. 1 for $t \approx 147.64$ years.

As can be seen in Fig. 2, the PBA is experiencing global deformations and shifts with time in the embedded phase space. The PBA's strangeness, with its foldings and stretchings, is also present in each of the snapshots shown in Fig. 2, but the mode of representation adopted here prevents one to display these fine-scale structures.[5]

As we will see in the next section, each PBA $\mathscr{A}(t)$ supports a complicated probability measure that describes the statistics of the dynamics and, in particular, that of the model's extremes events. The latter correspond to the PBA's filaments that meander with time in the embedded phase space. This meandering helps provide a useful physical interpretation of the model's dynamics.

For instance, due to the embedding used, at constant time, for each of the horizontal planes shown in Fig. 2, one can infer that a negative and large value of $h(t)$ followed by a negative and large value of $h(t+1)$, i.e., 1 year later, is less likely to occur in boreal winter (black and magenta curves in Fig. 2) than in boreal summer (blue and red curves in Fig. 2). This seasonal dependence of the extremes is well known in ENSO models, and it is reflected strikingly here by the global PBA's time evolution.

[5]Heavy curves have been used for a better visualization of the overall evolution in the three-dimensional representation used in Fig. 2.

We turn next to a natural class of probability measures supported by a strange PBA, such as the one shown in Fig. 1. These invariant measures will help complete our description of seasonal dependence of the extremes, as encoded by the time evolution of $\mathscr{A}(t)$, by attributing useful statistics to this dependence.

2.4 Pullback Statistical Equilibria of Periodically Forced Systems

In this section, we provide the theoretical underpinnings for the study of probability measures in periodically forced, infinite-dimensional systems like our ENSO model. Given a reference probability measure $\mathfrak{m}_0$ on the phase space X, one wishes to consider time-dependent probability measures μ_t on X that can be obtained as a weak limit of the measure $\mathfrak{m}_t := U(t,s) * \mathfrak{m}_0$, as $s \to -\infty$.

Equivalently, the probability measure $\mathfrak{m}_t$ is defined for any Borel set E of X by

$$\mathfrak{m}_t(E) = \mathfrak{m}_0(U(t,s)^{-1}(E)), \tag{19}$$

i.e., it gives the "$\mathfrak{m}_0$-volume" of points of X that fall into the set E when propagated by $U(t,s)$, and it characterizes therewith the evolution of the initial measure $\mathfrak{m}_0$ under the action of $U(t,s)$. A weak limit is understood here in the following sense: for all continuous and bounded real-valued function φ on X, we have

$$\lim_{s \to -\infty} \int_X \varphi(U(t,s)x)\, \mathrm{d}\mathfrak{m}_0(x) = \int_X \varphi(x)\, \mathrm{d}\mu_t(x). \tag{20}$$

In infinite dimensions, though, the existence of the limit on the left-hand side of (20) is not guaranteed in general, even in the autonomous case. By making, however, use of a generalized Banach limit[6] (Foias et al., 2001), a weaker version of (20) has been shown to hold in the autonomous setting[7] for a broad class of infinite-dimensional dissipative systems, as soon as they exhibit a global attractor; see Chekroun and Glatt-Holtz (2012, Theorem 2.2).

This result has been generalized to the non-autonomous setting by Lukaszewicz and Robinson (2014). In either case, autonomous or not, (Chekroun and Glatt-Holtz 2012, Theorem 2.2) or (Lukaszewicz and Robinson 2014, Theorem 4.1) ensures that such a limiting measure is necessarily invariant and supported by the global attractor. In the non-autonomous setting, the invariance property is

$$U(t,s) * \mu_s = \mu_t, \tag{21}$$

[6]Allowing, for instance, for a weighted combination over the possible accumulation points in X of the trajectory $s \mapsto U(t,s)x$.

[7]In this case, $U(t,s) = S(t-s)$ becomes a (semi-)flow and μ_t is time independent.

or, equivalently,

$$\int_{\mathscr{A}(s)} \varphi(U(t,s)x)\,\mathrm{d}\mu_s(x) = \int_{\mathscr{A}(t)} \varphi(x)\,\mathrm{d}\mu_t(x), \quad s \le t, \quad \varphi \in C_b(X), \tag{22}$$

with $C_b(X)$ the space of real-valued, continuous, and bounded functions on X.

In the periodically forced case of Eq. (6), the existence of a global PBA $\mathscr{A}(t)$ ensures thus that, starting from an initial probability measure $\mathfrak{m}_0$, an invariant measure μ_t supported by $\mathscr{A}(t)$ is reached at time t under the action of $U(t,s)$, as the initial time s is stretched into the past. Furthermore, recalling that $\mathscr{A}(t) = \mathscr{A}(t+T)$, cf. (18), one can prove from (21) and (16) that

$$\mu_{t+T} = \mu_t, \quad \text{with } T = 1 \text{ years.} \tag{23}$$

Obviously, the existence of a unique invariant probability measure μ_t that satisfies Eq. (20)—irrespectively of $\mathfrak{m}_0$—is not yet guaranteed at this stage. The difficulty does not come from the techniques underlying the aforementioned mathematical results, but rather from the infinite-dimensional nature of the phase space X. In finite dimensions, a unique measure μ_t satisfying (20), irrespective of any measure $\mathfrak{m}_0$ possessing a density with respect to the Lebesgue measure, has been shown to exist for several systems (Eckmann and Ruelle 1985; Ruelle 1999), giving rise often to a Sinai–Ruelle–Bowen (SRB) measure. In the non-autonomous setting, this measure describes the statistics of time evolutions of almost all solutions starting from the basin of attraction of a PBA $\mathscr{A}(t)$; see Ruelle (1999), Chekroun et al. (2011b), Young (2016).

There is, however, no direct generalization of the ideas surrounding SRB measures to infinite dimensions. This is due in part[8] to the absence of a notion of Lebesgue measure in function spaces such as X.

As mentioned earlier in Sect. 2.3, an important step towards the existence of an analogue of SRB measures in infinite dimension has been taken; see Young (2016) for a recent survey. It concerns the case of periodically forced systems that exhibit a limit cycle when the forcing is turned off. Loosely speaking, in the case of the origin losing its stability via a supercritical Hopf bifurcation, if the strong stable foliation W^{ss} originating from 0—and for which each W^{ss}-curve meets the limit cycle in exactly one point—has W^{ss}-curves twisted near the origin, then suitable periodic kicks in the vicinity of the supercritical limit cycle lead to folding and stretching of the phase space, and eventually to a strange attractor.

If the foliation is of finite codimension and sufficiently regular, e.g., Lipschitz continuous, then there is a well-defined Lebesgue measure class transversal to

[8]The other aspect of the problem that renders the analysis difficult is tied to the lack of smoothing of the flow in probability space by the Liouville equation (Chekroun et al., 2014)— in the present, deterministic setting—as compared to the Fokker–Planck equation, which is its counterpart in the presence of noise; see Chekroun et al. (submitted).

its leaves. If the codimension is two, for instance, and—for every embedded 2-D surface S transversal to the leaves of W^{ss}—a given property holds almost everywhere with respect to the Riemannian measure on S, then it holds almost everywhere transversal to W^{ss}. This way, a proper Lebesgue-like meaning to "almost all solutions" can be given, and the conclusion of Lu et al. (2013, Theorem 3.4) ensures the existence of SRB measures for a broad class of periodically kicked evolution equations in infinite dimension, whenever the kicks are suitable and the twist of the W^{ss}-curves sufficiently strong.

Such a theory of SRB measures provides a very useful and general geometric mechanism for the production of chaos from periodically kicked evolution equations, but its application to Eq. (6) requires a certain level of mathematical technicalities that go beyond the scope of this article. Instead, we will show in Sect. 3.2, by means of high-resolution numerics, that the singular nature of a statistical equilibrium μ_t satisfying (20), for an appropriately chosen initial probability measure $\mathfrak{m}_0$, strongly suggests the existence of such an SRB measure for Eq. (6), albeit without guaranteeing its uniqueness.

3 Chaos-to-Chaos Crisis and Pullback Symptoms

3.1 *Crisis Symptoms in the Kolmogorov–Smirnov Metric*

For simplicity, let us consider probability measures ρ and ν on the real line $\mathbb{R}$. We introduce the following abstract probability metric, subject to the choice of a set $\mathscr{F}$ of test functions:

$$d_{\mathscr{F}}(\rho, \nu) = \sup_{f \in \mathscr{F}} \left| \int f \, \mathrm{d}\rho - \int f \, \mathrm{d}\nu \right|. \tag{24}$$

If $\mathscr{F} = \{f : \|f\|_\infty \leq 1\}$, then $d_{\mathscr{F}}$ is the total variation metric TV. If

$$\mathscr{F} = \{\mathbb{1}_{(-\infty, x]}, \ x \in \mathbb{R}\}, \tag{25}$$

then $d_{\mathscr{F}}$ is the Kolmogorov–Smirnov (KS) metric d_{KS}.

It follows readily that, for any pair (ρ, ν) of probability measures,

$$d_{\mathrm{KS}}(\rho, \nu) \leq TV(\rho, \nu). \tag{26}$$

If one considers a family $\{\rho_\lambda\}$ of probability measures indexed by a parameter λ, a key property of the KS metric is that a discontinuity of the mapping $\lambda \mapsto d_{\mathrm{KS}}(\rho_{\lambda_0}, \rho_\lambda)$ at a point $\lambda = \lambda_*$ indicates a brutal change in the cumulative distribution function (CDF) of ρ_λ at that point. This change is given by

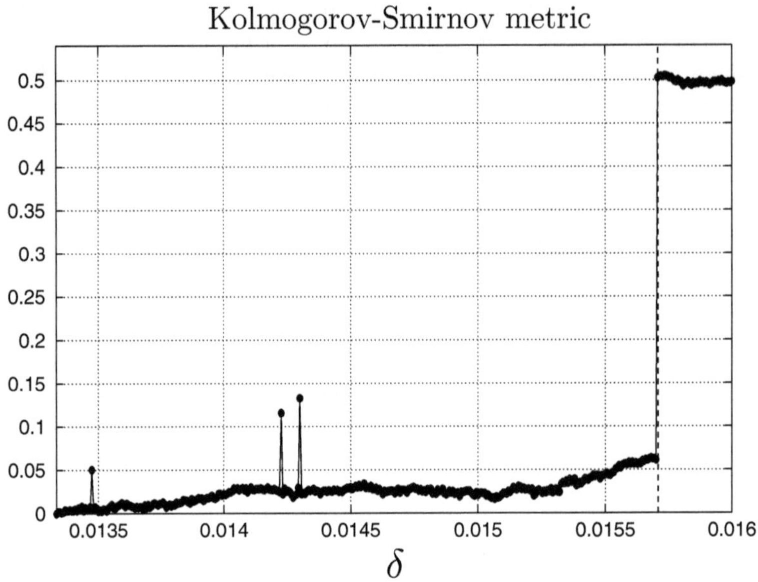

Fig. 3 Sharp transition in the Kolmogorov–Smirnov (KS) metric. The *vertical dashed line* emanates from the estimated value $15.7 \times 10^{-3} \leq \delta_* \leq 15.707 \times 10^{-3}$ at which a critical chaos-to-chaos transition occurs

$$d_{KS}(\rho_\lambda, \rho_{\lambda_*}) = \sup_x \left|\rho_\lambda((-\infty, x]) - \rho_{\lambda_*}((-\infty, x])\right|, \tag{27}$$

and there are standard statistical tests for its significance.

For Eq. (1) and for a given δ, we considered hereafter the probability distribution ρ_δ of a simulated time series $h(t)$ sampled every year. The support of this probability measure is contained in the real line, more exactly is contained in the projection of the global attractor of the time-T map (with $T = 1$ years) associated with Eq. (1). The simulations of $h(t)$ are each 85,000-years long and have been performed over a δ-grid of size 6.6667×10^{-6} from $\delta = 0$ to $\delta = 16 \times 10^{-3}$.

Given an arbitrary reference parameter δ_0, with $\delta_0 = 13.3 \times 10^{-3}$ here, we computed $d_{KS}(\rho_{\delta_0}, \rho_\delta)$, where we used the kernel estimation algorithm of Botev et al. (2010) to estimate each probability measure ρ_δ. The numerical results are reported in Fig. 3. From these results, a sharp discontinuity—up to the numerical accuracy of our experiments—can be reasonably conjectured to take place for a critical parameter value δ_* lying between $\delta = 15.7 \times 10^{-3}$ and $\delta = 15.707 \times 10^{-3}$. As a consequence, a discontinuity in the CDF of the probability measure ρ_δ occurs.

This approach based on the KS distance between one-dimensional CDFs is useful but it has its limitations. For instance, it does not allow one to distinguish what is happening dynamically right before and after the jump in the KS metric.

To get a better idea of the changes across δ_*, we examined carefully the time series $h(t)$ as the parameter δ is varied from $\delta < \delta_*$ in Fig. 4 to $\delta > \delta_*$ in Fig. 5. In Fig. 4, the most common year-to-year positive/negative excursions of h correspond to moderately warm (positive h anomaly, El Niño) and to moderately cold (negative h anomaly, La Niña) events. A small subset of these high and low excursions of h extend well beyond the typical range in Fig. 4, and are termed extreme El Niño and La Niña events.

As δ tends to δ_* from below, these extreme El Niño and La Niña events become less frequent; see the time series segments in panels (a) and (b) of Fig. 4. Such extremes disappear completely as δ_* is crossed, cf. Fig. 5.

The power spectral densities (PSDs) of the complete time series, though, as shown in Figs. 4 and 5, do not provide a clear imprint in the frequency domain of the increasing rarity of occurrence of these extreme events as one approaches δ_* from below (Fig. 4), nor about the disappearance of the latter as the critical parameter δ_* has been crossed (Fig. 5).

We show in the next section that more plentiful and reliable information regarding the nature of the dynamical transition occurring at $\delta = \delta_*$ is gained by visualizing the corresponding PBA, as well as by estimating a statistical equilibrium μ_t that this PBA supports and that satisfies Eq. (20) for an appropriate choice of initial probability measure $\mathfrak{m}_0$.

3.2 *Pullback Symptoms*

The high-resolution numerical experiments in this section are designed to shed light on the transition in the behavior of the PBA and of the pullback statistical equilibrium μ_t it supports, as the parameter δ crosses the critical value δ_*.

To estimate μ_t as per (20), the initial histories are drawn from an initial distribution $\mathfrak{m}_0$ and propagated according to the RFDE (6). The distribution $\mathfrak{m}_0$ is designed as follows. Over the interval $[-\tau, 0]$, with $\tau = \tau_2 \approx 3.3$ years, and for a grid resolution corresponding to n_g equally spaced points $\{\theta_j = -\tau(1 - j/n_g) : 1 \leq j \leq n_g\}$, the N initial histories ϕ_k are selected at random according to the formula

$$\phi_k(\theta_j) := -1 + \varepsilon k + (\varepsilon k)^{\frac{1}{p}} \xi_j, \quad 1 \leq k \leq N, \tag{28}$$

with $\varepsilon = 2/N$ and $p \geq 1$. Here the ξ_j's are n_g independent real-valued random normal variables of mean zero and unit variance.

The fractional exponent $1/p$ in (28) is chosen so that the initial distribution follows roughly a Gaussian shape in the embedded $(\phi(-1), \phi(0))$-plane[9]: the closer this exponent is to unity, the sharper the peak of the distribution, and the smaller it is, the more bell-shaped the distribution. Figure 6 shows a distribution of $N = 4 \times 10^5$ initial histories.

[9]Here $\phi(-1)$ corresponds to the value of the initial histories at -1 years.

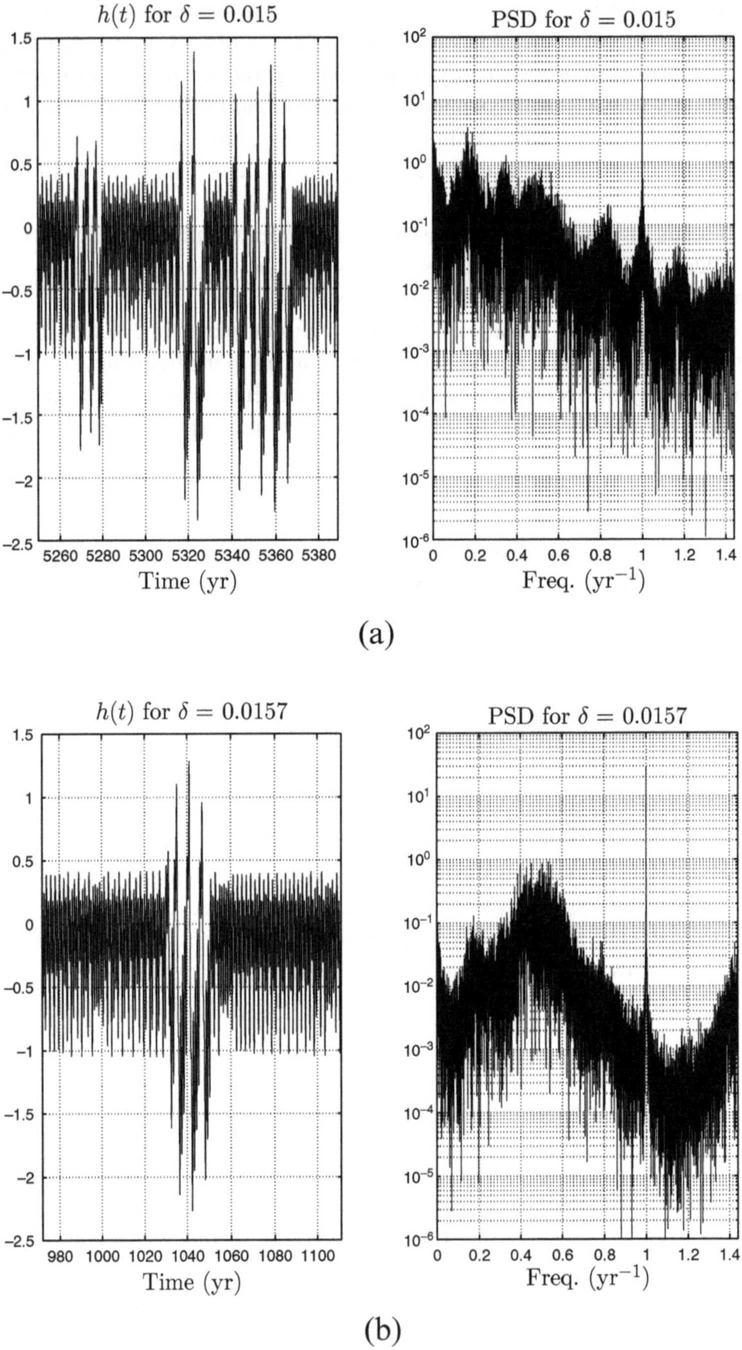

Fig. 4 Time series and PSDs for $\delta = 15 \times 10^{-3}$ and $\delta = 15.7 \times 10^{-3}$. Both these values of δ are strictly less than δ_*. (**a**) Time series and PSD. Here $a = 0.00630555$. (**b**) Time series and PSD. Here $a = 0.00630944$

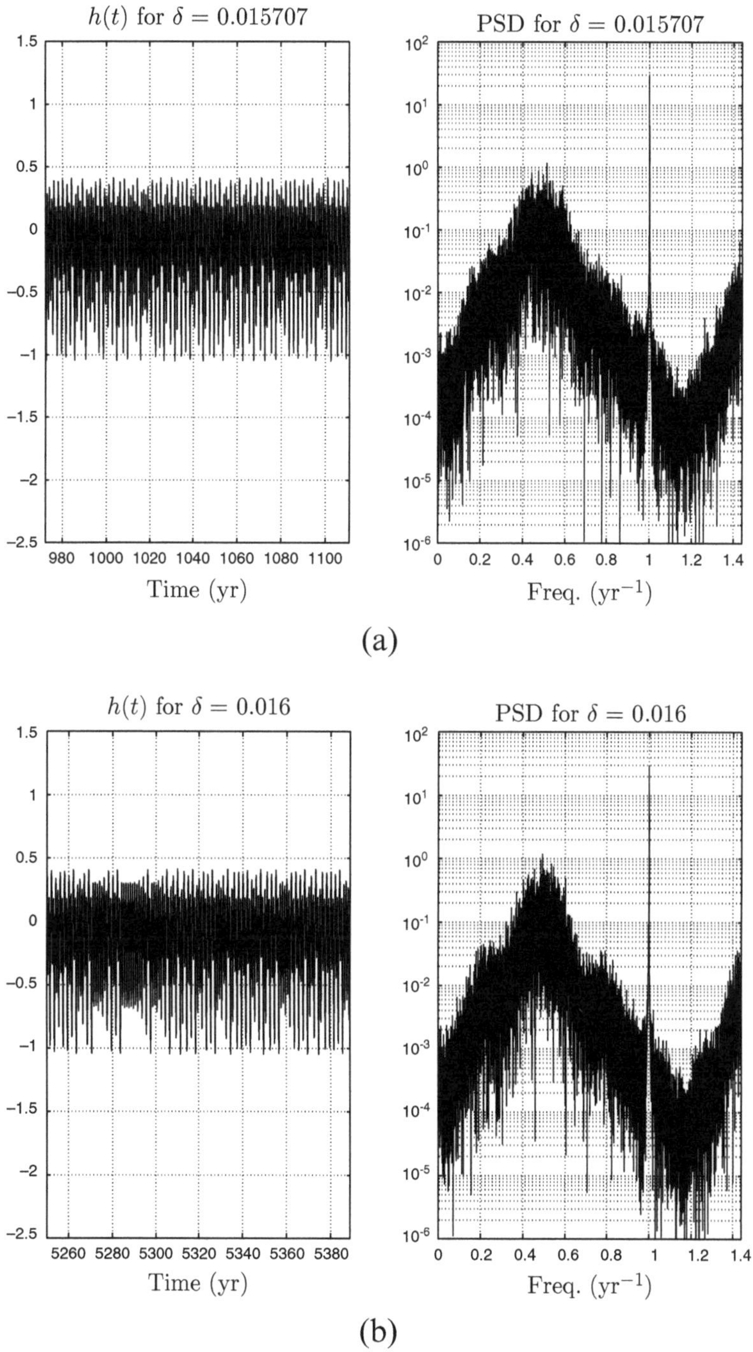

Fig. 5 Time series and PSDs for $\delta = 15.707 \times 10^{-3}$ and$\delta = 16 \times 10^{-3}$. Both these values of δ are strictly greater than δ_*. (**a**) Time series and PSD. Here $a = 0.00630948$. (**b**) Time series and PSD. Here $a = 0.00631111$

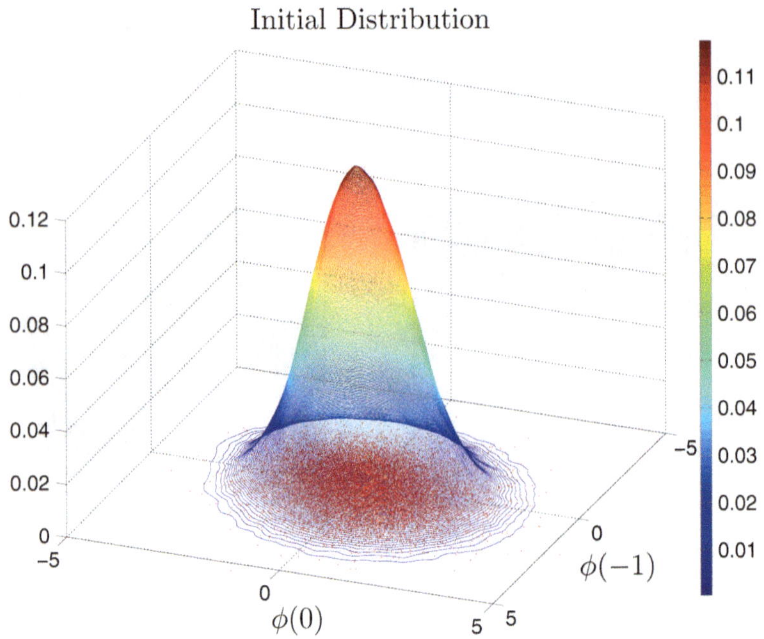

Fig. 6 A distribution m_0 of initial histories, embedded in the $(\phi(-1), \phi(0))$-plane. Here the initial histories are drawn according to (28) with $p = 12$ and $n_g = 400$

Due to the dissipative effects present in Eq. (6), one does not need to reach exactly the asymptotic limit in (20) in order to obtain a reliable approximation of μ_t. For instance, after flowing from $s = 0$ to $t = t_* \approx 147.64$ years, the $N = 4 \times 10^5$ initial histories whose distribution is shown in Fig. 6, one obtains the approximations of μ_t shown in Figs. 7 and 8. These approximations remain indistinguishable from those shown in these figures, when the same initial histories are flown from a time $s \ll 0$ up to the same t_* (not shown). Actually, even for some times $s > 0$, similar approximations (not shown) are obtained but we do not aim in this chapter to determine the minimum interval of time $t_* - s$ that ensures convergence in (20).

We focus here, as stated above, on the crisis of the global PBA and of its statistical equilibrium, when crossing the vertical dashed line in Fig. 3. For the sake of simplicity, we will no longer differentiate between μ_t and its approximations shown in Figs. 7 and 8.

Figures 7 and 8 clearly illustrate the singular nature of (the embedding of) μ_t with respect to the bell-shaped distribution m_0 shown in Fig. 6. This is not surprising, as the theory predicts that μ_t is supported by the strange PBA $\mathscr{A}_\delta(t)$, whose stretching and folding features were shown in Fig. 1 for $\delta = 15 \times 10^{-3}$. The PBA's strangeness is also manifest for the other values of δ in the interval $15.0 \times 10^{-3} \leq \delta \leq 15.707 \times 10^{-3}$ (not shown), and the singular support of the probability measures $\mu_t(\delta)$ is plotted as red curves in Figs. 7b and 8a, b.

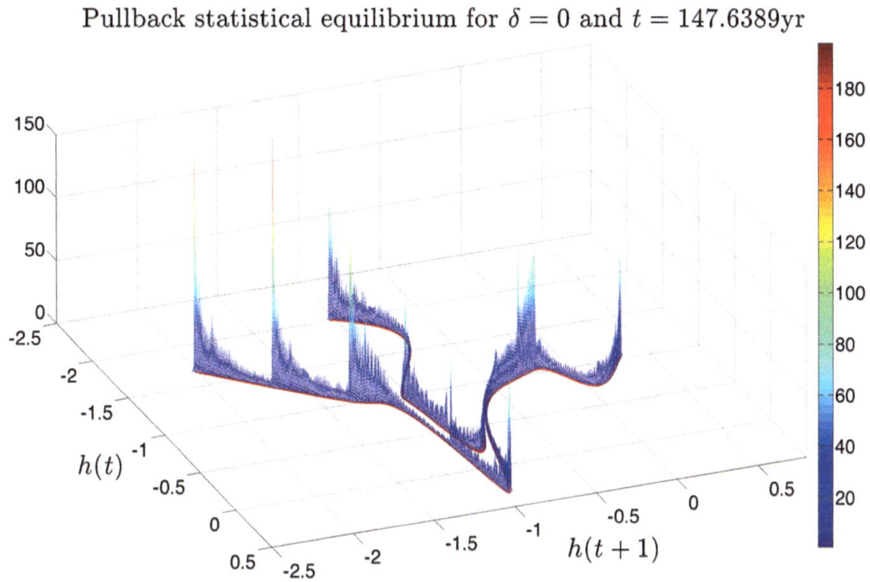

(a) $a = 0{:}00622222$.

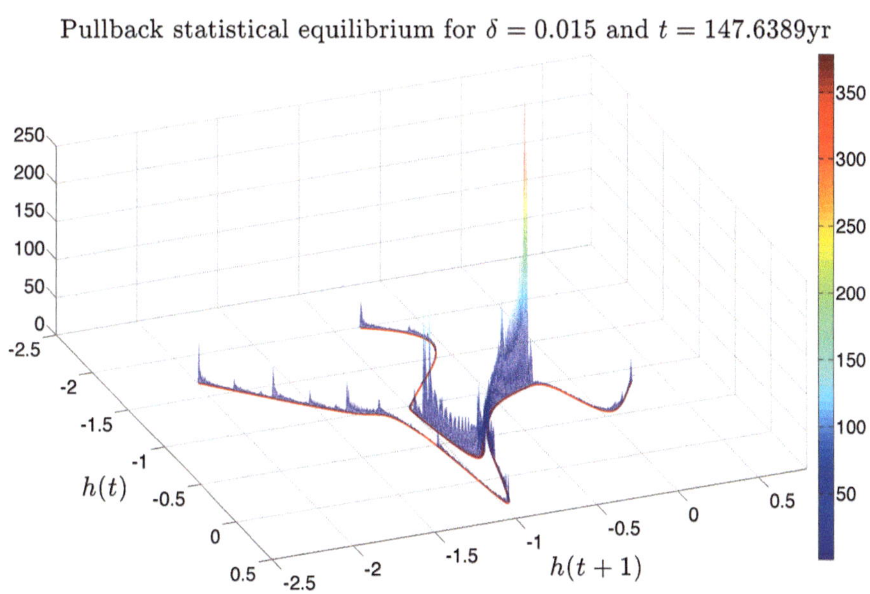

(b) $a = 0{:}00630555$:

Fig. 7 Embedding of the pullback statistical equilibrium $\mu_t(\delta)$ associated with the DDE (1). The embedding is shown within the $(h(t), h(t+1))$-plane, for $t \approx 147.64$ years and for (**a**) $\delta = 0 < \delta_*$ and (**b**) $\delta = 15 \times 10^{-3} \lesssim \delta_*$, respectively

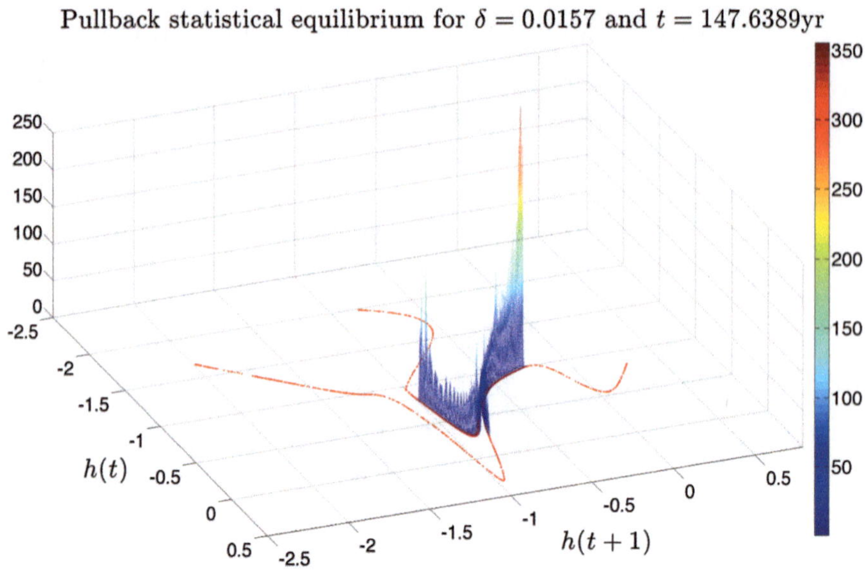

(a) a = 0:00630944:

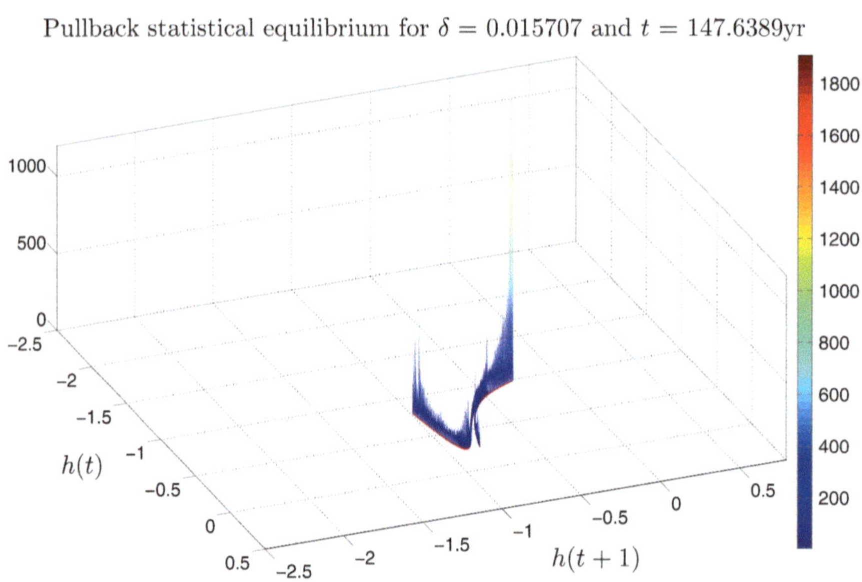

(b) a = 0:00630948:

Fig. 8 Same as Fig. 7 but for (**a**) $\delta = 15.7 \times 10^{-3} \lesssim \delta_*$ and (**b**) $\delta = 15.707 \times 10^{-3} \gtrsim \delta_*$

Recall that what is observed here is within the embedded phase space $(h(t), h(t+1))$, and one may ask to which extent we can rely on this embedding to reach conclusions about the true nature of μ_t in the full phase space X. Rigorous results from the dimension theory of Borel probability measures with compact support (Hunt and Kaloshi 1997) shed light on this issue. Denoting for a moment by $P\mu_t$ the projection of the measure μ_t onto the 2-D embedded phase space, like the one at hand, these results show that the correlation dimension D_2 (Grassberger and Procaccia 1983) of the measure $P\mu_t$ visualized herein after embedding and projection[10] is

$$D_2(P\mu_t) = \min(2, D_2(\mu_t)). \tag{29}$$

Therefore, if $D_2(P\mu_t)$ is strictly less than 2, we can conclude that the singular nature of this embedded measure—with respect to the Lebesgue measure of $\mathbb{R}^2$—reflects a genuinely singular nature of μ_t,[11] and it is not due to some numerical artifact.

We have estimated the correlation dimension $D_2(P\mu_t)$ following the algorithm of Grassberger and Procaccia (2004) and while taking into consideration the practical suggestions of Kostelich and Swinney (1989). We found that $D_2(P\mu_t) \approx 1.21$, which allows us to conclude that μ_t itself is singular, and not just its 2-D embedding shown in Figs. 7 or 8. For brevity's sake, we will not distinguish hereafter between μ_t and its 2-D embedding.

Another important point apparent from inspection of Figs. 7 and 8 concerns a key difference between the statistical equilibrium μ_t for $\delta = 0$ in Fig. 7a, located relatively far from the critical value δ_* at which the dynamical crisis occurs, and those shown for $\delta = 15 \times 10^{-3}$ in Fig. 7b and $\delta = 15.707 \times 10^{-3}$ in Fig. 8a, located both closer to and still below δ_*. The latter two statistical equilibria do exhibit elongated filaments, like those in Fig. 7a, but these filaments are much less populated by the nonlinear process than for the latter.

It follows that the statistical equilibrium μ_t supported by the PBA provides a global statistical description of the dynamics that is perfectly consistent with the observations reported at the end of Sect. 3.1 regarding the decrease in the rate of occurrence of extreme events as δ approaches δ_* from below. Indeed, the latter decrease is manifested here by a reduction of the mass of μ_t that populates the elongated filaments, until its total disappearance when δ_* has been crossed, in Fig. 8b.

In Fig. 8b, the bulk of μ_t survives the crossing of δ_*, while the elongated filaments have disappeared altogether, i.e., no more of the extreme class of events survive. These numerical results confirm that the regions of the strange PBA that survive the crisis are those that are populated by the system's most probable states. A closer look at these regions show that they correspond to regions in which the

[10]The "true" embedding dimension d given by the Takens embedding theorem may be greater than 2; see Robinson (2008) for a version of this theorem in the context of PBAs.

[11]With respect to the Lebesgue measure in $\mathbb{R}^d$.

PBA's foldings—like those shown in the inset of Fig. 1—are the most robust to perturbations.

3.3 Dynamical Interpretations

When the seasonal forcing is removed, i.e., $c = 0$ in Eq. (1), the ENSO model dynamics is periodic with a period T_δ, in years, that follows the empirical linear dependence

$$T_\delta = 8.7989 + 29.99(\delta - \delta_0), \tag{30}$$

throughout the interval $[\delta_0, 16 \times 10^{-3}]$ over which we performed the parameter-dependence experiments reported in this chapter.

The characteristics of the underlying frequency-locked regimes between the internal oscillatory frequency $\omega_i = T_\delta^{-1}$ and the driving annual frequency ω_a, i.e., the integers p and q for which $\omega_i = \omega_a p/q$, depend thus on δ. Such a frequency-locked behavior takes place—in parameter space—in a so-called p/q-Arnold tongue (Jensen et al. 1984; Jin et al. 1994; Tziperman et al. 1994) whose δ-dependence makes it a p_δ/q_δ-Arnold tongue.

The ENSO model of DDE (1), subject to seasonal forcing and over the entire range of δ-values considered here, exhibits chaotic behavior, as described in Sect. 2.3. One can thus infer that, for each δ, chaos results from overlapping of a p_δ/q_δ-Arnold tongue with another, p'_δ/q'_δ-Arnold tongue (Arnold 1988; Jensen et al. 1984).

The bifurcation theory of one-dimensional circle maps (e.g., Mori and Kuramoto 2013, Chap. 7.4) provides a possible explanation of the transition shown in Fig. 8, as δ is increased from $\delta = 15.7 \times 10^{-3}$ to $\delta = 15.707 \times 10^{-3}$. This theory addresses crises that occur within the overlap of two Arnold tongues, a region in which chaotic behavior occurs. Extrapolating to DDE models, such as the ENSO model investigated herein, and adopting the language of PBAs, one could argue that the transition observed in Figs. 3 and 8 results from the coexistence of two strange PBAs at each fixed δ.

If such were the case, the two coexisting PBAs would correspond here to the one that lies within the square $[-1, 0]^2$ of the $(h(t), h(t+1))$-plane and is shown by the red curve in Fig. 8b, along with the one that exhibits filament extending out of this box and shown by the red curve in Fig. 8a. In the present setting, the former is actually contained within the latter, but coexisting strange PBAs may, in general, be disjoint. If so, a crisis may still occur and manifest itself by a dynamics that hops between the two chaotic attractors, whether pullback or not, as one moves through parameter space; see Horita et al. (1988, Fig. 6). This phenomenon occurs whenever two Arnold tongues with nearby rotation numbers overlap, as certain crisis lines are crossed within the overlapping region; see Horita et al. (1988, Fig. 2).

A complementary explanation of the transition observed here is provided by the theory of attractor widening (Grebogi et al. 1987): see Grebogi et al. (1987, Figs. 5 & 6) for a similar crisis in the case of the Ikeda map. Adopting again the language of PBAs, a collision between the PBA shown in Fig. 8b and an unstable periodic orbit would be responsible for initiating the crisis. To get the attractor widening, the collision would have to occur as δ crosses δ_* from above.

Whatever the exact explanation of the crisis, our study provides—to the best of the authors' knowledge—the first identification of such a crisis occurring in a delay differential model, as well as its first characterization in terms of PBAs and the statistical equilibria they support. We leave the more detailed and mathematically rigorous dynamical characterization of this crisis for another investigation, and turn next to a discussion of the impact of the noise on such a chaos-to-chaos crisis.

3.4 Crisis Removal by Small Additive Noise

One could argue that similar characterizations of the dynamical crisis discussed so far could have been inferred from the system's Poincaré map and the corresponding forward attractor. This is actually a valid argument for periodically forced systems, like the non-autonomous DDE at hand. For the case of a T-periodic system, a relationship between PBAs and a notion of forward attractor is known to exist and it does not even rely on Poincaré maps.

More precisely, the set $\widetilde{\mathscr{A}} = \bigcup_{t\in[0,T]} U(t,0)\mathscr{A}(0)$, where $\mathscr{A}(0)$ denotes the global PBA at time $t = 0$, satisfies, for any bounded set B of X, the following forward attraction property:

$$\lim_{t\to+\infty} \sup_{\tau\in\mathbb{R}} \mathrm{dist}_X(U(\tau+t,t)B) = \widetilde{\mathscr{A}}. \tag{31}$$

We refer to Carvalho et al. (2013, Chap. 10.3) for a proof. The set $\widetilde{\mathscr{A}}$ is also known as a uniform forward global attractor, a concept introduced in Haraux (1991), cf. also Chepyzhov and Vishik (2002); it is the minimal compact set of X that attracts all the trajectories—uniformly with respect to the initial time—that start from a bounded set; see Haraux (1991, Chap. 8.3).

Nevertheless, the use of a standard Poincaré map or the concept of uniform attractor may hide, in the presence of noise, certain dynamical features that are revealed by a pullback approach that is not limited to the case of periodic or, more generally, deterministic non-autonomous forcing; see Ghil et al. (2008b), Chekroun et al. (2011b), Ghil (2017). We illustrate hereafter this point in the context of the DDE model (1).

Let us thus consider the following stochastic modification of Eq. (1):

$$\mathrm{d}h = \Big(aR\Big[h\Big(t-\frac{L}{2C_K}\Big)\Big]-bR\Big[h\Big(t-\frac{L}{C_K}-\frac{L}{2C_R}\Big)\Big]+c\cos(\omega_a t+\varphi)\Big)\,\mathrm{d}t+\sigma\,\mathrm{d}W_t, \tag{32}$$

where W_t denotes a one-dimensional Brownian motion and $\sigma \geq 0$. This noise term in Eq. (32) is motivated by the presence of atmospheric high-frequency variability in the coupled climate system (Blanke et al. 1997; Eckert and Latif 1997; Jin et al. 1996; Kleeman and Power 1994; Kleeman and Moore 1997; Roulston and Neelin 2000), a variability that is crudely represented herein by a white-noise process. A rigorous proof of the existence of random PBAs for such a non-autonomous stochastic DDE is beyond the scope of this chapter. We rely instead on numerical experiments to analyze the effects of noise on the inferred random PBA, as the parameter δ varies in a neighborhood of the critical value δ_* at which the chaos-to-chaos crisis occurs in the absence of noise.

Numerical results on the random PBA are shown in Fig. 9a, b for a noise intensity of $\sigma = 10^{-3}$, and the visual inspection of both panels strongly indicates that the chaos-to-chaos crisis did not survive the addition of small noise to the evolution equation. The results in Fig. 10a, b show, furthermore, that the corresponding pullback statistical equilibrium μ_t resembles the one obtained for $\delta = 0$ in the absence of noise, cf. Fig. 7a. Dynamically, the crisis in the deterministic version of the model, for $\sigma = 0$, was associated with the disappearance of extreme El Niño and La Niña events as the critical parameter value δ_* is crossed from below. The addition of noise in the system triggers once again these extreme events, manifested by the expansion of the PBA in the embedded phase space, as evident when comparing the panels (b) of Figs. 8 and 10.

Complementary experiments performed for smaller values of the noise intensity σ have been conducted and have shown that this phenomenon is robust, while reducing the noise does result in extreme events becoming less and less probable. This is noticeable, for instance, by comparing the panels of Fig. 9 with those of Fig. 10, in which the noise level is $\sigma = 10^{-4}$, while the same noise realization was used in both figures.

The statistical equilibria shown in Figs. 10a, b resemble those in Fig. 7b for $\delta = 15 \times 10^{-3}$, in which the main bulk of the density μ_t is located near the point $(-0.5, -1)$ in the $(h(t), h(t+1))$-plane, while the elongated PBA filaments—again like in Fig. 7b—are less populated by the dynamics than for $\sigma = 10^{-3}$, i.e., the extreme events are less likely to occur.

This removal of the crisis by the addition of a small additive noise is actually consistent with noise effects, as shown for the fundamental circle map in Ghil et al. (2008b). It was found there that a Devil's staircase step that corresponds to a rational rotation number can be "destroyed" by a sufficiently intense noise (Ghil et al. 2008b, Appendix B). In fact, the narrower a Devil's staircase steps is, the less robust is it to noise perturbations, while the wider ones are the most robust. Actually, the theory of topological equivalence in random dynamical systems—as analyzed in Arnold (2013), Cong (1996, 1997) and as explained in Ghil et al. (2008b, Appendix B)—implies that the elimination of a Devil's staircase step, for a sufficient amount of noise, is manifested by the disappearance of a p/q-Arnold tongue. As a consequence, the corresponding asymptotic dynamics is no longer a periodic random attractor but a random fixed point.

Given this understanding of the smoothing effect of noise on the circle map's fine-grained resonant landscape, and the universal character of the circle map, one

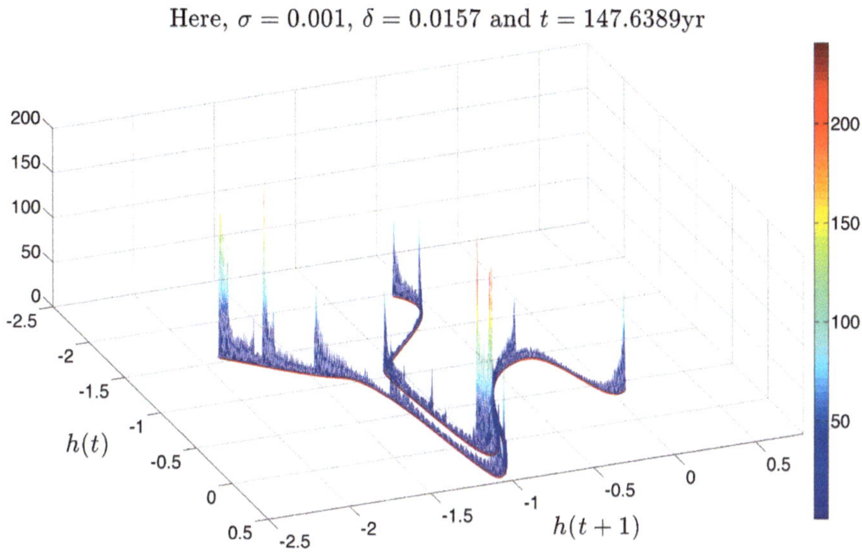

(a) a = 0:00630944:

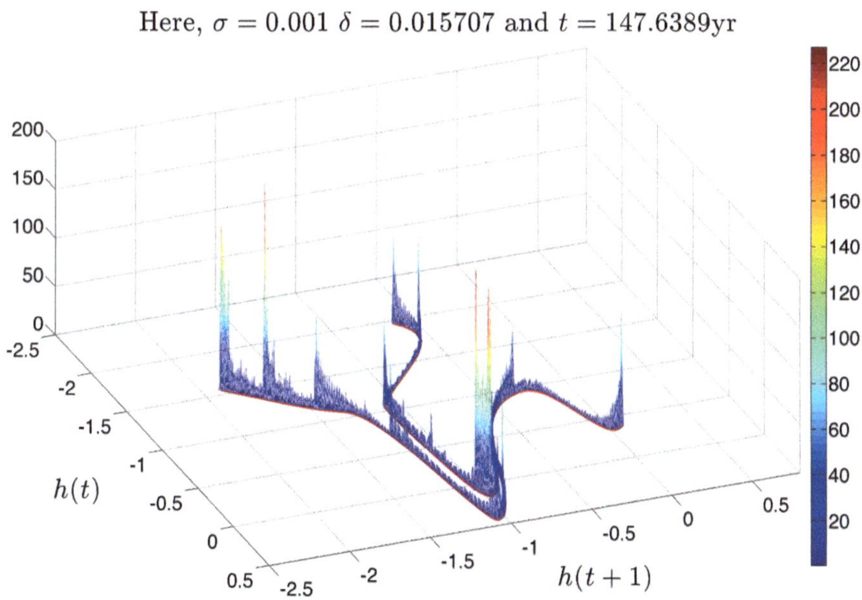

(b) a = 0:00630948.

Fig. 9 Same as Fig. 8 but for the stochastic DDE (32). The δ-values are again (**a**) $\delta = 15.7 \times 10^{-3}$ and (**b**) $\delta = 15.707 \times 10^{-3}$. The noise intensity $\sigma = 10^{-3}$ and the noise realization are the same for the panels (**a**) and (**b**), while the initial histories are again drawn from the distribution $\mathfrak{m}_0$ shown in Fig. 6

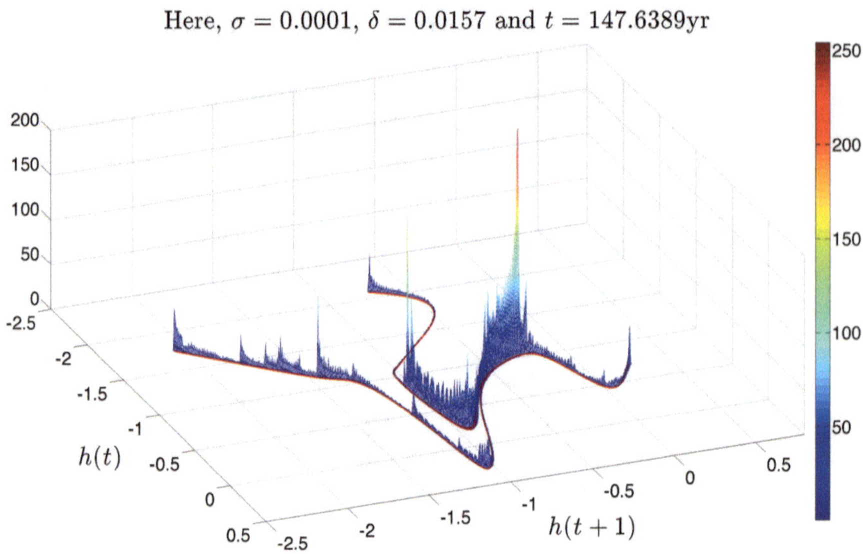

(a) $a = 0{:}00630944{:}$

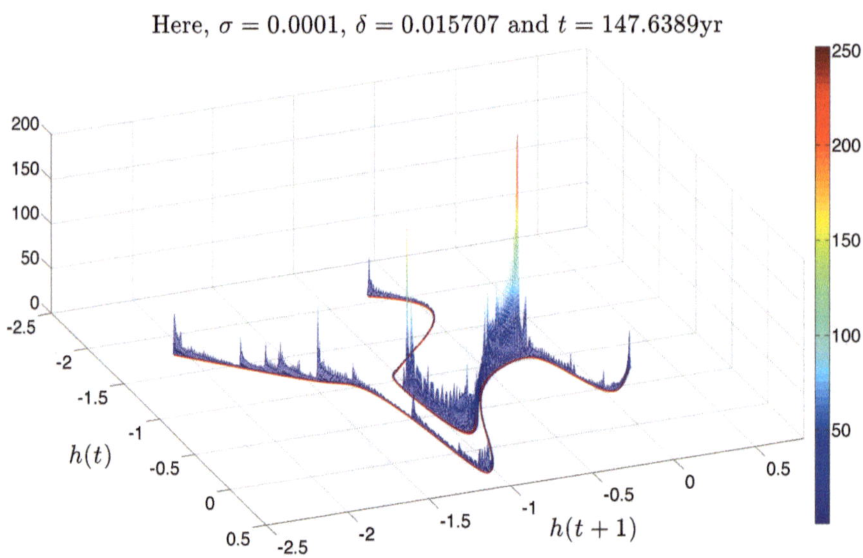

(b) $a = 0{:}00630948.$

Fig. 10 Same as Fig. 9 but with much smaller noise, $\sigma = 10^{-4}$. (**a**) $a = 0.00630944$. (**b**) $a = 0.00630948$

can deduce a heuristic result on the periodically forced DDE model considered here. To do so, recall the discussion of Sect. 3.3 about the dynamical origin of the chaos-to-chaos transition observed herein between $\delta = 15.7\times10^{-3}$ and $\delta = 15.707\times10^{-3}$, in the absence of noise; see again Figs. 3 and 8. Two dynamical mechanisms were proposed as potential causes of this transition.

In the case of the crossing of a crisis line within an overlap of two Arnold tongues (Mori and Kuramoto 2013), the removal of this crisis by the noise can be understood as the elimination of a nearby p/q-Arnold tongue; this elimination, in turn, induces the disappearance of the coexisting chaotic attractor, as discussed in Sect. 3.3. Such an explanation is consistent, furthermore, with the resemblance between the PBAs shown in Figs. 9 and 10, on the one hand, with those shown in Fig. 7, on the other.

In the case of an attractor widening scenario, according to Grebogi et al. (1987), the noise would be responsible for jiggling an unstable periodic orbit that lies near the PBA $\mathscr{A}(t)$ so as to collide with the latter. Such a collision can cause an attractor widening to occur even for parameter values for which this unstable periodic orbit does not collide with $\mathscr{A}(t)$ in the absence of noise.

Whatever the mechanisms behind the chaos-to-chaos crisis of interest here, the way the noise enters the governing equations is crucial in causing the removal of the crisis or not. Typically, certain state-dependent noises may preserve the ordering between stationary solutions or between more complicated invariant sets (Chekroun et al. 2016a). This ordering may, in turn, prevent the destruction of random periodic orbits and thus of p/q-Arnold tongues, as already pointed out in Ghil et al. (2008b, Appendix B); such is the case, for instance, in the circle map, if the noise enters nonlinearly into the phase of the rotation. Likewise, a random unstable periodic orbit may stay away from the PBA in the case of certain multiplicative noises, a situation that may prevent an attractor widening scenario à la Grebogi et al. (1987) to be realized. The rigorous reduction techniques of Chekroun et al. (2015a,b), along with the approximation techniques of Chekroun et al. (2016b), provide a natural framework for analyzing the effects of state-dependent noise on DDE models such as Eq. (1) and they will be pursued elsewhere.

Acknowledgements This work has been partially supported by the Office of Naval Research (ONR) Multidisciplinary University Research Initiative (MURI) grant N00014-12-1-0911 and N00014-16-1-2073 (MDC & MG), by the National Science Foundation grants OCE-1243175 (MDC & MG), DMS-1616981(MDC), and AGS-1540518 grant (JDN).

References

Arnold, V.I. 1988. *Geometrical methods in the theory of ordinary differential equations*, 2nd ed., Grundlehren der mathematischen Wissenschaften, vol. 250. Berlin: Springer.

Arnold, L. 2013. *Random dynamical systems*. Berlin: Springer Science & Business Media.

Battisti, D.S., and A.C. Hirst. 1989. Interannual variability in a tropical atmosphere–ocean model: Influence of the basic state, ocean geometry and nonlinearity. *Journal of the Atmospheric Sciences* 46(12): 1687–1712.

Bjerknes, J. 1969. Atmospheric teleconnections from the equatorial Pacific. *Monthly Weather Review* 97(3): 163–172.

Blanke, B., J.D. Neelin, and D. Gutzler. 1997. Estimating the effects of stochastic wind stress forcing on ENSO irregularity. *Journal of Climate* 10: 1473–1486.

Bódai, T., and T. Tél. 2012. Annual variability in a conceptual climate model: snapshot attractors, hysteresis in extreme events, and climate sensitivity. *Chaos: An Interdisciplinary Journal of Nonlinear Science* 22(2): 023110.

Bódai, T., G. Károlyi, and T. Tél. 2011. A chaotically driven model climate: extreme events and snapshot attractors. *Nonlinear Processes in Geophysics* 18(5): 573–580.

Bódai, T., G. Károlyi, and T. Tél. 2013. Driving a conceptual model climate by different processes: snapshot attractors and extreme events. *Physical Review E* 87(2): 022822.

Botev, Z.I., J.F. Grotowski, and D.P. Kroese. 2010. Kernel density estimation via diffusion. *The Annals of Statistics* 38(5): 2916–2957.

Cane, M.A. 1986. Experimental forecasts of el Niño. *Nature* 321: 827–832.

Caraballo, T., J.A. Langa, and J.C. Robinson. 2001. Attractors for differential equations with variable delays. *Journal of Mathematical Analysis and Applications* 260(2): 421–438.

Caraballo, T., P. Marın-Rubio, and J. Valero. 2005. Autonomous and non-autonomous attractors for differential equations with delays. *Journal of Differential Equations* 208(1): 9–41.

Carvalho, A., J.A. Langa, and J. Robinson. 2013. Attractors for infinite-dimensional non-autonomous dynamical systems In *Applied mathematical sciences*, vol. 182. Berlin: Springer.

Chekroun, M.D., and N.E. Glatt-Holtz. 2012. Invariant measures for dissipative dynamical systems: Abstract results and applications. *Communications in Mathematical Physics* 316(3): 723–761.

Chekroun, M.D., D. Kondrashov, and M. Ghil. 2011a. Predicting stochastic systems by noise sampling, and application to the El Niño-Southern Oscillation. *Proceedings of the National Academy of Sciences of the United States of America* 108: 11766–11771.

Chekroun, M.D., E. Simonnet, and M. Ghil. 2011b. Stochastic climate dynamics: random attractors and time-dependent invariant measures. *Physica D* 240(21): 1685–1700.

Chekroun, M.D., J.D. Neelin, D. Kondrashov, J.C. McWilliams, and M. Ghil. 2014. Rough parameter dependence in climate models: the role of Ruelle-Pollicott resonances. *Proceedings of the National Academy of Sciences of the United States of America* 111(5): 1684–1690.

Chekroun, M.D., H. Liu, and S. Wang. 2015a. *Approximation of stochastic invariant manifolds: stochastic manifolds for nonlinear SPDEs I*. Springer briefs in mathematics. New York: Springer.

Chekroun, M.D., H. Liu, and S. Wang. 2015b. *Parameterizing manifolds and non-Markovian reduced equations: stochastic manifolds for nonlinear SPDEs II*. Springer briefs in mathematics. New York: Springer.

Chekroun, M.D., E. Park, and R. Temam. 2016a. The Stampacchia maximum principle for stochastic partial differential equations and applications. *Journal of Differential Equations* 260(3): 2926–2972.

Chekroun, M.D., M. Ghil, H. Liu, and S. Wang. 2016b. Low-dimensional Galerkin approximations of nonlinear delay differential equations. *Discrete and Continuous Dynamical System A* 36(8): 4133–4177.

Chekroun, M.D., A. Tantet, H.A Dijkstra, and J.D. Neelin, Mixing Spectrum in reduced phase spaces of stochastic differential equations. Part I: theory (submitted)

Chepyzhov, V.V., and M.I. Vishik. 2002. *Attractors for equations of mathematical physics*, vol. 49 of Colloquium publications. Providence, RI: American Mathematical Society.

Cong, N.D. 1996. Topological classification of linear hyperbolic cocycles. *Journal of Dynamics and Differential Equations* 8(3): 427–467.

Cong, N.D. 1997. *Topological dynamics of random dynamical systems*. Oxford: Oxford University Press.

Crauel, H., and F. Flandoli. 1994. Attractors for random dynamical systems. *Probability Theory and Related Fields* 100(3): 365–393.

Crauel, H., A. Debussche, and F. Flandoli. 1997. Random attractors. *Journal of Dynamics and Differential Equations* 9(2): 307–341.

Diekmann, O., S.A. van Gils, S.M. Verduyn Lunel, and H.-O. Walther. 1995. *Delay equations: functional, complex, and nonlinear analysis*. Applied mathematical sciences, vol. 110. New York: Springer.

Drótos, G., T. Bódai, and T. Tél. 2015. Probabilistic concepts in a changing climate: a snapshot attractor picture. *Journal of Climate* 28(8): 3275–3288.
Eckert, C., and M. Latif. 1997. Predictability of a stochastically forced hybrid coupled model of El Niño. *Journal of Climate* 10(7): 1488–1504.
Eckmann, J.-P., and D. Ruelle. 1985. Ergodic theory of chaos and strange attractors. *Reviews of Modern Physics* 57: 617–656.
Foias, C., O. Manley, R. Rosa, and R. Temam. 2001. *Navier-Stokes equations and turbulence*. Encyclopedia of mathematics and its applications, vol. 83. Cambridge: Cambridge University Press.
Galanti, E., and E. Tziperman. 2000. ENSO's phase locking to the seasonal cycle in the fast-SST, fast-wave, and mixed-mode regimes. *Journal of the Atmospheric Sciences* 57(17): 2936–2950.
Ghil, M. 2017. The wind-driven ocean circulation: applying dynamical systems theory to a climate problem. *Discrete and Continuous Dynamical Systems-A* 37(1): 189–228.
Ghil, M., and N. Jiang. 1998. Recent forecast skill for the El Niño/Southern Oscillation. *Geophysical Research Letters* 25(2): 171–174.
Ghil, M., and I. Zaliapin. 2015. Understanding ENSO variability and its extrema: a delay differential equation approach. In *Observations, modeling and economics of extreme events*, ed. M. Ghil, M. Chavez, and J. Urrutia-Fucugauchi. Geophysical monographs, vol. 214, 63–78. Washington, DC: American Geophysical Union & Wiley.
Ghil, M., I. Zaliapin, and S. Thompson. 2008a. A delay differential model of ENSO variability: parametric instability and the distribution of extremes. *Nonlinear Processes in Geophysics* 15(3): 417–433.
Ghil, M., M.D. Chekroun, and E. Simonnet. 2008b. Climate dynamics and fluid mechanics: natural variability and related uncertainties. *Physica D* 237: 2111–2126.
Grassberger, P., and I. Procaccia. 1983. Characterization of strange attractors. *Physical Review Letters* 505: 346.
Grassberger, P., and I. Procaccia. 2004. Measuring the strangeness of strange attractors. In *The theory of chaotic attractors*, ed. B.R. Hunt, T.-Y. Tien-Yien Li, J.A. Kennedy, and H.E. Nusse, 170–189. Berlin: Springer.
Grebogi, C., E. Ott, F. Romeiras, and J.A. Yorke. 1987. Critical exponents for crisis-induced intermittency. *Physical Review A* 36(11): 5365.
Hale, J.K., and S.M. Verduyn-Lunel. 1993. *Introduction to functional-differential equations*. Applied mathematical sciences. vol. 99. New York: Springer.
Haraux, A. 1991. *Systèmes Dynamiques Dissipatifs et applications*, vol. 17. Paris: Masson.
Horita, T., H. Hata, H. Mori, T. Morita, K. Tomita, K. Shoichi, and H. Okamoto. 1988. Local structures of chaotic attractors and q-phase transitions at attractor-merging crises in the sine-circle maps. *Progress of Theoretical Physics* 80(5): 793–808.
Hunt, B.R., and V.Y. Kaloshin. 1997. How projections affect the dimension spectrum of fractal measures. *Nonlinearity* 10(5): 1031–1046.
Jensen, M.H., P. Bak, and T. Bohr. 1984. Transition to chaos by interaction of resonances in dissipative systems. I. Circle maps. *Physical Review A* 30(4): 1960.
Jiang, N., M. Ghil, and D. Neelin. 1995. Forecasts of equatorial Pacific SST anomalies by an autoregressive process using singular spectrum analysis. *Experimental Long-Lead Forecast Bulletin* 4(1): 24–27.
Jin, F.-F., and J.D. Neelin. 1993. Modes of interannual tropical ocean-atmosphere interaction-A unified view. Part III: analytical results in fully coupled cases. *Journal of the Atmospheric Sciences* 50(21): 3523–3540.
Jin, F.-F., J.D. Neelin, and M. Ghil. 1994. El Niño on the Devil's Staircase: annual subharmonic steps to chaos. *Science* 274: 70–72.
Jin, F.-F., J.D. Neelin, and M. Ghil. 1996. El Niño/Southern oscillation and the annual cycle: Subharmonic frequency locking and aperiodicity. *Physica D* 98: 442–465.
Keane, A., Krauskopf, B., and C. Postlethwaite. 2015. Delayed feedback versus seasonal forcing: resonance phenomena in an El Niño Southern Oscillation model. *SIAM Journal on Applied Dynamical Systems* 14(3): 1229–1257.

Keane, A., B. Krauskopf, and C. Postlethwaite. 2016. Investigating irregular behavior in a model for the El Niño Southern Oscillation with positive and negative delayed feedback. *SIAM Journal on Applied Dynamical Systems* 15(3): 1656–1689.

Kelley, J.L. 1975. *General topology*. Berlin: Springer Science & Business Media.

Kleeman, R., and S.B. Power. 1994. Limits to predictability in a coupled ocean-atmosphere model due to atmospheric noise. *Tellus A* 46(4): 529–540.

Kleeman, R., and A.M. Moore. 1997. A theory for the limitation of ENSO predictability due to stochastic atmospheric transients *Journal of the Atmospheric Sciences* 54(6): 753–767.

Kondrashov, D., M.D. Chekroun, A.W. Robertson, and M. Ghil. 2013. Low-order stochastic model and "past-noise forecasting" of the Madden-Julian oscillation. *Geophysical Research Letters* 40: 5305–5310. doi:10.1002/grl.50991.

Kondrashov, D., M.D. Chekroun, and M. Ghil. 2015. Data-driven non-Markovian closure models. *Physica D: Nonlinear Phenomena* 297: 33–55.

Kostelich, E.J., and H.L. Swinney. 1989. Practical considerations in estimating dimension from time series data. *Physica Scripta* 40(3): 436

Krauskopf, B., and J. Sieber. 2014. Bifurcation analysis of delay-induced resonances of the El-Niño Southern Oscillation. *Proc R Soc A* 470(2169). The Royal Society.

Lu, K., Q. Wang, and L.-S. Young. 2013. *Strange attractors for periodically forced parabolic equations*, vol. 224. Memoirs of the American mathematical society, vol. 1054. Providence, RI: American Mathematical Society.

Lukaszewicz, G., and J.C. Robinson. 2014. Invariant measures for non-autonomous dissipative dynamical systems. *Discrete and Continuous and Dynamical Systems A* 34(10): 4211–4222.

Mechoso, C.R., J.D. Neelin, and J.-Y. Yu. 2003. Testing simple models of ENSO. *Journal of the Atmospheric Sciences* 60: 305–318.

Mori, H., and Y. Kuramoto. 2013. *Dissipative structures and chaos*. New York: Springer.

Münnich, M., M.A. Cane, and S.E. Zebiak. 1991. A study of self-excited oscillations of the tropical ocean-atmosphere system. Part II: nonlinear cases. *Journal of the Atmospheric Sciences* 48(10): 1238–1248.

Neelin, J.D., D.S. Battisti, A.C. Hirst, F.-F. Jin, Y. Wakata, T. Yamagata, and S.E. Zebiak. 1998. ENSO theory. *Journal of Geophysical Research: Oceans (1978–2012)* 103(C7): 14261–14290.

Neelin, J.D., F.-F. Jin, and H.-H. Syu. 2000. Variations in ENSO phase-locking. *Journal of Climate* 13: 2570–2590.

Philander, S.G.H. 1992. *El Niño, La Niña, and the Southern oscillation.* San Diego: Academic.

Pierini, S., M. Ghil, and M.D. Chekroun. 2016. Exploring the pullback attractors of a low-order quasigeostrophic ocean model: the deterministic case. *Journal of Climate* 29(11): 4185–4202.

Robinson, J.C. 2008. A topological time-delay embedding theorem for infinite-dimensional cocycle dynamical systems. *Discrete and Continuous Dynamical Systems. Series B* 9(3–4): 731–741.

Romeiras, F.J., C. Grebogi, and E. Ott. 1990. Multifractal properties of snapshot attractors of random maps. *Physical Review A* 41(2): 784.

Roulston, M.S., and J.D. Neelin. 2000. The response of an ENSO model to climate noise, weather noise and intraseasonal forcing. *Geophysical Research Letters* 27: 3723–3726.

Ruelle, D. 1999. Smooth dynamics and new theoretical ideas in nonequilibrium statistical mechanics. *Journal of Statistical Physics* 95(1): 393–468.

Suarez, M.J., and P.S. Schopf. 1988. A delayed action oscillator for ENSO. *Journal of the Atmospheric Sciences* 45(21): 3283–3287.

Takens, F. 1981. Detecting strange attractors in turbulence. In *Dynamical systems and turbulence, Warwick 1980*, 366–381. New York: Springer.

Temam, R. 1997. *Infinite-dimensional dynamical systems in mechanics and physics*, 2nd ed., Applied Mathematical Sciences, vol. 68. New York: Springer.

Tziperman, E., L. Stone, M.A. Cane, and H. Jarosh. 1994. El Niño chaos: overlapping of resonances between the seasonal cycle and the Pacific ocean-atmosphere oscillator. *Science* 264(5155): 72–74.

Tziperman, E., M.A. Cane, and S.E. Zebiak. 1995. Irregularity and locking to the seasonal cycle in an ENSO prediction model as explained by the quasi-periodicity route to chaos. *Journal of the Atmospheric Sciences* 52(3): 293–306.
Wang, Q., and L.-S. Young. 2001. Strange attractors with one direction of instability. *Communications in Mathematical Physics* 218(1): 1–97.
Wang, Q., and L.-S. Young. 2003. Strange attractors in periodically-kicked limit cycles and Hopf bifurcations. *Communications in Mathematical Physics* 240(3): 509–529.
Wang, Q., and L.-S. Young. 2008. Toward a theory of rank one attractors. *Annals of Mathematics* 167: 349–480.
Young, L.-S. 2016. Generalizations of SRB measures to nonautonomous, random, and infinite dimensional systems. *Journal of Statistical Physics* 166: 494–515.

Shear-Wave Splitting Indicates Non-Linear Dynamic Deformation in the Crust and Upper Mantle

Stuart Crampin, Gulten Polat, Yuan Gao, David B. Taylor, and Nurcan Meral Ozel

Abstract We demonstrate that non-linear dynamic deformation exists throughout the crust and upper mantle of the Earth. Stress-aligned shear-wave splitting, seismic birefringence, is widely observed in the Earth's upper crust, lower-crust, and uppermost ~400 km of the mantle. Attributed to the effects of pervasive distributions of stress-aligned fluid-saturated microcracks in the crust (and controversially intergranular films of hydrated melt in the mantle), the degree splitting indicates that 'microcracks' are so closely spaced that they verge on failure in fracturing and earthquakes if there is any disturbance. Phenomena that verge on failure are critical systems with non-linear dynamics that impose a range of new properties on conventional sub-critical geophysics that we suggest is a *New Geophysics*. Consequently, shear-wave splitting provides directly interpretable information about the progress of non-linear dynamic deformation in the deep otherwise-inaccessible interior of the microcracked Earth. Possibly uniquely for non-linear dynamic phenomena, observation of shear-wave splitting allows the progress towards singularities to be monitored in deep in situ rock, so that earthquakes and volcanic eruptions can be predicted (we prefer *stress-forecast*). The response to other processes, such as hydraulic fracking, can be monitored, and in some cases calculated and effects predicted. Here, we review shear-wave splitting and demonstrate the prevalence of non-linear dynamic deformation of the New Geophysics in the crust and uppermost ~400 km of the mantle.

Keywords Monitoring fracking • *New Geophysics* • Non-linear dynamics • Shear-wave splitting • Stress-aligned microcracks • Stress-forecasting earthquakes • Stress-forecasting volcanic eruptions

---

Preamble
Schopenhauer 1788–1860: 'All truth passes through three stages: ridicule; violent-opposition; self-evident'.

S. Crampin (✉) • G. Polat • Y. Gao • D.B. Taylor • N.M. Ozel
British Geological Survey, The Lyell Centre, Edinburgh, EH14 4AP, Scotland, UK
e-mail: scrampin@ed.ac.uk

A.A. Tsonis (ed.), *Advances in Nonlinear Geosciences*,
DOI 10.1007/978-3-319-58895-7_2

1 Introduction

Since Crampin et al. (1980), shear-wave splitting (SWS) has been widely observed with stress-aligned polarizations and ~1.5 to ~4.5% shear-wave velocity anisotropy (SWVA) throughout the upper- and lower-crust (Crampin 1994), reviewed by Crampin and Peacock (2008). Similarly, since Ando et al. (1980), SWS has been widely observed with nearly identical stress-aligned polarizations and SWVA, throughout the uppermost ~400 km of the mantle, reviewed by Silver (1996) and Savage (1999).

The presence of SWS indicates some form of seismic anisotropy the effective elastic constants. The only anisotropic symmetry system that has the observed parallel SWS polarizations at a horizontal free-surface is hexagonal symmetry (transverse isotropy) with a horizontal axis of cylindrical symmetry (aka HTI-symmetry) (Crampin 1981; Crampin and Kirkwood 1981), and the only common geological phenomenon with HTI-symmetry is parallel vertical fluid-saturated microcracks. Hence the observed SWS at the surface indicates distributions of stress-aligned vertical fluid-saturated microcracks along almost all ray paths in the crust, and stress-aligned intergranular films of hydrated melt in the uppermost ~400 km of the mantle (Crampin 2003). Figure 1 is a schematic dimensionless illustration of SWS throughout the microcracked crust and upper mantle, where the (realistically imaged) closely spaced microcracks indicate sufficient compliance for non-linear dynamic (NLD) deformation.

Here, we briefly review SWS (and the evidence for extensive NLD deformation) in the Earth's interior, and demonstrate that, possibly uniquely for NLD studies, the progress towards singularities (fracture criticality and earthquakes in geophysics) can be monitored by analysing SWS time delays. We outline several applications of NLD.

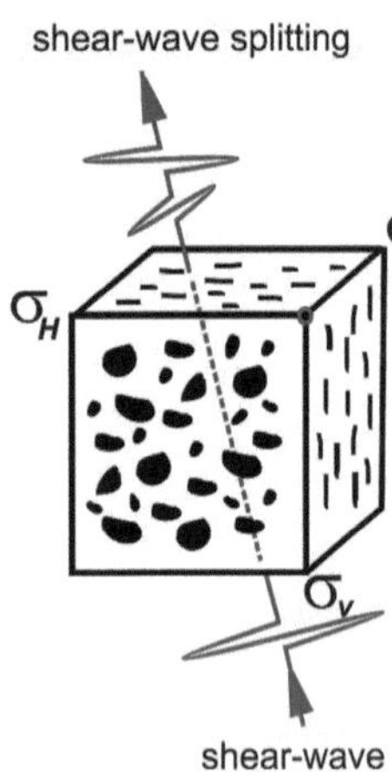

Fig. 1 Schematic illustration of shear-wave splitting on propagation through the fluid-saturated stress-aligned microcracks pervasive in almost all rocks in the Earth's crust (after Crampin 1994)

2 A Review of Shear-Wave Splitting

Azimuthally varying SWS is widely observed throughout the crust and uppermost ~400 km of the mantle (Crampin et al. 1980; Crampin 1994; Silver 1996; Savage 1999; Crampin and Peacock 2008; Crampin and Gao 2013). The SWS is generally interpreted as propagation through distributions of stress-aligned fluid-saturated microcracks in the upper- and lower-crust, where the fluid is typically water (possibly supercritical at depth), and in the upper-mantle, where the fluid is (more controversially) intergranular films of hydrated melt (Crampin 2003; Crampin and Gao 2016). Since crack density $\varepsilon \approx$ %SWVA/100 (Hudson 1981), the range of observed SWVA implies crack densities of $\varepsilon \approx 0.015$–0.045 (Crampin 1994). Percolation theory indicates through-going fluid pathways at $\varepsilon \approx 0.055$ (Crampin and Zatsepin 1997a, b; Crampin 1999). Associating fracture-criticality with the percolation-threshold, if there is any disturbance micro-cracks will fail at fracture-criticality of $\varepsilon \approx 0.055$ (SWVA $\approx 5.5\%$) (Crampin et al. 1999, 2008). Since SWVA of 1.5–4.5% is close to 5.5%, this implies that almost all in situ rocks throughout the crust and upper-mantle of the Earth verge on failure.

Figure 2 is a schematic illustration of crack distributions for a range of crack densities, where the two left-hand images are for observed SWS (Crampin 1994; Crampin and Gao 2013). Since phenomena verging on failure are critical-systems with NLD deformation. Observations of SWS indicate that much of the Earth above ~450 km depth has NLD deformation.

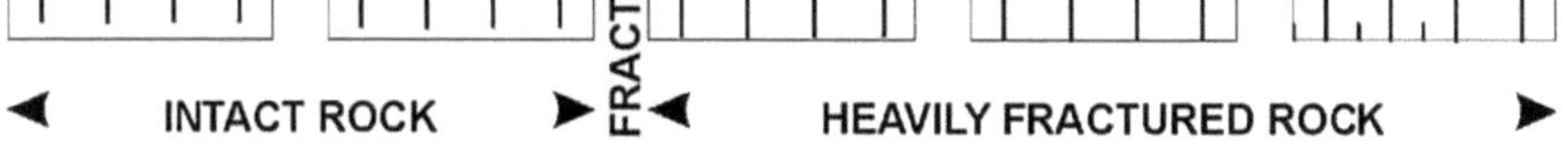

Fig. 2 Schematic (dimensionless) illustration of the observed percentages of shear-wave velocity-anisotropy interpreted as uniform distributions of equal-sized parallel penny-shaped cracks, where ε is crack density, and a is crack radius per unit cube. Images are 2D cross sections of 3D crack distributions. Fracture criticality is at the percolation threshold $\varepsilon = 0.055$ for stress-aligned microcracks, where cracks are so closely spaced they verge on fracturing if there is any disturbance (after Crampin 1994; Crampin and Zatsepin 1997a)

2.1 APE-Modelling

The suggested mechanism for low-level deformation is fluid-displacement by flow or dispersion between intergranular microcracks cracks at different orientations to the stress field in a process known as anisotropic poro-elasticity (APE) illustrated schematically in Fig. 3. Figure 3 shows APE-deformation of a dimensionless initially random distribution stress-aligned *vertical* cracks for four values of differential horizontal stress ($s_H = 0.0, 0.5, 1.0, 3.0$). Hexagons are elastically isotropic (Crampin and Kirkwood 1981) so that, under zero differential (horizontal) stress ($s_H = s_h = 0$) (top left) microcracks have equal aspect ratios (equal crack thicknesses), and the two solid hexagons are a small selection of randomly oriented vertical-microcracks with zero SWVA. Following a small increase in stress to $s_H = 0.5$ (top right), cracks perpendicular to s_H have greater pressure normal to the cracks than cracks parallel to s_H, and fluid migrates by flow or dispersion along the pressure gradients between neighbouring microcracks at different orientations to the stress field. Some cracks swell (increase in aspect-ratio) and some cracks thin, but without crack closure there is no significant SWVA and the overall elasticity is still essentially isotropic. As differential stress increases, there is a critical point

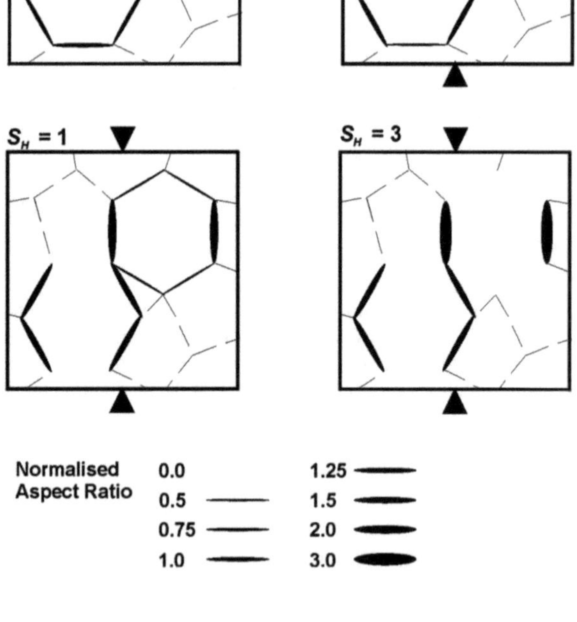

Fig. 3 Schematic (dimensionless, but spatially accurate), illustration of anisotropic poro-elasticity (APE) deformation of fluid-saturated microcracks. There is a more detailed description of the behaviour in Crampin (1999) (after Crampin 1999)

(normalized to $s_H = 1$), when cracks normal to s_H to first begin to close (bottom left), and SWVA jumps from zero to a minimum of approximately ~1.5%SWVA. This theoretical minimum is approximately the same as the minimum SWVA observed in the Earth, imaged in the left-hand diagram in Fig. 2, validating APE-deformation within the Earth (Crampin 1994). As s_H increases to $s_H \approx 3$, cracks begin to align (Fig. 3, bottom right), and at the percolation threshold at $s_H \approx 5.5$ (not modelled), fracture-criticality is reached and the rock fractures if there is any disturbance (Crampin and Zatsepin 1997a, b).

Note that the NLD of APE-deformation as illustrated in Fig. 3 is dimensionless and valid for any 3D distributions of crack and stress geometries. Figure 3 shows APE-deformation illustrated only for vertical cracks because: (1) parallel cracks can be drawn in orthogonal 2D planar diagrams as in Fig. 3; and (2), once below the depth, where increasing vertical stress σ_V equals the minimum horizontal stress σ_h, microcracks in the Earth do tend to be vertical, parallel and normal to σ_h.

The key effect of APE-deformation in the NLD deformation imaged in Fig. 3 is that increasing differential stress increases the aspect ratio (makes cracks swell) of cracks aligned perpendicular to the direction of minimum tectonic stress s_h. The NLD deformation of stress-accumulation before earthquakes, say, can be monitored by measuring the increasing *average* SWS time-delay in Band-1 directions in the shear-wave window (Crampin 1999; Crampin et al. 2008; Crampin and Gao 2016). The shear-wave window and Band-1 and Band-2 directions are described in Fig. 5, Appendix 1.

2.2 Measuring SWS Parameters on Seismograms

Various techniques have been used to automatically measure the polarizations (Φ) and time-delays (*dt*) of SWS on shear-wave seismograms. The results are generally unsatisfactory. The problem is that SWS, although simple in principle, in practice may be extremely complicated for seismograms in the crust, and it is difficult to assess the reliability of fully automated techniques (Crampin 2006). The preferred technique uses the semi-automatic Shear-Wave Analysis System (SWAS) for measuring shear-wave splitting parameters, where reliability of the results for each arrival displayed in 2D polarization diagrams is easy to visually assess (Hao et al. 2008). We show a typical example in Fig. 4 where SWAS is applied to a small $M = 1.43$ earthquake in SW Iceland.

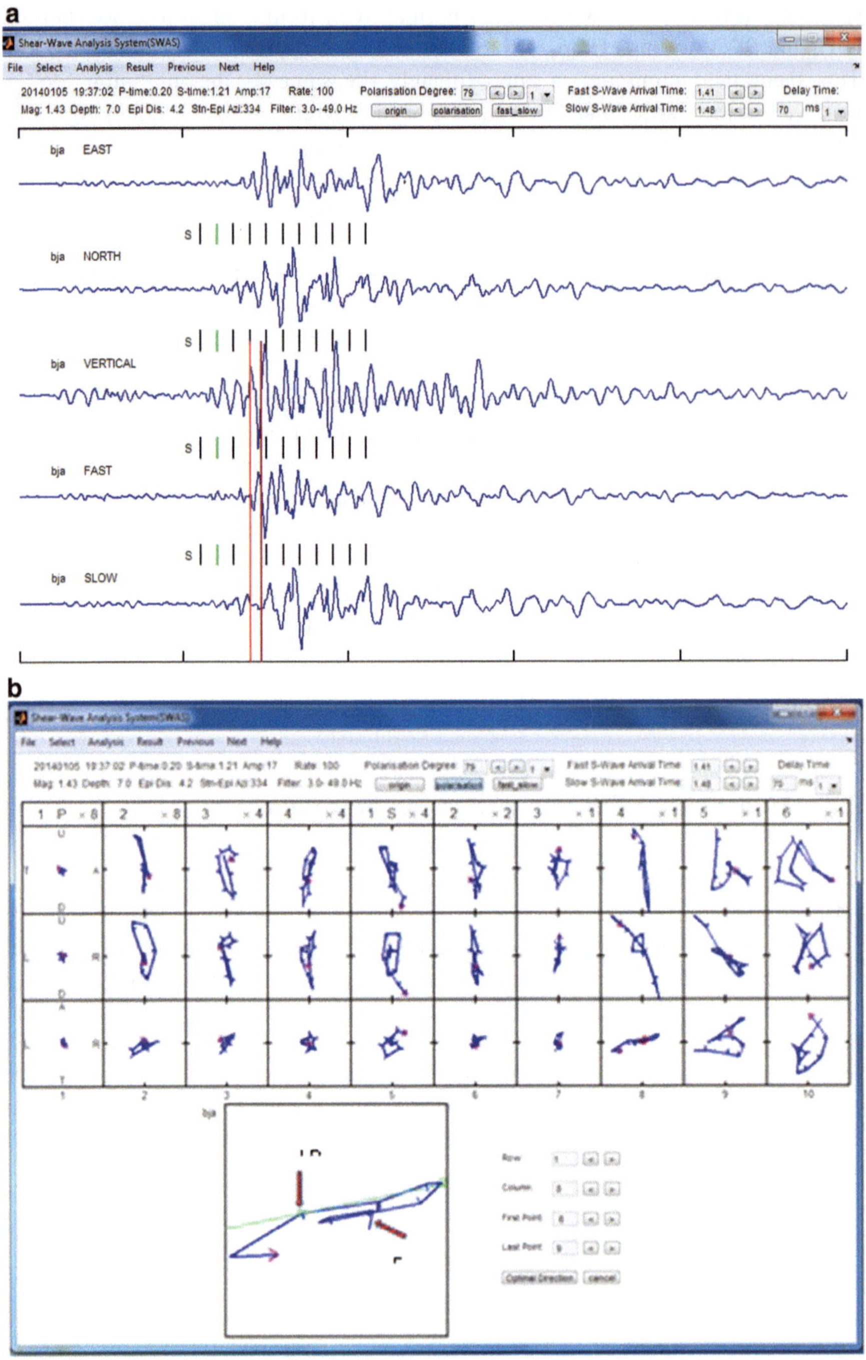

Fig. 4 (continued)

2.3 Shear-Wave Splitting (SWS) and Non-Linear Dynamic Deformation in the Earth

SWS is widely observed throughout the crust and uppermost ~400 km of the mantle with the configuration of Fig. 2 (Crampin and Peacock 2008) with the implied deformation of Fig. 3, which is clearly NLD deformation if there is any disturbance. This means that NLD deformation is prevalent throughout the crust and upper mantle of the Earth.

3 The New Geophysics

The NLD effects of APE-deformation are widely observed in extensive observations if SWS and imply a *New Geophysics* of wave propagation in critically microcracked rock with effects and properties that cannot be matched in conventional sub-critical geophysics without innumerable special cases (Crampin and Gao 2013). Table 1 lists observations of 19 different phenomena some along thousands to millions of shear-wave ray-paths (in industrial seismic-exploration), which match the effects of APE and support the critical New Geophysics. The list could easily be extended. This means that NLD deformation and New Geophysics are established throughout the crust and uppermost ~400 km of the mantle. There are no contrary observations known to us.

Seismic-wave propagation in such NLD critical-systems is necessarily different from wave propagation in conventional sub-critical geophysics and imposes the range of fundamentally new properties on sub-critical geophysics listed in Table 2. These new properties allow several new applications.

NLD is confirmed by the new properties of the *New Geophysics* of critical-systems of distributions of stress-aligned fluid-saturated microcracks that are not available in conventional sub-critical geophysics. With one exception the new properties in Table 2 have all been recognized in the Earth in some cases, thousands to millions of times, in oil company reflection seismics and other circumstances; however, these ideas are innovative and remain controversial (the Preamble is apposite). The exception, controllability, has not yet been tested. Observations of these prop-

◄

Fig. 4 Example of SWS measurements using SWAS (Hao et al. 2008). (**a**) Seismograms of small $M = 1.43$ earthquake recorded at station BJA in SW Iceland in recording geometry East, North, Vertical and rotated into orthogonal Fast and Slow shear-wave polarizations. *Red lines* mark the Fast and Slow SWS arrivals—note 'quiet' segment (the SWS time-delay, *dt*) between the *red lines* in the slow direction. (**b**) 3D polarization diagrams in the time intervals in (**a**) in directions (U)p, (D)own, and (T)owards, (A)way, and (L)eft and (R)ight from the source, with enlarged diagram showing semi-automated picks of fast (F) and slow (S) arrivals, where for illustration they are identified by manually positioned *arrows*. The Fast polarization is automatically picked in *green*

Table 1 Evidence for NLD supporting APE-deformation and the critical New Geophysics (Crampin and Gao 2013)

Evidence inexplicable in terms of conventional sub-critical geophysics[a]	
(1)	Shear-wave splitting is observed in almost all in situ rocks in the crust and upper mantle (Crampin 1994, 1999; Crampin and Peacock 2008; Crampin and Zatsepin 1997a)
(2)	There is a minimum shear-wave velocity anisotropy (SWVA) of $\sim$1.5% in almost all in situ rocks (Crampin 1994, 1999; Crampin and Peacock 2008; Crampin and Zatsepin 1997a)
(3)	There is a maximum SWVA of $\sim$5.5% in ostensibly unfractured rock (Crampin 1994, 1999; Crampin and Peacock 2008; Crampin and Zatsepin 1997a)
(4)	Fracture-criticality limit of SWVA is $\sim$5.5% in in situ rocks independent of rock-type, geology, tectonics, and porosity, etc., where SWVA of $\sim$5.5% is the percolation threshold for parallel cracks (Crampin 1994, 1999; Crampin and Peacock 2008; Crampin and Zatsepin 1997a)
(5)	High pore-fluid pressures induce 90°-flips in polarizations of the faster split shear-waves (Angerer et al. 2002; Crampin et al. 2002)
(6)	Explains the large ('±80%') scatter in shear-wave time-delays above small earthquakes (Crampin et al. 2002; Crampin et al. 2004b)
(7)	Effects of CO_2-injections on seismic reflection surveys modelled by APE (Angerer et al. 2002; Crampin et al. 2002; Crampin et al. 2004b)
(8)	Stress-accumulation observed before earthquakes (Volti and Crampin 2003b; Crampin et al. 1999; Crampin et al. 2008)
(9)	Time, magnitude and impending fault-break successfully stress-forecast in real time (Crampin et al. 1999; Crampin et al. 2008)
(10)	Stress-relaxation (crack-coalescence) observed before earthquakes (Crampin and Peacock 2008; Gao and Crampin 2004)
(11)	Stress-accumulation observed before volcanic eruptions (Crampin and Peacock 2008; Volti and Crampin 2003b; Crampin et al. 1999)
(12)	Extreme sensitivity: stress-variations observed in Iceland two and a half years before the Sumatra-Andaman EQ at the width of the Eurasian Plate from Indonesia (Crampin and Gao 2012)
(13)	Explains how a stressed rock differs from an unstressed rock
(14)	Explains how the enormous stress-energy before a large earthquake accumulates without inducing smaller earthquakes (Crampin et al. 2013)
(15)	Explains why initial stress drop at an earthquake is small (typically 2–4 MPa) and independent of earthquake magnitudes which may vary by over ten orders of magnitude (Crampin et al. 2013)
(16)	Explains how irregular fault-planes slip when constrained by enormous lithostatic stress (Crampin et al. 2013)
(17)	Explains why we cannot deterministically predict but can stress-forecast the time, magnitude and fault-break of impending earthquakes (Crampin et al. 2013)
(18)	Explains why the Gutenberg and Richter (1956) relationship between logarithms of cumulative frequencies of earthquakes and earthquake magnitudes is linear (Crampin et al. 2013)
(19)	Partly explains why, despite huge investments, average recovery is less than 40% of in-place oil (Crampin 2006)

[a]Without innumerable special cases

Table 2 Properties of *New Geophysics* of NLD in compliant critically microcracked in situ rock (after Crampin and Gao (2013))

(1)	Self-similarity:	Logarithmic plots of many properties are linear, such as Gutenberg-Richter relationship (Gutenberg and Richter 1956; Gao and Crampin 2004)
(2)	Monitorability:	Behaviour can be monitored with SWS (Crampin 1999; Crampin and Peacock 2005; Crampin and Peacock 2008)
(3)	Uniformity:	Statistical behaviour is more like other critical systems than it is to the underlying sub-critical physics (Crampin and Peacock 2005; Crampin and Peacock 2008)
(4)	Calculability:	Behaviour is more uniform (universal) than sub-critical behaviour and can be modelled or calculated with the equations of Anisotropic Poro-Elasticity, APE (Crampin and Peacock 2008; Angerer et al. 2002; Crampin and Zatsepin 1997a, b)
(5)	Predictability:	If impending changes can be quantified, behaviour can be predicted by APE (Crampin and Peacock 2008; Angerer et al. 2002; Crampin and Zatsepin 1997a, b)
(6)	Controllability:	If conditions can be monitored (Item 2), calculated (Item 4) and modified by injection pressures (Crampin and Peacock 2008; Angerer et al. 2002), in principle the behaviour of the in situ rock mass can be controlled by feedback (optimizing flow-directions by fluid-injection, say, in hydrocarbon production)
(7)	Universality:	Effects pervade all available space (Crampin and Gao 2013; Volti and Crampin 2003a, b; Crampin and Gao 2012)
(8)	Sensitivity:	Butterfly wings effect sensitivity to miniscule differences in initial conditions (Volti and Crampin 2003a, b; Crampin and Gao 2012)

erties of SWS allow us to monitor the approach of singularities in the interior of the medium to be monitored in detail. This property is believed to be unique in NLD and leads to several important applications. Three definitive applications, verifying NLD and New Geophysics, are briefly outlined in the Appendices: Appendix 2—stress-forecasting (predicting) earthquakes; Appendix 3—stress-forecasting (predicting) volcanic episodes; and Appendix 4—determining the response of a reservoir to critical and sub-critical fluid CO_2-injections (hydraulic fracking). These appendices are brief summaries, and more complete discussions are in the references.

4 Conclusions

We have shown that seismic shear-wave splitting monitors the non-linear dynamic deformation of the crust and uppermost ~400 km of the mantle. Deformation is by fluid movement by flow or dispersion between critical-systems of neighbouring 'microcracks' in the crust, and intergranular films of hydrolyzed melt in the mantle at different orientations to the stress-field. This can be imaged by anisotropic poro-elastic (APE) deformation of the stress-aligned fluid-saturated microcracks pervading almost all in situ rock. This behaviour has been classed as a New Geophysics and imposes many fundamentally new properties on conventional sub-critical geophysics and opens new applications for geophysics.

Table 3 lists proven and potential applications of NLD deformation. These ideas are controversial (see Preamble) and difficult get funded; hence, only three applications have currently been explored: stress-forecasting/predicting earthquakes (Appendix 2); stress-forecasting/predicting volcanic eruptions/episodes (Appendix 3); and modelling, calculating, and predicting the effects of fluid-injections (aka hydraulic fracking) (Appendix 4). Table 3 also lists other potential industrial and societal applications.

We conclude that the NLD deformation of New Geophysics exists throughout the crust and uppermost ~400 km of the mantle and has a number of possibly important proven and potential applications.

Table 3 Potential applications of NLD and SWS in the New Geophysics

Proven applications
(1) Using SWS for stress-forecasting/predicting impending earthquakes (Appendix 2; Crampin and Gao 2013)
(2) Using SWS for stress-forecasting/predicting impending volcanic eruptions (Appendix 2; Volti and Crampin 2003b)
(3) Using APE to monitor, calculate, predict hydraulic injection (hydraulic fracking) (Appendix 3; Angerer et al. 2002)
Potential applications in hydrocarbon recovery
(4) SWI: monitoring production with time-lapse seismics in Single-Well Imaging (Crampin and Gao 2013)
(5) SMORE: Slower production for More Oil REcovery (Crampin and Gao 2013)
(6) OWF: optimizing water-flooding with APE-modelling and feedback (Crampin and Gao 2013)
Other potential applications
(7) Monitoring seismic security of vulnerable locations: cities, nuclear power stations, dams, etc. (Crampin and Gao 2013)
(8) Monitoring, calculating and predicting the response of a reservoir to CO_2 sequestration (Crampin and Gao 2013)
(9) Monitoring nuclear waste deposition with an adjacent borehole stress-monitoring site (SMSs) (Crampin and Gao 2013)
(10) Monitoring rock failure in mining, landsides, tunnelling, etc. by adjacent SMSs (Crampin and Gao 2013)

Acknowledgements The authors thank Sheila Peacock and Peter Leary for their comments. Yuan Gao was partially supported by the National Natural Science Foundation of China, Project 41174042. We thank the Director of Science and Technology of the British Geological Survey (NERC) for approval to publish this paper.

Appendix 1: Ray-Path Geometry for Observing Undisturbed Shear Waves and SWS at the Shear-Wave Window at a Horizontal Free-Surface, and Identification of Band-1 and Band-2 Directions in Distributions of Parallel Vertical Fluid-Saturated Microcracks

Figure 5 shows ray-path geometry for observing undisturbed waveforms of *SV*-waves and SWS in stress-aligned fluid-saturated microcracks in the effective shear-wave window at a horizontal free-surface. (The wave-forms of *SH*-waves are preserved for all angles of incidence at a horizontal free-surface.) ABSCD is the crack-plane through distributions of parallel-vertical microcracks, and S is the recorder on a horizontal free-surface. The exact shear-wave window in an isotropic half-space is ray paths within the solid angle subtending $\sin^{-1}(Vs/Vp) \approx 35°$ marking the critical angle for *Vp* reflection (Booth and Crampin 1985). The effective

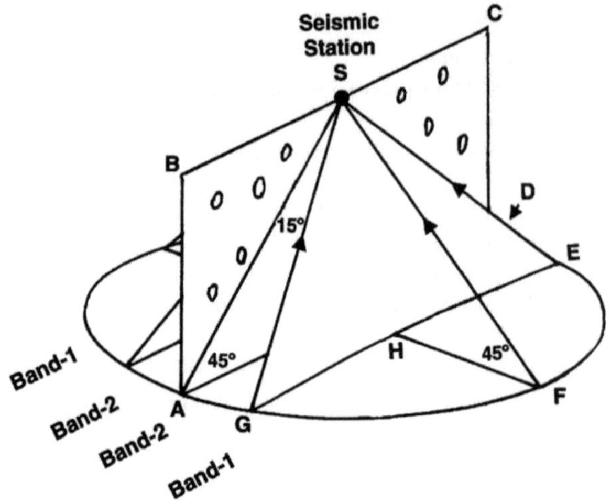

Fig. 5 Ray-path geometry for observing undisturbed waveforms of shear-waves and SWS in stress-aligned fluid-saturated microcracks in the shear-wave window at a horizontal free-surface (after Crampin and Gao 2013)

shear-wave window is ray paths within the solid angle AGFED-to-S and similar ray paths reflected in the crack-plane. However, near-surface low-velocity layers in the Earth bend rays upwards so that the effective shear-wave window may often be taken as straight-line ray paths out to 45° as in Fig. 5.

Band-1 directions to the free-surface, where time-delays are sensitive to crack aspect-ratio (Crampin 1999; Crampin and Gao 2016), are within the solid angle EFGH-to-S subtending 15–45° to the crack plane within the effective shear-wave window. Band-2 directions to the free-surface, where time-delays are dominated by crack-density (Crampin 1999), are within the solid angle ADEHG-to-S to the crack plane. Both Band-1 and Band-2 directions include equivalent solid-angle directions reflected in the far side of the imaged crack plane (After Crampin and Gao 2013).

Appendix 2: Monitoring NLD Deformation to Stress-Forecast Impending Earthquakes

The effects of changing stress on in situ rocks can be monitored by SWS imaging NLD changes in microcrack geometry (Crampin 1999; Crampin and Peacock 2008; Crampin and Gao 2016). Observations of SWS indicate that increases of stress in the Earth (typically originate from magma generation and subduction, and interactions at the margins of tectonic plates) can be monitored by measuring changes in SWS. Initially, such NLD stress-accumulation is widespread throughout tectonic plates and the stress-field does not initially identify the fault-planes where the stress will eventually be released by slippage in earthquakes. The accumulating stress modifies crack aspect ratios throughout the stressed rock-mass, until the microcrack geometry approaches levels of fracture-criticality (Crampin 1999). Only then does the stress-field concentrate on envelopes of weakness surrounding the impending fault-planes, and stress-relaxation occurs as microcracks coalesce onto the impending fault break in NLD deformation (Gao and Crampin 2004; Wu et al. 2006; Crampin and Peacock 2008; Crampin and Gao 2013; Crampin et al. 2013).

The Earth is highly heterogeneous and stress accumulates irregularly. If stress accumulates over a small rock volume, the increase will be rapid but the eventual earthquake will be small. If stress accumulates over a larger volume, the increase will be slower but the eventual earthquake will be larger. Consequently, durations of the changes and the magnitudes possess self-similarity, so that monitoring NLD changes in the surrounding rock mass allows the time, magnitude, and in some cases fault break, of the impending earthquake to be stress-forecast. Note that we refer to this phenomenon as *earthquake stress-forecasting*, rather than *earthquake forecasting* or *earthquake prediction*, to emphasize the different methodology.

New Geophysics demonstrates that stress-accumulation before earthquakes can be recognized by increasing *average* SWS time-delays in *Band-1* directions in the shear-wave window (Fig. 5), and corresponding decreases in *average* SWS time-delays for stress-relaxation (Crampin 1999; Crampin and Peacock 2008; Crampin and Gao 2013, 2016). NLD stress-accumulation was first positively identified in changes in SWS before a M 5 earthquake in Iceland with similar changes in SWS to those before a M 5.1 earthquake 6 months earlier (Fig. 6). A stress-forecast was emailed (10th Nov., 1998) to the Iceland Meteorological Office (IMO) '... an event could occur any time between now ($M \geq 5$) and the end of February ($M \geq 6$)' on a specified fault with continuing seismic activity. Three days later (13th Nov., 1998), a $M = 5$ earthquake occurred on the identified fault (Crampin et al. 1999, 2004a, 2008). We claim this as the first successful scientifically stress-forecast/predicted earthquake, as opposed to less-specific probabilistic estimates. Similar characteristic variations are seen retrospectively before 16 earthquakes elsewhere (Crampin and Gao 2015).

Later, it was recognized that the observed stress-accumulation stops abruptly before the impending earthquake occurs. There is stress-relaxation, average time-delays decrease, and the earthquake occurs at a comparatively low value of implied stress (Gao and Crampin 2004). Figure 7 shows stress-accumulation and stress-relaxation, before six earthquakes (and two laboratory experiments) ranging in magnitude from M 6 to M 1.7, in a normalized format convenient for displaying such characteristic changes. The successfully stress-forecast earthquake is Fig. 7c. All six earthquakes show similar behaviour despite orders of magnitude differences in released energy and durations of stress-accumulation ranging from 6 years to a few hours.

Logarithms of the durations of both the stress-accumulations and the stress-relaxations are both linear (self-similar) with the impending magnitudes (Crampin et al. 1999, 2008; Gao and Crampin 2004; Crampin and Peacock 2008). Stress-relaxation is interpreted as microcracks coalescing onto the impending fault-plane (Gao and Crampin 2004; Wu et al. 2006). Characteristic patterns of stress-accumulation increases and stress-relaxation (crack-coalescent) decreases have been recognized retrospectively before (currently) 15 earthquakes ranging from a M 1.7

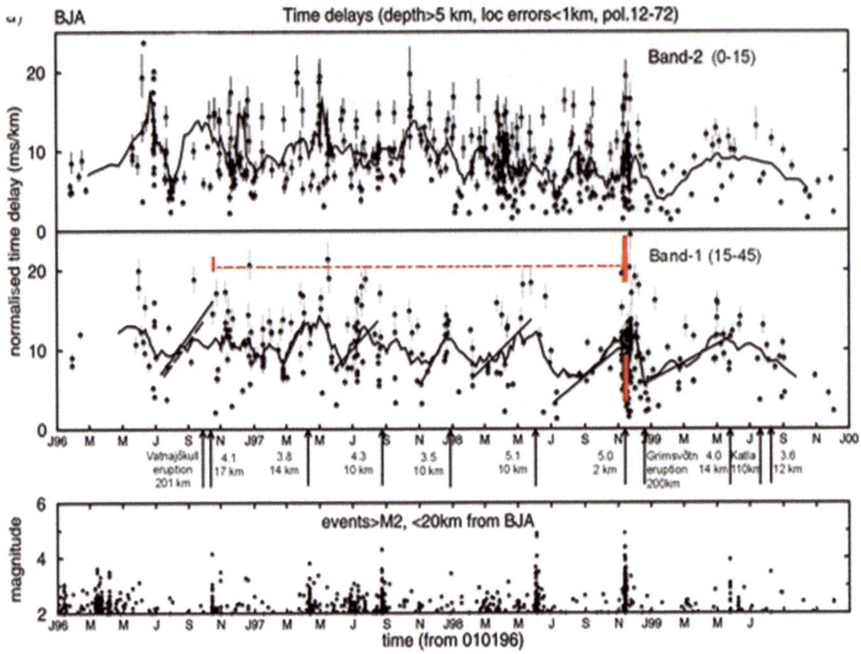

Fig. 6 Variations with time of SWS time-delays normalized to ms/km in Band-1 directions (*middle diagram*) and Band-2 directions (*Top diagram*) for 5 years at station BJA in SW Iceland showing, in Band-1, least-squares increases before larger earthquakes within 20 km of BJA in *lower diagram*. The curves in the time-delays in the *top* and *middle* diagrams are nine-point moving averages. The *red line* (Oct. 1996–Nov. 1998) marks a least-squares average of 2 ms/km/year relaxation interpreted as the Mid-Atlantic Ridge responds to the large Gjàlp, Vatnajökull eruption of Oct. 1996. The *vertical red bar* in Nov. 1998 marks the time of the successfully stress-forecast *M* 5 (Crampin et al. 1999, 2008). (Modified after Volti and Crampin 2003b)

swarm event in Iceland (Crampin et al. 2008) to the *M* 9.2, 2004, Sumatra-Andaman Earthquake (SAE), where changes in SWS were recognized in Iceland at the width of the Eurasian Plate ($\sim$10,500 km) from Indonesia (Crampin and Gao 2012). Before SAE, ten stress-forecasts were emailed to IMO (13th Sept., 2002 to 18th Feb., 2005) updated every few months, warning of an impending large earthquake (Crampin and Gao 2012). At that time the full NLD sensitivity of SWS had not been recognized, and a $M \approx 7$ earthquake in Iceland was stress-forecast. It was only in retrospect that it was recognized that the stress-forecasts were for the SAE (Crampin and Gao 2012).

Stress-forecasting is possible whenever SWS can be routinely monitored. Swarms of small earthquakes are generally far too scarce and irregular for routine monitoring of SWS. Only in Iceland where two transform faults of the Mid-Atlantic Ridge uniquely run onshore in SW Iceland and North-Central Iceland and provide the persistent low-level seismicity necessary for reliable routine stress-forecasting (Volti and Crampin 2003a, b).

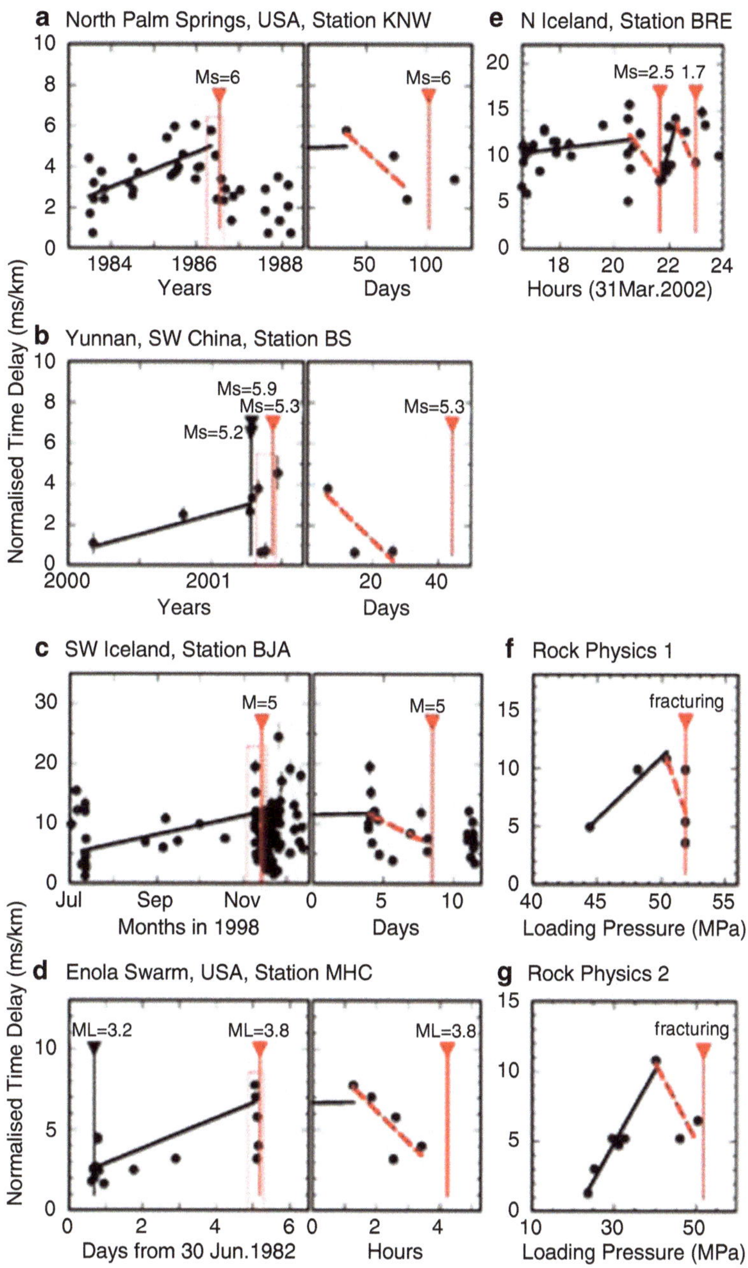

Fig. 7 Examples of stress-accumulation and stress-relaxation in field and in laboratory. Shear-wave time delays normalized to ms/km and plotted against time before six earthquakes ranging in magnitude from $M_S = 6$ to $M = 1.7$ and two laboratory experiments. More complete information is in Gao and Crampin (2004)

Note that New Geophysics implies that earthquakes cannot be predicted by monitoring effects at the source. Earthquakes are singularities which lead to deterministic chaos; thus, although the source effects may on occasions be modelled explicitly, they are essentially unrepeatable as they are likely to depend critically on otherwise negligible (butterfly effect) details of initial conditions. The only mechanism for stress-forecasting/prediction is using SWS to monitor stress-accumulation and stress-relaxation in the rock surrounding the impending earthquake (or volcanic eruption) by the conventional effects of changing stress on microcrack geometry in rocks surrounding the impending source (Crampin 1999; Crampin et al. 2008). The source of the shear-waves may either be the irregular and unreliable swarms of small earthquakes, or controlled-source Stress-Monitoring Sites (SMSs) (Crampin and Gao 2016). SMSs provide a mechanism for routinely monitoring stress accumulating before impending earthquakes and volcanic eruptions so that the earthquake or eruption can be stress-forecast.

Appendix 3: Monitoring NLD Deformation to Stress-Forecast Impending Volcanic Eruptions

Monitoring SWS before impending volcanic eruptions shows similar characteristic NLD deformation behaviour as that seen before earthquakes and can be similarly interpreted as stress-accumulation and stress-relaxation before the event.

Figure 8 compares stress-accumulation and stress-relaxation, in the normalized format of Fig. 7, before (a) the 2010 ash-cloud (flank) eruption of Eyjafjallajökull Volcano in SW Iceland (Liu et al. 1997) with (b) the successfully stress-forecast earthquake in Fig. 7c 90 km to the west (Gao and Crampin 2004). Both events show stress-accumulation increases, of 7 months and 4 months, respectively, and stress-relaxation decreases, of $\sim$40 days and four days, respectively. Considering the very different geophysical processes involved, the NLD behaviour of the variations of SWS time-delays seems remarkably similar and supports the existence of New Geophysics in the reservoir rock.

Appendix 4: Monitoring Fluid Injection (Aka Hydraulic Fracking)

Angerer et al. (2002) use APE to model the response of a cracked carbonate reservoir to critically high-pressure and low-pressure CO_2 injections (hydraulic fracking). Figure 9 shows seismograms of a multi-component 4-D (time-lapse 3-D) 3C reflection survey in Vacuum Field, New Mexico, in 1995, by the Reservoir Characterization Project (RCP), Colorado School of Mines (Roche et al. 1997).

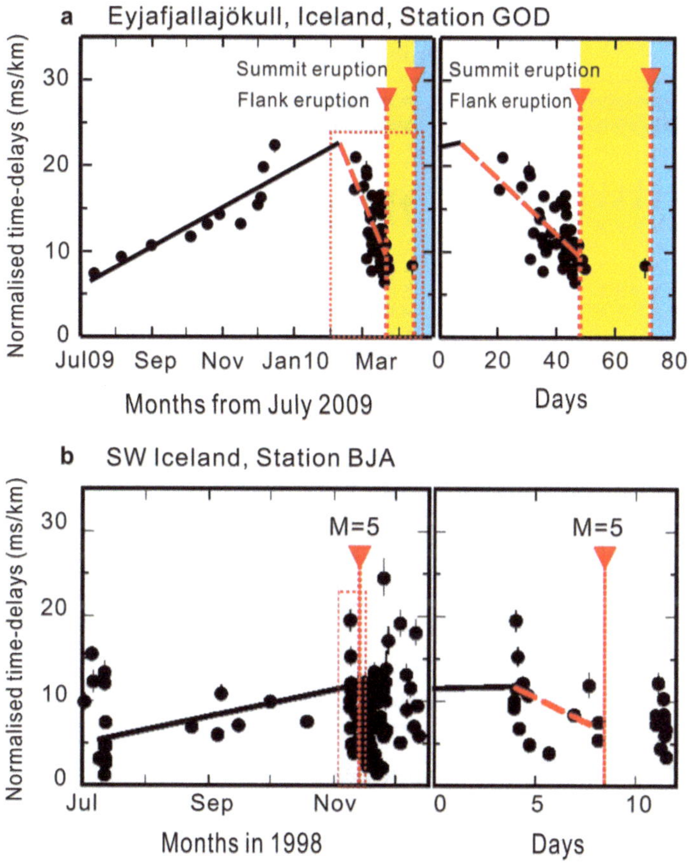

Fig. 8 Comparison of the behaviour of shear-wave splitting before (**a**) a volcanic eruption and before (**b**) an earthquake. The eruption is the ash cloud eruption of Vatnajökull, Iceland, March 2010 and the earthquake is in SW Iceland in Fig. 7c. Both show similar stress-accumulation increases and brief stress-relaxation (crack coalescence) decreases before both eruption and earthquake occurs

The record sections headed *S*1 and *S*2 are in the same orthogonal azimuthal directions. In (a) the pre-CO_2 injection: the arrowed arrivals at the top and bottom of the target zone are at 176 ms for *S*1 and 178 ms *S*2 so that *S*1 is the faster shear wave. In (b) the post-CO_2 injection, the target zone is at 204 ms for the *S*1-direction and 184 ms for *S*2. This means that the high-pore fluid pressure injection is critically high and has induced a 90°-*flip* in the orientation of the faster split shear-wave arrivals for both observed and calculated seismograms for shear waves travelling through the injection zone. Such 90°-flips have since been observed elsewhere in high-pressure reservoirs and near seismically active fault-planes where critically high pore-fluid pressures are encountered on all seismically active fault-planes (Crampin et al. 2002, 2004b).

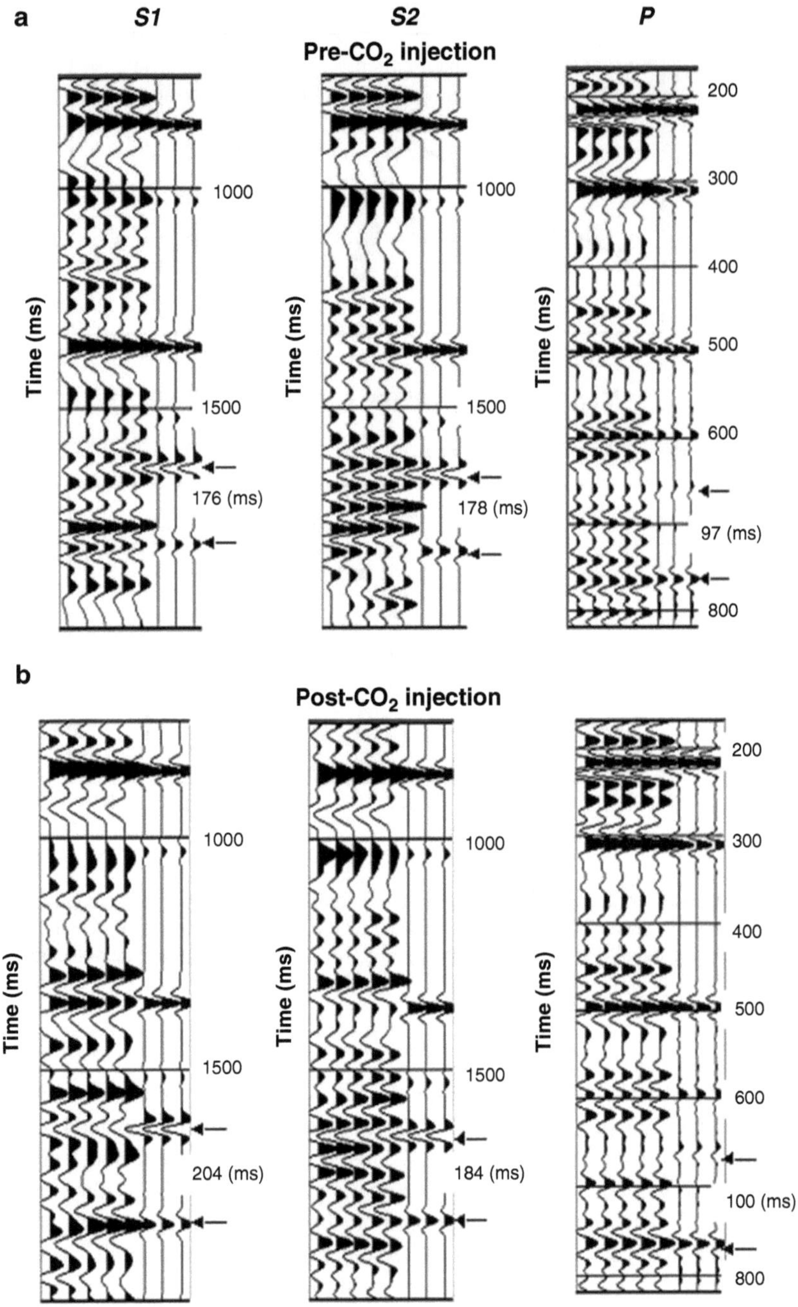

Fig. 9 (continued)

The 90°-flip was not expected, and the match of observations with APE is strong confirmation of the validity of APE and New Geophysics in crustal rock.

References

Ando, M., Y. Ishikawa, and H. Wada. 1980. S-wave anisotropy in the upper mantle under a volcanic area in Japan. *Nature* 286: 43–46.

Angerer, E., S. Crampin, X.-Y. Li, and T.L. Davis. 2002. Processing, modelling, and predicting time-lapse effects of overpressured fluid-injection in a fractured reservoir. *Geophysical Journal International* 149: 267–280.

Booth, D.C., and S. Crampin. 1985. Shear-wave polarizations on a curved wavefront at an isotropi free-surface. *Geophysical Journal of the Royal Astronomical Society* 83: 31–45.

Crampin, S. 1981. A review of wave motion in anisotropic and cracked elastic-media. *Wave Motion* 3: 343–391.

———. 1994. The fracture criticality of crustal rocks. *Geophysical Journal International* 118: 428–438.

———. 1999. Calculable fluid-rock interactions. *Journal of the Geological Society* 156: 501–514.

———. 2003. Aligned cracks not LPO as the cause of mantle anisotropy, EGS-AGU-EUG Joint Ass., Nice, 2003. *Geophysical Research Abstract* 5: 00205; with up-dated notes in online version.

———. 2006. The New Geophysics: a new understanding of fluid-rock deformation. In *Eurock 2006: multiphysics coupling and long term behaviour in rock mechanics*, ed. A. Van Cotthem, R. Charlier, J.-F. Thimus, and J.-P. Tshibangu, 539–544. London: Taylor & Francis.

Crampin, S., R. Evans, B. Üçer, M. Doyle, J.P. Davis, G.V. Yegorkina, and A. Miller. 1980. Observations of dilatancy-induced polarization anomalies and earthquake prediction. *Nature* 286: 874–877.

Crampin, S., and Y. Gao. 2012. Plate-wide deformation before the Sumatra-Andaman earthquake. *Journal of Asian Earth Sciences* 46: 61–19. doi:10.1016/j.jseaes.2011.1015.

———. 2013. The New Geophysics. *Terra Nova* 25: 173–180. doi:10.1111/ter.12030.

———. 2015. The physics underlying Gutenberg-Richter in the Earth and in the Moon. *Journal of Earth Science* 26: 134–139. doi:10.1007/s12583-015-0523-3.

———. 2016. Borehole Stress-Monitoring Sites (SMSs) for monitoring stress accumulation and predicting (stress-forecasting) impending earthquakes and eruptions. In *Workshop on earthquakes in North Iceland: proceedings WENI2 workshop*, Hüsavík Academic Center, Iceland, in press.

Fig. 9 (**a**) Pre-injection waveforms of a multi-component nearly vertical ray reflection survey near the centre of Vacuum Field, New Mexico, carbonate reservoir (Angerer et al. 2002). *S*1-, *S*2-, and *P*-waves are reflection sections with mutually orthogonal polarizations, where the horizontals *S*1, and *S*2, have been rotated into the split shear-wave polarizations parallel and perpendicular to the direction of maximum horizontal stress, respectively. Left-hand (LH) five traces are observed waveforms at adjacent recorders 17 m apart, and the right-hand (RH) three traces are synthetic seismograms modelled by APE to match the shear-wave and SWS arrivals. *Top* and *bottom* of injection zone for shear waves are marked by *arrows* with time-delays in ms/km. (**b**) Post-injection waveforms two-weeks after a high-pressure CO_2-injection (hydraulic fracking). Again, the LH traces are observations and RH traces are synthetic seismograms modelled by APE with the structure from (**a**) and an injection pressure of 6.4 MPa (after Angerer et al. 2002)

Crampin, S., Y. Gao, and A. De Santis. 2013. A few earthquake conundrums resolved. *Journal of Asian Earth Sciences* 62: 501–509. doi:10.1016/j.jseaes.1012.10.036,.
Crampin, S., Y. Gao, and S. Peacock. 2008. Stress-forecasting (not predicting) earthquakes: a paradigm shift. *Geology* 36: 427–430.
Crampin, S., and S.C. Kirkwood. 1981. Velocity variations in systems of anisotropic symmetry. *Journal of Geophysics* 49: 35–42.
Crampin, S., and S. Peacock. 2005. A review of shear-wave splitting in the compliant crack-critical anisotropic Earth. *Wave Motion* 41: 59–77.
———. 2008. A review of the current understanding of shear-wave splitting and common fallacies in interpretation. *Wave Motion* 45: 675–722.
Crampin, S., S. Peacock, Y. Gao, and S. Chastin. 2004b. The scatter of time-delays in shear-wave splitting above small earthquakes. *Geophysical Journal International* 156: 39–44.
Crampin, S., T. Volti, S. Chastin, A. Gudmundsson, and R. Stefánsson. 2002. Indication of high pore fluid pressures in a seismically-active fault zone. *Geophysical Journal International* 151: F1–F5.
Crampin, S., T. Volti, and R. Stefánsson. 1999. A successfully stress-forecast earthquake. *Geophysical Journal International* 138: F1–F5.
———. 2004a. Response to "A statistical evaluation of a 'stress forecast' earthquake" by T. Seher & I. G. Main. *Geophysical Journal International* 157: 194–199.
Crampin, S., and S.V. Zatsepin. 1997a. Changes of strain before earthquakes: the possibility of routine monitoring of both long-term and short-term precursors. *Journal of Physics of the Earth* 45: 1–26.
———. 1997b. Modelling the compliance of crustal rock: II – response to temporal changes before earthquakes. *Geophysical Journal International* 129: 495–506.
Gao, Y., and S. Crampin. 2004. Observations of stress relaxation before earthquakes. *Geophysical Journal International* 157: 578–582.
Gutenberg, B., and C.F. Richter. 1956. Magnitude and energy of earthquakes. *Annali di Geofisica* 9: 1–15.
Hao, P., Y. Gao, and S. Crampin. 2008. An expert system for measuring shear-wave splitting above small earthquakes. *Computeres & Geosciences* 34: 226–234.
Hudson, J.A. 1981. Wave speeds and attenuation of elastic waves in material containing cracks. *Geophysical Journal International* 64: 133–150.
Liu, Y., S. Crampin, and I. Main. 1997. Shear-wave anisotropy: spatial and temporal variations in time delays at Parkfield, Central California. *Geophysical Journal International* 130: 771–785.
Roche, S.L., T.L. Davis, and R.D. Benson. 1997. 4-D, 3-C seismic study at vacuum field, New Mexico. In *66th Annual international SEG meeting*, expanded abstracts, 886–889.
Savage, M.K. 1999. Seismic anisotropy and mantle deformation: what have we learned from shear wave splitting? *Reviews of Geophysics* 37: 65–106.
Silver, P.G. 1996. Seismic anisotropy beneath the continents: probing the depths of geology. *Annual Review of Earth and Planetary Sciences* 24: 385–432.
Volti, T., and S. Crampin. 2003a. A four-year study of shear-wave splitting in Iceland: 1. Background and preliminary analysis. In *New insights into structural interpretation and modelling*, ed. D.A. Nieuwland, vol. 212, 117–133. London: Geological Society, Special Publications.
———. 2003b. A four-year study of shear-wave splitting in Iceland: 2. Temporal changes before earthquakes and volcanic eruptions. In *New insights into structural interpretation and modelling*, ed. D.A. Nieuwland, vol. 212, 135–149. London: Geological Society, Special Publications.
Wu, J., S. Crampin, Y. Gao, P. Hao, and Y.-T. Chen. 2006. Smaller source earthquakes and improved measuring techniques allow the largest earthquakes in Iceland to be stress-forecast (with hindsight). *Geophysical Journal International* 166: 1293–1298.

Stochastic Parameterization of Subgrid-Scale Processes: A Review of Recent Physically Based Approaches

Jonathan Demaeyer and Stéphane Vannitsem

Abstract We review some recent methods of subgrid-scale parameterization used in the context of climate modeling. These methods are developed to take into account (subgrid) processes playing an important role in the correct representation of the atmospheric and climate variability. We illustrate these methods on a simple stochastic triad system relevant for the atmospheric and climate dynamics, and we show in particular that the stability properties of the underlying dynamics of the subgrid processes have a considerable impact on their performances.

Keywords Stochastic parameterization • Response theory • Multiscale system • Homogenization • Subgrid processes

1 Introduction

From a global point of view, the Earth system is composed of a myriad of different interacting components. These components can be regrouped in compartments like the atmosphere, the hydrosphere, the lithosphere, the biosphere, and the cryosphere (Olbers, 2001).[1] Those compartments play a role on different timescales from seconds to ice ages. In this perspective, the resulting Earth's climate is a "concert" at which each compartment seems to play its own partition with its own tempo. Their respective contribution to the total variability of an observable, say, e.g., the global temperature, is, however, the outcome of complex interactions between the different components, leading to an emergent dynamics far from the one that could be generated by a linear additive superposition principle (Nicolis and Nicolis, 1981, 2012).

[1]More recently, a new compartment has appeared, whose effect is not negligible at all and which is not predictable nor descriptive by evolution equations, namely the impact of the human activities.

J. Demaeyer (✉) • S. Vannitsem
Institut Royal Météorologique de Belgique, Avenue Circulaire, 3, 1180 Brussels, Belgium
e-mail: Jonathan.Demaeyer@meteo.be; Stephane.Vannitsem@meteo.be

A.A. Tsonis (ed.), *Advances in Nonlinear Geosciences*,
DOI 10.1007/978-3-319-58895-7_3

A paradigmatic example is provided in the work of Hasselmann, detailed in his seminal paper of 1976, which states precisely that the slowly evolving components of the climate system, besides their own dynamics due their own physical processes, also integrate the impact of the faster components (Hasselmann, 1976). Hasselmann describes this process using the analogy of the Brownian motion where a macro-particle in a liquid integrates the effect of the collisions with the fluid's micro-particles, leading to the erratic trajectory of the former. The interest of this framework is that it provides a natural description of the "red noise" spectral density observed in most climatic records and observations (Ghil et al., 2002; Lovejoy and Schertzer, 2013). Subsequently, during the following decade, stochastic modeling for meteorology and climatology became an important research topic (Frankignoul, 1979; Frankignoul and Hasselmann, 1977; Frankignoul and Müller, 1979; Lemke, 1977; Lemke et al., 1980; Nicolis, 1981, 1982; Nicolis and Nicolis, 1981; Penland, 1989) before falling into disuse in what has been described as a "lull" of work in this field (Arnold et al., 2003). However, during that period, the ideas that correct parameterizations of subgrid processes are important to improve climate and weather models gained popularity (Newman et al., 1997; Penland, 1996; Penland and Matrosova, 1994). Stochastic parameterizations for the "turbulent" closure in 2-D large-eddy simulations on the sphere have also been considered (Frederiksen, 1999; Frederiksen and Davies, 1997). It led recently to the implementation of stochastic schemes to correct the model errors (Nicolis, 2003, 2004) made in large numerical weather prediction (NWP) models (Buizza et al., 1999; Shutts, 2005), improving the reliability of probabilistic forecasts and correcting partially their variability (Doblas-Reyes et al., 2009; Nicolis, 2005). The relation between multiplicative noise and the non-Gaussian character observed in some geophysical variables has also been considered (Sardeshmukh and Penland, 2015; Sura et al., 2005), as well as stochastic models for the climate extremes (Sura, 2013).

Since the beginning of the twenty-first century, a revival of the interest in stochastic parameterization methods have occurred, due to the availability of new mathematical methods to perform the stochastic reduction of ODEs systems. Almost simultaneously, a rigorous mathematical framework for the Hasselmann "program" was devised (Arnold, 2001; Kifer, 2001, 2003) and a new method based on the singular perturbation theory of Markov processes (Majda et al., 2001) was proposed. The latter approach is currently known as the Majda–Timofeyev–Vanden-Eijnden (MTV) method. Both methods have been tested and implemented successfully in geophysical models (Arnold et al., 2003; Culina et al., 2011; Franzke et al., 2005; Vannitsem, 2014). The revival of the Hasselmann program has also stressed the need to consider the occurrence of very rare events triggered by the noise that allow for the solutions of the system to jump from one local attractor to another one (Arnold, 2001). Such events display recurrence timescales that are few orders greater than the timescale of the climate variables considered, and thus induce transitions between different climatic states. The statistics of these transitions is then given by the so-called *Large Deviations* theory (Freidlin and Wentzell, 1984) [for recent developments on this matter, see Bouchet et al. (2016)]. In addition to these two methods, the modeling of the effects of subgrid scale through conditional

Markov chain has been considered (Crommelin and Vanden-Eijnden, 2008) and recently, new stochastic parameterization techniques have been proposed, based on an expansion of the backward Kolmogorov equation (Abramov, 2015) and on the Ruelle response theory (Wouters and Lucarini, 2012). The latter has been tested on a simple coupled ocean-atmosphere model (Demaeyer and Vannitsem, 2016), on stochastic triads (Wouters et al., 2016), and on an adapted version of the Lorenz'96 model (Vissio and Lucarini, 2016).

This renewal of interest for stochastic modeling and reduction methods illustrates how fruitful was the original idea of Hasselmann. However, in view of the availability of several possible approaches, one might wonder about their efficiency in different situations. Indeed, depending on the specific purpose that it needs to fulfill, some parameterizations might perform better than others. The present review aims to shed some light on these questions and to illustrate some of the aforementioned parameterization methods on a simple model for which most of the calculations can be made analytically.

In Sect. 2, we will present the general framework in which the problem of stochastic parameterizations is posed. In Sect. 3, we present the different parameterizations that we shall consider for the analysis model. The stochastic triad model used here and the comparison are presented in Sect. 4. Finally, the conclusions are given in Sect. 5.

2 The Parameterization Problem

Consider the following system of ordinary differential equations (ODEs):

$$\dot{Z} = T(Z) \tag{1}$$

where $Z \in \mathscr{R}^d$ is a set of variables relevant for the problem under interest for which the tendencies $T(Z)$ are known. And suppose that one wants to separate this set of variables into two different subset $Z = (X, Y)$, with $X \in \mathscr{R}^m$ and $Y \in \mathscr{R}^n$. In general, such a decomposition is made such that the subset X and Y have strongly differing *response times* $\tau_Y \ll \tau_X$ (Arnold et al., 2003), but we will assume here that this constraint is not necessarily met. System (1) can then be expressed as:

$$\begin{cases} \dot{X} = F(X, Y) \\ \dot{Y} = H(X, Y) \end{cases} \tag{2}$$

The timescale of the X sub-system is typically (but not always) longer than the one of the Y sub-system, and it is often materialized by a parameter $\delta = \tau_Y/\tau_X \ll 1$ in front of the time derivative $\dot{Y}$. The X and the Y variables represent, respectively, the resolved and the unresolved sub-systems. The resolved sub-system is the part of the full system that we would like to simulate, i.e., generate explicitly and numerically its time-evolution. The general problem of model reduction consists thus to approximate the resolved component X as accurately as possible by obtaining a closed equation for the system X alone (Arnold et al., 2003). The term "accurately" here can have several meanings, depending on the kind of problem to solve. For instance, we can ask that the closed system for X has statistics that are very close to the ones of the X component of system (2). We can also ask that the closed system trajectories remain as close as possible to the trajectories of the full system for long times.

In general a parameterization of the sub-system Y is a relation Ξ between the two sub-systems:

$$Y = \Xi(X, t) \tag{3}$$

which allows to effectively close the equations for the sub-system X while retaining the effect of the coupling to the Y sub-system.

The problem of the model reduction is not new, and was considered first in celestial mechanics. Famous mathematicians have considered it and contributed to what is known nowadays as the theory of averaging (Sanders and Verhulst, 1985) and which forms the first step of the Hasselmann program (Arnold, 2001). The mathematical framework was set in the 1960s by the influential contribution of Bogoliubov and Mitropolski (1961). However, this averaging technique is a deterministic method which does not take into account the deviations from the average. The proposition of Hasselmann was thus to take into account these deviations by considering stochastic parameterization where the relation (3) can be considered in a statistical sense. In that framework, the Y sub-system and its effect on the sub-system X can be considered as a stochastic process, which possibly depends upon the state of the X sub-system. Different methods to achieve this program are now discussed in Sect. 3.

3 The Parameterization Methods

Let us now write system (2) as:

$$\begin{cases} \dot{X} = F_X(X) + \Psi_X(X, Y) \\ \dot{Y} = F_Y(Y) + \Psi_Y(X, Y) \end{cases} \tag{4}$$

where the coupling and the intrinsic dynamics are explicitly specified. In the present work, we shall focus on parameterizations that are defined in terms of stochastic processes. We will consider methods based

- on the Ruelle response theory (Demaeyer and Vannitsem, 2016; Wouters and Lucarini, 2012; Wouters et al., 2016).
- on the singular perturbation theory of Markov processes (Franzke et al., 2005; Majda et al., 2001).

- on the Hasselmann averaging methods (Arnold et al., 2003; Culina et al., 2011; Kifer, 2003; Vannitsem, 2014).
- on empirical methods (Arnold et al., 2013).

All these parameterizations can be written in the following form:

$$\dot{X} = F_X(X) + G(X,t) + \sigma(X) \cdot \tilde{\xi}(t) \tag{5}$$

where the matrix σ, the deterministic function G, and the random processes $\tilde{\xi}(t)$ have to be determined. The mathematical definition of these quantities obtained through averaging procedure and the measure being used to perform the averaging are usually both differing between the methods. These different choices are rooted in their different underlying hypothesis, as it will be discussed below. Specifically, the response theory method uses the measure of the uncoupled unresolved sub-system $\dot{Y} = F_Y(Y)$, the singular perturbation method uses the measure of the perturbation, and the averaging methods use the measure of the full unresolved sub-system $\dot{Y} = H(X, Y)$ with X considered as "frozen" (constant). Finally, the empirical methods use in general the output of the full unresolved Y sub-system, conditional or unconditional on the state X.

In the rest of the section, we shall describe more precisely each of the above methods.

3.1 The Method Based on Response Theory

This method is based on the Ruelle response theory (Ruelle, 1997, 2009) and was proposed by Wouters and Lucarini (2012, 2013). In this context, system (4) must be considered as two intrinsic sub-dynamics for X and Y that are weakly coupled. The response theory quantifies the contribution of the "perturbation" Ψ_X, Ψ_Y to the invariant measure[2] $\tilde{\rho}$ of the fully coupled system (4) as:

$$\tilde{\rho} = \rho_0 + \delta_\Psi \rho^{(1)} + \delta_{\Psi,\Psi} \rho^{(2)} + O(\Psi^3) \tag{6}$$

where ρ_0 is the invariant measure of the uncoupled system which is also supposed to be an existing, well-defined SRB measure. As shown in Wouters and Lucarini (2012), this theory gives the framework to parameterize the effect of the coupling on the component X. The parameterization is based on three different terms having a response similar, up to order two, to the couplings Ψ_X and Ψ_Y:

$$\dot{X} = F_X(X) + M_1(X) + M_2(X,t) + M_3(X,t) \tag{7}$$

[2] The theory assumes that for the system under consideration, a SRB measure (Young, 2002) exists (e.g., an Axiom-A system).

where

$$M_1(X) = \left\langle \Psi_X(X,Y) \right\rangle_{\rho_{0,Y}} \tag{8}$$

is an averaging term. $\rho_{0,Y}$ is the measure of the uncoupled system $\dot{Y} = F_Y(Y)$. The term $M_2(X,t) = \sigma_R(X,t)$ is a correlation term:

$$\left\langle \sigma_R(X,t) \otimes \sigma_R(X,t+s) \right\rangle = \mathrm{g}(X,s) = \left\langle \Psi'_X(X,Y) \otimes \Psi'_X\big(\phi^s_X(X), \phi^s_Y(Y)\big) \right\rangle_{\rho_{0,Y}} \tag{9}$$

where $\otimes$ is the outer product, $\Psi'_X(X,Y) = \Psi_X(X,Y) - M_1(X)$ is the centered perturbation, and ϕ^s_X, ϕ^s_Y. There are two flows and two systems $\dot{X} = F_X(X)$ and $\dot{Y} = F_Y(Y)$. The process σ_R is thus obtained by taking the square root of the matrix g, which is here accomplished by a decomposition of Ψ'_X on a proper basis (Wouters and Lucarini, 2012). The M_3 term is a memory term:

$$M_3(X,t) = \int_0^\infty \mathrm{d}s\, h(X(t-s), s). \tag{10}$$

involving the memory kernel

$$h(X,s) = \left\langle \Psi_Y(X,Y) \cdot \nabla_Y \Psi_X\big(\phi^s_X(X), \phi^s_Y(Y)\big) \right\rangle_{\rho_{0,Y}} \tag{11}$$

All the averages are thus taken with $\rho_{0,Y}$, the invariant measure of the unperturbed system $\dot{Y} = F_Y(Y)$. This particular choice of the measure is due to the perturbative nature of the method and simplifies the averaging procedure in many cases. The terms M_1, M_2 and M_3, are derived (Wouters and Lucarini, 2012) such that their responses up to order two match the response of the perturbation Ψ_X and Ψ_Y. Consequently, this ensures that for a *weak coupling*, the response of the parameterization (7) on the observables will be approximately the same as the coupling.

The advantages of this simplified averaging procedure (by using $\rho_{0,Y}$) should be tempered by the additional cost induced by the computation of the memory term, the latter implying that this parameterization is a non-Markovian one (Chekroun et al., 2015). However, the integral (10) in this memory term must only be evaluated from 0 up to the timescale τ_Y of the fast variable, due to the exponential decrease of the integrand. Moreover, in some cases, this non-Markovian parameterization can be effectively replaced by a Markovian one (Wouters et al., 2016).

3.2 Singular Perturbation Theory Method

Singular perturbation methods were developed in the 1970s for the analysis of the linear Boltzmann equation in some asymptotic limit (Ellis et al., 1975; Grad, 1969; Majda et al., 2001; Papanicolaou, 1976). Here, these methods are applicable if the problem can be cast into a Fokker–Planck equation. The procedure described in Majda et al. (2001) requires assumptions on the timescales of the different terms of system (4). In terms of the small parameter $\delta = \tau_Y/\tau_X$ defined in Sect. 2, the fast variability of the unresolved component Y is considered of order $O(\delta^{-2})$ and modeled as an Ornstein–Uhlenbeck process. The Markovian nature of the process defined by Eq. (4) and its singular behavior in the limit of an infinite timescale separation ($\delta \to 0$) allow then to apply the method.

More specifically, the parameter δ serves to distinguish terms with different timescales and is then used as a small perturbation parameter. In this setting, the backward Fokker–Planck equation reads (Majda et al., 2001):

$$-\frac{\partial \rho^{\delta}}{\partial s} = \left[\frac{1}{\delta^2}\mathscr{L}_1 + \frac{1}{\delta}\mathscr{L}_2 + \mathscr{L}_3\right]\rho^{\delta} \tag{12}$$

where the function $\rho^{\delta}(s, X, Y|t)$ is defined with the final value problem $f(X)$: $\rho^{\delta}(t, X, Y|t) = f(X)$. The function ρ^{δ} can be expanded in terms of δ and inserted in Eq. (12). The zeroth order of this equation ρ^0 can be shown to be independent of Y and its evolution given by a closed, averaged backward Fokker–Planck equation (Kurtz, 1973):

$$-\frac{\partial \rho^0}{\partial s} = \bar{\mathscr{L}}\rho^0 \tag{13}$$

This equation is obtained in the limit $\delta \to 0$ and gives the sought limiting, averaged process $X(t)$. Note that this procedure does not necessarily require the presence of the explicit small parameter δ in the original Eq. (4). Since δ disappears from Eq. (13), one can simply use the parameter to identify the fast terms to be considered, and eventually consider $\delta = 1$ (Franzke et al., 2005).

The parameterization obtained by this procedure is given by Franzke et al. (2005):

$$\dot{X} = F_X(X) + G(X) + \sqrt{2}\,\sigma_{\mathrm{MTV}}(X)\cdot\tilde{\xi}(t) \tag{14}$$

with

$$\begin{aligned} G(X) = \int_0^{\infty} \mathrm{d}s \Big[&\big\langle \Psi_Y(X,Y)\cdot\nabla_Y\Psi_X(X,\phi_Y^s(Y))\big\rangle_{\tilde{\rho}} \\ &+\big\langle \Psi_X(X,Y)\cdot\nabla_X\Psi_X(X,\phi_Y^s(Y))\big\rangle_{\tilde{\rho}}\Big] \end{aligned} \tag{15}$$

$$\sigma_{\mathrm{MTV}}(X) = \left(\int_0^\infty \mathrm{d}s \left\langle \Psi_X'(X,Y)\, \Psi_X'(X, \phi_Y^s(Y)) \right\rangle_{\tilde{\rho}} \right)^{1/2} \tag{16}$$

with the same notation as in the previous subsection. The measure $\tilde{\rho}$ is the measure of the $O(\delta^{-2})$ perturbation, i.e., the source of the fast variability of the unresolved Y component. This measure thus depends on which terms of the unresolved component are considered as "fast," and some assumptions should here be made. For instance, it is customary to consider as the fast terms the quadratic terms in Y and to replace them by Ornstein–Uhlenbeck processes whose measures are used to compute the averages (Franzke et al., 2005; Majda et al., 2001).

Finally, if one assumes that the source of the fast variability in the sub-system is given by the "intrinsic" term $F_Y(Y)$ (such that $\tilde{\rho} = \rho_{0,Y}$) and if the perturbation Ψ_X only depends on Y, this parameterization is simply given by the integration of the function $g(s)$ and $h(X,s)$ of the response theory parameterization given by Eqs. (9) and (11). This can be interpreted as an averaging of the latter parameterization when the timescale separation is infinite and X can thus be considered as constant over the timescale of the integrand. Therefore, M_2 can be modeled as a white noise and the memory term is Markovian.

3.3 Hasselmann Averaging Method

Since the initial work of Hasselmann in the 1970s (Hasselmann, 1976), various approaches have been considered to average directly the effects of the "fast" evolving variables on the "slow" ones. These methods assume in general a sufficient timescale separation between the resolved and unresolved components of the systems, and a direct average can be performed as,

$$\dot{X} = \bar{F}(X) = \left\langle F(X,Y) \right\rangle_{\rho_{Y|X}} \tag{17}$$

where $\rho_{Y|X}$ is the measure of the system

$$\dot{Y} = H(X,Y) \tag{18}$$

conditional on the value of X. In this approach, X is thus viewed as a constant parameter for the unresolved dynamics. In other words, this particular framework assumes that since X is slowly evolving with respect to the typical timescale of Y, it can be considered as "frozen" while Y evolves. With some rigorous assumptions, this approach has been mathematically justified (Kifer, 2003) and applied successfully to idealized geophysical models (Arnold et al., 2003) with non-trivial invariant measures. In the same vein, an approximation has been proposed in Abramov (2013) for the average (17), assuming that F is at most quadratic,

$$\left\langle F(X,Y)\right\rangle_{\rho_{Y|X}} = F(X,\bar{Y}(X)) + \frac{1}{2}\frac{\partial^2 F}{\partial Y^2}(X,\bar{Y}(X)) : \Sigma(X) \tag{19}$$

where ":" means the element-wise matrix product with summation and where

$$\bar{Y}(X) = \langle Y \rangle_{\rho_{Y|X}} \tag{20}$$

$$\Sigma(X) = \langle (Y - \bar{Y}(X) \otimes (Y - \bar{Y}(X)) \rangle_{\rho_{Y|X}}. \tag{21}$$

The approximation to the second order is particularly well suited for the application to atmospheric and climate flows for which the quadratic terms are usually the main non-linearities associated with the advection in the system.

In Abramov (2013), an approach based on the fluctuation–dissipation theorem is proposed to estimate the mean state $\bar{Y}(X)$ and the covariance matrix $\Sigma(X)$.

The deterministic parameterization (17) can be recast in a stochastic parameterization following the same principle. Such a parameterization is derived in Arnold et al. (2003), Abramov (2015) and reads

$$\dot{X} = \bar{F}(X) + \sigma_{\mathrm{A}}(X) \cdot \xi(t) \tag{22}$$

with

$$\sigma_{\mathrm{A}}(X) = \left(2\int_0^\infty \mathrm{d}s \left\langle \left(F(X,\phi^s_{Y|X}(Y)) - \bar{F}(X)\right)\left(F(X,Y) - \bar{F}(X)\right)\right\rangle_{\rho_{Y|X}}\right)^{1/2} \tag{23}$$

where $\phi^s_{Y|X}$ is the flow of the system (18) for X constant ("frozen"). A drawback of such an approach is that it requires that the measure $\rho_{Y|X}$ exists and is well-defined [ideally a SRB measure (Arnold et al., 2003)]. Such a requirement may not be always fulfilled, for instance, if the fast system conditional on the state X is unstable and does not possess any attractor (see Sect. 4 for an example).

3.4 Empirical Methods

The empirical methods are generally based on the statistical analysis of the timeseries Y of the full system (4). Many procedures exist as discussed in Sect. 1 but we will consider here a method based on state-dependent AR(1) processes proposed in Arnold et al. (2013). In this case, a timeserie $r(t)$ of the coupling part Ψ_X of the X tendency must first be computed with (4). The parameterization is then given by

$$\dot{X} = F_X(X) + \mathscr{U}(X) \tag{24}$$

with

$$\mathscr{U}(X) = \mathscr{U}_{\rm det}(X) + e(X(t),t). \tag{25}$$

The function $\mathscr{U}_{\rm det}(X)$ represents the deterministic part of the parameterization and is obtained by a least-squares fit of the timeserie $r(t)$ versus the timeserie $X(t)$ with the cubic function $\mathscr{U}_{\rm det}(X) = p_0 + p_1 X + p_2 X^2 + p_3 X^3$. The "stochastic" part $e(X(t),t)$ is then given by the following state-dependent AR(1) process:

$$e(X(t),t) = \phi \, \frac{\sigma_e(X(t))}{\sigma_e(X(t-\Delta t))} \, e(X(t-\Delta t), t-\Delta t) + \sigma_e(X(t)) \, (1-\phi^2)^{1/2} \, z(t) \tag{26}$$

where $z(t)$ is a standard Gaussian white noise process. The parameters of the process e are determined by considering the residual timeserie $r(t) - \mathscr{U}_{\rm det}(X(t))$ to compute the lag-1 autocorrelation ϕ and the state-dependent standard deviation $\sigma_e(X)$ which is modeled as $\sigma_e(X) = \sigma_0 + \sigma_1 |X|$ with the parameters σ_0 and σ_1 given by a binning procedure. The parameter Δt is the time step of integration of Eq. (24). Other empirical parameterizations have been proposed by Arnold et al. (2013), notably one with the function $\mathscr{U}(X) = (1 + e(t)) \, \mathscr{U}_{\rm det}(X)$ which resembles the SPPT[3] parameterization used in the ECMWF[4] Numerical Weather Prediction model (Buizza et al., 1999). However, the study shows no substantial differences with the parameterization (25).

4 Applications and Results

In this section, we will illustrate the various parameterizations described in Sect. 3 to the following example:

$$\begin{cases} \dot{X} = -DX + q\,\xi(t) + \frac{\varepsilon}{\delta} Y^{\mathsf{T}} \cdot \mathsf{C} \cdot Y \\ \dot{Y} = \frac{1}{\delta^2}\left(\mathsf{A} \cdot Y + \delta\, \mathsf{B}_Y \cdot \xi_Y(t)\right) + \frac{\varepsilon}{\delta} X \, \mathsf{V} \cdot Y \end{cases} \tag{27}$$

where $D > 0$, $q > 0$ and

$$Y = \begin{bmatrix} y_1 \\ y_2 \end{bmatrix}. \tag{28}$$

The matrices involved are defined as

[3]Acronym for *Stochastically Perturbed Parameterization Tendencies Scheme*.

[4]Acronym for *European Center for Medium-Range Weather Forecasts*.

$$\mathsf{C} = \begin{bmatrix} 0 & B \\ B & 0 \end{bmatrix} \quad , \quad \mathsf{A} = \begin{bmatrix} -a & \beta \\ -\beta & -a \end{bmatrix} \quad , \quad \mathsf{V} = \begin{bmatrix} 0 & B_1 \\ B_2 & 0 \end{bmatrix} \quad \text{and} \quad \mathsf{B}_Y = \begin{bmatrix} q_Y & 0 \\ 0 & q_Y \end{bmatrix} \tag{29}$$

with $a, \beta, q_Y > 0$. The process $\xi(t)$ and $\xi_Y(t)$ are uncorrelated standard Gaussian white noise processes.

The X and Y variables represent, respectively, the resolved and the unresolved sub-systems. The parameter $\delta > 0$ quantifies the timescale separation of the terms of the tendencies of the two components, with the three timescales $O(1)$,$O(\delta^{-1})$, and $O(\delta^{-2})$ as in Majda et al. (2001) (see also Sect. 3.2). Additionally, the parameter $\varepsilon > 0$ controls the coupling strength between the two sub-systems. In this setup the coupling is thus proportional to the ratio ε/δ, and therefore the characterization of the coupling as "weak" depends directly on the timescale separation.

The deterministic part of Eq. (27) is a well-known model called a triad encountered in fluid dynamics (Ohkitani and Kida, 1992; Smith and Waleffe, 1999; Waleffe, 1992), and in simplified geophysical flows, e.g. Majda et al. (2001), Wouters et al. (2016). Due to the presence of invariant manifolds, its mathematical structure can be found in higher-order model. See Demaeyer and Vannitsem (2016) for an example of such structure in the framework of a coupled ocean-atmosphere model. In the present study, the interest of the stochastic triad model (27) is that, $H(X, Y)$ being linear in Y, the measure $\rho_{0,Y}$ and $\rho_{Y|X}$ can be analytically computed since both $\dot{Y} = F_Y(Y)$ and $\dot{Y} = H(z, Y)|_{z=X}$ are two-dimensional Ornstein–Uhlenbeck processes. Therefore, for this simple case, the set of methods proposed in the previous section can be applied exactly without resorting to a binning procedure of the output of the Y sub-system.[5]

As energy conservation is a rule in physical systems in the absence of dissipation and fluctuations, we will adopt this rule for the current system. System (27) conserves the "energy" $(X^2 + y_1^2 + y_2^2)/2$ if the coefficient B, B_1, and B_2 are chosen such that (Majda et al., 2001; Smith and Waleffe, 1999)

$$2B + B_1 + B_2 = 0. \tag{30}$$

It allows for the following configurations of their signs: $(+, -, -)$, $(+, +, -)$, $(+, -, +)$, $(-, +, +)$, $(-, +, -)$, $(-, -, +)$. These different configurations are associated with different kinds of energy exchange scenarios and different stability properties (Waleffe, 1992).

We will focus on the two configurations $(-, -, +)$ and $(-, +, +)$, with parameters

1. $B = -0.0375$, $B_1 = -0.025$, $B_2 = 0.1$
2. $B = -0.0375$, $B_1 = 0.025$, $B_2 = 0.05$

[5]Except for the empirical methods which by definition use this kind of procedures.

and consider various values of the parameters δ and ε. The other parameters are fixed to $a = 0.01$, $D = 0.01$, and $\beta = 0.01/12$. Once the parameterizations have been developed, the different model versions have integrated over 4.5×10^5 timeunits with a timestep $\Delta t = 0.01$ after a transient period of 5.0×10^4 timeunits to let the system relax to its stationary state. The state X has been recorded every 0.1 timeunit, giving a dataset of 4.5×10^6 points for the analysis. The parameterizations given by Eqs. (7), (14), and (22) have been integrated with a second order Runge–Kutta (RK2) stochastic scheme which converges to the Stratonovich calculus (Hansen and Penland, 2006; Rüemelin, 1982). Equation (24) has been integrated with a deterministic RK2 scheme where the stochastic forcing $e(X, t)$ is considered constant during the timestep. The memory term M_3 appearing in the parameterization (7) and given by the integral (10) over the past of X has been computed numerically at each timestep. Although it increases considerably the integration time, this method is adopted in order to clarify the memory effect in Eq. (7). A Markovianization of this parameterization is possible (Wouters et al., 2016) but in the present case it would have required some assumptions that would blur the comparison of the methods.

The relative performances of the parameterizations can be tested in multiple ways, by comparing the climatology (the average state) or the variability (variance) of the systems (Nicolis, 2005). Another method is to look at the predictive skill score of the models, that is the ability of the parameterizations to provide skillful forecast compared to original system, as in Arnold et al. (2013), Wouters et al. (2016). On longer term, the good representation of the "climate" of a model by the parameterizations can be assessed by looking at the stationary probability densities and comparing them using some score (Abramov, 2012, 2013, 2015; Crommelin and Vanden-Eijnden, 2008; Franzke et al., 2005). The decorrelation properties of the models and the parameterizations can also be tested, to provide information about the correct representation of the timescales of the models. All those different aspects can be significant, depending on the purpose of the parameterization scheme. However, for the brevity of the present work, we shall focus on the probability densities and whether or not they are correctly reproduced by the parameterizations.

We present now the results obtained by with the proposed methods and consider first the different measures used for averaging in system (27).

4.1 Stability and Measures

All the ingredients needed to compute the parameterizations presented in Sect. 3 can be derived with the help of the covariance and the correlation of the Y variables in the framework of two different systems related to the unresolved dynamics, namely the unperturbed dynamics $\dot{Y} = F_Y(Y)$ and the unresolved dynamics $\dot{Y} = H(X, Y)$ with X frozen. The measure of the former is necessary to derive the response theory and the singular perturbation based parameterizations, while the latter is needed for the Hasselmann averaging method. These two systems are both two-dimensional

Ornstein–Uhlenbeck processes of the form

$$\dot{Y} = \mathsf{T} \cdot Y + \mathsf{B} \cdot \xi_Y(t) \tag{31}$$

for which, respectively, $\mathsf{T} = \mathsf{A}/\delta^2$ and $\mathsf{T} = \mathsf{A}/\delta^2 + (\varepsilon X/\delta)\, \mathsf{V} \cdot Y$. In both cases, we have $\mathsf{B} = \mathsf{B}_Y/\delta$. Their measure is then given by Wouters et al. (2016)

$$\rho(Y) = \frac{1}{\mathscr{Z}} \exp\left(-\frac{1}{2} Y^{\mathsf{T}} \cdot \Sigma^{-1} \cdot Y\right) \tag{32}$$

where $\mathscr{Z}$ is a normalization factor and where Σ is the covariance matrix solution of

$$\mathsf{T} \cdot \Sigma + \Sigma \cdot \mathsf{T}^{\mathsf{T}} = -\mathsf{B} \cdot \mathsf{B}^{\mathsf{T}}. \tag{33}$$

In order for theses processes to be stable, the real part of the eigenvalues of the matrix T must be negative (Gardiner, 2009) for every state X that the full coupled system (27) can possibly achieve. The eigenvalues of the system $\dot{Y} = F_Y(Y)$ are $\lambda_\pm = (-a \pm i\beta)/\delta^2$ and it is thus always stable (since $a > 0$ and $\beta \in \mathscr{R}$). On the other hand, the system $\dot{Y} = H(X, Y)$ has the eigenvalues $\lambda_\pm = (-a \pm \sqrt{\Delta(X)})/\delta^2$ with

$$\Delta(X) = -\left(B_1 X\delta\epsilon + \beta\right)\left(\beta - B_2 X\delta\epsilon\right) \tag{34}$$

Therefore, if

$$\mathrm{Re}\left(\sqrt{\Delta(X)}\right) > a \tag{35}$$

for some X, the Ornstein–Uhlenbeck process is unstable, and it is then called an *explosive* process. For any initial condition, the process diverges, and thus the only possible stationary measure is the trivial one. Consequently, Eq. (33) gives nonphysical solutions, the stationary covariance matrix does not exist, and the parameterizations depending upon cannot be derived.

For the system (27), if $\mathrm{sgn}(B_1\, B_2) = -1$, as in case 1, then the process is stable for every X if $a^2 > -\frac{(B_1+B_2)^2\beta^2}{4B_1B_2}$. For case 1, this inequality is satisfied, and thus the process (31) is stable for every X. Moreover, depending on the sign of $\Delta(X)$, the process for X fixed is a stochastic focus (if $\Delta(X) > 0$) or a stochastic damped oscillator (if $\Delta(X) < 0$). Here, it is a focus if

$$X \in [\min(-\beta/\delta\varepsilon B_1, \beta/\delta\varepsilon B_2), \max(-\beta/\delta\varepsilon B_1, \beta/\delta\varepsilon B_2)] \tag{36}$$

and an oscillator otherwise. That is, for the considered ε and δ parameters value, the system (27) is an oscillator for most of the X values.

If $\mathrm{sgn}(B_1\, B_2) = 1$, as in case 2, then the condition (35) must be satisfied for every state X. For case 2, this inequality was not satisfied for every state X for most of the values of the ε and δ parameters considered (see Sect. 4.3 below). The stability is therefore reversed as the system is non-oscillating for most of the X values.

To summarize, if B_1 and B_2 are of opposite sign, the dynamics of $\dot{Y} = H(X, Y)$ is stable and generally oscillatory. If B_1 and B_2 have the same sign, then the dynamics is unstable in most cases and generally hyperbolic. This is a consequence of the well-known difference of stability of the triads depending on their energy exchange properties (Waleffe, 1992).

For the interested reader, the exact calculation of the parameterization of Sect. 3 using the covariance and correlation matrices is detailed in the Appendix (see section "Appendix: Practical Computation of the Parameterizations").

4.2 The $(-, -, +)$ Stochastic Triad (Case 1)

Let us now consider case 1 corresponding to the $(-, -, +)$ stochastic triad for two different values of the timescale separation $\delta = 0.1$ and 0.4. For each of these timescale separation, we considered three values of the coupling strength ε: 0.05, 0.125, and 0.4. The probability densities associated with these different systems are represented in Figs. 1 and 2. For a timescale separation $\delta = 0.1$, the fully coupled dynamics given by Eq. (27) is quite well represented by all the proposed parameterizations. Since it is hard to distinguish the different density curves, a score such as the Hellinger distance (Arnold et al., 2013)

$$H(P, Q) = \frac{1}{2} \int \left(\sqrt{\mathrm{d}P} - \sqrt{\mathrm{d}Q} \right)^2 \tag{37}$$

between the distribution P of the full coupled system and the distribution Q of the parameterizations is worth computing to quantify the differences (the smaller the better). It is depicted for $\delta = 0.1$ on Fig. 3, and it shows that for a very small coupling parameter $\varepsilon = 0.05$, the best parameterization is the response theory given by Eq. (7). For larger values of ε, it is the Hasselmann averaging method which performs best. The empirical method gives a good correction of the uncoupled dynamics for $\varepsilon = 0.125$ but diverges for $\varepsilon = 0.4$. This may be due to instabilities introduced by the cubic deterministic parameterization $\mathscr{U}_{\mathrm{det}}(X)$ or to the inadequacy of the fitting function $\sigma_0 + \sigma_1 |X|$ for the standard deviation $\sigma_e(X)$ in the AR(1) process (26). Indeed, in general, this model fits quite well the statistics in the neighborhood of $X = 0$, but the standard deviation reaches a plateau for higher values of X. A more complicated fitting function would thus be necessary to get a stable dynamics. For a timescale separation $\delta = 0.4$, the same conclusions are reached, but the singular perturbation method performs not very well in all cases, as illustrated in Fig. 4 that for $\varepsilon = 0.125$ and 0.4. The response based and singular perturbation methods are even less effective than the uncoupled dynamics. It is not surprising for the latter since it is supposed to be valid in the limit $\delta \to 0$.

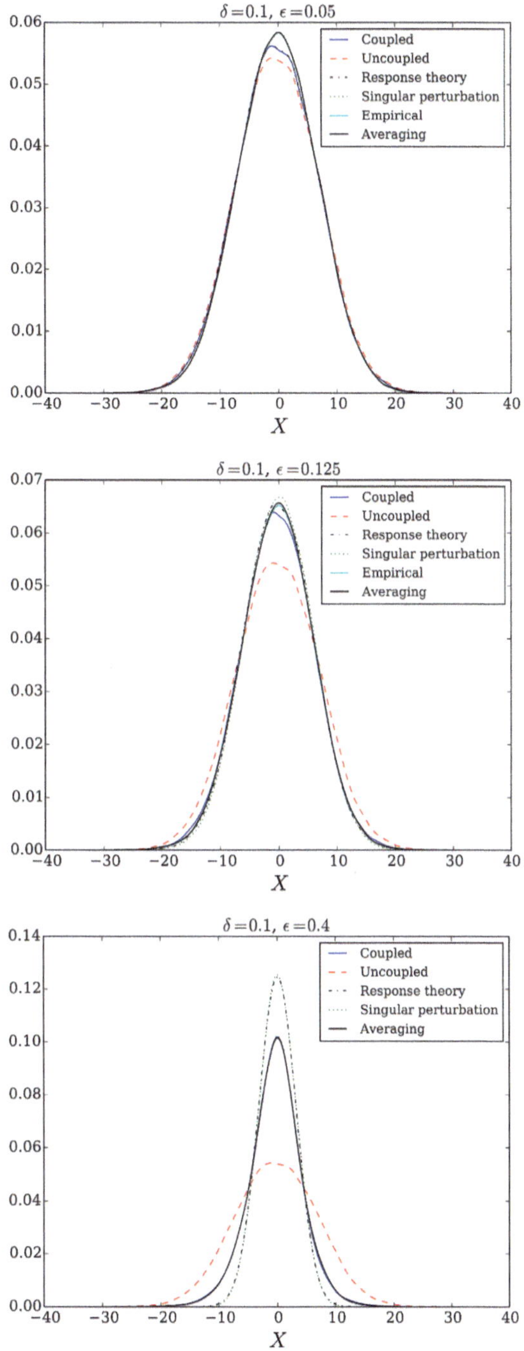

Fig. 1 Probability densities of the full coupled dynamics (27), the uncoupled dynamics $\dot{X} = -DX + q\xi(t)$, and the parameterized model versions for the timescale separation $\delta = 0.1$ and for the triad parameters of case 1. The empirical parameterization density is not represented for $\varepsilon = 0.4$ due to its divergence

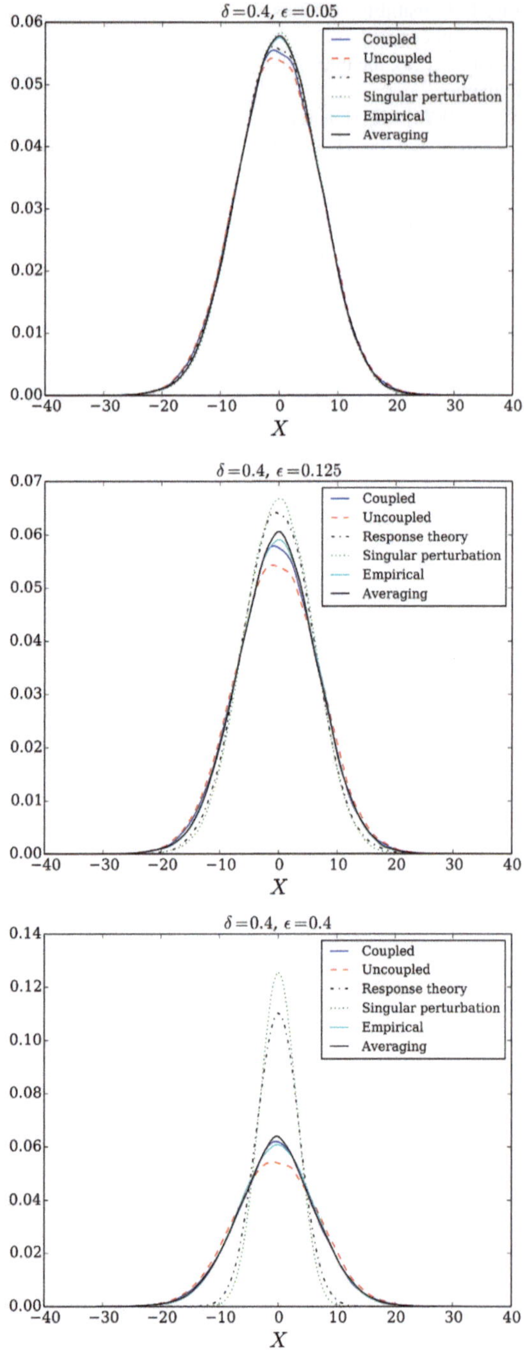

Fig. 2 Same as Fig. 1 but for the timescale separation $\delta = 0.4$

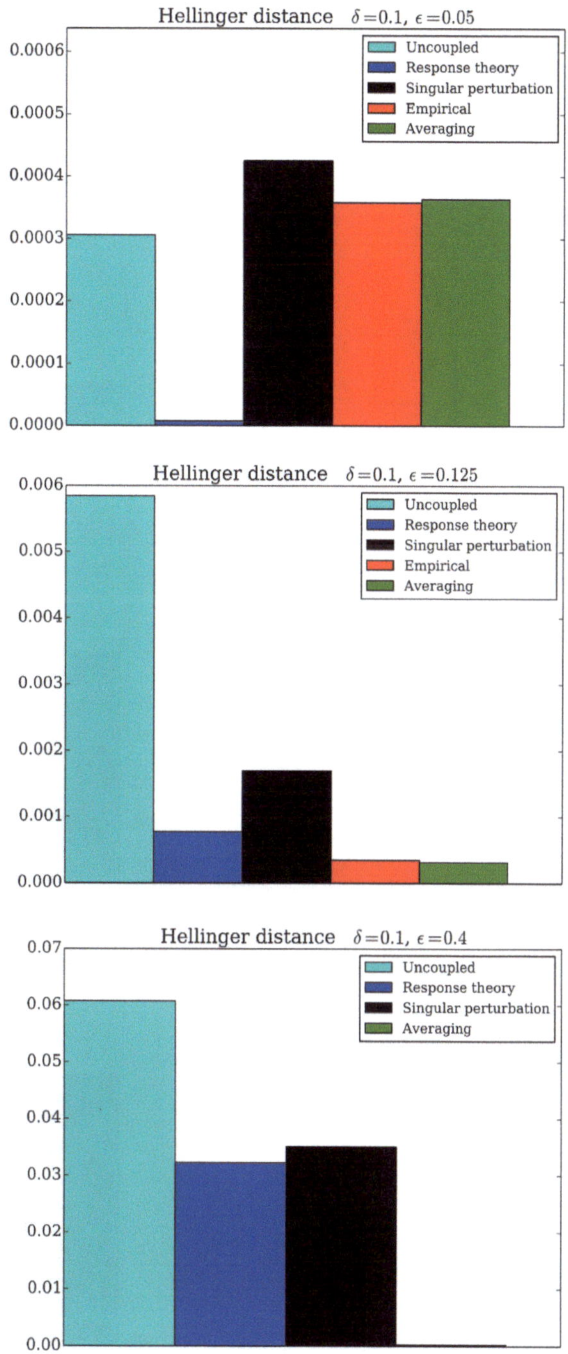

Fig. 3 Hellinger distance (37) between the densities of the different parameterized models and the full coupled system density for case 1. A small distance indicates that the two densities concerned are very similar. The Hellinger distance between the full coupled system and the uncoupled system distribution is depicted as reference. In case $\varepsilon = 0.4$, the empirical parameterization is not represented due to its divergence

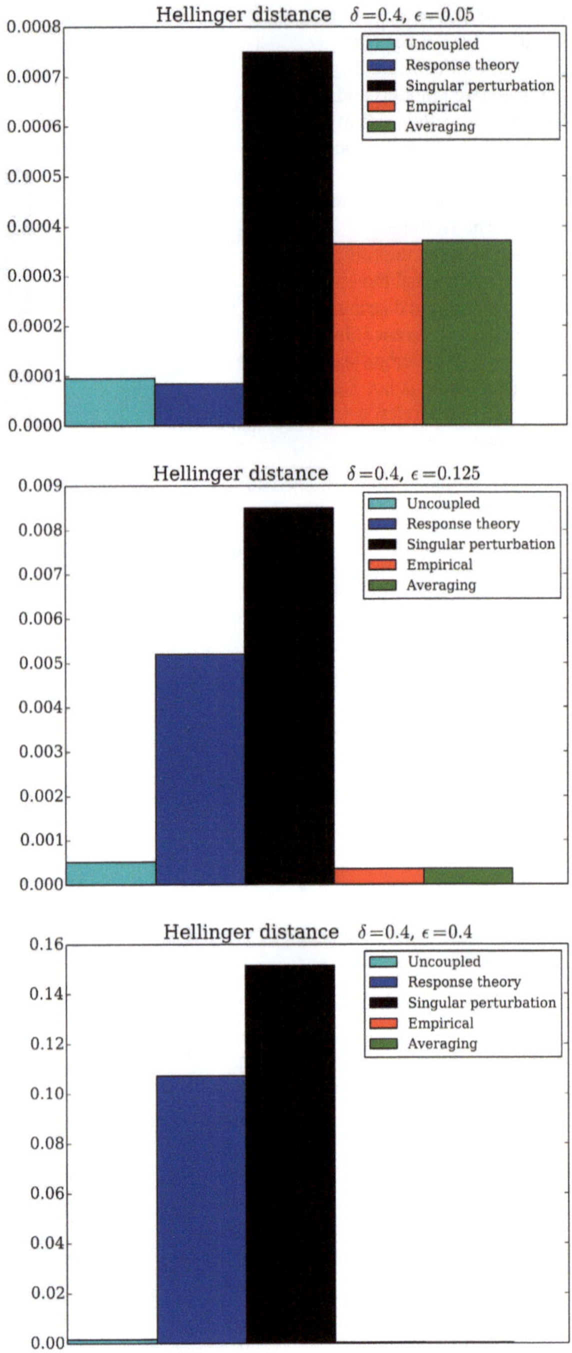

Fig. 4 Same as Fig. 3 but for the timescale separation $\delta = 0.4$

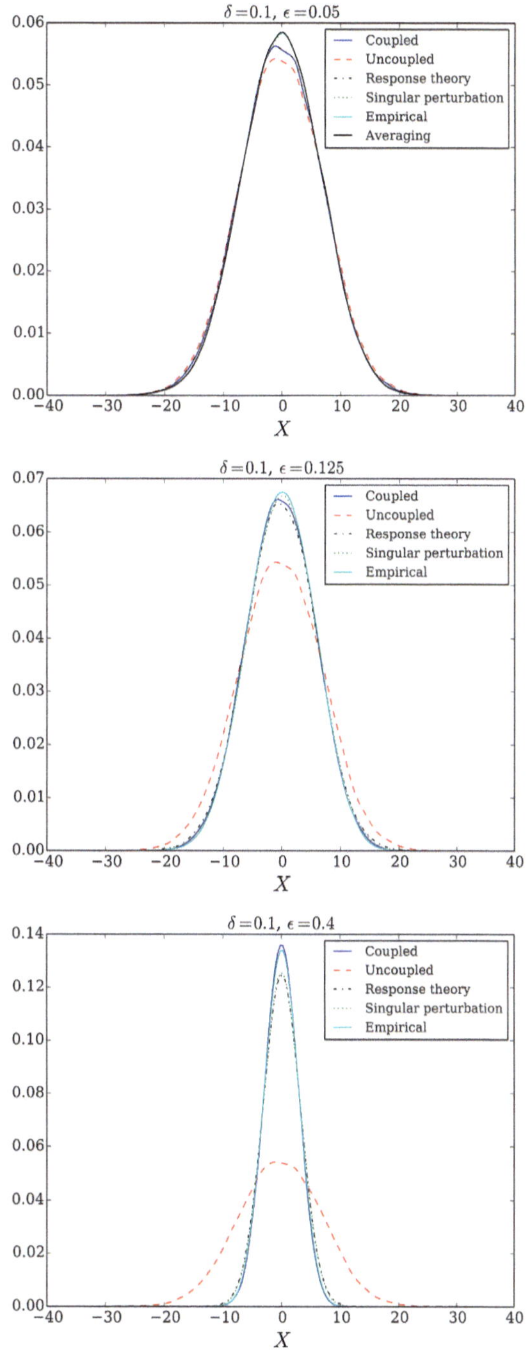

Fig. 5 Probability densities of the coupled full dynamics (27), the uncoupled dynamics $\dot{X} = -DX + q\xi(t)$, and the parameterized models for the timescale separation $\delta = 0.1$ and for the triad parameters of case 2. The direct averaging parameterization density is only represented for $\varepsilon = 0.05$ because the system diverges for the other values

Fig. 6 Same as Fig. 5 but for the timescale separation $\delta = 0.4$

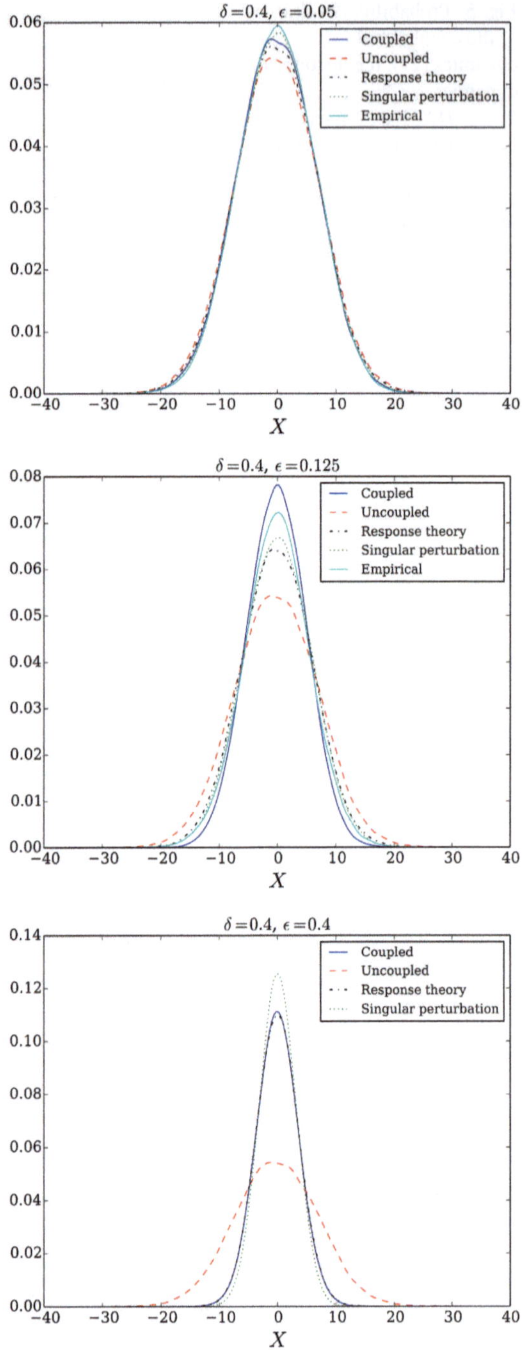

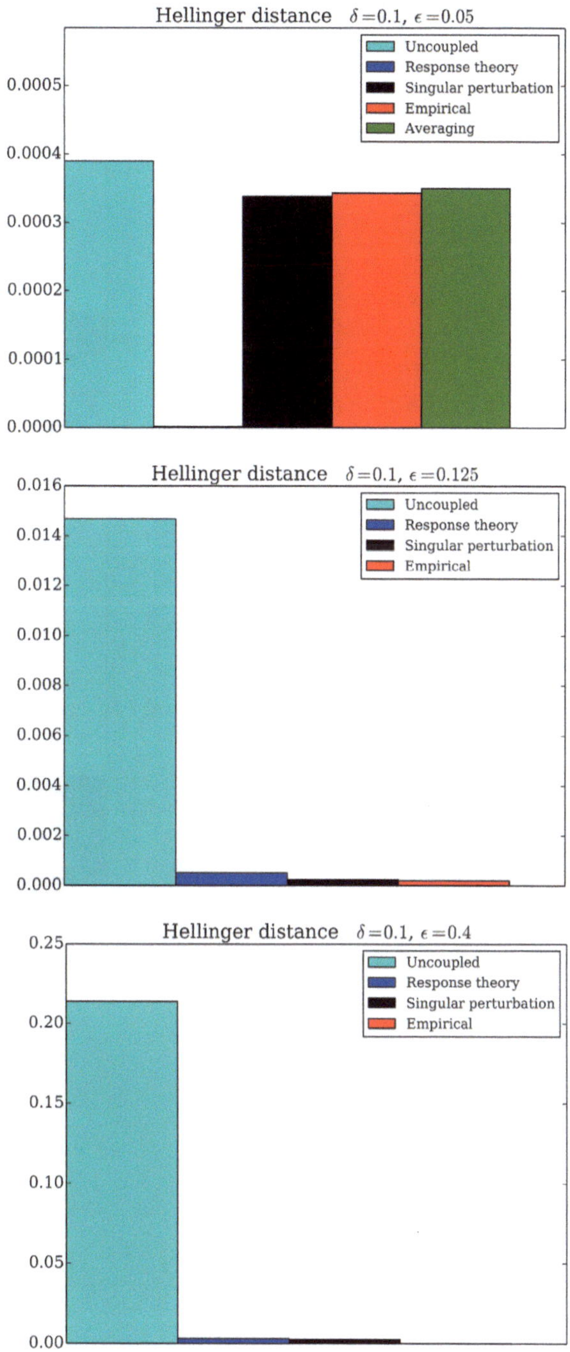

Fig. 7 Hellinger distance (37) between the densities of the different parameterized models and the full coupled system density for case 2. A small distance indicates that the two densities concerned are very similar. The Hellinger distance between the full coupled system and the uncoupled system distribution is depicted as reference. In case $\varepsilon = 0.4$, the empirical parameterization is not represented due to its divergence

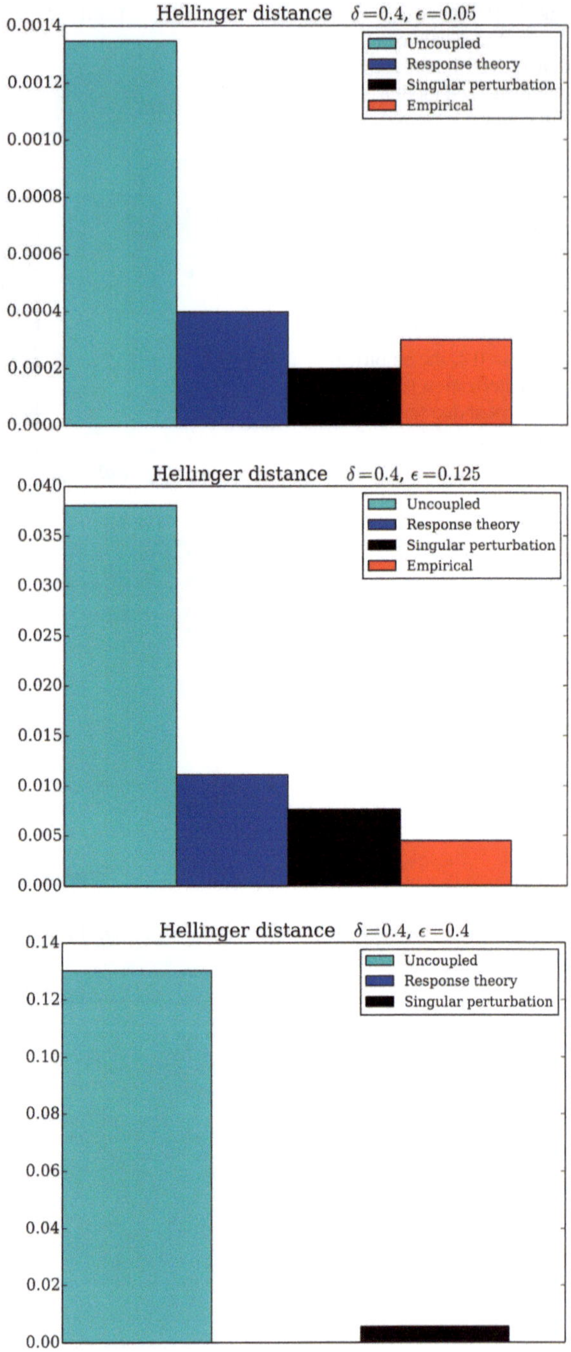

Fig. 8 Same as Fig. 7 but for the timescale separation $\delta = 0.4$

4.3 The $(-, +, +)$ Stochastic Triad (Case 2)

We now consider the parameters of case 2, for which the system (27) is a $(-, +, +)$ stochastic triad. The probability densities are depicted in Figs. 5 and 6, and the Hellinger distances are shown in Figs. 7 and 8. First, we must remark that the parameterization based on the Hasselmann's averaging method is not defined for most of the δ and ε parameters values. It is due to the fact that the dynamics of the unresolved component Y with X considered as a parameter is unstable, as shown in Sect. 4.1. Indeed, this linear system undergoes a bifurcation at some value $X^\star$ which destabilizes the dynamics $\dot{Y} = H(X, Y)$ with X frozen. Therefore, the measure $\rho_{Y|X}$ is not defined for some ranges of the full X dynamics and the method fails. The only case where this destabilization does not occur is for $\delta = 0.1$ and $\varepsilon = 0.05$, but the parameterization does not perform well. For these parameter values, the only parameterization that performs very well is the one based on response theory. For the other values of the parameters δ and ε, all the parameterizations have good performances. A particularly unexpected result is the very good correction provided by the response theory and singular perturbation based methods for the extreme case $\delta = 0.4$ and $\varepsilon = 0.4$ (see the bottom panel of Fig. 8). This have to be contrasted with their bad performances in the case of the other triad (see the bottom panel of Fig. 4). Note that for this extreme case, the direct averaging method fails and the empirical method is unstable and diverges.

4.4 Discussion

The results obtained so far with these two types of triads highlight the utility of the parameterization schemes discussed here. First, the empirical parameterization gives usually good results when it does not destabilize the dynamics. However, this method requires a case by case time-consuming statistical analysis whose complexity increases with the dimensionality of the problem considered. Physically based parameterizations do not require such an analysis, and the best approach in the present system is the Hasselmann averaging one, but it requires that the dynamics of the unresolved system be stable. It was thus very effective to correct the dynamics of the $(-, -, +)$ triad, but not the other triad $(-, +, +)$. In this latter case, the perturbative methods like the singular perturbation method or the response theory method give very good results. This difference is quite intriguing and interesting. It indicates that different physically-based parameterizations should be considered depending on the kind of problems encountered. In particular, the stability properties of the system considered seem to play an important role. This conclusion holds whatever the timescale separation and for the most realistic values of the coupling strength between the components ($\varepsilon = 0.125$ and 0.4). However, for very small values of the coupling strength, the response based method seems to be the best approach in all cases.

A question that is left open in the present work is to determine precisely which stability property is giving the contrasting observed result. More specifically, is it the hyperbolic instability of the $(-,+,+)$ triad which makes the perturbative approach and the response based parameterization perform so well, or is it simply the fact that it is unstable? On the other hand, is it the damped oscillatory behavior of the $(-,-,+)$ triad which makes the Hasselmann's method works well, or is it simply the fact that it is stable? Such questions should be addressed in the case of a more complex, globally stable system, which allows to have locally stable and unstable fast dynamics.

5 Conclusions

The parameterization of subgrid-scale processes is an important tool in model reduction, in order to improve the statistical properties of the forecasting systems. The variety of approaches available bear witness of the richness of the field but at the same time can also lead to questions on the best choice for the problem at hand. The purpose of the present review was to describe briefly some of the most recent methods and to illustrate them on a simple stochastic triad example. The methods covered include perturbative methods like the Ruelle response theory (Wouters and Lucarini, 2012), the singular perturbation theory (Majda et al., 2001), averaging methods like the Hasselmann method (Arnold et al., 2003; Hasselmann, 1976) and an empirical method (Arnold et al., 2013). As expected, these parameterizations provided contrasting results depending on the timescale separation and on the coupling between the resolved variables and the subgrid one. But more importantly, our results in the context of this simple triad stress the importance of the underlying stability properties of the unresolved system. It thus confirms a known result that the structure of the Jacobian and of the Hessian of a given system controls the behavior and performance of model error parameterizations (Nicolis, 2005).

Further comparisons of the different methods are needed in the context of more sophisticated systems in order to analyze the role of the stability properties of the subgrid scale processes on their performances. This type of analysis is currently under way in the context of a coupled ocean-atmosphere system (De Cruz et al., 2016).

Appendix: Practical Computation of the Parameterizations

In the following section, for illustrative purposes, we detail the computation that we have made to obtain the result of the present review. We start with the method based on response theory.

Response Theory Method

We consider the system (27) with the form (4) in mind. In this case, the influence of the Y sub-system on the X sub-system is parameterized as:

$$\dot{X} = -DX + q\,\xi(t) + M_1(X) + M_2(X,t) + M_3(X,t) \tag{38}$$

where then terms M_1, M_2, and M_3 are, respectively, given by Eqs. (8), (9, and (10). The average in these formula are performed with the measure $\rho_{0,Y}$ of the unperturbed Y dynamics $\dot{Y} = F_Y(Y)$. Since this latter is an Ornstein–Uhlenbeck process, its measure is the Wiener measure

$$\rho_{0,Y}(Y) = \frac{1}{\mathscr{Z}} \exp\left(-\frac{1}{2}\, Y^{\mathsf{T}} \cdot \Sigma^{-1} \cdot Y\right) \tag{39}$$

where Σ is the covariance matrix solution of

$$\mathsf{A} \cdot \Sigma + \Sigma \cdot \mathsf{A}^{\mathsf{T}} = -\mathsf{B}_Y \cdot \mathsf{B}_Y^{\mathsf{T}} \tag{40}$$

and $\mathscr{Z}$ is a normalization factor.

The covariance and correlation of the stationary process $\dot{Y} = F_Y(Y)$ are thus straightforward to compute (Gardiner, 2009):

$$\Sigma = \langle Y \otimes Y \rangle = \frac{q_Y^2}{2a}\,\mathsf{I} \tag{41}$$

$$\langle \phi_Y^t(Y) \otimes \phi_Y^s(Y) \rangle = \mathsf{E}(t-s) \cdot \Sigma, \quad t > s \tag{42}$$

$$\langle \phi_Y^t(Y) \otimes \phi_Y^s(Y) \rangle = \Sigma \cdot \mathsf{E}(s-t)^{\mathsf{T}}, \quad t < s \tag{43}$$

where I is the identity matrix, ϕ_Y^t is the flow of $\dot{Y} = F_Y(Y)$, and the matrix $\mathsf{E}(t)$ is the exponential

$$\mathsf{E}(t) = \exp(\mathsf{A}t/\delta^2) = e^{-at/\delta^2} \begin{bmatrix} \cos(\beta t/\delta^2) & \sin(\beta t/\delta^2) \\ -\sin(\beta t/\delta^2) & \cos(\beta t/\delta^2) \end{bmatrix} \tag{44}$$

The various terms M_i are then computed as follows.

The Term M_1

It is the average term:

$$M_1(X) = \langle \Psi_X(X,Y) \rangle \tag{45}$$

We have thus

$$M_1(X) = 2B\frac{\varepsilon}{\delta}\langle y_1\, y_2\rangle = 2B\frac{\varepsilon}{\delta}\,\Sigma_{12} = 0 \tag{46}$$

by using Eq. (41).

The Term M_2

It is the noise/correlation term which is defined here as:

$$M_2(t) = \sigma_R(t) \tag{47}$$

with

$$\langle\sigma_R(t)\sigma_R(t')\rangle = g(t-t') \tag{48}$$

and the correlation function

$$g(s) = \left\langle \Psi'_X(Y)\Psi'_X(\phi^s_Y(Y))\right\rangle \tag{49}$$

where $\Psi'_X(Y) = \Psi_X(Y) - M_1$. The result in the present case is given by the formula [see Demaeyer and Vannitsem (2016)]:

$$g(s) = \frac{\varepsilon^2}{\delta^2}\mathrm{Tr}\left(\left(\mathsf{C}+\mathsf{C}^{\mathsf{T}}\right)\cdot\Sigma\cdot\mathsf{E}(s)^{\mathsf{T}}\cdot\mathsf{C}^{\mathsf{T}}\cdot\mathsf{E}(s)\cdot\Sigma\right) = \frac{\varepsilon^2}{\delta^2}\frac{q_Y^4}{a^2}B^2e^{-2as/\delta^2}\cos(2\beta s/\delta^2) \tag{50}$$

The term M_2 must thus be devised as a process with the same correlation.

The Term M_3

This is the memory term, defined by

$$M_3(X,t) = \int_0^\infty \mathrm{d}s\; h(X(t-s),s) \tag{51}$$

with the memory kernel

$$h(X,s) = \langle\Psi_Y(X,Y)\cdot\nabla_Y\Psi_X(\phi^s_Y(Y))\rangle \tag{52}$$

which in the present case is given by the formula [see Demaeyer and Vannitsem (2016)]

$$h(X,s) = \frac{\varepsilon^2}{\delta^2}\mathrm{Tr}\left((X\,\mathsf{V}\cdot\Sigma)\cdot\left(\mathsf{E}(s)^{\mathsf{T}}\cdot\left(\mathsf{C}+\mathsf{C}^{\mathsf{T}}\right)\cdot\mathsf{E}(s)\right)\right) \tag{53}$$

$$= \frac{\varepsilon^2}{\delta^2}\frac{q_Y^2}{a}XB(B_1+B_2)e^{-2as/\delta^2}\cos(2\beta s/\delta^2) \tag{54}$$

The fact that the memory kernel (54) and the correlation function (50) present the same form implies that a Markovian parameterization is available (Wouters et al., 2016) even if by definition, Eq. (38) is a non-Markovian parameterization.

The Singular Perturbation Method

With this parameterization, the parameter δ serves to distinguish terms with different timescale and is then used as a small perturbation parameter (Franzke et al., 2005; Majda et al., 2001). The parameterization is given by:

$$\dot{X} = -D\,X + q\,\xi(t) + G(X) + \sqrt{2}\,\sigma_{\mathrm{MTV}}(X)\cdot\tilde{\xi}(t) \tag{55}$$

with notably $\langle\xi(t)\tilde{\xi}(t')\rangle = 0$ and

$$G(X) = \int_0^\infty \mathrm{d}s\left\langle \Psi_Y(X,Y)\cdot\nabla_Y\Psi_X(X,\phi_Y^s(Y))\right\rangle_{\tilde{\rho}} \tag{56}$$

$$\sigma_{\mathrm{MTV}}(X) = \left(\int_0^\infty \mathrm{d}s\left\langle \Psi_X'(X,Y)\,\Psi_X'(X,\phi_Y^s(Y))\right\rangle_{\tilde{\rho}}\right)^{1/2} \tag{57}$$

We see that the quantities appearing in this parameterization can easily be obtained from the functions h and g of section "Response Theory Method". Indeed we have

$$G(X) = \int_0^\infty \mathrm{d}s\; h(X,s) = \varepsilon^2 X q_Y^2 \frac{B(B_1+B_2)}{2(a^2+\beta^2)} \tag{58}$$

$$\mathsf{S}_{\mathrm{MTV}}(X) = \int_0^\infty \mathrm{d}s\; g(s) = \varepsilon^2\frac{q_Y^4 B^2}{2a(a^2+\beta^2)} \tag{59}$$

where we notice that the parameter δ has disappeared, since this parameterization is valid in the limit $\delta \to 0$.

Averaging Method

In this approach, we consider the system (2) and the parameterization (Abramov, 2013):

$$\dot{X} = \bar{F}(X) \tag{60}$$

with

$$\bar{F}(X) = \langle F(X,Y) \rangle_{\rho_{Y|X}} = F(X, \bar{Y}(X)) + \frac{1}{2} \frac{\partial^2 F}{\partial Y^2}(X, \bar{Y}(X)) : \Sigma(X) \tag{61}$$

and

$$\bar{Y}(X) = \langle Y \rangle_{\rho_{Y|X}} \tag{62}$$

$$\Sigma(X) = \langle (Y - \bar{Y}(X) \otimes (Y - \bar{Y}(X)) \rangle_{\rho_{Y|X}} \tag{63}$$

where $\rho_{Y|X}$ is the measure of the system $\dot{Y} = H(X, Y)$ with X "frozen." It is the measure of an Ornstein–Uhlenbeck process

$$\rho_{0,Y}(Y) = \frac{1}{\mathscr{Z}} \exp\left(-\frac{1}{2} Y^{\mathsf{T}} \cdot \Sigma^{-1}(X) \cdot Y \right) \tag{64}$$

where $\mathscr{Z}$ is a normalization factor and $\Sigma(X)$ is the stationary covariance matrix solution of

$$\mathsf{T}(X) \cdot \Sigma + \Sigma \cdot \mathsf{T}(X)^{\mathsf{T}} = -\frac{1}{\delta^2} \mathsf{B}_Y \cdot \mathsf{B}_Y^{\mathsf{T}} \tag{65}$$

with

$$\mathsf{T}(X) = \mathsf{A}/\delta^2 + \varepsilon X \, \mathsf{V}/\delta. \tag{66}$$

With the help of $\bar{Y}(X) = 0$ and $\Sigma(X)$, we can now rewrite Eq. (61) as

$$\begin{aligned} \bar{F}(X) &= F(X, 0) + \frac{\varepsilon}{\delta} \mathsf{C} : \Sigma(X) \\ &= -DX + q\xi(t) + \frac{B\,(B_1 + B_2)\, q_Y^2 X \epsilon^2}{2\,(a^2 + \beta^2 - X\beta\delta\varepsilon B_2 + X\delta\varepsilon B_1\,(\beta - X\delta\varepsilon B_2))} \end{aligned} \tag{67}$$

This forms a deterministic averaging parameterization. It can be extended into a stochastic parameterization (Abramov, 2015) as follows:

$$\dot{X} = \bar{F}(X) + \sigma_{\mathrm{A}}(X) \cdot \xi(t) \tag{68}$$

with

$$\sigma_{\mathrm{A}}(X) = \sqrt{S(X)} \tag{69}$$

and

$$S(X) = 2\int^{\infty}\left(2\int_0^{\infty} \mathrm{d}s \left\langle \left(F(X,\phi^s_{Y|X}(Y)) - \bar{F}(X)\right)\left(F(X,Y) - \bar{F}(X)\right)\right\rangle_{\rho_{Y|X}}\right)^{1/2} \tag{70}$$

We thus have

$$S(X) = 2\frac{\varepsilon^2}{\delta^2}\int_0^{\infty} \mathrm{d}s\ \mathrm{Tr}\left(\left(\mathsf{C}+\mathsf{C}^{\mathsf{T}}\right)\cdot\Sigma(X)\cdot\exp\left[\mathsf{T}(X)^{\mathsf{T}}s\right]\cdot\mathsf{C}^{\mathsf{T}}\cdot\exp\left[\mathsf{T}(X)s\right]\cdot\Sigma(X)\right)$$

where we have extended the result of Eq. (50) to the stationary Ornstein–Uhlenbeck process $\dot{Y} = H(X,Y)$ for X "frozen". The function $S(X)$ can be computed analytically using mathematical software but is a very complicate function that is not worth displaying in this review. This can, however, be provided upon query to the authors.

References

Abramov, R.V. 2012. *Multiscale Modeling & Simulation* 10(1): 28.
Abramov, R.V. 2013. *Multiscale Modeling & Simulation* 11(1): 134.
Abramov, R. 2015. *Fluids* 1(1): 2.
Arnold, L. 2001. *Stochastic climate models*, 141–157. New York: Springer.
Arnold, L., P. Imkeller, and Y. Wu. 2003. *Dynamical Systems* 18(4): 295.
Arnold, H., I. Moroz, and T. Palmer. 2013. *Philosophical Transactions of the Royal Society of London A: Mathematical, Physical and Engineering Sciences* 371(1991): 20110479.
Bogoliubov, N.N., and Y.A. Mitropolski. 1961. In *Asymptotic methods in the theory of non-linear oscillations*, ed. Bogoliubov, N.N., and Y.A. Mitropolski, vol. 1. New York: Gordon and Breach.
Bouchet, F., T. Grafke, T. Tangarife, and E. Vanden-Eijnden. 2016. *Journal of Statistical Physics* 162(4): 793.
Buizza, R., M. Milleer, and T. Palmer. 1999. *Quarterly Journal of the Royal Meteorological Society* 125(560): 2887
Chekroun, M.D., H. Liu, and S. Wang. 2015. *Approximation of stochastic invariant manifolds: stochastic manifolds for nonlinear SPDEs I*. Cham: Springer.
Crommelin, D., and E. Vanden-Eijnden. 2008. *Journal of the Atmospheric Sciences* 65(8): 2661.
Culina, J., S. Kravtsov, A.H. Monahan. 2011. *Journal of the Atmospheric Sciences* 68(2): 284.
De Cruz, L., J. Demaeyer, and S. Vannitsem. 2016. *Geoscientific Model Development* 9(8): 2793.
Demaeyer, J., and S. Vannitsem. 2017. Stochastic parametrization of subgrid-scale processes in coupled ocean-atmosphere systems: benefits and limitations of response theory. *Quarterly Journal of the Royal Meteorological Society*, 143(703), 881–896.
Doblas-Reyes, F., A. Weisheimer, M. Déqué, N. Keenlyside, M. McVean, J. Murphy, P. Rogel, D. Smith, and T. Palmer. 2009. *Quarterly Journal of the Royal Meteorological Society* 135(643): 1538.
Ellis, R.S., and M.A. Pinsky. 1975. *Journal de Mathématiques Pures et Appliqués* 54(9): 125.
Frankignoul, C. 1979. *Dynamics of Atmospheres and Oceans* 3(2–4): 465.
Frankignoul, C., and K. Hasselmann. 1977. *Tellus* 29(4): 289.
Frankignoul, C., and P. Müller. 1979. *Journal of Physical Oceanography* 9(1): 104.
Franzke, C., A.J. Majda, and E. Vanden-Eijnden. 2005. *Journal of the Atmospheric Sciences* 62(6): 1722.
Frederiksen, J.S. 1999. *Journal of the Atmospheric Sciences* 56(11): 1481.

Frederiksen, J.S., and A.G. Davies. 1997. *Journal of the Atmospheric Sciences* 54(20): 2475.
Freidlin, M.I., and A.D. Wentzell. 1984. *Random perturbations of dynamical systems*, 15–43. New York: Springer.
Gardiner, C.W. 2009. *Handbook of stochastic methods*, 4th ed. Berlin: Springer.
Ghil, M., M. Allen, M. Dettinger, K. Ide, D. Kondrashov, M. Mann, A.W. Robertson, A. Saunders, Y. Tian, F. Varadi, et al. 2002. *Reviews of Geophysics* 40(1): 3-1–3-41.
Grad, H. 1969. *Transport Theory* 1: 269.
Hansen, J.A., and C. Penland. 2006. *Monthly Weather Review* 134(10): 3006.
Hasselmann, K. 1976. *Tellus* A28(6): 473–485.
Kifer, Y. 2001. *Stochastic climate models*, 171–188. Berlin: Springer.
Kifer, Y. 2003. *Stochastics and Dynamics* 3(02): 213.
Kurtz, T.G. 1973. *Journal of Functional Analysis* 12(1): 55.
Lemke, P. 1977. *Tellus* 29(5): 385.
Lemke, P., E. Trinkl, and K. Hasselmann. 1980. *Journal of Physical Oceanography* 10(12): 2100.
Lovejoy, S., and D. Schertzer. 2013. *The Weather and climate: emergent laws and multifractal cascades*. Cambridge: Cambridge University Press.
Majda, A.J., I. Timofeyev, and E. Vanden Eijnden. 2001. *Communications on Pure and Applied Mathematics* 54(8): 891.
Newman, M., P.D. Sardeshmukh, and C. Penland. 1997. *Journal of the Atmospheric Sciences* 54(3): 435.
Nicolis, C. 1981. *Physics of solar variations*, 473–478. Berlin: Springer.
Nicolis, C. 1982. *Tellus* 34(1): 1.
Nicolis, C. 2003. *Journal of the Atmospheric Sciences* 60(17): 2208.
Nicolis, C. 2004. *Journal of the Atmospheric Sciences* 61(14): 1740.
Nicolis, C. 2005. *Quarterly Journal of the Royal Meteorological Society* 131(609): 2151.
Nicolis, C., and G. Nicolis. 1981. *Tellus* 33(3): 225.
Nicolis, G., and C. Nicolis. 2012. *Foundations of complex systems: emergence, information and prediction*. Singapore: World Scientific.
Ohkitani, K., and S. Kida. 1992. *Physics of Fluids A: Fluid Dynamics (1989–1993)* 4(4): 794.
Olbers, D. 2001. *Stochastic climate models*, 3–63. New York: Springer.
Papanicolaou, G.C. 1976. *Journal of Mathematics* 6(4): 653–674.
Penland, C. 1989. *Monthly Weather Review* 117(10): 2165.
Penland, C. 1996. *Physica D: Nonlinear Phenomena* 98(2): 534.
Penland, C., and L. Matrosova. 1994. *Journal of Climate* 7(9): 1352.
Ruelle, D. 1997. *Communications in Mathematical Physics* 187(1): 227.
Ruelle, D. 2009. *Nonlinearity* 22(4): 855.
Rüemelin, W. 1982. *SIAM Journal on Numerical Analysis* 19(3): 604.
Sanders, J., F. Verhulst, and J. Murdock (eds.). 2007. *Averaging Methods in Nonlinear Dynamical Systems. Collection: Applied Mathematical Sciences*, vol. 59, 2nd ed. New York: Springer.
Sardeshmukh, P.D., and C. Penland. 2015. *Chaos: An Interdisciplinary Journal of Nonlinear Science* 25(3): 036410.
Shutts, G. 2005. *Quarterly Journal of the Royal Meteorological Society* 131(612): 3079.
Smith, L.M., and F. Waleffe. 1999. *Physics of Fluids* 11: 1608.
Sura, P. 2013. *Extremes in a changing climate*, 181–222. Dordrecht: Springer.
Sura, P., M. Newman, C. Penland, and P. Sardeshmukh. 2005. *Journal of the Atmospheric Sciences* 62(5): 1391.
Vannitsem, S. 2014. *Philosophical Transactions of the Royal Society of London A: Mathematical, Physical and Engineering Sciences* 372: 20130282.
Vissio, G., and V. Lucarini. 2016. arXiv preprint arXiv:1612.07223.
Waleffe, F. 1992. *Physics of Fluids A: Fluid Dynamics (1989–1993)* 4(2): 350.

Wouters, J., and V. Lucarini. 2012. *Journal of Statistical Mechanics: Theory and Experiment* 2012(03): P03003.
Wouters, J., and V. Lucarini. 2013. *Journal of Statistical Physics* 151(5): 850.
Wouters, J., S.I. Dolaptchiev, V. Lucarini, and U. Achatz. 2016. *Nonlinear Processes in Geophysics* 23(6): 435.
Young, L.S. 2002. *Journal of Statistical Physics* 108(5): 733.

Large-Scale Atmospheric Phenomena Under the Lens of Ordinal Time-Series Analysis and Information Theory Measures

J.I. Deza, G. Tirabassi, M. Barreiro, and C. Masoller

Abstract This review presents a synthesis of our work done in the framework of the European project *Learning about Interacting Networks in Climate* (LINC, climatelinc.eu). We have applied tools of information theory and ordinal time series analysis to investigate large scale atmospheric phenomena from climatological datasets. Specifically, we considered monthly and daily Surface Air Temperature (SAT) time series (NCEP reanalysis) and used the climate network approach to represent statistical similarities and interdependencies between SAT time series in different geographical regions. Ordinal analysis uncovers how the structure of the climate network changes in different time scales (intra-season, intra-annual, and longer). We have also analyzed the directionally of the links of the network, and we have proposed novel approaches for uncovering communities formed by geographical regions with similar SAT properties.

Keywords Climate networks • Nonlinear time series analysis • Climate communities • Information transfer

1 Introduction

Complex networks constitute the huge revolution in nonlinear science in the twentieth century because it provides a unified framework for the study of a wide range of real-world complex systems, such as the Internet, social networks, transport networks, ecological and metabolic networks, and even the human brain (Albert and Barabasi 2002; Newman 2003; Boccaletti et al. 2006).

For understanding and extracting information from observed data, various methods for mapping statistical interdependencies between time series into "functional"

J.I. Deza • G. Tirabassi • M. Barreiro
Instituto de Física, Facultad de Ciencias, Universidad de la República, Igua 4225, Barcelona, Spain

C. Masoller (✉)
Departament de Fisica, Universitat Politecnica de Catalunya, Colom 11, Terrassa, 08222, Barcelona, Spain
e-mail: cristina.masoller@upc.edu

A.A. Tsonis (ed.), *Advances in Nonlinear Geosciences*,
DOI 10.1007/978-3-319-58895-7_4

networks have been proposed. These methods for constructing complex networks from data are complemented by a careful analysis of the inferred network, in order to detect fake links, missing links, hidden nodes, etc. (Timme 2007; Serrano et al. 2009; Shandilya and Timme 2011; Yu and Parlitz 2011; Rubido et al. 2014; Tirabassi et al. 2015a, b).

Considering the complexity of the inter-relations between the different elements that constitute the climate system, it is clear that the analysis of observed climatological data from a complex network perspective has great potential for yielding light into relevant, previously unknown features of our climate.

Indeed, in the last two decades the research field of climate networks has provided important insight into complex phenomena in our climate (Tsonis and Roebber 2004; Tsonis and Swanson 2008; Yamasaki et al. 2008; Donges et al. 2009; Barreiro et al. 2011; Fountalis et al. 2014; Hlinka et al. 2014; Tirabassi et al. 2015a, b). Nowadays climate networks are a research field located at the triple intersection of three active areas in nonlinear science: network theory, time series analysis, and climate dynamics.

The European project *Learning about Interacting Networks in Climate* (LINC, climatelinc.eu) brought together researchers from these communities with the goals of training the new generation of researchers, developing cutting-edge science, and promoting new collaborations. Here we present a summary of some of our results developed within the LINC project.

2 Time-Scale Analysis of Climate Interactions

The work by Barreiro et al. (2011) was a first approach to characterize the climate network by means of recurrent oscillatory patterns, with various time scales, as described by using symbolic *ordinal analysis* (Bandt and Pompe 2002). By mapping these processes into a climate network, we found that the structure of the network changes drastically at different time scales.

The symbolic method of ordinal analysis first divides a time series $x(t)$ of length M into $M - D$ overlapping vectors of dimension D. Then, each element of a vector is replaced by a number from 0 to $D - 1$, in accordance with its relative magnitude in the ordered sequence (0 corresponding to the smallest and $D - 1$ to the largest value in each vector). For example, with $D = 3$, the vector $(v_0, v_1, v_2) = (6.8, 11.5, 11)$ gives the "ordinal pattern" (OP) 201 because $v_2 < v_0 < v_1$. In this way, each vector has associated an OP composed by D symbols, and the symbol sequence comes from the comparison of neighboring values. With $D = 3$ the $3! = 6$ different patterns are (012, 021, 102, 120, 201, and 210). Last, the presence of recurrent oscillatory patterns in the time series is characterized by means of the probabilities of the ordinal patterns, computed from their frequency of occurrence in the time series.

A classical measure to investigate mutual interdependencies between time series is the mutual information (MI), which is computed from the probability distribution functions (PDFs) associated to the two time series, and the joint probability distribution. When using the ordinal probabilities to compute the MI, the PDF is computed with $D!$ bins, and the joint probability, with $D! \times D!$ bins. Therefore, to have a good statistics one must have enough data points in the time series, i.e., $M \gg D! \times D!$

A main advantage of ordinal analysis is that it allows selecting the time scale of the analysis by comparing L-lagged data points instead of consecutive data points. For example, in SAT reanalysis with monthly resolution, by comparing four data points separated by twelve months (i.e., $(v_0, v_{12}, v_{24}, v_{36})$) we can investigate recurrent oscillatory patterns with a characteristic time scale of 4 years.

When using an interdependency statistical measure, such as the MI, to define the links of the climate network, one must use an appropriate criterion to define which MI values are considered significant and represented as network links. Performing such significance analysis is a challenging task. A particularly important problem for climate networks is the fact that, due to physical proximity (i.e., due to the spatial embedding of the network), the strongest links are those between neighboring regions. Therefore, by using a high significance threshold, one ends up with a network in which long-distance links are scarce. On the other hand, by choosing a low significance threshold, a lot of "noise" is included in the network as fake links. Therefore, the challenge is how to select the threshold that provides the best compromise between the need to include relevant long-distance links that represent genuine atmospheric teleconnections, and the need to limit the proliferation of noisy links.

The networks obtained from Surface Air Temperature (SAT, NCEP/NCAR monthly reanalysis covering the period January 1949 to December 2006) with ordinal patterns formed by comparing SAT anomalies in the same month during four consecutive years (i.e., $D = 4$ and $L = 12$) are shown in Fig. 1 (Barreiro et al. 2011). In this figure, the networks obtained with different MI significance thresholds are shown. One can notice that in this "inter-annual" time scale the

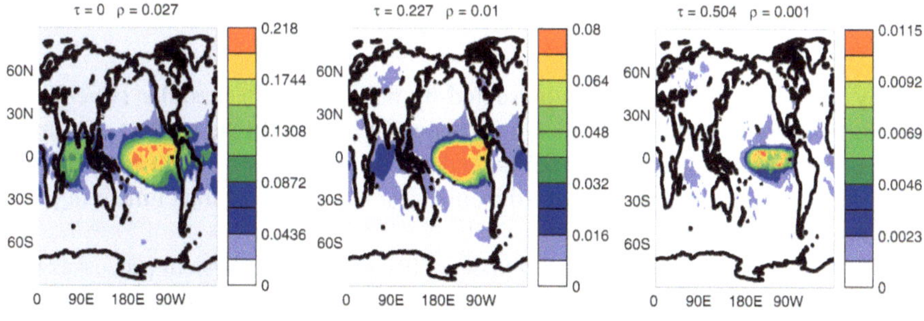

Fig. 1 Climate networks constructed by computing the mutual information (MI) from the probabilities of ordinal patterns of length $D = 4$ defined by comparing SAT anomalies (NCEP/NCAR monthly reanalysis) in consecutive years ($L = 12$). The color-code is such that the *white* (*red*) regions indicate the geographical areas with zero (largest) area weighted connectivity. The significance threshold increases from *left* to *right*. Adapted from Barreiro et al. (2011)

dominant atmospheric connections are located in the tropical Pacific and Indian Ocean areas, mainly associated with El Niño phenomenon. One can also notice that, as expected, the connectivity of the network decreases as the MI significance threshold is increased. For the highest threshold considered (shown in the right panel, here the threshold is selected such that the density of the network is 0.1% of the total possible links) the El Niño—Indian Ocean teleconnection is significantly weakened with respect to the lower threshold network (shown in the left panel, here the threshold is chosen equal to the maximum MI value obtained from surrogated data, which gives a network with 2.7% of the total links).

Figure 2 summarizes the effect of the lag L used to define the ordinal patterns (Deza et al. 2013). When the OPs are defined in terms of consecutive months (top row) the network links are mainly local. In the seasonal time scale (middle row) the tropical region becomes connected. Clearly, the extra-tropics become connected to the equatorial Pacific through atmospheric teleconnection processes only when considering inter-annual time scales (bottom row).

Figure 3 displays the climate network when the mutual information is computed with the classical approach, i.e., computing the PDFs from the histograms of values in the time series (i.e., without taking into account the ordering of the data points). We note that the network looks as a "superposition" of spatial structures which are present only in some of the maps shown in Fig. 2. See, for example, the highly connected green spot in the Labrador Sea, which is also seen in Fig. 2a and to a lesser extent in Fig. 2b; but is not present in Fig. 2c. The Labrador Sea is one of the most important regions of deep water formation in the north Atlantic. The formation of this water occurs in wintertime and depends on the passage of extratropical storms that cool the surface. The passage of storms is in turn related to the state of the North Atlantic Oscillation. As a result, there is a clear connection of the Labrador Sea with the rest of the north Atlantic mainly on seasonal time scales and is mostly independent of ENSO activity.

3 Climate Communities

Many natural systems can be represented by networks with modular structure in the form of communities of densely interconnected nodes. In the context of climate networks, climate communities can be understood as a set of geographical regions that share some common property (dynamical or statistical) of the climate in those regions.

The existence of such regions is expected because of the physical processes that govern our climate (ocean and atmospheric processes, solar forcing, vegetation, human activity, etc.), act in a similar way in distant regions (having similar effects), and therefore, distant regions can have similar climate. Examples include tropical rainforests, dry and arid regions, maritime regions, etc.

The methodology for constructing climate networks described in the previous section is not appropriate for detecting such community structure (i.e., regions which have similar climatic properties) because, as seen in Figs. 2 and 3, the short-

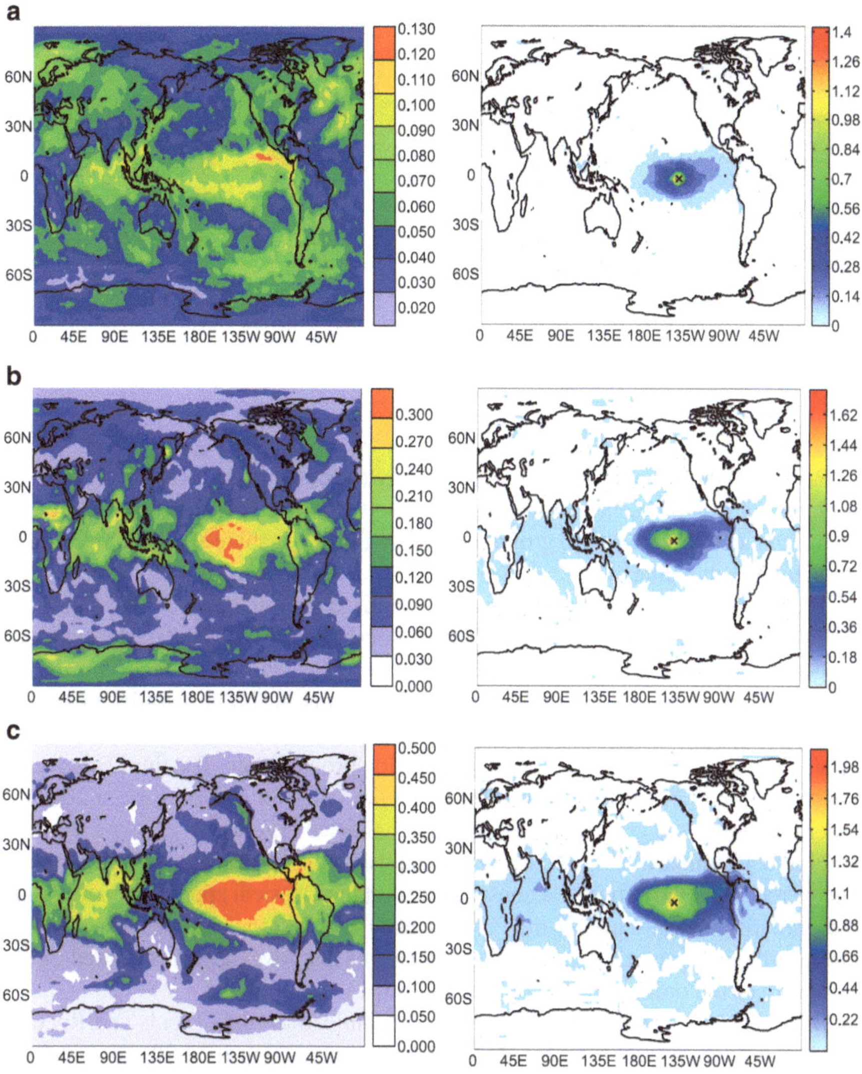

Fig. 2 Area weighted connectivity (*left column*) and connectivity maps (mutual information values of the significant links of the node indicated by an *x*, *right column*) using $D = 3$ OPs formed with three consecutive months ($L = 1$, *top row*), OPs formed with three equally spaced months covering a one-year period ($L = 3$, *middle row*); and OPs formed with 3 months in consecutive years ($L = 12$, *bottom row*). Adapted from Deza et al. (2013)

distance links between neighboring nodes dominate, and the northern and southern hemispheres are only indirectly or weakly connected. Therefore, in this network, areas of tropical rainforests, for example, which are located in different hemispheres won't be identified as belonging to the same community, because there won't be links that interconnect them.

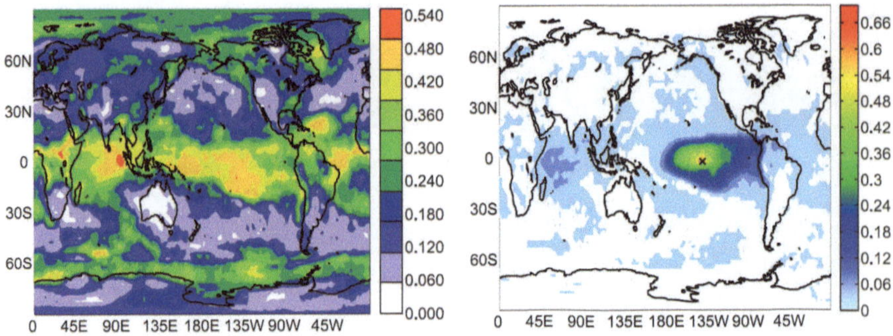

Fig. 3 As Fig. 2 but when the mutual information is computed from the PDFs of SAT anomaly data values. Adapted from Deza et al. (2013)

Recently we proposed a novel methodology to construct the network that allows overcoming this problem (Tirabassi and Masoller 2016). The methodology, based on ordinal analysis, allows to group together regions that share similar properties of the symbolic dynamics.

The main idea is to assign a high (low) weight to the link between two regions, if the ordinal transition probabilities (TPs) that describe the statistics of the symbolic sequence are very similar (very different) in the two regions. In other words, the symbolic sequences are mutually compared in terms of the probability of pattern "A" being followed by pattern "B." Then, a significance threshold is used to keep only the regions that have very similar transition probabilities. The third step was to run the Infomap community detection algorithm (Rosvall and Bergstrom 2007) in order to identify the groups of densely interconnected regions.

Figure 4 summarizes the results of the analysis. Panel a displays the communities uncovered when the network is constructed with the classical approach (in this case, the similarity measure used is the Pearson cross-correlation coefficient) and panel b displays the communities uncovered by means of the novel approach, based on the similarity of the ordinal transition probabilities.

By using the classical approach with a threshold $W = 0.5$ (Tsonis and Roebber 2004), Infomap algorithm uncovers 8604 communities, but only 20 are composed by more than two nodes. Figure 4a displays the largest 16 communities. The detected communities include the central-east equatorial Pacific dominated by El Niño, the tropical western Pacific, Indian Ocean, and tropical north Atlantic regions controlled mainly by the exchange of heat fluxes with the atmosphere, and the equatorial Atlantic cold tongue dominated by dynamical air–sea interaction. The other communities are small and some may be just noise.

In contrast, with the ordinal approach the community structure inferred, shown in Fig. 4b, divides the world in eight areas that share similar climatic properties, as measured by similar symbolic transition probabilities. There are five macro-communities: extratropical continents and southern ocean characterized by large SAT variability (indicated with number 0), northern oceans (2), regions of tropical

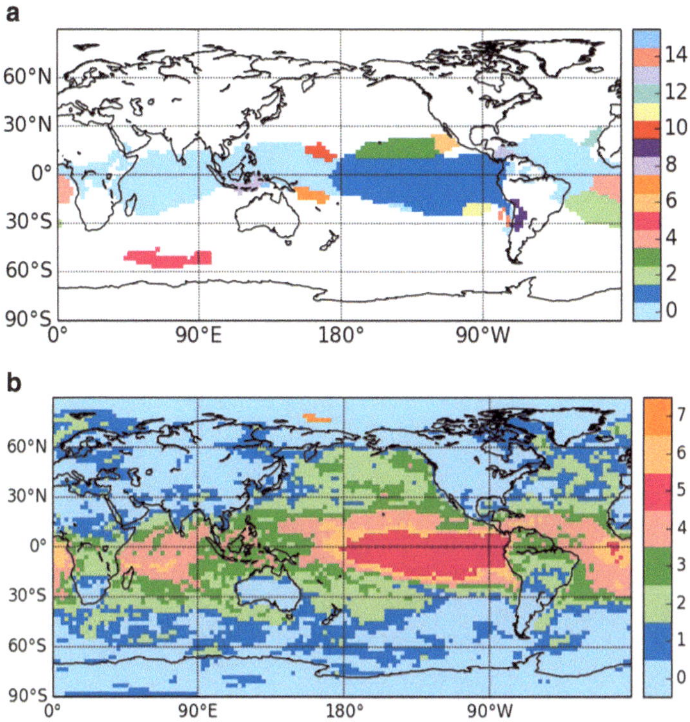

Fig. 4 Community structure uncovered by Infomap algorithm. The different communities are indicated with different colors. (**a**) The network is constructed by using the Pearson cross-correlation coefficient as a measure of dynamical similarity. (**b**) The network is constructed by calculating the similarity of the ordinal transition probabilities. In panel (**a**), for clarity, only the largest 16 communities are shown. Adapted from Tirabassi and Masoller (2016)

deep convection such as the western Pacific warm pool, Amazon and Congo basins (3), tropical oceans dominated air–sea heat fluxes (4) and ENSO basin (5). Then, there are also two boundary communities, indicated with numbers 1 and 6, which are placed at the communities interfaces.

Both methodologies identify the region dominated by the El Niño dynamics as a community, but there are differences in the rest. Compared to the communities calculated with the classical approach the new methodology is able to separate better in terms of processes dominating the SAT variability. For example, the new methodology (1) identifies the central equatorial Atlantic as having a similar behavior to El Niño, which is consistent with the literature (Zebiak 1993); (2) separates the behavior of SAT over the maritime continent from that of the Indian and tropical Atlantic oceans, consistent with a different rainfall regime, (3) considers the tropical north and south Atlantic as belonging to the same community, which is consistent because temperature is strongly controlled by air–sea heat fluxes.

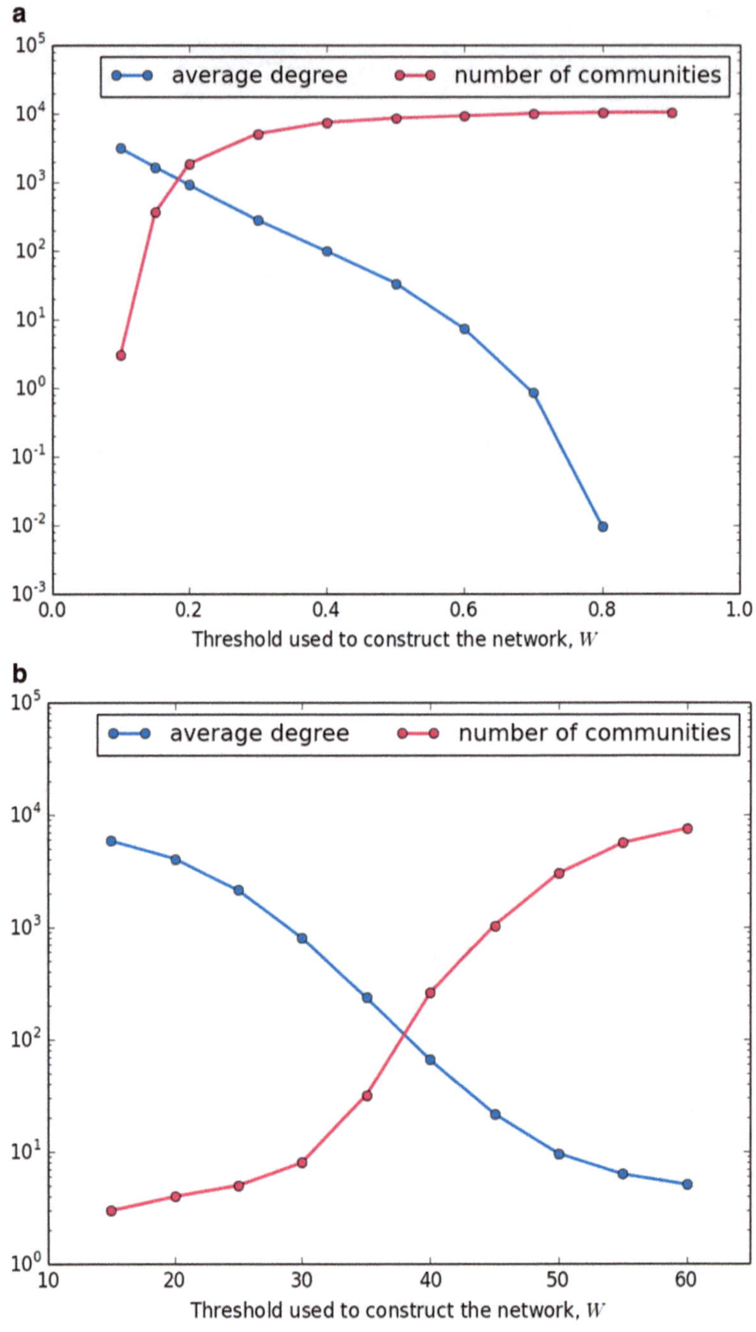

Fig. 5 Network average degree and number of communities vs. the threshold used to construct the network, W. In panel (**a**) the network is constructed by using the Pearson cross-correlation coefficient. For the community structure shown in Fig. 4a, the threshold used was $W = 0.5$ (as in

As discussed before, in order to construct a climate network, the links weights have to be pruned by using an adequate threshold. Decreasing the threshold leads to a more connected network, while increasing it results in a sparser one. The number of communities depends on the number of connections, which in turn depends on the threshold. In order to uncover a coherent, well-defined community structure, the threshold has to be carefully chosen. Figure 5 displays the number of communities and the average degree as a function of the threshold. It can be seen that there is a negative correlation between them. The fragmentation of the network into smaller communities (as community seven in Fig. 4b) can be due to the removal of relevant links that keep the bigger communities together. Thus, to obtain a meaningful community structure, we selected *ad hoc* a threshold that provided the best compromise between the need to limit the small-communities-proliferation and the need to include in the network only the relevant links.

4 Net Direction of Climate Interactions

A main drawback of the methodology discussed in the previous sections for inferring the climate network is that it uses a symmetric similarity measure (the mutual information or the Pearson correlation coefficient) that yield non-directed networks. In these networks the presence of a link indicates inter-dependency but the direction of the underlying interaction is not inferred. For improving the understanding of climate phenomena and its predictability, it is of foremost importance not only to be able to infer the presence of a link between two nodes but also to infer the direction of this interaction.

Deza et al. (2015) used a methodology that allows inferring directed interactions via an analysis of the net direction of information transfer. A measure was used—based on conditional mutual information—that quantifies the amount of information in a time-series $x(t)$, contained in τ time units in the past of another time series $y(t)$. The resulting network was found to be in full agreement with state-of-the-art knowledge in climate phenomena, validating in this way the methodology used. No assumptions about physical processes were made, except for the appropriate setting of the parameter τ.

Fig. 5 (continued) Tsonis and Roebber 2004). In panel (**b**) the network is constructed by calculating the similarity of the ordinal transition probabilities. For uncovering the communities shown in Fig. 4b, the threshold used was $W = 30$. It can be observed that with the first approach, a very low threshold needs to be used to uncover a small set of communities. However, using a low threshold has the strong disadvantage of including in the network many links which are not significant. In contrast, with the novel approach (by constructing the network considering the similarities of the transition probabilities), the variation of the number of communities with the threshold is more gradual, which allows uncovering a small set of communities by using a threshold that is not too low. Adapted from Tirabassi and Masoller (2016)

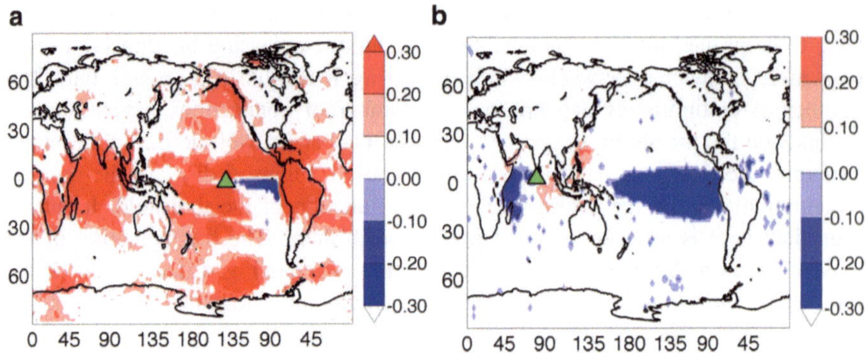

Fig. 6 Directionality of the links in a node in the central pacific (**a**) and in a node in the Indian Ocean (**b**) indicated with a *triangle*. The color code indicates the directionality index: outgoing links are shown in *red* while incoming links are shown in *blue*. The time scale of information transfer is $\tau = 30$ days. Adapted from Deza et al (2015)

The directionality measure and the statistical significance analysis are discussed in detail in Deza et al. (2015). Here we present two examples that illustrate the directional structure of the network. Figure 6 displays the directionality of the links of two nodes in the tropics (indicated with triangles) computed from SAT reanalysis data with daily resolution and parameter $\tau = 30$ days. The color code in this figure indicates the Directionality Index (DI): outgoing links are shown in red, while the incoming links are shown in blue.

Figure 6a shows, as expected, the central Pacific influenced by the eastern Pacific (in blue) and influencing the global network, with many regions in the tropics and in the extra-tropics in red. Reciprocally, Fig. 6b shows that the blue links come to the node in the Indian Ocean from a well-defined region in the central Pacific Ocean. In addition, few red outgoing links connect the node in the Indian Ocean to other regions. A main drawback of the directionality index used is that it does not distinguish indirect from direct information transfer. Therefore, the red areas influenced by the node in the Indian Ocean can be an artifact in the sense that these regions might be directly influenced by El Niño region.

Figure 7 displays the influence of the parameter τ that characterizes the time scale of the information transfer from one node to another. As an example, a region in southeastern South America is considered (indicated with a triangle). For synoptic time scales of a few days, the directionality index uncovers the existence of a wave train propagating with a southwest-northeast direction. Moreover, there is a clear separation line between regions with incoming and outgoing links. This configuration is characteristic of the progression of a front through the reference point. As the parameter τ increases, the extratropical wave train associated with synoptic time scales fades and only blue links to the tropics remain, consistent with an influence of the equatorial Pacific on the region on longer time scales, likely related to ENSO.

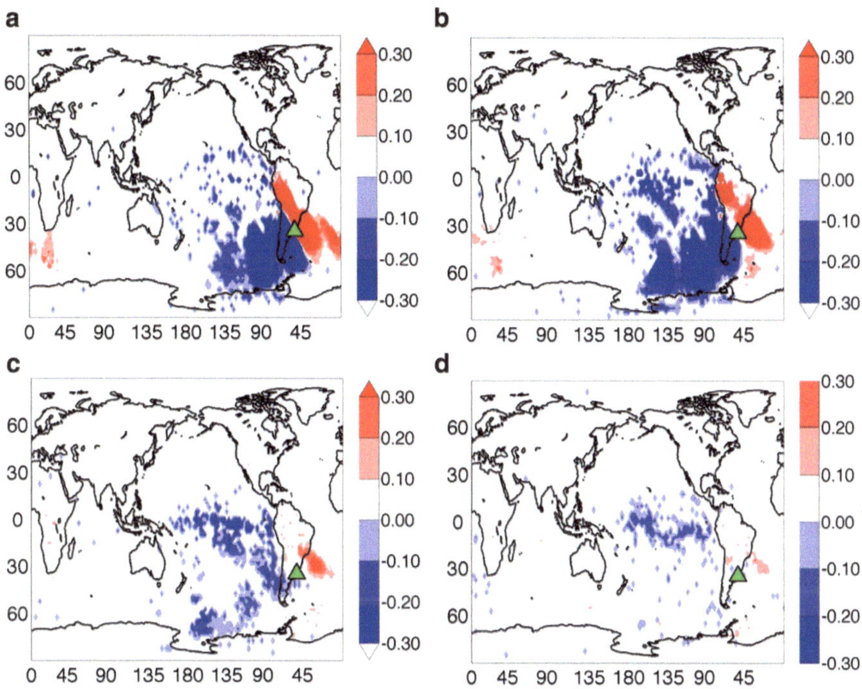

Fig. 7 Directionality of the links in a node in southeastern South America, indicated with a *triangle*. The color code indicates the directionality index: outgoing links are shown in *red* while incoming links are shown in *blue*. The time scale of information transfer τ is (**a**) 1 day, (**b**) 3 days, (**c**) 7 days, and (**d**) 30 days. Adapted from Deza (2015)

5 Conclusions

We have shown that symbolic time series analysis based on ordinal patterns and information theory measures, applied to surface air temperature anomalies (reanalysis data with monthly or daily resolution) are powerful tools for uncovering the large-scale structure of the climate network.

A main advantage of the ordinal methodology is that, by varying the dimension of the pattern and the year–month comparison, one can uncover memory processes with different time scales, and depending on the time scale considered, the climate network can change completely. Overall we found that on seasonal time scales the extra-tropical regions, mainly over Asia and North America, present strong connectivity, while in inter-annual time scales, the tropical Pacific clearly dominates.

A novel methodology for inferring the community structure of the climate network was proposed. Constructing the climate network by taking into account the similarity of the ordinal transition probabilities in different regions allowed to identify communities formed by geographical regions where the climate variability displays similar statistics of ordinal patterns. Five macro-communities were identified: extratropical continents, northern oceans, tropical convective regions, tropical oceans, and ENSO basin.

The analysis of the net directionality of the links revealed variability patterns consistent with well-known features of the global climate dynamics. For example, in the extra-tropics, the link direction revealed wave trains propagating from west to east, in both hemispheres. A drawback of the directionality index employed is that it does not distinguish direct from indirect interactions.

Ongoing and future work is aimed at exploring the suitability of other techniques of time series analysis, such as Hilbert analysis (Zappalà et al. 2016), other directionality measures (partial directed coherence and directed partial correlation, Tirabassi et al. 2017), and measures of distances between time series and entropy measures (Arizmendi et al. 2017) for gaining additional information from climatological datasets.

Acknowledgements This work was supported by the LINC project (FP7-PEOPLE-2011-ITN, Grant No. 289447), Spanish MINECO (FIS2015-66503-C3-2-P) and ICREA ACADEMIA.

References

Albert, R., and A.L. Barabasi. 2002. Statistical mechanics of complex networks. *Reviews of Modern Physics* 74: 47–97.

Arizmendi, F., M. Barreiro, and C. Masoller. 2017. Identifying large-scale patterns of nonlinearity and unpredictability in atmospheric data. *Scientific Reports* 7: 45676.

Bandt, C., and B. Pompe. 2002. Permutation entropy: a natural complexity measure for time series. *Physical Review Letters* 88: 174102.

Barreiro, M., A.C. Marti, and C. Masoller. 2011. Inferring long memory processes in the climate network via ordinal pattern analysis. *Chaos* 21: 013101.

Boccaletti, S., V. Latora, Y. Moreno, M. Chavez, and D.-U. Hwang. 2006. Complex networks: structure and dynamics. *Physics Reports* 424: 175–308.

Deza, J.I., M. Barreiro, and C. Masoller. 2013. Inferring interdependencies in climate networks constructed at inter-annual, intra-season and longer time scales. *The European Physical Journal Special Topics* 222: 511–523.

Deza, J.I., M. Barreiro, and C. Masoller. 2015. Assessing the direction of climate interactions by means of complex networks and information theoretic tools. *Chaos* 25: 033105.

Donges, J.F., Y. Zou, N. Marwan, and J. Kurths. 2009. The backbone of the climate network. *EPL* 87: 48007.

Fountalis, I., A. Bracco, and C. Dovrolis. 2014. Spatio-temporal network analysis for studying climate patterns. *Climate Dynamics* 42: 879–899.

Hlinka, J., D. Hartman, M. Vejmelka, D. Novotna, and M. Palus. 2014. Non-linear dependence and teleconnections in climate data: sources, relevance, nonstationarity. *Climate Dynamics* 42: 1873–1886.

Newman, M.E.J. 2003. The structure and function of complex networks. *SIAM Review* 45: 167–256.

Rosvall, R., and C.T. Bergstrom. 2007. An information-theoretic framework for resolving community structure in complex networks. *PNAS* 104: 7327–7331.

Rubido, N., A.C. Martí, E. Bianco-Martínez, C. Grebogi, M.S. Baptista, and C. Masoller. 2014. Exact detection of direct links in networks of interacting dynamical units. *New Journal of Physics* 16: 093010.

Serrano, M.A., M. Boguña, and A. Vespignani. 2009. Extracting the multiscale backbone of complex weighted networks. *PNAS* 106: 6483–6488.

Shandilya, S.G., and M. Timme. 2011. Inferring network topology from complex dynamics. *New Journal of Physics* 13: 013004.

Timme, M. 2007. Revealing network connectivity from response dynamics. *Physical Review Letters* 98: 224101.

———. 2016. Unravelling the community structure of the climate system by using lags and symbolic time-series analysis. *Scientific Reports* 6: 29804.

Tirabassi, G., C. Masoller, and M. Barreiro. 2015a. A study of the air–sea interaction in the South Atlantic Convergence Zone through Granger causality. *International Journal of Climatology* 35: 3440.

Tirabassi, G., R. Sevilla-Escoboza, J.M. Buldú, and C. Masoller. 2015b. Inferring the connectivity of coupled oscillators from time series statistical similarity analysis. *Scientific Reports* 5: 10829.

Tirabassi, G., L. Sommerlade, and C. Masoller. 2017. Inferring directed climatic interactions with renormalized partial directed coherence and directed partial correlation. *Chaos* 27: 035815.

Tsonis, A.A., and P. Roebber. 2004. The architecture of the climate network. *Physica A* 333: 497–504.

Tsonis, A.A., and K.L. Swanson. 2008. Topology and predictability of El Nino and La Nina networks. *Physical Review Letters* 100: 228502.

Yamasaki, K., A. Gozolchiani, and S. Havlin. 2008. Climate networks around the globe are significantly affected by El Nino. *Physical Review Letters* 100: 228501.

Yu, D., and U. Parlitz. 2011. Inferring network connectivity by delayed feedback control. *PloS One* 6: e24333.

Zappalà, D.A., M. Barreiro, and C. Masoller. 2016. Global atmospheric dynamics investigated by using Hilbert frequency analysis. *Entropy* 18: 408.

Zebiak, S.E. 1993. Air–sea interaction in the equatorial Atlantic region. *Journal of Climate* 6: 1567.

Supermodeling: Synchronization of Alternative Dynamical Models of a Single Objective Process

Gregory S. Duane, Wim Wiegerinck, Frank Selten, Mao-Lin Shen, and Noel Keenlyside

Abstract Imperfect models of the same objective process give an improved representation of that process, from which they assimilate data, if they are also coupled to one another. Inter-model coupling, through nudging, or more strongly through averaging of dynamical tendencies, typically gives synchronization or partial synchronization of models and hence formation of consensus. Previous studies of supermodels of interest for weather and climate prediction are here reviewed. The scheme has been applied to a hierarchy of models, ranging from simple systems of ordinary differential equations, to models based on the quasigeostrophic approximation to geophysical fluid dynamics, to primitive-equation fluid dynamical models, and finally to state-of-the-art climate models. Evidence is reviewed to test the claim that, in nonlinear systems, the synchronized-model scheme surpasses the usual procedure of averaging model outputs.

Keywords Synchronization • Data assimilation • Supermodel

1 Introduction

It has been established that a computational model that runs in parallel to the objective process being modeled can be conceived as synchronizing with that process through a one-way truth-model coupling (Duane et al. 2006; Yang et al.

G.S. Duane (✉)
Geophysical Institute, University of Bergen, Bergen, Norway

Department of Atmospheric and Oceanic Sciences, University of Colorado, Boulder, CO, USA
e-mail: gregory.duane@colorado.edu

M.-L. Shen • N. Keenlyside
Geophysical Institute, University of Bergen, Bergen, Norway

W. Wiegerinck
SNN Adaptive Intelligence, Nijmegen, The Netherlands

Department of Biophysics, Donders Institute for Brain, Cognition and Behaviour, Radboud University, Nijmegen, The Netherlands

F. Selten
Royal Netherlands Meteorological Institute (KNMI), De Bilt, The Netherlands

A.A. Tsonis (ed.), *Advances in Nonlinear Geosciences*,
DOI 10.1007/978-3-319-58895-7_5

2004). In numerical weather prediction, the repeated updates of the model based on new observations constitute the enterprise of *data assimilation*, methods for which are well developed in meteorology (Kalnay 2003). It can indeed be shown that Kalman filtering, the algorithm that provides the basis for the most common data assimilation methods, is also optimal for synchronization of truth and model under weak assumptions of local linearity (Duane et al. 2006).

Similarly, a biological organism perceives reality through a stream of incoming data and forms a prognostically useful perception that synchronizes with, but is distinct from objective reality. A conscious organism exhibits an additional feature: it perceives itself, focusing on its own thoughts in the same manner as it does the objective world. In this view, there must be semi-autonomous parts of a "conscious" mind that perceive one another. These components of the mind synchronize with one another, or in alternative language, they perform "data assimilation" from one another, with a limited exchange of information, lending an additional degree of objectivity to a conscious organism.

Such a scheme has actually been proposed in a computational science context, for the fusion of alternative computational models of the same objective process (Duane et al. 2009; Mirchev et al. 2012; van den Berge et al. 2011): different numerical models used to predict climate change in the twenty-first century differ by as much as a factor of two in the amount of globally averaged warming and differ completely in their projections for specific regions of the globe. Current practice is just to average the results of the different models. By synchronizing a small set of alternative models with each other, a more reliable and detailed consensus could be obtained.

The supermodel strategy is schematized in Fig. 1, for three constituent models. The three models perform data assimilation from (synchronization with) reality, through diffusive coupling with coefficient matrices K^i ("Kalman gains" in the language of data assimilation).

The lth variable in Model i is nudged to the lth variable in Model j with connection coefficient C_l^{ij}. The connections C_{ij} linking the three model systems can be chosen using yet a further extension of the synchronization paradigm: if

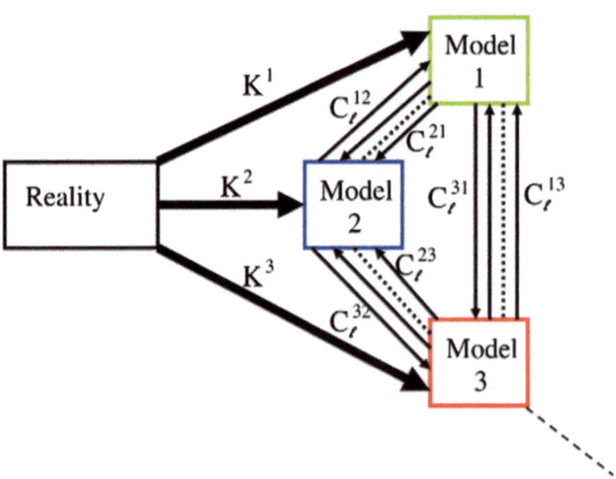

Fig. 1 In a supermodel, models are linked to each other, generally in both directions and to "reality" in one direction. Separate links between models, with distinct values of the connection coefficients C_l^{ij}, are introduced for different variables and for each direction of possible influence

two systems synchronize when their parameters match, then under some weak assumptions, as was proven in Duane et al. (2006), it is possible to prescribe a dynamical evolution law for general parameters in one of the systems, so that the parameters of the two systems, as well as the states will converge. In the present case, the tunable parameters are taken to be the connection coefficients (not the parameters of the separate models), and they are tuned under the peculiar assumption that reality itself is a similar suite of connected systems.

In the following sections, we present the results of the supermodeling approach in a hierarchy of increasingly complex models. Details of the learning algorithm are reviewed in the next section, for simple examples where the models are sets of a few ordinary differential equations. In Sect. 3, the strategy is applied to a partial differential equation model, the quasigeostrophic channel model, where the advantages of supermodeling can be clearly compared to ex post facto averaging. In Sect. 4 it is shown that the scheme can be applied to a fluid dynamical model capturing realistic features of the climate system. Preliminary efforts with state-of-the-art climate models are reviewed in Sect. 5, and the overall status of supermodeling is summarized in Sect. 6.

2 Supermodeling with Low-Order Models

A simple supermodel is constructed from a collection of Lorenz systems (Lorenz 1963) that each imperfectly represent a "true" Lorenz system. Three imperfect "model" Lorenz systems were generated by perturbing parameters in the differential equations for a given "real" Lorenz system and adding extra terms. The resulting suite is: $dx/dt = \sigma(y - z)$, $dy/dt = \rho x - y - xz$, $dz/dt = -\beta z + xy$, and

$$\begin{aligned} dx_i/dt &= \sigma_i\,(y_i - z_i) + \Sigma_{j\neq i}\; C^x_{ij}\,(x_j - x_i) + K^x\,(x - x_i) \\ dy_i/dt &= \rho x_i - y_i - x_i z_i + \mu_i + \Sigma_{j\neq i} C^y_{ij}\,(y_j - y_i) + K^y\,(y - y_i) \\ dz_i/dt &= -\beta_i z_i + x_i y_i + \Sigma_{j\neq i} C^z_{ij}\,(z_j - z_i) + K^z\,(z - z_i) \end{aligned} \tag{1}$$

where (x, y, z) is the real Lorenz system and (x_i, y_i, z_i) $i = 1, 2, 3$ are the three models. An extra term μ is present in the models, but not in the real system. Because of the relatively small number of variables available in this toy system, all possible directional couplings among corresponding variables in the three Lorenz systems were considered, giving 18 connection coefficients C^A_{ij} $A = x, y, z; i, j = 1, 2, 3, i \neq j$. The constants K^A $A = x, y, z$ are chosen arbitrarily so as to effect "data assimilation" from the "real" Lorenz system into the three coupled "model" systems.

It remains to determine connection coefficients C^A_{ij} that will define an optimal supermodel. The general method for parameter adaptation in any imperfect replica of any dynamical system with which the imperfect replica synchronizes (Duane et al. 2007), to be applied here, is the following: Consider a "real system" given by ODE's: $d\mathbf{x}/dt = \mathbf{f}(\mathbf{x}, \mathbf{p})$, $d\mathbf{p}/dt = 0$, where $\mathbf{x} \in \mathrm{R}^N$, $\mathbf{f}$: $\mathrm{R}^N \rightarrow \mathrm{R}^N$, and $\mathbf{p} \in \mathrm{R}^m$

is the vector of (unknown, constant) parameters of the system. Further assume that $\mathbf{s} = \mathbf{h}(\mathbf{x})$, where $\mathbf{h}$: $R^N \to R^n$, $n \le N$, is an n dimensional vector representing the experimental measurement output of the system. A "computational model" of the system is given by $d\mathbf{y}/dt = \mathbf{f}(\mathbf{y},\mathbf{q}) + \mathbf{u}(\mathbf{y},\mathbf{s})$, $d\mathbf{q}/dt = \mathbf{N}(\mathbf{y}, \mathbf{x} - \mathbf{y})$ where $\mathbf{N}(\mathbf{y}, 0) = 0$, and $\mathbf{u}$ is the control signal. Generally, the real system and its model are chaotic; for $\mathbf{u} = 0$ the simulation quickly diverges from the real system behavior. The problem is to find a parameter estimation law $\mathbf{N}$, so that $\mathbf{q} \to \mathbf{p}$, if we are given a control law $\mathbf{u}$ such that $\mathbf{y} \to \mathbf{x}$. Let $\mathbf{e} \equiv \mathbf{y} - \mathbf{x}$ and $\mathbf{r} \equiv \mathbf{q} - \mathbf{p}$. Consider a Lyapunov function $L_0(\mathbf{e})|_{\mathbf{q}=\mathbf{p}}$ that is positive definite and monotonically decreasing after some time, e.g., $L_0(\mathbf{e}) = e^2$. The recipe for the desired N, as proved in (Duane et al. 2007), is

$$N_j = -\delta_j \Sigma_i \left[(\partial L_0/\partial e_i) \left(\partial h_i/\partial r_j \right) \right] \tag{2}$$

where the δ_j are arbitrary positive constants, and $\mathbf{h} \equiv \mathbf{f}(\mathbf{y},\mathbf{r} + \mathbf{p}) - \mathbf{f}(\mathbf{y} - \mathbf{e},\mathbf{p})$. Typically, the first factor in brackets is simply e_i and the second factor is the cofactor of parameter p_j in the dynamical equation for x_i.

Letting the parameters to be estimated be the connection coefficients themselves (not the parameters of the separate models), the dynamical equation for these coefficients was chosen as:

$$dC_{ij}^X/dt = a\left(x_j - x_i\right)\left(x - \frac{1}{3}\Sigma_k x_k\right) - \varepsilon/\left(C_{ij}^X - C_{\max}\right)^2 + \varepsilon/\left(C_{ij}^X + \delta\right)^2 \tag{3}$$

with analogous equations for C^Y and C^Z, where the adaptation rate a is an arbitrary constant and the extra terms with coefficient ε dynamically constrain all couplings C^A to remain in the range $(-\delta, C_{\max})$ for some small number δ. The rule (3) has a simple interpretation: time integrals of the first terms on the right-hand side of each equation give the covariance between truth-model synchronization error, $x - 1/3\Sigma_k x_k$, and inter-model "nudging", $x_j - x_i$. We indeed want to increase or decrease the inter-model nudging, for a given pair of corresponding variables, depending on the sign and magnitude of this covariance. The procedure will produce a set of values for the connection coefficients that is at least locally optimal in the multidimensional space of connection values.

Figure 2a shows the results for a simple case in which each of the three model systems contains the "correct" equation for only one of the three variables and "incorrect" equations for the other two (Duane 2013; Duane et al. 2009). The couplings did not converge, but the coupled suite of "models" rapidly synchronized with the "real" system, even with the adaptation process turned off half-way through the simulation, so that the coupling coefficients C_{ij}^A subsequently held fixed values. (The three models also synchronized among themselves nearly identically.) The inter-model connections are needed, despite efforts, common in the modeling community (Tebaldi and Knutti 2007), to combine only the outputs of independently run models using Bayesian reasoning. The difference between corresponding variables in the "real" and coupled "model" systems was significantly less than the difference using the average outputs of the same suite of models, not coupled among themselves (Fig.

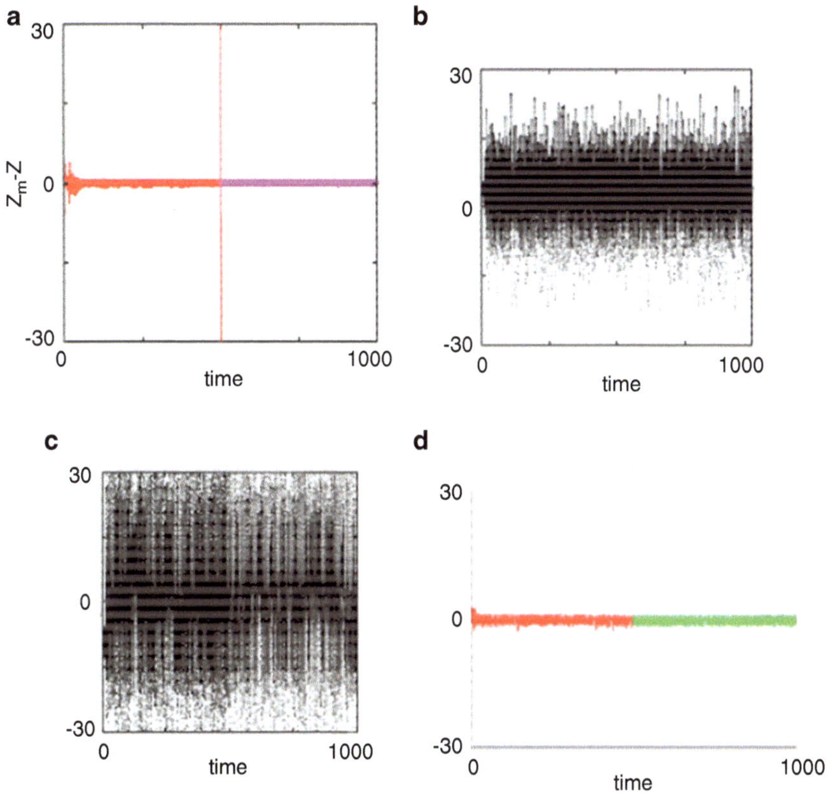

Fig. 2 Difference $z_m - z$ between "model" and "real" z vs. time for a Lorenz system with $\rho = 28$, $\beta = 8/3$, $\sigma = 10.0$ and an interconnected suite of models with $\rho_{1,2,3} = \rho$, $\beta_1 = \beta$, $\sigma_1 = 15.0$, $\mu_1 = 30.0$, $\beta_2 = 1.0$, $\sigma_2 = \sigma$, $\mu_2 = -30.0$, $\beta_3 = 4.0$, $\sigma_3 = 5.0$, $\mu_3 = 0$. The synchronization error is shown for (**a**) the average of the coupled suite $z_m = (z_1 + z_2 + z_3)/3$ with couplings C^A_{ij} adapted according to (3) for $0 < t < 500$ and held constant for $500 < t < 1000$; (**b**) the same average z_m, but with all $C^A_{ij} = 0$; (**c**) $z_m = z_1$, the output of the model with the best z equation, with $C^A_{ij} = 0$; (**d**) as in (**a**), but with $\beta_1 = 7/3$, $\sigma_2 = 13.0$, and $\mu_3 = 8.0$, so that no equation in any model is "correct"

2b). Further, without the model–model coupling, the output of the single model with the best equation for the given variable (in this case, z, modeled best by z_1 in Model 1) differed even more from "reality" than the average output of the three models (Fig. 2c). Therefore, it is unlikely that any ex post facto weighting scheme applied to the three outputs would give results equaling those of the synchronized suite. Internal synchronization within the multi-model "mind" is essential. The choice of semi-autonomous models to be combined is not essential—in a case where no model had the "correct" equation for any variable, results deteriorated only slightly (Fig. 2d).

The synchronization-based method for adapting the inter-model connections is only guaranteed to find a supermodel that is *locally* optimal in the space of connection coefficients. It is not yet known whether local optima are an impediment when such a space is high dimensional. However, Mirchev et al. (2012) obtained some improvement in another supermodel constructed from Lorenz systems by introducing stochasticity in the training procedure, a commonly used way to escape local optima.

Synchronization-based adaptation of coefficients is a form of machine learning on-the-fly, in which the coefficients typically oscillate wildly. A more stable procedure is to match entire segments of the supermodel trajectory to the real trajectory. One can introduce a cost function for mismatch, such as the one used by van den Berge et al. (2011):

$$F(\mathrm{C}) = \frac{1}{K\Delta}\sum_{i=1}^{K} \int_{t_i}^{t_i+\Delta} |x_s(\mathrm{C}, t) - x_0(t)|^2 \gamma^t dt \tag{4}$$

for a vector C of connection coefficients, defined as normalized sum over K short integrations of length Δ, with initial times t_i, of the squared error between the true trajectory x_0 and the supermodel trajectory x_s. The integration segments were chosen to overlap, so that $\Delta > t_{i+1} - t_i$. The factor γ^t with $0 < \gamma \leq 1$ is introduced to give stronger weight to the errors close to the initial conditions and discount the chaotic internal error growth that is not a result of model imperfections.

Results of trajectory-matching by minimizing (4) for a supermodel formed from imperfect replicas of a "true" Lorenz system are shown in Fig. 3. The algorithm is seen to be particularly useful for reproducing the true attractor, even where the attractors of the imperfect models are very different from truth.

3 Supermodeling vs. Output Averaging in Quasigeostrophic Models

3.1 Weighted Supermodeling

To investigate supermodeling with more complex models it is useful to consider a generalization arising from a limiting case of the connected supermodeling scheme described above. A class of supermodels is defined by defining the tendency for a given variable as a weighted average of the tendencies for that variable in the different models. That is, the parameters of the supermodel are weights w_i^l, with $w_i^l \geq 0$ and $\sum_i w_i^l = 1$, and the dynamics for the lth variable are given by:

$$dx^l/dt = \Sigma_i \; w_i^l f_i^l(\mathbf{x}) \tag{5}$$

Weighted supermodels can be considered as connected supermodels with infinitely strong connections, i.e., connections of the form κC_{ij}^i with $C_{ij}^i > 0$ and

Fig. 3 (**a–c**) Trajectories for three unconnected imperfect models (*black*) and for the "true" Lorenz system (*grey*). The trajectories include an initial transient as well as the attractor. (**d**) Trajectories for supermodel (*black*) trained by minimizing the cost function (4), and for the true Lorenz system (*grey*)

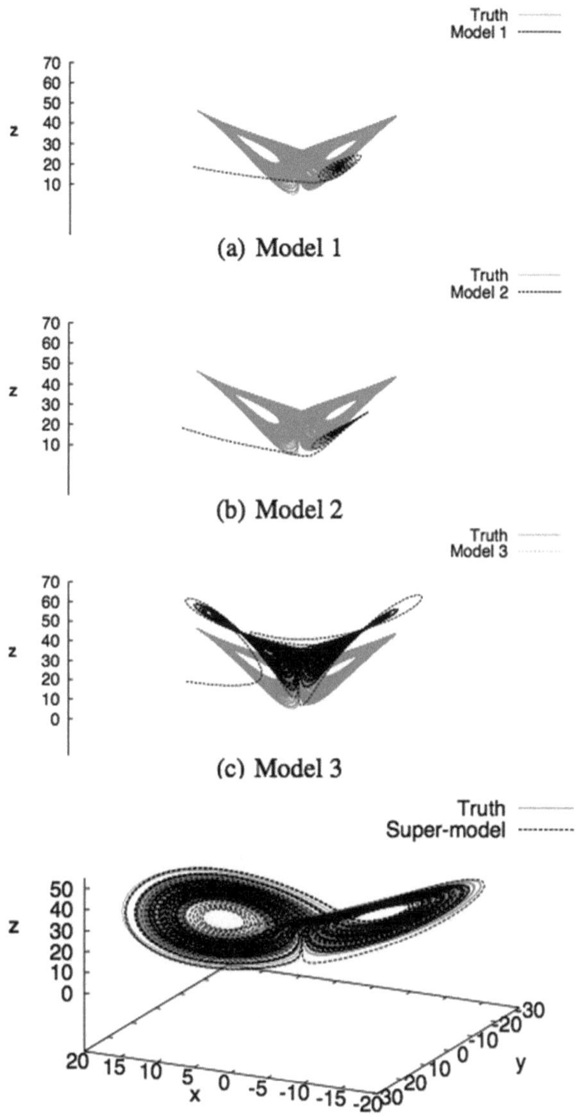

$\kappa \to \infty$. Thus the ratios of the large connections remain constant in the limit. In the limit it can be shown that all model states are completely synchronized $x_i^l = x_j^l$, and that the synchronized state follows the weighted averaged dynamics (5) (Wiegerinck et al. 2013).

3.2 Weighted Supermodels from Quasigeostrophic Models

The question of whether supermodels can exceed the performance of model output averages can now be addressed with models of more realistic complexity. If nonlinearities are strong enough so as to cause bifurcations in the climate systems as GHGs increase, it can be argued that output averaging will be insufficient to capture the effects and that supermodeling would be beneficial. However, there is little evidence for bifurcations of this type in model studies. But even without bifurcations, simple nonlinearity can still make the supermodel superior to an average of model outputs. This is perhaps most easily seen in the case where diagnostic properties depend non-monotonically on system parameters. Suppose we have two models of the form:

$$\begin{aligned} d\mathbf{x}/dt &= F(\mathbf{x}, p_1) \\ d\mathbf{x}/dt &= F(\mathbf{x}, p_2) \end{aligned} \tag{6}$$

where F is linear in the parameter p, and consider some diagnostic $P(p)$, e.g., mean temperature. Further suppose that $P(p_1) = P(p_2)$, but that for some intermediate value p_i, $p_1 < p_i < p_2$, $P(p_i) > P(p_1) = P(p_2)$. Then any weighted average of model outputs will only give the first value $P(p_1)$. A weighted supermodel, on the other hand, could readily reproduce the correct dynamics, that is $F(\mathbf{x}, p_i) = w_1 F(\mathbf{x}, p_1) + w_2 F(\mathbf{x}, p_2)$ for appropriately chosen weights w_1 and w_2, since F is linear in p. It is hypothesized that a connected supermodel could also give the correct result.

Consider specifically a quasigeostrophic model of a re-entrant channel on a β-plane given by:

$$Dq_i/Dt \equiv \partial q_i/\partial t + J(\psi_i, q_i) = F_i + D_i \tag{7}$$

where the layer $i = 1,2$, ψ is streamfunction, and the Jacobian $J(\psi_i, q_i)$ gives the advective contribution to the Lagrangian derivative D/Dt (Vautard and Legras 1988; Vautard et al. 1988). The forcing F is a relaxation term designed to induce a jet-like flow near the beginning of the channel:

$$F_i = \mu_0 \left(q_i^* - q_i \right) \tag{8}$$

for q_i^* corresponding to a streamfunction ψ^* that defines a jet. The dissipation terms D, boundary conditions, and other parameter values are given in Duane and Tribbia (2004).

The QG channel model vacillates between two dynamical regimes corresponding to "blocked" and "zonal" flow, as illustrated in Fig. 4. The response of the blocking activity to the forcing parameter μ_0 in (8) provides a simple example of non-monotonic behavior. For zero forcing, blocking frequency is zero due to damping by the dissipative terms. For large forcing, the flow is consistently jet-like, and again

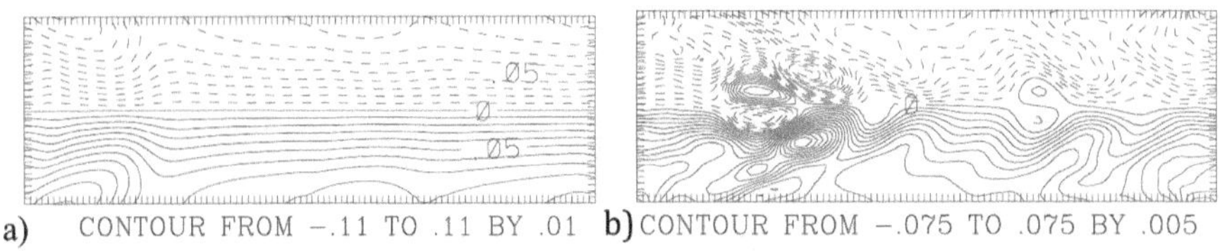

Fig. 4 Streamfunction (in dimensional units of $1.48 \times 10^9 \mathrm{m}^2\mathrm{s}^{-1}$) describing a typical zonal flow state (**a**), and a typical blocked flow state (**b**) in the two-layer quasigeostrophic channel model. Parameter values are as in Duane and Tribbia (2004). An average streamfunction for the two vertical layers $i = 1,2$ is shown

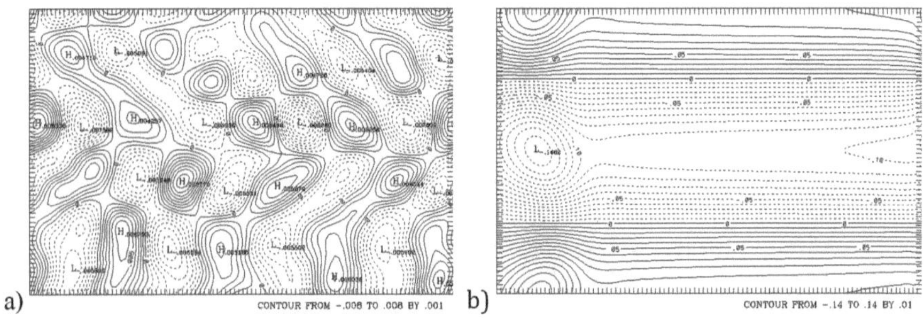

Fig. 5 Typical flows in the QG channel model with very small forcing coefficient ($\mu_0 = 0$) (**a**), and very large forcing coefficient ($\mu_0 = 3.0$) (**b**). (The spatial domain in each panel includes two channels with flows in opposite directions)

there is no blocking. Typical flow fields for these two cases are shown in Fig. 5a, b. (The zero-forcing flow in Fig. 5a is turbulent, but of low amplitude, and includes no blocks.)

A weighted supermodel formed from the two individual models illustrated in Fig. 5 can reproduce the true dynamics exactly for any value of the forcing coefficient μ_0 between $\mu_0 = 0$ and $\mu_0 = 3$, because μ_0 appears linearly in the tendency and so averaging tendencies effectively averages the μ_0 values (Duane 2015a). For the typical value $\mu_0 = 0.3$ used previously, the behavior is as illustrated in Fig. 6. The supermodel flow spends much time in the blocked regime, unlike the flows in the individual models or any weighted average thereof. (If the actual flow fields of the individual models are averaged, instead of the blocking frequencies, the same conclusion is reached.)

The learning task for the weights is equivalent to that for determining the single parameter μ_0 directly. The algorithm described in the previous section for parameter learning in models that synchronize with identical parameters (Duane et al. 2007), for instance, is effective in the present context. While the argument applies exactly to a weighted supermodel, it seems likely that a connected supermodel could also be formed from the two individual models illustrated in Fig. 5 that would approximate the "true" behavior for arbitrary forcing coefficient.

While a supermodel is clearly superior to an output average in the example given above, and in extreme cases generally, more linear behavior is expected for smaller

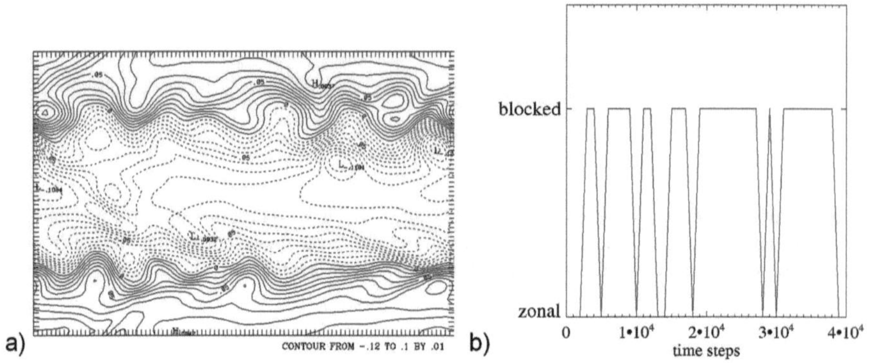

Fig. 6 Typical flow in the QG channel model with a "realistic" forcing coefficient ($\mu_0 = 0.3$) (**a**), and the history of vacillation of the flow in the bottom half of the domain between zonal and blocked regimes, sampled at low temporal resolution over the course of a simulation (**b**), using the blocking diagnostic defined in Duane and Tribbia (2004). The typical flow is also the exact solution to an appropriately weighted supermodel

inter-model differences as might occur in a realistic suite of models, such as the IPCC set. To construct a realistic experiment with toy models, a correspondence was established between parameter differences among the toy models, on the one hand, and differences among models or parameters used in actual climate projection, on the other. It was argued in Duane (2015a) that differences in the forcing coefficient μ_0 in the QG models are analogous to differences in climate model sensitivities to increased greenhouse gas levels. The latter sensitivities are known to vary among IPCC models by about $\pm 1/3$ of the average value. Considering proportional variations in μ_0 in the range $0.2 < \mu_0 < 0.4$, instead of the extreme range $0 < \mu_0 < 3.0$ used above, it was found that a weighted average of their blocking frequencies could reproduce the "true" behavior. At least in regard to blocking frequency, the advantage of supermodeling is lost in this less extreme case.

If one pays more attention to the detailed modes of variability, a subtle advantage remains. It is known that there is a very weak anticorrelation between blocking activity in the Atlantic and in the Pacific (Duane and Tribbia 2004). That effect could not possibly occur in an output average of models with Atlantic and Pacific forcing separately. It is thought that supermodeling will give improved predictions of other global multi-variable patterns of variability, where the relationships are stronger, as well.

4 Supermodeling with Primitive-Equation Models

A supermodel containing the main dynamical ingredients or real climate model was constructed from several versions of the intermediate complexity model SPEEDO (Severijns and Hazeleger 2009). The atmospheric component is the SPEEDY model that solves the primitive equations on a sphere using a spectral method. The spectral expansion is truncated at total wavenumber 30 which corresponds to a spatial resolution at the equator of about 500 km. It has eight vertical levels and simple parameterizations for radiation, convection, clouds, and precipitation. The solar radiation follows the seasonal cycle but the diurnal cycle is not imposed. Instead daily mean solar radiation fluxes are prescribed. The total number of degrees of freedom is 38,025:31,680 for the spectral coefficients of divergence, vorticity, temperature, specific humidity, and log of surface pressure plus 6345 to describe the land temperature, land moisture, and snow cover in the 2115 land points. The land component uses a simple bucket model to close the hydrological cycle over land and a heat budget equation that controls the land temperatures. The ocean component is the CLIO model (Goosse and Fichefet 1999). The CLIO model is a primitive-equation, free-surface ocean general circulation model coupled to a thermodynamic–dynamic sea-ice model. The ocean component includes a relatively sophisticated parameterization of vertical mixing. A three-layer sea-ice model, which takes into account sensible and latent heat storage in the snow-ice system, simulates the changes of snow and ice thickness in response to surface and bottom heat fluxes. In the computation of ice-dynamics, sea ice is considered to behave as a viscous-plastic continuum. The horizontal resolution of CLIO is 3° in latitude and longitude and there are 20 unevenly spaced vertical layers in the ocean. The CLIO model has a rotated grid over the North Atlantic Ocean in order to circumvent the singularity at the pole. The total number of degrees of freedom is on the order of 200,000.

Three SPEEDY atmospheres, with different parameters chosen to reflect the typical range of behavior of different atmospheric models, were coupled to the same ocean and the same land (see Fig. 7), and also to one another, by adding inter-atmosphere coupling terms to the dynamical equations for each atmosphere. The modified equation for the temperature field for model i ($i = 1 \dots 3$), for instance, is

$$\partial T_i/\partial t = \left(RT_i/c_p\right)\left(\dot{\sigma}_i/\sigma_i - \partial\dot{\sigma}_i/\partial\sigma_i - \nabla \cdot V_i\right) + \Sigma_j\left[C_s^{ij}\left(T_j - T_i\right)\ \delta\left(x - x_s\right)\right] \quad (9)$$

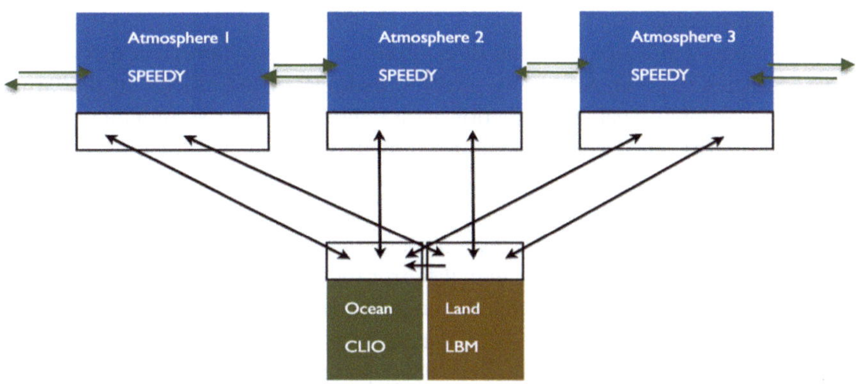

Fig. 7 Schematic representation of SPEEDO supermodel. The Ocean and Land models are the "true" Ocean and Land, respectively

where all variables are evaluated at position x and $\{x_s\}$ is a set of discrete coupling points. In (9), R is the gas constant, c_p is the specific heat at constant pressure, σ is a vertical pressure coordinate scaled with surface pressure, $\dot{\sigma}$ its time-derivative, V is the horizontal wind velocity, and C_s^{ij} is a connection coefficient linking the temperature fields between models i and j at position x_s. Dynamical equations for the other independent variables, u (east–west velocity), v (north–south velocity), and q (humidity) are similarly modified to include coupling terms linking the different models.

In the present situation, regarding the PDE as a very high-order ODE, the general rule for adaptation of parameters (2), as applied to the connection coefficients C^{ij}, gives

$$dC^{ij}/dt = a \int dx \, \left(T_j(x) - T_i(x)\right) \left(T^t(x) - \frac{1}{3} \Sigma_k \, T_k(x) \right) \tag{10}$$

where T^t is the true value of T, and a is an arbitrarily chosen learning rate. We assume spatially uniform connections C^{ij} that are independent of position s. Analogous rules are written to adapt the connections linking the other dynamical variables, with learning rates appropriate for their dynamics. The algorithm was tested by choosing one of the models to be a perfect replica of the "true" system; appropriate binary values for the connections did indeed result. All models are nudged to truth as the learning progresses; for the configuration studied, it was found that nudging to truth in the u field gave truth-model synchronization error rates that were useful in discriminating between good and bad models, so that the learning algorithm was effective.

Note that the last term in (9), connecting the models, tends to vanish as the models synchronize. This is desirable, so that each model satisfies its own physically motivated dynamical equation, without the influence of artificial coupling terms. Of, course, for each i, the parameters and hence the equations are different, so that the models cannot possibly synchronize completely. Typically, the differences in behavior are in small-scale processes that are not important for the large-scale behavior of interest.

The system was tested with the three arbitrarily chosen imperfect models of a "true" SPEEDO system, assuming ongoing nudging of the models to the "true" system, as if doing weather prediction with continuous data assimilation (Duane and Selten 2016). The "true" system also provided the land and ocean components for each of the imperfect models. Results are shown for the simple case of two identical models and a different third model in Fig. 8. It is seen that after 3 months, the truth-supermodel error, with adapted coefficients, is less than the error for each of the individual models, and less than the error for the supermodel with a choice of uniform connection coefficients that are not adapted.

Then the coefficients were frozen and atmospheric CO2 was doubled in the "true" system and in each of the models. Other parameters were also varied slightly. Results are shown in Fig. 9. It is seen that the supermodel gives reduced error after three months as compared to the weighted averages of the separate models,

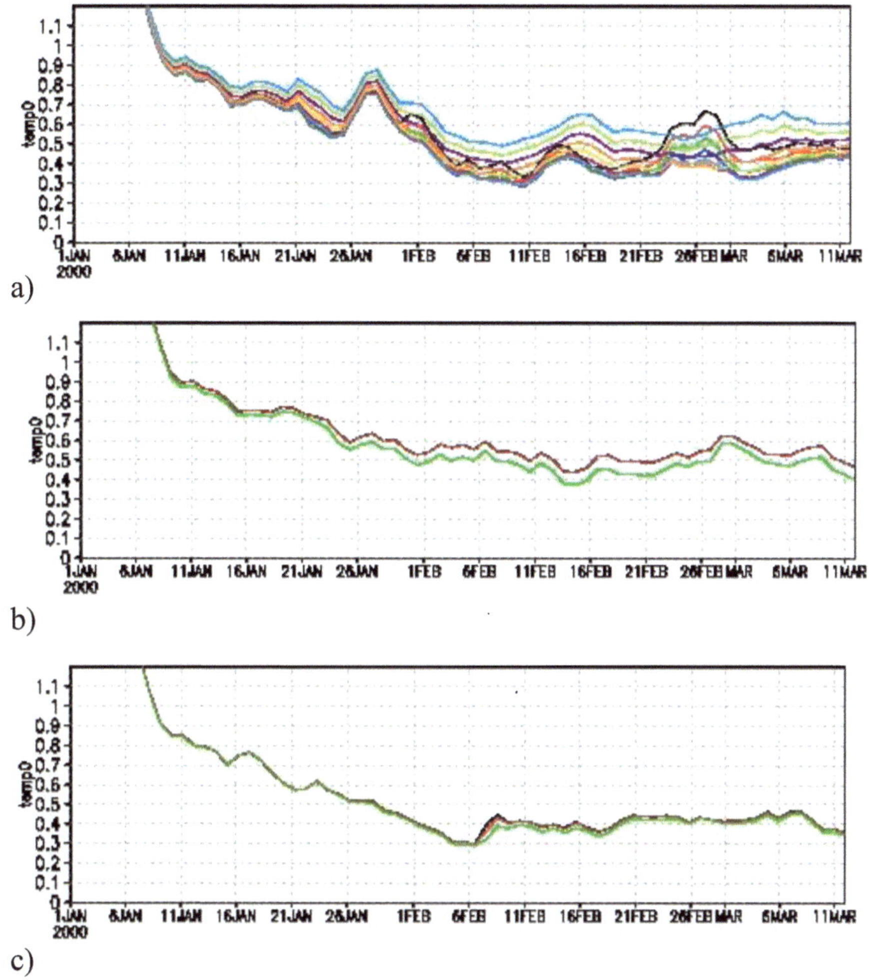

Fig. 8 Truth-model synchronization error in surface temperature (in °C) for (**a**) three SPEEDO models with parameters perturbed away from their values in a "true" SPEEDO model to which the imperfect models are nudged via the *u* variable (with two of the models identically perturbed) and various weighted combinations of their outputs; (**b**) a supermodel formed by connecting the three SPEEDO models through their dynamical equations according to Eq. (5) (for temperature) and analogous equations for *u*, *v*, and *q*, with constant and uniform connection coefficients C^{ij}; and (**c**) the same supermodel but with connections adapted according to (6) with analogous equations for the *u*, *v*, and *q* connections. (2 of the 3 constituent models in the supermodel were chosen to be the same, so there are only 2 dintinct lines in (b) and in (c))

but the coefficients learned from the single-CO2 runs are less than optimal. That is, a simple choice of uniform coefficients gives slightly better results than the learned coefficients (in this artificially constructed case where the imperfect models were about equally spaced around the true models), but the model with learned

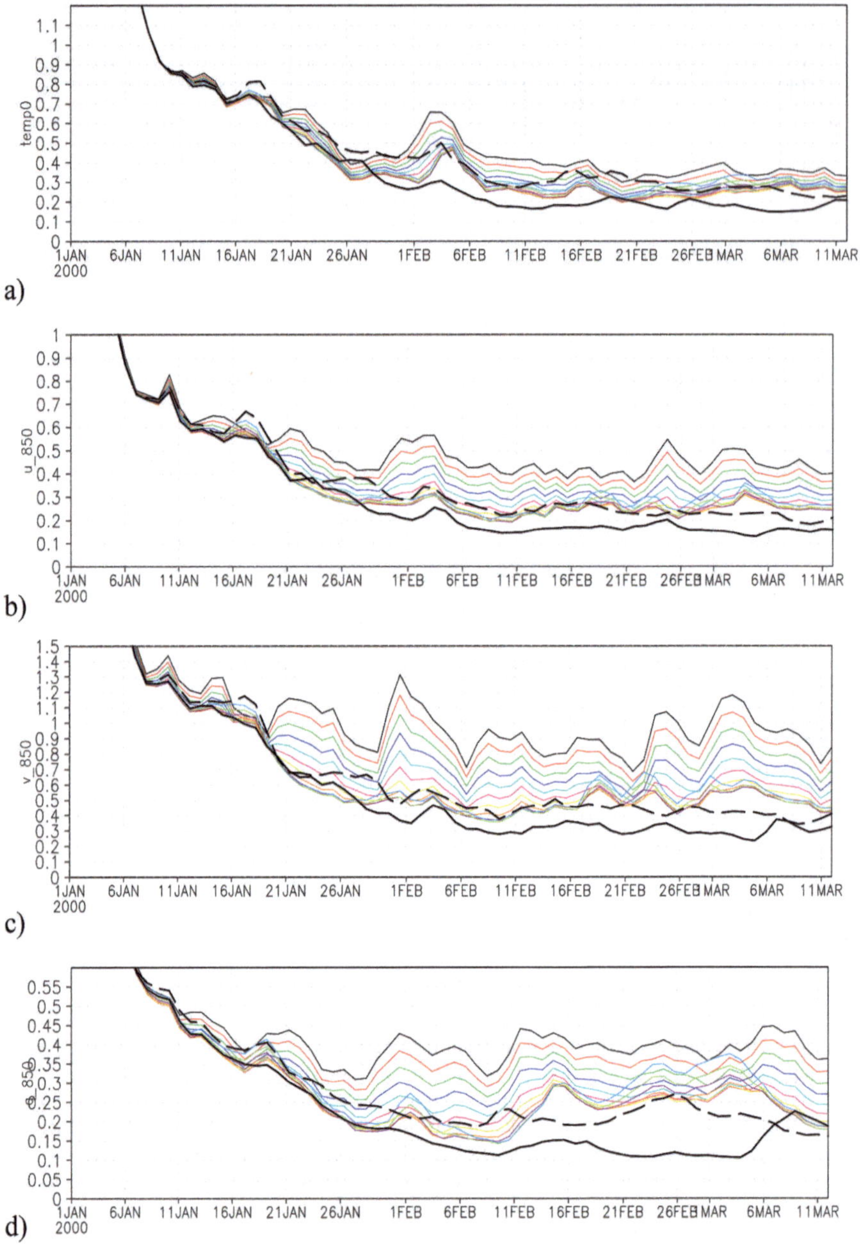

Fig. 9 Truth-model synchronization error in surface temperature (in °C) (**a**) for three SPEEDO models as in Fig. 8, but with doubled CO_2 in both truth and models, for various weighted combinations of model outputs (*colored lines*), a supermodel with uniform connections (*thick black line*), and a supermodel using the connection strengths from the present-CO_2 run (Fig. 8c) at final time (*dashed*). Correspondingly for error in zonal wind u at 850 mb (**b**), error in meridional wind v at 850 mb (**c**), and error in humidity q (**d**)

coefficients was still effective. Thus the supermodel is not only useful for exploring state space, but also for exploring an enlarged model space defined by variations in ancillary parameters.

5 A Weakly Connected Supermodel Formed From Full Climate Models Connected Only at the Ocean–Atmosphere Interface

Investigations with full climate models have thus far reached a stage in which different atmosphere models are connected to a common ocean, as in the early work of Kirtman et al. (2003) but not directly connected to each other. Yet even without the direct connections, the supermodel has been shown to be superior to any weighted combination of outputs of the individual models (Shen et al. 2016).

A climate model was built based on COSMOS (ECHAM5/MPIOM), developed at the Max-Planck Institut fur Meteorologie, Germany (Jungclaus et al. 2006), and involved two atmospheric general circulation models (AGCMs). The two models differed in their cumulus parameterization schemes, Nordeng (1994) and Tiedtke (1989), to represent typical model diversity because cumulus convection schemes normally have a strong impact on the climate state (Kim et al. 2011; Klocke et al. 2011; Mauritsen et al. 2012). The ocean model continuously interacts with the Nordeng atmosphere and Tiedtke atmosphere. AGCMs are problematic in representing real air-sea fluxes to different degrees of accuracy. Some may be better in representing momentum flux (i.e., wind stress on the ocean) and some in energy (heat) flux (Kirtman et al. 2003). Different weights were used for the energy, momentum, and mass (i.e., precipitation) fluxes felt by the common ocean, with the sum of the weights over the two models, for each type of flux, equal to unity. Each atmosphere feels only its own fluxes.

A machine learning technique, the Nelder–Mead method (Nelder and Mead 1965) was applied to optimize the weights for each of the fluxes. The Nelder–Mead method is also known as the simplex method, which is used to find a local minimum in multidimensional domain without having to compute gradients of a cost function. A performance index (Reichler and Kim 2008) computed over the Pacific region (160°E–90°W, 10°S–10°N) was used as a metric because there is partial synchronization over the tropical Pacific in this configuration; hence it is reasonable to expect that improvement can only be achieved over this area. The assessment was started from equal weights and followed the weights suggested by the simplex method. Each case was spun up for ten years and run for another 30 years to get a reasonable climatology. Over 300 cases were tested along the path to optimal weights, for which the performance index (error) was reduced and the correlation between zonal wind stress anomaly of two AGCMs is increased. Note that the variability of AGCMs tends to cancel over non-synchronized areas, thus reducing the ocean variability as well.

The behavior predicted by the supermodel was dramatically improved as shown in Fig. 10, in which both the SST and precipitation have better agreement with observations. The cold tongue is stopped around the International Date Line, which suggests that a west-Pacific warm pool was formed in the supermodel, unlike the situation in COSMOS(N), COSMOS(T), or their averaged output, COSMOS(E), in all of which the cold tongue crossed the International Date Line to the western Pacific and the variability of SST is much larger (not shown). The supermodel largely mitigates the double ITCZ error found in both COSMOS models and in most climate models.

The reduction of the SST bias in the supermodel implies that the whole dynamic is more realistic, suggesting that a much more realistic low level wind system exists in the supermodel, leading to a better latitudinal position of the Inter-tropical Convergence Zone (ITCZ). But it is still too wet in the South Pacific convergence zone.

The key to improved supermodel performance in this case appears to be in better representation of the air-sea feedbacks. In Fig. 11, we show the Bjerknes feedback and the thermodynamic feedback for the supermodel (SUMO), the individual models, and observations. The Bjerknes feedback in the supermodel is almost perfect and the thermodynamic feedback is much improved.

It can be shown that the supermodel is superior to any weighted combination of the two model outputs. In Fig. 12, we present a Taylor diagram that shows the correlation between model and observations, as well as the normalized standard deviation of the model field, for the various models. It is seen that the supermodel has almost the same standard deviation of SST as in the observed data, unlike any of the models, and the correlation coefficient is higher.

An objection to supermodeling in the meteorological community is that ensembles of model runs (where the models are the same or different) are usually used to estimate spread as an indication of error. One loses this information with supermodeling if the models synchronize nearly completely. However, the ensemble of models in the usual practice can be replaced by an ensemble of weights. One can examine the learning history, or simply look at the performance metric for a random sample of weights, to infer a plateau in weight space along which the performance is close to optimal. Then weights on that plateau can be used to define an ensemble of supermodels. Results of this procedure, shown in Fig. 13, give a plausible ensemble of SST fields. The models effectively "agree to disagree."

6 Conclusions

The supermodel scheme for the fusion of imperfect computational models is not limited to climate models. Supermodeling only requires that the constituent models come equipped with a procedure to assimilate new measurements from an objective process in real time and, hence, from one another. The scheme could thus also be applied to financial, physiological, or ecological models. It

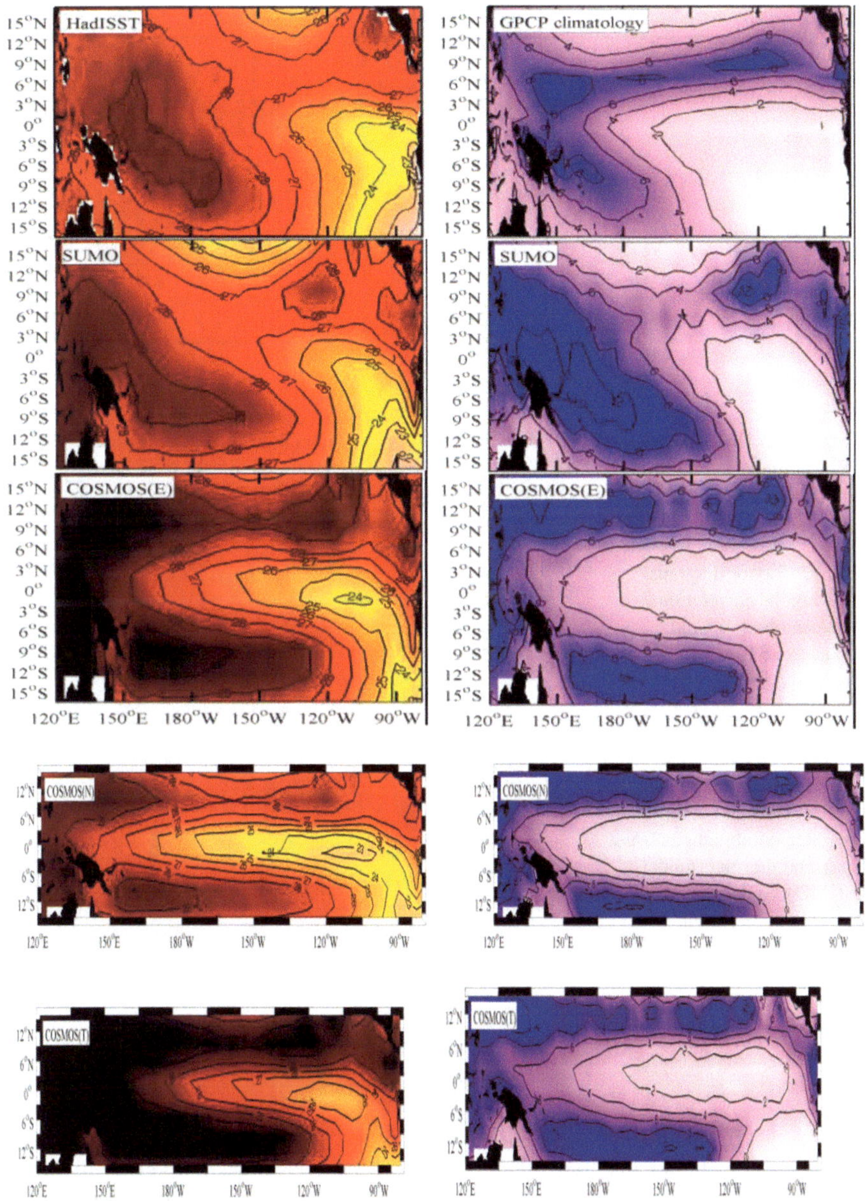

Fig. 10 The climatology sea-surface temperature (SST) (*left panel*, scale in °C) and precipitation (*right panel*, scale in mm/day) in the Tropical Pacific from observations, the trained supermodel (SUMO), the untrained, equal-weighted supermodel (COSMOS(E)), and the two constituent models, COSMOS(N) and COSMOS(T). Observed SST is from HadISST (1948–1979, the period used as a training set) while observed precipitation is from GPCP (1979–2012). Because the SST state over the equator is improved in the supermodel (SUMO), there is one ITCZ in SUMO

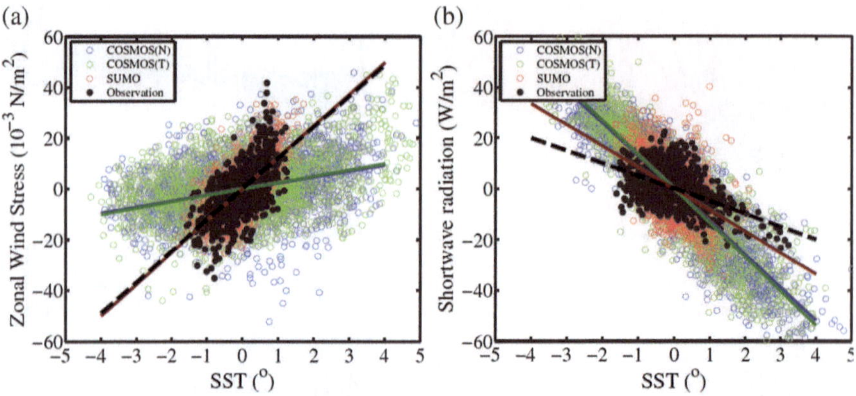

Fig. 11 (**a**) The Bjerknes feedback (*left panel*), describing the relationship between the east Pacific SST anomaly (over 5°S–5°N, 150°W–90°W, Niño 3 region) and the remote wind stress over the west Pacific (5°S–5°N, 160°E–150°W, Niño 4 region); (**b**) the thermodynamic damping (*right panel*) over the Niño 3 area.

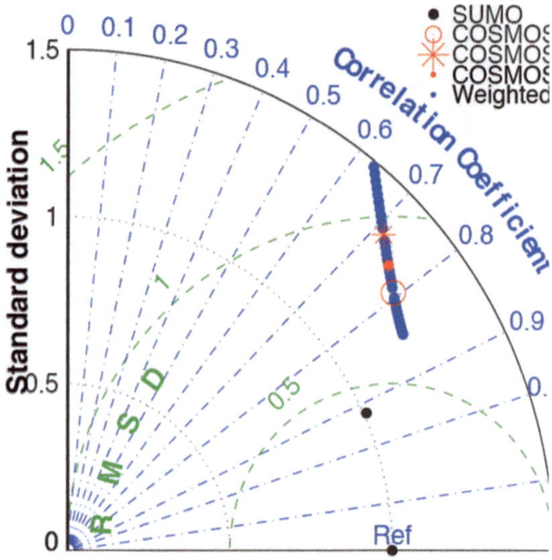

Fig. 12 Taylor diagram showing the correlation between observed and modeled SST over the Tropical Pacific, as well as the normalized standard deviation, for COSMOS(N), COSMOS(T), their equal-weighted combination COSMOS(E), all other weighted combinations (*thick line*), and the supermodel (SUMO). SUMO is clearly closer to observations (Ref) than any weighted average

has been speculated that the mind could also be conceived fundamentally as a supermodel, perceiving/synchronizing with the objective world, but also with a capacity for interaction among semi-autonomous components and resulting self-perception commonly experienced as consciousness (Duane 2015b).

Specific studies demonstrated that a wide range of coupling schemes and connection strengths will lead to inter-model synchronization and hence consensus. Conversely, in situations with a high degree of nonlinearity in the dynamics, synchronization is essential—the inter-model connections are needed to give results surpassing those of output averaging. Indeed the fact that a supermodel, in which the

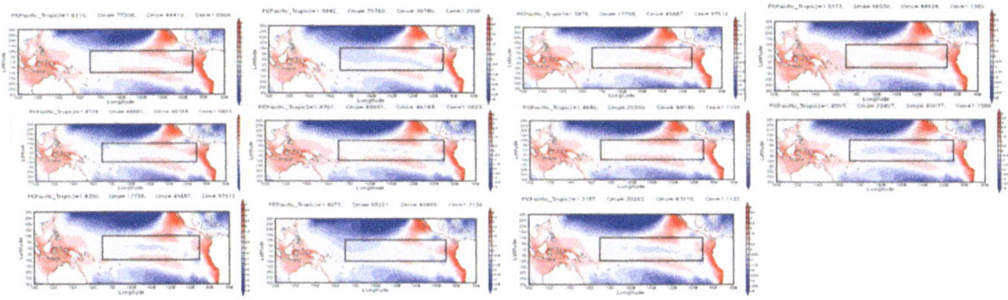

Fig. 13 SST fields for an ensemble of supermodels defined by examining the learning history to select combinations of weights that give near optimal performance, each of which defines a different supermodel, giving a plausible spread in results

constituent models are themselves synchronized, will in turn readily synchronize with an objective process, is an instance of a more general hypothesis about the relationship between internal and external synchronization (Duane 2009; Duane 2015b). The choice of semi-autonomous models to be combined is not essential, as long as the "gene pool" of models is diverse.

It is interesting that in both the quasigeostrophic supermodel described in Sect. 3 and the COSMOS supermodel described in Sect. 5, the constituent models err on the same side of reality, with an absence of blocking in the former case and an anomalous cold tongue in the latter one. Where there is such non-monotonic behavior, some type of weighted supermodel, and probably a connected supermodel, is guaranteed to outperform an output average. The commonality of such non-monotonic behavior is not yet clear. But perhaps a principle akin to that of self-organized criticality (Bak et al. 1987) is at work—when all scales are represented dynamically, the model naturally gravitates to some kind of critical state, a behavior that must be manually inserted in parameterized models or learned. The supermodel reduces the dimensionality of the learning problem by exploiting human experience to isolate the dimensions along which arbitrary choices tend to be made.

Synchronization, to whatever degree it is present, implies that the supermodel can be viewed more as a single model than as an ensemble of models. Thus detailed features will survive that would be washed out in an output average. However, in many applications one is only interested in statistical properties of these features, many of which are adequately represented by an average of the statistics of the separate systems. The degree of model nonlinearity in realistic situations will determine the advantage of supermodeling for capturing the structures of interest, or higher-order statistical properties thereof.

Acknowledgements Parts of the research reported here were performed under ERC Grant 648982, European Commission Grant 658602, and Dept. of Energy Grant DE-SC0005238

References

Bak, P., C. Tang, and K. Wiesenfeld. 1987. Self-organized criticality: an explanation of 1/f noise. *Physical Review Letters* 59: 381–384.

Duane, G.S. 2009. Synchronization of extended systems from internal coherence. *Physical Review E* 80: 015202.

———. 2013. Data assimilation as artificial perception and supermodeling as artificial consciousness. In *Consensus and synchronization in complex networks*, ed. Ljupco Kocarev. Berlin: Springer.

———. 2015a. Report on activities and findings under DOE grant "Collaborative research: an interactive multi-model for consensus on climate change" #DE-SC0005238.

———. 2015b. Synchronicity from synchronized chaos. *Entropy* 17: 1701–1733.

Duane, G.S., and F. Selten. 2016. Supermodeling by synchronization of alternative SPEEDO models. Paper presented at EGU General Assembly, No. 15945, Vienna, Austria.

Duane, G.S., and J.J. Tribbia. 2004. Weak Atlantic-Pacific teleconnections as synchronized chaos. *Journal of the Atmospheric Sciences* 61: 2149–2168.

Duane, G.S., J. Tribbia, and B. Kirtman. 2009. Consensus on long-range prediction by adaptive synchronization of models. Paper presented at EGU General Assembly, No. 13324, Vienna, Austria.

Duane, G.S., J.J. Tribbia, and J.B. Weiss. 2006. Synchronicity in predictive modeling: a new view of data assimilation. *Nonlinear Processes in Geophysics* 13: 601–612.

Duane, G.S., D.-C. Yu, and L. Kocarev. 2007. Identical synchronization, with translation invariance, implies parameter estimation. *Physics Letters A* 371: 416–420.

Goosse, H., and T. Fichefet. 1999. Importance of ice-ocean interactions for the global ocean circulation: a model study. *Journal of Geophysical Research* 104: 23337–23355.

Jungclaus, J.H., N. Keenlyside, M. Botzet, H. Haak, J.-J. Luo, M. Latif, J. Marotzke, U. Mikolalewicz, and E. Roeckner. 2006. Ocean circulation and tropical variability in the coupled model ECHAM5/MPI-OM. *Journal of Climate* 19: 3952–3972.

Kalnay, E. 2003. *Atmospheric modeling, data assimilation, and predictability*. Cambridge: Cambridge University Press.

Kim, D., Y.-S. Yang, D.-H. Kim, Y.-H. Kim, M. Watanabe, F.-F. Jin, and J.-S. Kug. 2011. El Nino-southern oscillation sensitivity to cumulus entrainment in a coupled general circulation model. *Journal of Geophysical Research* 116: D22112.

Kirtman, B.P., D. Min, P.S. Schopf, and E.K. Schneider. 2003. A new approach for coupled GCM sensitivity studies, COLA Technical Report No. 154.

Klocke, D., R. Pincus, and J. Quaas. 2011. On constraining estimates of climate sensitivity with present-day observations through model weighting. *Journal of Climate* 24: 6092–6099.

Lorenz, E.N. 1963. Deterministic non-periodic flow. *Journal of the Atmospheric Sciences* 20: 130–141.

Mauritsen, T., et al. 2012. Tuning the climate of a global model. *Journal of Advances in Modeling Earth Systems* 4: M00A01.

Mirchev, M., G.S. Duane, W.S. Tang, and L. Kocarev. 2012. Improved modeling by coupling imperfect models. *Communications in Nonlinear Science and Numerical Simulation* 17: 2471–2751.

Nelder, J.A., and R. Mead. 1965. A simplex method for function minimization. *The Computer Journal* 7: 308–313.

Nordeng, T.-E. 1994. Extended versions of the convective parametrization scheme at ECMWF and their impact on the mean and transient activity of the model in the tropics, Technical Memorandum No. 206, European Centre for Medium Range Weather Forecasts.

Reichler, T., and J. Kim. 2008. How well do coupled models simulate today's climate? *Bulletin of the American Meteorological Society* 89: 303–311.

Severijns, C., and W. Hazeleger. 2009. The efficient global primitive equation climate model Speedo. *Geoscientific Model Development Discussion* 2: 1115–1155.

Shen, M.-L., N. Keenlyside, F. Selten, W. Wiegerinck, and G.S. Duane. 2016. Dynamically combining climate models to "supermodel" the Tropical Pacific. *Geophysical Research Letters* 43: 359–366.

Tebaldi, C., and R. Knutti. 2007. The use of the multi-model ensemble in probabilistic climate projection. *Philosophical Transactions of the Royal Society of London A* 365: 2053–2075.

Tiedtke, M. 1989. A comprehensive mass flux scheme for cumulus parameterization in large-scale models. *Monthly Weather Review* 117: 1779–1800.

van den Berge, L.A., F.M. Selten, W. Wiegerinck, and G.S. Duane. 2011. A multi-model ensemble method that combines imperfect models through learning. *Earth System Dynamics* 2: 161–177.

Vautard, R., and B. Legras. 1988. On the source of mid-latitude low-frequency variability. Part II: nonlinear equilibration of weather regimes. *Journal of the Atmospheric Sciences* 45: 2845–2867.

Vautard, R., B. Legras, and M. Déqué. 1988. On the source of mid-latitude low frequency variability. Part I: a statistical approach to persistence. *Journal of the Atmospheric Sciences* 45: 2811–2843.

Wiegerinck, W., W. Burgers, and F. Selten. 2013. On the limit of large couplings and weighted averaged dynamics. In *Consensus and synchronization in complex networks*, ed. Ljupco Kocarev. Berlin: Springer.

Yang, S.-C., D. Baker, K. Cordes, M. Huff, G. Nagpal, E. Okereke, J. Villafane, and G.S. Duane. 2004. Data assimilation as synchronization of truth and model: experiments with the three-variable Lorenz system. *Journal of the Atmospheric Sciences* 63: 2340–2354.

Are We Measuring the Right Things for Climate?

Christopher Essex and Bjarne Andresen

Abstract If one could exist on climate scales would it make any more sense to measure laboratory-scale quantities to capture climate conditions than it does for us on the laboratory scale to compute wave functions to understand the weather? Clearly the quantum mechanical and the laboratory regime are constructed in terms of different physical variables. Why do we presume, then, that laboratory regime quantities like temperature continue to be the appropriate physical variables to measure in a climate regime? This paper suggests why we may not be measuring the right things and it will broach some alternatives in the context of a reformulation for relevant physics more natural to long timescales: slow time. Specifically it shows that fluctuating velocities can be "thermalized" in suitable averages suggesting that one might imagine climate in terms of a generalization of wind which may include persistent meteorological winds, or none at all. But it also shows that temperature cannot be "thermalized" on long time and space scales, making the notion of local equilibrium and simple generalizations of temperature problematic for climate.

Keywords Climate • Fundamental theory • Timescale • Thermodynamic variables • Closure

1 Introduction

We measure thermodynamic quantities like temperature, pressure, and humidity for weather—all strictly local and transient properties of a physical system out of global thermodynamic equilibrium. Should we measure the same things for climate? It is taken for granted that these things continue to have meaning for climate.

C. Essex (✉)
Department of Applied Mathematics, The University of Western Ontario, London, ON, Canada N6A 5B7
e-mail: essex@uwo.ca

B. Andresen
Niels Bohr Institute, University of Copenhagen, Universitetsparken 5, 2100 Copenhagen Ø, Denmark
e-mail: andresen@nbi.ku.dk

A.A. Tsonis (ed.), *Advances in Nonlinear Geosciences*,
DOI 10.1007/978-3-319-58895-7_6

Fig. 1 Two images of the same Niagara Falls downstream flow. The *left image* is an exposure of 0.4 s, while the *right-hand image* is exposed for 50 s. Note the flow features visible in the *right-hand image* (streamlines, bow waves, standing waves, vortices, etc.) that are not clearly visible or invisible in the *left image*

Moreover, the physical climate system is often viewed, to the contrary, as a stable thermodynamic system, only changeable through external influences, even though there is no physical reason to view it in that way. But perhaps there is something thermodynamic-like on large enough space and timescales. If climate can actually prove to have such a property, it must emerge from an unstable dynamical system where any direct thermodynamical connections are strictly local. Showing such a thing exists, if it even does, is a most challenging scientific problem.

In terms of thermodynamical quantities, there are few good analogs on the gravity-irrelevant, jiggling, and sticky kinetic-atomic scales, despite some interesting efforts to find thermodynamic-like analogues for those microscopic scales. Such conventional quantities remain tied to the laboratory regime. But studying climate is not unlike atomic physics upside down, where we are the atoms. While it is easy to mistake the appearance of, say, snow or palm trees for climate, these are only indirect manifestations of a grander physics. Trying to imagine that physics from a laboratory-scale perspective is like viruses trying to theorize about what the laboratory they are in looks like. This paper suggests that we may not be measuring the right things for climate, and it will broach some alternatives in the context of a reformulation for relevant physics more natural to long timescales: slow time.

To fix ideas, consider the images of Fig. 1. The left-hand image of Fig. 1 shows the turbulent water of the Niagara River downstream from Niagara Falls as the human eye sees it. The water flow is complex and turbulent as it self-interacts, and interacts with the shore and river bottom, not to mention surface interactions with the air. In contrast the right-hand image of the same scene shows phenomena previously only visible to the most educated eye, if visible at all. Streamlines, bow and standing waves, or downstream vortices are all plain in the right-hand image, which is a 50 s time exposure.

On the 50-s timescale physical phenomena reveal themselves that are invisible to the unaided eye. The reverse is also true. Things are visible to the eye that do not show up on the 50-s timescale. There is an old trick of architectural photographers that eliminates all traffic from an image by the use of long lenses, slow film, and time exposures. Some of what is visible to the human eye is thus made to disappear in the resulting images, not unlike the case of some turbulent water in the Niagara River images.

We have a sense of the appearance and disappearance of different physical phenomena between different physical regimes through the relationship between physics on the atomic and laboratory scales. But in that we also understand the physics of the laboratory scales stands independent of that of the atomic scales too, even though they are physically consistent and compatible (Essex 2011). We can in that sense "ignore" the atomic regime in studying laboratory-scale physics. That is we can make predictions of laboratory-scale phenomena in terms of laboratory-scale variables only, without explicitly referring directly to specific kinetic-scale variables.

Can we do this with the 50-s timescale fluid flow from Fig. 1? This is far from clear. Just because we see structure does not mean that there is a stand-alone physics, let alone dynamics for that regime. To see if there is dynamics one could generate a sequence of 50-s time exposures and then run the result as a video. Perhaps the streamlines and standing waves, etc., change and move in the resulting slow-time video, perhaps they do not. But if there is a dynamics of the 50-s regime that stands independent of the laboratory regime, one needs to be able to forecast what happens on the 50-s timescale video without requiring data from the laboratory regime. The resulting theory and its associated variables must be able to ignore the laboratory regime.

The closure problem of fluid mechanics is the famous failure to achieve independence for the physics of turbulent flows from the laboratory regime. Of course the theory, as realized in the Navier–Stokes differential equation, can be integrated to generate integrated variables, which (it was hoped) would stand as the measurables of a putative theory for turbulent flow, independent of the usual laboratory regime. But it failed.

Thus to this day not only can we not always accurately predict the flow in a pipe from first principles but we cannot accurately predict the lowest order statistic either from first principles. It failed because the integration of the equation creates more independent integrals over combinations of variables than original variables in the parent regime. Thus not all values are determined by the integrated equation within the integrated regime. It is always necessary to refer to the parent regime to evaluate them, and thus the integrated equation cannot forecast anything, except in (at best) an empirical manner. The integrated variables are not part of a stand-alone theory, but are subordinate to the laboratory regime. They do not represent the measurables of a stand-alone theory for turbulent flow.

The 50-s regime defined through Fig. 1 does not imply that there is anything special in comparison to, say, a 200-s regime, or a 4-h regime for that matter. All the issues of structure appearing and disappearing can still be in play between them, but none need to represent a regime with a stand-alone physical theory independent of the laboratory regime. The climate problem is simply a version of this problem, but on a much grander scale. But while there is no established on-going discussion of the 50-s regime, there is one for climate. While there are no putative variables for a putative 50-s theory being regularly measured, putative variables for climate have boldly been advanced without proof of their merit.

We do not yet know whether climate is simply a phenomenon subordinate to the meteorological regime (which only differs slightly from the laboratory regime as the parent regime for climate conceptualization), or a physically distinct regime with its own governing equations in terms of variables assembled in an as yet unknown manner from meteorological or kinetic primaries. If the answer to the existence question is yes, it is an open question as to whether what we measure or assemble from meteorological measurements today in the name of studying climate actually represents true climate measurables emerging from a stand-alone theory for climate.

While we cannot answer this question definitively we can use thinking from the beginnings of slow-time theory to look at aspects of this issue, assuming such a stand-alone climate regime exists. We can say that certain variables are not likely to help us with insight into a stand-alone theory for climate. In particular with previous work on the "slow-time Maxwellian" (Essex and Andresen 2015) we will show that local equilibrium will not likely survive in a climate regime, which makes any suppositions about an analog to meteorological local equilibrium problematic, suggesting that climatological measurables will not be simple averages over local thermodynamic states.

First we will address this by discussing how one might envision the thermalization of wind. We will find that the kinetic energy of wind is easily thermalized under particular conditions, making wind something that fits naturally into a climate picture where systematic winds survive averaging and random fluctuations can be envisioned as contributing to a long timescale version of temperature. Second we show that unlike wind, fluctuations in temperature cannot be thermalized, because they typically produce a distribution that does not have a Maxwellian shape.

This suggests that local thermodynamic states normal for meteorology cannot exist for a putative climate regime, and raises the question as to whether averages over local temperature will provide insight into climate.

2 No Wind

Let's start with a simple example and consider the effect of fluctuations in rest velocity, u, of a small volume of gas, i.e., wind. Without loss of generality we proceed in terms of fluctuations in one space dimension. Then the molecular velocity profile is the Maxwellian,

$$p(v; u, T) = \left(\frac{m}{2kT}\right)^{1/2} \frac{1}{\sqrt{\pi}} e^{-\frac{m}{2kT}(v-u)^2}. \tag{1}$$

Imagine that on a large timescale, e.g., the timescale of climate, winds experience reversals and ranges of magnitudes so that we may plausibly assume a normally distributed rest velocity u about $u = 0$ with σ_u being its standard deviation. If over the long timescale there is a prevalent velocity u_0, it is easy to translate this distribution to be around that u_0. Assuming that the central limit theorem holds, this convolution of the v and u distributions is itself a Gaussian,

$$p(v;\theta) = \left(\frac{m}{2k\theta}\right)^{1/2} \frac{1}{\sqrt{\pi}} e^{-\frac{m}{2k\theta}v^2} \tag{2}$$

but now with a revised effective temperature, θ,

$$\theta = \frac{\sigma_u^2 m}{k} + T \tag{3}$$

that contains the fluctuations of wind u. Suppose $\sigma_u \sim 5\,\mathrm{m/s}$, then for air at $T = 300\,\mathrm{K}$, $\sigma_u^2 m/k \sim 0.1\,\mathrm{K}$. This change of temperature of 0.1 K is for most practical purposes negligible. However, for other flows than the material wind, e.g., radiation, the ensuing revised effective temperature may be markedly changed. In any event, what is wind on the laboratory (meteorological) scale is still wind on the long timescale. But it has changed what is perceived as temperature.

The new temperature here, θ is an emergent feature of a well-defined underlying (small-scale) mechanism, not just a generalization. It is in all respects a legitimate temperature. As long as u is fluctuating in a Gaussian manner, all of the ideal gas relationships re-emerge, but in the temperature θ instead of T. For example, energy E along one axis is simply, $E = Nk\theta/2$, just as it is in T for the laboratory regime. Coarsening the timescale for fluctuations in u amounts to thermalizing the wind.

3 No Local Temperature

Next we turn to fluctuations in, temperature, over our long timescale as a more relevant quantity for climate predictions. Like before for u, we will assume that fluctuations in T, or some function of T, are normally distributed. This is speculation, but the aim is only to find a plausible slow-time scenario. Meanwhile we will not be working with T but θ defined in Eq. (3), where wind, u, has been thermalized. Actually, for mathematical convenience we will be working in the precision of a distribution rather than its standard deviation. The precision is 1/(standard deviation). For a Maxwellian velocity distribution like Eq. (1) we have that the standard deviation $\sigma_u \propto \sqrt{T}$ while the precision $\psi \propto \sqrt{\beta}$ where $\beta = 1/kT$. However, we will still refer to fluctuations in the precision as "temperature fluctuations." Thus larger precision means a tighter distribution.

Now the Gaussian precision, ψ, is defined by

$$\left(\frac{m}{2k\theta}\right)^{1/2} \frac{1}{\sqrt{\pi}} e^{-\frac{m}{2k\theta}v^2} = \frac{\psi}{\sqrt{\pi}} e^{-\psi^2 v^2}, \tag{4}$$

where $\psi = 1/(\sqrt{2}\sigma_\theta) = \sqrt{m/(2k\theta)} = \sqrt{m\beta_\theta/2}$ and has units of 1/velocity for the Maxwellian.

Let us now suppose that this precision itself is not constant but is normally distributed in a variable ξ about some reference value ψ_0 such that $\psi = \psi_0 + \xi$. Then Eq. (4) becomes

$$\frac{\psi}{\sqrt{\pi}} e^{-\psi^2 v^2} = \frac{\psi_0 + \xi}{\sqrt{\pi}} e^{-(\psi_0+\xi)^2 v^2} \equiv p_{v\xi}. \tag{5}$$

Since $\psi = \sqrt{m/2k\theta} > 0$ for finite θ, $\xi \in (-\psi_0, \infty)$ so that the normal distribution ought to be truncated. However, in typical statistical applications infinite domains are commonly used instead of semi-infinite ones. For example, the convention of spectroscopy is to integrate over spectral lines for frequencies, $\nu \in (-\infty, \infty)$, even though negative frequency makes little physical sense. In this case the inadmissible values contribute little to relevant integrals as well (Essex and Andresen 2015).

Taking this position we allow $\xi \in (-\infty, \infty)$ instead. The corresponding probability distribution function in ξ is

$$p_\xi = \frac{w}{\sqrt{\pi}} e^{-w^2 \xi^2}, \tag{6}$$

where w is the Gaussian precision for this ξ distribution with units of velocity. We will see that the resulting structure is such that w appears naturally in the expressions as a velocity, aiding interpretation of molecular velocity v regimes:

$$p(v; w, \psi_0) = \int_{-\infty}^{\infty} p_{v\xi} p_\xi d\xi = \frac{w^3 \psi_0}{\sqrt{\pi}(v^2 + w^2)^{3/2}} \exp\left(-\frac{w^2 \psi_0^2 v^2}{v^2 + w^2}\right). \tag{7}$$

This equation is the temperature counterpart of Eq. (2) for the wind average.

Two distinctive features emerge: This probability distribution function has polynomial (heavy) tails and a Gaussian core. The shift between these is controlled by the remarkable argument of the exponential, $-w^2\psi_0^2 v^2/(v^2 + w^2)$. Notice that Eq. (7) is almost symmetrical in v and w. For small velocities, when $v \ll w$, it becomes the classical Gaussian form $\exp(-\psi_0^2 v^2)$ since the denominator in the pre-factor, $(v^2 + w^2)^{3/2}$, behaves like a constant. For large velocities, $v \gg w$ the argument of the exponential approaches a constant leaving an asymptotic behavior of $\sim v^{-3}$. Figure 2 illustrates this mixed behavior.

Thus near the center of the probability distribution function it behaves like a Maxwellian with temperature θ while far from the core the simple notion of temperature is not sustainable. This Maxwellian is invalid for $|v| > |w|$, thus θ has no usable role in the sense of thermodynamics in that moments of the integral will not produce the traditional simple functions in terms of θ.

This is quite different from the result of letting the velocity u fluctuate, where the result was another Gaussian probability distribution function, but with a revised temperature, θ. The u fluctuations were naturally incorporated into the microscopic ones. This does not happen with the fluctuations in ξ since the microscopic quantity temperature or precision also appears in the normalization factor multiplying the

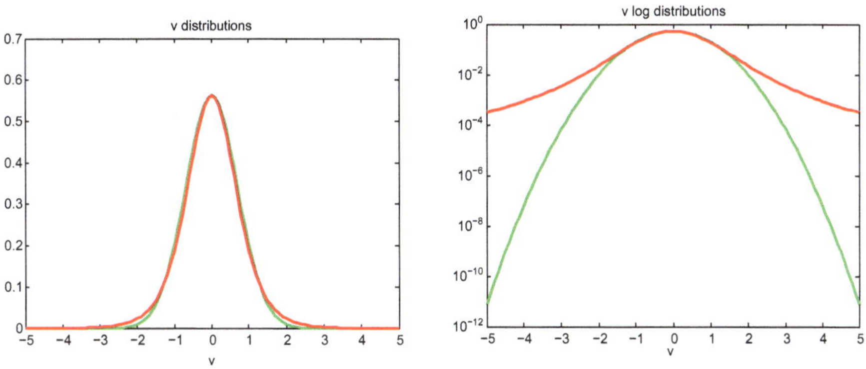

Fig. 2 The velocity distribution $p(v; w, \psi_0)$ of Eq. (7) for $w = 2.5$ and center precision $\psi_0 = 1$ (*red*). A pure Gaussian thermal distribution is shown in *green* for comparison. The *left frame* is a normal linear plot, the *right frame* a semilog plot where the agreement between the slow-time distribution (*red*) and a thermal distribution (*green*) for small velocities but large discrepancy at large velocities is even more evident

exponential in Eq. (1). Thus knowledge of short time quantities is needed for calculation of the longtime average of temperature. In other words, temperature cannot be part of a self-contained set of variables at long times.

4 Other Winds

The preceding makes two key points clear:

1. For finite w temperature cannot be thermalized like wind. Thus local equilibrium and all that it implies is tied to the laboratory regime, and not a property of large space and timescales.
2. Properties like wind can be formally thermalized as above, and mechanical pressure (distinct from thermodynamic pressure) continue to have meaning. Persistent winds on long timescales can be captured in the preceding by not assuming wind fluctuations are centered on zero.

Local equilibrium is tied entirely to the practical existence of intensive thermodynamic variables (Essex and Andresen 2013). Local conditions must then be characterized in a different manner in a putative climate regime. Unlike intensities, extensive thermodynamic variables can exist in such a regime. Thus we can still speak, for example, of energy and numbers of molecules. We can still imagine boundaries that such properties traverse, therefore fluxes still make sense. Vector flux densities divided by the corresponding volume densities of any extensive thermodynamic quantity of that slow regime will thus induce a local vector velocity field. This provides a way to distinguish between fixed conditions and evolution. When all vector velocity fields become identical, all processes stop. There is a rest frame in which there are no flows. The need for local equilibrium is thus circumvented. The various vector velocity fields are referred to as *generalized winds* (Essex 2013). Furthermore, all flows are put onto a common scale: velocity. A departure in velocities from each other is a measure of the vigor of processes.

5 Conclusion

This paper has contemplated the perspective of an observer who would regard the laboratory regime as jiggly and microscopic, much as we see the kinetic or nanoscales. We aimed to get beyond pure speculation by focusing on how the Maxwellian distribution might be seen by such a slow-time observer. The window of observation for this observer would be bounded by events that are too close in time to distinguish from his point of view (fast time), which would include our regime. We would regard the putative observer as experiencing slow time. Hence the resulting distribution is described as the slow-time Maxwellian.

The technique was to form compound distributions by fluctuating the wind, u, and temperature, T. Temperature and velocity emerge with a conjugate quality, which occurs explicitly in the case of thermalizing of wind. But it also appears in a more subtle manner in the precision picture of the Gaussian distribution because fluctuating precision led to a normal distribution with its own precision (i.e., the precision of the precision). The latter has units of velocity, and this velocity, w, plays a decisive role in the structure and behavior of the resulting compounded densities. It acts like a reference velocity separating regimes. It divides Gaussian-like structure from polynomial, heavy-tail structure.

An unusual hybrid of Gaussians with heavy tails emerges in this paper as a key feature. Heavy tails clearly can be expected to be a feature of the slow-time regime. This has some consequences. First, the notion of local equilibrium ceases to be strictly valid. There is no straightforward temperature, as there is in the Maxwellian case. There could be other qualities that might play such a role in the slow-time regime, but they would not be temperature strictly speaking. If w is large enough, the core would still behave Maxwellian, which would permit a limited return to temperature as long as the core of the probability distribution function is of importance. Second, the wings of the distribution need to be considered from a physical standpoint to avoid divergent moment integrals.

The slow-time observer is left with a rather different behavior for the ideal gas. There are heavy tails and a nearly Gaussian core, becoming more Gaussian with increasing w. But as the tails are heavy, we observe divergent second moments. Does this mean that energy becomes infinite? Not if there are only a finite number of particles and finite energy in the underlying system to begin with. The composition of probability distribution functions changes nothing in this regard.

The fundamental finding of this study is that while wind persists in slow time (the climate perspective), temperature does not. Hence any conclusions based on an extrapolation of short laboratory time measurements of temperature are ill founded: We are *not* measuring the right things.

References

Essex, C. 2011. Climate theory versus a theory for climate. *International Journal of Bifurcation and Chaos in Applied Sciences and Engineering* 21: 3477–3487.

Essex, C. 2013. Does laboratory-scale physics obstruct the development of a theory for climate? *Journal of Geophysical Research-Atmospheres* 118: 1218–1225.

Essex, C., and B. Andresen. 2013. The principal equations of state for classical particles, photons, and neutrinos. *Journal of Non-Equilibrium Thermodynamics* 38: 293–312.

Essex, C., and B. Andresen. 2015. Maxwellian velocity distributions in slow time. *Journal of Non-Equilibrium Thermodynamics* 40: 139–151.

What Have Complex Network Approaches Learned Us About El Niño?

Qing Yi Feng and Henk A. Dijkstra

Contribution to Nonlinear Advances in Geosciences

Abstract A short overview is given of recent work on the application of network techniques to the El Niño/Southern Oscillation phenomenon in the Tropical Pacific. Although several new and useful diagnostics have been developed, progress regarding the understanding of El Niño dynamics has been rather limited. Success has been claimed to forecast El Niño events 1 year ahead using network-based predictors, but tests are limited and the reason for this skill is still unclear.

Keywords Network approaches • El Niño • Diagnostics • Dynamics • Predictability

1 Introduction

Over the last decade, complex network-based approaches have been applied to tackle problems in climate dynamics (Tsonis et al., 2006). This is far from trivial as network theory deals with properties of graphs, while climate variability is associated with continuous fields (e.g. temperature) that evolve in time. A conversion from a continuous description to a discrete one can be easily made by considering the dependent quantities only on a grid (consisting either of observation locations or of model grid points), which define the 'nodes' of the graph. However, there are many ways to define the 'edges' or 'links' and, as will be discussed below, several suggestions for such a network 'inference' have been proposed.

In this short review, we will provide an overview what complex network approaches have learned us about the variability known as El Niño/Southern Oscillation (ENSO). The El Niño phase of this phenomenon appears about every 4 years

Q.Y. Feng • H.A. Dijkstra (✉)
Department of Physics and Astronomy, Institute for Marine and Atmospheric Research Utrecht, Utrecht University, Princetonplein 5, 3584 CC Utrecht, The Netherlands
e-mail: h.a.dijkstra@uu.nl

A.A. Tsonis (ed.), *Advances in Nonlinear Geosciences*,
DOI 10.1007/978-3-319-58895-7_7

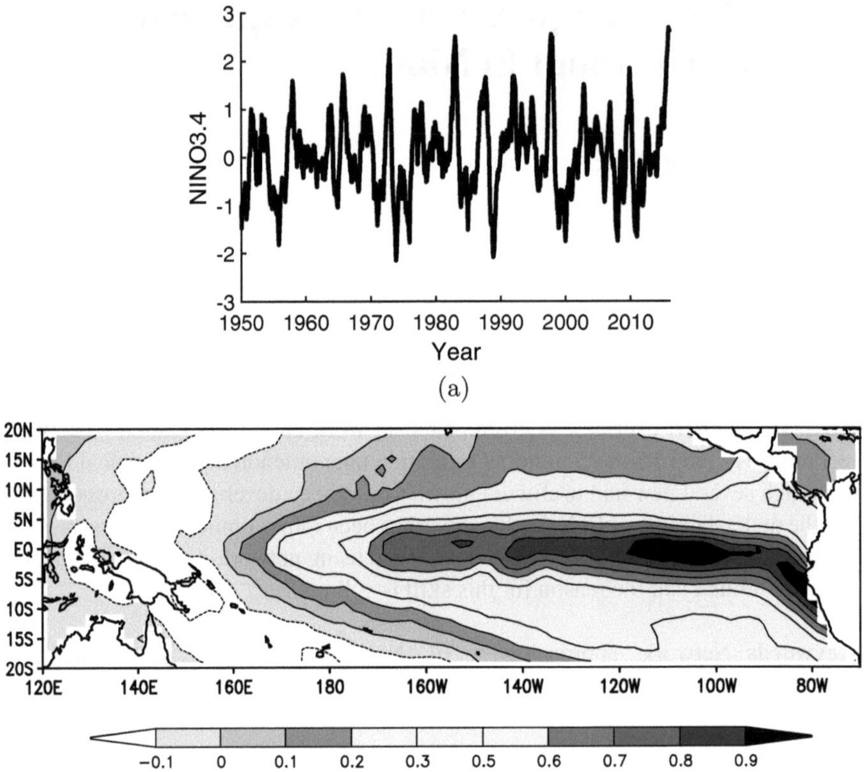

Fig. 1 (**a**) NINO3.4 index over the period 1950–2015. (**b**) First EOF of SST determined from HadISST data over the period 1950–2000. Data from the Climate Explorer (climexp.knmi.nl)

in the equatorial Pacific leading to a warming of the surface waters in the eastern equatorial Pacific up to 5 °C. The last El Niño had its maximum around December 2015 and was, in several measures, one of the strongest of the instrumental records. An often used ENSO index is NINO3.4, which is the area-averaged sea surface temperature (SST) anomaly over the region 120°W–170°W × 5°S–5°N (Fig. 1a). ENSO variability is traditionally analysed with linear, stationary statistical methods, such as principle component analysis. The first Empirical Orthogonal Function (EOF) of SST anomalies, for example, the Hadley Centre Sea Ice and Sea Surface Temperature (HadISST) data set, shows a pattern with largest amplitudes in the eastern Pacific (Fig. 1b) and strongly confined to the equatorial region.

As every El Niño event has substantial impact on climate worldwide, with typically droughts on the western part of the Pacific and flooding on the eastern part, it is important to develop skillful forecasts of the events, preferably with a 1 year lead time. Before such forecasts can be made, the system itself has to be well understood and the latter requires models that adequately capture the processes behind the phenomena. To test the output of these models, meaningful diagnostics from observations are required. In the results below, we address these three issues (diagnostics, dynamics, forecasting) and discuss how complex network theory has contributed to progress over the last decade.

2 Results

2.1 ENSO Diagnostics

Network methods have added several new diagnostics to the traditional statistical analyses techniques. The most straightforward diagnostic is the degree field of an unweighted network which arises by thresholding zero-lag cross (Pearson) correlations of SST time series at different locations in the equatorial Pacific (Feng and Dijkstra, 2016) over the whole period 1945–2010. This degree field shows very similar features as the first EOF (cf. Fig. 1b) that is determined directly from the eigenvectors of the covariance matrix. Indeed, there is a close relationship between principle component analysis and Pearson Correlation Climate Network (PCCN) analysis, with additional information on the higher-order statistical interrelationships provided by the network analysis (Donges et al., 2016).

More sophisticated network-based diagnostics were developed from surface atmospheric temperature (SAT) data by Gozolchiani et al. (2011) using optimal lag-τ correlations between different locations on the sphere. Using such link strengths leads to a weighted network, for which the in-degree (negative lag optimum) and out-degree (positive lag optimum) can be determined. Gozolchiani et al. (2011) identify a set of nodes, which they refer to as the El Niño Basin (ENB) nodes (Fig. 2a), that have a relatively low link strength during an El Niño event. The main result in Gozolchiani et al. (2011) is that the in- (out-) degree of these ENB nodes decreases (increases) substantially during an El Niño event and makes this set of nodes more autonomous (Fig. 2b). The same network reconstruction and analysis methods have been used in Wang et al. (2016) to detect equatorial Kelvin and Rossby waves in sea surface height data.

Community detection algorithms have also been applied to networks constructed from global data of SST, precipitation and SAT (Fountalis et al., 2015; Tantet and Dijkstra, 2014; Tsonis et al., 2010). Communities are groups of nodes tightly connected together and weakly connected to the rest of the network. As such, they can be regarded as subsystems which operate relatively independently of the other communities. An example of communities as deduced from a PCCN determined from SST data using the Infomap algorithm (Rosvall and Bergstrom, 2007) is shown in Fig. 3. Community #1 is by far the dominant community in terms of PageRank (69%) and size. Most of the nodes are located in the tropical Pacific (and related to ENSO) but remote patches, for example, in the extratropical Pacific and tropical Indian Ocean, are also part of this community. This shows the teleconnections that exist between the ENSO variability and remote regions over the globe (Tantet

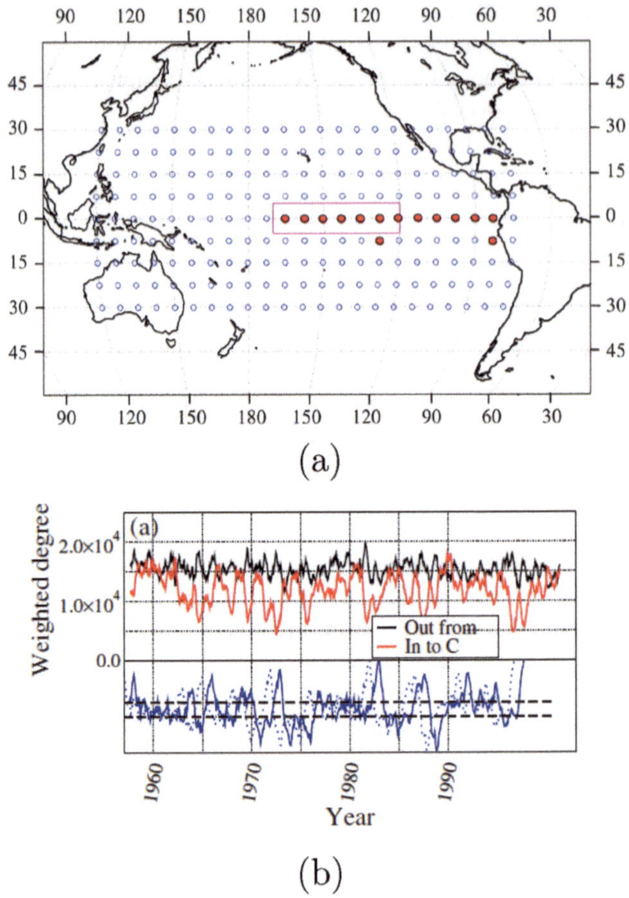

Fig. 2 (**a**) Nodes of the network within the El Niño Basin (ENB) indicated in *solid red symbols*, and the *red rectangle* denotes the NINO3.4 index area. (**b**) In and out-out links of the ENB nodes (the ENB nodes are indicated as group C). Figure from Gozolchiani et al. (2011) reproduced with permission from the American Physical Society (APS)

and Dijkstra, 2014). The community detection methods allow to bypass some shortcomings of EOF analysis (e.g. orthogonality) and hence provide additional information on global patterns of SST variability with respect to these classical analysis tools. The community analysis also gives new insight into the relationship between patterns of ENSO variability and the global mean surface temperature (Tantet and Dijkstra, 2014).

The disadvantages in using the thresholding in the network inference method have been shown in Fountalis et al. (2015) who use a cluster method to define connectedness in SST observational and model data. They also presented techniques, such as the Adjusted Rand Index (ARI) and the network distance *D*, to compare networks and efficiently used these to compare model results of

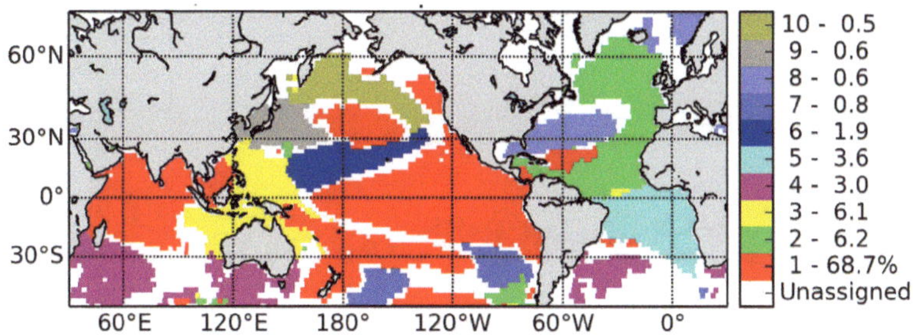

Fig. 3 Communities from a PCCN of SST, ordered by the total PageRank (Brin and Page, 1998) of their nodes. Figure from Tantet and Dijkstra (2014) reproduced with permission from the European Geosciences Union (EGU)

ENSO variability and those of observations. Network community techniques have also been applied to output of Global Climate Models (GCMs). For example, by analysing GCM output from CMIP5 models under the RCP8.5 forcing scenario, Fountalis et al. (2015) find the ENSO intensity will decrease after 2100 due to increase in greenhouse gas forcing.

2.2 *ENSO Dynamics*

The theory of ENSO has been developed through the Zebiak and Cane (ZC) model (Zebiak and Cane, 1987). The leading recharge–discharge oscillator view of ENSO consists (Jin, 1997) of the action of positive (Bjerknes') feedbacks that are responsible for the amplification of SST anomalies and ocean adjustment (i.e. through equatorial waves) providing a negative delayed feedback (Neelin et al., 1998). The strength of these feedbacks is measured by a coupling strength, usually indicated by μ, which is proportional to the change in wind stress due to a change in SST. The parameter μ also captures the strength of the ocean surface circulation response to changes in the surface wind stress.

In the ZC model, the (steady or seasonal) background Pacific climate, e.g., provided by observations, becomes unstable when the strength of the coupled processes exceeds a critical value. When $\mu > \mu_c$, oscillatory motion develops spontaneously (Fedorov and Philander, 2000) and the spatial pattern of the resulting variability is usually referred to as the ENSO mode. When conditions are such that $\mu < \mu_c$, the ENSO mode is damped and can only be excited by noise (Burgers, 1999; Penland, 1996). Hence, although the noise driven and sustained ENSO variability views are sometimes considered to be two different ENSO mechanisms, both are easily reconcilable (Dijkstra, 2013): it just depends on whether the background climate is stable ($\mu < \mu_c$) or unstable ($\mu > \mu_c$).

Both the background state and the growth/decay of the ENSO mode are controlled by similar coupled processes (Van der Vaart et al., 2000). In addition,

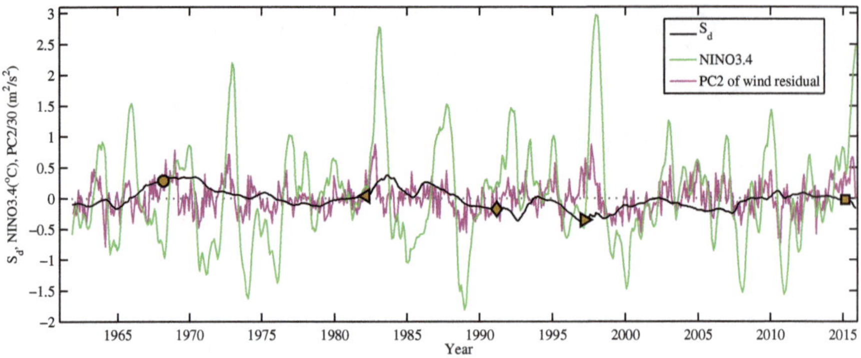

Fig. 4 The 10-year sliding window degree skewness index S_d (*black curve*), the 3-month running mean NINO3.4 index (*green curve*) from the observed SST, and the second principal component (PC2) of the wind-stress residual (*magenta curve*) from NCEP zonal wind-stress data of the same period. Figure from Feng and Dijkstra (2016) reproduced with permission from the American Physical Society (APS)

the background climate is also affected by processes outside of the Pacific basin such as those at midlatitudes and in the equatorial Indian Ocean and Atlantic Ocean (Wieners et al., 2016). Caused also by slow changes in the external radiative forcing, the background state has a strong non-stationary component on decadal-to-interdecadal time scales. The main challenging problem is whether one can determine if the feedbacks will amplify or damp SST anomalies in such a slow transient background state.

In Feng and Dijkstra (2016), this problem has been addressed using PCCNs and recurrence networks, both reconstructed from SST observations and SST output from the ZC model. From the ZC model results, the skewness of the degree distribution S_d of the PCCN was found to be a good indicator of the stability of the Pacific background state. Indeed, S_d is well anti-correlated with the Bjerknes stability (BJ) index (Kim and Jin, 2011) in several GCMs, an often used metric used to quantify the stability of the Pacific climate. The variation of S_d shown in Fig. 4 indicates periods of high background stability (high values of S_d) and ones with low background stability. For example, the relatively high value of S_d in early 1968 indicates that the Pacific background climate in 1968 was quite stable and the noise must have had a large influence on the development of the 1968 El Niño event. Indeed, the principle component of the wind stress residual from the National Centers for Environmental Prediction (NCEP) wind stress data (Kalnay et al., 1996) shows that high-noise variability occurred during early 1968 (the magenta curve in Fig. 4). On the contrary, the 1992 El Niño event would be considered as a sustained case, because of the relatively low value of S_d and low noise variability in early 1992. Actually, the value of S_d was overall low (less stable Pacific background state) during the early 1990s with a global minimum just before 1997.

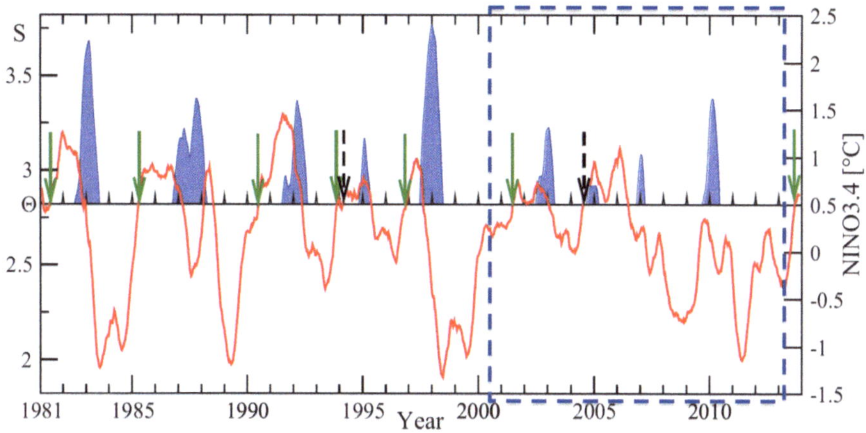

Fig. 5 Forecasting results in Ludescher et al. (2013) where the average link strength of the climate network S is plotted versus time. The value of Θ is the decision threshold. *Green arrows* indicate correct predictions and *dashed arrows* indicate false alarms. Figure from Ludescher et al. (2013) reproduced with permission from the US National Academy of Sciences (NAS)

2.3 ENSO Forecasting

The link strength concept proposed in Gozolchiani et al. (2011) was used in Ludescher et al. (2013) to use the average link strength as a predictor for El Niño events. The idea is that when the average link strength S of the ENB nodes (cf. Fig. 2) crosses a threshold while monotonically increasing, an El Niño will develop about 1 year later. A training SAT data set for the period 1950–1980 was used to determine the threshold Θ and then the period 1980–2011 was used as test data to evaluate the predictor (Fig. 5). The skill of the predictor over this test period is indeed remarkable (green arrows in Fig. 5 are correct predictions). Ludescher et al. (2014) used this method also to make a prediction of the El Niño event in 2014 (1 year ahead), but this turned out to be incorrect as a strong El Niño appeared more than 2 years later (at the end of 2015, see Fig. 1a).

In Feng et al. (2016), machine learning techniques are used to predict the occurrence of El Niño events. As attributes, several network quantities as in Gozolchiani et al. (2011) are used. The method used for supervised learning is an Artificial Neural Network with a 3×3 layer structure (three neurons per layer). The training set is from May 1949 to June 2001, the test set is from June 2001 to March 2014, and the prediction time τ is 12 months. Figure 6 shows the classification results on the test set, where 1 stands for the occurrence of an El Niño event and 0 means no event. By applying a specific time-series filter, which eliminates the isolated and transient events, and joins the adjacent events, Fig. 6 shows that this forecasting scheme gives accurate alarms 12 months ahead for the El Niño events at least in 2002 and 2006, without a false alarm in 2004. Hence, the machine learning toolbox also appears to provide skillful 1-year ahead predictions for the occurrence of El Niño events.

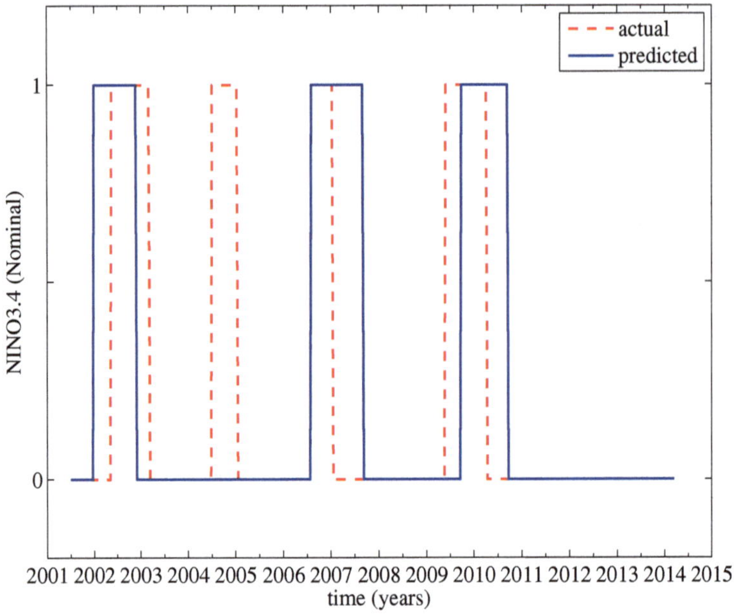

Fig. 6 Prediction results on the test set from June 2001 to March 2014. Figure from Feng et al. (2016) reproduced with permission from the European Geosciences Union (EGU)

3 Summary and Discussion

Through the example of the El Niño/Southern Oscillation (ENSO) variability, we have given a brief review of what complex network approaches have to offer regarding diagnostics, dynamics, and forecasting of this important phenomenon in climate research. Considering the amount of effort put into the application of network techniques to ENSO variability, the results are slightly disappointing as no real breakthroughs in either El Niño diagnostics, El Niño dynamics, or El Niño forecasting has occurred.

Network approaches definitely have led to new interesting diagnostics, in particular the average link strength of a weighted SAT network (Gozolchiani et al., 2011). The fact that the ENB nodes become more autonomous during the start of an El Niño can indeed be understood from ENSO theory. During the start of an El Niño, the SAT over the ENB region becomes more and more controlled by SST and hence the in-degree of these nodes is expected to decrease. On the other hand, SST is more and more controlling the winds over the equatorial Pacific and hence the out-degree of the ENB nodes increases.

Considering patterns of SST variability, information on higher-order statistical interrelations is provided by network analysis (Donges et al., 2016), compared to the traditional statistical analysis, such as EOFs analysis. In addition, with community analysis techniques one can isolate global SST patterns of variability, including ENSO teleconnections, much more clearly (Tantet and Dijkstra, 2014) than with EOF methods.

The results on determining properties of ENSO dynamics, such as the stability boundary, have been less successful. Certainly, there is a nice anti-correlation between the index S_d in Feng and Dijkstra (2016) and the Bjerknes stability index in Kim and Jin (2011), but S_d is only a nominal measure of stability and cannot solely determine the stability boundary from observations. The recurrence network measures used in Feng and Dijkstra (2016) indicate that there are different classes of El Niño's but give little connection to the underlying feedback processes causing these differences.

One of the more attractive results has been the claim that network- based predictors can provide skillful 1-year lead time forecasts of El Niño events (Ludescher et al., 2013). Indeed, it is not impossible that this could occur, as also the machine learning-based forecasts (Feng et al., 2016) indicate. However, the connection to the underlying physics is still lacking and needs to be clarified to put some confidence on these forecasts. This may then also provide reasons why the prediction for 2014 was not correct (Ludescher et al., 2014).

Acknowledgements We like to acknowledge the support of the LINC project (no. 289447) funded by the Marie-Curie ITN program (FP7-PEOPLE-2011-ITN) of the EC-FP7 framework.

References

Brin, S., and L. Page. 1998. The anatomy of a large-scale hypertextual Web search engine. *Computer Networks and ISDN Systems* 30(1–7): 107–117.
Burgers, G. 1999. The El Niño Stochastic oscillator. *Climate Dynamics* 15: 352–375.
Dijkstra, H.A. 2013. *Nonlinear climate dynamics*. Cambridge: Cambridge University Press.
Donges, J., I. Petrova, A. Loew, N. Marwan, and J. Kurths. 2016. How complex climate networks complement eigen techniques for the statistical analysis of climatological data, 1–18. arXiv:1305.6634v3 [physics.data-an] 9 Apr 2016.
Fedorov, A.V., and S.G. Philander. 2000. Is El Niño changing? *Science* 288: 1997–2002.
Feng, Q.Y., and H.A. Dijkstra. 2016. Climate network stability measures of El Niño variability. *Chaos: An Interdisciplinary Journal of Nonlinear Science* 27(3): 035801.
Feng, Q.Y., R. Vasile, M. Segond, A. Gozolchiani, Y. Wang, M. Abel, S. Havlin, A. Bunde, and H.A. Dijkstra. 2016. ClimateLearn: a machine-learning approach for climate prediction using network measures. *Geoscientific Model Development*. doi:10.5194/gmd-2015-273.
Fountalis, I., A. Bracco, and C. Dovrolis. 2015. ENSO in CMIP5 simulations: Network connectivity from the recent past to the twenty-third century. *Climate Dynamics* 45: 511–538.
Gozolchiani, A., S. Havlin, and K. Yamasaki. 2011. Emergence of El Niño as an autonomous component in the climate network. *Physical Review Letters* 107(14): 148501.
Jin, F.-F. 1997. An equatorial recharge paradigm for ENSO. II: A stripped-down coupled model. *Journal of the Atmospheric Sciences* 54: 830–8847.
Kalnay, E., M. Kanamitsu, R. Kistler, et al. 1996. The NCEP/NCAR 40-year reanalysis project. *Bulletin of the American Meteorological Society* 77(3): 437–471.
Kim, S.T., and F.-F. Jin. 2011. An ENSO stability analysis. Part I: Results from a hybrid coupled model. *Climate Dynamics* 36(7–8): 1593–1607.

Ludescher, J., A. Gozolchiani, M.I. Bogachev, A. Bunde, S. Havlin, and H.J. Schellnhuber. 2013. Improved El Niño forecasting by cooperativity detection. *Proceedings of the National Academy of Sciences* 110(29): 11742–11745.

Ludescher, J., A. Gozolchiani, M.I. Bogachev, A. Bunde, S. Havlin, and H.J. Schellnhuber. 2014. Very early warning of next El Niño. *Proceedings of the National Academy of Sciences* 111(6): 2064–2066.

Neelin, J., D.S. Battisti, A.C. Hirst, F.-F. Jin, Y. Wakata, T. Yamagata, and S.E. Zebiak. 1998. ENSO theory. *Journal of Geophysical Research* 103: 14261–14290.

Penland, C. 1996. A stochastic model of IndoPacific sea surface temperature anomalies. *Physica D: Nonlinear Phenomena* 98(2): 534–558.

Rosvall, M., and C. Bergstrom. 2007. An information-theoretic framework for resolving community structure in complex networks. *Proceedings of the National Academy of Sciences* 104(18): 7327–7331.

Tantet, A., and H.A. Dijkstra. 2014. An interaction network perspective on the relation between patterns of sea surface temperature variability and global mean surface temperature. *Earth System Dynamics Discussions* 4: 743–783.

Tsonis, A.A., K.L. Swanson, and P.J. Roebber. 2006. What do networks have to do with climate? *Bulletin of the American Meteorological Society* 87(5): 585–595.

Tsonis, A.A., G. Wang, K.L. Swanson, F.A. Rodrigues, and L.D.F. Costa. 2010. Community structure and dynamics in climate networks. *Climate Dynamics* 37(5–6): 933–940.

Van der Vaart, P.C.F., H.A. Dijkstra, and F.-F. Jin. 2000. The pacific cold tongue and the ENSO mode: Unified theory within the Zebiak-Cane model. *Journal of the Atmospheric Sciences* 57: 967–988.

Wang, Y., A. Gozolchiani, Y. Ashkenazy, and S. Havlin. 2016. Oceanic El-Niño wave dynamics and climate networks. *New Journal of Physics* 18(3): 1–10.

Wieners, C.E., W.P.M. De Ruijter, W. Ridderinkhof, A.S. von der Heydt, and H.A. Dijkstra. 2016. Coherent tropical Indo-Pacific interannual climate variability. *Journal of Climate* 29(11): 4269–4291.

Zebiak, S.E., and M.A. Cane. 1987. A model El Niño-Southern Oscillation. *Monthly Weather Review* 115: 2262–2278.

Late Quaternary Climate Response at 100 kyr: A Noise-Induced Cycle Suppression Mechanism

Ivan L'Heureux

Abstract Late quaternary climate proxies suggest the presence of a strong cycle at a period of about 100 kyr. It is thought that this cycle could be due to variations in the eccentricity of the Earth's orbit, as part of the Milankovitch forcing. However, based on simple energy balance arguments, the eccentricity variations are too small to explain the strength of the climatic response. Some amplification mechanisms based on ice sheet dynamics or ocean circulation models have been suggested to explain this paradox. But recently (Wallmann 2014), a different explanation was proposed. There, a non-linear biogeochemical model coupling seawater alkalinity, dissolved phosphate, dissolved inorganic carbon, and atmospheric carbon dioxide without any orbital forcing was developed. As the parameters vary, the system may undergo a Hopf bifurcation and exhibits self-organized oscillations with a period that has the appropriate order of magnitude but remains larger than 100 kyr. In this contribution, I revisit Wallmann's model by adding a weak stochastic periodic Milankovitch forcing at 100 kyr in the spirit of stochastic resonance phenomena. It is seen that for sufficiently high noise intensity, a noise-induced cycle suppression occurs, whereby the self-sustained oscillation of biogeochemical origin is destroyed and a strong signal persists at 100 kyr. This mechanism could thus provide an amplification mechanism for the presence of a strong response under the influence of a weak Milankovitch forcing.

Keywords Biogeochemical cycle • 100 kyr cycle • Milankovitch forcing • Noise-induced transitions • Stochastic resonance

1 Introduction

In the last million years, the Earth's climate has shown variations typically characterized by cycles of gradual cooling and glaciation followed by a sudden transition to a warm interglacial period. The climate proxies show strong signals at periods near 23, 41, and 100 kyr (Imbrie and Imbrie 1980; Petit et al. 1999).

I. L'Heureux (✉)
Department of Physics, University of Ottawa, Ottawa, ON, K1N6N5, Canada
e-mail: ilheureu@uottawa.ca

A.A. Tsonis (ed.), *Advances in Nonlinear Geosciences*,
DOI 10.1007/978-3-319-58895-7_8

The first two periods can be linked to the precession cycle of the Earth's orbit and the change in the obliquity of the Earth's axis (Milankovitch forcing). Although the variation of the eccentricity of the Earth's orbit exhibits a 100 kyr period, the resulting difference in insolation appears to be too small to have a direct effect on the climate. The cause of the strong signal at 100 kyr in climate proxies remains unclear.

Non-linear amplification mechanisms of the weak eccentricity forcing have been proposed to explain the 100 kyr cycle. For instance, ice sheet dynamics coupled with ocean circulation (Imbrie et al. 1993) could act as a non-linear amplifier of the precession and obliquity forcing terms. When the ice sheet becomes too large, internal dynamics would drive the climate in the 100 kyr spectral band. Other researchers have proposed that the 100 kyr signal is rather the manifestation of a self-organized limit-cycle solution in a non-linear climate system that is otherwise *not* forced by eccentricity variations. For instance, Gildor and Tziperman (2001) proposed an unforced non-linear box model coupling sea-ice and land-ice volume, air and sea-surface temperatures, and ocean salinity, in which limit-cycle solutions are obtained with a period of the order of 100 kyr. Saltzman and Maasch (1988) presented a three-variable model coupling global ice mass, North Atlantic Ocean circulation, and atmospheric CO_2 content. In that autonomous dynamical system, limit-cycle solutions with a period of the order of 100 kyr are also obtained. However, in his analysis of the Vostok ice core, Shackleton (2000) proposed that the 100 kyr signal does not arise from ice sheet dynamics, but that the global CO_2 cycle plays a determining role. Recently, Wallmann (2014) proposed an interesting biogeochemical model coupling the total ocean alkalinity, the dissolved phosphorus, the dissolved inorganic carbon, and the atmospheric CO_2 concentration, without any orbital forcing. In his model, the atmospheric CO_2 is the main driver of climate change. His model thus generates limit-cycle solutions in climate proxies that are broadly consistent with the sedimentary record. However, although the cycle period is of the order of 100 kyr, it remains slightly higher.

Another class of models considers the effect of randomness (noise) on a non-linear system (Horsthemke and Lefever 1984; Ridolfi et al. 2011). In Benzi et al. (1982), a simple bistable energy-balance climate model is considered, with a weak orbital forcing signal and an additive noise term. This forced bistable system may be thought of as double-well potential with a barrier height that is slightly modulated by the forcing. In the presence of noise, the system undergoes transitions from one stable state to the other. Another time scale appears in the problem: the inverse of the mean transition rate. This transition rate depends exponentially on the ratio of the barrier height to the noise intensity. Transitions are favorable when the noise intensity is such that the mean transition time is equal to the semi-period of the forcing. This signal amplification mechanism by the noise is termed "stochastic resonance." Another interesting effect of noise is "coherence resonance," whereas noise perturbs an unforced system in the neighborhood of a Hopf bifurcation point. In this case, noise-induced transitions occur between a stable point and a limit-cycle, without the need of any external forcing. Pelletier (2003) has applied this concept to a simple bistable energy-balance climate model subjected to a time-delay feedback due to lithospheric subsidence and ice sheet rebound.

In this contribution, I explore yet another effect of noise as it applies to Wallmann's biogeochemical climate model (Wallmann 2014). Parameter values are chosen such that—in absence of noise and orbital forcing—a limit-cycle solution is generated, with a period slightly larger than 100 kyr. I then add to the temperature a weak orbital forcing term at 100 kyr and a noise term. It is seen that, as the noise intensity increases, the stochastic limit-cycle becomes suppressed, so that only the orbital forcing signal survives. Yet, the dynamics feels the presence of the underlying deterministic limit-cycle and large variations in the climate proxies result. I call this mechanism "noise-induced cycle suppression."

The text is organized as follows. In Sect. 2, I offer a review of Wallmann's deterministic biogeochemical climate model. In Sect. 3, I illustrate the basic noise-induced cycle suppression mechanism by investigating a simple abstract model for a supercritical Hopf bifurcation subjected to a weak forcing and to multiplicative Gaussian white noise. In the next section, I present the numerical results pertaining to the weakly forced, noisy Wallmann's model. Section 5 offers concluding remarks.

2 Wallmann's Model: A Brief Review

In this section, I present a brief description of the Wallmann's deterministic model. The reader will find the details in Wallmann (2014). In this box model, the biosphere is reduced to one atmospheric compartment and three oceanic compartments: the shallow ocean *s* (depth smaller than 50 m), the intermediate thermocline region *i* (depth between 50 and 1200 m), and the deep ocean *d* (deeper than 1200 m). For each oceanic compartment, three dynamical variables are considered: the total alkalinity TA, the dissolved phosphorus concentration DP, and the dissolved inorganic carbon concentration DIC. In the atmospheric compartment, the dynamical variable of interest is the partial pressure of CO_2, pCO_2. Thus, ten variables are coupled together. In his approach, the solar forcing does not vary in time. The periodic behavior of the system rather results from biogeochemical self-organization coupled with sea-level changes parameterized by air temperature. Table 1 lists the ten ordinary differential equations defining the model. The various processes involved are listed in Table 2 and are now briefly introduced.

I first mention the biogeochemical and burial processes. Photosynthetic primary export production of particulate inorganic carbon (PIC), particulate organic phosphorus (POP), and particulate organic carbon (POC) is modeled by phosphorus-limited first-order kinetics in the shallow ocean. Microbial degradation of POP and POC in the thermocline and the deep oceans occurs at a rate proportional to the POP and POC production rate, the coefficient of proportionality being determined by mass balance between production, degradation, and burial. The kinetics of TA and DIC in the shallow ocean compartment is also directly affected by deposition of neritic carbonate shells (which represents a PIC burial term). In the deep oceans, PIC undergoes microbial degradation as well but is also affected by calcite dissolution or precipitation, characterized by the degree of saturation $\Omega = [Ca^{+2}][CO_3{}^{-2}]/K_{cal}$

Table 1 Wallmann's model: deterministic dynamical equations

Name of variable	Differential equation
DP_s (shallow ocean—μM):	$\frac{d(DP_s)}{dt} = V_s^{-1}\,(-F_{EPOP} + F_{IS} + F_{RDP})\,.$
DP_i (thermocline—μM):	$\frac{d(DP_i)}{dt} = V_i^{-1}\,(F_{DPOPI} + F_{DI} - F_{IS})\,.$
DP_d (deep ocean—μM):	$\frac{d(DP_d)}{dt} = V_d^{-1}\,(F_{DPOPD} - F_{DI} - F_{HY})\,.$
DIC_s (shallow ocean—μM):	$\frac{d(DIC_s)}{dt} = V_s^{-1}\,(-F_{EPOC} - F_{EPIC} - F_{BPICS} - F_{CO2SA} + F_{IS} + F_{RTA})\,.$
DIC_i (thermocline—μM):	$\frac{d(DIC_i)}{dt} = V_i^{-1}\,(F_{DPOCI} + F_{DI} - F_{IS})\,.$
DIC_d (deep ocean—μM):	$\frac{d(DIC_d)}{dt} = V_d^{-1}\,(F_{DPOCD} + F_{DPICD} - F_{DI} - F_{ALT} + F_{SP})\,.$
TA_s (shallow ocean—μM):	$\frac{d(TA_s)}{dt} = V_s^{-1}\,(-2F_{EPIC} - 2F_{BPICS} + F_{IS} + F_{RTA})\,.$
TA_i (thermocline—μM):	$\frac{d(TA_i)}{dt} = V_i^{-1}\,(F_{DI} - F_{IS})\,.$
TA_d (deep ocean—μM):	$\frac{d(TA_d)}{dt} = V_d^{-1}\,(2F_{DPICD} - F_{DI} - F_{ALT})\,.$
pCO_2 (μatm):	$\frac{d(pCO_2)}{dt} = M_a^{-1}\,(F_{CO2SA} - F_{WC} - F_{WS} + F_{VO} + F_{MC} + F_{WO})\,.$

with respect to calcite, where K_{cal} is the solubility of calcite in seawater. The calcium concentration $[Ca^{+2}]$ is considered constant, whereas K_{cal} is determined by using the thermodynamic expressions of Zeebe and Wolf-Gladrow (2001) from the knowledge of the deep ocean pH (which is a function of DIC and TA), deep ocean salinity, pressure, and temperature. The latter three parameters are considered constant.

Next, I mention the inter-compartment exchange processes. TA, DIC, and DP are affected by vertical eddy mixing from one oceanic compartment to the adjacent one. The mixing kinetics between two compartments is modeled as being proportional to their concentration difference. The CO_2 exchanged between the shallow ocean and the atmosphere is modeled as a term proportional to the CO_2 partial pressure difference between the sea surface and air, with a proportionality constant that depends on the gas transfer piston velocity and the solubility of CO_2 in seawater. The latter depends on sea-surface temperature.

I now list the external source and sink processes. Sources of atmospheric CO_2 include continental releases of CO_2 by volcanic activity, metamorphism, and organic carbon weathering. Continental weathering of carbonates and silicates constitutes sinks of atmospheric CO_2. The rate of these weathering processes is a function of air temperature. DP, TA, and DIC in the shallow oceans are affected by riverine inputs. Riverine input of DP is the result of apatite weathering. As such, it is equal to the sum of the weathering rates of carbonates, silicates, and organic carbon. Similarly, riverine inputs of TA and DIC depend on the weathering rates of silicates and continental carbonates. As far as the deep ocean is concerned, hydrothermal uptake of phosphorus constitutes a sink of DP with a rate proportional to its concentration, whereas hydrothermal release of H_2CO_3 at submarine spreading centers and carbonate formation resulting from the alteration of basaltic oceanic crust influence the dynamics of TA and DIC. The latter two processes are assumed to occur at a constant rate.

Table 2 Notation used in Table 1

Acronyms for the dynamical variables:	
DIC:	Dissolved inorganic carbon
DP:	Dissolved phosphorus
pCO_2:	Partial pressure of CO_2 in the atmosphere
PIC:	Particulate inorganic carbon
POC:	Particulate organic carbon
POP:	Particulate organic phosphorus
t:	Time
TA:	Total alkalinity
Mass currents (in Tmol/year) (Note: a subscript H refers to a modern Holocene value):	
F_{ALT}:	Uptake of dissolved carbon by alteration of basaltic oceanic crust
F_{BPICS}:	Burial of PIC in the shallow ocean by neritic shell deposits
F_{CO2SA}:	Exchange of CO_2 between the atmosphere and the shallow ocean
F_{DI}:	Transport of dissolved species by mixing from the deep ocean to the thermocline
F_{DPICD}:	Degradation of PIC in the deep ocean
F_{DPOCD}:	Degradation of POC in the deep ocean
F_{DPOCI}:	Degradation of POC in the thermocline
F_{DPOPD}:	Degradation of POP in the deep ocean
F_{DPOPI}:	Degradation of POP in the thermocline
F_{EPIC}:	Export production of PIC
F_{EPOC}:	Export production of POC
F_{EPOP}:	Export production of POP
F_{HY}:	Uptake of DP by hydrothermal activity
F_{IS}:	Transport of dissolved species by mixing from the thermocline to the shallow ocean
F_{MC}:	Source of atmospheric CO_2 from continental metamorphism
F_{RDP}:	Riverine DP input
F_{RTA}:	Riverine TA input
F_{SP}:	Source of dissolved carbon submarine spreading centers
F_{VO}:	Source of atmospheric CO_2 from continental volcanic activity
F_{WC}:	Sink of atmospheric CO_2 from weathering of continental carbonates
F_{WO}:	Source of atmospheric CO_2 from weathering of continental organic carbon
F_{WS}:	Sink of atmospheric CO_2 from weathering of continental silicates
Extensive parameters:	
M_a:	Number of moles in the atmosphere (1.773×10^8 Tmol)
V_d:	Volume of the deep ocean compartment (94.4×10^{16} m^3)
V_i:	Volume of the thermocline compartment (37.39×10^{16} m^3)
V_s:	Volume of the shallow ocean compartment (1.81×10^{16} m^3)

In the model, air temperature directly affects the silicates and continental carbonates weathering rates. Air temperature T_A is obtained by using a standard expression that relates global temperature to changes in the atmospheric CO_2 content:

$$T_A = \Gamma \log(pCO_2/280) + T_{AH} \tag{1}$$

where pCO_2 is in μatm, $\Gamma = 7.2$°C and $T_{AH} = 15$°C (*H* refers to "Holocene," the modern epoch). Similarly, the solubility of CO_2 in the shallow ocean depends on ocean surface pH (hence on its DIC and TA), salinity (considered constant), and sea-surface temperature, T_{SS}. By analogy with Eq. (1), the latter is taken as

$$T_{SS} = \Gamma_{SS} \log(pCO_2/280) + T_{SSH} \tag{2}$$

where $\Gamma_{SS} = 4.1$ °C and $T_{SSH} = 17.88$°C. An equilibrium thermodynamic calculation of the solubility is then performed using the expression of Zeebe and Wolf-Gladrow (2001).

Finally, an important feature of Wallmann's model is the consideration of sea-level falls SLF. As pCO_2 changes, air temperature changes, and sea-level is affected. The lower is the air temperature, the higher the sea-level fall. Reports by Grant et al. (2012) and Foster and Rohling (2013) suggest a linear correlation between sea-level falls and air temperature:

$$\text{SLF} = \text{SLF}_{\text{LGM}} \frac{T_{\text{AH}} - T_A}{T_{\text{AH}} - T_{\text{ALGM}}} \tag{3}$$

where the calibration is taken at the last glacial maximum (LGM) with $\text{SLF}_{\text{LGM}} = 120$ m and $T_{\text{ALGM}} = 12$°C. From hypsographic data (Eakins and Sharman 2012), the sea-level changes are then used to calculate the change in exposed shelf area A_{EX}, the seafloor area covered by the shallow oceans A_{SS}, and the margin (depth < 1200 m) seafloor area A_M. These changes in areas directly affect some rate processes. Thus, the total rates of organic carbon and continental carbonates weathering are proportional to the exposed shelf area A_{EX}. The total burial rate of PIC in the shallow ocean (neritic carbonate deposition) is proportional to seafloor area under the shallow oceans, A_{SS}. The total burial rates of POC and POP in the shallow and thermocline ocean compartments are proportional to the margin seafloor area, A_M. And the burial of POC in the deep ocean is inversely proportional to A_M. The latter effect reflects the fact that, for high sea-level falls (low A_M), larger volumes of fine-grained terrigenous sediments from the exposed continental shelf are deposited, which in turn enhances the burial rates of POC.

In fact, Wallmann explored various amended versions of his base model. In the first one, the rate coefficients for ocean ventilation (mixing terms) are no longer constant but change linearly with sea-surface temperature T_{SS}. This reflects the fact that mixing is not as efficient under glacial conditions, when T_{SS} is smaller. This version of the model generates limit-cycle solutions for a significantly larger range of parameter values. This is the version that I have adopted here.

Other model versions proposed by Wallmann include (1) an increase in phosphorus utilization due to dust-driven iron fertilization under glacial conditions; (2) a decrease in the rate of POP and POC microbial degradation in the thermocline resulting from the increase in sinking velocity of particulate matter as more dust provides more ballast material under glacial conditions; and (3) a stepwise change in the mixing rate between the thermocline and the deep ocean during cooling trends while keeping the mixing rate between the thermocline and the shallow ocean unchanged. These models produce response curves that are more asymmetric and (in the third version) a limit-cycle period closer to 100 kyr. However, they use a large number of fitting parameters. I will not consider these variants here.

Using reasonable estimates of the modern parameters, Wallmann was able to reproduce a steady-state solution consistent with modern values of the TA, DIC, PD, and pCO_2. Some parameters are less well constrained, however. So, it is estimated that the riverine phosphorus input current F_{RDPH} varies from 0.1 to 0.18 Tmol/year and the total atmospheric carbon source current $F_{\mathrm{VO}} + F_{\mathrm{MC}} + F_{\mathrm{WOH}}$ can vary between 10 and 16 Tmol/year.

Typically, the actual riverine inputs of phosphorus and carbon are smaller than their steady-state burial rates, so that DP, TA, and DIC initially decline. At a very early stage, the decrease in DP decreases photosynthetic production and the atmospheric CO_2 increases briefly. But, soon, the decline in TA and DIC generates a decrease in atmospheric CO_2 and a temperature decrease. As a result, the sea-level falls. Many positive feedback mechanisms then enhance the decrease in atmospheric CO_2. (1) As the exposed shelf area increases, the total weathering of continental carbonates increases, thus removing more CO_2 from the atmosphere. (2) As the submerged ocean area decreases, the total burial of particulate matter decreases, thus leaving more DP in the ocean, which enhances photosynthetic production and decreases atmospheric CO_2. (3) As the sea-surface temperature decreases, the CO_2 solubility in seawater increases, thus contributing to a further decrease in atmospheric CO_2. However, there are negative feedbacks that limit the decrease in CO_2. (1) As temperature decreases, the margin oceanic area decreases and less dissolved carbon is removed by burial of neritic shells. Thus, DIC and TA increase, contributing to an increase in atmospheric CO_2. (2) As temperature decreases, the weathering of continental silicates (an atmospheric CO_2 sink) is not as efficient. (3) As the sea-level falls, the exposed shelf area increases and the total weathering of organic carbon increases, directly contributing to atmospheric CO_2. (4) As the submerged area decreases, the total burial of organic carbon decreases, removing less CO_2 from the ocean, which contributes to an increase in atmospheric CO_2. Thus, the potential exists for self-oscillatory solutions. The balance between these mechanisms (modulated by a temperature-dependent mixing when ventilation is included) will determine whether a fixed point or a limit-cycle solution is obtained.

To illustrate the model, Fig. 1 shows the air temperature, the sea-surface temperature, the CO_2 atmospheric concentration, the sea-level and the oceanic pH as a function of time using $F_{\mathrm{RDPH}} = 0.15$ Tmol/year, $F_{\mathrm{VO}} + F_{\mathrm{MC}} + F_{\mathrm{WOH}} = 11.15$ Tmol/year, and reasonable values of the other parameters. This solution clearly exhibits a limit-cycle behavior with a cycle period equal to 162 kyr. Figure 2 shows the phase diagram in (F_{VO}, F_{RDPH}) space with $F_{\mathrm{MC}} = 3$ Tmol/year and $F_{\mathrm{WOH}} = 7$ Tmol/year. It demonstrates the existence of a stable limit-cycle solution between two stable fixed points regions. Figure 3 shows the bifurcation diagram corresponding to the vertical dashed line at $F_{\mathrm{RDPH}} = 0.15$ Tmol/year in Fig. 2 (the value used in

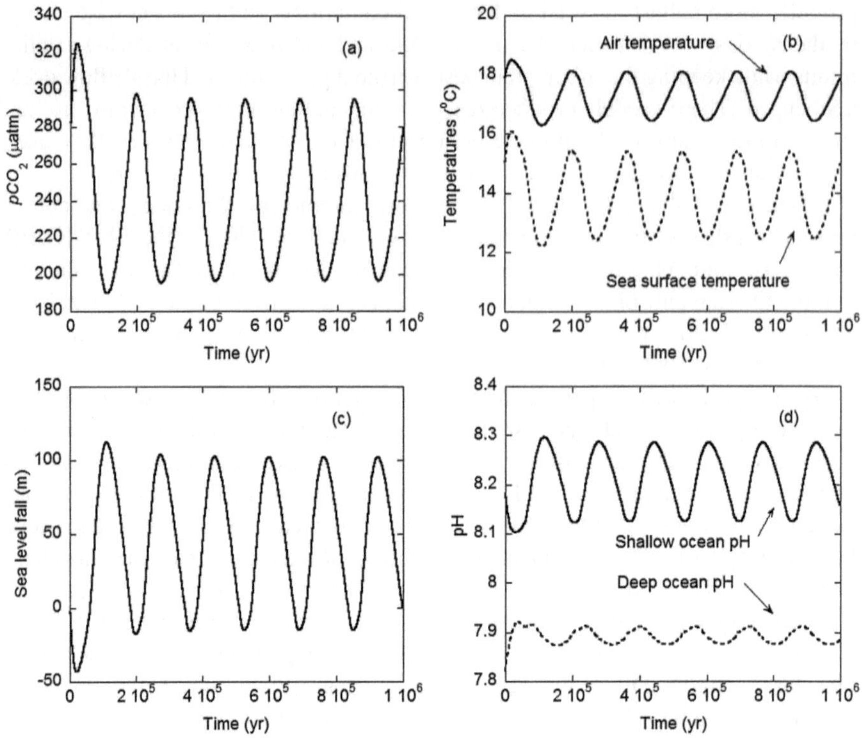

Fig. 1 Time series for the deterministic Wallmann's model with ocean ventilation. $F_{VO} = 1.15$ Tmol/year, $F_{MC} = 3$ Tmol/year, $F_{WOH} = 7$ Tmol/year, $F_{RDPH} = 0.15$ Tmol/year. The other parameter values are taken from Wallmann (2014). (**a**) Atmospheric CO_2 partial pressure; (**b**) air and sea-surface temperatures; (**c**) sea-level fall; (**d**) shallow and deep oceans pH

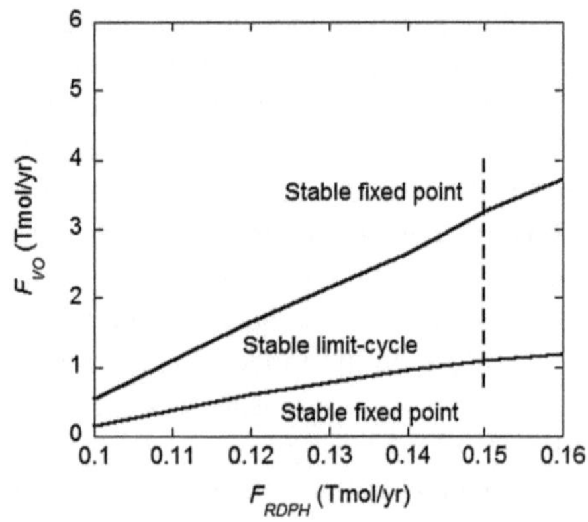

Fig. 2 Phase diagram in (F_{VO}, F_{RDPH}) space for the deterministic Wallmann's model with ocean ventilation. $F_{MC} = 3$ Tmol/year, $F_{WOH} = 7$ Tmol/year. The other parameter values are taken from Wallmann (2014). The vertical dashed line is a cut at $F_{RDPH} = 0.15$ Tmol/year corresponding to Fig. 1

Fig. 1). It shows that the limit-cycle solution issues from a stable focus through a supercritical Hopf bifurcation at a high value of F_{VO}. On the other hand, at a lower value of F_{VO}, the system undergoes a subcritical Hopf bifurcation, generating an

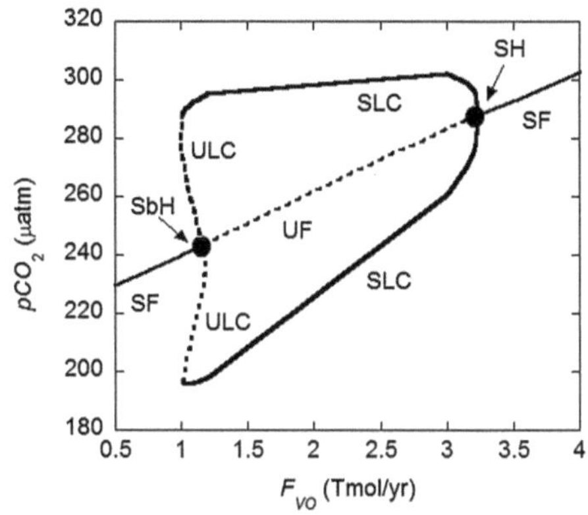

Fig. 3 Bifurcation diagram as F_{VO} varies for the deterministic Wallmann's model with ocean ventilation. $F_{\mathrm{MC}} = 3$ Tmol/year, $F_{\mathrm{WOH}} = 7$ Tmol/year, $F_{\mathrm{RDPH}} = 0.15$ Tmol/year (*vertical dashed line* in Fig. 2). The other parameter values are taken from Wallmann (2014). *SF* stable focus, *SLC* stable limit-cycle, *UF* unstable focus, *ULC* unstable limit-cycle. The two *black dots* indicate the position of the Hopf bifurcations, subcritical (SbH) and supercritical (SH)

unstable limit-cycle that coexists with the stable one. Eventually, at a smaller value of F_{VO}, the two limit-cycles collide and cease to exist. Only a stable focus remains. Thus, there exists a small bistability window where a stable focus coexists with a stable limit-cycle.

3 Noise-Induced Cycle Suppression: A Toy Model

Before considering a stochastic version of the Wallmann's biogeochemical model, I wish to explore the dynamics of a simple system that undergoes a Hopf bifurcation from a stable focus to a stable limit-cycle solution under the influence of multiplicative parametric noise and of a weak external periodic driving term. This simple but relevant toy model will illustrate the main features of a "noise-induced cycle suppression" mechanism whereby only a relatively strong signal at the driving frequency persists for sufficiently high noise intensity.

The toy model is described by two coupled dimensionless dynamical variables x and y as follows.

$$\begin{aligned} \dot{x} &= \mu x - y - \left(x^2 + y^2\right) x + x\xi(t) + A \sin\left(2\pi f_{\mathrm{ex}} t\right); \\ \dot{y} &= \mu y + x - \left(x^2 + y^2\right) y + y\xi(t) + A \sin\left(2\pi f_{\mathrm{ex}} t\right). \end{aligned} \tag{4}$$

Here, μ is a bifurcation parameter and $\xi(t)$ is a Gaussian white noise term characterized by

$$< \xi(t) >= 0; \quad < \xi(t)\xi\left(t'\right) >= 2\sigma\delta\left(t - t'\right) \tag{5}$$

where the average <> is taken over the realizations of the process and σ is the noise intensity. The noise term in Eq. (4) can be interpreted as rendering the bifurcation parameter stochastic $\mu \to \mu + \xi(t)$. The external driving force in Eq. (4) is characterized by an amplitude A and a frequency f_{ex}.

In the deterministic ($\sigma = 0$) undriven ($A = 0$) case, the system exhibits a supercritical Hopf bifurcation at $\mu = 0$. For $\mu < 0$, only a stable focus ($x = 0$, $y = 0$) exists, whereas for $\mu > 0$ the focus loses its stability but a stable limit-cycle solution of radius $R_{LC} = \sqrt{x_{LC}^2 + y_{LC}^2} = \sqrt{\mu}$ and period 2π develops.

The stochastic but undriven case was previously investigated (Bashkirtseva et al. 2007). It is easier to describe the system in terms of the radius in dynamical space, $R = (x^2 + y^2)^{1/2}$, and the angle $\varphi = \tan^{-1}(y/x)$ between the direction of the vector (x,y) and the x-axis. The dynamical system reduces to

$$\dot{R} = \mu R - R^3 + R\xi(t); \quad \dot{\phi} = 1. \tag{6}$$

The dynamics of R and $\dot{\phi}$ is uncoupled, and only the former is driven by a random term. Let $p(R, t)$ be the radial distribution, defined such that $p(R,t)\mathrm{d}R$ is the probability of finding the system with a value of R between R and $R + \mathrm{d}R$ at time t, for any angle. It obeys the following Fokker–Planck equation with the Stratonovich interpretation (Horsthemke and Lefever 1984; Gardiner 1983):

$$\frac{\partial p}{\partial t} = -\frac{\partial}{\partial R}\left(\mu R - R^3\right)p + \sigma\frac{\partial}{\partial R}\left(R\frac{\partial}{\partial R}Rp\right). \tag{7}$$

The steady-state solution p_s reads:

$$p_s = CR^{\mu/\sigma - 1}\exp\left(-R^2/2\sigma\right) \tag{8}$$

where C is a normalization constant. For a fixed value of the control parameter $\mu > 0$, the shape of the distribution changes as the noise intensity σ varies (Fig. 4). For $\sigma = 0$, the system is deterministic and the distribution reduces to a Dirac delta-function at $R = R_{LC}$. For $0 < \sigma < \mu$, the distribution exhibits a maximum at $R_{max} = (\mu - \sigma)^{1/2}$. Noise-induced states are defined by the position of the maxima in the steady-state distribution (Horsthemke and Lefever 1984). Thus, R_{max} corresponds to a stochastic limit-cycle whose radius is smaller than its deterministic counterpart $\sqrt{\mu}$. For $\sigma = \mu$, the maximum reaches $R_{max} = 0$ with a distribution that stays finite there. Finally, for $\sigma > \mu$, the maximum is at $R_{max} = 0$ but the distribution diverges there (while remaining integrable). Thus, as the noise intensity increases, the stochastic limit-cycle becomes smaller and eventually disappears at the critical value $\sigma^* = \mu$. One can say that the noise suppresses the stochastic limit-cycle at σ^*. In terms of the variables x or y, the distribution $p_s(x)$ and $p_s(y)$ will be bimodal for $\sigma < \sigma^*$ with a maximum in x (and y) at $\pm A$ and monomodal for $\sigma \geq \sigma^*$, with a maximum at 0. These regimes are illustrated in Fig. 4. Here, normalized simulated distributions $p_s(R)$ are shown, using a first-order Euler

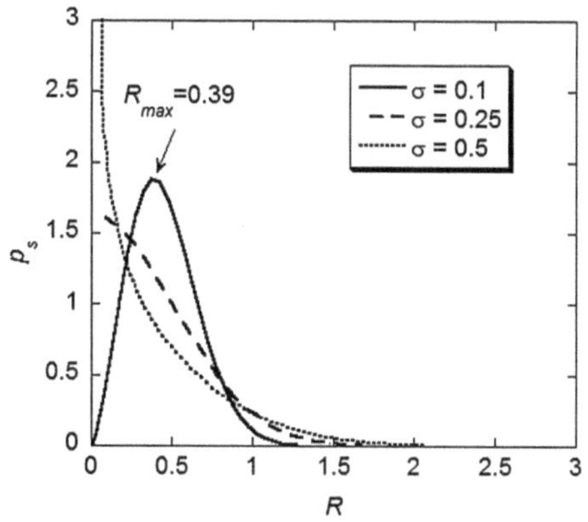

Fig. 4 Steady-state distribution $p_s(R)$ for the undriven ($A = 0$) stochastic toy model of Sect. 3 with $\mu = 0.25$ and various values of σ. The curves show normalized simulation results obtained from the numerical solution of Eqs. (4) and (5). The analytical solutions of Eq. (8) are virtually undistinguishable from these curves. For $\sigma = 0.1$, the maximum in p_s is at $R_{\text{max}} = (\mu - \sigma)^{1/2} = 0.387$ in agreement with the simulation result

stochastic numerical algorithm (Sancho et al. 1982). The analytical curves (not shown) are practically undistinguishable from the simulations.

I now consider the deterministic ($\sigma = 0$) driven ($A \neq 0$) case. Figure 5a illustrates the power spectrum P associated with the variable x as a function of frequency f, for $\mu = 0.25, A = 0.1$, and $f_{\text{ex}} = 0.1$. The peak P_{LC} at $f = 1/(2\pi) \approx 0.16$ corresponds to the limit-cycle signal, whereas the peak P_{ex} at $f = 0.1$ is associated with the external signal. The amplitude A is chosen small enough so that the limit-cycle signal dominates over the external one. The power spectrum ratio $r \equiv P_{\text{ex}}/P_{\text{LC}} = 0.09$. I now switch on the noise intensity. Figure 5b and c illustrate the power spectra for $\sigma = 0.1$ ($r = 0.34$) and $\sigma = 1.0$ ($r = 2.14$), respectively. It is seen that, as the noise intensity becomes larger, the ratio r increases and, for $\sigma > 0.37$, it becomes larger than one. In summary, for a sufficiently large noise intensity, the dominant signal is rather associated with the external driving frequency. As the stochastic limit-cycle ceases to exist through the noise-induced suppression mechanism, the external signal persists as the dominant organizing clock. Yet, the variance in R remains comparable to the deterministic limit-cycle case. Actually, the same qualitative behavior is observed when a colored Ornstein–Uhlenbeck noise is used instead of a Gaussian white noise (results not shown).

In fact, this mechanism could be applied to a wide range of situations in which a system exhibiting a deterministic limit-cycle is subjected to multiplicative noise and a weak forcing. I will now apply this approach to the Wallmann's model.

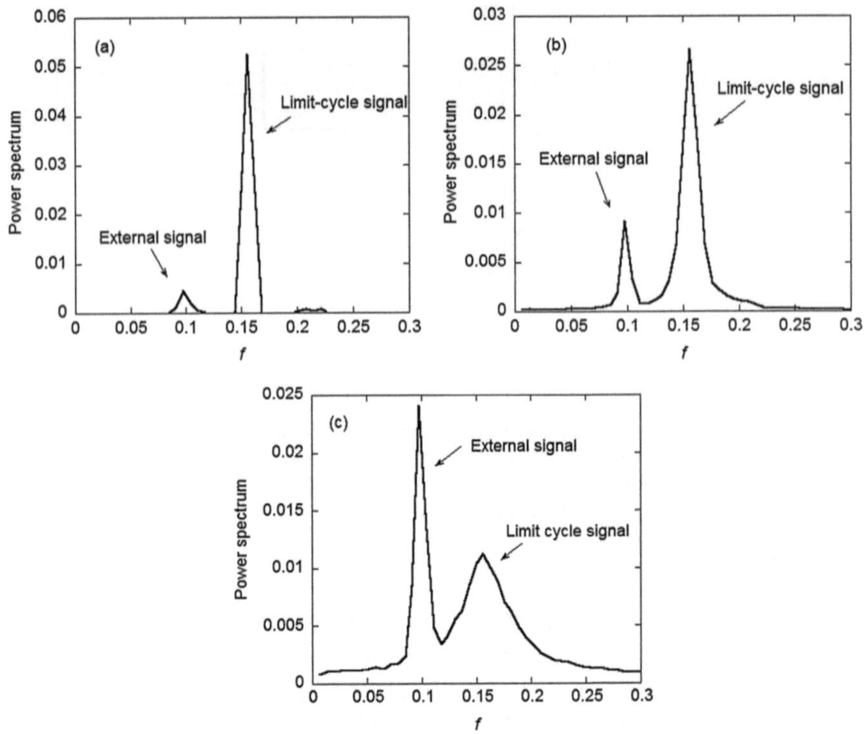

Fig. 5 Power spectrum associated with the variable x of the toy model of Sect. 3 as a function of frequency f for the driven case with $A = 0.1$, $f_{ex} = 0.1$, and $\mu = 0.25$. **(a)** Deterministic case, $\sigma = 0$. The small signal at $f \approx 0.21$ is due to non-linear coupling between the external driving and the limit-cycle. **(b)** $\sigma = 0.1$. **(c)** $\sigma = 1.0$

4 Forced Stochastic Wallmann's Model

I consider an analogous stochastic, weakly driven version of the Wallmann's model. The weak forcing results from the Milankovitch orbital eccentricity variation. The so-called climate sensitivity factor k describes the amplitude of the temperature change ΔT resulting from a change in solar insolation ΔQ in the spirit of linear response theory (Douglass and Clader 2002). In this case, the relative variation in solar insolation due to the eccentricity variations is small, $a = 5 \times 10^{-4}$ (Berger 1978; Benzi et al. 1983). One can write

$$\Delta T = k\Delta Q = kQa \tag{9}$$

where Q is the solar constant (1367 W/m^2). Using simple energy balance equation (Benzi et al. 1983), it is possible to relate k to other climate parameters with the result

$$k = E\tau_0/CQ \tag{10}$$

where E is the energy flux re-emitted in the infrared, τ_0 is a relaxation time to equilibrium, and C is an effective heat capacity per area. Many authors have used estimates for these factors or have estimated k from the observations (Ghil 1976; Douglass and Clader 2002; Schwartz 2007; Schwartz et al. 2010) with the result that ΔT may vary from 0.02 K to 0.12 K. I have adopted the conservative value $A = 0.04$ K for the forcing amplitude here. The following contribution from the solar forcing is thus added to the air temperature equation, Eq. (1):

$$\Delta T_f = A \sin(2\pi f t) \tag{11}$$

where, for simplicity, I use a driving frequency $f = (100 \text{ kyr})^{-1}$. The sea-surface temperature T_{SS} (Eq. (2)) will also have a similar contribution. For simplicity, I have assumed a forcing amplitude in proportion to the CO_2 warming response coefficients:

$$\Delta T_{fSS} = A(\Gamma_{SS}/\Gamma)\sin(2\pi f t) \equiv Ae \sin(2\pi f t) \tag{12}$$

where $e = \Gamma_{SS}/\Gamma$.

So far, the air temperature is determined by the CO_2 concentration with the addition of a small forcing term due to the eccentricity variations of the Earth's orbit. To these, in the spirit of Benzi et al. (1981, 1983), I add to T_A a random term $\eta(t)$ that represents the various unknown other factors that may influence the global temperature. A similar term is added to the sea-surface temperature in which the effect of the noise is also modulated by the factor $e = \Gamma_{SS}/\Gamma$. The dynamics of the non-linear system is then described by a set of forced, stochastic ordinary differential equations in which the noise is multiplicative and non-linear.

For simplicity, I choose an Ornstein–Uhlenbeck colored noise characterized by a correlation time τ. It is itself described by the following dynamics:

$$\frac{d\eta}{dt} = -\frac{\eta}{\tau} + \frac{\xi(t)}{\tau}. \tag{13}$$

Here, $\xi(t)$ is a Gaussian white noise of intensity σ, which is in turn characterized by

$$< \xi(t) >= 0; \quad < \xi(t)\xi(t') >= 2\sigma\delta(t - t') \tag{14}$$

in which the average <> is taken over the realizations of the stochastic process. It turns out that the system (13) and (14) can be solved exactly through the recurrence relation:

$$\eta(t + \Delta t) = \eta(t)\exp(-\Delta t/\tau) + \alpha(t)\sqrt{\sigma/\tau}[1 - \exp(-2\Delta t/\tau)]^{1/2}; \tag{15}$$
$$\eta(0) = \alpha(0)\sqrt{\sigma/\tau}.$$

Here, Δt is an arbitrary time increment and $\alpha(t)$ is a random number chosen from a normal Gaussian distribution of zero average and unit variance. A straightforward numerical algorithm then ensues. The statistical properties of the η noise are

$$\langle \eta(t) \rangle = 0; \quad \langle \eta(t + \Delta t)\, \eta(t) \rangle = (\sigma/\tau) \exp(-\Delta t/\tau). \tag{16}$$

Schwartz (2007) has estimated the current correlation time of air temperature from observed time series. He obtains $\tau = 5 \pm 1$ year. I will fix τ at 5 year. In their simulations of the 100 kyr climate cycle undergoing stochastic resonance, Benzi et al. (1981, 1982, 1983) considered noise intensities from 0 to 0.22 K^2/year. However, they added the stochastic term on an equation for dT/dt, not on T itself. To compare the noise intensities, I need to multiply their noise intensities by τ^2. With $\tau = 5$ year, this corresponds to a range $0 < \sigma < 5.5$ K^2 year. I will actually consider $\sigma < 6$ K^2 year here.

Including all terms, the air temperature (Eq. (1)) becomes

$$T_A = \Gamma \log(pCO_2/280) + T_{AH} + A \sin(2\pi f t) + \eta(t), \tag{17}$$

whereas the sea-surface temperature (Eq. (2)) is amended to

$$T_{SS} = \Gamma_{SS} \log(pCO_2/280) + T_{SSH} + e\,(A \sin(2\pi f t) + \eta(t)). \tag{18}$$

The graphs in Fig. 6 illustrate the pCO_2 steady-state distribution obtained from numerical simulations of the stochastic Wallmann's model without orbital forcing ($A = 0$). The parameter values are the same as in Fig. 1 except that $\sigma \neq 0$. In the deterministic case, there is a limit-cycle solution and the distribution is bimodal with sharp peaks at the turning points of the cycle. For small σ ($\sigma < 1.5$ K^2 year), the distribution remains bimodal. However, for $\sigma > 1.5$ K^2 year, the stochastic limit-

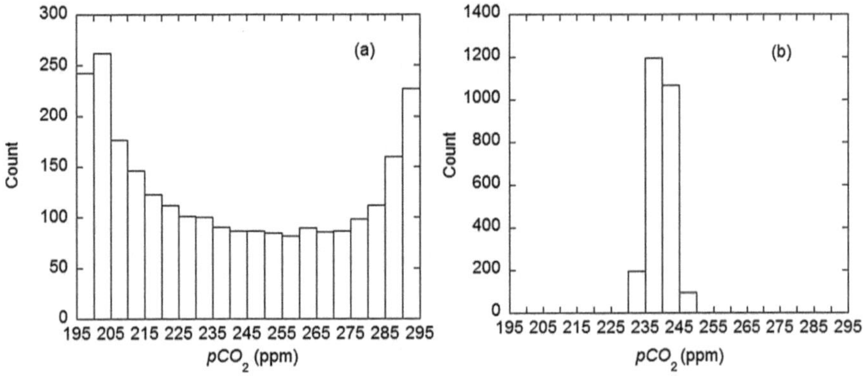

Fig. 6 Steady-state p_s of the CO_2 distribution for the unforced ($A = 0$) stochastic Wallmann's model with $\tau = 5$ year. The other parameter values are as in Fig. 1. (**a**) $\sigma = 0.1$ K^2 year; (**b**) $\sigma = 4$ K^2 year

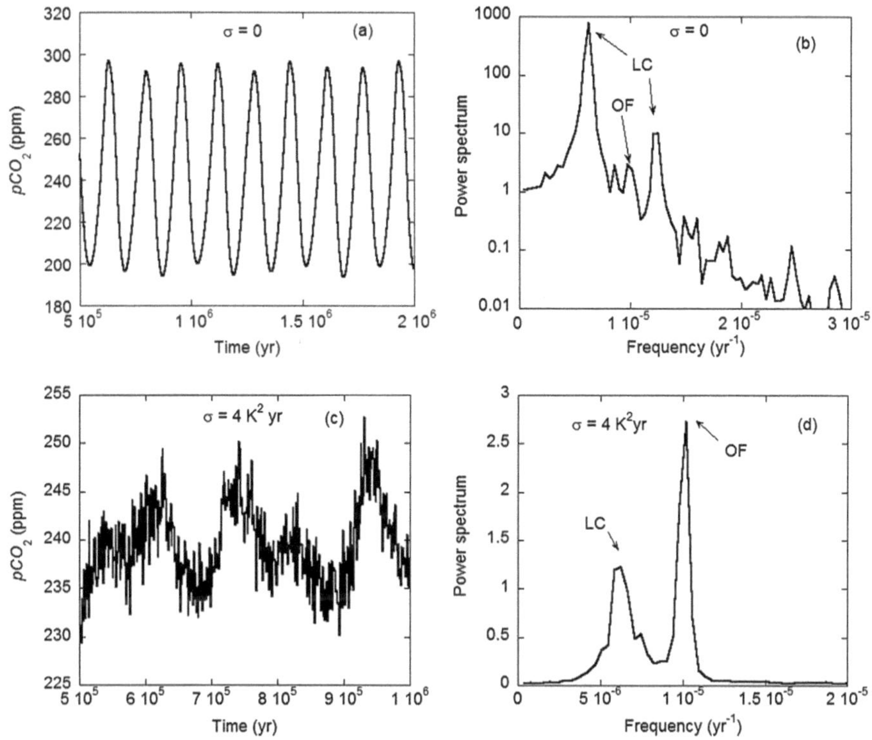

Fig. 7 Wallmann's model for the orbital forcing case ($A = 0.04$ K). pCO_2 times series and power spectrum in the deterministic case, $\sigma = 0$ (**a**, **b**) and for $\sigma = 4$ K^2 year and $\tau = 5$ year (**c**, **d**). The other parameter values are as in Fig. 1. LC indicates the power spectrum peak at the limit-cycle frequency and its first harmonic; OF shows the power spectrum peak at the orbiting forcing frequency

cycle is suppressed by the noise and the distribution becomes monomodal. There is a noise-induced cycle suppression at $\sigma \approx 1.5$ K^2 year analogous to what happens in the toy model (Fig. 4).

Figure 7 illustrates the pCO_2 time series and its power spectrum for the same situation as Fig. 1, except that the orbital forcing is active with $A = 0.04$ K, for the deterministic case ($\sigma = 0$) and a stochastically driven example ($\sigma = 4$ K^2 year). In the deterministic case, the power spectrum exhibits a main peak at the limit-cycle frequency, as well as some of its harmonics. The peak corresponding to the orbital forcing frequency is also present, but remains small compared to the limit-cycle signal. In the stochastic case of Fig. 7d, the situation is reversed: it is the orbital forcing signal that dominates. This is analogous to what happens in the toy model through the noise-induced cycle suppression mechanism (Fig. 5).

Figure 8 shows various properties of the stochastic Wallmann's model with fixed orbital forcing as the noise intensity varies. The deterministic unforced case corresponds to that of Fig. 1. Figure 8a shows the ratio P_{OF}/P_{LC} of the peak P_{OF} in

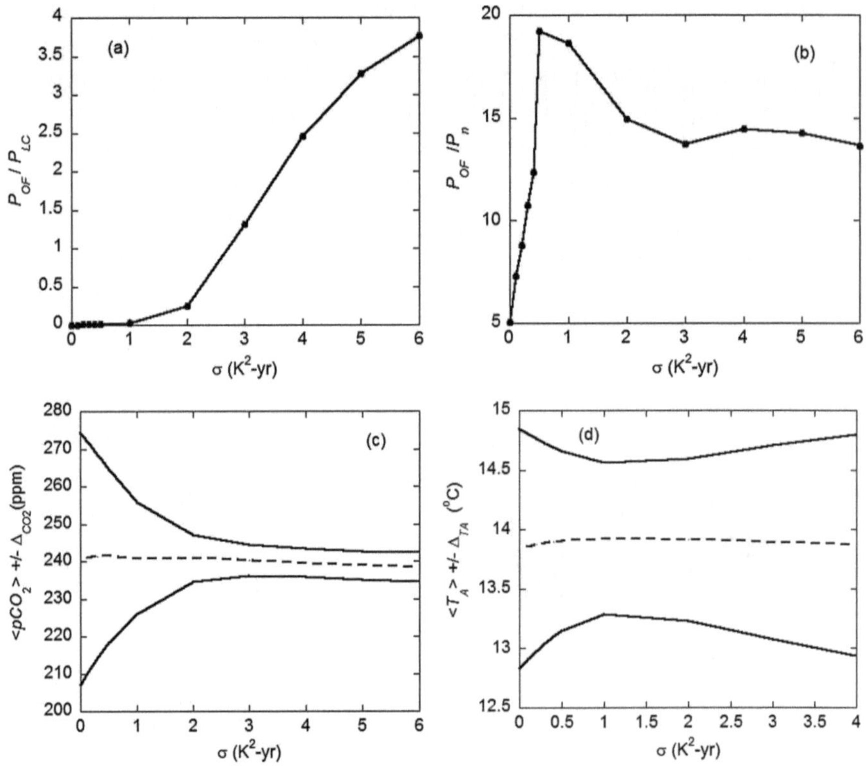

Fig. 8 Various statistical properties of the forced ($A = 0.04$ K) stochastic Wallmann's model as a function of the noise intensity σ with $\tau = 5$ year. The other parameters are as in Fig. 1. (**a**) Ratio $P_{\rm OF}/P_{\rm LC}$ of the OF power spectrum peak first to the first LC power spectrum peak. (**b**) Ratio $P_{\rm OF}/P_n$ of the OF power spectrum peak to the background power spectrum value at the OF frequency (noise signal). (**c**) Ensemble average of pCO_2 (*dotted line*) $\pm$ its standard deviation (*continuous lines*). (**d**) Ensemble average of the air temperature T_A (*dotted line*) $\pm$ its standard deviation (*continuous lines*)

the power spectrum at the orbital forcing frequency to the peak $P_{\rm LC}$ at the limit-cycle frequency. The ratio varies monotonously with σ and reaches a value of one at a critical $\sigma \approx 2.7$ K^2 year. In other words, as the noise intensity increases from 0, the signal is first stronger at the deterministic limit-cycle frequency. At $\sigma \approx 1.5$ K^2 year, the system undergoes a noise-induced cycle suppression. The signal is still dominating at the limit-cycle frequency but the contribution from the orbital forcing frequency becomes non-negligible. At $\sigma > 2.7$ K^2 year, the latter dominates the overall signal. Figure 8b shows the ratio $P_{\rm OF}/P_n$ where P_n is the noise signal at the orbital forcing frequency, defined as the interpolated background level. For smaller value of σ, $P_{\rm OF}$ is too small, whereas the background noise level becomes larger for large σ. Thus, the plot of Fig. 8b exhibits a stochastic-resonance-like peak at $\sigma \approx 0.5$ K^2 year. Figure 8c, d shows the ensemble average and the statistical standard deviation for pCO_2 and the air temperature T_A for various σ. For small σ, the standard deviation reflects nothing more than the amplitude of the deterministic limit-cycle. Although the average value of these variables does not change much with σ, the standard deviations remain significant as the limit-cycle gets suppressed by the noise. For example, the overall air temperature variation has a minimum value of 1.3 °C and becomes even larger as the noise intensity increases.

In summary, the model indicates that, as the noise intensity becomes sufficiently large, the climatic signal is stronger at the orbital forcing frequency than at the limit-cycle frequency, whereas the temperature variations (and other climatic variables) remain significantly large. In contrast, the deterministic system would also exhibit large temperature variations but at a different frequency.

5 Conclusion

In this contribution, the starting point was Wallmann's model (Wallmann 2014), a simple box model of the biogeochemical cycling of phosphorus, alkalinity, and dissolved inorganic carbon in the ocean, coupled with sea-level changes, without orbital forcing. In that model, self-organized oscillating solutions are found for a wide range of parameter values. (Interestingly, there is also a narrow range of parameter values for which the model exhibits bistability between a limit-cycle and a fixed point, but this feature is not exploited here.) I have then extended Wallmann's model by driving the system with a weak orbital forcing and with a non-linear multiplicative colored noise process. For sufficiently large noise intensities, the signal at the orbital forcing frequency dominates over the limit-cycle signal. In spite of the smallness of the orbital forcing amplitude, the fluctuations in the dynamical variables about their averages remain significant. On the basis of a comparison with a simple "toy" model of a noisy, weakly forced, Hopf-bifurcating system, I propose that this new orbital forcing signal amplification mechanism results from a noise-induced cycle suppression.

Of course, a more complete model should include the precession and obliquity effects of the orbital forcing, and a more detailed physics of the ocean circulation and ice dynamics. But the biogeochemical feedback mechanism explored here should not be ignored.

References

Bashkirtseva, I., L. Ryashko, and H. Schurz. 2007. Stochastic bifurcations for random force oscillations. *IFAC Proceedings* 3: 213–217.

Benzi, R., A. Sutera, and A. Vulpiani. 1981. The mechanism of stochastic resonance. *Journal of Physics A* 14: L453–L457.

Benzi, R., G. Parisi, A. Sutera, and A. Vulpiani. 1982. Stochastic resonance in climatic change. *Tellus* 34: 10–16.

———. 1983. A theory of stochastic resonance in climatic change. *SIAM Journal on Applied Mathematics* 43: 565–578.
Berger, A. 1978. Long-term variations of daily insolation and Quaternary climatic changes. *Journal of the Atmospheric Sciences* 35: 2362–2367.
Douglass, D.H., and B.D. Clader. 2002. Climate sensitivity of the Earth to solar irradiance. *Geophysical Research Letters* 29. doi:10.1029/2002GL015345.
Eakins, B.W., and G.F. Sharman. 2012. *Hypsographic curve of Earth's surface from ETOPO1*. Boulder, CO: NOAA National Geophysical Data Center.
Foster, G.L., and E.J. Rohling. 2013. Relationship between sea level and climate forcing by CO_2 on geological timescales. *PNAS* 110: 1209–1214.
Gardiner, C.W. 1983. *Handbook of stochastic methods*. Berlin: Springer, 442 p.
Ghil, M. 1976. Climate stability for a Sellers-type model. *Journal of the Atmospheric Sciences* 33: 3–20.
Gildor, H., and E. Tziperman. 2001. A sea ice climate switch mechanism for the 100-kyr glacial cycles. *Journal of Geophysical Research* 106: 9117–9133.
Grant, K.M., E.J. Rohling, M. Bar-Matthews, A. Ayalon, M. Medina-Elizalde, C. Bronk Ramsey, C. Satow, and A.P. Roberts. 2012. Rapid coupling between ice volume and polar temperature over the past 150,000 years. *Nature* 491: 744–747.
Horsthemke, W., and R. Lefever. 1984. *Noise-induced transitions: theory and applications in physics, chemistry and biology*. Berlin: Springer, 322 p.
Imbrie, J., and J.Z. Imbrie. 1980. Modeling the climatic response to orbital variations. *Science* 207: 943–953.
Imbrie, J., et al. 1993. On the structure and origin of major glaciation cycles. 2. The 100,000-year cycle. *Paleoceanography* 8: 699–735.
Pelletier, J.D. 2003. Coherence resonance in ice ages. *Journal of Geophysical Research* 108: 4645.
Petit, J.R., et al. 1999. Climate and atmospheric history of the past 420,000 years from the Vostok ice core, Antarctica. *Nature* 399: 429–436.
Ridolfi, L., P. D'Odorico, and F. Laio. 2011. *Noise-induced phenomena in the environmental sciences*. New York: Cambridge University Press, 313 p.
Saltzman, B., and D.K. Maasch. 1988. Carbon cycle instability as a cause of the late Pleistocene ice age oscillations: modeling the asymmetric response. *Global Biogeochemical Cycles* 2: 177–185.
Sancho, J.M., M. San Miguel, S.L. Katz, and J.D. Gunton. 1982. Analytical and numerical studies of multiplicative noise. *Physical Review A* 26: 1589–1609.
Shackleton, N.J. 2000. The 100,000-year ice age cycle identified and found to lag temperature, carbon dioxide and orbital eccentricity. *Science* 289: 1897–1902.
Schwartz, S.E. 2007. Heat capacity, time constant and sensitivity of Earth's climate system. *Journal of Geophysical Research* 112: D24S05.
Schwartz, S.E., R.J. Charlson, R.A. Kahn, J.A. Ogren, and H. Rodhe. 2010. Why hasn't Earth warmed as much as expected? *Journal of Climate* 23: 2453–2464.
Wallmann, K. 2014. Is late Quaternary climate change governed by self-sustained oscillations in atmospheric CO_2? *Geochimica et Cosmochimica Acta* 132: 413–439.
Zeebe, R., and D. Wolf-Gladrow. 2001. *CO_2in seawater: equilibrium, kinetics and isotopes*. Amsterdam: Elsevier, 360 p.

Role of Nonlinear Eddy Forcing in the Dynamics of Multiple Zonal Jets

Igor Kamenkovich and Pavel Berloff

Abstract Turbulent oceanic and atmospheric flows are characterized by persistent nearly zonal alternating currents, referred to as multiple zonal jets. These jets emerge from self-organization of the flow, and are maintained by persistent action of transient fluctuations ("eddies"). The action of these eddies takes a form of internally generated eddy forcing that can either resist the jets or support them against dissipation and other processes. This review chapter is concerned with the role of the eddy forcing in the dynamics of the multiple zonal jets in the ocean, but the results are also applicable to atmospheric flows and some isolated jets in the ocean.

Keywords Zonal Jets • Mesoscale Eddies • Eddy Forcing • Nonlinear Dynamics

1 Introduction

Motions in the atmospheres and oceans have strong preference for the zonal direction, that is, along the latitude circles. This preference can be understood through conservation of angular momentum: purely zonal motions can conserve this quantity without changing their velocity, whereas any displacement in the meridional (north–south) direction results in changes in the zonal velocity and vorticity. The associated "β-effect" can act as a restoring force, resulting in the westward-propagating planetary waves (Rossby waves) and a fundamental asymmetry between the south–north and west–east directions. These fundamental properties explain the existence of persistent multiple, alternating zonal currents (Rhines 1975), often simply referred to as zonal jets.

The most striking example of zonal jets can be observed in the atmospheres of giant gaseous planets, Jupiter and Saturn, where they are manifested by alternating

I. Kamenkovich (✉)
RSMAS, University of Miami, Miami, FL, 33149, USA
e-mail: ikamenkovich@miami.edu

P. Berloff
Imperial College London, London, UK

A.A. Tsonis (ed.), *Advances in Nonlinear Geosciences*,
DOI 10.1007/978-3-319-58895-7_9

bands of high and low albedo (Sanchez-Lavega et al. in press). In the Earth's troposphere, nearly zonal jets are found in mid-latitudes (around 45°N and 50°S) and subtropics (30°N–35°N and 30°S); the stratospheric circulation is characterized by polar vortices in both hemispheres with strong westward jets (e.g., Wallace and Hobbs 2006). Similar westerly jets are found in the atmospheres of other terrestrial planets, such as Mars, Venus, and Titan (Mitchell et al. in press). Well-pronounced zonal jets are also observed at the Equator and in the tropics in all Earth's oceans (e.g., Godfrey et al. 2001). The existence of stationary oceanic jets in subtropics and mid-latitudes is more controversial, because the magnitudes of transient velocity anomalies (so-called mesoscale eddies) often exceed the magnitude of the jets. In other words, the oceanic zonal jets are "latent" (Berloff et al. 2009; Kamenkovich et al. 2009), and some spatio-temporal filtering is usually required for their detection. These latent zonal jets have been observed in the time-averaged anomalies of the geostrophic velocity estimated from altimeter data (Maximenko et al. 2005, 2008; Sokolov and Rintoul 2009; Huang et al. 2007; Buckingham et al. 2014). Identification of these jets with in situ observations is challenging due to often poor horizontal resolution and extent in time; narrow zonal currents, with direction alternating in latitude, were nevertheless detected in float measurements in the Brazil basin of the deep South Atlantic (Treguier et al. 2003; Hogg and Owens 1999); North Atlantic (van Sebille et al. 2011; who combined altimetry with data from profiling Argo floats), and in the Southern Ocean (Nowlin and Klinck 1986; Orsi et al. 1995).

The typical meridional width of these jets is several Rossby deformation radii; outside the tropics, the latter length scale varies between approximately 10 and 50 km in the oceans, but is considerably longer in the atmosphere. For this reason, Earth's atmosphere is simply not large enough to "fit" multiple zonal jets, and this phenomenon is mostly characteristic for the oceans and giant planets. The discussion in this chapter is concerned primarily with oceanic jets, although results are expected to be applicable to atmospheric jets, on both terrestrial and giant gas planets. The mechanisms of emergence of the jets in various turbulent flows are still under debate, and various theories have been proposed (see Zonal Jets, eds. Galperin and Read, in press). This review chapter focuses instead on how stationary zonal currents can persist despite dissipation and other processes that draw energy away from them. Nonlinear jet dynamics is explored using a concept of the eddy forcing, first in an idealized quasi-geostrophic model and then in General Circulation Models (GCMs).

2 Quasi-Geostrophic Dynamics

Under the assumptions of quasi-geostrophic (QG) dynamics, the evolution of the flow is described by conservation of the potential vorticity (PV) (Pedlosky 1987); if the ocean or atmosphere is described by a set of N stacked moving isopycnal (constant density) layers, the conservation law takes the following form:

$$\frac{\partial Q_n}{\partial t} + J(\psi_n, Q_n) = \upsilon \nabla^4 \psi_n + \delta_{nN} k_{\mathrm{bot}} \nabla^2 \psi_N, n = 1, \ldots, N, \tag{1}$$

where subscript "n" denotes the nth layer, J is the Jacobian operator, υ is the lateral viscosity, k_{bot} is the bottom friction coefficient, and δ_{nN} is the Kronecker delta. ψ is the streamfunction that determines circulation in each layer:

$$u_n = -\frac{\partial \psi_n}{\partial y}, v_n = \frac{\partial \psi_n}{\partial x} \tag{2}$$

PV in each layer is given by

$$Q_n = \beta y + \nabla^2 \psi_n - (1 - \delta_{n1}) \frac{1}{R_{n1}^2} (\psi_n - \psi_{n-1}) - (1 - \delta_{nN}) \frac{1}{R_{n2}^2} (\psi_n - \psi_{n+1}) \tag{3}$$

where R_{n1} and R_{n2} are the stratification parameters and β is the Coriolis parameter.

We next consider the flow in a periodic zonal domain, most relevant to the atmospheres and the Antarctic Circumpolar Current, and assume a persistent source of available potential energy (Haidvogel and Held 1980), due to, for example, differential heating from the sun; this external energy source sustains a broad background zonal flow U_n. Evolution of eddies in this system is then described by the following set of equations:

$$\left(\frac{\partial}{\partial t} + U_n \frac{\partial}{\partial x}\right) q_n + \beta \frac{\partial \varphi_n}{\partial x} + J(\varphi_n, q_n) = \upsilon \nabla^4 \varphi_n + \delta_{nN} k_{\mathrm{bot}} \nabla^2 \varphi_N, \mathrm{n} = 1, \ldots, N \tag{4}$$

where we took $\psi_n = -U_n y + \varphi_n$, and the eddy PV is

$$q_n = \nabla^2 \varphi_n - (1 - \delta_{n1}) \frac{1}{R_{n1}^2} (\varphi_n - \varphi_{n-1}) - (1 - \delta_{nN}) \frac{1}{R_{n2}^2} (\varphi_n - \varphi_{n+1}) \tag{5}$$

If the vertical shear in the background flow is sufficiently strong, the associated meridional tilt of the mean isopycnals enables baroclinic instability and growing linear disturbances, which later reach finite amplitudes and decay due to dissipation and nonlinear interactions. This constantly evolving flow can reach a statistically steady state, with stationary zonal jets and residual "eddy" field:

$$\varphi_n = \overline{\varphi_n}(y) + \varphi_n'(x, y, t) \tag{6}$$

where the overbar represents time- and zonal averaging. Williams (1979) was the first to discuss QG simulations of QG jets in Jupiter, by using a two-layer ($N = 2$) model. An example of stationary oceanic jets is shown in Fig. 1 for a two-layer model and an eastward flow with $U = 0.05\ \mathrm{ms}^{-1}$.

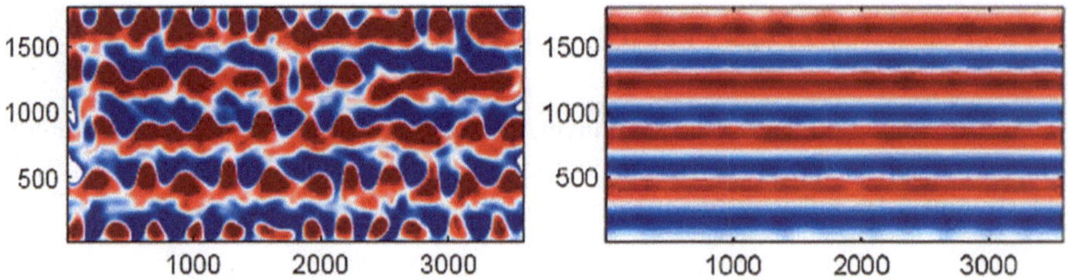

Fig. 1 Flow with multiple zonal jets in a two-layer QG simulation with an eastward background flow. Streamfunction in the *top layer*: *left panel*—instantaneous; *right panel*—time averaged

2.1 Eddy Forcing

Substitution of (6) into (4)–(5) and zonally and time-averaging the results lead to the dynamical balance for the jets in each isopycnal layer:

$$0 = -\frac{\partial}{\partial y}\overline{\left(\frac{\partial \varphi_n'}{\partial x} q_n'\right)} + \upsilon \frac{\partial^4}{\partial y^4}\overline{\varphi_n} + \delta_{nN} k_{\mathrm{bot}} \frac{\partial^2}{\partial y^2}\overline{\varphi_n} \tag{7}$$

$$\overline{-\frac{\partial}{\partial y}\left(\frac{\partial \varphi_n'}{\partial x} q_n'\right)} = \overbrace{\overline{-\frac{\partial}{\partial y}\left(\frac{\partial \varphi_n'}{\partial x} \nabla^2 \varphi_n'\right)}}^{\text{Reynolds Stress Term}} \tag{8}$$

$$\overbrace{\overline{-\frac{\partial}{\partial y}\left(\frac{\partial \varphi_n'}{\partial x}\left\{(1-\delta_{n1})\frac{1}{R_{\mathrm{n1}}^2}\left(\varphi_n' - \varphi_{n-1}'\right) - (1-\delta_{nN})\frac{1}{R_{\mathrm{n2}}^2}\left(\varphi_n' - \varphi_{n+1}'\right)\right\}\right)}}^{\text{Form Stress Term}}$$

Equations (7) and (8) describe a balance between dissipation and two nonlinear terms: convergence of the meridional flux of relative vorticity ("Reynolds Stress Term," or RST) and convergence of the meridional flux of buoyancy anomalies ("Form Stress Term," or FST). Both terms represent correlation between eddy velocities and components of q, and can be interpreted as internally generated "eddy forcing." The eddy forcing is the result of self-organization of the turbulent flow, leading to systematic input (removal) of relative vorticity and buoyancy into (from) zonal jets.

2.2 Analysis of the Role of Eddy Forcing in QG Flows

Useful insights into the jet dynamics can be derived from the meridional structure of various terms in (7–8). For example, positive spatial correlation of the eddy forcing with the zonal-mean relative vorticity $\overline{\partial^2\varphi/\partial y^2}$ implies negative correlation with the viscous term (second term on the right-hand side of Eq. (7)), and can be interpreted as the tendency to balance dissipation and support the jets. More generally, positive correlation of the eddy forcing with the PV of the jets implies that the forcing supports the jets, because a sudden reduction in the forcing magnitude would induce PV time tendency that acts to destroy the jets (Kamenkovich et al. 2009).

Panetta (1993) analyzed PV balance for jets in the simplest version of a baroclinic system—a two-layer version of Eq. (1) ($N = 2$) with the eastward background flow U in the upper layer only. His analysis demonstrated that RST acts to sustain the jets, because their meridional structure is out of phase with the dissipation terms in Eq. (7). Similarly, Lee (1997) reported the importance of RST in the formation, splitting, and maintenance of multiple atmospheric jets. These results imply that the action of eddies can be interpreted as "negative viscosity"—upgradient fluxes of relative vorticity and momentum. In contrast, Panetta (1993) found FST resisting the jets, because their meridional structure is in-phase with the dissipation terms. This property is apparently consistent with baroclinic instability of the jets, since in the linear regime FST is expected to draw energy from the mean currents (Pedlosky 1987), but, as we will see below, is not universal for all flows with jets.

Further insights into the jest dynamics can be gained from the decomposing of the streamfunction into its barotropic (vertically averaged, angle brackets) and baroclinic (depth-dependent residual, stars) components, for example:

$$\langle \varphi \rangle = \frac{1}{H} \int_{-H}^{0} \varphi dz, \varphi^{*} = \varphi - \langle \varphi \rangle , \tag{9}$$

where H is the total depth of the fluid. The barotropic component of the eddy forcing can then be split into two terms, one due to the advection of the barotropic component of PV by the barotropic meridional velocity (the "BRT-BRT" term), and another—the advection of the baroclinic component of PV by the baroclinic meridional velocity. Similarly, the baroclinic components of the eddy forcing can be written as a sum of two barotropic–baroclinic (BRT-BCL) and one baroclinic–baroclinic (BCL-BCL) terms:

$$\begin{aligned} \frac{\partial}{\partial y}\left\langle \overline{\frac{\partial \varphi_n'}{\partial x} q_n'} \right\rangle &= \frac{\partial}{\partial y} \overline{\left\langle \frac{\partial \varphi_n'}{\partial x} \right\rangle \langle q_n' \rangle} + \frac{\partial}{\partial y} \left\langle \overline{\left(\frac{\partial \varphi_n'}{\partial x} \right)^{*} q_n'^{*}} \right\rangle \\ \frac{\partial}{\partial y} \left(\overline{\frac{\partial \varphi_n'}{\partial x} q_n'} \right)^{*} &= \frac{\partial}{\partial y} \overline{\left\langle \frac{\partial \varphi_n'}{\partial x} \right\rangle q_n'^{*}} + \frac{\partial}{\partial y} \overline{\left(\frac{\partial \varphi_n'}{\partial x} \right)^{*} \langle q_n' \rangle} + \frac{\partial}{\partial y} \left(\overline{\left(\frac{\partial \varphi_n'}{\partial x} \right)^{*} q_n'^{*}} \right)^{*} \end{aligned} \tag{10}$$

Berloff et al. (2009) analyzed the dynamical balances for the barotropic and baroclinic components of the jets, extending the analysis by Panetta (1993) to both eastward and westward background flows and to both the two-layer ($N = 2$) and the three-layer ($N = 3$) systems. They concluded that the BRT-BRT and BCL-BCL components act to support barotropic jets, and this is true in both the eastward and westward background flows (Fig. 2); barotropic and first baroclinic component of the eddy fields play the leading roles in the process. This result demonstrates the importance of the baroclinic eddies for the barotropic jets and suggests that

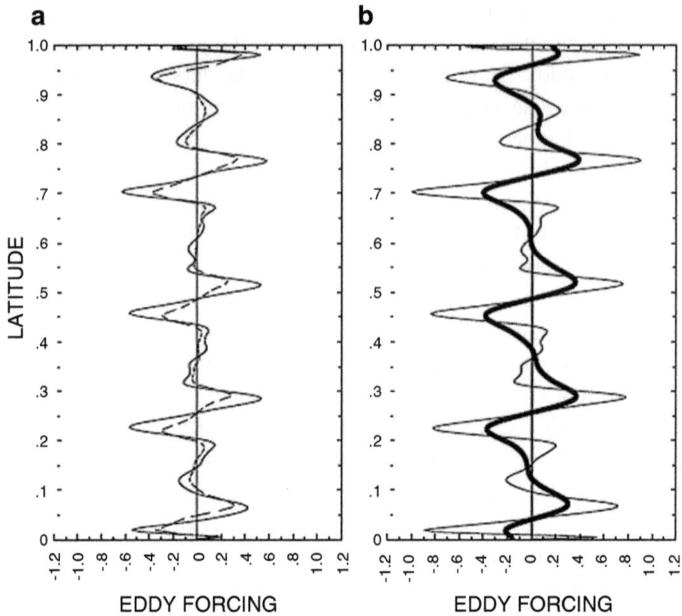

Fig. 2 Barotropic eddy forcing and its components. (**a**) BRT-BRT (*solid line*) and BCL-BCL (*dashed line*) eddy forcing components of the two-layer flow in the eastward background flow. (**b**) Full eddy forcing (*thin*) is shown along with the time-mean barotropic PV component (*thick line*). The eddy forcing itself and its components are normalized by the maximum value of the eddy forcing; the barotropic PV is shown with arbitrary units. Adapted from Berloff et al. (2009). ©American Meteorological Society. Used with permission

purely barotropic models will inadequately describe barotropic nonlinear processes. Similarly, barotropic components of the eddy field are important for the dynamics of the baroclinic jets, and their interactions with barotropic eddies act to maintain the jets via the BRT-BCL eddy forcing (Fig. 3). In agreement with Panetta (1993), FST acts to destroy the baroclinic jets whereas RST maintains them, but only when the background flow is eastward. For the westward flow, the roles are reversed, and the jets are resisted by RST ("positive viscosity") and supported by FST.

The value of the correlation between the eddy forcing and PV can be interpreted as the efficiency of the eddy forcing in supporting/resisting the jets and used to understand jet dynamics. Berloff et al. (2011) demonstrated that the decrease in the eddy forcing efficiency can be associated with weakening of the jets and strengthening of eddies. The resulting increased latency is typically associated with higher bottom friction (k_{bot}); see also Panetta (1993). The authors further speculate that this property can potentially explain the difference between latent jets in the Earth's oceans and manifest jets in Jupiter's atmosphere. Berloff and Kamenkovich (2013a, b) used the eddy forcing and its efficiency to gain insights into the relative importance of the eddy–eddy and eddy–jet interactions in the dynamics of the flow. They demonstrated that normal modes that are derived from a reduced model

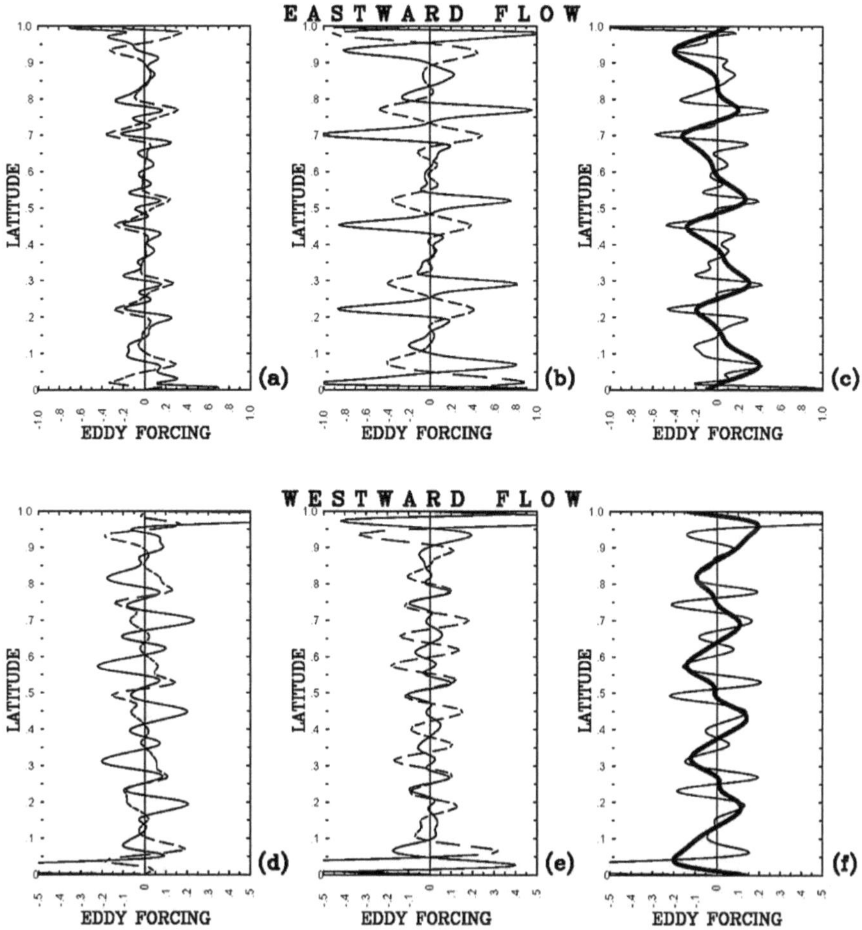

Fig. 3 Baroclinic eddy forcing and its components in the two-layer eastward background flow: (**a**) BRT-BCL (*solid line*) and BCL-BCL (*dashed line*); (**b**) RST (*solid line*) and FST (*dashed line*); (**c**) full eddy forcing (*thin line*) and the time-mean baroclinic PV component (*thick line*). The eddy forcing itself and its components are normalized by the maximum value of the eddy forcing; the baroclinic PV is shown with arbitrary units. (**d–f**) The same quantities as in (**a–c**), but for the westward background flow. Adapted from Berloff et al. (2009). ©American Meteorological Society. Used with permission

linearized around the mean state with zonal jets can explain several properties of the spectrum and eddy forcing in the full nonlinear solution. For example, several distinct parts of the full nonlinear spectrum can be associated with these normal modes, which signifies the importance of the eddy–jet interactions in determining the eddy field; the authors refer to this property as the "linear control." These linear modes also differ in the efficiency of their eddy forcing, which helps to classify these modes according to their contribution to the jet dynamics.

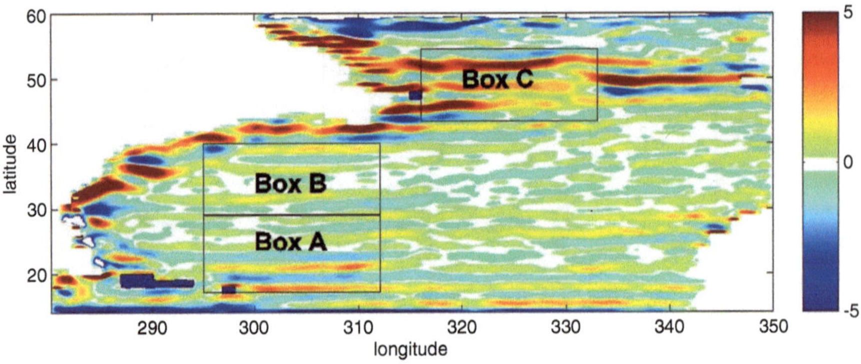

Fig. 4 Zonal velocities at the depth of 550 m in GCM simulations, time averaged over 9 years; the units are 10^{-2} ms^{-1}: Three box regions, A, B and C, are shown by the black lines. Adapted from Kamenkovich et al. (2009). ©American Meteorological Society. Used with permission

3 Jets and Eddy Forcing in GCMs

General Circulation Models (GCMs) aim to provide the most realistic simulations of the ocean circulation. Unlike the QG system whose dynamics is determined by a single variable (PV), primitive equations in a GCM describe evolution of momentum, temperature, and salinity. The discussion here is focused on properties that can be diagnosed from these three variables: relative vorticity, potential density, and PV.

At high spatial resolution (grid size less than approximately 25 km), GCM simulations exhibit latent zonal jets in most of the World Ocean (e.g., Nakano and Hasumi 2005; Richards et al. 2006; Melnichenko et al. 2010). Kamenkovich et al. (2009) studied these jets in a regional GCM of the North Atlantic. The jets are visible in time-averaged fields at all depth levels and dominate zonal circulation in most of the deep ocean; they are most pronounced in the subpolar region (north of the Gulf Stream) and in the western part of the subtropical region (south of the Gulf Stream; Fig. 4). The analysis in Kamenkovich et al. (2009) is focused on the gyre interior (box regions A, B, and C) where the large-scale background flow is nearly zonal and comparison with QG-channel studies is more appropriate. Melnichenko et al. (2010) analyzed quasi-zonal jets in the Eastern Pacific (Fig. 5), both in satellite altimetry data (at the surface) and high-resolution GCM simulations (at the surface and 600 m depth). In these regions, the background flow is nonzonal—south-eastward in the Northern Hemisphere and north-westward in the Southern Hemisphere, which complicates the direct comparison with studies in a QG channel, but introduces additional interesting dynamical features.

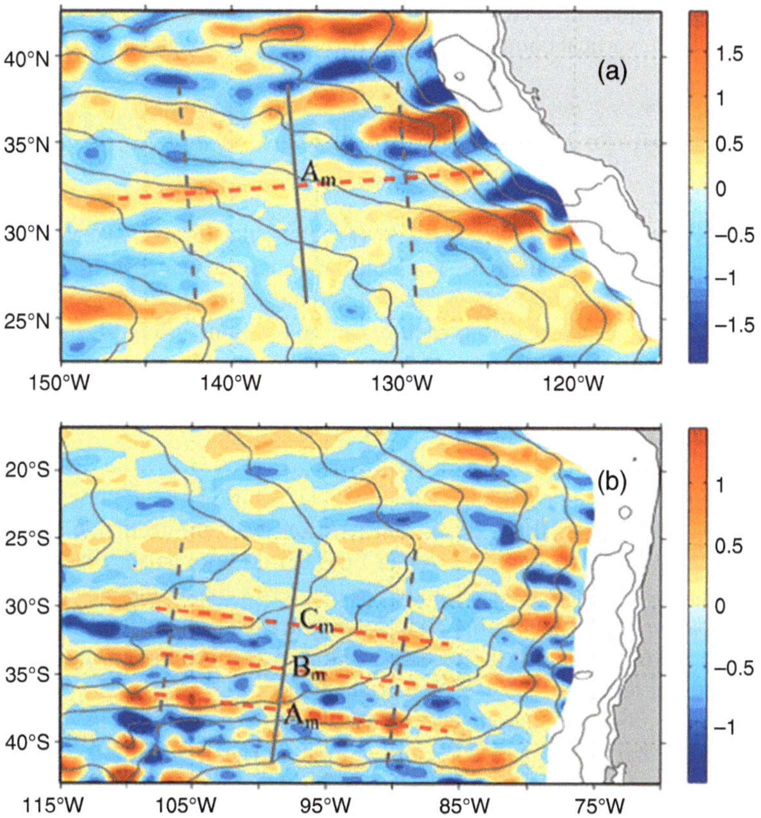

Fig. 5 Quasi-zonal jets in the eastern North Pacific (**a**) and South Pacific (**b**) from a GCM simulation. Contours of the 1993–2002 mean sea surface height (*gray solid lines*, contour interval is 5 cm) are shown in gray solid lines. Red dashed lines approximate crests of the selected jets. Adapted from Melnichenko et al. (2010). ©Springer. Used with permission

3.1 Relative Vorticity

Kamenkovich et al. (2009) analyzed the balance for the relative vorticity, $\zeta = \frac{\partial v}{\partial x} - \frac{\partial u}{\partial y}$, for simplicity written here in Cartesian coordinates (e.g., Vallis 2006):

$$\frac{\partial \zeta}{\partial t} = -\nabla \cdot (\boldsymbol{u}\boldsymbol{\zeta}) - \beta v + f\frac{\partial w}{\partial z} + \Phi_{\mathrm{STR}} + \Phi_{\mathrm{BCL}} + \Phi_{\mathrm{DIS}} \quad (11)$$

where the first three terms on the right-hand side of Eq. (11) are, from left to right, the advection of the relative vorticity by the three-dimensional velocity $u = (u.v.w)$, advection of the planetary vorticity, and the linear stretching term. In the QG dynamics, the latter term is closely related to FST in the PV equation, through the buoyancy equation (Pedlosky 1987). The remaining terms are the nonlinear stretching term, vertical component of the baroclinic vector, and dissipation:

$$\Phi_{\mathrm{STR}}=\frac{\partial u}{\partial z}\frac{\partial w}{\partial y}-\frac{\partial v}{\partial z}\frac{\partial w}{\partial x}+\varsigma\frac{\partial w}{\partial z},\ \Phi_{\mathrm{BCL}}=\frac{1}{\rho^2}\left(\frac{\partial P}{\partial y}\frac{\partial \rho}{\partial x}-\frac{\partial P}{\partial x}\frac{\partial \rho}{\partial y}\right),\ \Phi_{\mathrm{DIS}}=\frac{\partial F_y}{\partial x}-\frac{\partial F_x}{\partial y} \tag{12}$$

where P is pressure, ρ—density, and $F = (Fx, Fy)$—the dissipative term. Since the first two terms were verified to be small, the dynamics can be expected to be similar to a QG flow. After decomposition (6), with the mean defined as an average in time and in the zonal direction within each of the box regions in Fig. 4, the nonlinear advection term splits into the two components:

$$\overline{\nabla\cdot(\boldsymbol{u}\zeta)}=\overline{\nabla\cdot(\boldsymbol{u}'\zeta')}+\nabla\cdot(\overline{\boldsymbol{u}}\overline{\zeta}) \tag{13}$$

The first term on the right-hand side of Eq. (13) is RST of the previous section. Unlike the QG channel flow, the second term on the right-hand side of Eq. (13) and the planetary vorticity term $-\beta\overline{v}$ in Eq. (11) do not vanish. As a result, the relative vorticity balance involves eddy forcing, terms due to the mean circulation and dissipation. In particular, the eddy forcing acts to support the barotropic jets in regions A and C, against the resisting action of the mean advection of relative vorticity, linear terms $-\beta\overline{v}+f\frac{\partial\overline{w}}{\partial z}$, and dissipation.

As in the previous section, decomposition (9) can be used to separate the effects of the barotropic and baroclinic currents. The BCL-BCL component of the RST supports the barotropic jets everywhere in the domain (Fig. 6a). This supporting role of the eddy forcing and its BCL-BCL component is in agreement with the results from the QG channel flow. In contrast, the BRT-BRT term has a weak resisting effect in the subtropical gyre, but supports barotropic jets in the subpolar region; only the latter conclusion is in agreement with QG results. The dynamical balance for the first baroclinic mode in the subpolar gyre is consistent with conclusions from Berloff et al. (2009): RST, BCL-BCL, and BRT-BCL eddy forcing all acting to support the jets (Fig. 6b). Subtropical region is more complicated, and the correlation of the BCL-BCL eddy forcing is low. The misalignment of the eddy forcing and relative vorticity profiles can be interpreted as a tendency of the eddy forcing to shift the jets; and the jets indeed tend to drift in the meridional direction. Similarly, Melnichenko et al. (2010) demonstrated that RST acts to resist and shift jets in the upper ocean of the eastern Pacific.

3.2 Density

The effects of eddies on potential density σ are analyzed within the framework of the Transformed Eulerian Mean (Andrews and McIntire 1976), where the eddy density fluxes are expressed in terms of the eddy-induced velocities (u_e, v_e, w_e). Taking the time average and ignoring the time evolution term in the approximate density

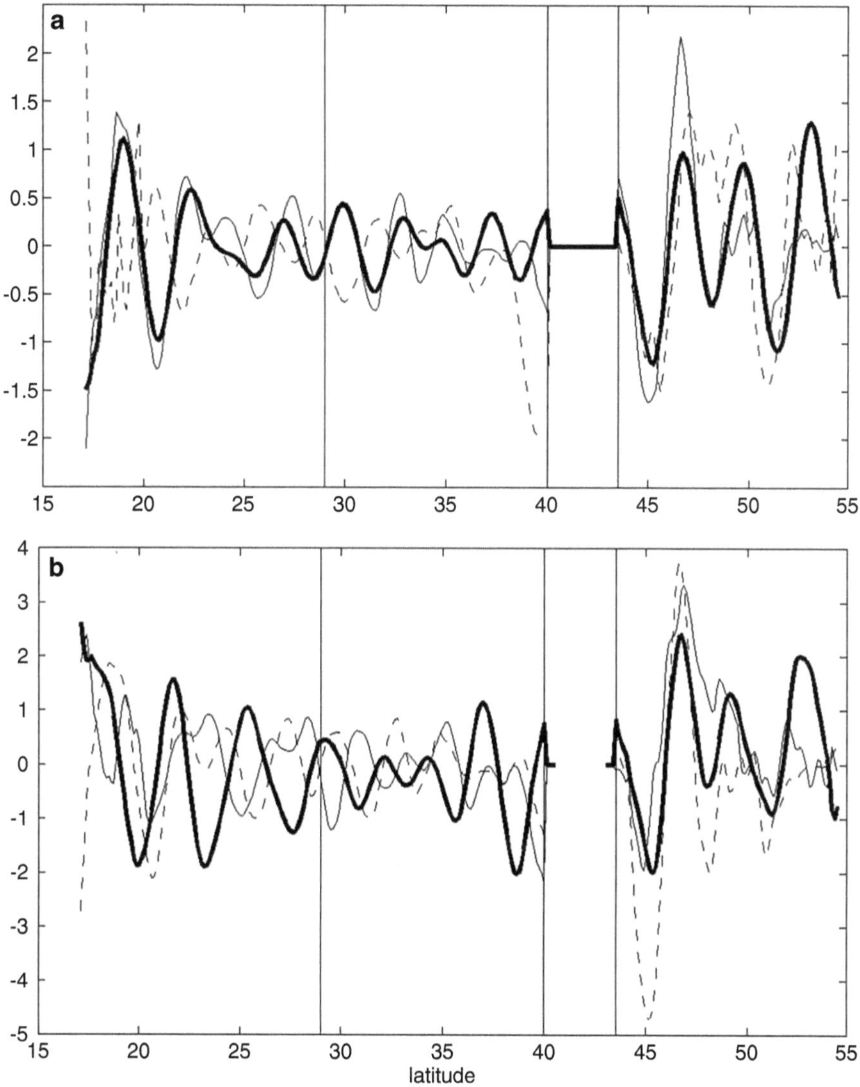

Fig. 6 Role of eddy forcing in the dynamics of jets in GCM simulations. Panel (**a**): BCL-BCL (*thin solid*) and BRT-BRT (*dashed*) eddy forcing and the barotropic relative vorticity (*heavy solid*). Panel (**b**): BRT-BCL (*thin solid*) and BCL-BCL (*dashed*) eddy forcing and the baroclinic relative vorticity (heavy solid); all terms are projected on the first baroclinic mode in zonal velocity (Pedlosky 1987). Quantities are zonally averaged within each of the box regions shown in Fig. 4; the latitudinal boundaries of the regions are marked with vertical lines. Relative vorticity is scaled to be comparable in magnitude to the advective terms. Adapted from Kamenkovich et al. (2009). ©American Meteorological Society. Used with permission

equation, one can derive a balance between the advection of the density flux by the residual (mean plus eddy-induced) circulation, divergence of the projection of the eddy density flux on the cross-isopycnal direction G, and diffusion Φ_{DIFF}:

$$\nabla \cdot \{(\overline{\boldsymbol{u}} + \boldsymbol{u}_e)\overline{\sigma}\} = -\frac{\partial \overline{G}}{\partial z} + \overline{\Phi_{\mathrm{DIFF}}} \tag{14}$$

where the eddy-induced velocities are defined through the eddy density fluxes and the mean stratification:

$$u_e = -\frac{\partial}{\partial z}\left(\frac{\overline{u'\sigma'}}{\partial\overline{\sigma}/\partial z}\right), v_e = -\frac{\partial}{\partial z}\left(\frac{\overline{v'\sigma'}}{\partial\overline{\sigma}/\partial z}\right), \tag{15}$$

$$w_e = \frac{\partial}{\partial x}\left(\frac{\overline{v'\sigma'}}{\partial\overline{\sigma}/\partial z}\right) + \frac{\partial}{\partial y}\left(\frac{\overline{v'\sigma'}}{\partial\overline{\sigma}/\partial z}\right), G = \frac{\overline{\boldsymbol{u}'\boldsymbol{\sigma}'} \cdot \nabla\overline{\sigma}}{\partial\overline{\sigma}/\partial z}.$$

To infer the role each term in Eq. (14) in supporting the density anomalies associated with the jets, Kamenkovich et al. (2009) calculated spatial correlations between the zonal-means of these terms and isopycnal height anomalies. The latter quantity is defined as the difference between the actual isopycnal height $Z(\sigma)$ and its value smoothed with the 5-degree running-mean filter. Since the convergence of the eddy-induced density flux (eddy forcing) acts to shoal isopycnals, a positive correlation between the eddy forcing and height anomalies implies that the eddies act to maintain the "banded" structure in density. The density balance is dominated by the mean and eddy-induced advection, whereas the cross-isopycnal term (first term on the right-hand side of Eq. (14)) plays a secondary role. The eddy-induced advection term acts to support isopycnal height anomalies on most isopycnals in the subtropical gyre (Fig. 7), which is in disagreement with the common assumption that mesoscale eddies act to "iron out" isopycnals and to remove the mean available potential energy from the flow. On the other hand, the eddy forcing acts to destroy the banded structure in the subpolar region (Box C). The convergence of the cross-isopycnal eddy forcing resists the jets almost everywhere.

3.3 Potential Vorticity

The equation for PV, $q = (\nabla \times \boldsymbol{u} + f\boldsymbol{k}) \cdot \nabla\rho$, can be written as (Pedlosky 1987):

$$\frac{\partial q}{\partial t} + \boldsymbol{u} \cdot \nabla q = (\nabla \times \boldsymbol{u} + f\boldsymbol{k}) \cdot \nabla\left(\frac{\partial \rho}{\partial t} + \boldsymbol{u} \cdot \nabla\rho\right) + \frac{\nabla\rho}{\rho} \cdot (\nabla \times \boldsymbol{F}) \tag{16}$$

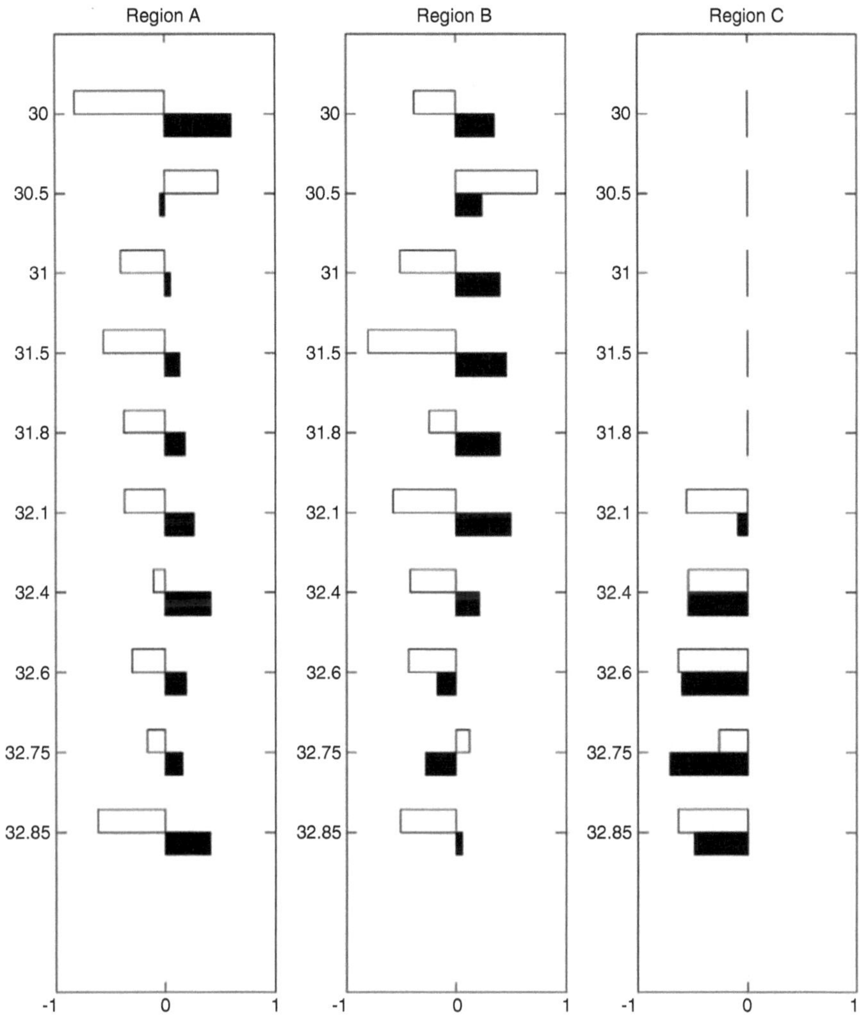

Fig. 7 Role of eddies in the potential density balance. Spatial correlation between zonally averaged isopycnal height anomalies and eddy density evolution terms: along-isopycnal eddy-induced advection term (*black*) and cross-isopycnal eddy advection (*white*). The correlation is computed separately in each of the three regions: region A (*left panel*), B (*middle panel*), and C (*right panel*), and on ten isopycnal surfaces. Positive correlation implies that a corresponding density evolution term acts to sustain the banded structure in the isopycnal height. Adapted from Kamenkovich et al. (2009). ©American Meteorological Society. Used with permission

where k is the unit vector in the vertical direction. The two terms on the right-hand side of Eq. (16) represent sources/sinks of the PV: the diabatic term and the dissipation term. Both terms are of the same order of magnitude as the advection on the left-hand side. After transformation (6) and collection of only the leading order terms, the time-mean Eq. (16) reduces to the balance between the group of terms that depend on the time-mean fields, the dissipative term, and the eddy forcing (Kamenkovich et al. 2009), written here in the flux divergence form:

$$\nabla \cdot \left\{ \overline{\boldsymbol{u}} \left(f + \overline{\xi} \right) \frac{\partial \overline{\rho}}{\partial z} \right\} - f \frac{\partial}{\partial z} \left(\nabla \cdot \overline{\boldsymbol{u}} \overline{\rho} \right) - \overline{\frac{\nabla \rho}{\rho} \cdot \left(\nabla \times \boldsymbol{F} \right)} \approx -\nabla \cdot \left(\overline{\boldsymbol{u}' \varsigma'} \frac{\partial \overline{\rho}}{\partial z} \right)$$
$$-\nabla \cdot \left(\overline{f \boldsymbol{u}' \frac{\partial \rho'}{\partial z}} \right) + f \frac{\partial}{\partial z} \left(\nabla \cdot \overline{\boldsymbol{u}' \rho'} \right) \quad (17)$$

The eddy forcing (the right-hand side of Eq. (17)) consists of three terms, from left to right: (1) generalized RST, which is advection of relative vorticity multiplied by the mean stratification term; (2) generalized FST, which is advection of stratification anomalies multiplied by the planetary vorticity; and (3) "density flux term" (DFT). Note that the generalized FST and DFT tend to compensate each other because the geostrophic components of velocities are in thermal wind balance. Furthermore, the sum of the second term on the left-hand side of Eq. (16) (mean density flux term) and DST is balanced by the explicit diffusion and sources/sinks in the density balance.

The mean PV anomalies, defined as the difference between the full and large-scale mean PV field, exhibit banded structure reflecting the presence of the jets (not shown). The spatial correlation between the zonal average of these PV anomalies and the three different components of the eddy forcing is reported in Fig. 8. On most isopycnal surfaces, the banded PV anomalies are supported by the DFT and resisted by the generalized RST and FST. This result is consistent with the supporting role of eddy forcing in the density balance (previous section), because PV anomalies in this study are dominated by density fluctuations, rather than relative vorticity.

Melnichenko et al. (2010) demonstrated that the eastern Pacific jets cross the mean PV contours, which implies divergence of the mean PV advection. The eddy forcing balances the mean PV divergence, which can be interpreted as a tendency of eddies to drive zonal jets across the mean PV contours (Fig. 9). The dominant component of the PV eddy forcing is the generalized FST.

4 Summary and Discussion

This review chapter discusses nonlinear dynamics of multiple zonal jets that result from self-organization of turbulent atmospheric and oceanic flows. Transient eddies in these flows grow because of an external input of energy and, upon reaching finite amplitudes, play a key role in the dynamics by transporting density, and relative and potential vorticity. Convergence of the corresponding eddy fluxes can correspond to a systematic source of these properties, or internally generated, nonlinear eddy forcing. Time-averaged eddy forcing plays a key role in the dynamics of the

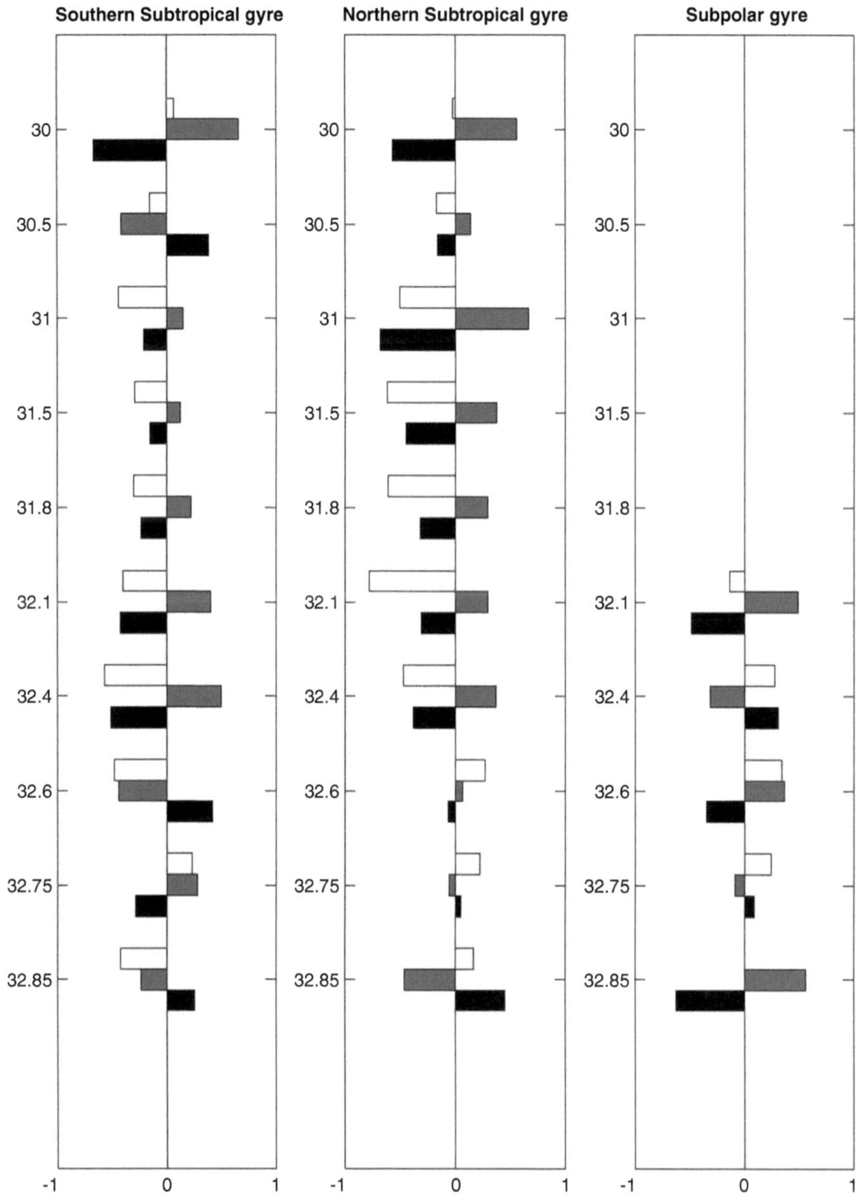

Fig. 8 Role of eddy forcing in PV balance. Spatial correlation between zonally averaged PV anomaly and eddy forcings: generalized RST (*white*), generalized FST (*black*), and DFT (*gray*). The correlation is computed separately in each of the three regions: region A (*left panel*), B (*middle panel*), and C (*right panel*) and on ten isopycnal surfaces. Positive correlation implies that the corresponding eddy forcing acts to sustain the banded structure in PV. Adapted from Kamenkovich et al. (2009). ©American Meteorological Society. Used with permission

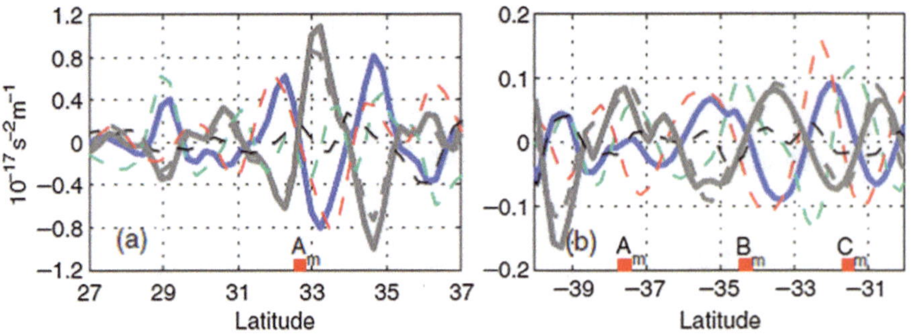

Fig. 9 Time-mean PV balance on a selected isopycnal in the Northeastern (*left*) and Southeastern (*right*) Pacific, from a high-resolution GCM. Thick blue curves show the total mean PV advection; thick gray curves—total eddy forcing, thick dashed gray curves—generalized FST. Thin dashed curves show components of the mean PV advection term. All terms are averaged along the quasi-zonal jets in Fig. 5. Adapted from Melnichenko et al. (2010). ©Springer. Used with permission

stationary zonal jets, by balancing dissipation and advection of the time-mean properties by the mean currents. The nonlinear nature of the process implies that the description of jet dynamics cannot be complete unless all flow components are accounted for in the analysis. First, numerical simulations that do not resolve eddies cannot simulate these jets, despite their relatively large size. Second, nonlinear coupling between barotropic and baroclinic components of the eddy fields is critical for the jet dynamics. For example, barotropic–barotropic and baroclinic–baroclinic interactions are equally important in the dynamics of the barotropic jets, which implies that purely barotropic models of jets are incomplete.

Jets in a flat-bottom QG channel flow are aligned with the mean PV contours and the mean circulation conserves PV in the absence of dissipation. In contrast, quasi-zonal jets in two-dimensional oceanic flows cross nonzonal-mean PV contours, which leads to convergence/divergence of the mean PV fluxes, which is in turn balanced by the eddy forcing. In this case, jets are often non-stationary and drift in the meridional direction (Boland et al. 2012; Chen et al. 2015, 2016), and the drift can be caused by a systematic meridional shift of the eddy forcing relative to the jets. The reason for the jets crossing the PV contours is not clear. The mean PV contours change their orientation with depth; the barotropic PV gradient is nearly zonal over the regions with flat bottom. Given the importance of interaction between the barotropic and baroclinic components of the eddies, it is possible that the direction of the jets is determined by the barotropic dynamics, but this hypothesis needs to be verified by additional analysis.

This chapter discusses multiple zonal jets, but eddy forcing also plays a similarly important role in the dynamics of oceanic isolated jets, such as the Gulf Stream and Kuroshio extensions (e.g., Waterman et al. 2011; Shevchenko and Berloff 2015). This property explains why ocean models that do not resolve eddies cannot simulate a path of these western boundary currents after their separation from the coast. Momentum diffusion used in these models to parameterize the effects of eddies cannot represent upgradient momentum and PV fluxes associated with eddy forcing. Different approaches to parameterization of these effects have been suggested (Berloff 2015, 2016), but their universal applicability still needs to be demonstrated.

This review was concerned with only the time-averaged eddy forcing in the dynamical balances for the stationary jets. Time dependence in the eddy forcing is significant, and fluctuations often exceed the time-mean value by several orders of magnitude (Berloff 2016); these fluctuations can have an indirect impact on the mean circulation and lead to rectified flows. Multiple zonal jets are clearly a non-linear phenomenon, but the importance of the nonlinear dynamics in maintaining the jets does not imply that linear arguments cannot be used to interpret other components of the flow. For example, properties of eddies can in some cases be determined by eddy–jet interactions, in a linearized system with the background flow that includes jets. Accurate assessment of the importance of nonlinearity in oceanic processes is a challenging task, but is essential for our understanding of ocean dynamics.

References

Andrews, D.G., and M.E. McIntire. 1976. Planetary waves in horizontal and vertical shear: The generalized Eliassen-Palm relations and the mean zonal acceleration. *Journal of the Atmospheric Sciences* 33: 2031–2048.

Berloff, P. 2015. Dynamically consistent parameterization of mesoscale eddies. Part I: Simple model. *Ocean Modelling* 87: 1–19.

———. 2016. Dynamically consistent parameterization of mesoscale eddies. Part II: Eddy fluxes and diffusivity from transient impulses. *Fluids* 1: 22. doi:10.3390/fluids1030022.

Berloff, P., and I. Kamenkovich. 2013a. On spectral analysis of mesoscale eddies. Part I: Linear analysis. *Journal of Physical Oceanography* 43: 2505–2527.

———. 2013b. On spectral analysis of mesoscale eddies. Part II: Nonlinear analysis. *Journal of Physical Oceanography* 43: 2528–2544.

Berloff, P., I. Kamenkovich, and J. Pedlosky. 2009. Model of multiple zonal jets in the oceans: Dynamical and kinematical analysis. *Journal of Physical Oceanography* 39: 2711–2734.

Berloff, P., S. Karabasov, T. Farrar, and I. Kamenkovich. 2011. On latency of multiple zonal jets in the oceans. *Journal of Fluid Mechanics* 686: 534–567.

Boland, E.J.D., A.F. Thompson, E. Shuckburgh, and P.H. Haynes. 2012. The formation of nonzonal jets over sloped topography. *Journal of Physical Oceanography* 42: 1635–1651.

Buckingham, C.E., P.C. Cornillon, F. Schloesser, and K.M. Obenour. 2014. Global observations of quasi-zonal bands in microwave sea surface temperature. *Journal of Geophysical Research* 119 (8): 4840–4866.

Chen, C., I. Kamenkovich, and P. Berloff. 2015. On the dynamics of flows induced by topographic ridges. *Journal of Physical Oceanography* 45: 927–940.

———. 2016. Eddy trains and striations in quasi-geostrophic simulations and the ocean. *Journal of Physical Oceanography* 46: 2827–2850.

Godfrey, J.S., G.C. Johnson, M.J. McPhaden, G. Reverdin, and S.E. Wijffels, 2001: The tropical ocean circulation. Chap. 4.3, pages 215–246 of: Siedler, Gerold, Church, John A, and Gould, John (eds), Ocean circulation and climate: Observing and modelling the global ocean. International geophysics series, vol. 77. Academic San Diego, CA.

Haidvogel, D., and I. Held. 1980: Homogeneous quasi-geostrophic turbulence driven by a uniform temperature gradient. *Jounal of the Atmospheric Sciences***37**: 2644–2660.

Hogg, N., and B. Owens. 1999. Direct measurement of deep circulation within the Brazil basin. *Deep Sea Research* 46: 335–353.

Huang, H.-P., A. Kaplan, E. Curchitser, and N. Maximenko. 2007. The degree of anisotropy for mid-ocean currents from satellite observations and an eddy-permitting model simulation. *Journal of Geophysical Research* 112: C09005.

Kamenkovich, I., P. Berloff, and J. Pedlosky. 2009. Role of eddy forcing in the dynamics of multiple zonal jets in a model of the North Atlantic. *Journal of Physical Oceanography* 39: 1361–1379.

Lee, S. 1997. Maintenance of multiple jets in a baroclinic flow. *Journal of the Atmospheric Sciences* 54: 1726–1738.

Maximenko, N., B. Bang, and H. Sasaki. 2005. Observational evidence of alternating zonal jets in the world ocean. *Geophysical Research Letters* 32: L12607.

Maximenko, N.A., O.V. Melnichenko, P.P. Niiler, and H. Sasaki. 2008. Stationary mesoscale jet-like features in the ocean. *Geophysical Research Letters* 35 (8): L08603.

Melnichenko, O.V., N.A. Maximenko, N. Schneider, and H. Sasaki. 2010. Quasi-stationary striations in basin-scale oceanic circulation: Vorticity balance from observations and eddy-resolving model. *Ocean Dynamics* 60: 653–666.

Mitchell, J.L., T. Birner, G. Lapeyre, N. Nakamura, P.L. Read, G. Riviere, A. Sanchez-Lavega, and G.K. Vallis. in press. Terrestrial atmospheres. In *Zonal jets*, ed. B. Galperin and A. Read.

Nakano, H., and H. Hasumi. 2005. A series of zonal jets embedded in the broad zonal flows in the Pacific obtained in eddy-permitting ocean general circulation models. *Journal of Physical Oceanography* 35: 474–488.

Nowlin, W., and J. Klinck. 1986. The physics of the Antarctic Circumpolar Current. *Reviews of Geophysics* 24: 469–491.

Orsi, A.H., T. Whitworth, and W. Nowlin. 1995. On the meridional extent and fronts of the Antarctic Circumpolar Current. *Deep Sea Research* 42: 641–673.

Panetta, L. 1993. Zonal jets in wide baroclinically unstable regions: Persistence and scale selection. *Journal of the Atmospheric Sciences* 50: 2073–2106.

Pedlosky, J. 1987. *Geophysical fluid dynamics*. 2nd ed. New York, NY: Springer-Verlag. 710pp.

Rhines, P. 1975. Waves and turbulence on a beta-plane. *Journal of Fluid Mechanics* 69: 417–443.

Richards, K., N. Maximenko, F. Bryan, and H. Sasaki. 2006. Zonal jets in the Pacific Ocean. *Geophysical Research Letters* 33: L03605.

Sanchez-Lavega, A., L.A. Sromovcky, A.P. Showman, A.D. Del Genio, R.M.B. Young, R. Hueso Alonso, E. Garicia-Melendo, Y. Kaspi, G.S. Orotn, N. Barrado-Izagirre, D.S. Choi, and J.M. Barbara. in press. Gas giants. In *Zonal jets*, ed. B. Galperin and A. Read.

Shevchenko, I., and P. Berloff. 2015. Multi-layer quasi-geostrophic ocean dynamics in eddy-resolving regimes. *Ocean Modelling* 94: 1–14.

Sokolov, S., and S. Rintoul. 2009. Circumpolar structure and distribution of the Antarctic Circumpolar Current fronts: 1. Mean circumpolar paths. *Journal of Geophysical Research* 114: C11018. doi:10.1029/2008JC005108.

Treguier, A., N. Hogg, M. Maltrud, K. Speer, and V. Thierry. 2003. The origin of deep zonal flows in the Brazil basin. *Journal of Physical Oceanography* 33: 580–599.

Vallis, G.K. 2006. *Atmospheric and oceanic fluid dynamics*. Cambridge: Cambridge University Press.

Van Sebille, E., I. Kamenkovich, and J. Willis. 2011. Quasi-zonal jets in 3D Argo data of the northeast Atlantic. *Geophysical Research Letters* 38: L02606. doi:10.1029/2010GL046267.

Wallace, J.M., and P.V. Hobbs. 2006. Atmospheric science. In *An introductory survey*, 2nd ed. New York, NY: Academic.

Waterman, S., N.G. Hogg, and S.R. Jayne. 2011. Eddy-mean flow interaction in the Kuroshio extension region. *Journal of Physical Oceanography* 41 (6): 1182–1208. doi:10.1175/2010JPO4564.1.

Williams, G. 1979. Planetary circulations: 2. The jovian quasigeostrophic regime. *Journal of the Atmospheric Sciences* 36: 932–968.

Data-Adaptive Harmonic Decomposition and Stochastic Modeling of Arctic Sea Ice

Dmitri Kondrashov, Mickaël D. Chekroun, Xiaojun Yuan, and Michael Ghil

Abstract We present and apply a novel method of describing and modeling complex multivariate datasets in the geosciences and elsewhere. Data-adaptive harmonic (DAH) decomposition identifies narrow-banded, spatio-temporal modes (DAHMs) whose frequencies are not necessarily integer multiples of each other. The evolution in time of the DAH coefficients (DAHCs) of these modes can be modeled using a set of coupled Stuart-Landau stochastic differential equations that capture the modes' frequencies and amplitude modulation in time and space. This methodology is applied first to a challenging synthetic dataset and then to Arctic sea ice concentration (SIC) data from the US National Snow and Ice Data Center (NSIDC). The 36-year (1979–2014) dataset is parsimoniously and accurately described by our DAHMs. Preliminary results indicate that simulations using our multilayer Stuart-Landau model (MSLM) of SICs are stable for much longer time intervals, beyond the end of the twenty-first century, and exhibit interdecadal variability consistent with past historical records. Preliminary results indicate that this MSLM is quite skillful in predicting September sea ice extent.

D. Kondrashov (✉)
Department of Atmospheric and Oceanic Sciences, University of California, Los Angeles, CA, USA

Institute of Applied Physics of the Russian Academy of Sciences, Nizhny Novgorod, Russia
e-mail: dkondras@atmos.ucla.edu

M.D. Chekroun
Department of Atmospheric and Oceanic Sciences, University of California, Los Angeles, CA, USA
e-mail: mchekroun@atmos.ucla.edu

X. Yuan
Lamont-Doherty Earth Observatory of Columbia University, Palisades, CA, USA
e-mail: xyuan@ldeo.columbia.edu

M. Ghil
Geosciences Department, Ecole Normale Supérieure and PSL Research University, Paris, France

Department of Atmospheric and Oceanic Sciences, University of California, Los Angeles, CA, USA
e-mail: ghil@atmos.ucla.edu

A.A. Tsonis (ed.), *Advances in Nonlinear Geosciences*,
DOI 10.1007/978-3-319-58895-7_10

Keywords Inverse modeling • Multilayer stochastic models • Power and phase spectra • Cross-correlations • Hankel matrices

1 Data-Adaptive Harmonic Decomposition

The data-adaptive harmonic (DAH) decomposition introduced in Chekroun and Kondrashov (2017) is a signal processing methodology that allows for a data-adaptive decomposition of power and phase spectra by adapting the time embedding approach to the study of time series introduced in Broomhead and King (1986), Vautard and Ghil (1989), Elsner and Tsonis (1996), and its multivariate extensions. However, unlike other methodologies that rely on time embedding—such as Multichannel Singular Spectrum Analysis (M-SSA) (Ghil et al., 2002) or Laplacian spectral analysis (Giannakis and Majda, 2012)—DAH uses integral-operator techniques that help decompose the original signal into narrow-banded signals; while data-adaptive, these elementary signals remain narrow-banded for each separate, discrete Fourier frequency.

At a practical level, the key feature of the DAH method is that it relies on the construction of matrices that exploit cross-correlations in a different way than found in standard statistical methods, such as in Principal Component Analysis (PCA) (Preisendorfer, 1988). As explained in Chekroun and Kondrashov (2017) and discussed below, the eigenmodes associated with the matrices constructed by DAH exhibit a data-adaptive feature that shows up in their phase rather than in their shape. To wit, these modes form an orthogonal set of oscillating functions within the embedding window that is characterized by an interlacing of their zeros, as is the case for the eigenfunctions of Sturm–Liouville boundary-value problems for ordinary differential equations (e.g., Hartman 1986). While this interlacing property is intrinsic to the modes obtained by the DAH approach, the location of their zeros depends on the dataset at hand.

It is for this reason, that these modes are referred hereafter as *data-adaptive harmonic modes* (DAHMs). As a result, the elementary signals come in pairs, which are composed—as far as permitted by the available information and resolution—by such modes in exact phase quadrature. This property allows one to extract the aforementioned narrow-banded but amplitude-modulated time series, whose sum represents the original signal, as time series of DAH coefficients (DAHCs) obtained by projecting the input dataset onto the DAHMs. These features are at the core of identifying spatio-temporal oscillatory modes in the noisy synthetic dataset introduced in Sect. 2, as well as in the DAH analysis of a dataset of Arctic Sea Ice extent (Fetterer et al., 2010) performed in Sect. 3; finally they permit the DAH-enabled nonlinear stochastic modeling of Sect. 4. Numerical details appear in Appendices 1 and 2.

2 DAH Identification of Spatio-Temporal Oscillatory Modes

Here we evaluate our DAH methodology by applying it to a synthetic dataset designed as a testbed for the classical Prony problem of identifying "hidden periodicities" in a noisy environment (e.g., Marple 1987). Pisarenko harmonic decomposition (Pisarenko, 1973) is a well-known method of frequency estimation by using time-lagged correlations, and it assumes that a signal $x(n)$ consists of p complex exponentials superimposed on white noise. However, the algorithm is restricted to the univariate case, and its practical usefulness is somewhat limited due to the white-noise assumption and to the fact that p must be known a priori.

The M-SSA methodology (Ghil et al., 2002) also relies on time-lagged correlations, and it can be applied for identifying oscillatory modes without the limitations inherent in Pisarenko (1973). A challenge for M-SSA, however, is the degeneracy problem in discriminating between oscillatory modes having similar energy but distinct temporal frequencies and spatial patterns; Groth and Ghil (2011) introduced a suitably modified varimax rotation of the M-SSA modes that helps to deal with this shortcoming. To demonstrate the DAH capabilities for mode identification, we will rely on the synthetic dataset provided at http://www.atmos.ucla.edu/tcd/ssa/guide/mssa/mssarot.html, as part of the SSA-MTM Toolkit for time series analysis, https://dept.atmos.ucla.edu/tcd/ssa-mtm-toolkit; this dataset is used in the freeware Toolkit to illustrate the varimax-rotated M-SSA algorithm introduced in Groth and Ghil (2011).

We thus consider a short and noisy spatio-temporal dataset describing the time evolution of a d-dimensional vector $\mathbf{y}(t_n) := (y_1(t_n), \ldots, y_d(t_n))$ over the interval $n = 1, \ldots, N$; here $d = 6$ and $N = 130$. The full dataset shown in Fig. 1f consists of a coherent component $\mathbf{s}(t)$ embedded into *temporally correlated*, albeit spatially uncorrelated noise $\mathbf{r}(t)$:

$$\mathbf{y}(t_n) = (1 - \nu)^{1/2}\, \mathbf{s}(t_n) + \nu^{1/2}\, \mathbf{r}(t_n)\,. \tag{1}$$

The coherent component $\mathbf{s}(t)$ in Fig. 1e is the sum of the four oscillatory modes $x_k^i(t)$ with varying amplitude and phase across the six spatial channels, as shown in Fig. 1a–d:

$$s_k(t_n) = \sum_{i=1}^{4} x_k^i(t_n), \quad k = 1, \ldots, 6; \tag{2}$$

these modes are given by

$$x_k^i(t) = \left(\frac{\alpha_k^i}{2}\right)^{1/2} \sin(2\pi f_i t + \Phi_k^i), \quad k = 1, \ldots, 6, \;\; i = 1, \ldots, 4, \tag{3}$$

and each phase Φ_k^i is obtained independently as a random variable uniformly distributed in $[0, 2\pi]$.

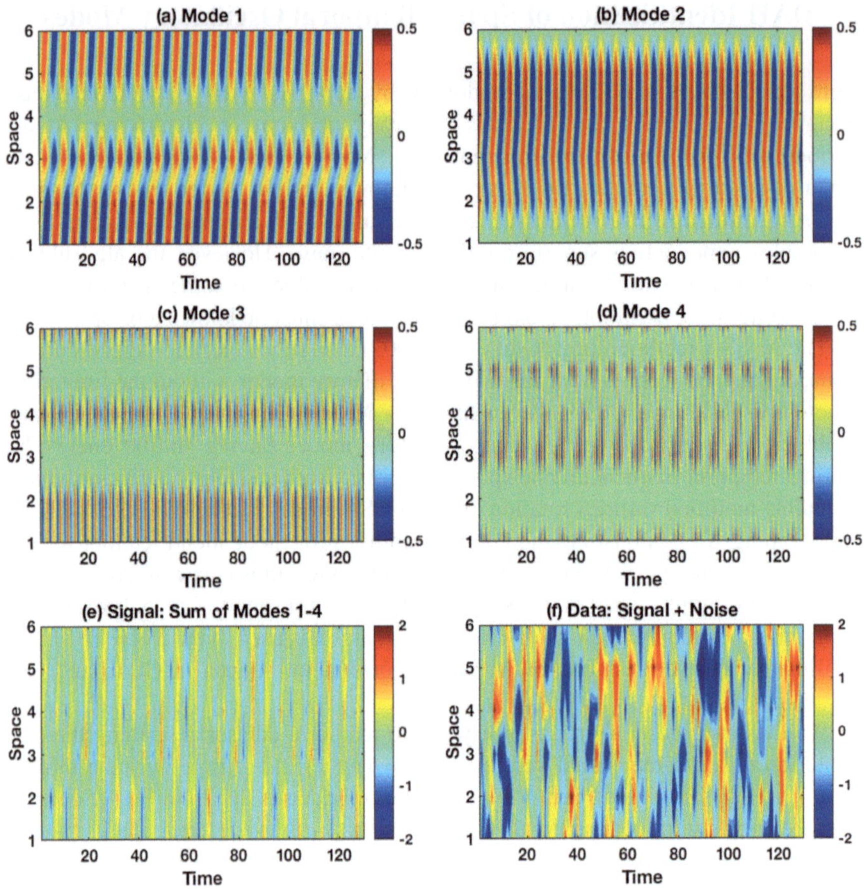

Fig. 1 Multivariate spatio-temporal dataset representing six channels in space and 130 points in time: (**a–d**) four harmonic modes $\{\mathbf{x}^i(t) : i = 1, \ldots, 4\}$ having fixed temporal frequencies but different amplitudes and phases in each of the six channels; see Eq. (3). Their sum $\mathbf{s}(t)$ defines the coherent component given by Eq. (2) shown in panel (**e**); (**f**) total dataset representing the sum of the coherent component $\mathbf{s}(t)$ and of the temporal *red* noise $r_k(t)$ in each of the $\{k = 1, \ldots, d\}$ channels; see text for details

The periodicities of the four oscillatory modes are not integer multiples of the sampling time nor of each other, while the respective frequencies $f_1 = 1/7.5, f_2 = 1/6, f_3 = 1/2.8$ and $f_4 = 1/2.3$ (in sampling units) are located in both the low-frequency and high-frequency part of the power spectrum. The amplitudes α_i^j are prescribed across the spatial channels so that 3 distinct modes contribute to each channel, albeit with different amplitudes; see Table 1. The random choice of the phases Φ_k^i in Eq. (3) results in arbitrary phase shifts across the spatial channels; see Fig. 1. The coefficient $\nu = 0.7$ in Eq. (1) guarantees that the noise component has larger variance than the signal; this fact is obvious from a comparison of the

Table 1 Amplitude modulation of the four oscillatory modes across six spatial channels; see Eq. (3)

α_k^i	$i = 1$	$i = 2$	$i = 3$	$i = 4$
$k = 1$	0.4	0.0	0.3	0.3
$k = 2$	0.4	0.2	0.4	0.0
$k = 3$	0.3	0.3	0.0	0.4
$k = 4$	0.0	0.4	0.4	0.2
$k = 5$	0.2	0.4	0.0	0.4
$k = 6$	0.3	0.0	0.4	0.3

The index k is for the channels, while the index i is for the modes

"clean" (Fig. 1e) with the "noisy" (Fig. 1f), and it makes the identification problem that much more challenging.

The block-Hankel matrix $\mathfrak{C}$ of the DAH decomposition (see Appendix 1) has $d = 6$ blocks of dimension $M' \times M'$, where M' is the embedding dimension. The choice of M' is based on two competing goals: (1) to obtain reliable estimates of autocorrelations from noisy and short datasets; and (2) to resolve the dataset's frequency domain for identification purposes with sufficient accuracy. We chose $M' = 119$, which results in a total number $dM' = 714$ of DAH eigenvalues λ_j and eigenvectors $\mathbf{E}_j$, i.e., $1 \leq j \leq dM'$.

Each of the DAH eigenvectors represents a data-adaptive spatio-temporal pattern associated with a fixed temporal frequency; the latter are equally spaced at intervals of $1/(M' - 1)$ in the Nyquist interval $[0, 0.5]$. Moreover, each temporal frequency is associated with d pairs of DAH eigenvalues that are opposite in sign but equal in absolute value, except at $f = 0$, where there is only one eigenvector per eigenvalue.

Figure 2 shows the DAH spectrum composed of the values $|\lambda_j|$ (red full circles), and obtained here for the synthetic dataset in Fig. 1f. The frequencies of the oscillatory modes that make up the coherent component are identified by eigenpairs located above the noisy background, and marked by the black arrows.

The time-embedded structure of these eigenvectors is shown in Fig. 3, with each pair (E_j, E_j') plotted by red and blue lines, respectively. This structure conveys information about the amplitude modulation across spatial channels, and the figure demonstrates that indeed the eigenvectors for each pair, except at zero frequency, are in phase quadrature, i.e., shifted by one quarter of the associated period.

The latter property is reminiscent of Fourier decomposition, based on sine and cosine pairs with the same periodicity, as well as of the similar property of oscillatory SSA eigenpairs (Ghil et al., 2002). The kth spatial channel $\mathbf{E}_k^j$ of a particular multivariate DAHM—i.e., for a DAH with $d \geq 2$—that is associated with a frequency

$$\omega_\ell = \frac{2\pi(\ell - 1)}{M' - 1}, \quad \ell = 1, \ldots, \frac{M' + 1}{2}. \tag{4}$$

can be analytically expressed—for each $1 \leq j \leq dM'$—as an oscillatory function in the embedding time-window variable τ as follows:

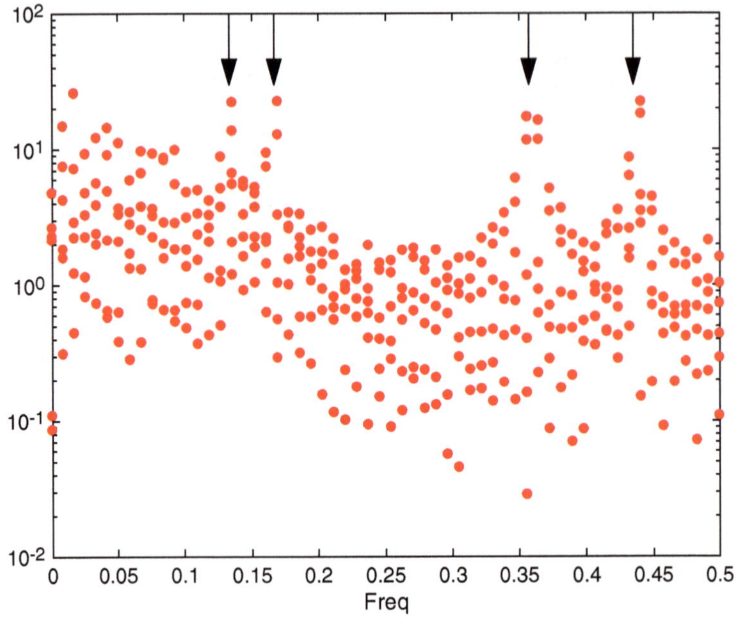

Fig. 2 DAH spectrum of the noisy dataset in Fig. 1f. Each *red full circle* corresponds to a pair $\pm|\lambda_j|$ with distinct eigenvectors $(\mathbf{E}_j, \mathbf{E}_j')$; the latter represent the same temporal frequency f_j but are time-shifted so as to be in phase quadrature, cf. Fig. 3 below. *Arrows point* to the temporal frequencies of four oscillatory modes that do correspond to those shown in Fig. 1a–d. The frequencies of the DAH eigenvectors are equally spaced between 0 and 0.5, and the total number of DAH pairs in each frequency bin is equal to the number of channels $d = 6$ in the dataset. The data-adaptive DAH modes describe amplitude and phase modulation between the spatial channels and are shown in Fig. 3; they do permit the faithful reconstruction of the reference modes in Fig. 1a–d, as shown in Fig. 4a–d below

$$\mathbf{E}_k^j(\tau) = B_k^j(\omega_\ell)\sin(\omega_\ell\tau + \phi_k^j(\omega_\ell)), \quad 1 \le k \le d,\ 1 \le \tau \le M'; \tag{5}$$

here both amplitudes $B_k^j(\omega_\ell)$ and phases $\phi_k^j(\omega_\ell)$ are *data-adaptive* (Chekroun and Kondrashov, 2017).

Moreover, the theory shows that the phases $\phi_k^j(\omega_l)$ for the modes in each pair are shifted by one fourth of the period, i.e., DAHMs are in exact phase quadrature, as for sine–cosine pairs, but in a data-adaptive fashion, encapsulated into the phase. Indeed, as proved in Chekroun and Kondrashov (2017), in the case of univariate time series, the DAH modes provide the phase spectrum contained in each frequency ω_ℓ [given in (4)] via the analytical formula:

$$\Phi(\omega_\ell) = \arg(\lambda_j\widehat{\mathbf{E}}^j(\omega_\ell)) - \arg(\overline{\widehat{\mathbf{E}}^j}(\omega_\ell)),\ 1 \le j \le dM', \tag{6}$$

where $\widehat{\mathbf{E}}^j$ and $\overline{\widehat{\mathbf{E}}^j}$ denote, respectively, the Fourier transform of $\mathbf{E}^j$ and its complex conjugate.

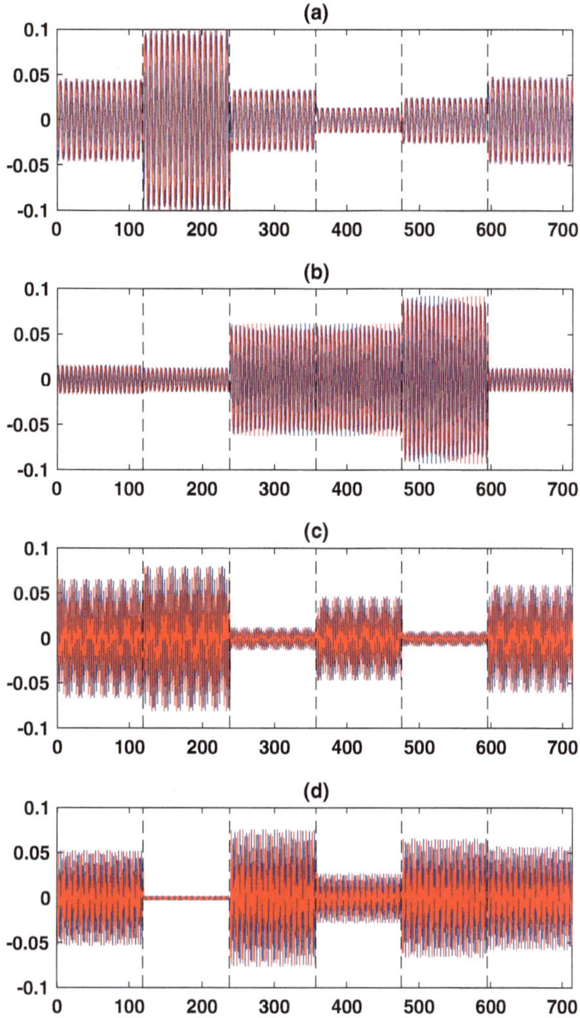

Fig. 3 (**a-d**): Eigenvectors $(\mathbf{E}_j, \mathbf{E}_j')$ of the leading spectral DAH pairs for the four frequencies that are closest to those of the four spatio-temporal oscillatory modes in Fig. 1a–d, i.e., f_1,f_2,f_3, and f_4, respectively; see Eq. (3). The x-axis represents the embedding dimension dM', while the *vertical dashed lines* mark six M'-long segments that correspond to $d = 6$ spatial channels. For each spatial channel, the eigenvectors of a given frequency convey different phases and are shifted by a quarter of the associated period, i.e., they are in exact phase quadrature

The precise information about amplitude and phase modulation of the oscillatory modes captured by the DAHMs allows one to perform highly accurate reconstructions in the space-time domain, cf. Eq. (14) in Appendix 1 below. Figure 4 shows the space-time patterns of the harmonic reconstruction components (HRCs) given by Eq. (15); these patterns are obtained using all the DAH pairs in the frequency bins that contain the target periodicities f_1,f_2,f_3, and f_4. These patterns match quite well in frequency and phase those of the reference coherent components in Fig. 1a–d, although they do underestimate their amplitude as a consequence of the large noise level. In fact, the normalized root-mean-square (rmse) error, averaged over time and space, is roughly 0.5 for all four modes.

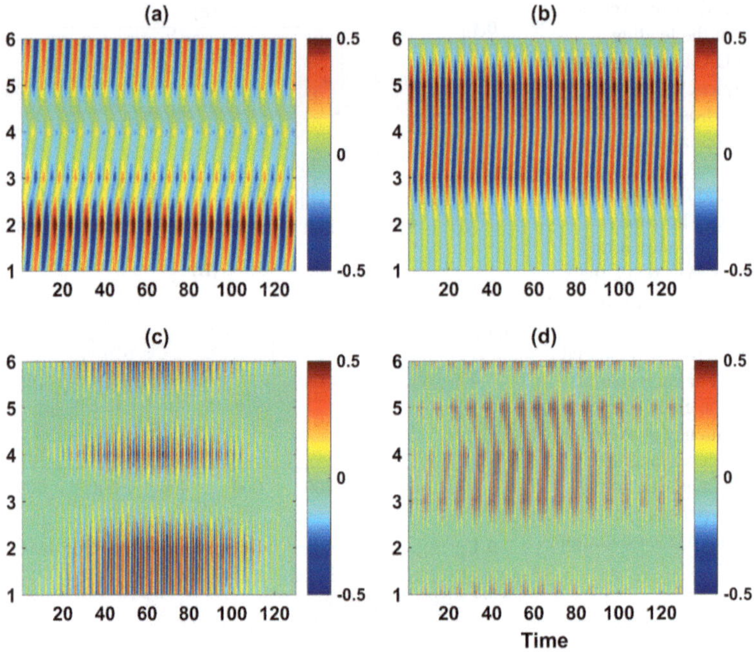

Fig. 4 DAH reconstruction associated with the frequencies of the four dominant oscillatory DAH pairs, as marked by the *arrows* in Fig. 2, and obtained by using the DAH pairs in the corresponding frequency bins. The resulting patterns match reasonably well the reference patterns shown in Fig. 1a–d. (**a**) Mode 1. (**b**) Mode 2. (**c**) Mode 3. (**d**) Mode 4

These results show that DAH does correctly detect the temporal frequencies of distinct oscillatory modes in a very noisy multichannel dataset. Moreover, it also captures fairly well their distinct phase and amplitude across the spatial channels.

3 DAH Decomposition of Arctic Sea Ice Concentrations

Decline in Arctic Sea ice extent is an area of active scientific research with profound climatic and socio-economic implications, both negative—on global temperatures—and positive—by facilitating navigation in polar waters (Sigmond et al., 2016). The key variable of interest to study Arctic Sea ice dynamics is the so-called sea ice concentration (SIC), which measures the relative amount of reference area covered by ice at a given location; SIC is given in percentage points (0–100%). An important indicator of Arctic sea ice conditions is the so-called Sea Ice Extent (SIE), defined as the total surface area of the Arctic region having SIC greater than 15%.

The widely used Sea Ice Index (SII) from the National Snow and Ice Data Center (NSIDC) relies exclusively on passive microwave measurements, which provide a 35-year-long dataset of daily SICs from 1979 to the present. The satellite observations are automatically processed by the National Aeronautics and Space Administration (NASA) Team (Cavalieri et al., 1996) and Bootstrap (Comiso, 2014) algorithms to create daily SIC maps; both algorithms have their own biases and limitations.

We have used the monthly NSIDC dataset for SIC over the January 1979–December 2014 interval, available on a 25 km $\times$ 25 km polar stereographic grid; this dataset is based on the Bootstrap algorithm (Comiso, 2014). The data version used has been coarse-grained onto a $2^\circ \times 0.5^\circ$ grid, representing 7400 spatial degrees of freedom each month and $N = 432$ monthly maps.

First, we removed the seasonal cycle by computing SIC anomalies with respect to each calendar month. Figure 5 shows that the dynamics of SIC anomalies is very different in key Arctic regions, namely the Bering Sea, Baffin Bay, Barents Sea, and Chuckhi Sea. In particular, SIC anomalies in the Baffin Bay and Chuckhi Sea are dominated by the seasonal cycle and a strong downward trend, while internal dynamics is more prominent in the Bering and Barents Seas.

Figure 6a shows that the variability of SIC anomalies is mostly concentrated in the marginal seas of the Arctic Ocean, while it is very small over the North Pole, where the sea remains ice-covered at all times. To extract the dominant modes of SIC variability, empirical orthogonal function (EOF) decomposition (Preisendorfer, 1988) was applied to the dataset. The 12 leading EOFs account for 82% of SIC anomaly variance: excluding the Bering Sea, which is only in very limited contact with the Arctic Ocean, these EOFs capture most of the variance in the marginal seas, cf. Fig. 6b.

Figure 7 shows the corresponding time series of principal components (PCs). The trend component is most prominent in the leading pair of PCs, although it is present, to a lesser extent, in other PCs as well. Moreover, the trend component strongly depends on the calendar month, being more pronounced in fall than in winter; hence there is also strong annual variability in the 1st and 2nd PC, superimposed on the trend. To summarize, SIC PCs exhibit a complex mixture of annual cycle, intraseasonal, interannual, and long-term time scales; this complexity represents a serious challenge for data-driven analysis and modeling techniques, but will be successfully addressed by DAH decomposition.

Figure 8 shows the multivariate DAH spectrum of $d = 12$ PCs for the SIC dataset, with an embedding dimension of $M' = 59$ months. Each full circle in this figure is associated with a pair of DAHMs, except at zero frequency, where the modes are not paired, cf. Eq. (5). The seasonally dependent trend is clearly isolated by the pairs associated with annual-cycle harmonics and located well above the continuous background.

The spatio-temporal patterns of the DAH modes shown in the left and center panels of Fig. 9 reveal useful dynamical information on the combined evolution and mutual influence of SIC's PCs in particular frequency bands. For example, the dominant variability patterns—i.e., those corresponding to the pair having the

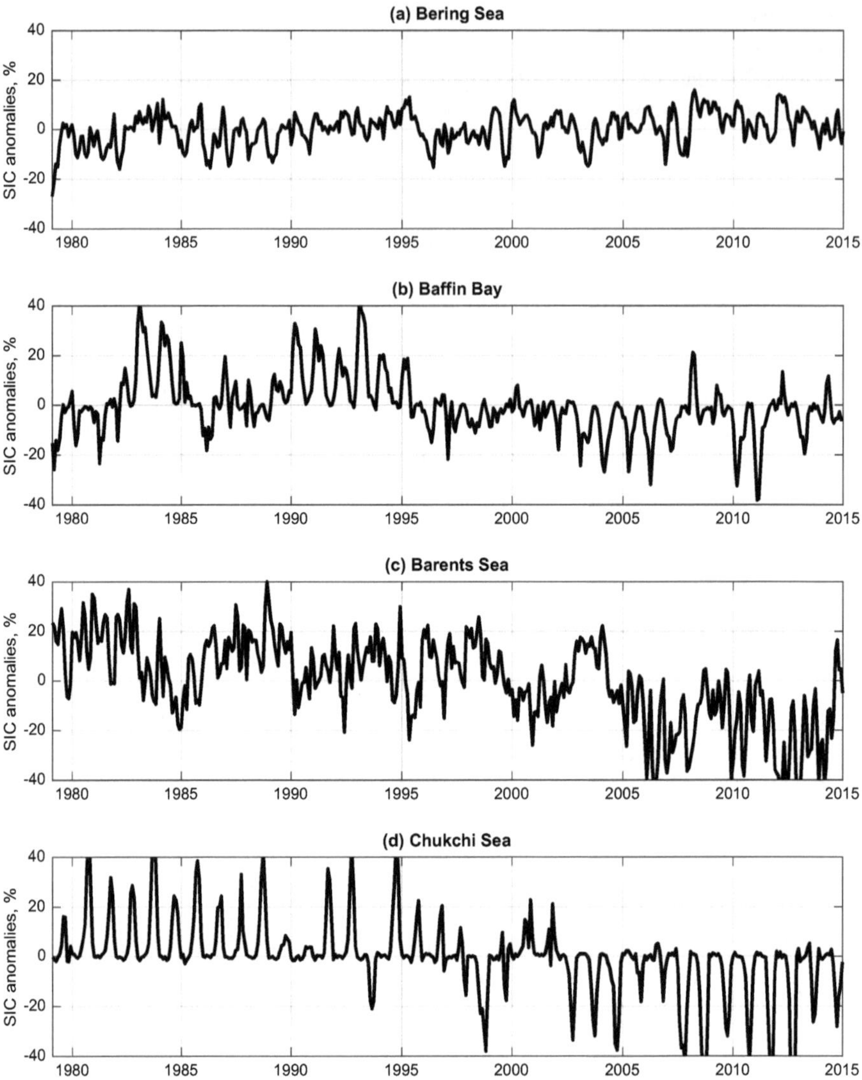

Fig. 5 Monthly time series for sea ice concentration (SIC) anomalies in key Arctic regions; see text for details. (**a–d**) Bering Sea (182°E–192°E, 58°N–62°N); Baffin Bay (298°E–304°E, 61°N–66°N); Barents Sea (34°E–54°E, 76°N–80°N); and Chuckhi Sea (190°E–210°E, 72°N–76°N)

largest $|\lambda_j|$ at a particular frequency—convey in-phase, out-of-phase, and time-lagged influences between different PCs. The DAHMs associated with the *same* frequency and ranked top-to-bottom by their DAH spectral value behave in a similar fashion, as shown in Fig. 10 for the 12-month periodicity. Note that the DAHMs are always in phase quadrature, except at zero frequency.

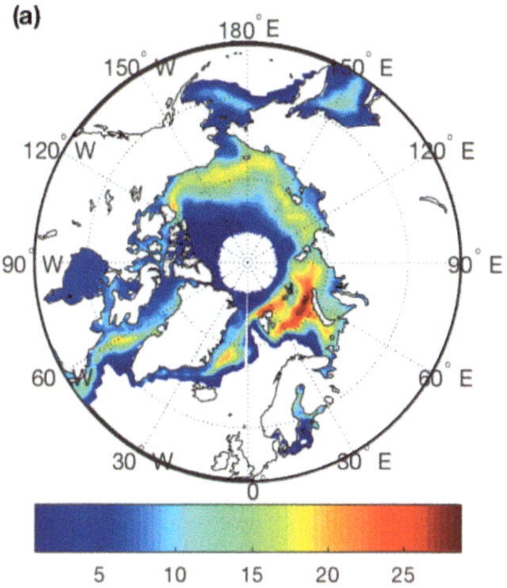

Fig. 6 Spatial distribution of SIC variability. (**a**) Standard deviation of SIC anomalies; and (**b**) fraction of SIC variance captured by the 12 leading EOFs of SIC anomalies. *Color bars* are in percentage units and nondimensional, in (**a**) and (**b**), respectively

On the other hand, although the DAH coefficients A_j are not formally orthogonal in time—see Eq. (13) and its discussion in Appendix 1—they also exhibit a certain phase–quadrature relationship that depends on whether the window M is sufficiently large to resolve the decay of temporal correlations of a given dataset. Typically, the larger M (subject to the length of the record), the more apparent is the phase quadrature between a pair of DAHCs associated with the same frequency.

Shown in the right panels of Figs. 9 and 10, the DAHCs constituting a given $\mathbf{A}_j$-pair account for narrow-band temporal information contained at the characteristic frequency associated with the respective $\mathbf{E}_j$-pair. The latter pairs are shown in the left and center panels of these two figures, respectively, and they reflect differences in amplitude and a shift of, approximately, a quarter of a period. As we can see, the phase-quadrature property of the DAHCs is satisfied to a reasonable degree, which bodes well for the success of the stochastic-modeling approach described in the next section.

4 Stochastic Modeling of Arctic SICs

The recent *Multilayer Stochastic Model* (MSM) framework introduced in Kondrashov et al. (2015) emphasizes the key role of nonlinear, stochastic, and non-Markovian effects in deriving data-driven closure models. Such models have been shown to posses considerable skill in simulating and predicting the main dynamical features of a targeted spatio-temporal field, given either as the output of a high-end geophysical model or as a set of observations. The MSM approach generalizes various multilevel inverse models, including Empirical Model Reduction (EMR) (Kravtsov et al., 2005, 2009): it allows for greater flexibility in the choice of

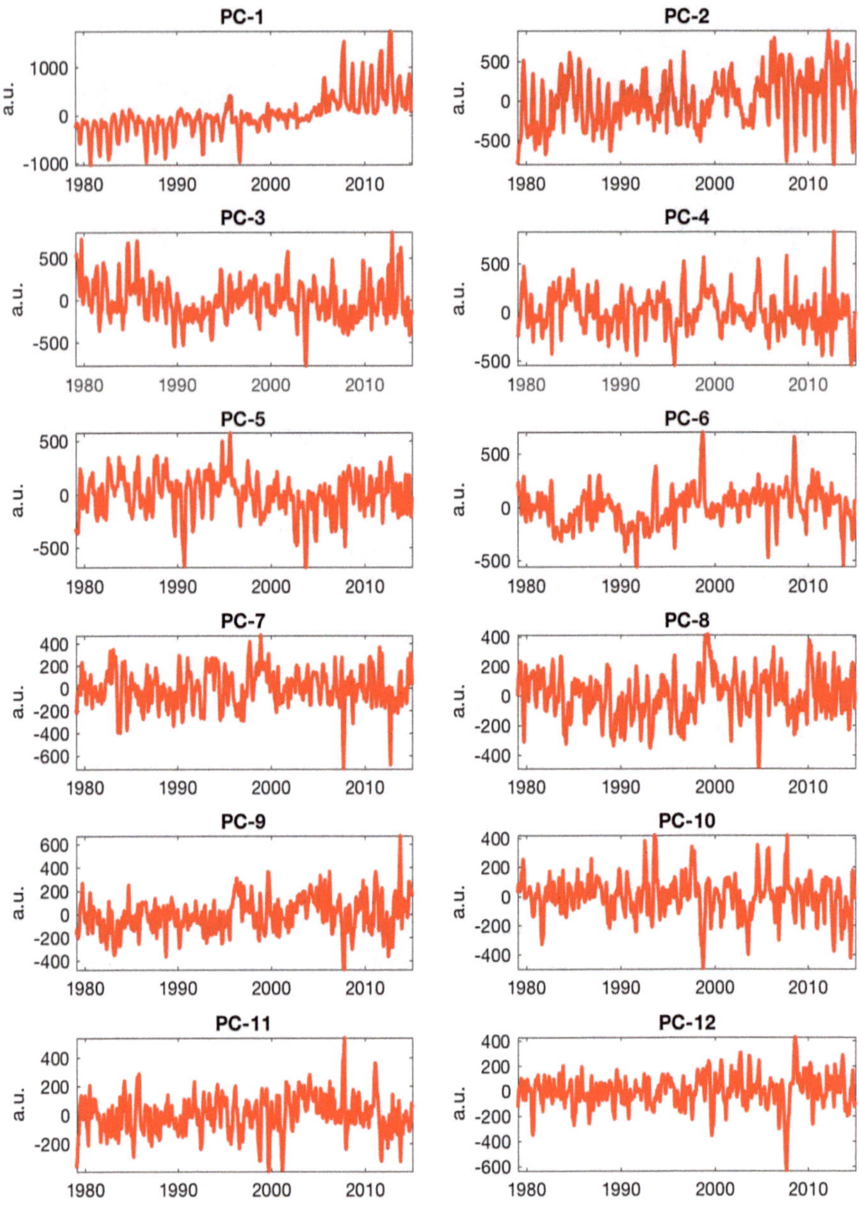

Fig. 7 Time series of the 12 leading principal components (PCs) of SIC anomalies. The seasonally dependent trend component is very prominent in the 1st and 2nd PC

the nonlinear predictors, while ensuring stable asymptotic behavior, such as the existence of a global random attractor (Chekroun et al., 2011); see Theorem 3.1 and Corollary 3.2 in Kondrashov et al. (2015).

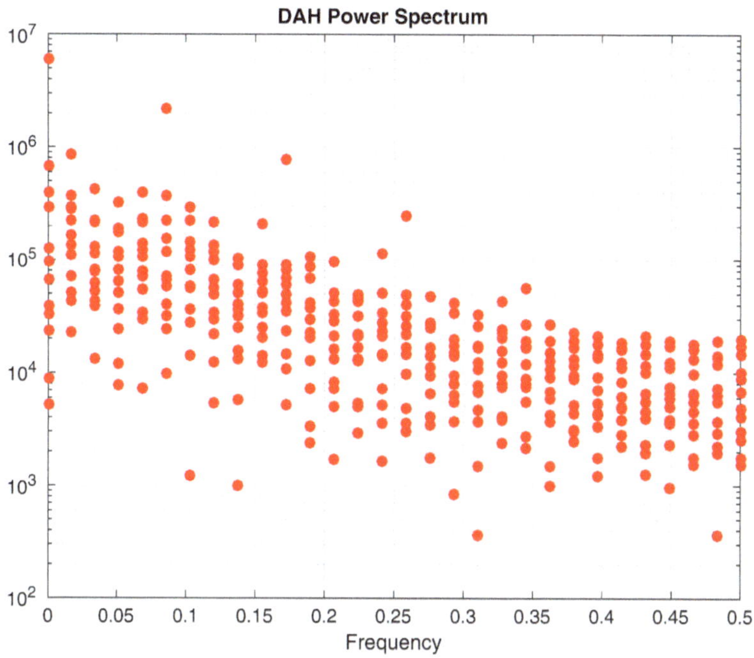

Fig. 8 DAH spectrum of the 12 leading PCs of the SIC dataset, using an embedding window of $M' = 59$ months

However, if the input dataset is not large enough and exhibits a mixture of several time scales, this approach may propose numerous predictors that require one to estimate too many model coefficients, a situation that makes accurate and stable estimates quite difficult. Alternative algorithms are thus called for, and DAH decomposition provides such an alternative. We show here, in the context of Arctic sea ice modeling, that an appropriate change of the basis—in a data-adaptive manner—reduces the data-driven modeling effort to elemental MSMs stacked by frequency, and requires only estimating a fixed and much smaller number of coefficients.

These elemental models fall into the class of networks of linearly coupled Stuart-Landau oscillators (Zakharova et al., 2016), which may include memory terms (Selivanov et al., 2012) and are described below. Given a sequence of partial observations of a dynamical-model simulation, the DAHCs allow one to recast these observations so that they can be reproduced by a simple stochastic model. Such a model can be inferred within a universal parametric family, provided, roughly speaking, that the window whether the window M is sufficiently large to resolve the decay of temporal correlations of a given dataset, as discussed in Appendix 1.

Stuart-Landau (SL) models with additive noise form a generic class of models that capture (1) the frequency f and (2) the amplitude modulations of the A_j's corresponding to a given narrow-band DAHC pair, denoted by $(x(t), y(t))$:

$$\dot{z} = (\mu + i\gamma)z - (1 + i\beta)|z|^2 z + \varepsilon_t, \quad z \in \mathbb{C}; \tag{7}$$

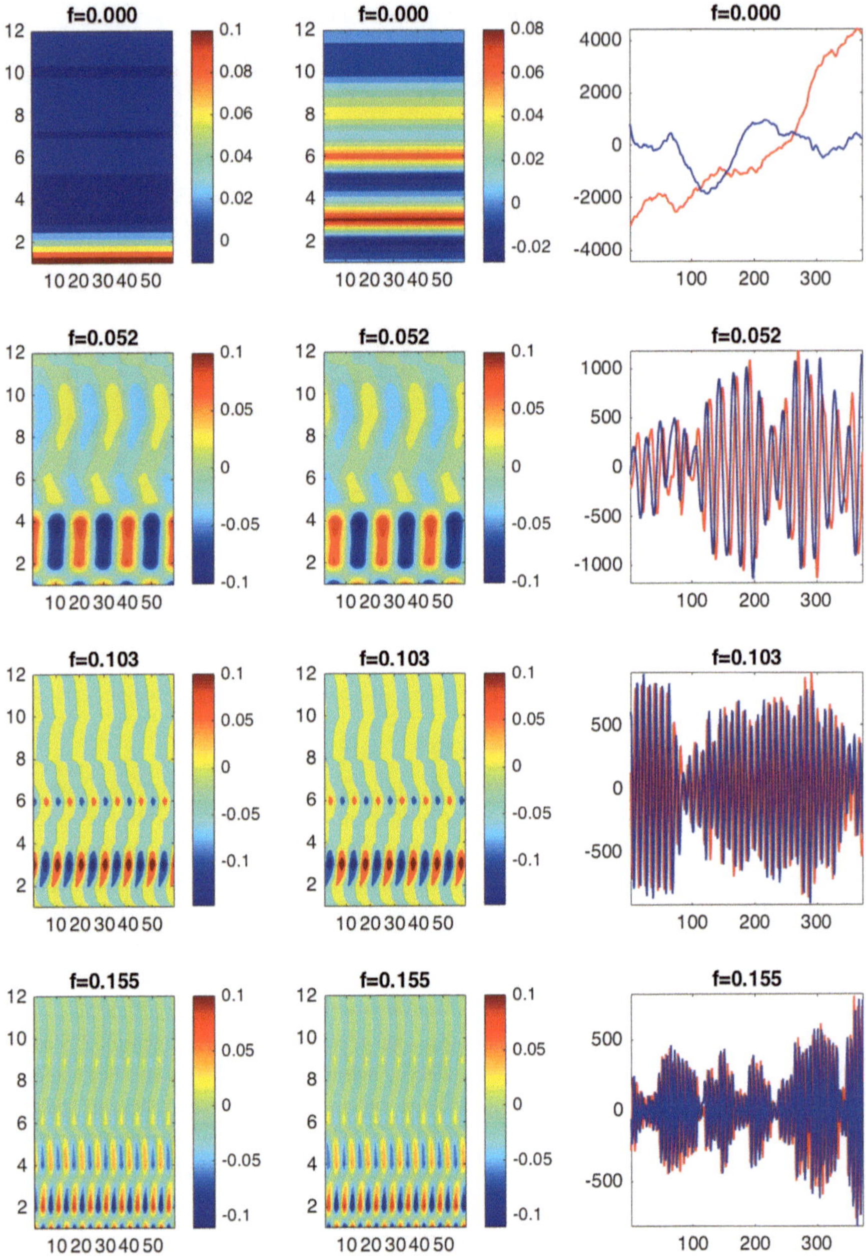

Fig. 9 *Left and center columns:* Spatio-temporal DAH modes (DAHMs) that correspond to the leading DAH pair (1,2) in the SIC dataset's spectrum at selected frequencies: x-axis—embedding dimension, y-axis—PC index. *Right column:* Corresponding temporal DAH coefficients (DAHCs). The four selected frequencies, $f = 0.0, 0.052, 0.103$ and $f = 0.155$, appear in the caption of each panel

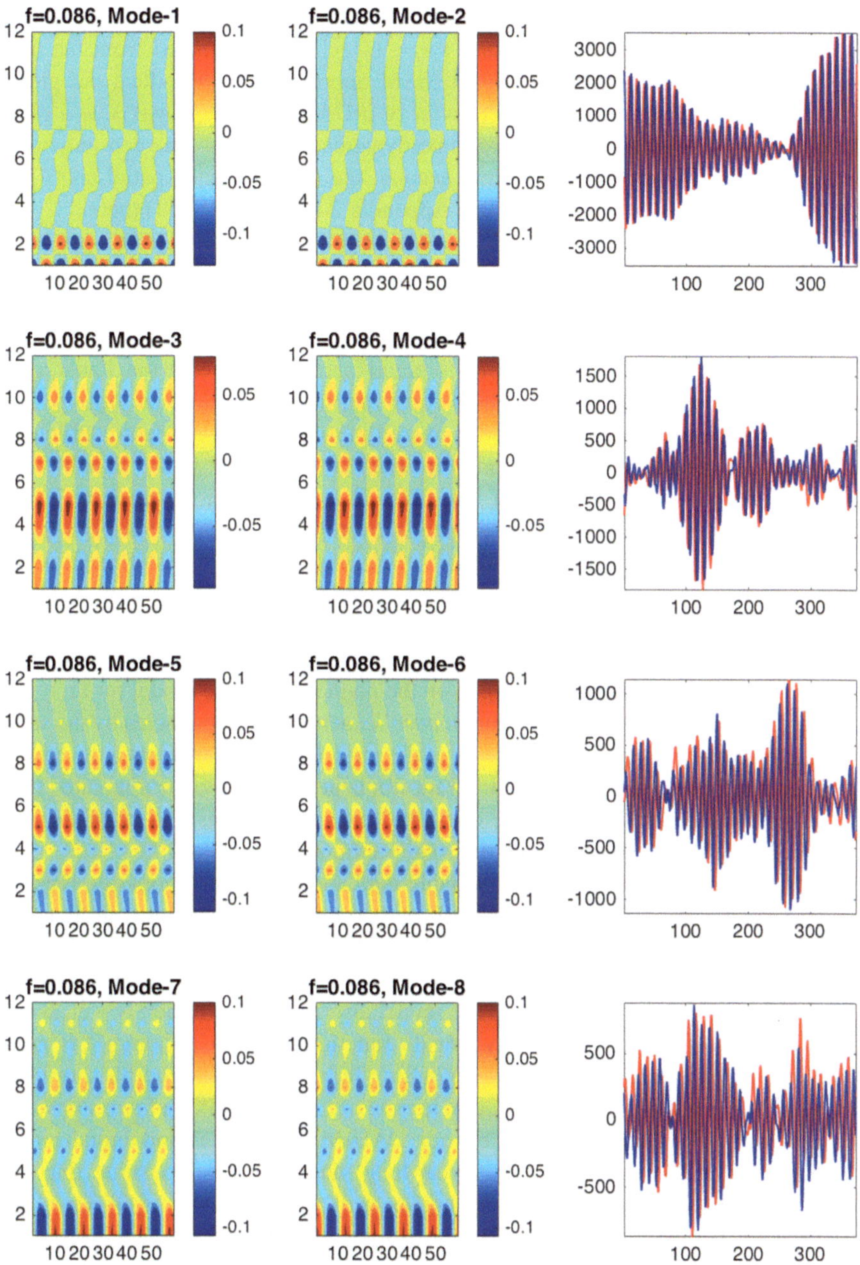

Fig. 10 Same as Fig. 9, except for showing the four leading pairs at the 12-month periodicity, $f = 0.086$. The DAHMs (1,2), (3,4), (5,6), and (7,8) appear in the caption of each panel

here $z(t) = x(t) + iy(t)$ $(i^2 = -1)$ and the real parameters μ, γ, and β, as well as the properties of the driving noise $\varepsilon_t = (\varepsilon_j^x, \varepsilon_j^y)$, are estimated from the time history of $z(t)$ by the aforementioned MSM approach. To reproduce the global phase coherence of the collective behavior of d DAH pairs $(x_j(t), y_j(t))$, at a given frequency $f \neq 0$, requires an appropriate dynamical coupling between individual SL oscillators, along with taking into account the temporal and spatial cross-pair correlations in the driving noise ε_t; see Appendix 2 and Eq. (MSLM) there.

Thus, for each frequency f, the 12 associated pairs of temporal DAHCs are modeled by Eq. (MSLM). First, the model coefficients can be estimated *in parallel* for each frequency, i.e., by successive pairwise regressions, subject to linear constraints on $\beta_j(f), \alpha_j(f)$, and $\sigma_j(f)$ that impose the necessary model structure in Eq. (MSLM) for each (x_j, y_j) pair; these constraints entail antisymmetry for the linear part, without the coupling terms, as well as equal and nonpositive values $\sigma_j(f) \leq 0$ to ensure asymptotic stability. Hence the overall number of independent coefficients to estimate is fixed and relatively small for each (x_j, y_j) pair; e.g., the main layer of Eq. (MSLM) involves estimation of $3 + 4(d - 1) = 47$ coefficients from the $2N' = 748$ DAH-processed Arctic SIC observations, over the full time interval 1979–2014; see Appendix 1 for the definition of $N' = N - M' + 1$, with the window width $M' = 59$ months. Extra layers are added as needed until the regression residuals for the last layer can be approximated by white noise, according to the stopping test described in Kondrashov et al. (2015, Appendix A); these layers convey temporal correlations in the stochastic forcing ε_t on the main layer of the model for (x_j, y_j).

Second, the DAH-MSLMs are run *in parallel* across the frequencies by the same white-noise realization in the last layer of the model, which represents a dynamical mechanism for coupling between different frequencies. Finally, the simulated time series of the temporal DAHCs are converted back to the phase space of the SIC dataset's PCs, by convolution with the spatio-temporal DAHM's.

Despite the limited amount of available data and their nonstationarity, Figs. 11 and 12 show very good modeling skill in reproducing the complex structure of the autocorrelation functions (ACFs) of the SIC dataset's PCs, as simulated by the optimal DAH-MSLM model with $M' = 59$ and having three additional layers in Eq. (MSLM) to model the noise ε_t. The model also captures sufficiently well skewness and kurtosis of the probability density functions (PDFs), although it is more challenging to capture the bumps in the PDFs' "tails," due to the record's shortness.

Figures 13 and 14 show the evolution in time of the leading PCs of two stochastic ensemble members, as simulated by our DAH-MSLM model and initialized in January 1979. These extended, 129-year-long simulations demonstrate that our optimized stochastic-dynamic model agrees well with the existing 36-year-long SIC record, is numerically stable for much longer times, and displays interesting dynamical behavior such as multidecadal variability in PC-1. Such variability has been documented by Walsh and Chapman (2015) in their reconstruction of sea ice extent anomalies from historical records.

One reason for the success of our model's simulations relies on the ability of the DAH approach to extract modulated time series of DAHCs that are narrow-banded

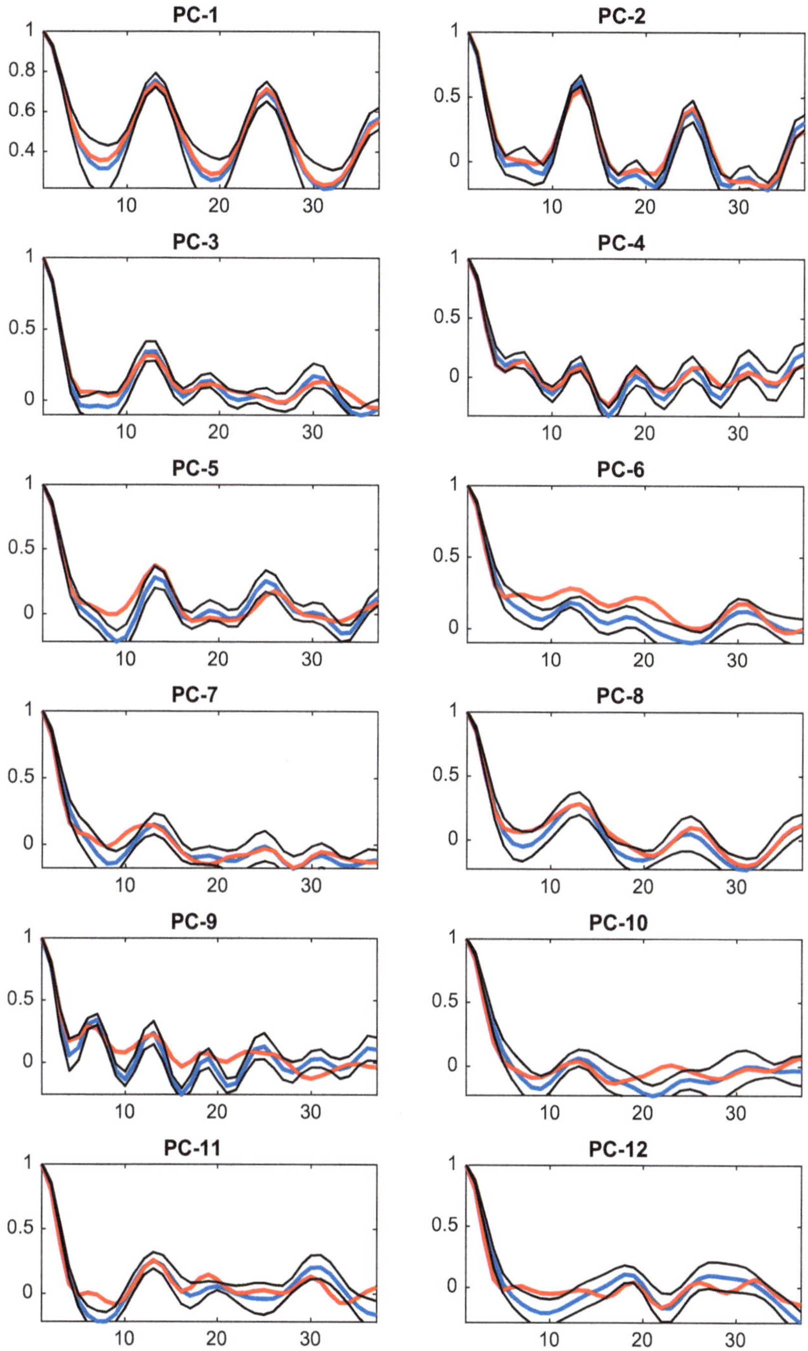

Fig. 11 The autocorrelation functions (ACFs) of the SIC dataset's PCs: *red*—observations, *black*—ensemble mean of stochastic-dynamic simulations by the DAH-MSLM approach; *blue*—standard deviation of the ensemble

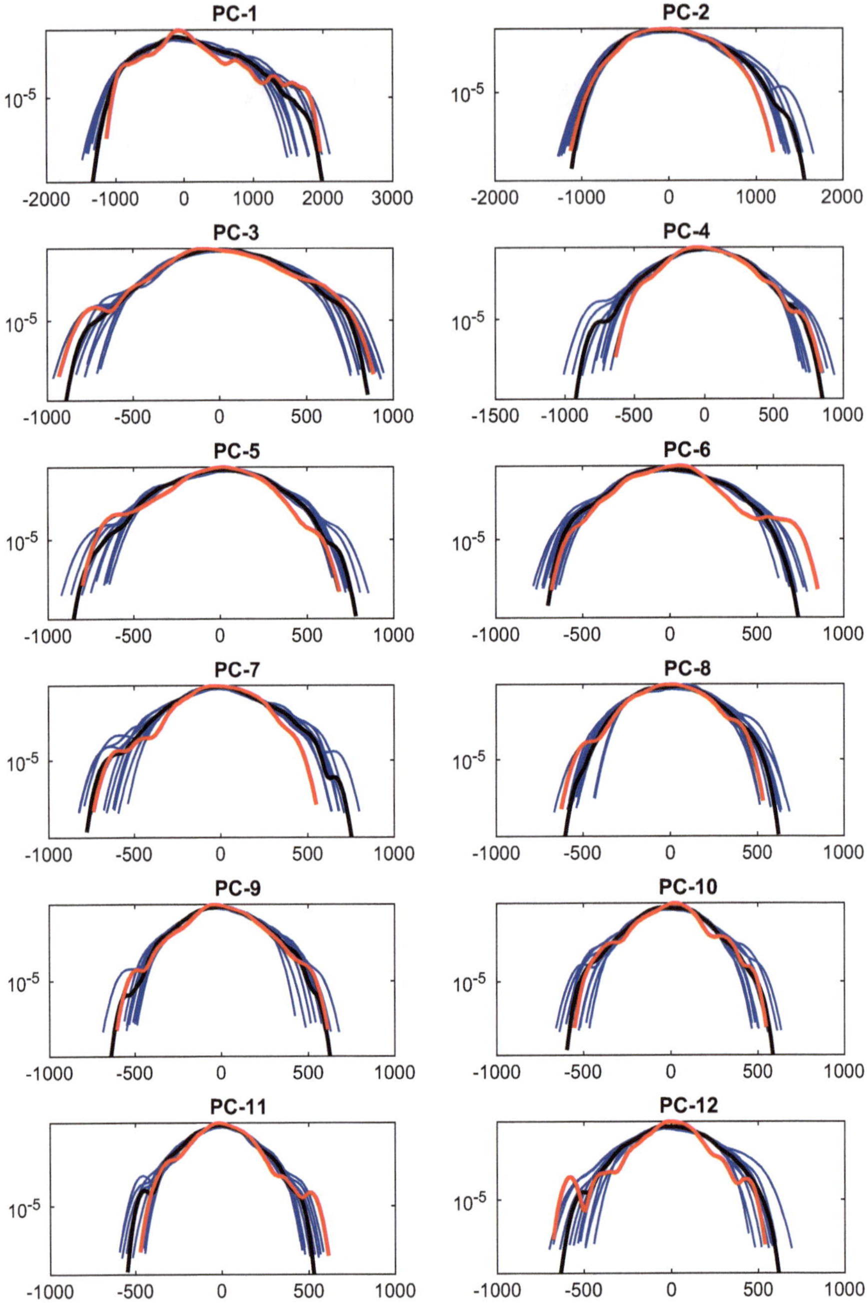

Fig. 12 Same as Fig. 11, except for the probability density functions (PDFs): the *blue lines* now represent individual ensemble members

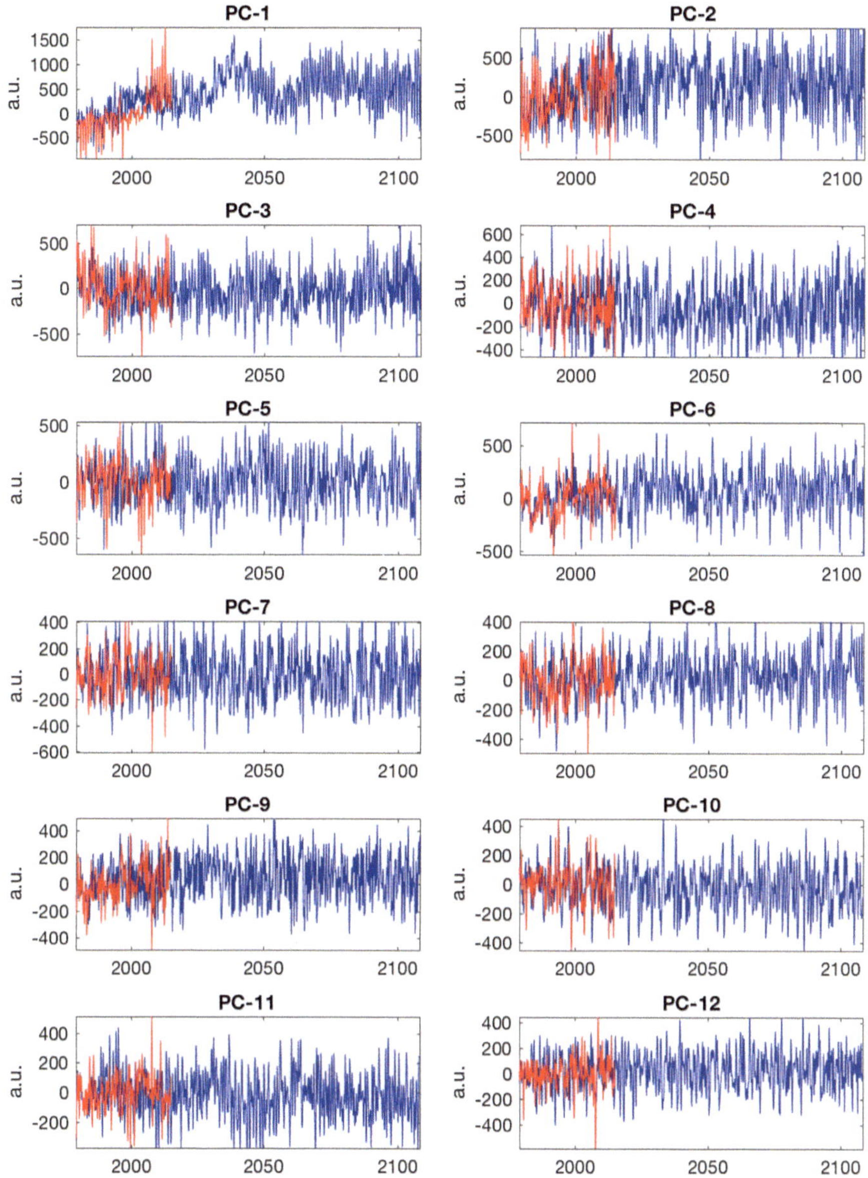

Fig. 13 Extended simulation of the Arctic SIC conditions. *Red*—observational dataset of the 12 leading PCs for 1979–2014 (36 years); *blue*—129-year-long stochastic simulation by the DAH-MSLM approach

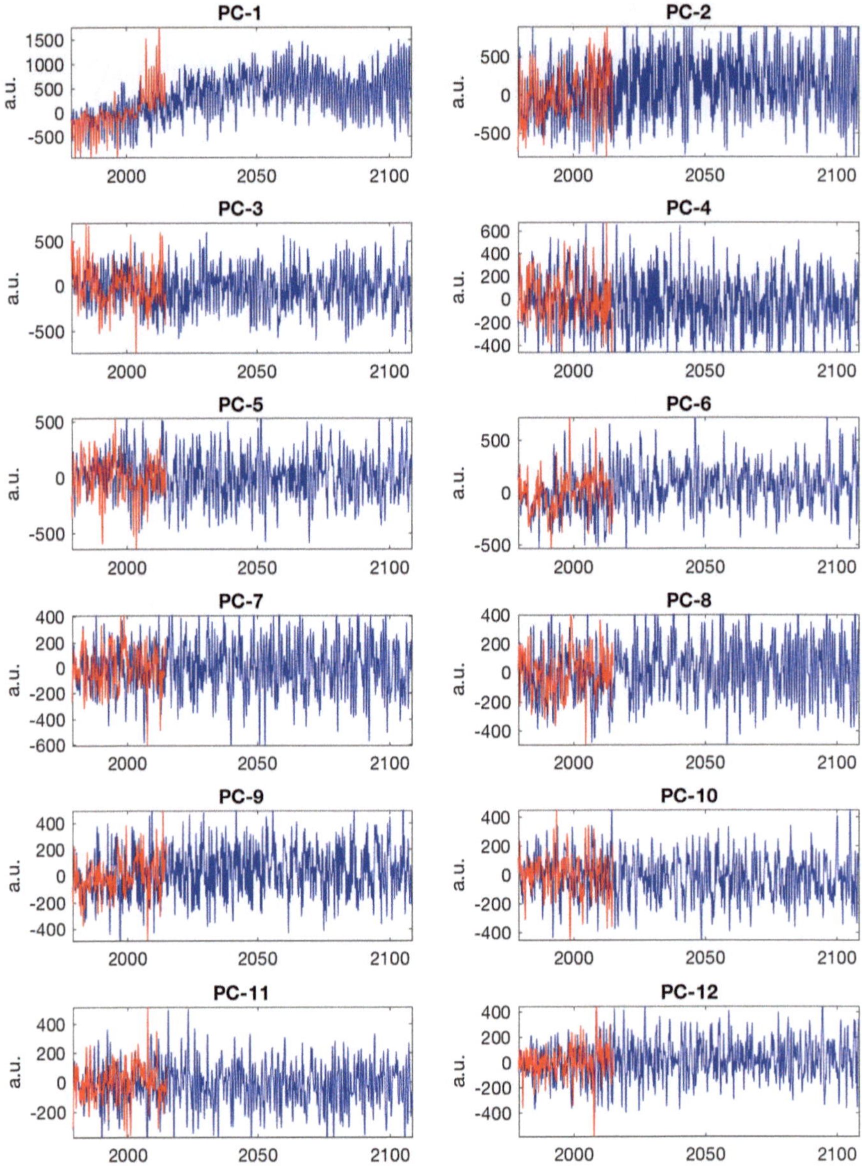

Fig. 14 Same as in Fig. 13, but for another stochastic realization

in the frequency domain and exhibit phase quadrature in the time domain. Another important reason is that the class of MSLMs introduced herein is intrinsically well adapted to the modeling of such time series.

It is worth mentioning that the less narrow-banded the DAHCs, the worse their modeling using MSLM. For the Arctic Sea Ice dataset of Comiso (2014), as repre-

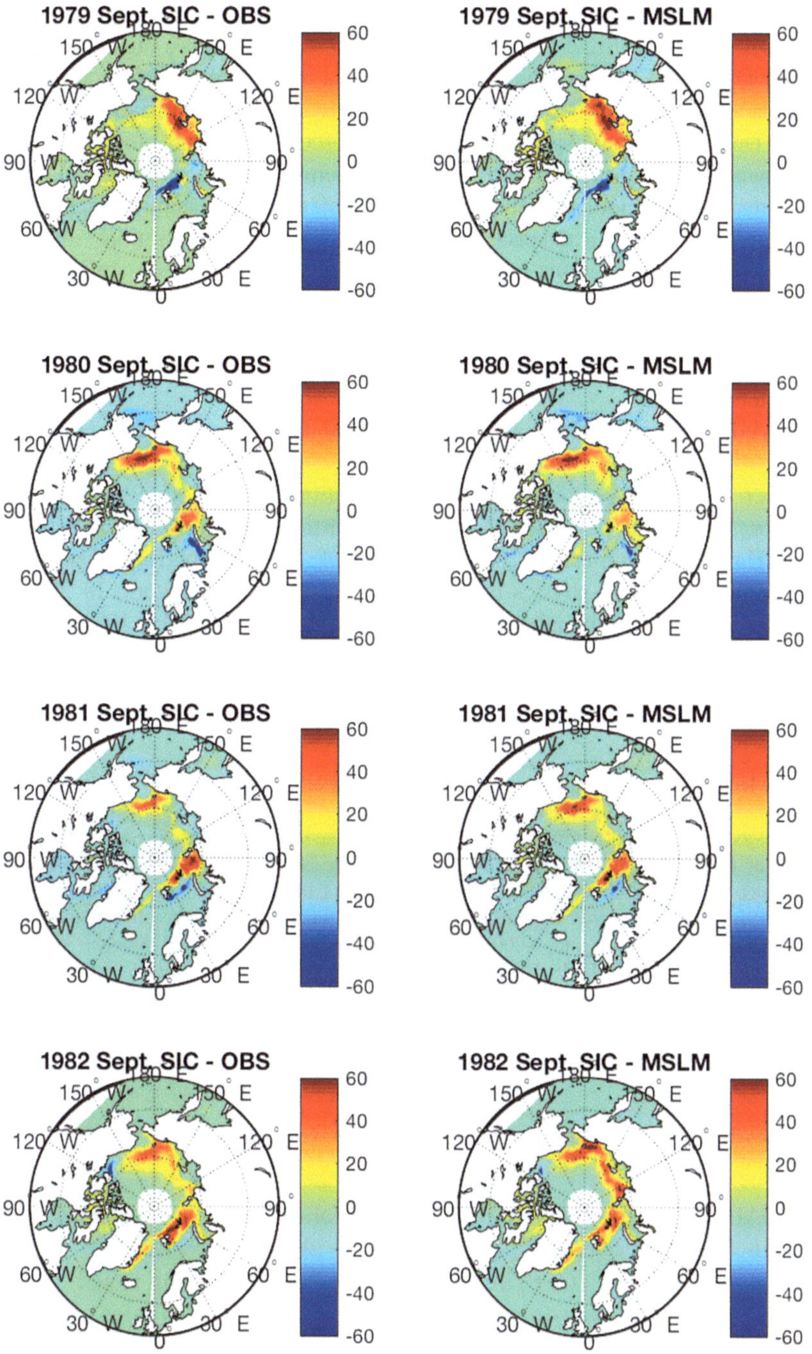

Fig. 15 Simulations of September SICs by using our DAH-MSLM approach. *Left*—observed September SIC anomalies; *right*—hindcast of the DAH-MSLM model, initialized in January 1979. *Caption of each panel* indicates the particular September being compared, OBS vs. MSLM, for 1979, 1980, 1981, and 1982

sented by the SIC PCs, the DAH decomposition provides just the right time series of DAHCs for the MSLM modeling approach to be efficient; see Figs. 9 and 10.

Our DAH-MSLM model is able to produce a remarkable near-synchronization of the simulations with observations during the first 4 years that start with January 1979. This approximate synchronization holds for almost every noise realization, as shown, for instance, in Fig. 15 for one ensemble member, using a particular noise realization: plotted in the figure are September SIC anomalies for 1979–1982 in gridded physical space, with the maps of the observations in the left column and the simulations in the right one. The match between simulation and observation is visually excellent and only starts deteriorating in September 1982. The potential predictive skill of our DAH-MSLM model suggested by these plots implies highly promising potential of developed approach for real-time forecasting of September SIE.

Indeed, this potential forecast skill has been tentatively confirmed by the present authors by using the Multisensor Analyzed Sea Ice Extent (MASIE) dataset (Fetterer et al., 2010) for the Sea Ice Prediction Network (SIPN, http://www.arcus.org/sipn). Our DAH-MSLM model's real-time SIE forecast for September 2016 (Hamilton and Stroeve, 2016; Stroeve et al., 2015) outperformed most other statistical models and physics-based models in the SIPN network. In 2016, the multimodel-median September SIPN estimate in August was 4.4×10^6 km^2, with a quartile range of 4.2–4.7×10^6 km^2, vs. the actual observed value of 4.72×10^6 km^2. The real-time DAH-MSLM August prediction for SIPN's 2016 September Outlook was 4.79×10^6 km^2.

Acknowledgements The authors would like to acknowledge Andreas Groth for developing the synthetic dataset in the SSA-MTM Toolkit example of varimax-rotated M-SSA (http://www.atmos.ucla.edu/tcd/ssa/guide/mssa/mssarot.html); it is this dataset that was utilized in Sect. 2. Preliminary results of this research were reported at "30 Years of Nonlinear Dynamics in Geosciences" conference in Rhodes, Greece, July 2017. The design of this study and the development of the DAH-MSLM techniques were supported by ONR's Multidisciplinary Research Initiative (MURI) grants N00014-12-1-0911 and N00014-16-1-2073, and by the National Science Foundation grants OCE-1243175 and DMS-1616981. Analysis of Arctic sea ice data was also supported by Government of Russian Federation (Agreement No. 14.Z50.31.0033 with the Institute of Applied Physics of RAS).

Appendix 1: Details on the DAH Decomposition

The DAH modes (DAHMs) are obtained as follows. First, we estimate from a given d-channel time series $\mathbf{X}(t_n) = (X_1(t_n), \ldots, X_d(t_n))$, $n = 1, \ldots, N$, the *cross-correlation coefficient* (CCF) $\rho_\tau^{(p,q)}$ at lag τ between channels p and q, where $-M + 1 \leq \tau \leq M - 1$. In spectral analysis, it is common to refer to M as the window width.

Next, we form the following Hankel matrix:

$$\mathbf{H}^{(p,q)} = \begin{pmatrix} \rho_{-M+1}^{(p,q)} & \rho_{-M+2}^{(p,q)} & \cdots & \rho_{0}^{(p,q)} & \rho_{1}^{(p,q)} & \cdots & \rho_{M-1}^{(p,q)} \\ \rho_{-M+2}^{(p,q)} & \iddots & \iddots & \iddots & \iddots & \iddots & \rho_{-M+1}^{(p,q)} \\ \vdots & \iddots & \iddots & \iddots & \iddots & \iddots & \rho_{-M+2}^{(p,q)} \\ \rho_{0}^{(p,q)} & \iddots & \iddots & \iddots & \rho_{-M+1}^{(p,q)} & \iddots & \vdots \\ \rho_{1}^{(p,q)} & \iddots & \iddots & \iddots & \rho_{-M+2}^{(p,q)} & \iddots & \rho_{0}^{(p,q)} \\ \vdots & \rho_{M-1}^{(p,q)} & \rho_{-M+1}^{(p,q)} & \iddots & \iddots & \iddots & \vdots \\ \rho_{M-1}^{(p,q)} & \rho_{-M+1}^{(p,q)} & \rho_{-M+2}^{(p,q)} & \cdots & \rho_{0}^{(p,q)} & \cdots & \rho_{M-2}^{(p,q)} \end{pmatrix}. \tag{8}$$

Equivalently, this matrix can be viewed as a left circulant matrix formed from the $(2M-1)$-dimensional row $r = (\rho_{-M+1}^{(p,q)}, \ldots, \rho_0^{(p,q)}, \ldots, \rho_{M-1}^{(p,q)})$, i.e.:

$$\mathbf{H}^{(p,q)} = l\text{-circ}(\rho_{-M+1}^{(p,q)}, \ldots, \rho_{-1}^{k,k')}, \rho_0^{(p,q)}, \rho_1^{(p,q)} \ldots, \rho_{M-1}^{(p,q)})\,; \tag{9}$$

in other words, the rows of $\mathbf{H}^{(p,q)}$ are obtained by successive shifts to the left by one position, starting from r as a first row. Finally, we consider the block-Hankel matrix $\mathfrak{C}$ formed by d^2 blocks of size $(2M-1) \times (2M-1)$, each given according to

$$\begin{aligned} \mathfrak{C}^{(p,q)} &= \mathbf{H}^{(p,q)}, \text{ if } 1 \le p \le q \le d, \\ \mathfrak{C}^{(p,q)} &= \left(\mathbf{H}^{(q,p)}\right), \text{ otherwise.} \end{aligned} \tag{10}$$

Note that $\mathfrak{C}$ is symmetric by construction due to symmetry of its building blocks $\mathbf{H}^{(p,q)}$, i.e., $\mathfrak{C}^{(p,q)} = \mathfrak{C}^{(q,p)}$, and hereafter we use $M' = 2M - 1$ for concision, reindexing the string $\{-M+1, \ldots, M-1\}$ from 1 to M' as necessary.

The DAH eigenpairs $(\lambda_j, \mathbf{E}^j)$, with $1 \le j \le dM'$, reveal useful information about the variability contained in the multivariate time series. In contrast to other data-adaptive methods built from cross-correlations, each of the DAH eigenvectors $\mathbf{E}^j$ represents a data-adaptive spatio-temporal pattern naturally associated with a Fourier frequency ω_l given by

$$\omega_\ell = \frac{2\pi(\ell-1)}{M'-1}, \quad \ell = 1, \ldots, \frac{M'+1}{2}. \tag{11}$$

These frequencies are equally spaced within the Nyquist interval $[0, 0.5]$ with a resolution of $1/(M'-1)$, essentially given by the embedding dimension M.

Each temporal frequency ω_ℓ is associated with d pairs of DAH eigenvalues $\pm\lambda_j$ that are opposite in sign but equal in absolute value, except at zero frequency, where there is only one eigenvector per eigenvalue, for a total of $2d(M-1)+d$ eigenvalues.

The association between a particular frequency and a given DAHM is obtained by counting zero-crossings δ_j across the window width M for all channels:

$$\delta_j = \sum_{k=1}^{d} \sum_{\tau=1}^{M'-1} \left(1 - \mathrm{sign}(\mathbf{E}_k^j(\tau)\mathbf{E}_k^j(\tau+1))\right), \quad 1 \le j \le dM' . \tag{12}$$

One can thus assign a frequency that is in one-to-one correspondence to δ_j. In Eq. (12), $\mathbf{E}_k^j$ denotes the kth spatial component of the DAHM, $\mathbf{E}^j$. One can then rank the DAHMs from the lowest to the highest frequency by simply looking at their number of sign changes. As shown in Chekroun and Kondrashov (2017), the corresponding fraction of the energy they capture is given by $|\lambda_j|$, up to a scaling factor.

By analogy with M-SSA (Ghil et al., 2002), the multivariate dataset $\mathbf{X}$ can be projected onto the orthogonal set formed by the $\mathbf{E}^j$'s, to obtain the DAH expansion coefficients (DAHCs):

$$A_j(t) = \sum_{\tau=1}^{M'} \sum_{k=1}^{d} X_k(t+\tau-1)\mathbf{E}_k^j(\tau), \tag{13}$$

where t varies from 1 to $N' = N - M' + 1$.

Although the DAHCs are not formally orthogonal in time, they also exhibit a phase–quadrature relationship that depends on whether the window M is sufficiently large to resolve the decay of temporal correlations of a given dataset. Typically, the larger M (subject to the length of the record), the more apparent is the phase quadrature between a pair of DAHCs associated with the same frequency.

Furthermore, any subset $\mathbf{B} \subset \mathbf{A}$ of DAHCs, as well as the full set $\mathbf{A}$, can be convolved with associated $\mathbf{E}_j$'s, for partial or full reconstruction of the original data, respectively. The transformation between $\mathbf{X}$ and $\mathbf{A}$ is unitary, i.e., there is no loss of variance. Thus, the jth RC at time t for channel k is given by:

$$R_k^j(t) = \frac{1}{M_t} \sum_{\tau=L_t}^{U_t} A_j(t-\tau+1)\mathbf{E}_k^j(\tau). \tag{14}$$

The normalization factor M_t equals M', except near the ends of the time series (Ghil et al., 2002), and the sum of all the RCs recovers the original time series.

It is also useful to consider harmonic reconstruction components (HRCs), namely a sum of d RC pairs corresponding to a particular frequency $\omega_\ell \neq 0$:

$$R_k^{\omega_\ell}(t) = \sum_{j \in \mathscr{J}_\ell} R_k^j(t), \tag{15}$$

where $\mathscr{J}_\ell$ denotes the set of all the indices j associated with the frequency ω_ℓ. By construction, for each nonzero frequency, this set is constituted by $2d$ elements.

Appendix 2: Details on the MSLM Modeling

As discussed in Sect. 4, the DAHMs extract harmonic components of variability that allow for a reduction of the data-driven modeling effort to a simple class of elemental multilayer stochastic models [MSMs: Kondrashov et al. (2015)]; these MSMs are stacked by frequency and only coupled at different frequencies by the same noise realization.

In the simplest case of one layer for the modeled noise, this construction leads to stochastic models of the form:

$$
\begin{aligned}
\dot{x}_j &= \beta_j(f)x_j - \alpha_j(f)y_j + \sigma_j(f)x_j(x_j^2 + y_j^2) + \sum_{i \neq j}^{d} b_{ij}^x(f)x_i + \sum_{i \neq j}^{d} a_{ij}^x(f)y_i + \varepsilon_j^x, \\
\dot{y}_j &= \alpha_j(f)x_j + \beta_j(f)y_j + \sigma_j(f)y_j(x_j^2 + y_j^2) + \sum_{i \neq j}^{d} a_{ij}^y(f)x_i + \sum_{i \neq j}^{d} b_{ij}^y(f)y_i + \varepsilon_j^y, \\
\dot{\varepsilon}_j^x &= L_{11}^j(f)x_j + L_{12}^j(f)y_j + M_{11}^j(f)\varepsilon_j^x + M_{12}^j(f)\varepsilon_j^y + Q_{11}^j(f)\dot{W}_1^j \\
&\quad + Q_{12}^j(f)\dot{W}_2^j + \sum_{i \neq j}^{d}\sum_{k=1}^{2} Q_{1k}^i(f)\dot{W}_k^i, \\
\dot{\varepsilon}_j^y &= L_{21}^j(f)x_j + L_{22}^j(f)y_j + M_{21}^j(f)\varepsilon_j^x + M_{22}^j(f)\varepsilon_j^y + Q_{21}^j(f)\dot{W}_1^j \\
&\quad + Q_{22}^j(f)\dot{W}_2^j + \sum_{i \neq j}^{d}\sum_{k=1}^{2} Q_{2k}^i(f)\dot{W}_k^i.
\end{aligned}
\tag{MSLM}
$$

In (MSLM), the index j varies in the set of indices $\mathcal{J}_f$ associated with a single frequency f, determined by the zero-crossings of the corresponding $\mathbf{E}^j$'s. When $f \neq 0$, this set consists of d elements. In practice $f = \omega_\ell/(2\pi)$ is determined by a Fourier frequency ω_ℓ given in Eq. (11). The W_k^j's with k in $\{1, 2\}$ and j in $\{1, \ldots, d\}$ form $2d$ independent Brownian motions.

We call these models *multilayer stochastic Stuart-Landau models* (MSLM). At a given frequency f, the d pairs are linearly coupled as indicated by the terms in the sums apparent in the x_j- and y_j-equations. In (MSLM) and for a given pair indexed by j, the noise term $(\varepsilon_j^x, \varepsilon_j^y)$ is modeled by means of linear dependencies involving only $(\varepsilon_j^x, \varepsilon_j^y)$, on the one hand, and the jth pair (x_j, y_j), on the other.

Obviously, for a given pair, and following Kondrashov et al. (2015), more layers can be added as needed to (MSLM), when the noise term $(\varepsilon_j^x, \varepsilon_j^y)$ at the first level is not white. In this case, the extra layers will depend linearly on the jth pair (x_j, y_j), and on the noise residuals from the previous layers. The sums in the ε_j^x- and ε_j^y-equations take into account "spatial" correlations between the pairs, at the level of the noise. Note that for the null frequency, $f \equiv 0$, there are exactly d modes that are not paired, and they are modeled by a linear multilayer stochastic model as in Kondrashov et al. (2015).

Note that Eq. (MSLM) can be generalized further by allowing coupling of (x_j, y_j) pairs at neighboring frequencies, which can be useful for certain applications where cross-frequency interactions are important. Equations (MSLM) are discretized in time and integrated numerically forward from initial conditions that respect the initialization procedure described in Kondrashov et al. (2015, Appendix B).

References

Broomhead, D.S., and G.P. King. 1986. Extracting qualitative dynamics from experimental data. *Physica D: Nonlinear Phenomena* 20(2): 217–236.

Cavalieri, D., C. Parkinson, P. Gloersen, and H.J. Zwally. 1996. Updated Yearly Sea Ice Concentrations from Nimbus-7 SMMR and DMSP SSM/I-SSMIS Passive Microwave Data, 1979–2010, Digital media, National Snow and Ice Data Center, Boulder, CO.

Chekroun, M.D., and D. Kondrashov. 2017. Data-adaptive harmonic spectra and multilayer Stuart-Landau models. HAL preprint, hal-01537797.

Chekroun, M.D., E. Simonnet, and M. Ghil. 2011. Stochastic climate dynamics: Random attractors and time-dependent invariant measures. *Physica D* 240: 1685–1700.

Comiso, J.C. 2014. Bootstrap Sea Ice Concentrations from Nimbus-7 SMMR and DMSP SSM/I-SSMIS. Version 2 [Northern Hemisphere daily data]. Digital media, NASA National Snow and Ice Data Center, Distributed Active Archive Center, Boulder, CO.

Elsner, J.B., and A.A. Tsonis. 1996. *Singular spectrum analysis: a new tool in time series analysis*. Berlin: Springer Science & Business Media.

Fetterer, F., M. Savoie, S. Helfrich, and P. Clemente-Colón. 2010. *Multisensor analyzed sea ice extent - Northern Hemisphere*. Digital media. Boulder, CO: National Snow and Ice Data Center.

Ghil, M., M.R. Allen, M.D. Dettinger, K. Ide, D. Kondrashov, M.E. Mann, A.W. Robertson, A. Saunders, Y. Tian, F. Varadi, and P. Yiou. 2002. Advanced spectral methods for climatic time series. *Reviews of Geophysics* 40: 3-1–3-41.

Giannakis, D., and A.J. Majda. 2012. Nonlinear Laplacian spectral analysis for time series with intermittency and low-frequency variability. *Proceedings of the National Academy of Sciences of the United States of America* 109(7): 2222–2227.

Groth, A., and M. Ghil. 2011. Multivariate singular spectrum analysis and the road to phase synchronization. *Physical Review E* 84: 036206.

Hamilton, L.C., and J. Stroeve. 2016. 400 predictions: the SEARCH Sea Ice Outlook 2008–2015. *Polar Geography* 39(4): 274–287.

Hartman, P. 1986. *Ordinary differential equations*, 2nd ed. Classics in Applied Mathematics, vol. 38. Philadelphia: SIAM.

Kondrashov, D., M.D. Chekroun, and M. Ghil. 2015. Data-driven non-Markovian closure models. *Physica D* 297: 33–55.

Kravtsov, S., D. Kondrashov, and M. Ghil. 2005. Multi-level regression modeling of nonlinear processes: Derivation and applications to climatic variability. *Journal of Climate* 18(21): 4404–4424.

Kravtsov, S., D. Kondrashov, and M. Ghil. 2009. Empirical model reduction and the modeling hierarchy in climate dynamics and the geosciences. In *Stochastic physics and climate modeling*, ed. Palmer, T.N., and P. Williams, 35–72. Cambridge: Cambridge University Press.

Marple, S.L. 1987. *Digital spectral analysis with applications*. Englewood Cliffs, NJ: Prentice-Hall.

Pisarenko, V.F. 1973. The retrieval of harmonics from a covariance function. *Geophysical Journal International* 33(3): 347–366.
Preisendorfer, R.W. 1988. *Principal component analysis in meteorology and oceanography*, 425 pp. New York: Elsevier.
Selivanov, A.A., J. Lehnert, T. Dahms, P. Hövel, A.L. Fradkov, and E. Schöll. 2012. Adaptive synchronization in delay-coupled networks of Stuart-Landau oscillators. *Physical Review E* 85: 016201.
Sigmond, M., M.C. Reader, G.M. Flato, W.J. Merryfield, and A. Tivy. 2016. Skillful seasonal forecasts of Arctic sea ice retreat and advance dates in a dynamical forecast system. *Geophysical Research Letters* 43(24): 12457–12465.
Stroeve, J., E. Blanchard-Wrigglesworth, V. Guemas, S. Howell, F. Massonnet, and S. Tietsche. 2015. Improving predictions of Arctic sea ice extent. *Eos, Transactions of the American Geophysical Union*, 96. doi:10.1029/2015EO031431. https://eos.org/features/improving-predictions-of-arctic-sea-ice-extent.
Vautard, R., and M. Ghil. 1989. Singular spectrum analysis in nonlinear dynamics, with applications to paleoclimatic time series. *Physica D: Nonlinear Phenomena* 35(3): 395–424.
Walsh, J., and W. Chapman. 2015. Variability of sea ice extent over decadal and longer timescales. In *Climate change: multidecadal and beyond*, ed. Chang, C.P., M. Ghil, M. Latif, and J.M. Wallace, 203–217. Singapore/London: World Scientific/Imperial College Press.
Zakharova, A., S. Loos, J. Siebert, A. Gjurchinovski, J.C. Claussen, and E. Schöll. 2016. Controlling chimera patterns in networks: Interplay of structure, noise, and delay in control of self-organizing nonlinear systems. In *Control of self-organizing nonlinear systems*, ed. Hövel, P., E. Schöll, and S.H.L. Klapp. Berlin: Springer.

Pisarenko, V.F. 1973. The retrieval of harmonics from a covariance function. *Geophysical Journal International* 33 (3): 347–366.
Preisendorfer, R.W. 1988. *Principal component analysis in meteorology and oceanography*. 425 pp. New York: Elsevier.
[illegible]
[illegible]
[illegible]
[illegible]
Vautard, R., and M. Ghil. 1989. Singular spectrum analysis in nonlinear dynamics, [illegible]
[illegible]
[illegible]

Cautionary Remarks on the Auto-Correlation Analysis of Self-Similar Time Series

Sung Yong Kim

Abstract As the time-domain analysis of non-linear time series in geosciences, the auto-correlations of the self-similar time series are examined to identify spurious decorrelation structures in terms of the number of independent pulses and the shape of decay patterns. The self-similar time series is defined as a continuous time series having similar shapes of disturbance or amplitudes of which statistics is non-Gaussian, such as records of river flows, rainfall, wind speed, concentration of Chlorophyll, and inertial amplitudes in geosciences. In this chapter, the auto-correlations of the modeled self-similar time series are evaluated and the relevant cautionary remarks are discussed.

Keywords Self-similar time series • Correlation • Decorrelation scale • Non-gaussian data

1 Introduction

Signals and noise in nature are considered as Gaussian random processes, which are typically represented by their mean and standard deviation. In particular, a covariance function (called as "kernels") has a primary role to determine the shape of prior and posterior of Gaussian random processes (e.g., Rasmussen and Williams 2006). Moreover, a correlation function is the covariance function normalized by standard deviation and maintains the de-correlated structure and characteristics in the domain where the process is defined. In the time-domain analysis, auto-correlations of time series under Gaussian statistics have been used to identify the decorrelation time scale. Similarly, the auto-correlations of the spatial data quantify the de-correlated structures (e.g., Bendat and Piersol 2000; Priestley 1981). For instance, the cross-correlation of the ambient noise field received from two

S.Y. Kim (✉)
Environmental Fluid Mechanics Laboratory, Department of Mechanical Engineering, Korea Advanced Institute of Science and Technology, 291 Daehak-ro, Yuseong-gu, Daejeon 34141, Republic of Korea
e-mail: syongkim@kaist.ac.kr

A.A. Tsonis (ed.), *Advances in Nonlinear Geosciences*,
DOI 10.1007/978-3-319-58895-7_11

points can recover the time-domain Green's function (or empirical noise function), between two locations (e.g., Courtland 2008; Roux et al. 2005).

However, since non-Gaussian variables can be described with chi-squared or Poisson statistics, the classic correlation analysis may not be applicable to data under non-Gaussian statistics. As an example of non-Gaussian variables, a self-similar time series is considered as a continuous time series having similar shapes of disturbance or amplitudes of which statistics is non-Gaussian, such as records of river flows, rainfall, wind speed, concentration of Chlorophyll, and inertial amplitudes in geosciences. In characterizing the non-Gaussian variable, a careful treatment may be required, which is different from typical data analysis techniques used in the data under Gaussian statistics. For instance, Park et al. (2009) estimated the decorrelation time scales of the pure-inertial amplitude using an auto-correlation analysis of their time series as a self-similar time series (e.g., a series of saw-teeth). They concluded that the estimate of the decorrelation scales depends on the decay pattern of individual pulses (e.g., exponential or Gaussian shape) and the decorrelation scale of a self-similar time series decaying with a Gaussian shape is equal to $\sqrt{2}$ times of the actual length scale [see Appendix of Park et al. (2009) for more details].

In this chapter, experiments to quantify the decorrelation scales of a self-similar series are conducted with an evaluation of whether the decorrelation scales depend on the decay patterns and the number of individual pulses. Some portion of descriptions is excerpted from Kim et al. (2014).

2 Data Analysis

Three sets of synthetic self-similar time series [$d(t)$] having multiple double-sided pulses that decay in the exponential, Gaussian, and linear manner, respectively [Eqs. (1)–(4)] are generated. Primary parameters [e.g., amplitudes (a_n), decorrelation time scales (λ_n), and timings of individual pulses (t_n)] in each time series are chosen as random variables:

$$d^e(t) = \sum_{n=1}^{N} a_n^e b_n^e(t) = \sum_{n=1}^{N} a_n^e \exp\left[-\frac{|t-t_n|}{\lambda_n}\right] , \tag{1}$$

$$d^g(t) = \sum_{n=1}^{N} a_n^g b_n^g(t) = \sum_{n=1}^{N} a_n^g \exp\left[-\frac{(t-t_n)^2}{\lambda_n^2}\right] , \tag{2}$$

$$d^l(t) = \sum_{n=1}^{N} a_n^l b_n^l(t) = \sum_{n=1}^{N} a_n^l \left(\frac{t-t_n}{t_n-\beta} + 1\right) , \tag{3}$$

where a_n, λ_n, and β denote the independent amplitude at each t_n. the decorrelation scale, and the slope parameter of the linear, respectively. Superscripts of e, g, and l indicate the exponential, Gaussian, and linear functions, respectively.

Moreover, the self-similar time series having multiple single-sided pulses (positive only) can be simulated with a constraint of

$$d_n(t) = \begin{cases} d_n(t), & \text{if } t \geq t_n \\ 0, & \text{if } t < t_n \end{cases}. \tag{4}$$

The auto-correlations (ρ) of self-similar time series are evaluated with

$$\rho(\Delta t) = \frac{\langle d(t+\Delta t) d(t)^{\dagger} \rangle}{\sqrt{\langle d(t+\Delta t)^2 \rangle} \sqrt{\langle d(t)^2 \rangle}}, \tag{5}$$

where Δt denotes the time lag.

The auto-correlations of three sets of time series show the exponentially decaying shape regardless of the decaying pattern of the original time series (Fig. 1). Moreover, the number of pulses, the resolution of the time axis, decorrelation time scales, timings of pulses, and amplitudes did not change the shape of auto-correlation. The decorrelation time scale obtained from the auto-correlation does not have any relevance with decay scales of individual pulses. Thus, the decay scale of the inertial amplitudes is nothing to do with $\sqrt{2}$ times of the actual length scale as described in Park et al. (2009). In addition, the auto-correlation of the double-sided pulses shows a Gaussian shape regardless of the decay pattern of the time series (Fig. 2).

3 Conclusion

The self-similar time series requires careful analysis because its auto-correlations may have spuriously de-correlated structures as they are independent of the decay patterns (e.g., exponential, Gaussian, and linear), the number of pulses, the types of pulses (e.g., single- and double-sided pluses), and the number of realizations. Thus, it would be more appropriate that a set of the self-similar time series is considered individually or averaged, then, is approximated with any functions with a decay pattern. The individuals or the composite mean of self-similar time series be fitted with any functions having a decay pattern in order to quantify the decorrelation (time) scales is suggested. If other dynamic variables (e.g., wind stress) are available, the statistically computed response function can provide the decorrelation scales as presented in Kim and Kosro (2013).

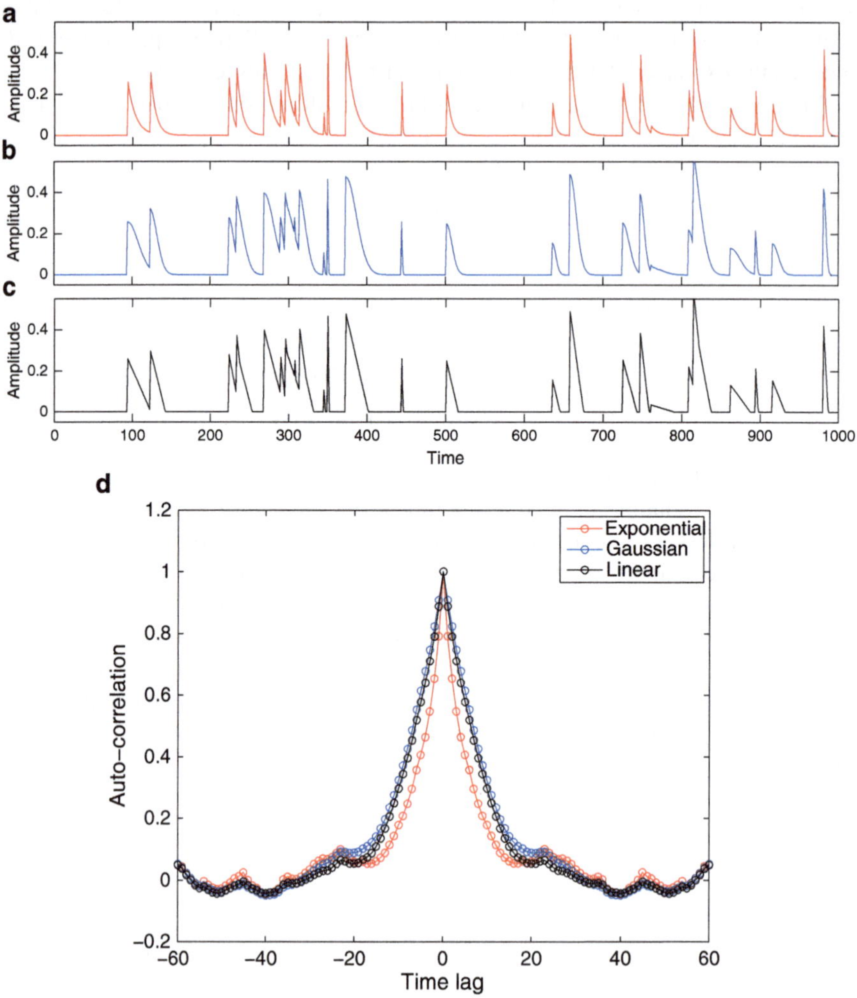

Fig. 1 (**a**)–(**c**): Three sets of self-similar time series with multiple single-sided pulses which decay with the exponential, Gaussian, and linear shapes. The amplitude, decay coefficients, and timings are chosen randomly [Eqs. (1)–(3)]. (**d**) Auto-correlations of self-similar time series given in (**a**)–(**c**)

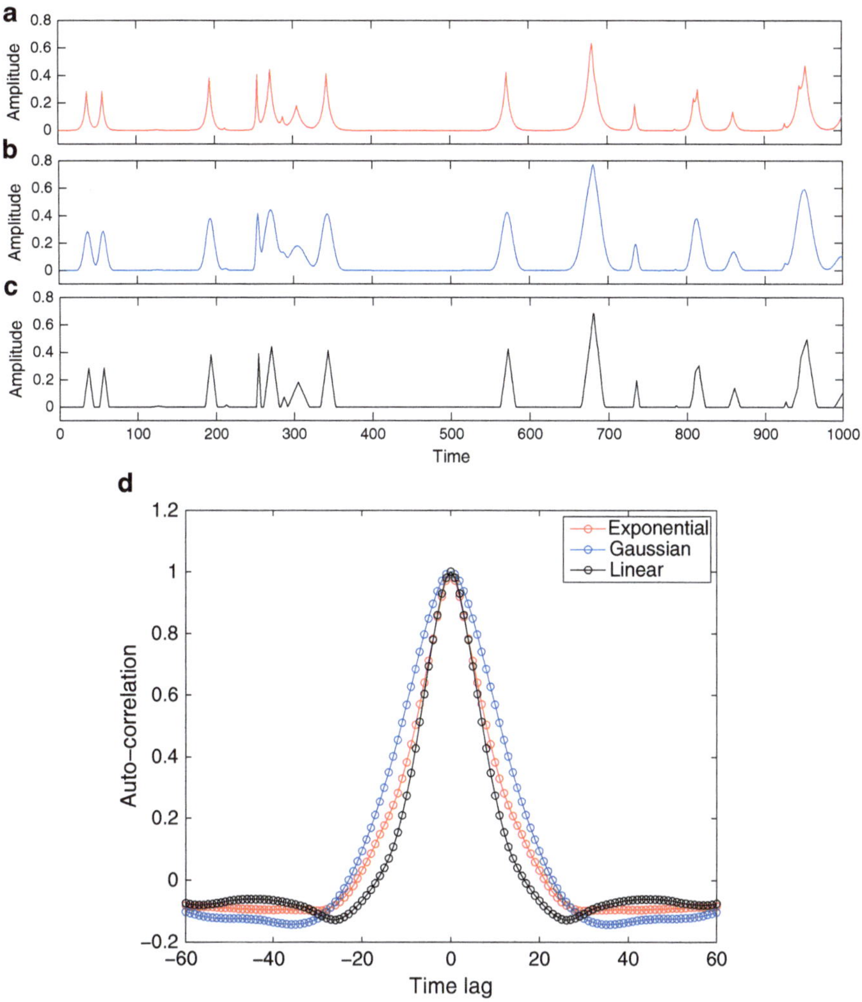

Fig. 2 (**a**)–(**c**): Three sets of self-similar time series with multiple double-sided pulses which decay with exponential, Gaussian, and linear shapes, respectively. The amplitude, decay coefficients, and timings are chosen as randomly [Eqs. (1)–(3)]. (**d**) Auto-correlations of self-similar time series given in (**a**)–(**c**)

Acknowledgements This work is supported by a grant through the Disaster and Safety Management Institute, Ministry of Public Safety and Security (MPSS-CG-2016-05), Republic of Korea as an excerpt of Kim et al. (2014).

References

Bendat, J., and A.G. Piersol. 2000. *Random data analysis and measurement procedures*, 3rd ed., 594 pp. New York: Wiley.
Courtland, R. 2008. Earth science: harnessing the hum. *Nature* 453(7192): 146–148.
Kim, S.Y., and P.M. Kosro. 2013. Observations of near-inertial surface currents off Oregon: decorrelation time and length scales. *Journal of Geophysical Research* 118(7): 3723–3736. doi:10.1002/jgrc.20235.
Kim, S.Y., P.M. Kosro, and A.L. Kurapov. 2014. Evaluation of directly wind-coherent near-inertial surface currents off Oregon using a statistical parameterization and analytical and numerical models. *Journal of Geophysical Research* 119(10): 6631–6654. doi:10.1002/2014JC010115.
Park, J., K. Kim, and R. Schmitt. 2009. Global distribution of the decay timescale of mixed layer inertial motions observed by satellite-tracked drifters. *Journal of Geophysical Research* 114(C11): C11010. doi:10.1029/2008JC005216.
Priestley, M.B. 1981. *Spectral analysis and time series*, 890 pp. London: Academic.
Rasmussen, C., and C. Williams. 2006. *Gaussian processes for machine learning*. Cambridge: MIT.
Roux, P., K. Sabra, W. Kuperman, and A. Roux. 2005. Ambient noise cross correlation in free space: theoretical approach. *The Journal of the Acoustical Society of America* 117: 79–84.

Emergence of Coherent Clusters in the Ocean

A.D. Kirwan Jr., H.S. Huntley, and H. Chang

Abstract Why does material tend to congregate in long coherent clusters at the surface of the ocean when it is well known that the ocean is dispersive? Here we review some recent research that addresses this question. A standard diagnostic for discerning transport pathways in incompressible 2D flows is the finite time Lyapunov exponent (FTLE). The FTLE can be expressed as the average of two rarely evaluated Lagrangian objects: the dilation and stretch rates. The stretch rate accounts for the ability of fluid shear to change the shape of fluid blobs, and for incompressible fluids it is the FTLE. However, in the real ocean and especially at submesoscales, the horizontal divergence is not negligible. This is quantified by the dilation rate, which is identically zero in 2D incompressible flow. Our analysis demonstrates that the combination of fluid dilation and stretch enhances accumulation of buoyant material along thin clusters in an otherwise dispersing ocean.

Keywords Lyapunov exponents • Clustering • Dilation • Stretch • Singular value decomposition • Deformation • Mixing • Transport boundaries

1 Introduction

Dispersive mechanisms operating in the world's oceans span six or seven orders of magnitude in space and time scales. For mesoscale and large-scale phenomena these range from the Rossby deformation radius to basin scales and weeks to years, respectively. Submesoscale processes operate in the range of meters to a few kilometers and hours to a few days. The low-end extreme is given by turbulent processes that act on centimeter and minute scales. Two noteworthy examples of dispersion at the largest scales are cargo falling off freighters and being transported across ocean basins (Ebbesmeyer and Ingraham, 1992, 1994) and radioactive isotopes released from the 2011 Fukushima Daiichi nuclear reactor accident in

A.D. Kirwan Jr. (✉) • H.S. Huntley • H. Chang
School of Marine Science and Policy, University of Delaware, Robinson Hall, Newark, DE 19716, USA
e-mail: adk@udel.edu; helgah@udel.edu; changh@udel.edu

A.A. Tsonis (ed.), *Advances in Nonlinear Geosciences*,
DOI 10.1007/978-3-319-58895-7_12

Japan making their way to the North American continental shelf (Smith et al., 2015). On smaller scales, dye experiments and drifter releases in various parts of the world have shown dispersive characteristics (Okubo, 1971; Poje et al., 2014, 2016).

These and many other observations demonstrate that the ocean is, on average, dispersive at virtually all scales. In view of this, it seems somewhat paradoxical that material also is observed to form coherent clusters in certain regions. The two phenomena are clearly related, yet call for separate diagnostics. Here we review some recent results relevant to the distinction between dispersion and clustering patterns.

The discussion is organized as follows. Section 2 reviews some examples of coherent clusters in geophysical fluid settings. Section 3 summarizes recent theoretical results that bear on the issue of formation of coherent clusters in otherwise dispersive fluids. Section 4 reviews recent numerical experiments exhibiting both dispersion and clustering of material at the ocean surface. Section 5 summarizes the principal findings and offers some speculations about future research.

2 Examples of the Emergence of Coherent Clusters

One of the earliest examples of the formation of coherent clusters in fluid flow was provided by Aref (1984). He considered the 2D flow of an incompressible and barotropic fluid in a cylinder with stirring provided by two vortices. Steady rotation of the vortices produces a simple stream function (Fig. 1, reproduced from his paper). However, when the vortices oscillate in their positions, the flow becomes chaotic. This was illustrated by following the evolution of a rectangular blob of particles, initially located approximately midway and slightly off the axis connecting the two stirrers, as shown in panel Fig. 1b. Instead of dispersing isotropically, the blob evolves into intricate thin structures before reaching the final state of near uniform distribution.

What about natural flows? Images of a variety of floating substances on the ocean surface show strikingly similar patterns of concentrated clustering. Satellite images of chlorophyll have documented intricate interwoven lines reminiscent of Aref's simple mixing model (Fig. 2a). Distinctive lines of sargassum have also been observed from ships; Fig. 2b shows drifting buoys having congregated in the same convergence zone as the seaweed.

Clustering was also observed in the aftermath of the Deepwater Horizon oil spill in the northeastern Gulf of Mexico in 2010. The oil slick displayed sharp fronts and lengthy striations, both on small scales observable from ships (Fig. 3a, b) and on large scales as seen from satellites (Fig. 3c, d). Figure 3c, d was taken just over 2 weeks apart. Note the huge increase in the elongation of the cluster over this relatively short time.

A striking aspect of these images is the vast range of spatial scales over which string-like clusters of material are observed. Such accumulations are well documented in virtually all parts of the world ocean. Figure 4a shows a concentration

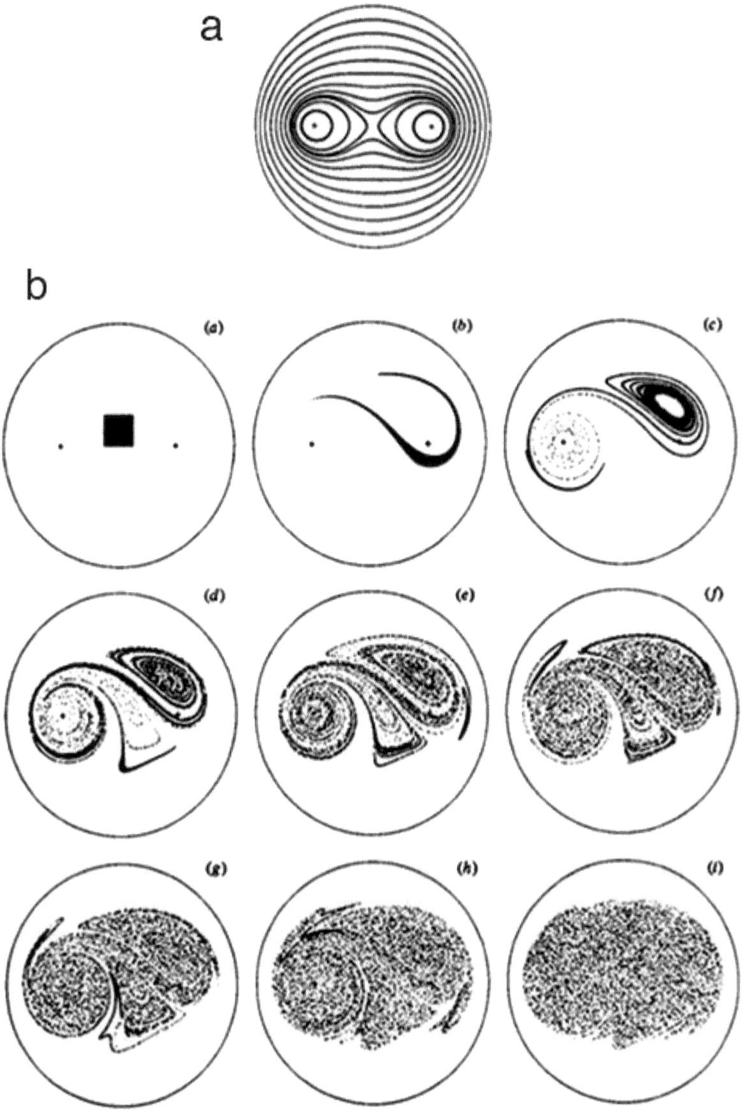

Fig. 1 (**a**) Streamlines when the fluid in a cylinder is stirred by two fixed vortices located at the crosses rotating at a constant rate. (**b**) The evolution of a grid of particles when stirred by vortices oscillating with a period of 1 and amplitude 0.5 [see Aref (1984) for more details]. Snapshots are taken at times 1, 2, 3, 4, 5, 6, 9, and 12. [Reproduced from Aref (1984)]

of debris floating in the Pacific Ocean originating from the tsunami that hit the east coast of Japan in 2011. Figure 4b illustrates that such clusters are even observed in the Arctic Ocean, in this case made up of ice floes.

(a) (b)

Fig. 2 (**a**) Chlorophyll in the western Gulf of Mexico, as observed by MERIS. (Image courtesy of ESA.) (**b**) Surface drifters and sargassum clustering along a front in the northeastern Gulf of Mexico. (Image courtesy of Tamay Özgöomen)

3 Theory

The flow fields producing the structures shown in Figs. 2, 3, and 4 differ in important ways from those responsible for Fig. 1. The ocean flows are not necessarily 2D incompressible. Moreover, they are governed by geophysical fluid dynamics and so include rotation and stratification along with turbulent dispersive processes. The theory we review here sheds some light on how to quantify the effects of the 2D compressibility and separate them from the deformation that is common to both types of flows.

Although our analysis is restricted to 2D velocity fields, this is not unduly restrictive as the clustered material depicted in Figs. 2, 3, and 4 is buoyant and thus confined to the ocean surface. Even though horizontal divergence there is generally not zero, the vertical velocities typically are negligible relative to horizontal velocities.

Any material blob in a 2D compressible flow is subject to two types of deformations, those that change its area—which we will call *dilation*—and those that change shape while preserving area—which we will refer to as *stretch*. Nonlinear combinations of these mechanisms can result in complex shapes. The cartoon in Fig. 5 illustrates these concepts.

The first order approximation to the map taking a blob in its initial state to a later state is given by the deformation tensor:

$$\mathbf{F} = \frac{\partial \mathbf{x}}{\partial \mathbf{x}_0} = \begin{bmatrix} \frac{\partial x}{\partial x_0} & \frac{\partial x}{\partial y_0} \\ \frac{\partial y}{\partial x_0} & \frac{\partial y}{\partial y_0} \end{bmatrix}. \quad (1)$$

Here $\mathbf{x}$ is the current position of a particle as a function of its initial position $\mathbf{x}_0$ and time t.

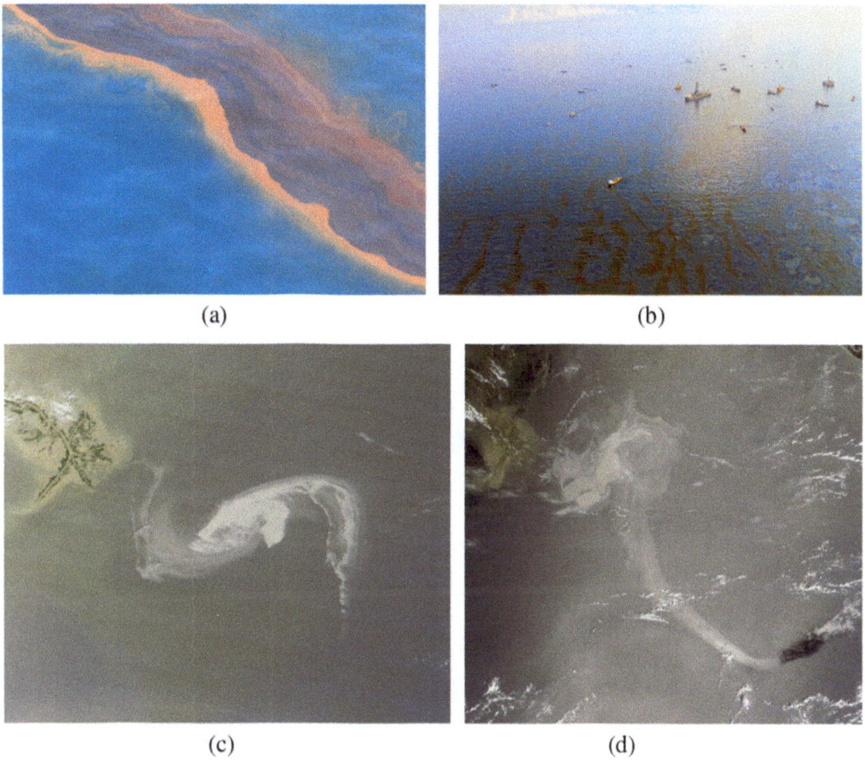

Fig. 3 Aerial (*top*) and satellite (*bottom*) images of the 2010 Deepwater Horizon oil spill in the Gulf of Mexico. (**a**) A sharp front has developed in the oil slick, 12 May 2010. (Image courtesy of NOAA.) (**b**) Striations are visible with oil sheen near the clean up activities, 1 June 2010. (Image courtesy of Green Fire Productions.) (**c**) MODIS image of the oil slick on 29 April 2010. (Image courtesy of NASA.) (**d**) MODIS image of the oil slick on 17 May 2010. (Image courtesy of NASA)

The deformation tensor quantifies both stretch and dilation through its singular value decomposition (SVD):

$$\mathbf{F} = \mathbf{U}\Sigma\mathbf{V}^* = \mathbf{U}\begin{bmatrix}\mu_1 & 0\\ 0 & \mu_2\end{bmatrix}\mathbf{V}^*. \tag{2}$$

The singular values of $\mathbf{F}$ are $\mu_1 \geq \mu_2$; $\mathbf{U}$ and $\mathbf{V}$ are real unitary matrices. The * superscript is the transpose operator. In theoretical mechanics, $\mathbf{F}$ is often decomposed into a right stretch tensor $\mathbf{R}$ or a left stretch tensor $\mathbf{L}$. Their relationship to the SVD is as follows:

$$\begin{aligned}\mathbf{RR} &= \mathbf{V}\Sigma^2\mathbf{V}^*\\ \mathbf{LL} &= \mathbf{U}\Sigma^2\mathbf{U}^*.\end{aligned} \tag{3}$$

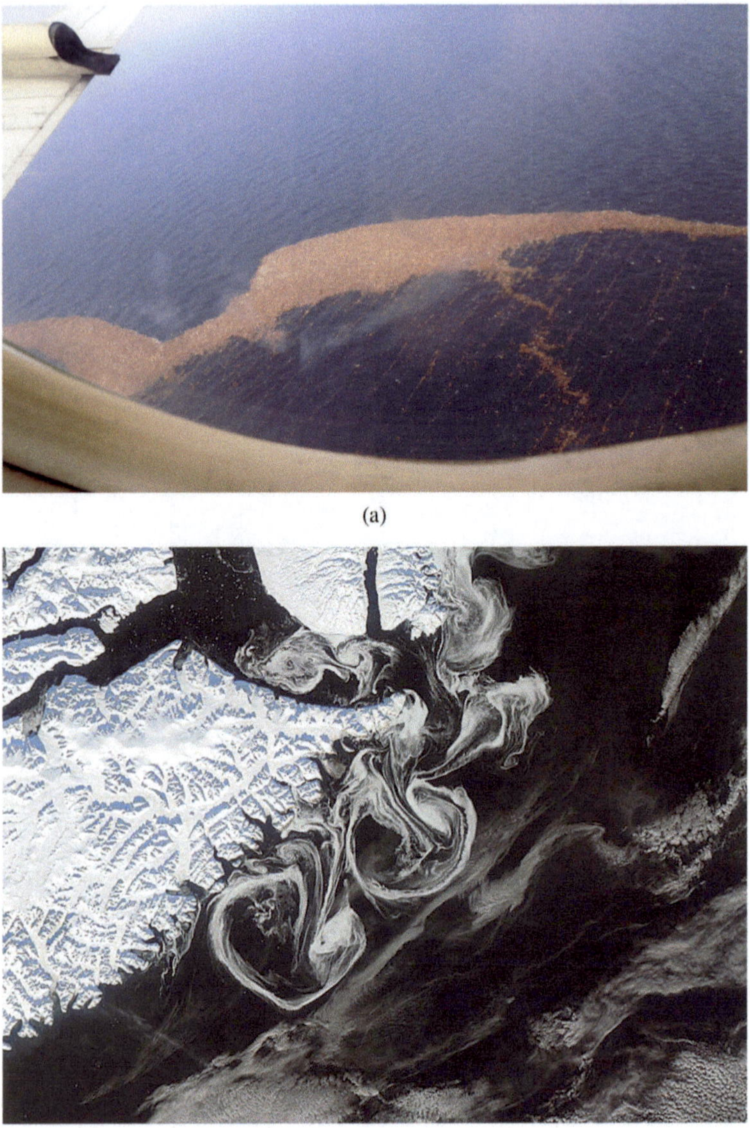

(a)

(b)

Fig. 4 (**a**) Debris off the east coast of Japan, a few days after the nation was struck by a tsunami. (Image courtesy of the US Navy.) (**b**) MODIS image of sea ice off eastern Greenland, 16 October 2012. (Image courtesy of NASA)

As shown by Huntley et al. (2015), the μ_i have useful physical interpretations. Their product $\mu_1\mu_2$ characterizes the change in area of an incremental blob, while the ratio $\mu_1\mu_2^{-1}$ characterizes the blob's stretch or change in aspect ratio.

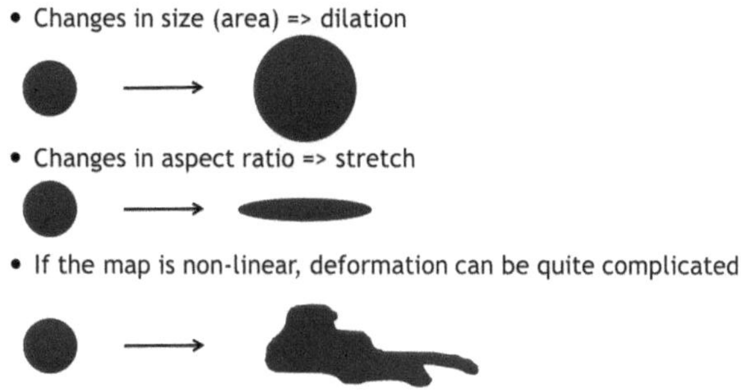

Fig. 5 Cartoon depicting types of deformation for a blob of fluid, including dilation and stretch effects

For integration time T, the rates of dilation and stretch are given by

$$\Delta = \frac{\log(\mu_1\mu_2)}{T}$$

$$\Sigma = \frac{\log(\mu_1\mu_2^{-1})}{T}. \tag{4}$$

Here T is the integration time. These can be nicely related to the finite time Lyapunov exponent (FTLE), which is the classic tool for identifying coherent Lagrangian structures. The FTLE is defined as $\Lambda = \log(\mu_1)/T$. From (4) it follows that this is simply the average of Δ and Σ:

$$\Lambda = \frac{\log \mu_1}{T} = \frac{\Delta + \Sigma}{2}. \tag{5}$$

In the dynamical systems literature, ridges in the FTLE field evaluated over the interval $[t_0, t_0 + T]$ are associated with stable or inflowing manifolds, while those in the FTLE field evaluated over the interval $[t_0 - T, t_0]$ are associated with unstable or outflowing manifolds.

Figure 6 is an example of the application of traditional FTLE methodology to a current field from the Gulf of Mexico. In this figure the red and blue curves are, respectively, the backward and forward in time Λ. Each is a transport boundary over the time period it was calculated. Their intersection is at a critical trajectory. See Kirwan (2006) for a discussion of the connection between FTLE intersections and critical trajectories. Panel (a), on 11 October, shows the intersection of two strong inflowing and outflowing Λ ridges near $24N$, $93.5W$. Also seen in the figure are several circular blobs: three green, one yellow, one black, and one orange. Over the next 20 days the yellow and orange blobs flow along the inflowing manifold and collapse onto the outflowing manifold. The black blob, initially centered on the intersection, simply collapses along the outflowing curve. In contrast, the green blobs, initially located at the centers of eddies, are merely distorted by the shears within the eddies. A purple blob is initialized as a string-like feature along the inflowing manifold. It collapses onto the manifold intersection. This example shows the potential power of FTLE to identify transport boundaries.

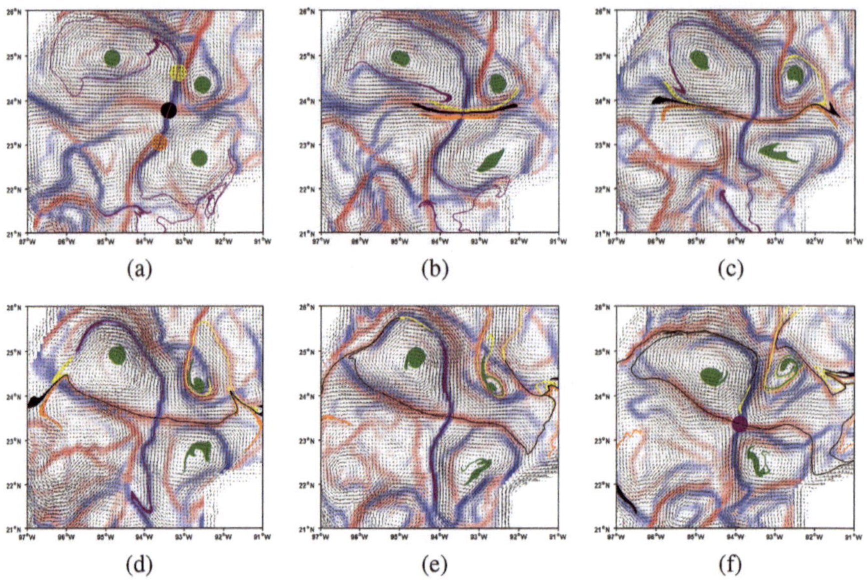

Fig. 6 Example of the inflowing (*blue*) and outflowing (*red*) FTLE ridges from a simulation of currents in the Gulf of Mexico at 50 m depth. Strategically placed blobs illustrate the impact of these structures on nearby material. *Green blobs* are initialized inside eddies. The *black blob* is positioned over the intersection of the inflowing and outflowing manifolds at the initial time, whereas the *purple blob* ends up centered at the intersection. *Yellow and orange blobs* begin on opposite sides of the outflowing manifold, centered on the inflowing one. (**a**) 11 October 1998. (**b**) 15 October 1998. (**c**) 19 October 1998. (**d**) 23 October 1998. (**e**) 27 October 1998. (**f**) 31 October 1998

4 Results

Not all transport boundaries are associated with clustering behavior, however. To tease out what relative separation is due to dilation—and hence indicative of clustering or dispersion—and what is due to stretch—and hence not related to clustering phenomena, it is necessary to split the FTLE into its components of dilation and stretch rates. To address this issue we now present some results from the application of the theory outlined in Sect. 3 to flows in the Gulf of Mexico. Further details of related applications are given in Huntley et al. (2015) and Jacobs et al. (2016).

How important is the horizontal divergence at the surface of the ocean in the formation of clusters? Huntley et al. (2015) considered the full model velocity field and its geostrophic approximation from a fully non-linear data-assimilating high resolution hydrostatic model hindcast near a Loop Current Ring in the Gulf of Mexico. The two velocity fields for this flow were nearly indistinguishable, yet the resulting particle distributions differed wildly. Here we repeat this experiment with a flow in the northeastern Gulf of Mexico that has a somewhat greater non-geostrophic contribution. Figure 7 compares the congregation of surface material from the full and divergence-free geostrophic velocity fields. The top left panel (a) shows the sea surface height anomaly field and the model hindcast of the velocity

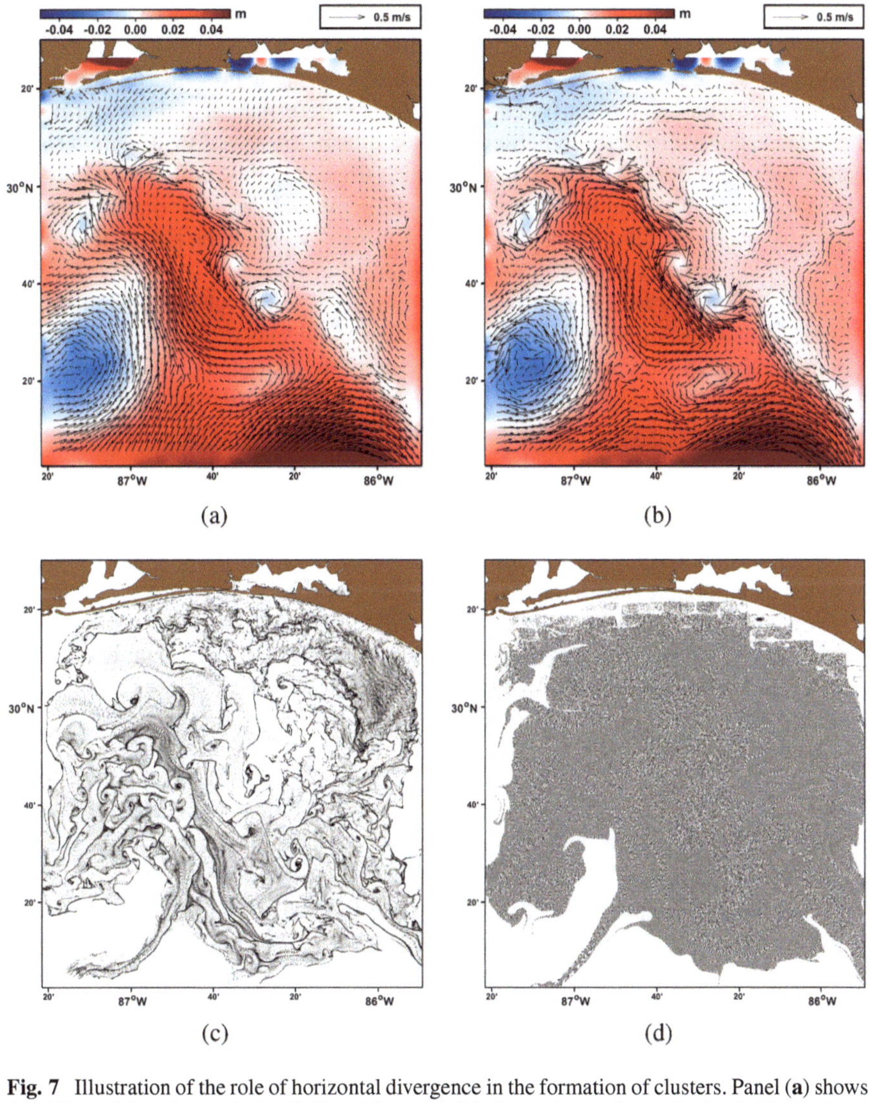

Fig. 7 Illustration of the role of horizontal divergence in the formation of clusters. Panel (**a**) shows the full model surface velocity field, while panel (**b**) shows the surface velocity field calculated from the sea surface height field using the geostrophic assumption on 16 December 2013. The *colored field* in the background is the sea surface height anomaly in the model. Panels (**c**) and (**d**) show the end positions of an initially uniform grid of particles for the respective velocity fields after 3 days

field on 16 December 2013. See Jacobs et al. (2016) for model details. The top right panel (b) shows the same sea surface height anomaly field and the geostrophic velocity calculated from that height field. Particles were started on a regular grid with approximately 0.0022-deg spacing, and their trajectories were evaluated from the respective velocity fields. The lower two panels of Fig. 7 show the particle distributions after 3 days. Panel (c) indicates significant clustering of particles along long thin lines, whereas there is no such effect in the geostrophic flow distribution shown in panel (d), although that plot does show organized deformation of the uniform grid.

What can the difference in particle distributions be attributed to? Figure 8 shows Δ, Σ, and Λ for the full velocity field over the same 3-day time period used in Fig. 7. All three fields show similar features. Dilation and stretch rates are similarly strong, with fine-scale structure. Since the features are slightly offset, the FTLE field, being their average, appears generally smoother with softer ridges.

The dilation rate field in Fig. 8a illustrates the competing roles of dispersion and cluster formation. Much of this field is red, which indicates regions where particles are dispersing. In contrast the blue cluster regions are restricted to long filaments.

For the geostrophic velocity field, as shown in Fig. 7, the dilation rate vanishes, of course, since there is no divergence. The stretch rate is shown in Fig. 8d. The FTLE would simply be half of this field. Note that the geostrophic stretch rate shows some similarities with that of the full velocity field, but the structures generally have a larger scale, and the distinct eddies have a much more dominant signature in ridges wrapping around them. (The scatter at the northern end of the domain should be ignored as unreliable due to its proximity to the coastline.)

In the case analyzed by Huntley et al. (2015), where the ageostrophic velocity component is weaker, the FTLE fields for the two sets of velocities are much more similar, indicating that it is not the divergence per se that leads to the distinction in the case presented in Fig. 8.

5 Discussion

We have summarized a theory that decomposes the FTLE into dilation and stretch components. In purely geostrophic flow the dilation is 0, and there is no bunching of particles along thin curves (Fig. 7). Although the full model velocity field is similar, the particles tend to congregate along curves of strong negative dilation. The dilation rate Δ alone, however, does not fully account for the deformation of the patch of particles: Transport pathways are delineated by Λ.

The distinction between regions of buoyant particle accumulation characterized by Δ and transport boundaries characterized by Λ seems important to us. As shown by Poje et al. (2014, 2016) this region of the Gulf of Mexico obeys classic turbulent dispersive scaling. At the same time horizontal convergence produced by submesoscale dynamics will cause some particles to accumulate along lines. Averaged separation behavior, thus, is only part of the picture.

The complete story is still evolving. In a related study, Jacobs et al. (2016) analyzed the horizontal divergence along trajectories in a series of numerical experiments in both a deep portion of the Gulf of Mexico and the adjacent continental shelf. The clustering patterns between these regions were generally

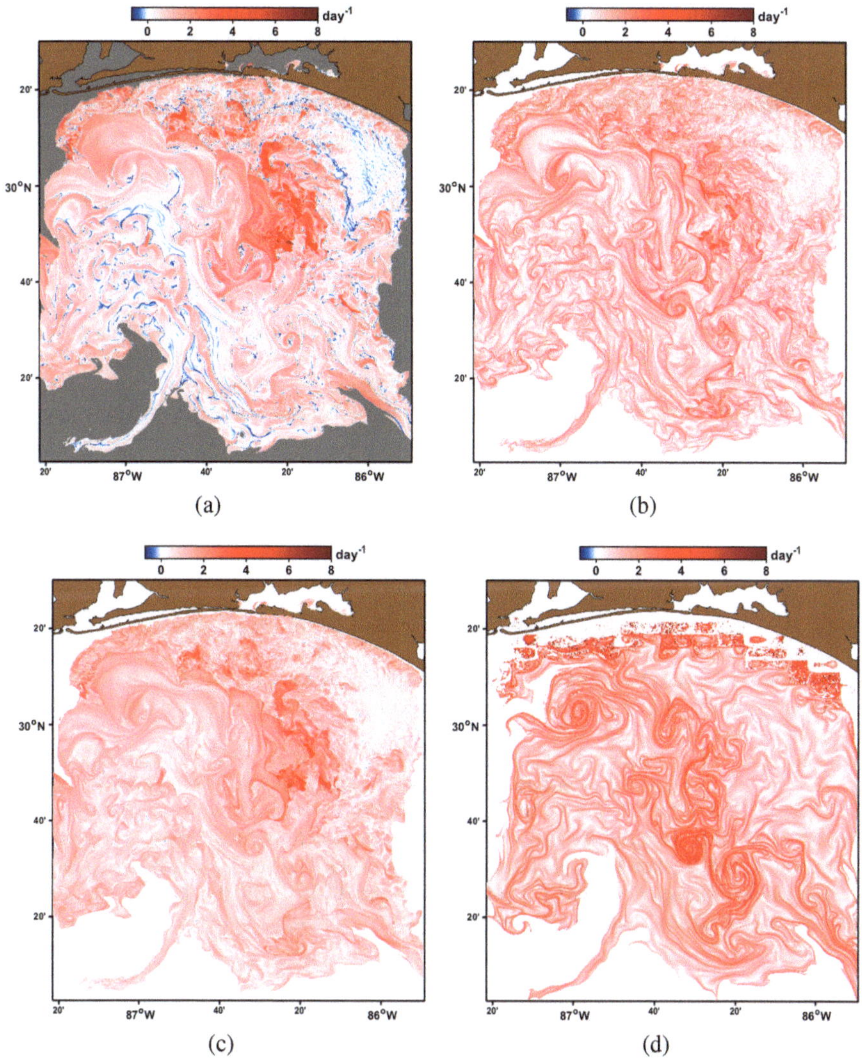

Fig. 8 The (**a**) dilation rate, (**b**) stretch rate, and (**c**) the FTLE for the full velocity field and the same period shown in Fig. 7. (**d**) The stretch rate for the geostrophic velocity field over the same time

similar, but with different characteristic temporal and spatial scales. Initially, small-scale ocean processes were found to produce small-scale cluster patterns. Over time, the small clusters broke up but larger scale clusters arose from lower-frequency flow phenomena. This suggests that clustering phenomena and the diagnostics developed here are critically dependent on phenomenological scales and consequently on model resolution.

Two recent field experiments in the Gulf of Mexico should shed more light on the matter. The LAgrangian Submesoscale ExpeRiment (LASER) was conducted in the DeSoto Canyon, near the Deepwater Horizon spill site, in January–February 2016. This experiment monitored Lagrangian properties down to scales of meters. A forthcoming experiment, Submesoscale Processes and Lagrangian Analysis on the SHelf (SPLASH), scheduled for spring 2017, will acquire similar data on the continental shelf off Port Fourchon. A primary goal of these experiments is to quantify the competition between dispersion and clustering down to these scales, complementing the model results.

Acknowledgements This work was funded in part by grant N00014-11-1-0087 from the Office of Naval Research for MURI *OCEAN 3D+1* and in part by a grant from The Gulf of Mexico Research Initiative to the Consortium for Advanced Research on Transport of Hydrocarbon in the Environment. The authors thank Karal Gregory for technical assistance in preparation of the manuscript.

References

Aref, H. 1984. Stirring by chaotic advection. *Journal of Fluid Mechanics* 143(1–21). doi:10.1017/S0022112084001233.

Ebbesmeyer, C.C., and W.J. Ingraham. 1992. Shoe spill in the north pacific. *Eos* 73(34): 361–365.

Ebbesmeyer, C.C., and W.J. Ingraham. 1994. Pacific toy spill fuels ocean current pathways research. *Eos* 75(37): 425–430.

Huntley, H.S., B.L. Lipphardt Jr., G. Jacobs, and A.D. Kirwan Jr. 2015. Clusters, deformation, and dilation: Diagnostics for material accumulation regions. *Journal of Geophysical Research, Oceans* 120: 6622–6636. doi:10.1002/2015JC011036.

Jacobs, G.A., H.S. Huntley, A.D. Kirwan Jr., B.L. Lipphardt Jr., T. Campbell, T. Smith, K. Edwards, and B. Bartels. 2016. Ocean processes underlying surface clustering. *Journal of Geophysical Research, Oceans* 121: 180–197. doi:10.1002/2015JC011140.

Kirwan, A.D. Jr. 2006. Dynamics of "critical" trajectories. *Progress in Oceanography* 70: 448–465.

Okubo, A. 1971. Oceanic diffusion diagrams. *Deep Sea Research* 18(8): 789–802. doi:10.1016/0011-7471(71)90046-5.

Poje, A.C., T.M. Özgökmen, B.L. Lipphardt Jr., B.K. Haus, E.H. Ryan, A.C. Haza, G.A. Jacobs, A.J.H.M. Reniers, M.J. Olascoaga, G. Novelli, A. Griffa, F.J. Beron-Vera, S.S. Chen, E. Coelho, P.J. Jogan, A.D. Kirwan Jr., H.S. Huntley, and A.J. Mariano. 2014. Submesoscale dispersion in the vicinity of the deepwater horizon spill. *Proceedings of the National Academy of Sciences of the United States of America* 111(35): 12693–12698.

Poje, A.C., T.M. Özgökmen, D.J. Bogucki, and A.D. Kirwan Jr. 2016. Evidence of a forward energy cascade and Kolmogorov self-similarity in submesoscale ocean surface drifter observations. *Physics of Fluids*: 020701-1–020701-10. Special issue for Prof. John Lumley

Smith, J.N., R.M. Brown, W.J. Williams, M. Robert, R. Nelson, and S.B. Moran. 2015. Arrival of the Fukushima radioactivity plume in North American continental waters. *Proceedings of the National Academy of Sciences of the United States of America* 112(5): 1310–1315. doi:10.1073/pnas.1412814112.

The Rise and Fall of Thermodynamic Complexity and the Arrow of Time

A.D. Kirwan Jr. and William Seitz

Abstract Complexity Theory is an eclectic collection of theoretical approaches to a wide variety of nonlinear problems that typically involve many degrees of freedom. Despite numerous claims, there does not appear to be a universal basis for the various approaches. Here we report on recent attempts to provide such a basis. Our approach is based on the *partial order* of Boltzmann states under majorization and thus is grounded in the Second Law of Thermodynamics. However, here we do not appeal to any energetic constraints. By majorizing the Boltzmann states we identify a new statistical mechanical entity, namely a multivalued function that maps Boltzmann entropy to the size or order of sets of incomparable Boltzmann entropy states. We call this *thermodynamic complexity*. This is a concave function of entropy, peaking near mid-entropy and falling to zero at maximum and minimum entropies. It remains to be seen if this approach can be rigorously applied to other areas, but heuristic arguments given here indicate broad applicability.

Keywords Boltzmann entropy • Second law of thermodynamics • Majorization • Thermodynamic complexity • Incomparability • Posets • Young diagram lattice • Arrow of time

1 Introduction

The notion that all things evolve irreversibly towards equilibrium is deeply imbedded in science and experience. This is codified in the Second Law of Thermodynamics, where the final equilibrium state achieves maximum entropy; this state in turn is often characterized as being maximally disordered or chaotic. Although

A.D. Kirwan Jr. (✉)
School of Marine Science and Policy, University of Delaware, Robinson Hall, Newark, DE 19716, USA
e-mail: adk@udel.edu

W. Seitz
Department of Marine Sciences, Galveston Campus of Texas A&M University, Galveston, TX 77553, USA
e-mail: seitzw@tamug.edu

A.A. Tsonis (ed.), *Advances in Nonlinear Geosciences*,
DOI 10.1007/978-3-319-58895-7_13

this principle was originally formulated for inanimate thermodynamic systems, it is generally accepted as a universal truth. Recent developments in the nonlinear sciences have shown that the evolution to a final equilibrium of maximum entropy is not simple. In fluid dynamics, for example, considerable effort is directed towards understanding the dynamics of "coherent structures" in otherwise turbulent flows. See Haller (2015) for a recent topical review. The reaction–diffusion equations and subsequent laboratory studies have shown that intricate patterns can arise from complex chemical reactions. Nonlinear models with many degrees of freedom also arise in the biological sciences and economics. There too, identifiable structures appear to emerge from seemingly incoherent backgrounds far from equilibrium. The proceedings of a conference on coherent structures in complex systems (Reguera et al., 2001) beautifully document the huge variety of disciplines in which the notion of coherent structures plays a fundamental role.

These developments have led to the concept that complex structures can emerge far from equilibrium. At the risk of fomenting confusion we shall use the term *complex* to describe such structures. Although widely recognized in many disciplines, there seems to be no universal quantitative definition of complexity. Weaver (1948) posited two types of complexity. In this categorization disorganized complexity is characterized randomness while organized complexity referred to large systems with many interacting components. Lloyd (2001) listed 31 measures. Johnson (2009) defined complexity science as "the study of the phenomena which emerge from a collection of interacting objects." Starting in 1971 with Ruelle and Takens' work on turbulence (Ruelle and Takens, 1971) a number of investigators May (1975), Feigenbaum (1978, 1979), Feigenbaum et al. (1982), and Rand et al. (1982) have attempted to relate complexity to the onset of chaos.

These approaches have provided useful and important insight into many disciplines. But we are struck by an apparent lack of universality. For example, the "universality" of the approach to chaos advocated in Feigenbaum (1978, 1979), Feigenbaum et al. (1982), and Rand et al. (1982) is specific to their definition of "chaos." Nor is it clear to us what the final equilibrium state should be for these systems. The approach taken here is different. Our hypothesis is that complex systems typically undergo a systematic evolution from simple configurations to those that exhibit considerable complexity, with noticeable structure, before reaching some final equilibrium configuration typically characterized by total loss of coherence. We attempt to quantify this evolution using standard, but somewhat obscure, mathematical tools.

A critical question is what standard should be used for the evolution of a complex system. Here we select Boltzmann entropy, as this is the "mother" entropy for nearly all scientific applications. Variants are used in quantum statistical mechanics, economics, and diversity studies in ecology and sociology. Because of this we expect the analysis below may have significant impact in a number of disciplines.

Our report is organized as follows. Section 2 gives a short review of Boltzmann entropy. In Sect. 3 we review recent work relevant to our thesis about the evolution of systems. Section 4 introduces the notion of majorization and incomparability and applies this to a Boltzmann system of modest dimension. In Sect. 5 we speculate how our approach might fit into a general view of the evolution of many complex systems.

2 The Mother of All Entropies: Boltzmann

The Boltzmann entropy, $S = k \ln \Omega$, was the original statistical mechanical model of a gas. It was developed by the Austrian physicist L. Boltzmann in the early 1870s. In this equation k is the Boltzmann constant and Ω is the number of ways N distinguishable particles can be distributed among N states with λ_j particles in state j. From elementary considerations this is

$$\Omega = \frac{N!}{\prod_{j=1}^{N} \lambda_j!}. \tag{1}$$

It is appropriate here to introduce some special notation. We characterize entropy configurations by the partition notation

$$S(\boldsymbol{\lambda}; N) \leftrightarrow [\lambda_1, \ldots, \lambda_N]. \tag{2}$$

We will refer to $[\lambda_j]$ as a partition of S. In this notation the ordering of the partition is such that $\lambda_j \geq \lambda_{j+1}$. By way of illustration consider the state $N = 5$. The minimum and maximum entropy partitions are simply denoted by [5] and $[1^5]$, respectively. Two possible intermediate partitions are [3, 2] and $[3, 1^2]$.

Clearly the minimum entropy occurs when all the particles are in one state and the maximum entropy is when each particle occupies its own unique state. The minimum entropy configuration is often regarded as the most organized condition while the maximum entropy configuration characterizes total randomness or chaos since it represents the maximum number of ways the particles can be distributed amongst all the states. We argue that the minimum and maximum entropy states are uninteresting. In the former everything is the same while in the latter state everything is different. It is much more interesting to ponder what happens when there is a mixture of sameness and differentness.

To illustrate how this works consider a configuration of N particles given by specific values of $[\lambda_j]$. S characterizes the macrostate while Ω gives the number of ways that value of S can be achieved by rearrangements of the particles between the $[\lambda_j]$, keeping the value of each $[\lambda_j]$ fixed. For example, for three particles one might have a distribution of particles 1 and 2 in state λ_1, particle 3 in state λ_2, and no particles in state λ_3. The same value of entropy would be achieved if particle 1 swapped positions with particle 3. The different combinations of particles are called microstates. Generally microstates are unobservable while macrostates are observables.

We will be concerned with the number of partitions available for a given state size N. It is well known that for the Boltzmann system this is exactly the number of integer partitions of N. Hardy and Ramanujan (1918) showed that for large N this was approximately

$$P(N) \approx \left(\frac{1}{4N\sqrt{3}}\right) \exp\sqrt{\frac{2N}{3}}. \tag{3}$$

Evidently for even modest state sizes the number of available entropy partitions is huge.

3 Surprising Properties of *S*

Because of its long history in science, mathematical properties of the Boltzmann entropy are generally taken for granted. Here we review some recent research that reveal new facets of this marvelous function.

We start with an example of the complete evolution of a Boltzmann system with ten particles from its minimum to maximum entropy configurations. That is we start with a configuration [10] and follow its evolution to the configuration $[1^{10}]$. This is conveniently illustrated by a Young Diagram Lattice (YDL) as shown in Fig. 1.

For $N = 10$ there are 42 entropy states. They are indicated in the YDL by the rows and columns in the diagram. The lowest entropy state is at the top when all ten objects are in one row. The first evolutionary step is to move one object to the adjacent row. Now there are two choices for the second step. One can either move a second object to the second row or move that object to a third row. The former move produces a partition $[8, 2]$ while the latter produces $[8, 1^2]$. This situation illustrates two possible strategies. The "diversity" strategy has no constraint to what row a particle can be moved. The "mixing" strategy requires that objects move the minimum number of rows. So in this example the partition $[8, 2]$ resulted from the mixing paradigm while the partition $[8, 1^2]$ was a possible example of the diversity paradigm. Obviously the mixing paradigm is a special case of the diversity paradigm.

Casual inspection of Fig. 1 reveals that there are many possible routes to the final equilibrium partition $[1^{10}]$ at the bottom of the ladder where there is one object in every row for either paradigm. Hence the question: Are there preferred routes to equilibrium? Seitz and Kirwan (2014) ran Monte Carlo simulations for a variety of state sizes N for both the mixing and diversity paradigms. Figure 2, taken from that paper, summarizes the results. The diversity paradigm path length scales nearly linearly with N for over two decades while the mixing paradigm scales as $N^{1.375}$. These results are close to limiting asymptotic values for these two paradigms. It is straightforward to show that for the diversity paradigm the shortest route is simply the state size N. The longest route for the mixing paradigm is a bit more complicated to analyze but it approaches $N^{4/3}$ for large N. Shorter routes go down the sides of the YDL in Fig. 1 while longer routes are those that run through the middle of the

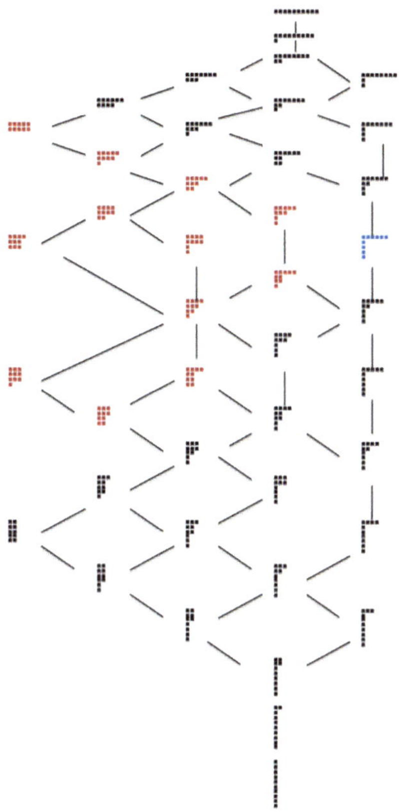

Fig. 1 Young Diagram Lattice for $N = 10$ showing partitions incomparable to $[6, 1, 1, 1, 1]$

lattice. It is noteworthy that both paradigms are much different from the classic thermodynamic paradigm, which requires that the route to equilibrium pass through every partition.

Since the routes scale close to the asymptotic limits for both cases the question arises whether some partitions are visited more frequently than others. As noted by Seitz and Kirwan (2014), the Monte Carlo simulations revealed a surprising variation in partition visitations. For example, rectangular shaped partitions such as $[6^4, \ldots]$ for $N = 36$ are rarely visited by either paradigm. Such partitions tend to be located on the extreme left side of YDLs. Partitions located on the extreme right side such as $[5, 2, 1^3]$ are visited more frequently by the diversity paradigm. The large variation in visitation frequencies as reported by Seitz and Kirwan (2014) was unexpected and is counter to the notion in classical thermodynamics regarding the approach to equilibrium. Surely further inquiry is warranted into this phenomenon.

The most surprising property of S was the preponderance of degenerate or doppelgänger entropy partitions reported by Kirwan and Seitz (2016). The first example was found at $N = 7$ where it is easy to see that the partitions $[4, 1^3]$ and $[3, 2^2]$ have the same entropies. Previously doppelgänger entropy partitions were reported in Seitz and Kirwan (2014), but were regarded as pathological since the

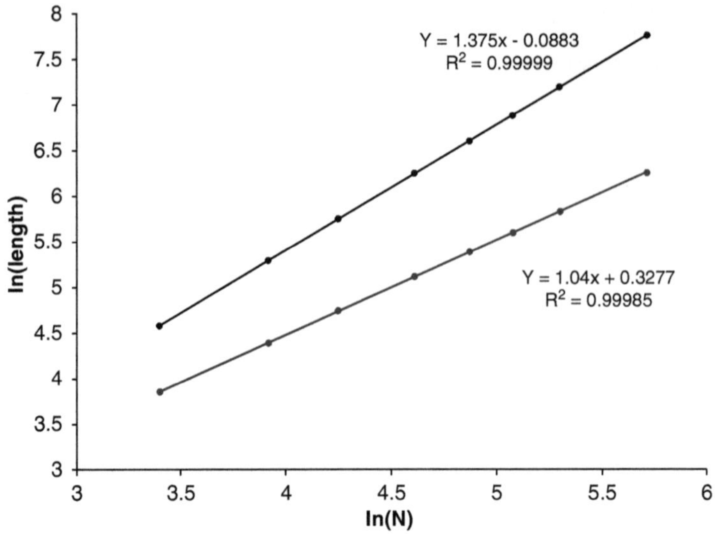

Fig. 2 Log-log plots of the average path length vs. N. The top curve is for the mixing lattice. Note slope is 1 for diversity but fractional for mixing

number of states grows as the integer partition of N. However, this proved not to be the case. For a modest state size of $N = 50$ where there are 204,226 states only 5% were *not* doppelgängers.

Motivated by this finding Kirwan and Seitz (2016) developed a theory for doppelgänger entropies. They noted that doppelgänger entropies occurred when the number of objects in a row in the YDL could be expressed as the product of two or more factorials. The case $N = 7$ is the simplest example. Since $4 = 2 \cdot 2$ then $4! = 3!2!2!$, thus the partitions $[4, 1^3]$ and $[3, 2^2]$ have the same entropy. This seed occurs first at $N = 7$ but the combination will be repeated for increasing values of N. In fact they showed that this seed grows as the integer partition of $N - 7$. Of course for $N > 7$ more seeds are available which also grow as integer partitions. For sufficiently large N the first seeds can intersect to produce vielgänger entropy states. The first occurrence of this is at $N = 12$ where the $4! = 3!2!2!$ seed intersects the $6! = 5!3!$ seed to produce the isoentropy partitions $[5, 4, 1^3]$, $[5, 3, 2^2]$, and $[6, 2^2, 1^2]$. At $N = 50$ Kirwan and Seitz (2016) found one entropy value produced by 86 partitions. On the other hand, there were over 5000 examples of just doppelgänger entropy partitions. Figure 3, taken from their paper, shows the frequency of the degenerate entropy partitions.

What does this mean for state systems of large N? The theory developed in Kirwan and Seitz (2016) shows that the fraction of unique entropy partitions relative to the state size goes as $N^{-1/2}$. Thus for even modest state sizes typical in many applications there are very few unique entropy states.

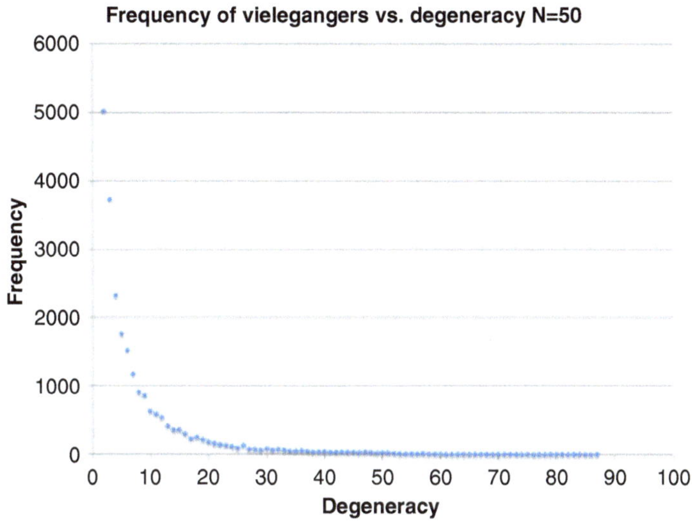

Fig. 3 Frequency decrease with degeneracy

4 Majorization, Partial Orders, and Incomparability

The concept of partially ordering partitions was introduced in combinatorics by Muirhead (1903). The partial order is determined by a mathematical operation called majorization. This is a way of comparing different multidimensional vectors whose elements are positive numbers. The convention used here is that the vector elements are arranged in descending order. More specifically consider two entropy partitions, both of size N given by $[\lambda_i]$ and $[\mu_i]$. If

$$\sum_i^m \lambda_i \geq \sum_i^m \mu_i \text{ for } m = 1, \ldots, N$$

$$\sum_i^N \lambda_i = \sum_i^N \mu_i \tag{4}$$

then $[\lambda_i]$ majorizes $[\mu_i]$. If $[\lambda_i]$ neither $[\mu_i]$ majorize each other, then these partitions are said to be *incomparable*. When applied to the evolution of entropy incomparability has a physical consequence. If a lower entropy partition is incomparable with a higher entropy partition, then any path from the lower partition to equilibrium cannot pass through the higher partition. In other words incomparability is a powerful diagnostic for identifying forbidden states.

As an example consider YDL in Fig. 1 and the sequence of partitions arranged in order of increasing entropy $[8, 2]$, $[8, 1^2]$, $[7, 3]$, and $[7, 2, 1]$. Clearly $[8, 2]$ majorizes the other three partitions and both $[8, 1^2]$ and $[7, 3]$ majorize $[7, 2, 1]$. But $[8, 1^2]$ does not majorize [7, 3]! Hence the state [8, 1^2] cannot evolve to the next higher entropy state, which is [7, 3]. Even in this small state size the number of incomparable partitions can be quite large. The blue partition [5, 1^5] in that figure is incomparable with the 12 red partitions.

A curious feature of the partitions of degenerate entropy states is that they are mutually incomparable (Kirwan and Seitz, 2016). Apparently there is no isentropic trajectory that will connect two partitions with the same entropy.

For these reasons incomparable entropy partitions seem important to us. It is the mechanism that produced the huge disparity in partition visitations noted in Sect. 3. It is a distinctive feature of degenerate entropy states. It also runs counter to the classical thermodynamic view that as a system evolves to equilibrium it passes through a continuum of all possible intermediate states.

Can this be used to quantify incomparability for the Boltzmann system? Here is a synopsis of the approach used by Seitz and Kirwan (2016). Suppose there is a set X with a partial order **P**. They defined C_P of any element of X as the number of elements of X that it is incomparable with. C_P is readily computed for modest state sizes but the calculation is exponentially hard as $\sqrt{N}$ increases because of the exponential increase in the number of states. We call C_P *thermodynamic complexity*.

Seitz and Kirwan (2016) calculated C_P for $N = 50$. This is shown in Fig. 4. In this figure the entropy was normalized by the maximum entropy 50! and the incomparability by the maximum value of C_P.

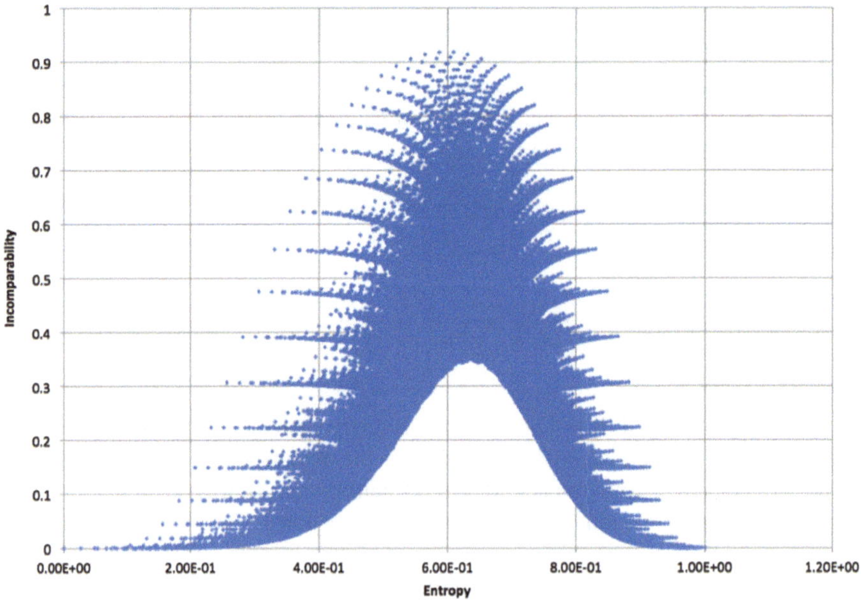

Fig. 4 Normalized incomparability/$IP(N)$ vs. normalized entropy for $N = 50$

We note three intriguing features of this figure. One is the spines of nearly constant incomparability but rapidly increasing entropy. The outermost points on the spines are states with partitions $[50 - m, 1^m]$. In the YDL shown in Fig. 1 these states arise from evolutionary paths that go down the right side. The states that make up the minimum C_P arise from paths that favor the left side. The partition with the maximum C_P is $[25, 1^{25}]$. The second feature is the asymmetry of the maximum C_P relative to the normalized entropy. This partition occurs at a normalized entropy of

$$\bar{S}(25, 1^{25}) = \frac{S(25, 1^{25})}{S(1^{50})} \approx \left(\frac{1}{2}\right)\left[1 + \left(\frac{\ln 2}{\ln 25 - 1}\right)\right] \approx 0.619017 \tag{5}$$

and not at $\bar{S} = 0.5$ as one might expect. As Seitz and Kirwan (2016) show, the offset from $\bar{S} = 0.5$ is simply a finite state size effect. They show that as $N \to \infty$, $\bar{S} \to 0.5$. The third feature is the remarkable absence of '"simple" states for intermediate normalized entropy values. Except for the tails near the minimum and maximum entropies, all entropy partitions have significant incomparability, or are thermodynamically complex. Evidentially incomparability is a fundamental characteristic of entropy partitions.

5 Discussion

What might be the broader significance of these results? First we emphasize that all results presented here are quantitative. They are based just on fundamental classical thermodynamics and statistical mechanics concepts and the mathematical operation of majorization. Moreover the results are independent of any energetic considerations. Since incomparability or thermodynamic complexity is distinct from entropy and yet consistent with the evolution of complex systems one might expect from the Second Law, it should be considered as a new thermodynamic state variable.

Since classical thermodynamic and statistical mechanics concepts have been adapted by many other disciplines it is tempting to speculate about the impact thermodynamic complexity might have on complexity studies in other disciplines. As discussed in Sect. 1 other complexity paradigms are concerned with an abrupt transition to chaos through period doubling or metrics arising from Johnson's (2009) definition of complexity. Thermodynamic complexity is not compatible with either paradigm. It is a measure of systems that are inaccessible and hence do not interact directly. Moreover as seen in Fig. 4 there is a distinct rise and fall of thermodynamic complexity as the Boltzmann system evolves towards final equilibrium.

Finally consider one of the cherished views of entropy as the "arrow of time." It is often used in qualitative arguments as to why complex systems such as individuals and civilizations age and die. Figures 5 and 6 show two examples. Figure 5 is an image of a famous woodcutting by Baltasar Talamantes depicting the life span of

Fig. 5 The complexity of individuals is greatest at middle age and less at birth and death. (The female Steps of Life, a woodcut by Baltasar Talamantes, late eighteenth century)

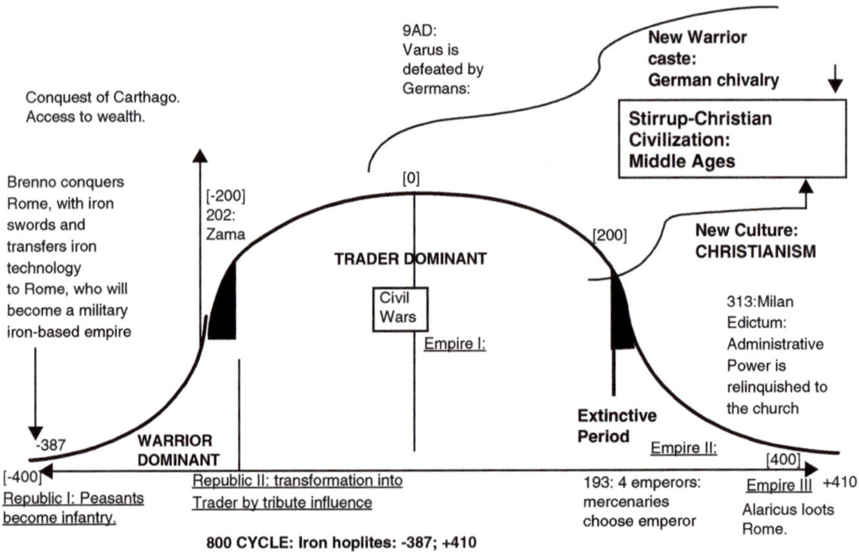

Fig. 6 Cartoon of the rise and fall of the Roman empire

a medieval woman. Figure 6 is a cartoon of the rise and fall of the Roman empire. The abscissa in both figures is time but the ordinates are unspecified. It has been argued that both individuals and empires ultimately die and that this process is a consequence of the Second Law. However, the unstated ordinates in both figures indicate that the artists sensed that something else was going on during the evolution. Near the mid-life period in both cases there is a maximum in "complexity." The woman's life is complicated by family commitments and likely too many other factors too numerous to be depicted. In Fig. 6 life at the apex of the Roman empire involved many factors not present during the warrior dominant phase. The similarity of these figures with Fig. 4 is striking. Were the artists attempting to portray in the ordinates the complexity of systems far from final equilibrium?

Thermodynamic complexity as developed here is completely consistent with the arrow of time property of entropy. It qualitatively depicts the rise and fall of complexity in the lifetime of any organism and also societal structures. The other complexity metrics noted previously do not. It will be both stimulating and challenging to apply majorization and incomparability to non-thermodynamic complex systems.

Acknowledgements We thank Karal Gregory for technical assistance in preparation of this manuscript.

References

Feigenbaum, M.J. 1978. Quantitative universality for a class of nonlinear transformations. *Journal of Statistical Physics* 19: 25–52.

Feigenbaum, M.J. 1979. The universal metric properties of nonlinear transformations. *Journal of Statistical Physics* 21: 669–702.

Feigenbaum, M.J., L.P. Kadanoff, and S.J. Shenker. 1982. Quasiperiodicity in dissipative systems: A renormalization group analysis. *Physica D* 5: 370–386.

Haller, G. 2015. Lagrangian coherent structures. *Annual Review of Fluid Mechanics* 47: 137–162.

Hardy, G., and S. Ramanujan. 1918. Asymptotic formulae in combinatory analysis. *Proceedings of the London Mathematical Society* 17: 75–115.

Johnson, N.F. 2009. *Simply complexity: a clear guide to complexity theory.* Oxford: Oneworld Publications.

Kirwan, A.D. Jr., and W. Seitz. 2016. Doppelgänger entropies. *Journal of Mathematical Chemistry* 54: 1942–1951. doi:10.1007/s10910-016-0658-z. http://dx.doi.org/10.1007/s10910-016-0658-z

Lloyd, S. 2001. Measures of complexity: a nonexhaustive list. *IEEE Control Systems* 21(4): 7–8.

May, R.M. 1975. Deterministic models with chaotic dynamics. Nature 356(55141): 165–166.

Muirhead, R. 1903. Some methods applicable to identities and inequalities of symmetric algebraic functions of n letters. *Proceedings of the Edinburgh Mathematical Society* 21: 144–157.

Rand, D., S. Ostlund, J. Sethna, and E.D. Siggia. 1982. Universal transition from quasiperiodicity to chaos in dissipative systems. *Physica D* 49: 132–135.

Reguera, D., J.M. Rubí, and L.L. Bonilla, eds. 2001. *Coherent structures in complex systems.* Lecture Notes in Physics, vol. 567. Berlin: Springer.

Ruelle, D., and F. Takens. 1971. On the nature of turbulence. *Communications in Mathematical Physics* Addendum 23, 343–344.

Seitz, W., and A.D. Kirwan Jr. 2014. Entropy vs. majorization: What determines complexity? *Entropy* 16(7): 3793–3807. doi:10.3390/e16073793. http://www.mdpi.com/1099-4300/16/7/3793

Seitz, W., and A.D. Kirwan Jr. 2016. Boltzmann complexity: An emergent property of the majorization partial order. *Entropy* 18(10): 347. doi:10.3390/e18100347. http://www.mdpi.com/1099-4300/18/10/347

Weaver, W. 1948. Science and complexity. *American Scientist* 36: 536–544.

From Fractals to Stochastics: Seeking Theoretical Consistency in Analysis of Geophysical Data

Demetris Koutsoyiannis, Panayiotis Dimitriadis, Federico Lombardo, and Spencer Stevens

Abstract Fractal-based techniques have opened new avenues in the analysis of geophysical data. On the other hand, there is often a lack of appreciation of both the statistical uncertainty in the results and the theoretical properties of the stochastic concepts associated with these techniques. Several examples are presented which illustrate suspect results of fractal techniques. It is proposed that concepts used in fractal analyses are stochastic concepts and the fractal techniques can readily be incorporated into the theory of stochastic processes. This would be beneficial in studying biases and uncertainties of results in a theoretically consistent framework, and in avoiding unfounded conclusions. In this respect, a general methodology for theoretically justified stochastic processes, which evolve in continuous time and stem from maximum entropy production considerations, is proposed. Some important modelling issues are discussed with focus on model identification and fitting often made using inappropriate methods. The theoretical framework is applied to several processes, including turbulent velocities measured every several microseconds, and wind and temperature measurements. The applications show that several peculiar behaviours observed in these processes are easily explained and reproduced by stochastic techniques.

Keywords Fractal techniques • Multifractals • Stochastics • Power spectrum • Hurst-Kolmogorov dynamics • Grid turbulence • Wind speed • Air temperature

D. Koutsoyiannis (✉) • P. Dimitriadis
Department of Water Resources and Environmental Engineering, School of Civil Engineering, National Technical University of Athens, Heroon Polytechneiou 5, 15780, Zographou, Greece
e-mail: dk@ntua.gr

F. Lombardo
Dipartimento di Ingegneria Civile, Edile e Ambientale, Sapienza Università di Roma, Via Eudossiana, 18, 00184, Rome, Italy

S. Stevens
Independent Researcher, London, UK

A.A. Tsonis (ed.), *Advances in Nonlinear Geosciences*,
DOI 10.1007/978-3-319-58895-7_14

I regard intuition and imagination as immensely important: we need them to invent a theory. But intuition, just because it may persuade and convince us of the truth of what we have intuited, may badly mislead us: it is an invaluable helper, but also a dangerous helper, for it tends to make us uncritical. We must always meet it with respect, with gratitude, and with an effort to be severely critical of it. (Karl Popper, preface to "The Open Universe: An Argument for Indeterminism", 1982).

1 Introduction

Over the past 30 years or more, considerable literature highlighted the fractal (self-similar, self-affine, multifractal) characteristics of many complex patterns that characterize geophysical processes. Fractal literature provides a framework in which a simple process, involving a basic operation repeated many times, can represent natural patterns that can be of extraordinary complexity (Falconer 2014; Scholz and Mandelbrot 1989). In a variety of applications, geophysical systems are viewed as fractals that follow certain scaling rules over a broad (even unlimited) range of scales, implying that the degree of their irregularity and/or fragmentation is identical at all those scales. Mathematically, these rules are power laws with exponents being related to a fractal dimension. Roughly speaking, the fractal dimension is a measure of the prominence of complexity of a pattern when viewed at very small scales. Therefore, the fractal dimension is originally a local property, notwithstanding the fact that in fractal literature the local properties are reflected in the global ones (Mandelbrot 1982).

Finding that a complex system is characterized by fractal (or multifractal) behaviour with particular scaling exponents represents a desideratum for many practicing geophysicists and engineers (von Kármán 1940), because this finding will help in describing the system dynamics with very simple formulae and few parameters, in order to obtain predictions on the future behaviour of the system. Such dynamics is usually denoted as fractal or multifractal, depending on whether it is characterized by one scaling exponent or by a multitude of scaling exponents.

However, if we agree that scientific theories are mental constructs rather than the physical reality per se, then we should also agree that there are no true fractals in nature. Although there are natural phenomena that have been explained in terms of fractal mathematics, "natural fractals" (such as coastlines, turbulence in fluids, cloud boundaries, etc.) can usefully be regarded as such only over an appropriate range of scales, with the fractal description inevitably ceasing to be valid if they are viewed out of this range of scales.

Since asymptotic properties of geophysical processes are crucial for the quantification of future uncertainty, as well as for planning and design purposes, many applications of fractal theory tend to be descriptive rather than predictive (Falconer 2014; Kantelhardt 2009). In the foundational treatise on fractals, Mandelbrot (1982) made such a distinction clear, but it has become somewhat blurred in recent literature.

We maintain and show in the following that careful use of stochastics (which includes probability theory, statistics and stochastic processes) can deal with all problems about complex geophysical processes in a more rigorous manner and more effectively than fractals can do.

2 Why not to Prefer Fractals over Stochastics

In spite of the difficulty even mathematicians have in formally defining fractals (Falconer 2014; Mandelbrot 1982), their wide popularity stems from the concept of symmetry—in particular, expanding symmetry. From the birth of science and philosophy, symmetry has been closely related to harmony and beauty, and this was to prove decisive for its role in theories of nature. Both ancients and moderns often believed indeed that there is a close association in mathematics between beauty and truth.

A common research theme in the study of complex systems is the pursuit of universal properties that transcend specific system details. In this way, fractal-based techniques have opened new avenues in the analysis of geophysical data. According to Scholz and Mandelbrot (1989):

> *One possible broad explanation of the role of fractals in geophysics may be found in probabilistic limit theorems, and in the existence of classical "universality classes" related to them. The reason is illustrated by the following fact. Wiener's scalar Brownian motion process W(t) is the limit of the linearly rescaled random processes that belong to its very wide domain of attraction. Therefore, it is itself the fixed point of the rescaling process. That is, its graph is a self-affine fractal set, a curve. The argument suggests that geometric shapes relative to probabilistic limit theorems can be expected to be fractal sets.*

On the other hand, the concept of fractals has been closely associated from the outset with mathematical constructions involving infinite operations on simple, deterministically defined, objects. Simple nonlinear dynamical systems were also enrolled in illustrating the emergence of fractal structures. This association with determinism and simplicity has been prominent and shaped the evolution of the fractal literature.

Even when studying more complex systems, such as the evolution of geophysical processes, the intuitive zeal was to make them comply with the simplicity of the archetypal fractal mathematical objects. Thus, several studies attempted to demonstrate that irregular fluctuations observed in natural processes are *au fond* manifestations of underlying deterministic dynamics with low dimensionality, hence rendering probabilistic descriptions unnecessary. If we assume, for example, that the evolutions of all temporal and spatial patterns of geophysics result from deterministic chaos, then we may derive the underlying deterministic rules on the basis of their strange attractors, which have a fractal structure (Grassberger and Procaccia 1983). However, such an approach is questionable in geophysics (Koutsoyiannis 2006).

The opposite reading of the same finding would be more sensible, in our view. Specifically, if simple underlying dynamics can produce irregular fluctuations and eventually, unpredictable trajectories, then, *a fortiori*, more complex systems are even more unpredictable. In this line of thought, Koutsoyiannis (2010b) used a caricature geophysical system, which is low dimensional deterministic by construction, and showed that we cannot get rid of uncertainty. Hence, probability theory and its extension, stochastics, become absolutely necessary even for the simplest systems. This argument may also be used in order to criticize the determinist point of view that probability considerations enter into science only if our knowledge is insufficient to enable us to make predictions with certainty (Popper 1982).

Stochastics has its own rules of calculations and estimations, which go far beyond classical calculus in order to deal with uncertain quantities represented as random variables and stochastic processes. Fractal studies often fail to appreciate this and apply algorithms referring to uncertain quantities with standard mathematical calculations. They do so even when using stochastic concepts, such as statistical moments, (auto)correlations and power spectra. Thus, they produce results which not only fail to recognize the statistical uncertainty but may be fundamentally flawed, i.e., inconsistent with theory. In the subsections below, we summarize some of the problems often characterizing fractal studies which make us advocate the dedication to proper theoretical concepts, offered by the theory of stochastics.

2.1 Ambiguity

Even the very terms *fractal* and *multifractal* remain without an agreed mathematical definition. This is a severe drawback, as without proper definitions we cannot build a scientific theory. The importance of definitions in science has been emphasized in the following philosophical note by the great Russian mathematician Nikolai Luzin:

> *Each definition is a piece of secret ripped from Nature by the human spirit. I insist on this: any complicated thing, being illumined by definitions, being laid out in them, being broken up into pieces, will be separated into pieces completely transparent even to a child, excluding foggy and dark parts that our intuition whispers to us while acting, separating into logical pieces, then only can we move further, towards new successes due to definitions* (from Graham and Kantor 2009).

This is not the case with fractals. Instead, fractals are usually identified intuitively; for example, Falconer (2014) refers to a set F as a fractal, when:

1. F has a fine structure, i.e., detail on arbitrary small scales;
2. F is too irregular to be described in traditional geometrical language, both locally and globally;
3. F has some form of self-similarity, perhaps approximate or statistical;
4. usually, the "fractal dimension" of F (defined in some way) is greater than its topological dimension;
5. in most cases of interest, F is defined in a very simple way, perhaps recursively.

Mandelbrot, who coined the term *fractal* in 1975, tried to theorize about the absence of a definition, arguing just opposite of Luzin:

> *Let me argue that this situation ought not create concern and steal time from useful work. Entire fields of mathematics thrive for centuries with a clear but evolving self-image, and nothing resembling a definition* (Mandelbrot 1999, p. 14).

One may indeed recall cases where mathematical concepts did not have proper definitions for centuries; *probability* is a characteristic example. However, the expression "nothing resembling a definition" may be a gross exaggeration. In the example of probability there never was lack of definitions; the problem was that the definitions were problematic (e.g., suffering from circular logic, like in the previous sentence). Once Kolmogorov (1933) gave a proper definition to probability, he opened new avenues. Certainly, absence of a definition entails domination of intuition over logic, dark over light, or uncritical acting over critical thinking (cf. the excerpt by Luzin above and that by Popper in the opening motto of the paper).

Nevertheless, Mandelbrot's aversion from defining concepts, which he does not regard as "useful work" to do, has influenced the entire field of fractals. Even in cases where clear definitions exist, Mandelbrot encourages neglecting them and preferring intuitive notions. The following excerpt provides an example for the well-defined concept of stationarity, which is central in stochastics (see Koutsoyiannis and Montanari 2015):

> *[Mandelbrot* 1982*] observes that* "Ordinary words used in scientific discourse combine (a) diverse intuitive meanings, dependent on the user, and (b) formal definitions, each of which singles out one special meaning and enshrines it mathematically. The terms stationary and ergodic are fortunate in that mathematicians agree on them. However, experience indicates that many engineers, physicists, and practical statisticians pay lip service to the mathematical definition, but hold narrower views." *That is, many mathematically stationary processes are not intuitively stationary. By and large, those processes exemplify wild randomness, a circumstance that provides genuine justification for distinguishing a narrower and a wider view of stationarity* (Mandelbrot 1999, p. 7).

Even when Mandelbrot attempts to provide a definition for the central concept of a *multifractal*, he bases that definition on the intuitive concept of a "multibox cartoon":

> *Definition. The term multifractal denotes the most general category of multibox cartoons. It allows the generator to combine axial boxes and diagonal boxes with non-identical values of H_i from $H_{\min} > 0$ to $H_{\max} < \infty$* (Mandelbrot 1999, p. 45; see Sect. 3 below about the meaning of *H*).

The ambiguity does not concern merely definitions. "Peaceful coexistence" of different numerical values for the same mathematical concept has also been advocated:

> *We are done now with explaining the peaceful coexistence of two values of D: the dimension $D = 1/H = 2$ applies to that three-dimensional curve, as well as to the trail obtained by projecting on the plane (X, Y). However, the projections of the three dimensional curve on the planes (t, X) and (t, Y) are of dimension $D = 2 - H = 1.5$* (Mandelbrot 1999, p. 45).

In fact, when dealing with geophysical processes, one can easily get rid of ambiguity through stochastics. Careful use of stochastics can deal with all problems involving fractals of non-deterministic type in a more rigorous manner and more effectively.

2.2 *Confusion Between Local and Global Properties of Processes*

Indeed, attempts to remove ambiguity based on stochastics are not rare, as indicated by the following excerpt:

> *There is no "official" consensus on the definition of a fractal. However, what is generally agreed on is that the Hausdorff measure and Hausdorff dimension play a key role. One possible definition of a fractal is then for example that it is a set* $A \subseteq R^k$ *whose Hausdorff dimension* $\dim_{\text{Haus}} A$ *is not an integer* (Beran et al. 2013, p. 178).

Other researchers who seek for clarity also agree on this; for example, Veneziano and Langousis (2010, p. 4) state that the most general and mathematically satisfactory definition of fractal dimension is the Hausdorff dimension. Here it is important to note that the Hausdorff dimension expresses a local property, an asymptotic measure as a radius δ for covering the set A tends to zero. This is more evident in the so-called *box-counting dimension*, which is an upper bound for D_{Haus} (Beran et al. 2013, p. 181–182) and is defined as $\dim_{\text{Box}} A = \lim_{\delta \to 0} \log N_\delta / \log \delta$where N_δ is the minimal number of sets U_i needed for a δ-cover of A.

However, as in the fractal literature it is intuitively believed that the local properties repeat themselves at bigger and bigger scales, and given the general frame of ambiguity, the local properties have been confused with global ones, such as the long-range dependence. Indeed:

> *In the context of time series analysis, fractal behaviour is often mentioned as synonym for long-range dependence. Though there are strong connections between the two notions, they are also in some sense completely different* (Beran et al. 2013, p. 178).

Even Mandelbrot (1999, p. 3) referred to the difference of locality and globality, but in a rather obscure way:

> *The importance of the contrast between mildness and wildness is in part due to its links with a contrast between locality and globality.*

However, this was not enough to hinder the fractal literature from confusing fractal behaviour with long-range dependence.

Gneiting and Schlather (2004) were perhaps the first to clarify the issue and highlight the fact that fractal properties and long-range dependence are independent of each other. They used a process with Cauchy-type autocovariance function, which was first proposed by Yaglom (1987, p. 365) and also referred to by Wackernagel (1995, p. 219; 1998, p. 246), while a similar one was used by Koutsoyiannis (2000) in discrete time. Using this process, they demonstrated that the fractal and Hurst properties (long-range dependence) are two different things, independent to each other: The fractal parameter determines the local properties (the roughness) of the process (as time $t \to 0$) while the Hurst parameter determines the global properties of the process (as $t \to \infty$).

2.3 *Use of the Abstract Mathematical Objects as if They are Natural Objects*

In mathematical processes the local and global properties can be the same. The obvious example is the Hurst-Kolmogorov (HK) process (see below), also known as fractional Gaussian noise (Mandelbrot and Van Ness 1968), which is described by a single scaling exponent applicable to all scales. Scale independence or absence of characteristic scales in a process or a phenomenon is mathematically and intuitively attractive. Indeed, it would imply that simple physical dynamics could produce complex phenomena that exhibit startling similarities over all scales. However, in Nature complex phenomena are influenced by different mechanisms and agents, each one acting at a different characteristic scale, and therefore absence of characteristic scales is only a dream. Besides, the assumption of absence of characteristic time scales would have consequences that would be absurd. Some examples follow:

- The speculation that rivers are fractals with fractal dimension >1 (e.g., 1.2) has been very popular. However, if that were the case, it would mean that the number of sets of its δ-cover would be a power law of δ with exponent >1 for arbitrary low δ. As a direct consequence, the geometrical length of the river would be infinite (a curve with dimension >1 has infinite length; Falconer 2014) and any particle of water would take infinite time to reach the sea.
- If a Hurst-Kolmogorov process (whose variance is a power law of time scale; Eq. (16) below) were applicable for arbitrary short time scales, it would entail infinite variance of the instantaneous, continuous-time process which would imply infinite energy.
- If an antipersistent Hurst-Kolmogorov process (with Hurst exponent $H < 0.5$; see below) were applicable for arbitrary short time scales, it would entail negative autocovariance (anti-correlation) for arbitrary small lags which is absurd. For in a natural process, the autocorrelation should tend to 1 as lag tends to 0.

All these paradoxes are easily resolved if we abandon the idea of absence of characteristic scales and admit that below (or above) a certain characteristic scale the respective power laws cease to hold.

2.4 Hasty Use of Stochastic Concepts

Stochastic concepts such as statistical moments (marginal or joint, e.g., covariances), and spectral densities have been widely used in the fractal literature, usually by making calculations using data and at the same time ignoring the theoretical properties of those concepts. A typical example is the power spectrum, $s(w)$, where w denotes frequency (inverse time scale), and its log-log slope $s^{\#}(w)$. The latter represents the log–log derivative, which for any function $f(x)$ is defined as:

$$f^{\#}(x) := \frac{\mathrm{d}\,(\ln f(x))}{\mathrm{d}\,(\ln x)} = \frac{xf'(x)}{f(x)} \tag{1}$$

The HK process is used as a benchmark as it has a power spectrum with constant slope, i.e., $s(w) \propto w^{\beta}$, where the constant slope $\beta = s^{\#}(w)$ is related to the Hurst parameter H (Eq. (16) below) by $\beta = 1 - 2H$. The special case $H = 0.5$, which signifies the white noise, corresponds to $\beta = 0$, thus complying with the fact that the white noise spectrum is flat ($s(w) =$ constant). As $H \to 1$, which is the highest possible value, $\beta \to -1$, which is the lowest possible value for a stationary and ergodic process.

However, a huge number of studies exploring several data sets have reported steeper constant slopes, i.e., $\beta < -1$, also suggesting $H > 1$, which is absurd. Other studies assume that slopes $\beta < -1$ are theoretically consistent, also claiming that the particular value $\beta = -2$ corresponds to the power spectrum of the Brownian motion (the integral over time of white noise), which is a nonstationary process. This line of thought is extended further, in the characterization of processes. Specifically, the power spectrum has been often regarded as a tool to identify whether a process is stationary or nonstationary: values $\beta > -1$ are thought to suggest a stationary process while values $\beta < -1$ are thought to confirm the nonstationarity of the process. The fact is, however, that the entire line of thought is theoretically inconsistent and such numerical results, usually reported, are artefacts due to insufficient data or inadequate estimation algorithms.

Before we describe the details for recovering from the incorrect application of the power spectrum, it would be informative to trace how incorrect results can appear.In the example of Fig. 1, 1024 data points have been generated from a stationary stochastic process and the empirical power spectrum, calculated from these data, has been plotted. To apply some smoothing (as per Bartlett's (1948) method), the empirical power spectrum was constructed by averaging from 8 segments, in which the data were separated (since without smoothing the power spectrum would be exceptionally rough). The stochastic process has the theoretical power spectrum with the indicated varying slope (specifically, it is an HHK process, defined in Eq. (17) below, with parameters $M = 0.5$, $H = 0.8$, $\alpha = \lambda = 1$; see also Koutsoyiannis 2014). On its right tail the power spectrum has an asymptotic slope of -2, which is not inconsistent nor does it indicate nonstationarity (actually, a right-tail slope of -2 is precisely the slope of a stationary Markov model; see below). In contrast, on

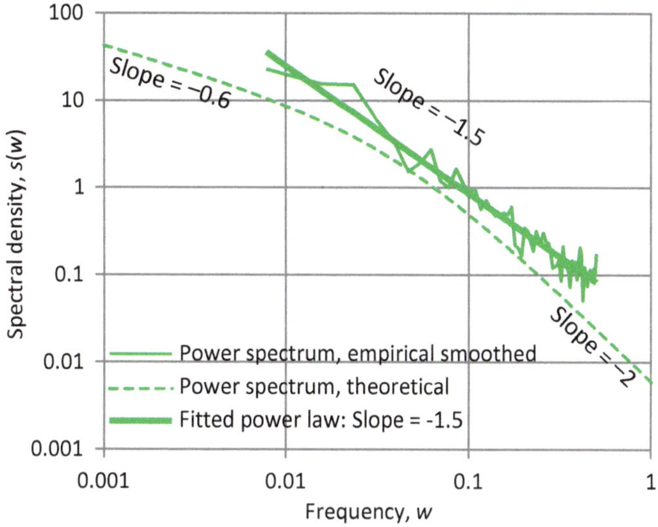

Fig. 1 Illustration of inconsistent results derived by hasty use of the power spectrum

its left tail the power spectrum has an asymptotic slope of −0.6, which is strictly >−1 (were it not, it would be inconsistent with theory, as will be detailed below).

From the shape of the theoretical power spectrum it can be imagined that if the time step and length of the data set were such that we could "see" only at frequencies >0.1, then we would conclude that we have a constant slope of −2 and, if we followed the standard fractal line of thought, we would claim that the process is (nonstationary) Brownian motion. Of course, all these would be incorrect as the model is purely stationary and not at all related to Brownian motion.

Even with the given data set, which allows us to "see" frequencies much lower than 0.1 (by an order of magnitude or more), the empirical power spectrum may again mislead us. For, even after the aforementioned smoothing, the empirical power spectrum is too rough to recover the underlying model and its parameters. Furthermore, it involves high bias and it suggests a misleading constant slope of −1.5. Just knowing the theoretical properties, as well as the uncertainty and bias of the power spectrum as a stochastic tool, we would avoid making erroneous claims, even though it is doubtful if this would help us to identify the correct model (see Dimitriadis and Koutsoyiannis 2015). Nonetheless, identifying the model from data and recovering the theoretically consistent asymptotic slopes (−0.6 and −2) are possible but need other methods (CS—see below).

The theoretical properties of the power spectrum which we need to know to avoid false claims include the following:

- Once we make the power spectrum of a process as a function of frequency, we have tacitly assumed a stationary process. In a nonstationary process, both the autocovariance and the spectral density, i.e., the Fourier transform of the autocovariance, are functions of two variables, one being related to "absolute" time (see, e.g., Dechant and Lutz 2015). Thus, there is no meaning in using a stationary representation (setting the power spectrum as a function of frequency only) and, at the same time, claiming nonstationarity. Even though this tactic has been very common, it is inconsistent. Furthermore, we should be aware that the customary Wiener-Khinchin theorem relating autocovariance and power spectrum pertains to stationary processes. This theoretical knowledge will prevent us from making claims of nonstationarity while using formulations and tools pertaining to stationary stochastic processes. In addition, we should be aware that claiming nonstationarity based solely on inductive reasoning is inconsistent (Koutsoyiannis and Montanari 2015).

- Once we use the power spectrum of a process for inference, as we always do, we should be aware that inference from data is only possible when the process is ergodic. As shown in Appendix 1, in an ergodic process, the asymptotic slope on the left tail of the power spectrum cannot be steeper than −1. Thus, there is no meaning in reporting slopes in empirical power spectra $s^{\#} < -1$ (e.g., $s^{\#} = -1.5$, as in the example of Fig. 1) and at the same time making any claim about the process properties (e.g., of nonstationarity) based on the power spectrum. Actually, such a steep slope, when emerging from processing of data, does not suggest that a process is non-ergodic, it rather identifies inconsistent estimation.
- We should be aware of the close relationship of ergodicity and stationarity (Koutsoyiannis and Montanari 2015). In particular, a nonstationary process is nonergodic and thus any estimates from data (including those of the power spectrum) are meaningless when we claim nonstationarity.
- As a result of the above listed theoretical points, constant slopes $\beta < -1$ of the power spectrum are invalid and indicate either inadequate length of data or inconsistent estimation algorithm. Likewise, non-constant slopes of power spectrum steeper than −1 ($s^{\#}(w) < -1$) for small frequencies ($w \to 0$) are equally invalid. We note that steep slopes ($s^{\#}(w) < -1$) are mathematically and physically possible for medium and large w—actually they are quite frequent in geophysical processes (see also Koutsoyiannis 2013a, b; Koutsoyiannis et al. 2013; Dimitriadis and Koutsoyiannis 2015).

2.5 *Misspecification/Misinterpretation of Scaling Laws*

The applicability of fractal analyses to complex phenomena of the real world essentially relies on the empirical detection of power-law relationships in observational data. Therefore, such analyses heavily rely on available data series and their statistical processing; and since they ask statistical questions, they must rely on probability theory (Stumpf and Porter 2012).

However, as the inference from data obeys statistical laws and is affected by statistical uncertainty and bias, we should respect these laws in making inference. Some examples can demonstrate that such respect is often not paid in fractal studies. The interested reader could perform a Google search with related terms (e.g.,

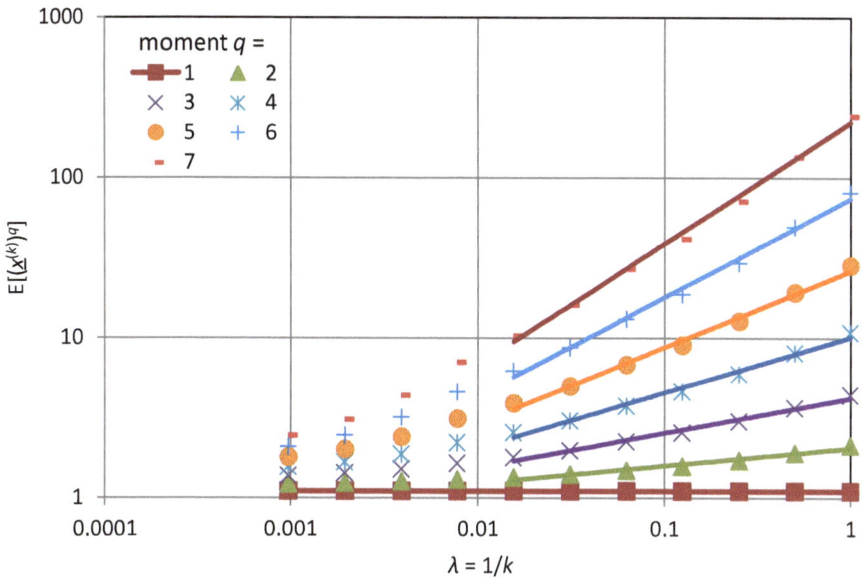

Fig. 2 Illustration of spurious scaling laws between raw moments and inverse time scale

universal multifractal rainfall—see also Koutsoyiannis 2010a) and several studies will be listed that identify multifractal behaviour of rainfall. This is usually done in terms of scaling relationships between raw moments of the averaged process $\underline{x}^{(k)}$ at time scale k, i.e., $\mathrm{E}[(\underline{x}^{(k)})^q]$ (or inverse time scale $\lambda := 1/k$), for several orders of moments q. Such scaling relationships are graphically identified on log-log plots and then the relationship of the scaling exponent (slope) K as a function q (the function $K(q)$) is empirically constructed (even though, according to universal multifractals, there exists a theoretical model for $K(q)$ that one can fit to empirical data; cf. Eq. (2.12) in Tessier et al. 1993).

A graphical example is provided in Fig. 2 to illustrate that the entire procedure is problematic from the outset. A time series with length $N = 2^{13} = 8192$ was generated from the HK process with Hurst parameter $H = 0.8$ and Gaussian distribution $N(1,1)$. Some scaling laws seem to appear at a range of time scales. One could be led to assume a multifractal behaviour and specify a $K(q)$ function. All these, however, are spurious. The truth is that there is no multifractal behaviour here. As shown theoretically by Lombardo et al. (2014) for $q = 2$, there is no constant slope K but, as $\lambda \to 0$ (or $k \to \infty$), the slope decreases to $K(q) = 0$. Also the slope empirically estimated for small k (large λ) is too low compared to its theoretical value.

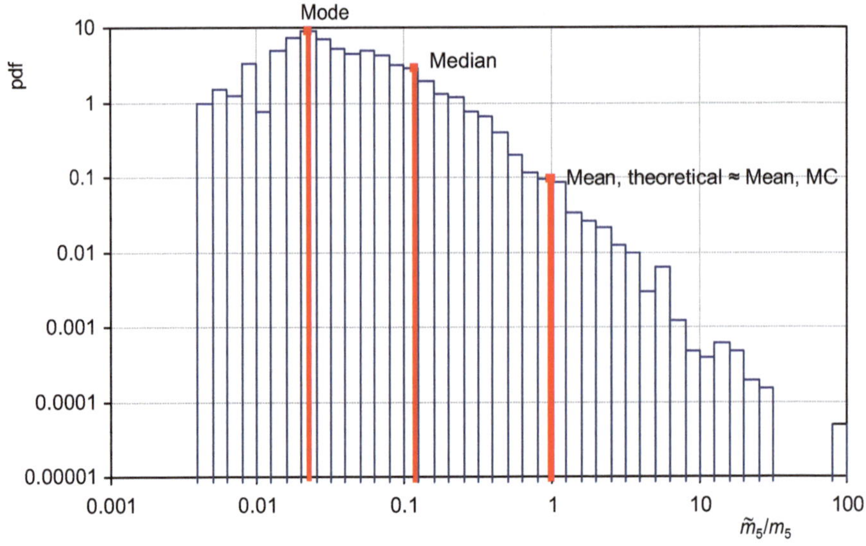

Fig. 3 Illustration of the statistical distribution of the estimate $\tilde{m}_5$ of the fifth moment m_5 of the Pareto distribution (pdf stands for probability density function)

2.6 *Neglect of Statistical Bias and Variation*

The above example illustrates a symptom of a more general tendency in the fractal literature to treat observations (time series) deterministically, confusing random variables with their realizations and ignoring statistical bias and variation. In the example of Fig. 2, high-order moments up to $q = 7$ have been used, as actually happens is several multifractal studies (this can be verified in studies that could be located with the Google search mentioned above).

However, high-order moments, which have been popular in multifractal studies, are well known in statistics to have minimal information content and therefore are avoided. This is further illustrated in Fig. 3, constructed after Monte Carlo simulation of the fifth moment of a Pareto distribution with shape parameter 0.15 and for sample size $n = 100$ (Papalexiou et al. 2010; see also Lombardo et al. 2014).

Here the theory guarantees that there is no estimation bias, but the distribution function is enormously skewed. The mode is nearly two orders of magnitude less than the mean and the probability that a calculation, based on data, will reach the mean is two orders of magnitude lower than the probability of obtaining the mode. Therefore, there is no meaning in using such uncertain quantity, with so skewed distribution, in any type of inference.

2.7 Confusion Between Different Scaling Behaviours

Scaling relationships, expressed as power laws between involved quantities, have been central in fractal studies. Yet their meaning has been obscure, while quite different scaling laws with different meanings are confused and regarded to be of the same nature. This is like regarding the different physical laws that involve the product of two quantities (e.g., $F = m\ a$, $W = F\ s$, $m = \rho\ V$, where F, m, a, W, s, ρ and V denote force, mass, acceleration, work, displacement, density and volume, respectively) as a manifestation of the same magical law of multiplicative quantities.

It is thus important to differentiate the unlike types of scaling met in geophysical processes and clarify their meaning. We can distinguish the following types of scaling (where the formal definitions of the various terms are given in Sect. 3):

- *Temporal scaling* indicates dependence in time and is expressed as a power law of some second-order property (marginal or joint second central moment) of a process with respect to a quantity related to time. We can further subdivide temporal scaling into:
  - Hurst behaviour, which is expressed as a power function of autocorrelation vs. time lag or climacogram vs. time scale;
  - fractal (local) behaviour, which is expressed as a power function of structure function vs. time lag or climacogram-based structure function (see below) vs. time scale.
- *Spatial scaling* is similar to temporal scaling but indicating dependence in space.
- *State scaling* is totally irrelevant to temporal and spatial scaling; it is related to the marginal distribution of the process and indicates a heavy-tailed distribution (a power law of probability of exceedance vs. state).
- *Scaling of* (high-order) *moments* with time scale; while in theory this cannot be excluded, in most empirical studies it perhaps is an artefact related to other types of scaling and, as explained above, it is usually spurious because high-order moments are not reliably estimated from data.

As already mentioned, in real world systems scaling laws never extend to the entire range of scales. Usually they are asymptotic laws, with different exponents at each edge. Asymptotic scaling laws abound because, in our view, they are a mathematical necessity (Koutsoyiannis 2014). The asymptotic behaviour of stochastic properties of processes (such as survival function, autocovariance, structure function, climacogram, etc.) should necessarily tend to zero at one edge (e.g., at infinity) and the decay to zero can be exponential (fast decay) or of power-type (slow decay). In the latter case, the emergence of an asymptotic power law is obvious, whether it holds in the form of scaling in state (heavy-tailed distributions) or in time (long-term persistence). Both cases have been verified in geophysical time series (e.g., O'Connell et al. 2016; Markonis and Koutsoyiannis 2016, 2013; Dimitriadis and Koutsoyiannis 2017). According to this view, scaling behaviours are just manifestations of enhanced uncertainty and are consistent with the principle of maximum entropy (Koutsoyiannis 2011; see also below). The connection of scaling with maximum entropy constitutes also a connection of stochastic representations of natural processes with statistical physics.

3 Fundamentals of Stochastics for Geophysics

In this section, we give a very brief presentation of the most fundamental concepts of stochastics. Later, in Sect. 5 we will show that these concepts suffice to model complex phenomena without making any use of the fractal nomenclature, even though some of these phenomena are thought to belong to the preferential domain of the fractal literature (e.g., turbulence).

3.1 The Meaning of Randomness and Stochastics

A deterministic world view is founded on a concept of sharp exactness. A deterministic mathematical description of a system uses regular variables (e.g., x) which are represented as numbers. The change of the system state is represented as a *trajectory* $x(t)$, which is the sequence of a system's states x as time t changes.

In an indeterministic world view there is uncertainty or randomness, where the latter term does not mean anything more than unpredictability or intrinsic uncertainty. A system's description is done in terms of random variables. A random variable $\underline{x}$ is an abstract mathematical entity whose realizations x belong to a set of possible numerical values. A random variable $\underline{x}$ is associated with a probability density (or mass) function $f(x)$. Notice the different notation of random variables (underlined, according to the Dutch notation; Hemelrijk 1966) from regular ones. The evolution of a system over time is no longer sufficient to be represented as a trajectory but as a stochastic process $\underline{x}(t)$, which is a collection of (usually infinitely many) random variables $\underline{x}$ indexed by t (typically representing time). A realization (sample) $x(t)$ of $\underline{x}(t)$ is a trajectory; if it is known at certain points t_i, $i = 1, 2, \ldots$, it is a time series.

The mathematics of random variables and stochastic processes is termed stochastics, and is composed of probability theory, statistics and stochastic processes. Most natural processes evolve in continuous time but they are observed in discrete time, instantaneously or by averaging. Accordingly, the stochastic processes devised to represent the natural processes should evolve in continuous time and be converted into discrete time, as illustrated in Fig. 4.

While a stochastic process denotes, by conception, change (process = change), there should be some properties that are unchanged in time. This implies the concept of stationarity (Koutsoyiannis and Montanari 2015), which is central in stochastics. For the remaining part of this article, the processes are assumed to be stationary, noting that nonstationary processes should be converted to stationary before their

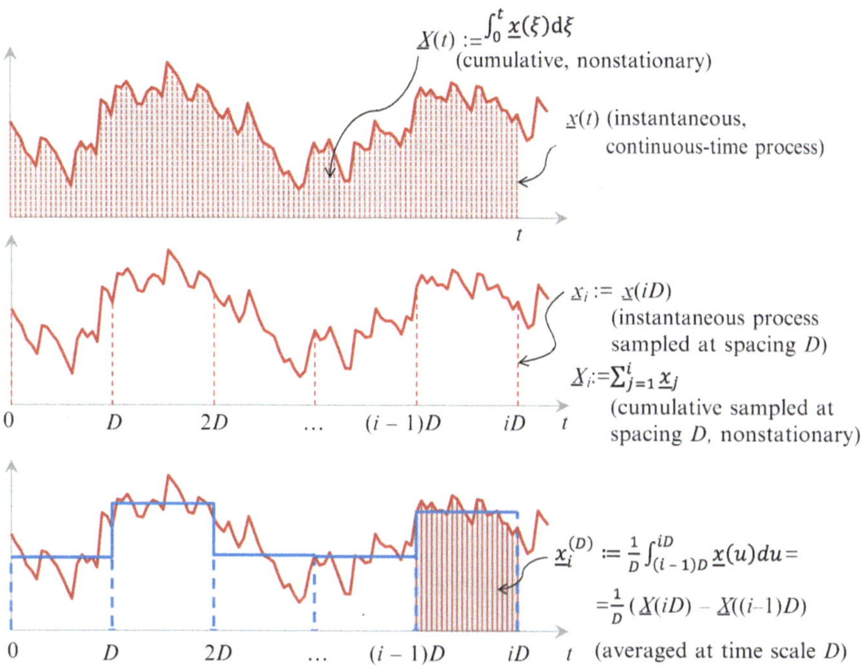

Fig. 4 Explanatory sketch for a stochastic process in continuous time and two different representations in discrete time. Note that the graphs display a realization of the process (it is impossible to display the process as such) while the notation is for the process per se

study (for example, the cumulative process $\underline{X}(t)$ in Fig. 4 is nonstationary, but by differentiating it in time we obtain the stationary process $\underline{x}(\mathrm{t})$). The most customary properties of a stationary stochastic process are its second-order properties:

- *Autocovariance function*, $c(h) := \mathrm{Cov}[\underline{x}(t), \underline{x}(t+h)]$.
- *Power spectrum* (also known as *spectral density*), $s(w)$; it is defined as the Fourier transform of the autocovariance function, i.e., by Eq. (2).
- *Structure function* (also known as *semivariogram* or *variogram*), $v(h) := (1/2)\,\mathrm{Var}\,[\underline{x}(t) - \underline{x}\,(t+h)]$.
- *Climacogram*, $\gamma(k) := \mathrm{Var}[\underline{x}_i^{(k)}]$, where $\underline{x}_i^{(k)}$ is the averaged process over time scale k (see Fig. 4 and substitute a varying time scale k for the constant time interval D).

For time-related quantities, in the above notation and in the next part of this article, we use the following symbols, where Latin letters denote dimensional quantities and Greek letters dimensionless ones, where the latter are convenient when using the discrete-time variants of a process:

- *Time unit* (time step in case of sampling or time scale in case of aggregating or averaging), D.

- *Time*, $t = \tau D$ (alternatively for strictly integer $i = 1, 2, \ldots, t = i D$ where t is continuous time and i discrete time).
- *Time lag*, $h = \eta D$.
- *Time scale* $k = \kappa D$.
- *Frequency*, $w = \omega / D$, related to time scale by $w = 1/k$, $\omega = 1/\kappa$.

All these properties are transformations of one another, i.e.:

$$s(w) = 4\int_0^\infty c(h)\cos(2\pi w h)\,\mathrm{d}h, \quad c(h) = \int_0^\infty s(w)\cos(2\pi w h)\,\mathrm{d}w \tag{2}$$

$$v(h) = c(0) - c(h), \quad c(h) = c(0) - v(h) \tag{3}$$

$$\gamma(k) = 2\int_0^1 (1 - v)\,c\,(vk)\,\mathrm{d}v, \quad c(h) = \frac{1}{2}\frac{\mathrm{d}^2\left(h^2\gamma(h)\right)}{\mathrm{d}h^2} \tag{4}$$

where Eq. (3) is valid when the variance of the instantaneous process is finite ($\gamma_0 := \gamma(0) \equiv c(0) \neq \infty$).

The climacogram is not as popular as the other tools but it has several good properties due to its simplicity, close relationship to entropy (see below) and more stable behaviour, which is an advantage in model identification and fitting from data. In particular, when estimated from data, the climacogram behaves better than all other tools, which involve high bias and statistical variation (Dimitriadis and Koutsoyiannis 2015; Koutsoyiannis 2016). The climacogram involves bias too, but this can be determined analytically and included in the estimation. Furthermore, it enables the definition of additional useful tools as shown in Table 1.

The CSF, $\xi(k)$, behaves like the structure function $v(h)$ and is related to the latter by the same way as the climacogram $\gamma(k)$ is related to the autocovariance function $c(h)$:

Table 1 Climacogram-based metrics of stochastic processes

Metric/usefulness	Definition	Comments
Climacogram Useful for the global asymptotic behaviour ($k \to \infty$)	$\gamma(k) := \mathrm{Var}[\underline{x}_i^{(k)}]$	For an ergodic process for $k \to \infty$, $\gamma(k) \to 0$ necessarily
Climacogram-based structure function (CSF) Useful for the local asymptotic behaviour ($k \to 0$)	$\xi(k) := \gamma_0 - \gamma(k)$	The definition presupposes that the variance γ_0 is finite
Climacogram-based spectrum (CS) Useful for both the global and local asymptotic behaviour	$\psi(w) := \frac{2}{w\gamma_0}\gamma\,(1/w)\,\xi\,(1/w) = \frac{2\,\gamma(1/w)}{w}\left(1 - \frac{\gamma(1/w)}{\gamma_0}\right)$	It combines the climacogram and the CSF; valid even for infinite variance

$$c(h) = \frac{1}{2}\frac{\mathrm{d}^2\left(h^2\gamma(h)\right)}{\mathrm{d}h^2}, \quad v(h) = \frac{1}{2}\frac{\mathrm{d}^2\left(h^2\xi(h)\right)}{\mathrm{d}h^2} \tag{5}$$

The CS, $\psi(w)$, behaves like the power spectrum; it has same dimensions, and in most cases has precisely the same asymptotic behaviour as the power spectrum, but it is smoother and more convenient in model identification and fitting (see Sect. 5).

3.2 Second-Order Properties at Discrete Time

Once the continuous-time properties are known, the discrete-time ones can be readily calculated. For example, and assuming a time interval D for discretization, as in Fig. 4, the autocovariance of the averaged process is:

$$c_\eta^{(D)} = \mathrm{Cov}\left[\underline{x}_\tau^{(D)}, \underline{x}_{\tau+\eta}^{(D)}\right] = \frac{1}{D^2}\left(\frac{\Gamma\left(|\eta+1|D\right) + \Gamma\left((|\eta-1\,|D\right)}{2} - \Gamma\left(|\eta|D\right)\right) \tag{6}$$

where $\Gamma(D) := \mathrm{Var}[\underline{X}(D)] = D^2\gamma(D)$. Also, the power spectrum of the averaged process can be calculated from:

$$s_\mathrm{d}^{(D)}(\omega) = 2c_0^{(D)} + 4\sum_{\eta=1}^{\infty} c_\eta^{(D)}\cos\left(2\pi\eta\omega\right) \tag{7}$$

where $s_\mathrm{d}^{(D)}(\omega) := s^{(D)}(w)/D$ (nondimensionalized spectral density), whereas the discrete-time power spectrum $s^{(D)}(w)$ is related to the continuous-time one by (Koutsoyiannis 2016)

$$s^{(D)}(\omega) = \sum_{j=-\infty}^{\infty} s\left(w + \frac{j}{D}\right)\mathrm{sin\,c}^2\left(\pi\left(wD + j\right)\right) \tag{8}$$

More details and additional cases can be found in Koutsoyiannis (2013b, 2016).

3.3 Cautionary Notes for Model Fitting

Model identification and fitting is much more important than commonly thought. Even the statistical literature has paid little attention to the fact that direct estimation of *any statistic* of a process (except perhaps for the mean) is not possible merely from the data. We always need to assume a model to estimate statistics.

Any statistical estimator $\hat{\underline{s}}$ of a true parameter s is biased either strictly (meaning: $\mathrm{E}[\hat{\underline{s}}] \neq s$) or loosely (meaning: mode $[\hat{\underline{s}}] \neq s$). Model fitting is necessarily based on discrete-time data and needs to consider the effects of (a) discretization and (b) bias.

It is commonly thought that the standard estimator of the variance from a sample of size n is unbiased if we divide the sum of squared deviations from mean by $n-1$ instead of n (Eq. 9). This is correct only if the assumed model is the white noise. Otherwise, the estimation is biased and, if the process has long-range dependence, the bias can be substantial. The climacogram, which is none other than the variance, needs to consider this bias. Actually, it is easy to analytically estimate the bias and the effect of discretization, once a model has been assumed in continuous time.

Let us consider a process with climacogram $\gamma(k)$, from which we have a time series for an observation period T (multiple of the time step D), each one giving the averaged process $\underline{x}_i^{(D)}$ at a time step D. We form time series for scales that are multiples of D, i.e., $k = \kappa D$, $\kappa = 1, 2, \ldots$, and we wish to estimate the variance at any such scale (including that at scale D, for $\kappa = 1$). The standard estimator $\hat{\underline{\gamma}}(k)$ of the variance $\gamma(k)$ is

$$\hat{\underline{\gamma}}(k) := \frac{1}{n-1}\sum_{i=1}^{n}\left(\underline{x}_i^{(k)} - \underline{x}_1^{(T)}\right)^2 = \frac{1}{T/k-1}\sum_{i=1}^{T/k}\left(\underline{x}_i^{(k)} - \underline{x}_1^{(T)}\right)^2 \tag{9}$$

where by inspection it is seen that $\underline{x}_1^{(T)}$ is the sample mean, while it was assumed that T is a multiple of k so that the sample size is $n = T/D$ (if not, we should replace T with $\lfloor T/k \rfloor k$, where $\lfloor . \rfloor$ denotes the floor of a real number). It can be then shown (Koutsoyiannis 2011, 2016) that the bias can be calculated from

$$\mathrm{E}\left[\hat{\underline{\gamma}}(k)\right] = \chi(k,T)\,\gamma(k),\ \chi(k,T) = \frac{1-\gamma(T)/\gamma(k)}{1-k/T} = \frac{1-(k/T)^2\Gamma(T)/\Gamma(k)}{1-k/T} \tag{10}$$

3.4 Entropy and Entropy Production

As already mentioned, the emergence of scaling from maximum entropy considerations may provide the theoretical background in modelling complex natural processes by scaling laws.

The Boltzmann-Gibbs-Shannon entropy of a cumulative process $\underline{X}(t)$ with probability density function $f(X;t)$ is a dimensionless quantity defined as:

$$\Phi[\underline{X}(t)] := \mathrm{E}\left[-\ln\frac{f(\underline{X};t)}{m(\underline{X})}\right] = -\int_{-\infty}^{\infty}\ln\frac{f(X;t)}{m(X)}f(X;t)\,dX \tag{11}$$

where $m(X)$ is the density of a background measure (typically Lebesgue). The entropy production in logarithmic time (EPLT) is a dimensionless quantity, the derivative of entropy in logarithmic time (Koutsoyiannis 2011):

$$\phi(t) \equiv \phi[\underline{X}(t)] := \Phi'[\underline{X}(t)]\,t \equiv \mathrm{d}\Phi[\underline{X}(t)]/\mathrm{d}(\ln t) \tag{12}$$

For a Gaussian process with constant density of background measure, $m(X) \equiv m$, the entropy depends on its variance $\Gamma(t)$ only and is:

$$\Phi[\underline{X}(t)] = (1/2)\ \ln\left(2ne\Gamma(t)/m^2\right), \phi(t) = \Gamma'(t)\,t/2\Gamma(t) \tag{13}$$

When the past ($t < 0$) and the present ($t = 0$) are observed, instead of the unconditional variance $\Gamma(t)$ we should use a variance $\Gamma_{\mathrm{C}}(t)$ conditional on the past and present:

$$\Gamma_{\mathrm{C}}(t) \approx 2\Gamma(t) - \frac{\Gamma(2t)}{2}, \quad \varphi_{\mathrm{C}}(t) = \frac{\Gamma_{\mathrm{C}}'(t)t}{2\Gamma_{\mathrm{C}}(t)} \approx \frac{\left(2\Gamma'^{(t)} - \Gamma'^{(2t)}\right)t}{4\Gamma(t) - \Gamma(2t)} \tag{14}$$

3.5 *Resulting Processes from Maximizing Entropy Production*

Koutsoyiannis (2011) assumed that the behaviour seen in natural processes is consistent with extremization of entropy production and provided a framework to derive processes maximizing entropy production. Using simple constraints in maximization, such as known variance at the scale $k = D = 1$, and lag one autocovariance for the same time scale, the following processes extremizing the EPLT $\varphi(t)$ and $\varphi_{\mathrm{C}}(t)$ can be derived, which are also depicted in Fig. 5 in terms of their EPLT and climacograms.

- A *Markov* process:

$$c(h) = \lambda \mathrm{e}^{-h/\alpha}, \gamma(k) = \frac{2\lambda}{k/\alpha}\left(1 - \frac{1 - \mathrm{e}^{-k/\alpha}}{k/\alpha}\right) \tag{15}$$

maximizes entropy production for small times but minimizes it for large times.

- A *Hurst-Kolmogorov* (HK) process:

$$\gamma(k) = \lambda(\alpha/k)^{2-2H} \tag{16}$$

maximizes entropy production for large times but minimizes it for small times.

- A *Hybrid Hurst-Kolmogorov* (HHK) process:

$$\gamma(k) = \lambda\left(1 + (k/\alpha)^{2M}\right)^{\frac{H-1}{M}} \tag{17}$$

maximizes entropy production both at small and large time scales.

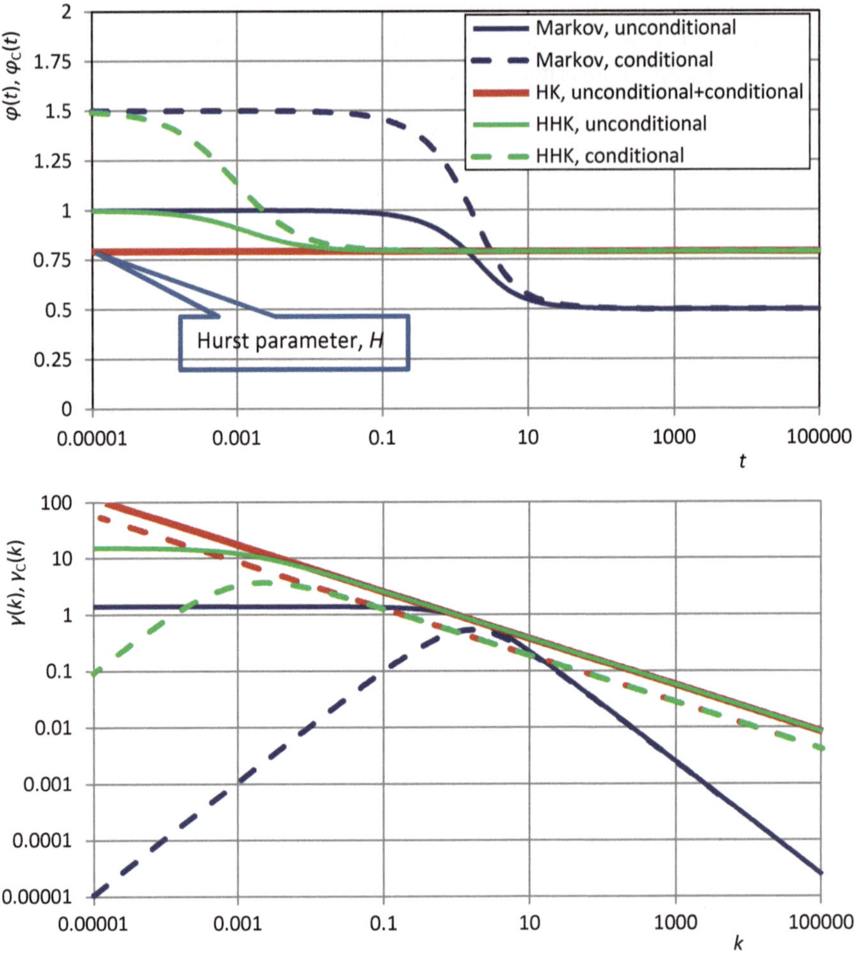

Fig. 5 EPLTs (*upper*) and climacograms (*lower*) of the three processes extremizing entropy production. At time scale $k = 1$ all three processes have the same variance, $\gamma(1) = 1$, and the same autocovariance for lag 1, $c_1^{(1)} = 0.5$. Their parameters are (see text for their definitions): for the Markov process $\alpha = 0.8686$, $\lambda = 1.4176$; for the HK process $a = 0.0013539$, $\lambda = 15.5032$, $H = 0.7925$; for the HHK process $a = 0.0013539$, $\lambda = 15.5093$, $M = 0.5$, $H = 0.7925$ (adapted from Koutsoyiannis 2016).

In these definitions α and λ are scale parameters with dimensions of $[t]$ and $[x^2]$, respectively. The parameter H (in honour of Hurst) is the Hurst parameter which determines the global properties of the process (as $k \to \infty$). The parameter M (in honour of Mandelbrot) is the fractal parameter which determines the local properties (as $k \to 0$). Both H and M are dimensionless numbers in the interval $(0, 1)$. In the HHK process, locality and globality are clearly independent of each other, each one characterized by an asymptotic power law. Hence, it allows explicit control of both asymptotic logarithmic slopes of the CS $\psi^{\#}(k)$ and the power spectrum $s^{\#}(w)$. In the special case where $H = M = 0.5$, HHK is practically indistinguishable from a Markov process, even though not precisely identical. Furthermore, as $\alpha \to 0$, the process tends to a pure HK process with the same Hurst parameter H. Also, for any specific parameter set, HHK exhibits Markov behaviour for small time scales (if $M = 0.5$, or similar to Markov if $M \neq 0.5$) and Hurst behaviour for large time scales, as seen in Fig. 5.

The HHK process is consistent with natural behaviours and remedies known inconsistencies of the HK process (discussed in subsection "2.3"), while retaining the persistence or antipersistence properties. Specifically, the variance of the instantaneous process is always finite ($\gamma_0 = \gamma(0) = \lambda$), while even for $0 < H < 0.5$ the initial part of the autocovariance function for small lags is positive for all variants of the process (continuous time, discrete time, either sampled or averaged, for a small time interval D).

4 Simulation of Stochastic Processes Respecting Their Fractal Properties

Monte Carlo (stochastic) simulation is an important numerical method for resolving problems that have no analytical solution. Obviously, simulation is performed in discrete time, at a convenient discretization step. The following method based on the so-called symmetric moving average (SMA) scheme (Koutsoyiannis 2000, 2016) can be used to exactly simulate any Gaussian process, with any arbitrary autocovariance function (provided that it is mathematically feasible). It can also approximate, with controlled accuracy, any non-Gaussian process with any arbitrary autocovariance function and any marginal distribution function.

4.1 The Symmetric Moving Average Scheme

The SMA scheme can directly generate time series x_i (where for simplicity we have omitted the time interval D in the notation) from any process $\underline{x}_i$ with any type of dependence by:

$$\underline{x}_i = \sum_{l=-\infty}^{\infty} a_{|l|} \underline{v}_{i+l} \tag{18}$$

where a_l are coefficients calculated from the autocovariance function and $\underline{v}_i$ is white noise averaged in discrete time. Assuming that the power spectrum $s_{\mathrm{d}}^{(D)}(\omega)$ of the averaged discrete-time process is known (from the equations listed above), it has been shown (Koutsoyiannis 2000) that the Fourier transform $s_\mathrm{d}^a(\omega)$ of the a_l series of coefficients is related to the power spectrum of the discrete time process as

$$s_\mathrm{d}^a(\omega) = \sqrt{2 s_\mathrm{d}^{(D)}(\omega)} \tag{19}$$

Thus, to calculate a_l we first determine $s_\mathrm{d}^a(\omega)$ from the power spectrum of the process and then we invert the Fourier transform to estimate all a_l.

4.2 Handling of Truncation Error

It is expected that the coefficients a_l will decrease with increasing l and will be negligible beyond some q ($l > q$), so that we can truncate (18) to

$$\underline{x}_i = \sum_{l=-q}^{q} a_{|l|} \underline{v}_{i+l} \tag{20}$$

This introduces some truncation error in the resulting autocovariance function. To adjust for this on the variance, we calculate the a_l from

$$a_l = a_l' + a'' \tag{21}$$

where the coefficients a_l' are calculated from inverting the Fourier transform of either $s_\mathrm{d}^a(\omega)$ or $s_\mathrm{d}^a(\omega)\,(1 - \mathrm{sinc}\,(2\pi\omega q))$ (two options; Koutsoyiannis 2016).

The constant a'' is determined so that the variance is exactly preserved:

$$\gamma(D) = \sum_{l=-q}^{q} a_{|l|}^2 = \sum_{l=-q}^{q} \left(a_{|l|}' + a''\right)^2 \tag{22}$$

Solving for a'', this yields:

$$a'' = \sqrt{\frac{\gamma(D) - \Sigma a'^2}{2q+1} + \left(\frac{\Sigma a'}{2q+1}\right)^2} - \frac{\Sigma a'}{2q+1} \tag{23}$$

where $\Sigma a' := \sum_{l=-q}^{q} a_{|l|}'$ and $\Sigma a'^2 := \sum_{l=-q}^{q} a_{|l|}'^2$.

4.3 Handling of Moments Higher than Second-Order

In addition to being general for any second-order properties (autocovariance function), the SMA method can explicitly preserve higher order marginal moments.

Here it should be made clear that, while, as already mentioned, high-order moments cannot be estimated reliably from data, non-Gaussianity is very commonly verified empirically and also derived by theoretical reasoning (Koutsoyiannis 2014, 2005). An easy manner to simulate non-Gaussian (e.g., skewed) distributions is to calculate theoretically (not from the data) their moments and then explicitly preserve these moments in simulation. Preservation of three or four central moments usually provides good approximations to the theoretical distributions. Apparently, by preserving four moments, a non-Gaussian distribution is not precisely preserved. What can be assumed to be preserved is a Maximum Entropy (ME) approximation of the distribution constrained by the known moments. For four known moments of the variable $\underline{x}$ this approximation should be an exponentiated fourth-order polynomial of x (Jaynes 1957; Papoulis 1991), which can be written as

$$f(x) := \frac{1}{\lambda_0} \mathrm{e}^{-\left(\frac{x}{\lambda_1} + \mathrm{sign}(\lambda_2)\left(\frac{x}{\lambda_2}\right)^2 + \left(\frac{x}{\lambda_3}\right)^3 + \left(\frac{x}{\lambda_4}\right)^4\right)} \tag{24}$$

where λ_i are parameters, all with dimensions $[x]$ (with $\lambda_4 \geq 0$).

The third and fourth moments are more conveniently expressed in terms of the coefficients of skewness and kurtosis, respectively. To produce a discrete-time process $\underline{x}_i$ with coefficient of skewness $C_{\mathrm{s},x}$ we need to use a white-noise process $\underline{v}_i$ with coefficient of skewness (Koutsoyiannis 2000):

$$C_{s,v} = C_{s,x} \frac{\left(\sum_{l=-q}^{q} a_{|l|}^2\right)^{3/2}}{\sum_{l=-q}^{q} a_{|l|}^3} \tag{25}$$

Likewise, to produce a process $\underline{x}_i$ with coefficient of kurtosis $C_{\mathrm{k},x}$ the process $\underline{v}_i$ should have coefficient of kurtosis (Dimitriadis and Koutsoyiannis 2017):

$$C_{k,v} = \frac{C_{k,x}\left(\sum_{l=-q}^{q} a_{|l|}^2\right)^2 - 6\sum_{l=-q}^{q-1}\sum_{k=l+1}^{q} a_{|l|}^2 a_{|k|}^2}{\sum_{l=-q}^{q} a_{|l|}^4} \tag{26}$$

Four-parameter distributions are needed to preserve skewness and kurtosis; details are provided by Dimitriadis and Koutsoyiannis (2017). Illustration of the very good performance of the method in the generation of non-Gaussian white noise is provided in Fig. 6 for popular distribution functions such as Weibull, gamma, lognormal and Pareto.

It is finally noted that the method can also be used in multivariate processes, represented by vectors of random variables (Koutsoyiannis 2000).

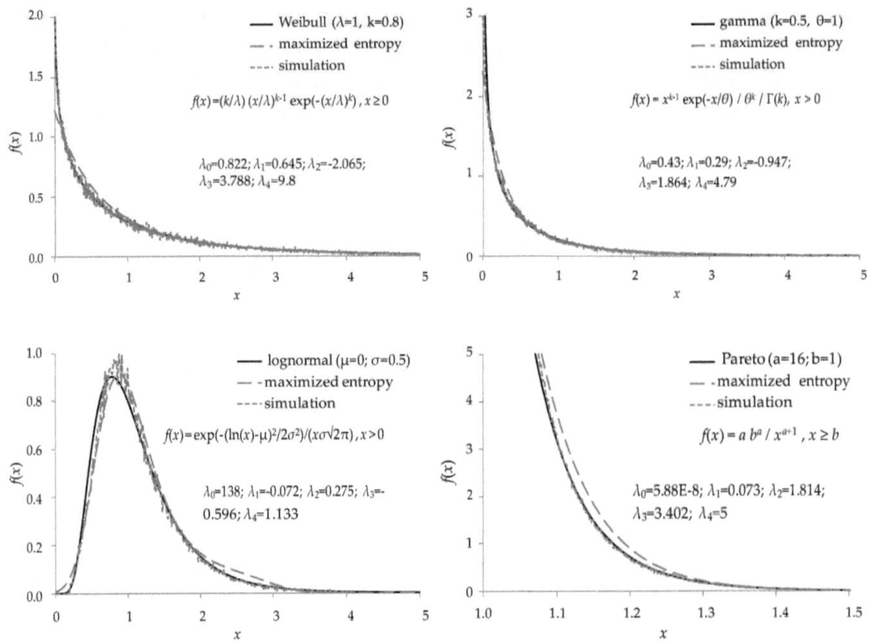

Fig. 6 Various two-parameter probability density functions along with their fitted ME approximations and the empirical probability density from a single synthetic time series with $n = 10^5$ (from Dimitriadis and Koutsoyiannis 2017)

5 Applications

5.1 Application 1: Turbulence

Estimation of high-order moments involves large uncertainty and cannot be reliable in the typically short time series of geophysical processes. However, in laboratory experiments at sampling intervals of μs, very large samples can be formed which can support the reliable estimation of high-order moments. Here we use grid-turbulence data made available on the Internet by the Johns Hopkins University (http://www.me.jhu.edu/meneveau/datasets/datamap.html). This dataset consists of 40 time series with $n = 36 \times 10^6$ data points of longitudinal wind velocity along the flow direction, all measured at a sampling time interval of 25 μs by X-wire probes placed downstream of the grid (Kang et al. 2003).

By standardizing all series (see Dimitriadis et al. 2016; Dimitriadis and Koutsoyiannis 2017) we formed a sample of $40 \times 36 \times 10^6 = 1.44 \times 10^9$ values to estimate the marginal distribution, and an ensemble of 40 series, each with 36×10^6 values to estimate the dependence structure through the climacogram. Based on this dataset we built a stochastic model of turbulence, which to verify we performed

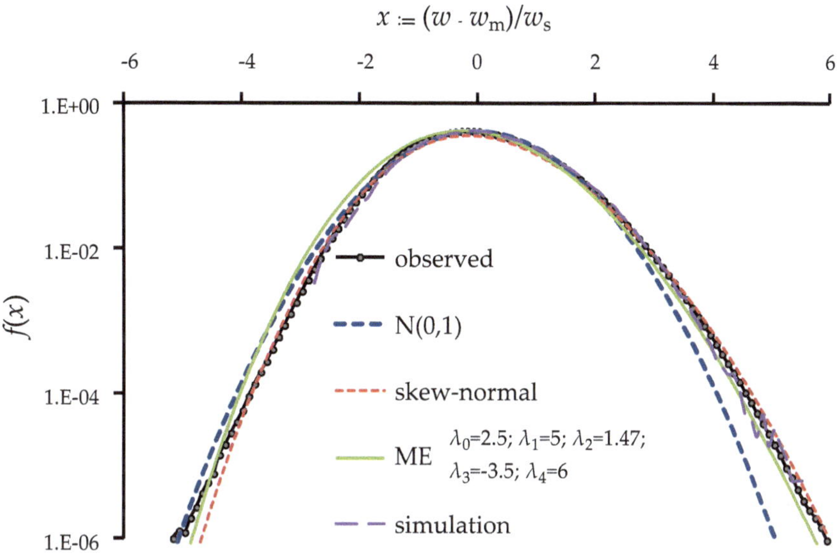

Fig. 7 Probability density function of the measured turbulent velocity w standardized, in each time series, by the mean w_m and standard deviation w_s, compared to that of a single simulation using the SMA scheme preserving the first four moments; the standard normal distribution $N(0,1)$ and the skew normal (both not used in simulation) are also shown. The ME approximation, also shown in the figure, is the one used in simulations

stochastic simulation using the SMA framework with $n = 10^6$ values and compared the synthetic data with the measurements using several tools.

In terms of the marginal distribution, the time series are nearly Gaussian but not exactly Gaussian. There are slight deviations from normality toward positive skewness, as indicated by the coefficient of skewness, which is 0.2 instead of 0, and that of kurtosis, which is 3.1 instead of 3, as well as from the plot of the probability function shown in Fig. 7. This divergence of fully developed turbulent processes from normality has been also justified theoretically (Wilczek et al. 2011). Interestingly, these slight differences from normality result in highly non-normal distribution of the white noise $\underline{v}_i$ of the SMA model (skewness $C_{s,v} = 3.26$; kurtosis $C_{k,v} = 12.30$!); this should have substantial effects in some aspects of turbulence.

For the stochastic dependence of the turbulent velocity process, after some exploratory analysis, we assumed a model consisting of the sum of two equally weighted processes, an HHK and a Markovian:

$$\gamma(k) = \frac{\lambda}{2}\left(1 + (k/\alpha)^{2M}\right)^{\frac{H-1}{M}} + \frac{\lambda}{k/\alpha}\left(1 - \frac{1 - e^{-k/\alpha}}{k/\alpha}\right) \tag{27}$$

We fitted the model to the climacogram, the structure function, the CS and the power spectrum, calculated as the average of the 40 series. The fitting is shown

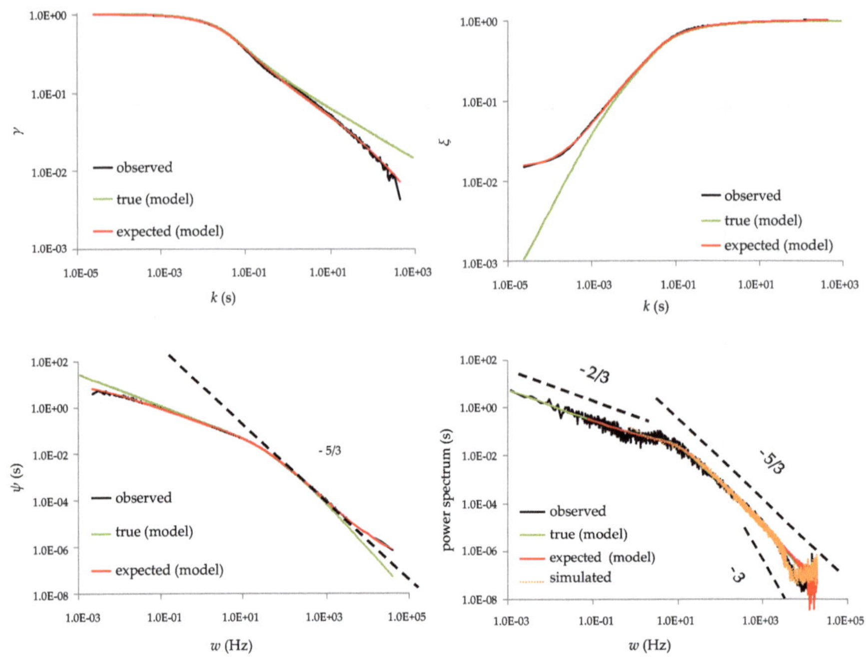

Fig. 8 Empirical, true and expected values of the climacogram (*upper left*), CSF (*upper right*), CS (*lower left*) and power spectrum (*lower right*). The "observed" is the average from the 40 time series

in Fig. 8; the four parameters of the model are estimated as: $\lambda = 1$, $\alpha = 14$ ms, $M = 1/3$, $H = 5/6$. As seen in Fig. 8, the model is indistinguishable from the data, measured or synthesized, when the climacogram or its derivatives CSF and CS are used. Note that the comparison of the empirical quantities is not made with the true ones but with the expected, in order to take account of the bias.

The power spectrum is much rougher than the other tools, yet a good model fit can be clearly seen. Kolmogorov's "5/3" law of turbulence (K41 self-similar model; Kolmogorov 1941) is also evident in the power spectrum for $w > 10$ Hz. Steepening of the power spectrum slope for even larger frequencies ($w > 1000$ Hz), which has also reported in several studies, is also apparent in Fig. 8. This, however, seems to be a numerical effect (resulting from discretization and bias), as the same behaviour appears also in the simulated data from a model whose structure (Eq. 27) does not include anything that would justify steepening of the slope.

It is extremely insightful to investigate the high-order properties of the velocity increments, i.e., differences of velocities at adjacent times with a certain time distance (lag) h. In particular, the variation of high-order moments of the velocity increments with increasing h (i.e., the moments $v_p := \mathrm{E}\left[|\underline{x}(t) - \underline{x}(t+h)|^p\right]$ for $p > 2$) has been associated with the intermittent behaviour of turbulence and has

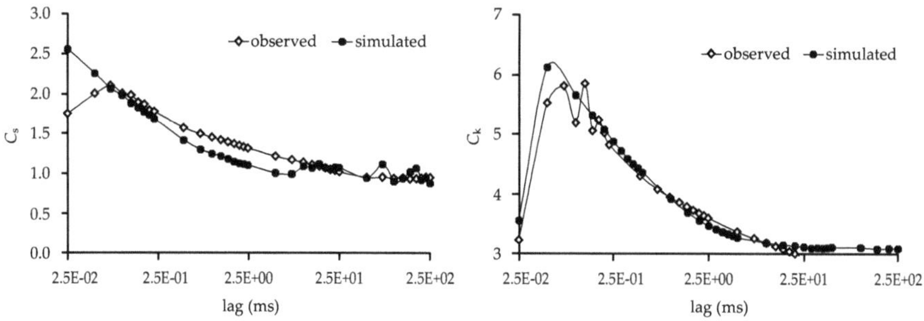

Fig. 9 Empirical and simulated coefficients of skewness (*left*) and kurtosis (*right*) of the velocity increments vs. lag

been mentioned as the intermittent effect (Frisch 2006, Sect. 8.3), first discovered in turbulence by Batchelor and Townsend (1949). Therefore it is important to preserve this variation. The model in Eq. (27) does not make any effort for such preservation. However, as seen in Fig. 9, these are preserved well and effortless. Therefore, it is no longer puzzling to have large kurtosis (even > 5) in velocity increments, even though the velocity is almost normal. No additional assumption, model component, or even model parameter is necessary. Similar good preservation appears also for the skewness of velocity increments (Fig. 9).

The huge data size in this application allows evaluation of even higher moments and construction of a plot (Fig. 10) of the exponent ζ_p vs. moment order p of an assumed scaling relationship

$$v_p := \mathrm{E}\left[\left|\underline{x}(t) - \underline{x}(t+h)\right|^p\right] \approx h^{\zeta_p} \tag{28}$$

which has been very common in the literature. Again the agreement between the simulated and measured data is impressive, particularly if we bear in mind the fact that no provision has been made to this aim. Some more simulations have been used to investigate this further and a number of additional curves have been plotted in Fig. 10. It is thus seen that the HHK model alone fails to preserve this actual behaviour if a Gaussian distribution is assumed; it rather approached the K41 self-similar model (Kolmogorov 1941) as reproduced by Frisch (2006, Fig. 8.8). Similar results are obtained if a Markov dependence structure is assumed along with the modelled marginal distribution based on the empirical moments (Fig. 7). Interestingly, if we combine the modelled distribution (Fig. 7) and the modelled climacogram (Eq. 27), then we adequately preserve the intermittent effect without the need for any other mono-fractal (such as the β-model) nor multifractal models (cf. Frisch 2006, Sect. 8.5) and not even the She–Leveque model (She and Leveque 1994), which is also plotted in Fig. 10 (Frisch 2006, Sects. 8.6.4 and 8.6.5) and behaves also well against the empirical data.

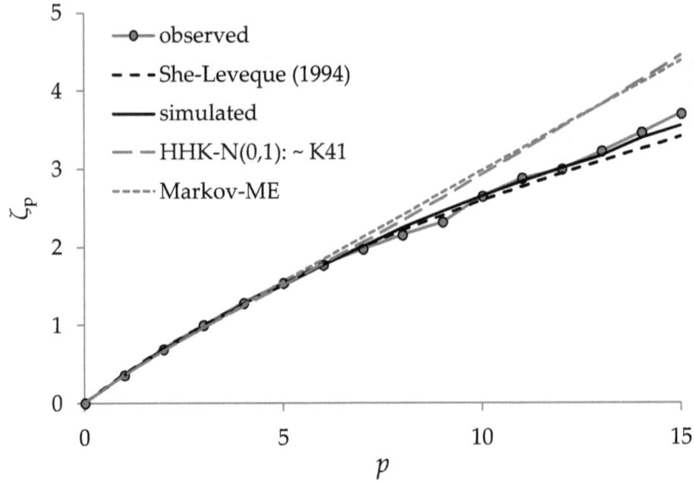

Fig. 10 Empirical values of the scaling exponent ζ_p vs. moment order p of the scaling relationship (28)

In conclusion, this application shows that all important properties of turbulence, including its short- and long-term characteristics, as well as intermittency, can be very well modelled without any mystery but using a parsimonious stochastic model, theoretically justified on the basis of the maximization of entropy production (Koutsoyiannis 2011), with both Hurst and fractal behaviours and slightly non-Gaussian distribution (with skewness of 0.2 and kurtosis of just 3.1).

5.2 Application 2: Wind

Understanding atmospheric motion in the form of wind is essential to many fields in geophysics. Wind is considered one of the most important processes in hydrometeorology since, along with temperature, it drives climate dynamics. Currently, the interest for modelling and forecasting of wind has increased due to the importance of wind power production in the frame of renewable energy resources development.

For the investigation of the large scale of atmospheric wind speed, we use over 15000 meteorological stations around the globe (Fig. 11, upper) recorded mostly by anemometers and with hourly resolution (www.noaa.gov; GHCN database). In total, we analyse almost 4000 stations from different sites and climatic regimes by selecting time series that are still operational, with at least one year length of data, at least one non-zero measurement per three hours on average and at least 80% of non-zero values for the whole time series (Fig. 11, middle). This data set is referred to below as "global".

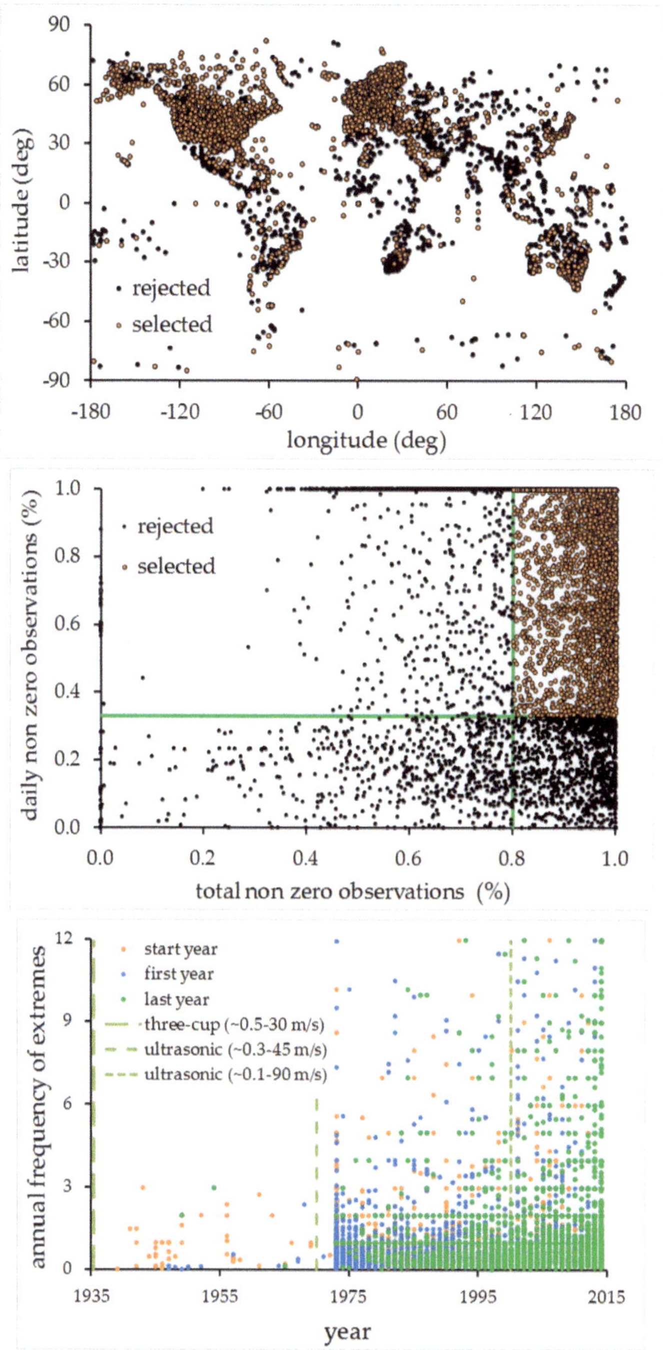

Fig. 11 (*Upper*) Distribution of the wind speed stations over the globe; (*middle*) sketch about the selection of the stations in the analysis; (*lower*) evolution of the frequency of measured extremes in the stations (where the 'start' year denotes the first operational year of the station and the 'first' and 'last' year denote the first and last year that an extreme value was recorded, respectively)

By standardizing all series we formed a sample of $\sim 0.5 \times 10^9$ values to estimate the marginal distribution, and an ensemble of 3886 series, each with $\sim 10^5$ values on average, to estimate the dependence structure through the climacogram. A known problem of field measurements of wind (particularly those originating from over 70 years ago) is that the technology of measuring devices has been rapidly changed (Manwell et al. 2010, Sect. 2.8.3). For example, in Fig. 11 (lower) we illustrate a rather virtual increase of extreme wind events after the 1970s which is mainly due to the inability of older devices to properly measure wind speeds over 30 m/s (i.e., category I of Saffir–Simpson hurricane wind scale). Furthermore, in common anemometer instrumentation there is a lower threshold of speed that could be measured, usually within the range 0.1–0.5 m/s (e.g., www.pce-instruments.com). It should be noted that, as the recorded wind speed decreases, so does the instrumental accuracy and it may be a good practice to always set the minimum threshold to 0.5 m/s to avoid measuring the errors of the instrument (e.g., zero or extremely low values) in place of the actual wind speed that can never reach an exact zero value.

In an attempt to incorporate smaller scales, starting from the microscale of turbulence, we include again the dataset of the previous application of turbulence, using it as an indicator of the similar statistical properties of small scale wind (Castaing et al. 1990). In addition to the 40 time series of the longitudinal turbulent velocity, here we also use another 40 time series of transverse velocity, measured at the same points with the longitudinal one; again each time series has $n = 36 \times 10^6$ data points with a sampling interval of 25 μs. The coefficients of skewness and kurtosis are estimated as 0.1 and 3.1 for the transverse velocity, respectively. Stochastic similarities between small-scale atmospheric wind and turbulent processes abound in the literature as, for example, in terms of the marginal distribution (Monahan 2013 and the references therein), of the distribution of fluctuations (Bottcher et al. 2007 and the references therein), of the second-order dependence structure (Dimitriadis et al. 2016 and the references therein) and of higher-order behaviour such as intermittency (e.g., Mahrt 1989).

Finally, to link the large and small scale of atmospheric wind we analyse an additional time series, referred to as "medium", provided by NCAR/EOL of one-month length and with a 10 Hz resolution. This time series has been recorded by a sonic anemometer on a meteorological tower located at Beaumont KS and it includes over 25×10^6 longitudinal and transverse wind speed measurements (http://data.eol.ucar.edu/; Doran 2011).

The statistical characteristics based on moments up to fourth order are shown in Fig. 12; interestingly, there appears to be a rather well-defined relationship between mean and standard deviation. The plot of coefficient of kurtosis vs. coefficient of skewness indicates that Weibull distribution falls close to the lower bound of the scatter of empirical points.

Numerous works have been conducted for the distribution of the surface wind speed (see Appendix 2 for a sample of recent studies). The Weibull distribution has proven very useful in describing the wind magnitude distribution for over three decades (Monahan 2006 and the references therein). However, various studies illus-

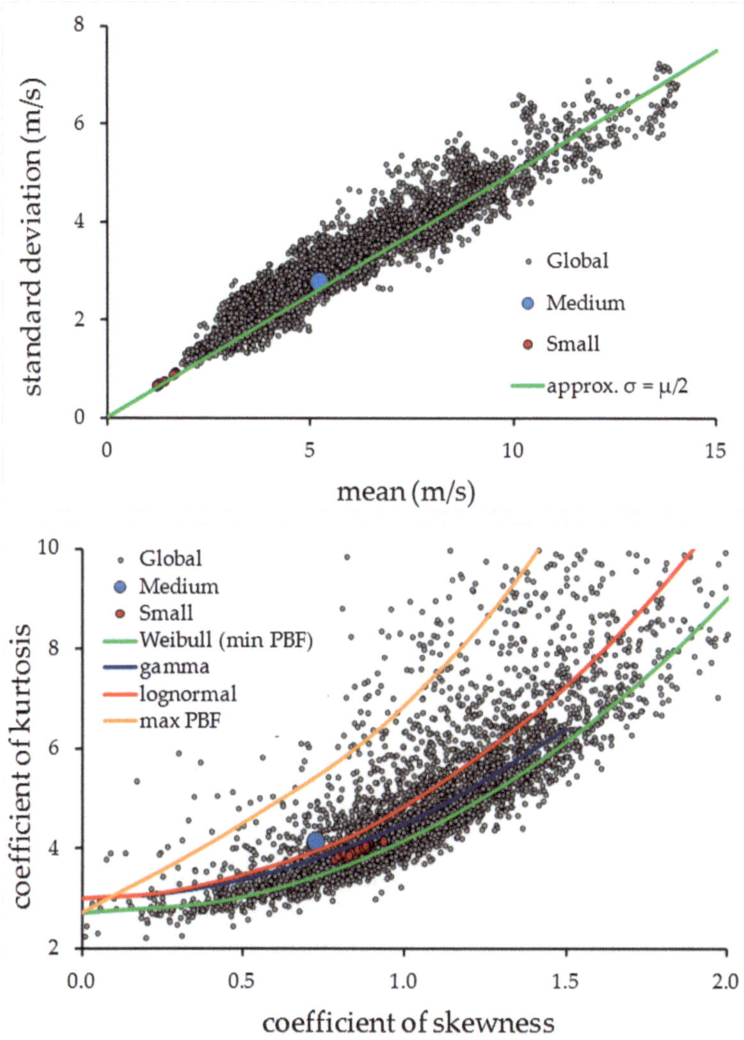

Fig. 12 Standard deviation vs. mean (*upper*) and coefficient of kurtosis vs. coefficient of skewness of all time series used in Application 2

trate empirical as well as physically based deviations from the Weibull distribution (Drobinski and Coulais 2012 and the references therein). Due to the discussed limitations of properly measuring wind speed most studies have focused on a local or small scale. In such cases where there is limited empirical evidence, we could search for a physical justification for the left and right tail of the probability function.

It can easily be proven that the length of a vector of uncorrelated Gaussian distributions with zero mean and equal variance follows the Rayleigh distribution.

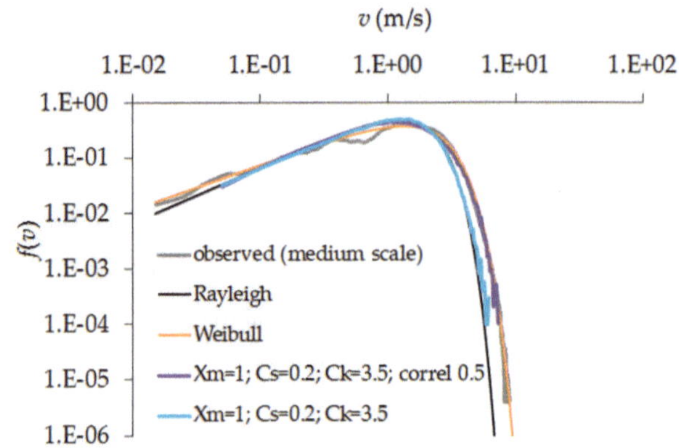

Fig. 13 Probability density function of the medium scale time series along with theoretical and Monte Carlo generated distributions

However, there is empirical and theoretical evidence (Application 1) that the small-scale distribution of turbulence is not Gaussian and it is expected that this should also be the case for the components of wind speed. Through Monte Carlo experiments we illustrate that correlated non-Gaussian components result in a distribution close to Weibull and are in agreement with small and medium scale observations (an example is shown in Fig. 13).

The distribution of the "global" time series appears to deviate from Weibull, gamma, and lognormal distributions, and is closer to a distribution with a much heavier tail:

$$F(v) = 1 - \left(1 + \left(\frac{v}{\alpha v_s}\right)^b\right)^{-c/b} \tag{29}$$

where $v > 0$ is the wind speed, v_s is the standard deviation of the wind speed process; α is a scale parameter, and b and c are the shape parameters of the marginal distribution, all three dimensionless. For this distribution we use the name Pareto-Burr-Feller (PBF) to give credit to the engineer V. Pareto, who discovered a family of power-type distributions for the investigation of the size distribution of incomes in a society (Singh and Maddala 1976), to Burr (1942) who identified and analysed (but without giving a justification) a function first proposed as an algebraic form by Bierens de Haan, and to Feller (1970) who linked it to the Beta function and distribution. Other names such as Pareto type IV or Burr type VII are also in use for the same distribution. Interestingly, the PBF distribution has two different asymptotic properties, i.e., the Weibull distribution for low wind speeds and the Pareto distribution for large ones. The derivation of PBF from maximum entropy has been studied in Yari and Borzadaran (2010). The PBF has been used in a variety of independent fields (see Brouers 2015). Therefore, it seems that there is a strong physical as well as empirical justification for applying the PBF to the analysis of the wind process.

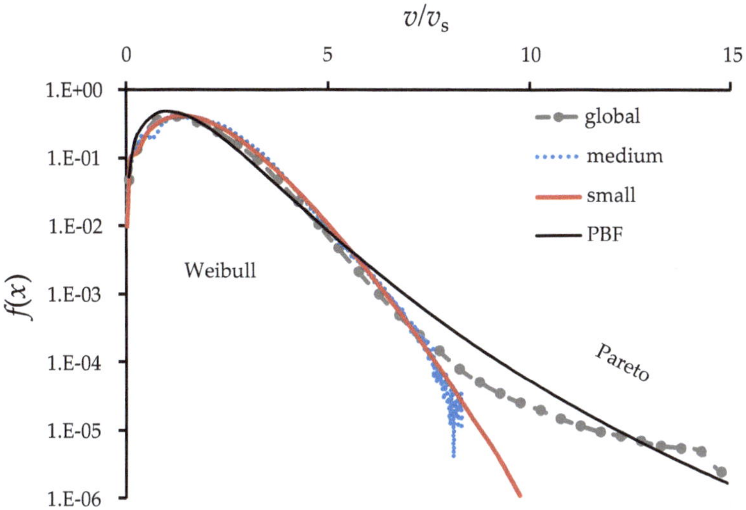

Fig. 14 Probability density function of the velocity of grid-turbulent data (small) and of the wind speed of the medium and global scale time series along with fitted theoretical distributions

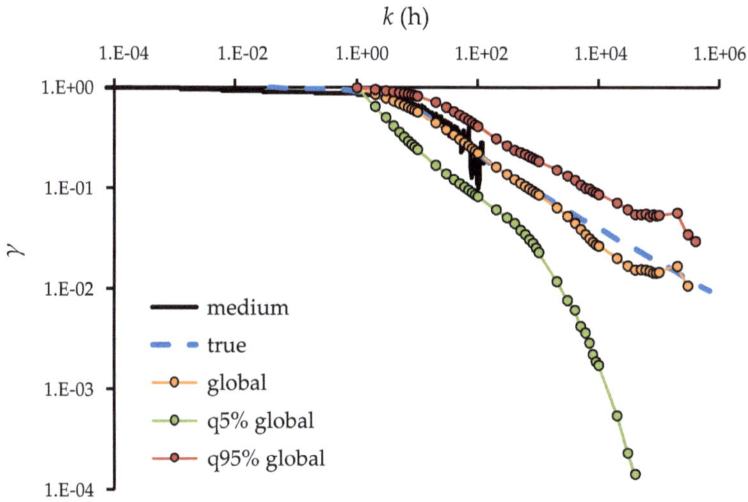

Fig. 15 Climacogram of the wind speed process estimated from the medium and global series

The distribution fitted to all data sets is shown in Fig. 14 and the fitted parameters are $\alpha = 3.5$, $b = 1.9$, $c = 8.5$. The mean estimated climacograms from the data (Fig. 15) indicate that the model of Eq. (27) is also applicable for the wind speed at all scales with parameters estimated as $\lambda \approx 1$, $M = 1/3$, $H = 5/6$ and $\alpha = 6$ h.

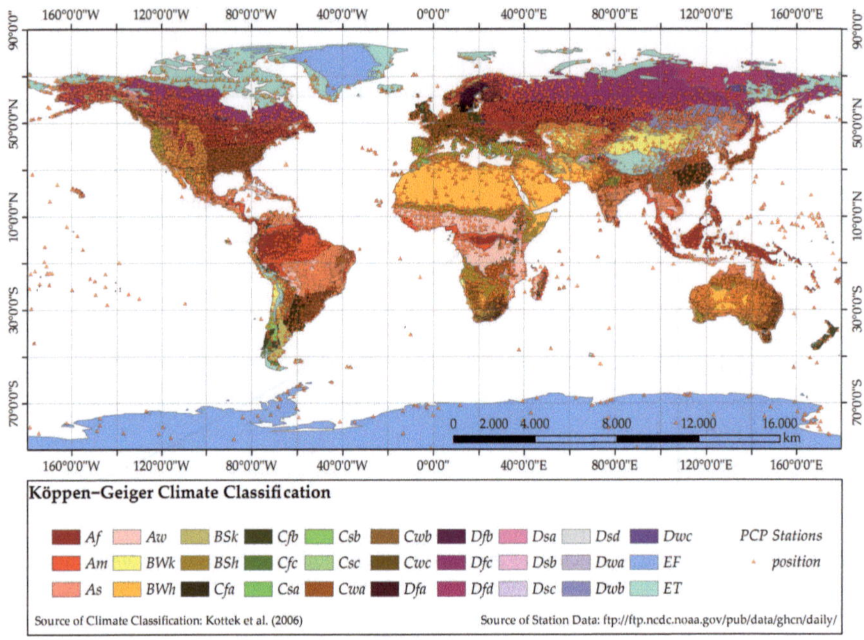

Fig. 16 Locations of the selected hourly time series of air temperature

5.3 Application 3: Temperature

In this last application we analyse the dependence structure of the air temperature process close to surface. For the microscale structure, we use a 10 Hz resolution time series recorded for a two-month period via a sonic anemometer at Beaumont, USA (https://data.eol.ucar.edu/dataset/45.910). For the macro-scale structure, we use a global database of hourly air temperature (https://www.ncdc.noaa.gov/data-access/land-based-station-data). In total, we analyse over 5000 stations from different sites and climatic regimes by selecting time series with at least 1 year length and at least one measurement per three hours (Fig. 16).

It can be assumed that the air temperature process follows a Gaussian distribution (Koutsoyiannis 2005). Indeed, Fig. 17 shows that the 90% of the time series have skewness around 0 ± 1 and kurtosis around 3 ± 1. We normalize all time series and we estimate the dependence structure through the climacogram, autocovariance, and power spectrum.

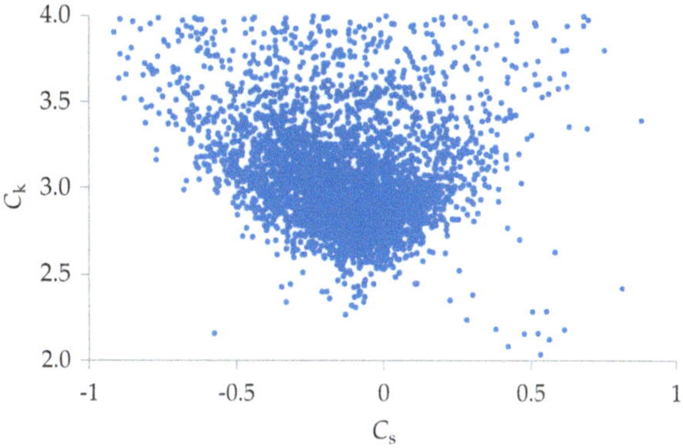

Fig. 17 Coefficient of skewness vs. coefficient of kurtosis for ~90% of the macro-scale temperature time series

The mean estimated climacograms from the data (Fig. 18) and the CS (Fig. 19) indicate that, interestingly, the model of Eq. (27) is also applicable here with parameters estimated as $\lambda \approx 1$, $M = 1/3$, $H = 5/6$ and $\alpha = 3.3$ d.

6 Concluding Remarks

Stochastics offers a strong basis for modelling and interpretation of natural behaviours and can directly incorporate, in a rigorous manner, useful concepts from the fractal literature, removing the ambiguity characterizing many fractal studies. Stochastics offers all tools for data analysis, inductive inference and prediction with quantified uncertainty, but above all it offers the basis for a logical world view.

We owe the well-founded and rigorous mathematical theory of stochastics to Kolmogorov (1931, 1933, 1938), including the foundation of scaling processes (Kolmogorov 1940). This theory has often been distorted but there exist textbooks consistent with it (e.g., Papoulis 1991).

Calculating values of sample statistics without considering their statistical properties (bias and statistical variation) can yield misleading results. Without proper attention to the underlying stochastics, we can even "identify" phenomena that do not exist and take statistical sampling effects as natural behaviours.

A general methodology for data analysis and construction of synthetic time series is possible provided that we have a good understanding of stochastics. In particular, the applications presented here suggest a promising characterization of different geophysical processes in a unified manner and with a simple and parsimonious stochastic model, appropriate for a range of scales spanning several orders of magnitude.

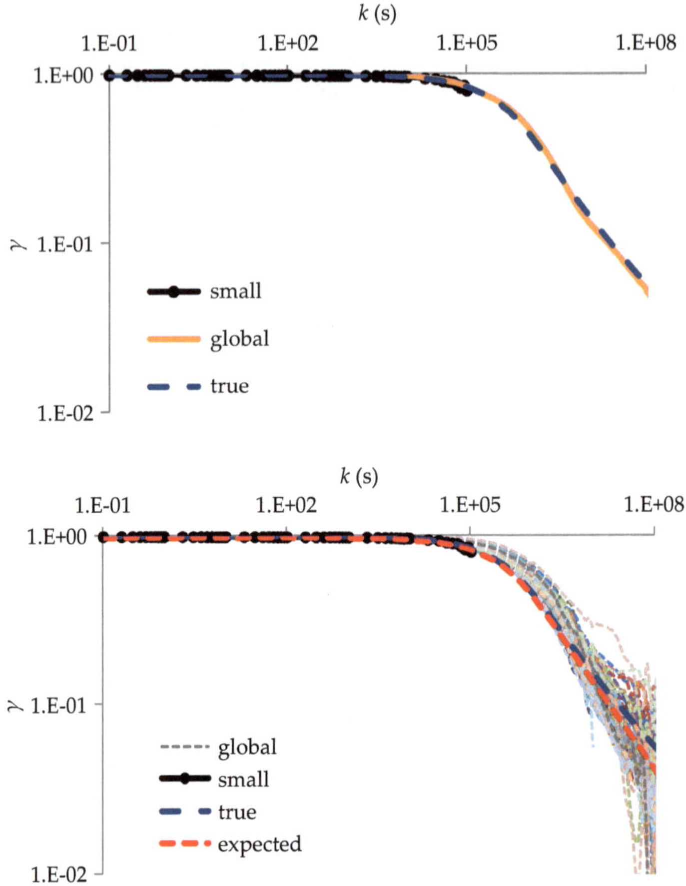

Fig. 18 Climacogram of the normalized temperature for the micro-scale time series (small) and the set of hourly air temperature time series (global; *upper*: average climacogram; *lower*: climacograms of 100 different time series), compared to the fitted model of Eq. (27) (true and expected)

Appendix A: Proof of Infeasibility of Too Steep Slopes in Power Spectrum at Low Frequencies

This proof is summarized here from Koutsoyiannis (2013b) and Koutsoyiannis et al. (2013).

Let us assume the contrary, i.e., that for frequency range $0 \leq w \leq \varepsilon$ (with ε however small) the log-log derivative is $s^{\#}(w) = \beta$, or else $s(w) = \alpha\, w^{\beta}$ where α and β are constants, with $\beta < -1$. As a result of (2) and (4) the climacogram is related to power spectrum by:

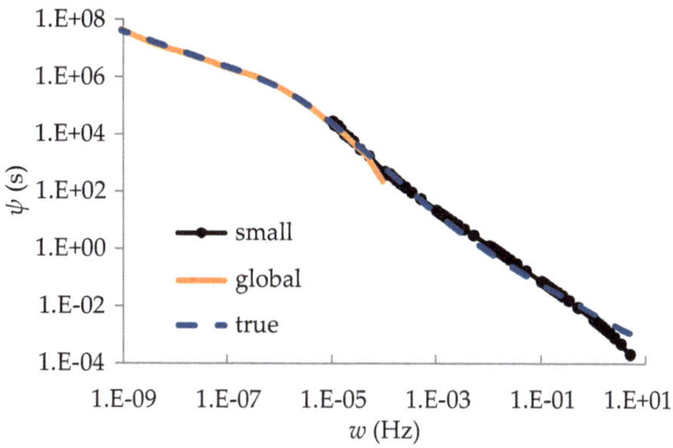

Fig. 19 CS of the normalized temperature for the micro-scale time series (small) and the set of hourly air temperature time series (global; average from all time series), compared to the fitted model of Eq. (27) (true)

$$\gamma(k) = \int_0^\infty s(w)\,\mathrm{sinc}^2\,(\pi wk)\,\mathrm{d}w \tag{30}$$

The sinc^2 function within the integral takes significant values only for $w < 1/k$ (cf. Papoulis 1991, p. 433). Assuming a scale $k \gg 1/\varepsilon$,

$$\gamma(k) = \int_0^\infty s(w)\mathrm{sinc}^2\,(\pi wk)\,\mathrm{d}w \approx \int_0^\varepsilon \alpha w^\beta\,\mathrm{sinc}^2\,(\pi wk)\,\mathrm{d}w \tag{31}$$

On the other hand, it can be easily seen that, for $0 < w < 1/k$, the following inequality holds:

$$\mathrm{sinc}\,(\pi wk) \geq 1 - wk \geq 0 \tag{32}$$

Since $\varepsilon \gg 1/k$, while the function in the integral (31) is nonnegative,

$$\gamma(k) \approx \int_0^\varepsilon \alpha w^\beta \mathrm{sinc}^2\,(\pi wk)\,\mathrm{d}w \geq \int_0^{1/k} \alpha w^\beta \mathrm{sinc}^2\,(\pi wk)\,\mathrm{d}w \geq \int_0^{1/k} \alpha w^\beta (1 - wk)^2 \mathrm{d}w \tag{33}$$

By substituting $\omega = wk$ into Eq. (33), we find:

$$\gamma(k) \geq ak^{-\beta-1} \int_0^1 \omega^\beta (1 - \omega)^2 \mathrm{d}\omega \tag{34}$$

To evaluate the integral in (34) we take the limit for $q \to \infty$ of the integral:

$$B(q) := \int_{1/q}^{1} \omega^{\beta}(1-\omega)^2 \mathrm{d}\upsilon = \frac{1-q^{-1-\beta}}{1+\beta} - 2\frac{1-q^{-2-\beta}}{2+\beta} + \frac{1-q^{-3-\beta}}{3+\beta} \quad (35)$$

Clearly, the limit of $B(q)$ as $q \to \infty$ depends on that of the term with the highest exponent, i.e., $q^{-1-\beta}$. For $\beta < -1$ this term diverges and thus, $B(0) = +\infty$. Then, by virtue of the inequality (34), $\gamma(k)= \infty$. For a (mean) ergodic processes $\gamma(k)$ should necessary tend to 0 for $k \to \infty$ (Papoulis 1991, p. 429). Therefore, the process is nonergodic.

It is interesting to note here that, when $|\beta| < 1$, the integral in (31) can be evaluated to give:

$$\gamma(k) \approx \alpha \int_{0}^{\infty} w^{\beta} \operatorname{sinc}^2(\pi w \Delta)\, \mathrm{d}w = \frac{\alpha \Gamma(1+\beta) \operatorname{sinc}(\pi\beta/2)}{2(1-\beta)(2\pi)^{\beta} k^{1+\beta}} \quad (36)$$

Clearly, for $k \to \infty$, the last expression gives $\gamma(k) \to 0$ and thus for $|\beta| < 1$ the process is mean ergodic.

This analysis for $\beta < -1$ generalizes a result by Papoulis (1991, p. 434) who shows that an impulse at $w = 0$ corresponds to a non-ergodic process.

Appendix B: Literature Review on the Distribution Function of Wind Speed

A large variety of distributions in the literature (with the most common to be Gaussian, gamma, Weibull, lognormal, Pareto and generalizations thereof as well as mixtures with each other) show equally good agreement with atmospheric wind measurements recorded at different sites around the globe with different climatic conditions.

A sample of recent publications is listed in Table 2 along with the proposed distributions. However, some distributions seem to exhibit good agreement with data at the left or right tail mostly due to different lengths of the examined time series, while arguably most distributions do not exhibit good agreement for the whole range.

Table 2 Recent publications on the distribution function of wind speed

Reference	General characteristics	Proposed distribution	Comments
Aksoy et al. (2004)	1 station; 4 years	Weibull	Markov chain
Monahan (2006)	Global; sea-surface; wind speed	Weibull	Non Rayleigh
Bottcher et al. (2007)	Laboratory; 4 atmospheric stations; wind components	Castaing et al. (1990)	Standard deviation with a lognormal model for intermittency
Kiss and Janosi (2008)	Reanalysis data over Europe	Generalized gamma	Non-Rayleigh; non-Weibull
Morgan et al. (2011)	178 offshore time series; 10-min wind speed	Kappa	14 distribution tested; non-Weibull; non Rayleigh
Lo Brano et al. (2011)	Wind speed over Palermo	Burr	Tested: Weibull, Rayleigh, lognormal, gamma, inverse-Gaussian, Pearson V
Drobinski and Coulais (2012)	3 stations; high altitude; wind components	Rayleigh-Rice	Non-Weibull, Elliptical distribution to model skewness
Wu et al. (2013)	Inner Mongolia region	Lognormal	Weibull; logistic
Ouarda et al. (2015)	9 stations in United Arab Emirates	Kappa, generalized gamma	18 distributions tested with mixture properties

References

Aksoy, H.Z., T. Fuad, A. Aytek, and N. Erdem. 2004. Stochastic generation of hourly mean wind speed data. *Renewable Energy* 29: 2111–2131.

Arnold, B.C. 1983. *Pareto distributions*. Fairland, MD: International Co-operative Publishing House.

Bartlett, M.S. 1948. Smoothing periodograms from time series with continuous spectra. *Nature* 161 (4096): 686–687. doi:10.1038/161686a0.

Batchelor, G.K., and A.A. Townsend. 1949. The nature of turbulent motion at large wave-numbers. *Proceedings of the Royal Society of London A* 199: 238–255.

Beran, J., Y. Feng, S. Ghosh, and R. Kulik. 2013. *Long-memory processes: Probabilistic properties and statistical methods*. Berlin: Springer.

Bottcher, F., S. Barth, and J. Peinke. 2007. Small and large scale fluctuations in atmospheric wind speeds. *Stochastic Environmental Research and Risk Assessment* 21: 299–308.

Burr, I.W. 1942. Cumulative frequency functions. *Annals of Mathematical Statistics* 13: 215–235.

Brouers, F. 2015. The Burr XII distribution family and the maximum entropy principle: Power-law phenomena are not necessarily nonextensive. *Open Journal of Statistics* 5: 730–741.

Castaing, B., Y. Gagne, and E.J. Hopfinger. 1990. Velocity probability density functions of high Reynolds number turbulence. *Physica D* 46: 177–200.

Dechant, A., and E. Lutz. 2015. Wiener-Khinchin theorem for nonstationary scale-invariant processes. *Physical Review Letters* 115 (8): 080603.

Dimitriadis, P., and D. Koutsoyiannis. 2015. Climacogram versus autocovariance and power spectrum in stochastic modelling for Markovian and Hurst–Kolmogorov processes. *Stochastic Environmental Research and Risk Assessment* 29 (6): 1649–1669. doi:10.1007/s00477-015-1023-7.

———. 2017. Stochastic synthesis approximating any process dependence and distribution. *Stochastic Environmental Research and Risk Assessment*. (in review).

Dimitriadis, P., D. Koutsoyiannis, and P. Papanicolaou. 2016. Stochastic similarities between the microscale of turbulence and hydrometeorological processes. *Hydrological Sciences Journal* 61 (9): 1623–1640.

Doran, C. (2011). Anemometer–Sonic at ABLE Beaumont Site Data. Version 1.0. UCAR/NCAR–Earth Observing Laboratory. http://data.eol.ucar.edu/dataset/45.910. Accessed 07 Jan 2017.

Drobinski, P., and C. Coulais. 2012. Is the Weibull distribution really suited for wind statistic modelling and wind power evaluation. *Journal of Physics Conference Series* 753: 5–8.

Falconer, K. 2014. *Fractal geometry: Mathematical foundations and applications*. 3rd ed. Chichester: Wiley.

Feller, W. 1970. *An introduction to probability and its applications*. Vol. II. 2nd ed. New York, NY: Wiley.

Frisch, U. 2006. *Turbulence: The legacy of A. N. Kolmogorov*. Cambridge: Cambridge University Press.

Gneiting, T., and M. Schlather. 2004. Stochastic models that separate fractal dimension and the Hurst effect. *Society for Industrial and Applied Mathematics Review* 46 (2): 269–282.

Graham, L., and J.-M. Kantor. 2009. *Naming infinity: A true story of religious mysticism and mathematical creativity*. Cambridge: Harvard University Press.

Grassberger, P., and I. Procaccia. 1983. Characterization of strange attractors. *Physical Review Letters* 50 (5): 346–349.

Hemelrijk, J. 1966. Underlining random variables. *Statistica Neerlandica* 20 (1): 1–7.

Jaynes, E.T. 1957. Information theory and statistical mechanics. *Physics Review* 106: 620.

Kang, H.S., S. Chester, and C. Meneveau. 2003. Decaying turbulence in an active-grid-generated flow and comparisons with large-eddy simulation. *Journal of Fluid Mechanics* 480: 129–160.

Kantelhardt, J.W. 2009. Fractal and multifractal time series. In *Encyclopedia of complexity and systems science*, ed. R.A. Meyers, vol. LXXX, 3754–3778. Berlin: Springer.

Kiss, P., and I.M. Janosi. 2008. Comprehensive empirical analysis of ERA-40 surface wind speed distribution over Europe. *Energy Conversion and Management* 49 (8): 2142–2151. doi:10.1016/j.enconman.2008.02.003.

Kolmogorov, A.N. 1931. Uber die analytischen Methoden in der Wahrscheinlichkcitsrechnung. *Mathematische Annalen, English translation: On analytical methods in probability theory, In: Kolmogorov, A.N., 1992. Selected Works of A. N. Kolmogorov—Volume 2, Probability Theory and Mathematical Statistics A. N. Shiryayev, ed., Kluwer, Dordrecht, The Netherlands, pp. 62–108* 104: 415–458.

———. (1933). *Grundbegrijfe der Wahrscheinlichkeitsrechnung*, Ergebnisseder Math. (2), Berlin. (2nd English Edition: Foundations of the theory of probability, 84 pp. Chelsea Publishing Company, New York, 1956).

———. 1938. A simplified proof of the Birkhoff-Khinchin ergodic theorem. *Uspekhi Matematicheskikh Nauk* 5: 52–56. (English edition: Kolmogorov, A.N., 1991, Selected Works of A. N. Kolmogorov - Volume 1, Mathematics and Mechanics, Tikhomirov, V. M. ed., Kluwer, Dordrecht, The Netherlands, pp. 271—276).

———. 1940. Wiener spirals and some other interesting curves in a Hilbert space. *Doklady Akademii Nauk SSSR* 26: 115–118. (English translation in: V.M. Tikhomirov, ed., 1991, Selected works of A.N. Kolmogorov, Volume I: Mathematics and mechanics, 324—326. Springer, Berlin).

———. 1941. Dissipation energy in locally isotropic turbulence. *Doklady Akademii Nauk SSSR* 32: 16–18.

Koutsoyiannis, D. 2000. A generalized mathematical framework for stochastic simulation and forecast of hydrologic time series. *Water Resources Research* 36 (6): 1519–1533.

Koutsoyiannis, D. 2005. Uncertainty, entropy, scaling and hydrological stochastics. 1. Marginal distributional properties of hydrological processes and state scaling. *Hydrological Sciences Journal* 50 (3): 381–404. doi:10.1623/hysj.50.3.381.65031.

———. 2006. On the quest for chaotic attractors in hydrological processes. *Hydrological Sciences Journal* 51 (6): 1065–1091.

———. 2010a. Some problems in inference from time series of geophysical processes (solicited). In *European Geosciences Union General Assembly*, Geophysical research abstracts, EGU2010-14229, vol. 12. Vienna: European Geosciences Union. (http://www.itia.ntua.gr/973/).

———. 2010b. A random walk on water. *Hydrology and Earth System Sciences* 14: 585–601.

———. 2011. Hurst-Kolmogorov dynamics as a result of extremal entropy production. *Physica A* 390 (8): 1424–1432.

———. 2013a. Climacogram-based pseudospectrum: a simple tool to assess scaling properties. In *European Geosciences Union General Assembly*, Geophysical research abstracts, EGU2013-4209, vol. 15. Vienna: European Geosciences Union. (http://itia.ntua.gr/1328).

———. 2013b. *Encolpion of stochastics: Fundamentals of stochastic processes*. Athens: Department of Water Resources and Environmental Engineering, National Technical University of Athens. (http://www.itia.ntua.gr/1317/).

———. 2014. *Random musings on stochastics (Lorenz Lecture), AGU 2014 Fall Meeting*. San Francisco, USA: American Geophysical Union. doi:10.13140/RG.2.1.2852.8804. (http://www.itia.ntua.gr/en/docinfo/1500/).

———. 2016. Generic and parsimonious stochastic modelling for hydrology and beyond. *Hydrological Sciences Journal* 61 (2): 225–244. doi:10.1080/02626667.2015.1016950.

Koutsoyiannis, D., and A. Montanari. 2015. Negligent killing of scientific concepts: the stationarity case. *Hydrological Sciences Journal* 60 (7–8): 1174–1183.

Koutsoyiannis, D., F. Lombardo, E. Volpi, and S.M. Papalexiou. 2013. Is consistency a limitation?—Reply to "Further (monofractal) limitations of climactograms" by Lovejoy et al., Comment in the review of "Just two moments! A cautionary note against use of high-order moments in multifractal models in hydrology" by Lombardo et al. *Hydrology and Earth System Sciences Discussions* 10: C5397. (http://www.hydrol-earth-syst-sci-discuss.net/10/C5397/2013/hessd-10-C5397-2013-supplement.pdf).

Lo Brano, V., A. Orioli, G. Ciulla, and S. Culotta. 2011. Quality of wind speed fitting distributions for the urban area of Palermo, Italy. *Renewable Energy* 36: 1026–1039.

Lombardo, F., E. Volpi, D. Koutsoyiannis, and S.M. Papalexiou. 2014. Just two moments! A cautionary note against use of high-order moments in multifractal models in hydrology. *Hydrology and Earth System Sciences* 18: 243–255.

Mahrt, L. 1989. Intermittency of atmospheric turbulence. *Journal of the Atmospheric Sciences* 46: 79–95.

Mandelbrot, B.B. 1982. *The fractal geometry of nature*. New York, NY: W. H. Freeman.

———. 1999. *Multifractals and 1/f noise: Wild self-affinity in physics (1963–1976)*. New York, NY: Springer.

Mandelbrot, B.B., and J.W. Van Ness. 1968. Fractional Brownian motions, fractional noises and applications. *SIAM Review* 10: 422–437.

Manwell, J.F., J.G. McGowan, and A.L. Rogers. 2010. *Wind energy explained*. 2nd ed. Amherst, MA: Wiley.

Markonis, Y., and D. Koutsoyiannis. 2013. Climatic variability over time scales spanning nine orders of magnitude: Connecting Milankovitch cycles with Hurst–Kolmogorov dynamics. *Surveys in Geophysics* 34 (2): 181–207.

———. 2016. Scale-dependence of persistence in precipitation records. *Nature Climate Change* 6: 399–401.

Monahan, A.H. 2006. The probability distribution of sea surface wind speeds. Part I. Theory and sea winds observations. *Journal of Climate* 19: 497–520.

———. 2013. The Gaussian statistical predictability of wind speeds. *Journal of Climate* 26: 5563–5577.

Morgan, E.C., M. Lackner, R.M. Vogel, and L.G. Baise. 2011. Probability distributions for offshore wind speeds. *Energy Conversion and Management* 52 (1): 15–26.

O'Connell, P.E., D. Koutsoyiannis, H.F. Lins, Y. Markonis, A. Montanari, and T.A. Cohn. 2016. The scientific legacy of Harold Edwin Hurst (1880–1978). *Hydrological Sciences Journal* 61 (9): 1571–1590.

Ouarda, T.B.M.J., C. Charron, J.Y. Shin, P.R. Marpu, A.H. Al-Mandoos, M.H. Al-Tamimi, et al. 2015. Probability distributions of wind speed in the UAE. *Energy Conversion and Management* 93: 414–434.

Papalexiou, S.M., D. Koutsoyiannis, and A. Montanari. 2010. *Mind the bias!* STAHY Official Workshop: Advances in statistical hydrology. Taormina, Italy: International Association of Hydrological Sciences.

Papoulis, A. 1991. *Probability, random variables, and stochastic processes*. 3rd ed. New York, NY: McGraw-Hill.

Popper, K.R. 1982. *The open universe: An argument for indeterminism*. London: Hutchinson.

Scholz, C.H., and B.B. Mandelbrot. 1989. *Fractals in geophysics*. Basel: Birkhäuser Verlag.

She, Z.S., and E. Leveque. 1994. Universal scaling laws in fully developed turbulence. *Physical Review Letters* 72: 336.

Singh, S.K., and G.S. Maddala. 1976. A function for size distribution of incomes. *Econometrica* 44: 963–970.

Stumpf, M.P.H., and M.A. Porter. 2012. Critical truths about power laws. *Science* 335: 665–666.

Tessier, Y., S. Lovejoy, and D. Schertzer. 1993. Universal multifractals: theory and observations for rain and clouds. *Journal of Applied Meteorology* 32 (2): 223–250.

Veneziano, D., and A. Langousis. 2010. Scaling and fractals in hydrology. In *Advances in data-based approaches for hydrologic modeling and forecasting*, ed. B. Sivakumar and R. Berndtsson . Singapore: World Scientific.145 pages

von Kármán, T. 1940. The engineer grapples with nonlinear problems. *Bulletin of the American Mathematical Society* 46: 615–683. doi:10.1090/S0002-9904-1940-07266-0.

Wackernagel, H. 1995. *Multivariate geostatistics*. Berlin: Springer.

———. 1998. *Multivariate geostatistics, 2nd completely revised edition*. Berlin: Springer.

Wilczek, M., A. Daitche, and R. Friedrich. 2011. On the velocity distribution in homogeneous isotropic turbulence: Correlations and deviations from Gaussianity. *Journal of Fluid Mechanics* 676: 191–217.

Wu, J., J. Wanga, and D. Chib. 2013. Wind energy potential assessment for the site of Inner Mongolia in China. *Renewable and Sustainable Energy Reviews* 21: 215–228.

Yaglom, A.M. 1987. *Correlation theory of stationary and related random functions*. New York, NY: Springer.

Yari, G.H., and G.R.M. Borzadaran. 2010. Entropy for Pareto-types and its order statistics distributions. *Communications in Information and Systems* 10 (3): 193–201.

Role of Nonlinear Dynamics in Accelerated Warming of Great Lakes

Sergey Kravtsov, Noriyuki Sugiyama, and Paul Roebber

Abstract In recent decades, the Laurentian Great Lakes have undergone rapid surface warming with the summertime trends substantially exceeding the warming rates of surrounding land. Warming of the deepest Lake Superior was the strongest, and that of the shallowest Lake Erie—the weakest of all lakes. We investigate the dynamics of accelerated lake warming in idealized coupled thermodynamic lake–ice–atmosphere models. These models are shown to exhibit, under identical seasonally varying forcing, multiple possible stable equilibrium cycles, or regimes, with different maximum summertime temperatures and varying degrees of wintertime ice cover. The simulated lake response to linear climate change in the presence of the atmospheric noise rationalizes the observed accelerated warming of the lakes, the correlation between wintertime ice cover and next summer's lake-surface temperature, as well as higher warming trends of the (occasionally wintertime ice-covered) deep-lake vs. shallow-lake regions, in terms of the corresponding characteristics of the forced transitions between colder and warmer lake regimes. Since the regime behavior in the models considered arises due to nonlinear dynamics rooted in the ice–albedo feedback, this feedback is also the root cause of the accelerated lake warming simulated by these models.

Keywords Great Lakes • Regional warming trends • Multiple climate regimes • Ice–albedo feedback

1 Introduction

In recent decades, a large number of lakes all over the globe have been undergoing rapid increase in surface water temperature (Schneider et al. 2009; Schneider and Hook 2010; O'Reilly et al. 2015). Furthermore, many of the lakes exhibited summertime warming trends exceeding the globally averaged surface temperature trend over land. This is in sharp contrast with the observed oceanic surface warming,

S. Kravtsov (✉) • N. Sugiyama • P. Roebber
Department of Mathematical Sciences, Atmospheric Sciences Group, University of Wisconsin-Milwaukee, Milwaukee, WI, 53201, USA
e-mail: kravtsov@uwm.edu

A.A. Tsonis (ed.), *Advances in Nonlinear Geosciences*,
DOI 10.1007/978-3-319-58895-7_15

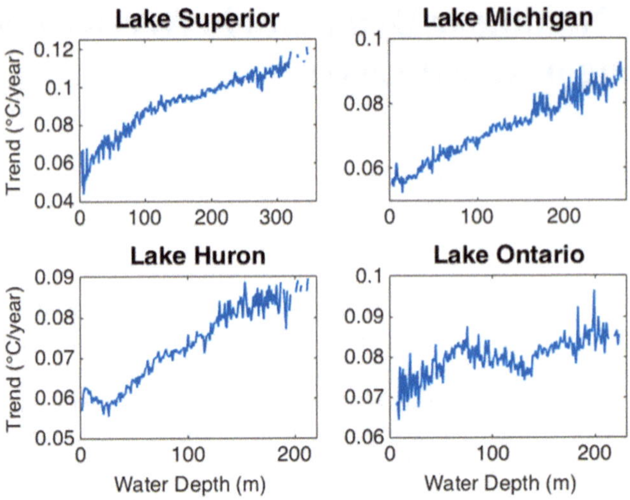

Fig. 1 The 1995–2012 warming trend in the annual-mean surface water temperature versus water depth for Great Lakes. The trends were spatially averaged over the areas of a given depth within each lake (see panel captions). These results are based on the satellite observations of surface temperature from the Great Lakes Surface Environmental Analysis (GLSEA) operated by NOAA's Great Lakes Environmental Research Laboratory (GLERL)

which was generally smaller than the warming over land (see, for example, Manabe et al. 1991; Sutton et al. 2007; Joshi et al. 2008; Joshi and Gregory 2008; Byrne and O'Gorman 2012, among others).

Despite that some of the ice-free lakes were exhibiting rapid warming (Schneider et al. 2009), the wintertime ice-covered lakes were found, on average, to warm significantly faster than the ice-free lakes (O'Reilly et al. 2015), at the rates also exceeding those of ambient air temperatures. This accelerated warming appears to be associated with a variety of climatic drivers and, interestingly, depends on the lakes morphology, with the deepest lakes exhibiting the largest warming trends (Austin and Colman 2007; Hampton et al. 2008; Zhong et al. 2016). Among Great Lakes, for example, Lake Superior (the deepest) has the strongest, and Lake Erie (the shallowest)—the weakest surface warming trend (Austin and Colman 2007). This dependence of the warming rates on depth is also found within individual lakes (Fig. 1). Yet another interesting aspect of the Great Lakes' recent evolution, which also appears to depend on the depth of the lake, is an apparent discontinuous jump in the time series of their summertime surface temperature, lake's heat content, and some other lake properties at around 1997–1998. This discontinuity was, once again, most pronounced in Lake Superior, and least pronounced in Lake Erie (Van Cleave et al. 2014; Gronewold et al. 2015; Zhong et al. 2016).

The dynamical causes of the accelerated warming of mid-latitude lakes are still a subject of debate. A combination of explanatory factors have been considered, such as increases in incoming shortwave radiation and air temperature (Arvola et al. 2010; Ackerman et al. 2013; Foster and Heidinger 2013; Fink et al. 2014; Gronewold et al. 2015), shorter lake-ice durations (Magnuson 2000), as well as an earlier onset and longer duration of the summer stratification (Austin and Colman 2007; Austin and Allen 2011; Piccolroaz et al. 2015; Zhong et al. 2016). Of the processes mentioned above, the direct response to surface air temperature trends appears to dominate the surface warming of small, shallow lakes (Toffolon et al. 2014), but other processes may be equally or more important in determining the response of deeper, larger lakes (Zhong et al. 2016).

Since the seasonal presence of lake ice is clearly a factor characterizing the majority of the most rapidly warming lakes, the ice–albedo feedback has been suggested as a root dynamical cause of accelerated lake warming (Austin and Colman 2007). In support of this idea, Hanrahan et al. (2010) found a correlation between the amount of winter ice cover and the summer surface water temperature of Lake Michigan. By contrast, Vavrus et al. (1996), Gerbush et al. (2008), and Zhong et al. (2016) argue that the net influence of lake ice on the lake's response to ambient warming is limited due to compensation between ice–albedo and insulating effects of the ice.

Here we address the multi-faceted problem of the accelerated lake warming using an idealized lake–ice–atmosphere coupled model. The central result of this study is an identification of multiple stable equilibrium seasonal cycles of the lakes (hereafter, the lakes' regional climate regimes) in our coupled model. These nonlinear regimes occur throughout the range of model geometries we considered, from one-column lakes of uniform depth to three-column lakes mimicking the geometry of individual Great Lakes, and derive their existence from the lake–ice–albedo feedback. Global-warming experiments with our coupled model rationalize many qualitative and quantitative aspects of the observed accelerated lake warming, including the dependence of the warming trends on lake depth, the association between wintertime ice cover and next summer's surface temperatures, and abrupt regional climate change associated with transitions between warm and cold lake-climate regimes.

2 Coupled Lake–Ice–Atmosphere Model

Adequately addressing dynamics of the Great Lakes' regional climate variability requires faithful simulation of the lake/lake-ice seasonal cycle. Typically, lake temperatures remain vertically homogenous throughout a substantial portion of the spring and fall seasons, and the lakes become stratified in winter and summer (Fig. 2); the lake ice appears when lake-surface temperatures cool below 0°C. The previous formulations of the one-dimensional lake models (e.g., Hostetler and Bartlein 1990) exhibited substantial biases in the duration of both the stratified and lake-ice seasons of deep lakes (Martynov et al. 2010). We introduced improvements in the lake-model vertical mixing scheme to alleviate these biases and developed a coupled configuration of the model with an interactive atmosphere to address lakes' regional climate change.

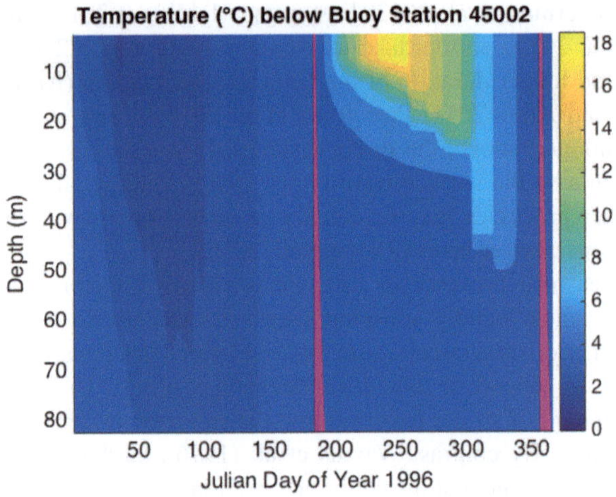

Fig. 2 A simulation of temperature (°C) below buoy station 45,002 in northern Lake Michigan, for the year 1996. The purple stripes correspond to the regions where temperature is 3.98 ± 0.05°C. The vertical grid spacing is 5 m

a. *Model geometry and experimental setup.* We considered an idealized lake that has n lake columns characterized by a variable time- and depth-dependent temperature. If $n = 1$, the lake has a uniform depth; we also considered the case with $n = 3$ to model the lakes with non-trivial bathymetry. The lake is surrounded by land and overlaid by two atmospheric layers (Fig. 3). Lake columns do not exchange heat horizontally, and we assume no heat transport through the bottom of the lake. The lake absorbs and emits radiation and exchanges heat with the lower atmospheric layer at the surface. The lower atmospheric layer, nominally the atmospheric boundary layer, is divided into parts whose boundaries coincide with those of lake columns or land; each part has a distinct variable temperature predicted by the coupled model equations, and we allow lateral heat transport between adjacent parts of this layer. On the other hand, the uppermost layer represents the lower free atmosphere and has a specified variable temperature $T_{a,u}$, which enters the formulation of the model's forcing in both stationary and global-warming experiments (see below). The model behavior is a function of a number of free parameters. Two such parameters are the relative size of land surrounding the lake and the efficiency of heat transport within the atmospheric boundary layer; both parameters affect the magnitude of the lake's simulated warming trend.
b. *Lake model.* The individual columns of the lake model are governed by the one-dimensional model formulation of Hostetler and Bartlein (1990), with empirical improvements in its so-called enhanced minimum diffusion scheme on top of the modifications suggested by Fang and Stefan (1998) and Bennington et al. (2014). These further modifications were designed to achieve better modeling of the lakes' seasonal cycle, in particular in conjunction with the correct simulation

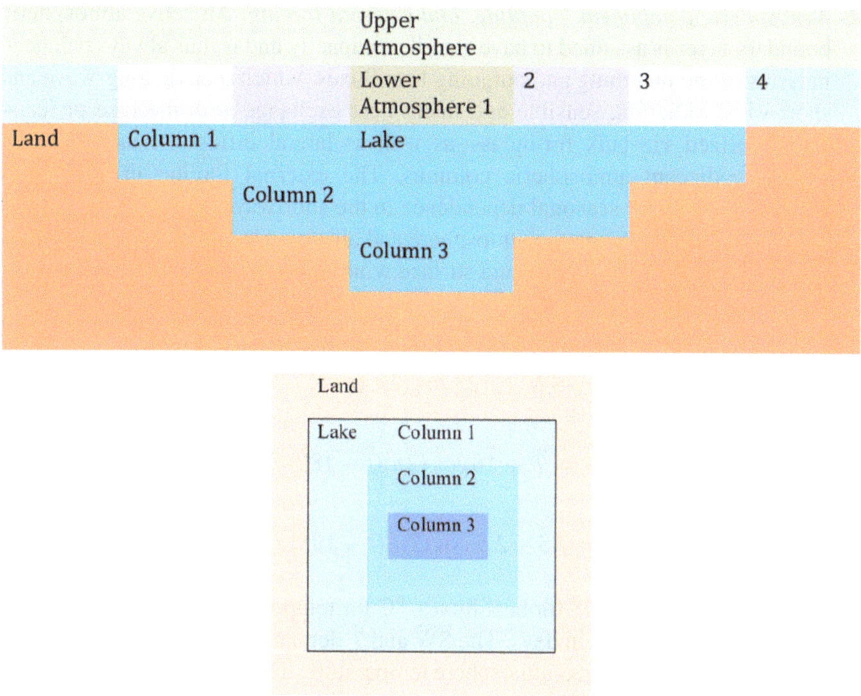

Fig. 3 Geometry of the coupled model with three-column lake component. *Top panel*: cross-section; *bottom panel*: plan view

of the onset date and duration of the summer stratification season, which is notoriously difficult to simulate in one-dimensional deep-lake models (see Martynov et al. 2010; Subin et al. 2012; Bennington et al. 2014; Zhong et al. 2016). Simulations with our modified uncoupled lake–lake–ice model driven by the observed variable atmospheric radiation, surface temperature, and wind result in a fairly good match between the simulated lake temperatures (Fig. 2) and observations thereof (not shown here).

c. *Ice model*. The lake ice is simulated using one-dimensional thermodynamic sea-ice model of Semtner (1976) modified to exclude the effects of brine pockets and explicit representation of the snow cover. To account for the latter, we instead set the surface albedo of the ice exceeding the 10 cm thickness to 0.45, which is between the typical ice and snow albedo; the surface albedo for the ice thickness h between 0 and 10 cm in our model changes linearly from 0.05 (open water value) to 0.45. For simplicity, we ignore the insulating effects of snow. The type of the ice model we used was also different depending on the ice thickness. In particular, we used what Semtner (1976) called the 0-layer model for the thin ice ($h \leq 10$ cm), and the four-layer ice model otherwise. We find that the multi-layer ice model leads to simulating a more realistic—shorter—ice-season duration compared to the 0-layer models described in Hostetler and Bartlein (1990), especially for deep lakes.

d. *Atmospheric component, coupling, and external forcing.* An active atmospheric boundary layer is assumed to have zero heat capacity and is thus always balanced in terms of the incoming and outgoing heat fluxes, which include long-wave and short-wave radiation, sensible and latent heat exchange with the lake or ice—parameterized via bulk formulas—as well as lateral diffusive heat transports between adjacent atmospheric columns. The external forcing in the model reflects the periodic seasonal dependence in the shortwave radiation SW (which is all transmitted through the atmosphere and absorbed by the lake or land), free-atmosphere temperature $T_{a,u}$, and surface wind speed u:

$$\mathrm{SW}(t) = \overline{\mathrm{SW}} + 125\cos\left(2\pi\left(t - 172\right)/365\right), \tag{1}$$

$$T_{a,u}(t) = \overline{T} + 16\cos\left(2\pi\left(t - 180\right)/365\right), \tag{2}$$

$$u(t) = 7.5 - 2.5\cos\left(2\pi\left(t - 195\right)/365\right). \tag{3}$$

The units here are Wm^{-2} for heat fluxes, °C for temperatures, and ms^{-2} for wind speeds; time t is measured in days. The $\overline{\mathrm{SW}}$ and $\overline{T}$ denote the annual-mean values of short-wave radiation and free-atmosphere temperature. The amplitude of $T_{a,u}(t)$ was chosen so that the simulated surface water temperature seasonal variation roughly matched that of Great Lakes. The seasonal variation of $u(t)$ was chosen based on the fact that the climatological surface wind speed over land surrounding the Great Lakes is roughly 5 or 6 ms^{-1} in winter and 3 or 4 ms^{-1} in summer, and that surface wind speed is generally greater over the Great Lakes than over land. For the single-column lake experiments, we set $\overline{\mathrm{SW}} = 175\ \mathrm{Wm}^{-2}$, which is comparable to the amount of downward shortwave radiation in the Great Lakes region. The phase shift in the formula of downward shortwave radiation was chosen so that the radiation reaches its maximum value on June 21st, but the phase shifts in the formulas of the other two quantities are somewhat arbitrary, except to ensure that the free-atmosphere temperature reaches its maximum value in summer and surface wind speed in winter.

Below, we will extensively analyze the numerical experiments in which the free-atmosphere annual-mean temperature $\overline{T}$ exhibits a linear trend of ±0.04°C per year and/or quasi-periodic or stochastic interannual variability.

3 Multiple Regimes in Lakes of Uniform Depth

a. *Hysteresis behavior.* To identify multiple stable equilibrium seasonal cycles of the lakes, we computed the hysteresis curves in the phase plane of the lake's maximum (summertime) surface water temperature and the concurrent annual-

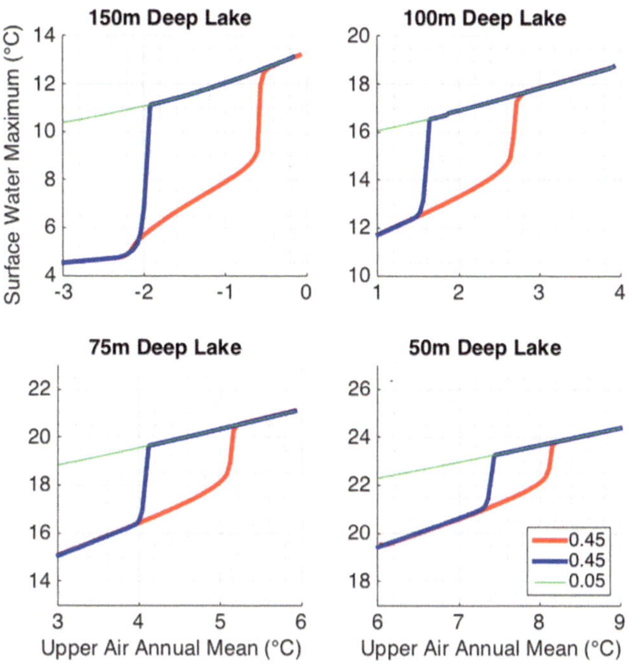

Fig. 4 The hysteresis diagrams in the phase plane of maximum (summertime) surface water temperature for the lakes of different depths (see panel captions) and the concurrent annual-mean lower free atmosphere's temperature forcing $\overline{T}$. The blue curves indicate the evolution of maximum surface water temperature when $\overline{T}$ slowly warms at the rate of +0.04°C per year. The red curves indicate the evolution of maximum surface water temperature when T slowly cools at the rate of +0.04°C per year. The ice albedo is at the default value of 0.45. For comparison, the green curves show the results of simulations with the ice albedo set to the water albedo of 0.05, which exhibit no difference between the slow warming and slow cooling results. Units are °C

mean free-atmosphere temperature (Fig. 4). We first used a steady (seasonally periodic) forcing with the low value of the annual-mean free-atmosphere temperature $\overline{T}$ to reach a seasonally varying lake equilibrium characterized by abundant wintertime ice cover and low summertime surface temperatures. We then added a linear trend of +0.04°C per year to $\overline{T}$ and followed the evolution of the lake's seasonal cycle (red curves in Fig. 4). This trend is slow enough that the resulting forcing is essentially quasi-stationary, and leads to the lake seasonal cycle initially exhibiting gradual changes, with progressively less wintertime ice cover (not shown) and progressively warmer summertime temperature. This behavior ends when the lake abruptly transitions, at some value of the free-atmosphere temperature T_{max}, to the wintertime ice-free state (not shown), which has a higher maximum (summertime) lake-surface temperature. Upon this transition, the ice-free warm state resumes gradual linear changes under a continued free-atmosphere temperature trend. Starting from the rightmost part of the hysteresis diagrams in Fig. 4, we now reverse the sign of the annual-mean free-atmosphere temperature trend, making it equal to −0.04°C per year (blue curves). The ice-free state gradually cools down until it reaches another threshold value of the free-atmosphere temperature $T_{\min}$ and transitions abruptly back to the cold regime with substantial wintertime lake-ice cover. Further slow decrease of the $\overline{T}$ forcing results in the quasi-stationary and linear lake-temperature changes along the original line of the experiment with the warming trend.

In the range of the annual-mean upper air temperature forcing between $T_{\min}$ and $T_{\max}$, the seasonally varying warm and cold climate regimes described above coexist. The occurrence of multiple equilibrium seasonal cycles of the lakes crucially depends on the lake-ice–albedo feedback—we obtained no evidence of multiple regimes in any simulations in which the ice-surface albedo was made equal to that of water (green curves).

b. *Multiple regimes in one-column lakes of different depths*. Different panels of Fig. 6 correspond to the hysteresis diagrams computed for the one-column coupled lake models of different depths. We observe that: (i) the multiple regimes of deeper lakes occur at colder values of $\overline{T}$ forcing compared to the multiple regimes of shallower lakes; (ii) the range $T_{\max}$—$T_{\min}$ of $\overline{T}$ in which the two regimes exist simultaneously is larger for deeper lakes; and (iii) the difference in the maximum summertime temperature between the two regimes is also larger for deeper lakes.

All of these properties can be rationalized by studying seasonal cycles of the shallow and deep lakes in their cold and warm regimes (Fig. 5). Throughout most of the cold season, the lake water remains vertically mixed throughout the whole column for shallower lakes and over the depths exceeding 100 m or more for deeper lakes (see, for example, Assel 1986). Hence, a deeper lake has a larger thermal inertia and it takes more forcing (and colder free-atmosphere temperatures) to cool it down to freezing temperature and form ice in winter, explaining the property (i) above.

Properties (ii) and (iii) also have to do with a larger effective thermal inertia of deep lakes vs. that of the shallow lakes, albeit not quite as directly as the property (i). The ultimate reason behind (ii) and (iii) is that shallower lakes exhibit a longer stratified season in summer than deeper lakes (see Fig. 5). This is due, in turn, to an earlier onset of the spring overturn (which happens when the surface temperature reaches the value of 3.98°C corresponding to the largest density of water) in shallow regions of the lakes. By contrast, deeper lake columns have more water to mix, so the vertical density profile of a deeper lake remains nearly homogeneous and its surface temperature remains just below the maximum density threshold of 3.98°C longer than that of a shallower lake. An earlier spring overturn and an earlier formation of the summertime surface mixed layer in shallow-lake areas is also a feature of the observed seasonal cycles of the lakes (not shown).

A typical depth of the summertime surface mixed layer of the Great Lakes is 10–20 m, so this layer's thermal inertia is really small (see, for example, Assel 1986; McCormick and Meadows 1988), and it responds to the atmospheric forcing

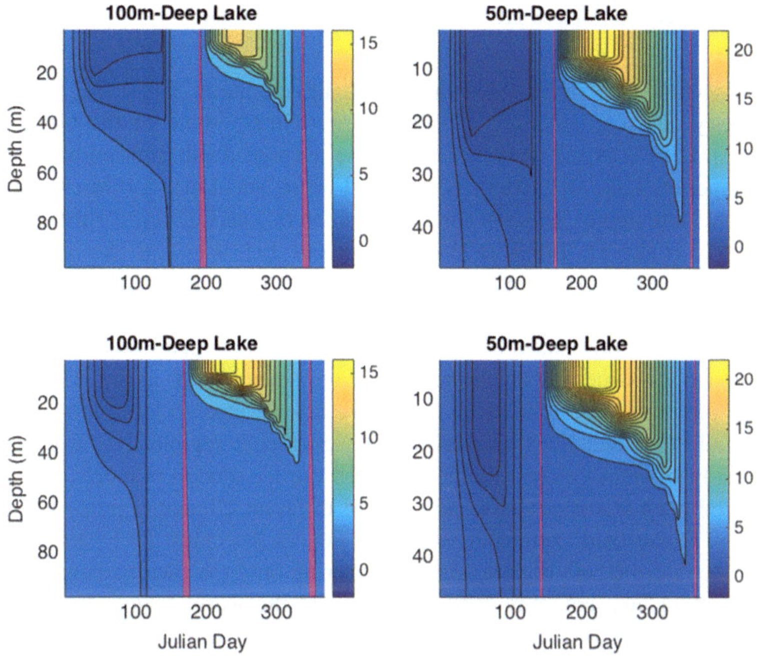

Fig. 5 Seasonal cycle of lake temperature in °C for the 100-m-deep lake (*left column*) and 50-m-deep lake (*right column*). Top row: cold regime; *bottom row*: warm regime. The purple stripes correspond to the regions where temperature is 3.98 ± 0.05°C

fairly quickly. A longer duration of the summertime stratified season in shallower lakes thus translates to a longer time to efficiently alleviate the differences in the maximum summertime temperature between the cold and warm regimes—which dynamically originate earlier, in the cold season, due to workings of the ice–albedo feedback—via radiation and sensible/latent heat loss to the atmosphere [property (iii)]. Property (ii) is a byproduct of property (iii): the smaller the temperature "gap" between the two regimes is, the smaller is the range of free-atmosphere temperature $T_{max}-T_{min}$ is in which these regimes coexist.

The consequence of properties (ii) and (iii) is that shallower lakes transition from one regime to the other more easily than deeper lakes in response to forcing. In particular, under the action of atmospheric noise with amplitude between the temperature "gap" values $T_{max}-T_{min}$ characterizing a shallow lake and a deep lake, the regime behavior of the shallow lake may not be immediately apparent as the temperature "trajectory" would wander chaotically between the two regimes. On the other hand, the deep lake in this case would be characterized by quasi-stable regime behavior, possibly with occasional and easily identifiable transitions between the two regimes. These properties help explain amplification of the surface warming trends of deeper lakes vs. shallower lakes in the presence of global warming and atmospheric noise (see Sect. 5).

4 Multiple Regimes in Three-Column Lakes

In this section, we study the behavior of three-column lakes (Fig. 3) whose bathymetry characteristics are chosen to approximate some of the Great Lakes (Table 1). Lake 1 is the deepest lake whose average depth approximates that of Lake Superior, Lake 3 ("Erie") is the shallowest, and Lake 2 ("Michigan") has an intermediate depth. The $\overline{SW}$ forcing parameters [see Eq. (1)] for these lakes—175 Wm^{-2} (Lake 1), 190 Wm^{-2} (Lake 2), and 195 Wm^{-2} (Lake 3)—are also roughly comparable to the amounts of long-term mean shortwave radiation over Lakes Superior, Michigan, and Erie.

Figure 6 (left) presents the hysteresis curves or the deepest column of Lake 2 computed in the same way as for the one-column lake of Sect. 3 (Fig. 4). Similar to the case of single-column lakes in Fig. 4, the Lake-2 three-column model without the ice–albedo effect does not have multiple climate regimes. By contrast, the full version of three-column model in which the ice albedo is much higher than that of the open water can have up to three different climate regimes for certain values of $\overline{T}$: the cold regime in which ice covers the entire lake surface during winter, the intermediate regime with ice covering only the intermediate-depth and the shallowest lake columns during winter, and the warm regime with only the shallowest lake column covered with ice in winter. The three sets of hysteresis curves in Fig. 6 (right) show the maximum (summertime) temperature for the three columns: the deepest (black), intermediate-depth (cyan), and the shallowest column (red), as a function of $\overline{T}$.

Table 1 Geometry of three-column lake models

	Depth (m)	Fractional area
Lake 1		
Column 1	50	0.1
Column 2	150	0.5
Column 3	225	0.3
Land	–	0.1
Lake 2		
Column 1	30	0.2
Column 2	80	0.5
Column 3	140	0.2
Land	–	0.1
Lake 3		
Column 1	15	0.4
Column 2	20	0.4
Column 3	40	0.1
Land	–	0.1

Lake 1 mimics Lake Superior, Lake 2—Michigan, and Lake 3—Erie

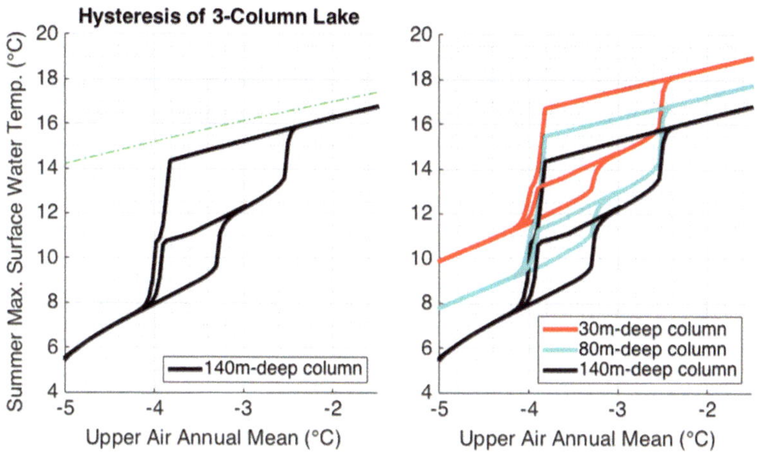

Fig. 6 *Left*: The hysteresis diagram in the phase plane of maximum (summertime) surface water temperature for the deepest column of Lake 2 and the concurrent annual-mean free-atmosphere temperature forcing $\overline{T}$. We consider two cases: simulations with the ice albedo at the default value of 0.45 (*black*) and simulations with the ice albedo set to the water albedo of 0.05 (*green*). Units are °C. This figure is analogous to Fig. 4 for the one-column lake model. Right: The hysteresis diagrams of maximum (summertime) surface water temperature for Lake-2 model's deep (*black*), intermediate (*cyan*), and shallow (*red*) columns, shown together. The ice albedo here is at the default value of 0.45

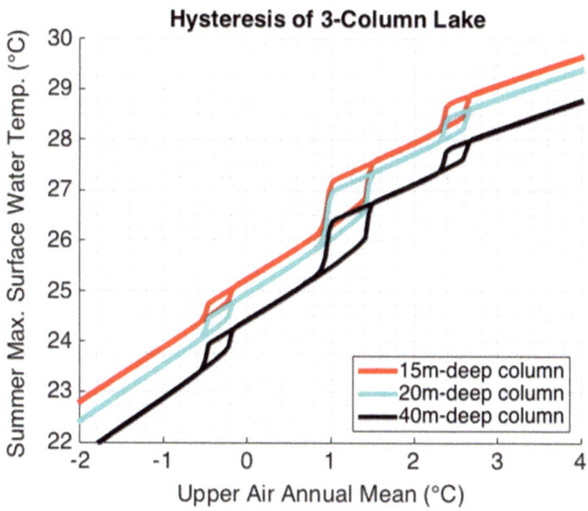

Fig. 7 *Top*: The hysteresis diagrams of maximum (summertime) surface water temperature for Lake-3 model's deep (*black*), intermediate (*cyan*), and shallow (*red*) columns. This figure is analogous to Fig. 6 (*right*)

The bifurcation diagram for the shallowest Lake-3 model (Fig. 7) is qualitatively different from that for Lake 2 in that the regime transitions are spread out along the $\overline{T}$ axis, and we do not find a range of $\overline{T}$ in which more than two regimes coexist. The three sets of regime transitions in this case correspond to the transitions of the deepest, intermediate-depth, and shallowest lake columns from their respective wintertime ice-covered to perennially ice-free states, respectively. The $\overline{T}$ thresholds for this transition are the lowest (−0.4°C, −0.2°C) for the deepest lake column, intermediate (1.4°C, 2.4°C) for the intermediate-depth column, and the highest (2.4°C, 2.6°C) for the shallowest lake column; this is consistent with the property (i) of Sect. 3. Furthermore, the gaps between maximum (summertime) temperature of warmer-vs.-colder regimes within each regime pair are in general much smaller than for the Lake-2 regimes, in accord with property (iii) of the one-column models in Sect. 3. Note, however, that the relative sizes of these gaps are not merely the function of the lake depth, as in one-column models, but also depend on the relative areas of the lake columns (Table 1) and the efficiency of the horizontal atmospheric heat transport (not shown).

In summary, while the regime structure of the three-column lakes is more complex than that of flat-bottom lakes, the properties of the regimes and, in particular, their dependence on the lake depth in the two cases, are consistent.

5 Response of Lakes to Global Warming

a. *Lacustrine regional amplification of global warming*. The bifurcation diagrams of the previous section were obtained by adding linear trends to the annual-mean free-atmosphere temperature $\overline{T}$. We now examine the evolution of three-column lake models under such warming trend (of 0.04°C per year) to gain insight into how the lake dynamics may amplify global warming on a regional scale. In the experiments of this section, we also added an idealized interannual variability on top of the linear global-warming signal in $\overline{T}$, by introducing alternating biennial anomalies of ±2°C to the $\overline{T}$ time series. The standard deviation of the resulting interannual variability is similar to the observed variations (not shown).

For each of our three idealized lake models, we started with atmospheric conditions cold enough to freeze the entire lake in winter, and followed the evolution of the lakes' seasonal cycle in a long global-warming simulation setup as described above. We then computed the slopes of linear trends in the annual-mean lake-surface temperature for each lake column over the 20 years sliding window. The resulting values of the maximum warming rates are listed in Table 2. Note that these warming trends all exceed the global-warming rate of 0.04°C per year, are largest for the deepest Lake 1 and smallest for the shallowest Lake 3, in accordance with observations (Sect. 1). We also recover the observed correlation between lake-column depth and surface warming rates within each lake, with the deepest lake columns exhibiting the largest warming rates.

Table 2 The peak warming trends (°C per year) in the annual-mean surface-water temperature for the three idealized lake models from Table 1 subjected to the linear trend of 0.04°C per year plus a periodic interannual variability: see text for details

Warming-trend	Shallow column	Intermediate column	Deep column	Overall
Lake 1	0.137	0.171	0.183	0.171
Lake 2	0.090	0.119	0.129	0.116
Lake 3	0.063	0.085	0.092	0.076

The warming trends were computed using the 20 years moving-window linear least-square trends of lake-surface temperature

The latter properties in our global-warming experiments stem from the fact that the peak warming rates of the lakes arise due to transitions between the lake regimes, as cold regimes are gradually becoming less and less likely under the global warming. The peak differences between cold and warm regimes of deep lakes are larger than those between the regimes of shallow lakes (see Sects. 3 and 4); hence, deep lakes tend to exhibit larger warming rates. Furthermore, since the dynamical inertia of the shallow lakes is smaller, they are more likely to transition back and forth between their cold and warm regimes due interannual atmospheric variability compared to the deep lakes. These multiple transitions smear out the peak warming rates of shallow lakes even further.

b. *Discontinuous behavior of deep lakes*. Finally, we present, in Fig. 8, an example of simulation with our Lake 2 forced by a combination of linear global warming trend in $\overline{T}$ and random Gaussian noise in both $\overline{T}$ and $\overline{\text{SW}}$, with the standard deviations of 2°C and 6 Wm^{-2}, respectively; these values are consistent with observations of atmospheric interannual variability. The lake starts from the cold regime at low values of $\overline{T}$ and. As $\overline{T}$ gradually warms, the lake starts to transition back and forth between its colder and warmer regimes before arriving permanently to its final warm state. In the first half of the time series, the lake's cold regime is preferred, with the lake only experiencing occasional transitions to the warm regime for 1 or 2 years (where the minimum temperature remains above freezing throughout the year). In the second half of the time series though, the situation is completely reversed, with the warm regime being clearly dominant (ice only reappears in this column twice after the simulation year 90). This simulation qualitatively mimics the behavior of Lake Superior. Prior to 1997, this lake's climate was dominated by cold regime with an extensive wintertime ice cover (maximum >80%) and low summertime temperatures. After 1998, the lakes switched to the warm regime with maximum ice cover <60% and warm lake temperature in summer: during this period, the lake's cold regime appeared three times, in years 2003, 2009, and 2014–2015, but neither of these occurrences lasted more than 2 years.

The multiple stochastically forced transitions introduce an apparent decadal variability in the lake-temperature time series, consistent with the interannual memory of deep lakes, and "diffuse" the lake warming to occupy a longer time

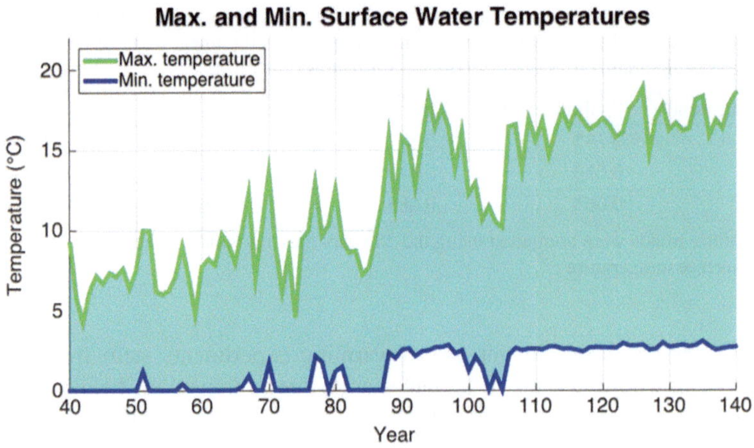

Fig. 8 A realization of the "global warming" experiment (forced by the steady 0.04°C per year trend in the upper-air annual-mean temperature $\overline{T}$) with superimposed atmospheric stochastic forcing for Lake 2. Shown are the time series of the maximum (*green*) and minimum (*blue*) surface water temperatures of the deepest column of Lake 2

interval. Still, note the jump-like character of the lake-temperature time series in Fig. 8, with a clear step-like increase in maximum summertime water temperature around year 88 of simulation (equivalent to year 1998 in the case of the observed Lake Superior transition to its warm regime). By contrast, the time series of surface-water temperatures of shallower lakes exhibit less clear regime transitions (not shown), due to their smaller thermal and dynamical inertia. Hence, we expect the discontinuous regime behavior to be most pronounced for the deepest lakes like Lake Superior, and less so for shallower lakes.

6 Conclusions

The main result of the present study is that nonlinear dynamics operating in our coupled lake-ice–atmosphere model allows us to faithfully simulate large amplification of the global-warming signal in deep-lake areas, as was observed in the Great Lake region during recent decades. These dynamics manifest in the existence of multiple regional climate regimes of the lakes—that is, distinct seasonal cycles of the lakes, with warmer or colder summertime temperatures and less or more extensive wintertime ice cover—arising under the identical seasonally varying forcing. The persistence characteristics and sheer differences between the regimes depend on the depth of the lake. Deep lakes, which have a large thermal/dynamical inertia, exhibit large differences between the regimes and are resilient to external perturbations, whereas the differences between shallow-lake regimes are less pronounced, and the transitions between them under interannual atmospheric variability are easier to achieve. Hence, the deep lakes exhibit a stronger—often jump-like—response to global warming forcing as they undergo changes toward a state in which their warmer regimes gradually become progressively more likely, consistent with observations.

The regimes in our model only occur in the presence of the ice–albedo feedback nonlinearity; therefore, our results corroborate Austin and Colman's (2007) original hypothesis about the central role of this feedback in the accelerated warming of Lake Superior. Our hypothesis of nonlinear regime dynamics behind the lacustrine regional amplification of global warming is, however, novel, and complements a rich spectrum of existing theories (see Sect. 1). Sorting out relative contributions to the lake warming from a large suite of possible linear and nonlinear mechanisms will require further work.

Nonlinear regimes due to ice–albedo feedback have been studied before in a variety of climatic problems, including that of glacial-to-interglacial transitions which involve land-ice and sea-ice feedbacks, as well as in addressing a possibility of abrupt changes in Arctic sea ice under climate change (see Merryfield et al. 2008 for a review). Our present study revisits this concept in a novel context of the regional climate change and provides a new framework for assessing and understanding climatic effects of mid-latitude lakes.

Acknowledgements We thank Michael Notaro, Stephen Vavrus, and Yafang Zhong for numerous valuable discussions. This work was supported by the NSF grant 1236620.

References

Ackerman, S., A. Heidinger, M. Foster, and B. Maddux. 2013. Satellite regional cloud climatology over the Great Lakes. *Remote Sensing* 5: 6223–6240.

Arvola, L., G. George, D.M. Livingstone, M. Jarvinen, T. Blenckner, M.T. Dokulil, E. Jennings, C.N. Aonghusa, P. Ñoges, T. Noges, and G.A. Weyhenmeyer. 2010. The impact of the changing climate on the thermal characteristics of lakes. In *The impact of climate change on european lakes. Aquatic ecology series*, ed. G. George, 85–101. Netherlands: Springer.

Assel, R. 1986. Fall and winter thermal structure of Lake Superior. *Journal of Great Lakes Research* 12 (4): 251–262.

Austin, J., and J. Allen. 2011. Sensitivity of summer Lake Superior thermal structure to meteorological forcing. *Limnology and Oceanography* 56: 1141–1154.

Austin, J.A., and S.M. Colman. 2007. Lake Superior summer water temperatures are increasing more rapidly than regional air temperatures: a positive ice-albedo feedback. *Geophysical Research Letters* 34: L06604.

Bennington, V., M. Notaro, and K.D. Holman. 2014. Improving climate sensitivity of deep lakes within a regional climate model and its impact on simulated climate. *Journal of Climate* 27: 2886–2911.

Byrne, M.P., and P.S. O'Gorman. 2012. Land–ocean warming contrast over a wide range of climates: convective quasi-equlibrium theory and idealized simulations. *Journal of Climate* 26: 4000–4016.

Fang, X., and H.G. Stefan. 1998. Temperature variability in lake sediments. *Water Resources Research* 34: 717–729. doi:10.1029/97WR03517.

Fink, G., M. Schmid, B. Wahl, T. Wolf, and A. Wuest. 2014. Heat flux modifications related to climate-induced warming of large European lakes. *Water Resources Research* 50: 2072–2085.

Foster, M., and A. Heidinger. 2013. PATMOS-x: results from a diurnally corrected 30-yr satellite cloud climatology. *Journal of Climate* 26: 414–425.

Gerbush, M., D. Kristovich, and N. Laird. 2008. Mesoscale boundary layer and heat flux variations over pack ice-covered Lake Erie. *Journal of Applied Meteorology and Climatology* 47: 668–682.

Gronewold, A.D., E.J. Anderson, B. Lofgren, P.D. Blanken, J. Wang, J. Smith, T. Hunter, G. Lang, C.A. Stow, D. Beletsky, and J. Bratton. 2015. Impacts of extreme 2013–2014 winter conditions on Lake Michigan's fall heat content, surface temperature, and evaporation. *Geophysical Research Letters* 42: 3364–3370.

Hampton, S.E., L.R. Izmest'eva, M.V. Moore, S.L. Katz, B. Dennis, and E.A. Silow. 2008. Sixty years of environmental change in the world's largest freshwater lake—Lake Baikal, Siberia. *Global Change Biology* 14 (8): 1947–1958. doi:10.1111/j.1365-2486.2008.01616.x.

Hanrahan, J.L., S. Kravtsov, and P.J. Roebber. 2010. Connecting past and present climate variability to the water levels of Lake Michigan and Huron. *Geophysical Research Letters* 37: L01701.

Hostetler, S., and P.J. Bartlein. 1990. Simulation of lake evaporation with application to modeling lake-level variations at Harney-Malheur Lake, Oregon. *Water Resources Research* 26: 2603–2612.

Joshi, M.M., and J.M. Gregory. 2008. Dependence of the land–sea contrast in surface climate response on the nature of the forcing. *Geophysical Research Letters* 35: L24802.

Joshi, M.M., J.M. Gregory, M.J. Webb, D.M.H. Sexton, and T.C. John. 2008. Mechanism for the land/seawarming contrast exhibited by simulations of climate change. *Climate Dynamics* 30: 455–465.

Magnuson, J.J. 2000. Historical trends in lake and river ice cover in the Northern Hemisphere. *Science* 289 (5485): 1743–1746. doi:10.1126/science.289.5485.1743.

Manabe, S., R.J. Stouffer, M.J. Spelman, and K. Bryan. 1991. Transit responses of a coupled ocean–atmosphere model to gradual changes of atmospheric CO2. Part I: Annual mean response. *Journal of Climate* 4: 785–818.

Martynov, A., L. Sushama, and R. Laprise. 2010. Simulation of temperate freezing lakes by one- dimensional lake models: performance assessment for interactive coupling with regional climate models. *Boreal Environment Research* 15: 143–164.

McCormick, M.J., and G.A. Meadows. 1988. An intercomparison of four mixed layer models in a shallow inland sea. *Journal of Geophysical Research* 93: 6774–6788.

Merryfield, W.J., M.M. Holland, and A.H. Monahan. 2008. Multiple equilibria and abrupt transitions in Arctic summer sea ice extent. In *Arctic sea ice decline: observations, projections, mechanisms, and implications*, ed. E.T. DeWeaver, C.M. Bitz, and L.-B. Tremblay, 151–174. doi:10.1029/180GM11.

O'Reilly, C.M., S. Sharma, D.K. Gray, S.E. Hampton, et al. 2015. Rapid and highly variable warming of lake surface waters around the globe. *Geophysical Research Letters* 42: 10773–10781. doi:10.1002/2015GL066235.

Piccolroaz, S., M. Toffolon, and B. Majone. 2015. The role of stratification on lakes' thermal response: the case of Lake Superior. *Water Resources Research* 51: 7878–7894. doi:10.1002/2014WR016555.

Schneider, P., and S.J. Hook. 2010. Space observations of inland water bodies show rapid surface warming since 1985. *Geophysical Research Letters* 37: L22405.

Schneider, P., S.J. Hook, R.G. Radocinski, G.K. Corlett, G.C. Hulley, S.G. Schladow, and T.E. Steissberg. 2009. Satellite observations indicate rapid warming trend for lakes in California and Nevada. *Geophysical Research Letters* 36: L22402. doi:10.1029/2009GL040846.

Semtner, A.J. 1976. A model for the thermodynamic growth of sea ice in numerical investigations of climate. *Journal of Physical Oceanography* 6: 379–389.

Subin, Z.M., W.J. Riley, and D. Mironov. 2012. An improved lake model for climate simulations: Model structure, evaluation, and sensitivity analyses in CESM1. *Journal of Advances in Modeling Earth Systems* 4: M02001. doi:10.1029/2011MS000072.

Sutton, R.T., B. Dong, and M.G. Gregory. 2007. Land/sea warming ratio in response to climate change: IPCC AR4 model results and comparison with observations. *Geophysical Research Letters* 34: L02701.

Toffolon, M., S. Piccolroaz, B. Majone, A. Soja, F. Peeters, M. Schmid, and A. Wüest. 2014. Prediction of surface temperature in lakes with different morphology using air temperature. *Limnology and Oceanography* 59 (6): 2185–2202.

Van Cleave, K., J.D. Lenters, J. Wang, and E.M. Verhamme. 2014. A regime shift in Lake Superior ice cover, evaporation, and water temperature following the warm El Niño winter of 1997–1998. *Limnology and Oceanography* 59 (6): 1889–1898. doi:10.4319/lo.2014.59.6.1889.

Vavrus, S., R. Wynne, and J. Foley. 1996. Measuring the sensitivity of southern Wisconsin lake ice to climate variations and lake depth using a numerical model. *Limnology and Oceanography* 41: 822–831.

Zhong, Y., M. Notaro, and S.J. Vavrus. 2016. Recent accelerated warming of the Laurentian Great Lakes: physical drivers. *Limnology and Oceanography* 61 (5): 1762–1786. doi:10.1002/lno.10331.

The Prediction of Nonlinear Polar Motion Based on Artificial Neural Network (ANN) and Fuzzy Inference System (FIS)

Ramazan Alper Kuçak, Raşit Uluğ, and Orhan Akyılmaz

Abstract The Earth rotation movement characterizes the situation of the whole Earth movement, as well as the interaction between the Earth's various layers such as the Earth's core, mantle, crust, and atmosphere. Prediction of the Earth rotation parameters (ERPs) is important for near real-time applications including navigation, precise positioning, remote sensing and landslide monitoring, etc. In such studies, the analysis of time series is also important for highly accurate and reliable predictions. Therefore, prediction of ERPs at least over a few days in the future is necessary. At present, there are two major forecasting methods for ERP: linear and nonlinear models. The nonlinear models include: sequence of artificial neural network (ANN), fuzzy inference system, and other methods. Fuzzy inference system (FIS) and traditional artificial neural networks (ANN) provide good predictions of polar motion (PM). In this study, for the prediction of Earth rotation parameters, International Earth Rotation and Reference System Service (IERS) C04 daily time series data from 1990 to 2015 was used for training. From 1 to 120 days in future of ERPs values were predicted by using the data of 5, 15, and 25 years in ANN. The results of ANN and ANFIS were compared with observed values. The results indicate that the longer training data are used in ANN and ANFIS, the more accurate prediction can be obtained.

Keywords Polar motion • Artificial neural network • Fuzzy inference system • Earth rotation parameters

1 Introduction

With the development of high precision space geodetic techniques, Earth Orientation Parameters (EOPs) which is very important for many geodetic applications can accurately be enhanced. Their exact values are very important for many investiga-

---

R.A. Kuçak (✉) • R. Uluğ • O. Akyılmaz
Faculty of Civil Engineering, Department of Geomatics Engineering, Istanbul Technical University, Maslak, 34469, Istanbul, Turkey
e-mail: kucak15@itu.edu.tr

A.A. Tsonis (ed.), *Advances in Nonlinear Geosciences*,
DOI 10.1007/978-3-319-58895-7_16

tions in geodesy and astronomy, e.g., for high precision terrestrial navigation by use of the Global Navigation Satellite System (GNSS), for navigation of Earth satellites and interplanetary spacecrafts, and for laser ranging to satellites and to the Moon (Schuh et al. 2002).

As these parameters are determined by space geodetic techniques such as GNSS, Very Long Baseline Interferometry (VLBI), and Satellite Laser Ranging (SLR), they are not available in real time (Akyilmaz and Kutterer 2004). For many geodetic application, it is necessary to predict this value at least over a few days. Until now, various algorithms and prediction methods have been used by Akyilmaz and Kutterer (2004), McCarthy and Luzum (1991), Freedman et al. (1994), Ulrich (2000), Schuh et al. (2002). In this study, by using data with different temporal lengths (5, 15, 25 years), polar motion components were predicted by using artificial neural network (ANN) and adaptive network based fuzzy inference system (ANFIS) for 120 days in future and differences between predicted values and real values were compared.

ANN and ANFIS have already been successfully applied in many scientific applications however; they have different algorithms for analyzing time series. ANN used here was examined by Egger (1992) for prediction of EOPs, simply it is a computational model based on the structure and functions of biological neural networks and it will be briefly explained in Sect. 2. Information that flows through the network affects the structure of the ANN because a neural network learns from data. On the other hand, ANFIS used here is primarily based on fuzzy set theory which was introduced by Zadeh (1965) (Akyilmaz and Kutterer 2004) and it will be briefly explained in Sect. 3.

2 Artificial Neural Network

An ANN is a mathematical model that tries to simulate the structure and functionalities of biological neural networks (Krenker et al. 2009). Warren McCulloch and Walter Pitts (1943) created a computational model for neural networks based on mathematics and algorithms called threshold logic. This model paved the way for neural network research to split into two distinct approaches. One approach focused on biological processes in the brain and the other focused on the application of neural networks to artificial intelligence. ANN composes of three simple rules: multiplication, summation, and activation. At the first stage inputs are weighted what means that every input value is multiplied with individual weight. The second section is a summation of function that sums all weighted inputs and bias (Krenker et al. 2009). In the third and the last section of ANN, the sum of the previously weighted inputs and bias is passing through an activation function that is also called transfer function.

Neural networks are typically organized in layers. Layers are consisting of a number of interconnected nodes which contain an activation function. Patterns are presented to the network via the input layer, which communicates to one or more hidden layers, where the actual processing is done via a system of weighted connections. The hidden layers link to output layer, where final answer is delivered. Although there are many different kinds of learning rules used by neural network, feed forward back propagation method was used in this study. In feed forward back propagation method information moves only in one direction, forward, from the input nodes through the hidden nodes and to the output nodes. There are no cycles or loops in the network (Auer et al. 2008).

3 Fuzzy Inference System

Fuzzy inference system transforms fuzzy information from an input space ("premises") to an output space ("consequents") by means of fuzzy if-then rules (Akyilmaz and Kutterer 2004). These if-then rule statements are used to formulate the conditional statements that comprise fuzzy logic. A single fuzzy if-then rule assumes the form "if x is A then y is B," where A and B are linguistic values defined by fuzzy sets on the variable spaces X and Y, respectively. They may differ only in the types of membership functions in the consequent parts (Akyilmaz and Kutterer 2004).

Adaptive network based fuzzy inference system (ANFIS) has been developed by Jang (1993). ANFIS is a type of neural network that is focused on Takagi–Sugeno fuzzy inference system. ANFIS is a well-known artificial intelligence technique (Bisht and Jangid 2011). In our study first-order Takagi and Sugeno (1983) ANFIS was used in which the consequent part is linear function of the input variables and supervised learning algorithm is based on a hybrid algorithm. The advantage of ANFIS is that it is not complicated as much as ANN, and hybrid algorithm provides fast convergence time.

4 Data Reduction and Generation of Training Patterns

Daily values of PM were obtained from International Earth Rotation and Reference System Service (IERS), daily time series between 1990 and 2015 were used for training and validation of two different model. In order to clearly detect the differences between ANN and ANFIS, data were divided into three parts as 5, 15, and 25 years. For all three cases, 80% of the data were used for training and remaining 20% were used for validation. In order to make a prediction, validation set to be covered by the range of training set. Therefore, a linear trend using training data was estimated and removed from the original polar motion series (including both training and validation data). The residual data after linear trend reduction were used in predictions. To generate training patterns following formulation was adopted for both x and y components:

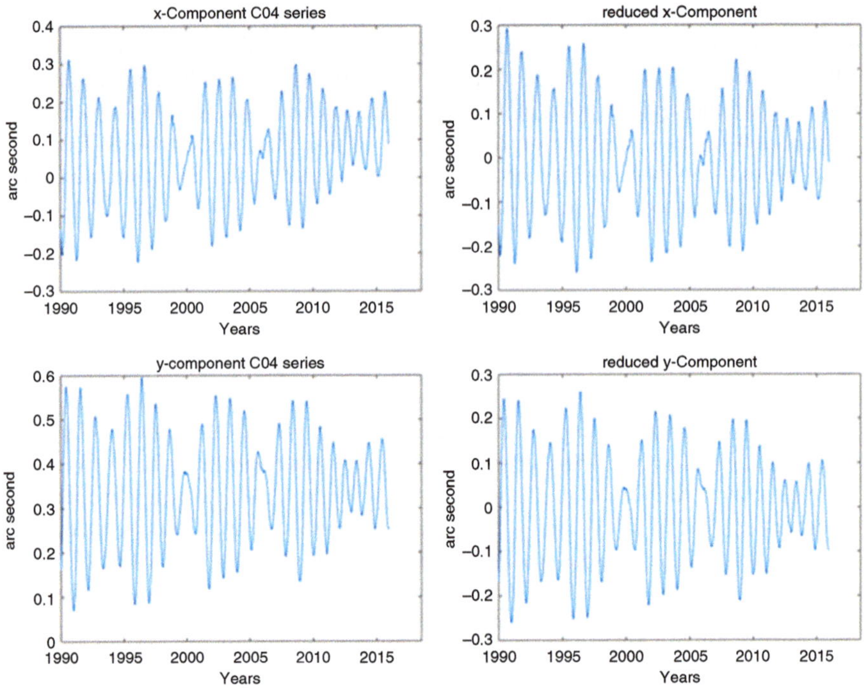

Fig. 1 Polar motion components and reduction process

$$\{x\,(t-5k)\,,x\,(t-4k)\,,x\,(t-3k)\,,x\,(t-2k)\,,x\,(t-k)\} \rightarrow \{x(t)\}$$

where $x(t)$ is values of the time series which is predicted, and k is the number indicating the day in future to be predicted (Fig. 1 middle panel).

These patterns are shifted along the whole time series of both reduced polar motion components. After the generation of the training patterns, data were divided into two parts as input and output. To make a good prediction Levenberg–Marquardt learning algorithm (LMA) which is also known as damped least square method (DLM) was used in ANN. This learning algorithm provides a numerical solution to the problem of minimizing a nonlinear function. It is fast and has stable convergence but can be preferred as long as the ANN has a single output which is the case in our PM prediction models. In ANN applications, this algorithm is suitable for training small- and medium-sized problems (Yu and Wilamowski 2011). On the other hand, the number of fuzzy if-then rule is directly related to the number of membership functions in each variable space. For example, in this study five variables in input space and each one is represented by three membership function were used; this means that the number of fuzzy if-then rules is equal to $3^5 = 243$. Membership function type was chosen triangular membership function (trimf) and supervised learning algorithm is based on a hybrid algorithm, which is a combination of gradient descent and Kalman filter. The advantage of this update algorithm is the very fast convergence and guarantee for reaching the global optimum (Akyilmaz and Kutterer 2005) of the objective/cost function of the problem.

For predictions, each ANN and ANFIS model were composed for prediction of 1 day in future and afterwards the predicted values were used as inputs for already existing models to calculate the corresponding values for the day 2, 3, ...30. Because of RMS errors increased rapidly after 30 days with a linear trend, prediction was not extrapolated. To calculate RMS error following equation was used:

$$\mathrm{RMS}_d = \sqrt{\frac{1}{n}\sum_{i=1}^{n}\left(\widehat{i}_d^i - \iota_d^i\right)^2}$$

where $\widehat{i}_d^i$ is the predicted value of ANN and ANFIS network for day d, ι_d^i is the corresponding actual value from IERS C04 series, and n is the number of predictions.

5 Prediction Results and Comparison

The results predicted from 5, 15, and 25 years of data by ANN and ANFIS were compared in Tables 1 and 2.

As seen from Tables 1 and 2, there is a big difference between the RMS errors of x and y components; RMS errors of x components are much higher than y components. This is because the x component has different character from y components and possibly more sensitive to the geophysical phenomena. Because

Table 1 RMS error values of ANN prediction

	RMSE of ANN (mas)					
	5 years		15 years		25 years	
Prediction day	*x*	*y*	*x*	*y*	*x*	*y*
1	0,75	0,04	0,67	0,03	0,67	0,02
2	1,69	0,18	1,50	0,17	1,50	0,15
3	2,54	0,59	2,19	0,56	1,87	0,53
4	3,13	1,04	2,61	0,98	2,20	0,93
5	3,53	1,47	2,82	1,39	2,63	1,30
10	4,83	2,46	2,89	2,19	2,85	1,88
15	5,21	3,13	2,45	2,57	2,99	1,93
30	12,00	7,01	1,94	4,84	2,51	2,15
60	37,75	22,18	8,26	16,08	3,88	5,64
90	68,08	35,32	26,69	27,68	11,00	7,42
120	99,01	39,52	53,74	34,60	23,65	8,37

mas milliarcsecond

Table 2 RMS error values of ANFIS prediction

	RMSE of ANFIS (mas)					
	5 years		15 years		25 years	
Prediction day	*x*	*y*	*x*	*y*	*x*	*y*
1	0,71	0,02	0,70	0,01	0,69	0,01
2	1,59	0,22	1,58	0,21	1,55	0,22
3	2,36	0,63	2,33	0,62	2,29	0,64
4	2,86	1,08	2,81	1,07	2,76	1,10
5	3,18	1,51	3,10	1,50	3,03	1,53
10	3,92	2,53	3,62	2,46	3,47	2,45
15	3,72	3,19	3,19	3,02	3,01	2,90
30	7,25	6,58	4,87	5,71	4,13	4,75
60	24,34	19,03	17,62	15,29	14,93	10,83
90	47,45	29,20	36,65	21,88	31,36	12,85
120	73,70	33,31	58,98	23,79	51,53	13,57

mas milliarcsecond

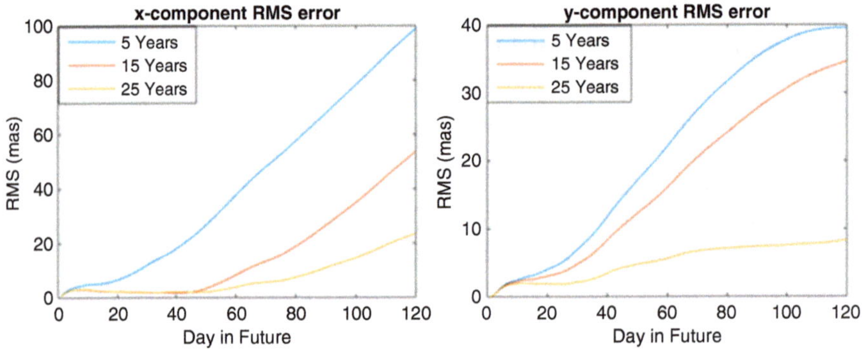

Fig. 2 RMS errors of prediction obtained from ANN

of predicted values were used as inputs for the next day, RMS error increases rapidly after the 30th day in future predictions. The values in the table clearly show that more input data can provide more accuracy, however, more data also contain disruptive values. In such a case, more complicated process of data reduction and generating training patterns may help to reduce the effect of disruptive values, but this is beyond the scope of this paper and left for a future study.

A graphical comparison of RMS errors using 5, 15, and 25 years of data for ANN and ANFIS prediction models is given in Figs. 2 and 3, respectively. The first 5 days in future predictions both by ANN and ANFIS is interesting. In particular, the x component from ANN predictions look very close each other, however, after the 5th day, the RMS errors improve with the increasing number of input data. Although first 5 days seem very close each other for all cases, the RMS error which uses 25 years of input C04 data is lower (Figs. 2 and 3 middle panel).

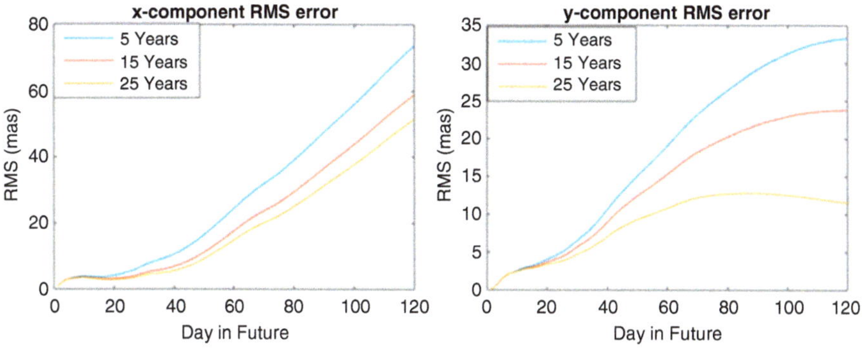

Fig. 3 RMS errors of prediction obtained form ANFIS

6 Conclusions

In this study, to predict polar motion component with different time span inputs, artificial neural network and adaptive neuro fuzzy inference system were used. The results clearly indicate that ANN and ANFIS are appropriate tools to predict polar motion components. ANFIS provides more accurate predictions using as short as 5 years of observation data; however, RMS errors of ANN using 15 and 25 years of observation data are lower than ANFIS. This does not mean ANN is better than ANFIS because some researchers, e.g., Akyilmaz and Kutterer (2004) found out total opposite results. This is due to different time span data, different learning algorithm, number of membership functions etc. can change the results. On the other hand, ANN offers very good prediction but it is very complex to handle. In addition, one has to calculate a refined a priori model before training network. However, ANFIS is less complex than ANN and depending on our experience, training of ANN takes much longer time than that of ANFIS. Although we think that older data could contain disruptive values, the results also indicate that more input data can provide better prediction for polar motion component. This is because either ANN or ANFIS can thus learn low frequency behavior of the PM and use this information during the prediction processes.

References

Akyilmaz, O., and H. Kutterer. 2004. Prediction of earth rotation parameters by fuzzy inference systems. *Journal of Geodesy* 78 (1–2): 82–93.

———. 2005. Fuzzy inference systems for the prediction of earth rotation parameters. *Iag Symp* 128: 582–587.

Auer, P., H. Burgsteiner, and W. Maass. 2008. A learning rule for very simple universal approximators consisting of a single layer of perceptrons. *Neural Networks* 21 (5): 786–795.

Bisht, D.C., and A. Jangid. 2011. Discharge modelling using adaptive neuro-fuzzy inference system. *International Journal of Advanced Science and Technology* 31: 99–114.

Egger, D. 1992. Neuronales Netz prädiziert Erdrotation. *Allgemeine Vermessungsnachrichten (AVN)* 11/12: 517–524.

Freedman, A.P., J.A. Steppe, J.O. Dickey, T.M. Eubanks, and L.Y. Sung. 1994. The short-term prediction of universal time and length of day using atmospheric angular momentum. *Journal of Geophysical Research* 99(B4): 6981–6996.

Jang, J.S.R. 1993. ANFIS: adaptive-network-based fuzzy inference system. *IEEE Transactions on Systems, Man, and Cybernetics* 23: 665–685.

Krenker, A., M. Volk, U. Sedlar, J. Bester, and A. Kos. 2009. Bidirectional artificial neural networks for mobile-phone fraud detection. *Etri J* 31 (1): 92–94.

McCarthy, D.D., and B.J. Luzum. 1991. Prediction of Earth orientation. *Bulletin géodésique* 65 (1): 18–21.

McCulloch, W., and W. Pitts. 1943. A logical calculus of the ideas immanent in nervous activity. *The bulletin of mathematical biophysics* 5(4): 115–133.

Schuh, H., M. Ulrich, D. Egger, J. Muller, and W. Schwegmann. 2002. Prediction of Earth orientation parameters by artificial neural networks. *J Geodesy* 76 (5): 247–258.

Takagi T, and Sugeno M. "Derivation of fuzzy control rules from human operator's control actions." In *Proceedings of the IFAC symposium on fuzzy information, knowledge representation and decision analysis*, vol. 6, pp. 55–60. 1983.

Ulrich, M. 2000. *Vorhersage der Erdrotationsparameter mit Hilfe Neuronaler Netze*, IAPG/FESG No. 9. München: Institut für Astronomische und Physikalische Geodisie, Technische Universittit München.

Yu, H., and B.M. Wilamowski. 2011. Levenberg–marquardt training. *Industrial Electronics Handbook* 5 (12): 1.

Zadeh, L.A. 1965. Fuzzy sets. *Information and Control* 8 (3): 338–353.

Harnessing Butterflies: Theory and Practice of the Stochastic Seasonal to Interannual Prediction System (StocSIPS)

S. Lovejoy, L. Del Rio Amador, and R. Hébert

Abstract The atmosphere is governed by continuum mechanics and thermodynamics yet simultaneously obeys statistical turbulence laws. Up until its deterministic predictability limit ($\tau_w \approx 10$ days), only general circulation models (GCMs) have been used for prediction; the turbulent laws being still too difficult to exploit. However, beyond τ_w—in macroweather—the GCMs effectively become stochastic with internal variability fluctuating about the model—not the real world—climate and their predictions are poor. In contrast, the turbulent macroweather laws become advantageously notable due to (a) low macroweather intermittency that allows for a Gaussian approximation, and (b) thanks to a statistical space-time factorization symmetry that (for predictions) allows much decoupling of the strongly correlated spatial degrees of freedom. The laws imply new stochastic predictability limits. We show that pure macroweather—such as in GCMs without external forcings (control runs)—can be forecast nearly to these limits by the ScaLIng Macroweather Model (SLIMM) that exploits huge system memory that forces the forecasts to converge to the real world climate.

To apply SLIMM to the real world requires pre-processing to take into account anthropogenic and other low frequency external forcings. We compare the overall Stochastic Seasonal to Interannual Prediction System (StocSIPS, operational since April 2016) with a classical GCM (CanSIPS) showing that StocSIPS is superior for forecasts 2 months and further in the future, particularly over land. In addition, the relative advantage of StocSIPS increases with forecast lead time.

In this chapter we review the science behind StocSIPS and give some details of its implementation and we evaluate its skill both absolute and relative to CanSIPS.

Keywords Scaling • Forecasting • Prediction • Weather • Macro weather • Climate • Stochastic • Fractals • Multifractals

S. Lovejoy (✉) • L. Del Rio Amador • R. Hébert
Department of Physics, McGill University, 3600 University St., Montreal, QC, H3A 2T8, Canada
e-mail: lovejoy@physics.mcgill.ca

A.A. Tsonis (ed.), *Advances in Nonlinear Geosciences*,
DOI 10.1007/978-3-319-58895-7_17

1 Introduction

1.1 Deterministic, Stochastic, Low Level, High Level Laws

L. F. Richardson's "Weather forecasting by numerical process" (1922) opened the era numerical weather prediction. Richardson not only wrote down the modern equations of atmospheric dynamics, but he also pioneered numerical techniques for their solution and even laboriously attempted a manual integration. Yet this work also contained the seed of an alternative: buried in the middle of a paragraph, he slyly inserted the now iconic poem: "Big whirls have little whirls that feed on their velocity, little whirls have smaller whirls and so on to viscosity (in the molecular sense)". Soon afterwards, this was followed by the Richardson 4/3 law of turbulent diffusion (Richardson 1926), which today is celebrated as the starting point for modern theories of turbulence including the key idea of cascades and scale invariance. Unencumbered by later notions of meso-scale, with remarkable prescience, he even proposed that his scaling law could hold from dissipation up to planetary scales, a hypothesis that has been increasingly confirmed in recent years. Today, he is simultaneously honoured by the Royal Meteorological Society's Richardson prize as the father of numerical weather prediction, and by the Nonlinear Processes division the European Geosciences Union's Richardson medal as the grandfather of turbulence approaches.

Richardson was not alone in believing that in the limit of strong nonlinearity (high Reynolds number, *Re*), that fluids would obey new high level turbulent laws. Since then, Kolmogorov, Corrsin, Obhukhov, Bolgiano and others proposed analogous laws, the most famous of which is the Kolmogorov law for velocity fluctuations (it is nearly equivalent to Richardson's law). While the laws of continuum mechanics and thermodynamics are deterministic, the classical turbulent laws characterize the statistics of fluctuations as a function of space-time scale; they are stochastic. Just as the laws of statistical mechanics are presumed to be compatible with those of continuum mechanics—and even though no proof (yet) exists—the latter are also presumed to be compatible with the higher level turbulence laws, see the comprehensive review (Lovejoy and Schertzer 2013).

If both continuum mechanics and turbulent laws are valid, then both are potentially exploitable for making forecasts. Yet for reasons that we describe below, for forecasting, only the brute force integration of the equations of continuum mechanics—general circulation models (GCMs)—have been developed to any degree. In this paper we review an early attempt to directly exploit the turbulent laws for macroweather forecasting, i.e. for forecasts beyond the deterministic predictability limit ($\approx$10 days).

1.2 The Status of the Turbulent Laws

The classical turbulent laws are of the scaling form: fluctuation $\approx$ (turbulent flux) $\times$ (scale)^H where H is the fluctuation exponent (for the Kolmogorov law, $H = 1/3$, see below). The scaling form is a consequence of the scale invariance of the governing laws; symbolically, *(laws)* $\longrightarrow$ *(scale change by factor* λ*)* $\longrightarrow$ λ^H *(laws)*, (note that the scale change must be anisotropic, see Schertzer et al. (2012)). The atmosphere has structures spanning the range of scales from planetary to submillimetric with $Re \approx 10^{12}$: making it in principle an ideal place to test such high Re theories. However, the classical laws were based on very restrictive assumptions, they used unrealistic notions of turbulent flux and scale. In particular, the fluxes (which are actually in Fourier space and typically go from small to large wavenumbers) were assumed to be homogeneous or at least quasi-Gaussian. However a basic feature of atmospheric dynamics is that almost all of the energy and other fluxes are sparsely distributed in storms—and in their centres—and this enormous turbulent intermittency was not taken into account. In addition, the classical notion of scale was naïve: it was taken to be the usual Euclidean distance between two points, i.e. it was isotropic, the same in all directions.

To be realistic, Schertzer and Lovejoy (1985) argued that the classical laws needed to be generalized precisely to take into account intermittency and anisotropy (especially stratification) and they introduced the main tools: multifractal cascade processes and Generalized Scale Invariance. Profiting from the golden age of geophysical data (remotely sensed, in situ and airborne), models and reanalyses (model–data hybrids), a growing body of work has largely vindicated this view, and has resulted in a quantitative characterization of the relevant multifractal hierarchy of exponents over wide ranges of space and time scales. While the laws are indeed of the (generalized) scaling form indicated above, with only a few exceptions the values of the exponents still have not been derived theoretically. They are nevertheless robust with quite similar values being found in diverse empirical data sets as well as in GCM outputs.

While large scale boundary conditions clearly affect the largest scales of flows, at small enough scales, the latter become unimportant so that, for example, in the atmosphere for scales below about 5000 km, the predictions of turbulent cascade theories are accurate to within typically $\pm 0.5\%$ (see, e.g., Chap. 4 of Lovejoy and Schertzer (2013), although at larger scales, deviations are important. If the turbulent laws are insensitive to driving mechanisms and boundary conditions, then they should be "universal", operating, for example, in other planetary atmospheres. This prediction was largely confirmed in a quantitative comparison of turbulent laws on Earth and on Mars. It turns out with the exception of the largest factor of five or so in scale that statistically, we are twins with our sister planet (Chen et al. 2016), see Fig. 1a, b!

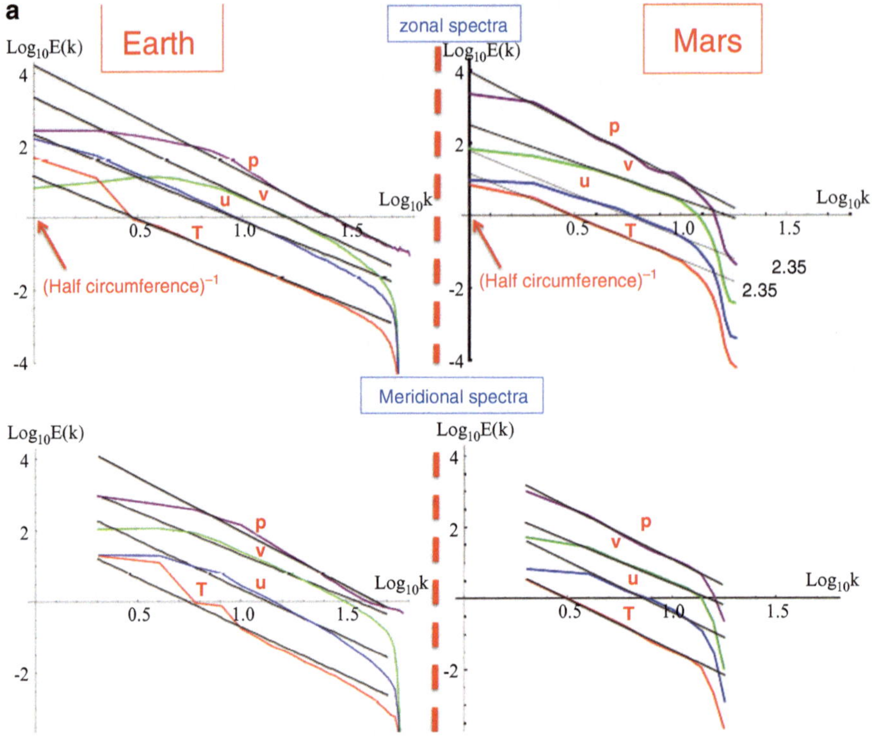

Fig. 1 (**a**) (*Top row*): The zonal spectra of Earth (*top left*) and Mars (*top right*) as functions of the nondimensional wave numbers for the pressure (*p*, *purple*), meridional wind (*v*, *green*), zonal wind (*u*, *blue*) and temperature (T, *red*) lines. The data for Earth were taken at 69% atmospheric pressure for 2006 between latitudes ±45o. The data for Mars were taken at 83% atmospheric pressure for Martian Year 24 to 26 between latitudes ±45o. The reference lines (*top left*, Earth) have absolute slopes, from top to bottom: 3.00, 2.40, 2.40, and 2.75 (for *p*, *v*, *u*, and *T*, respectively). Top right (Mars) have reference lines with absolute slopes, from top to bottom: 3.00, 2.05, 2.35 and 2.35 (for *p*, *v*, *u* and *T*, respectively). The spectra have been rescaled to add a vertical offset for clarity and wavenumber $k = 1$ corresponds to the half circumference of the respective planets. (*Bottom row*): The same as top row except for the meridional spectra of Earth (*left*) and Mars (*right*). The reference lines (*left*, Earth) have absolute slopes, from top to bottom: 3.00, 2.75, 2.75 and 2.40 (for *p*, *v*, *u* and *T*, respectively). The reference lines (right, Mars) have absolute slopes, from top to bottom: 3.00, 2.40, 2.80 and 2.80 (for *p*, *v*, *u* and *T*, respectively). The spectra have been rescaled to add a vertical offset for clarity. Adapted from (Chen et al. 2016). (**b**) The three known weather–macroweather transitions: air over the Earth (*black* and *upper purple*), the Sea Surface Temperature (SST, ocean) at 5° resolution (*lower blue*) and air over Mars (*Green* and *orange*). The air over earth curve is from 30 years of daily data from a French station (*Macon*, *black*) and from air temps for last 100 years (5° × 5° resolution NOAA NCDC), the spectrum of monthly averaged SST is from the same database (*blue*, *bottom*). The Mars spectra are from Viking lander data (*orange*) as well as MACDA Mars reanalysis data (*Green*) based on thermal infrared retrievals from the Thermal Emission Spectrometer (TES) for the Mars Global Surveyor satellite. The strong green and orange "spikes" at the right are the Martian diurnal cycle and its harmonics. Adapted from Lovejoy et al. (2014). (**c**) Spectra from the 20CR reanalysis (1871–2008) at 45°N for temperature (*T*), zonal and meridional wind (*u*, *v*) and specific humidity (h_s). The reference lines have correspond to $\beta_{mw} = 0.2$, $\beta_w = 2$ left to right, respectively. Adapted from Lovejoy and Schertzer (2013)

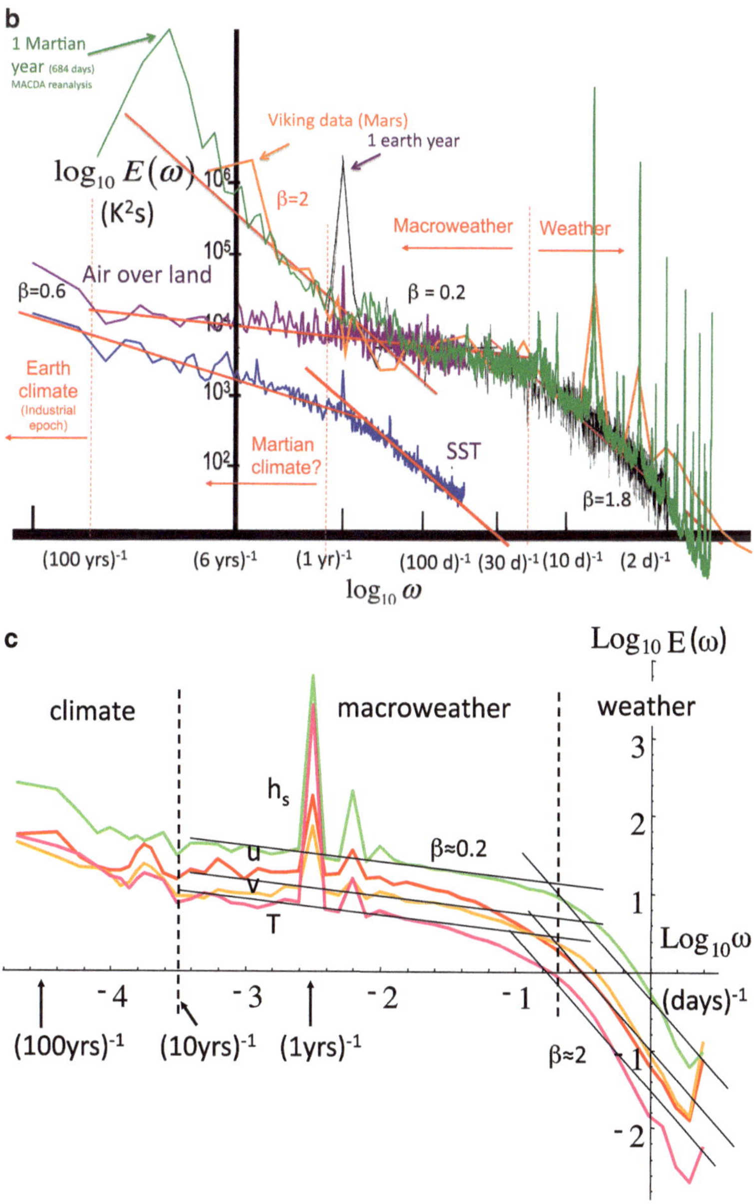

Fig. 1 (continued)

1.3 Status of Forecasts Based on the Classical Laws and their Prospects with Turbulence Laws

Over the last decades, conventional numerical approaches have developed to the point where they are now skilful up until nearly their theoretical (deterministic) predictability limits—itself close to the lifetimes of planetary structures (about 10 days, see below). Actually—due to stochastic parametrizations—state of the art ensemble GCM forecasts are stochastic–deterministic hybrids, but this limit is still fundamental. At the same time, the strong intermittency (multifractality) over this range has meant that stochastic forecasts based on the turbulent laws must be mathematically treated as (state) *vector* anisotropic multifractal cascade processes, the mathematical understanding of which is still in its infancy (see, e.g., Schertzer and Lovejoy (1995)), GCMs are the only alternative. However, if we consider scales of many lifetimes of planetary structures—the macroweather regime—then the situation is quite different. On the one hand, because of the butterfly effect (sensitive dependence on initial conditions), in macroweather even fully deterministic GCMs become stochastic. On the other hand, as pointed out in Lovejoy and Schertzer (2013) (Lovejoy and de Lima 2015; Lovejoy et al. 2015) in their macroweather limit, the turbulence laws become much simpler and—as we review below—can already be used to yield monthly, seasonal, annual and decadal forecasts that are comparable or better than the GCM alternatives. The stochastic forecasts that we describe here thus effectively harness the butterfly effect. Significantly, their forecasts already appear to be close to new—stochastic—predictability limits.

As we review below, there are two principal reasons that macroweather turbulent laws are tractable for forecasts. The first is that macroweather intermittency is generally low enough that a Gaussian model is a workable approximation (although not for the extremes)—and the corresponding prediction problem has been mathematically solved. This is the basis of the ScaLIng Macroweather Model (SLIMM (Lovejoy et al. 2015)) that is the core of the Stochastic Seasonal and Annual Prediction System (StocSIPS) that we describe in this review paper. The second macroweather simplification is that the usual size-lifetime relations breakdown, being replaced by new ones and an important new property called "statistical space-time factorization" (SSTF) holds (at least approximately). It turns out that the SSTF effectively transforms the forecast problem from a familiar deterministic nonlinear PDE *initial value* problem into a stochastic, fractional order linear ODE *past value* problem. In contrast at macroweather time scales, a fundamental GCM limitation comes to the fore: each GCM converges to its own model climate, not to the real world climate. While this was not important at shorter weather scales, now it becomes a fundamental obstacle. We conclude that for macroweather forecasting, the turbulent approach becomes attractive while the GCM approach becomes unattractive. Below, we compare the skills of the two different approaches and underline the advantages of exploiting the turbulent laws.

This review is structured as follows: we first discuss and summarize macroweather statistics (Sect. 2). In Sect. 3, we describe the forecast model and its skill, and in Sect. 4, we compare stochastic hindcasts with GCMs both with and without external forcings. In Sect. 5 we conclude.

2 Macroweather Statistics

2.1 *The Transition from Weather to Macroweather*

Ever since the first atmospheric spectra (Panofsky and Van der Hoven 1955; Van der Hoven 1957), it has been known that there is a drastic change in atmospheric statistics at time scales of several days. At first ascribed to “migratory pressure systems”, termed a “synoptic maximum” (Kolesnikov and Monin 1965), it was eventually theorized as baroclinic instability (Vallis 2010). However, its presence in all the atmospheric fields (Fig. 1c), its true origin and its fundamental implications could not be appreciated until the turbulent laws were extended to planetary scales.

The key point is that the horizontal dynamics are controlled by the energy flux ε to smaller scales (units W/Kg, also known as the “energy rate density”). Although this is the same dimensional quantity upon which the Kolmogorov law is based ($\Delta v = \varepsilon^{1/3}L^{1/3}$ for the velocity difference Δv across a distance *L)*, it had not been suggested that it hold up to planetary scales; Kolmogorov himself believed that it would not hold to more than several hundred metres (Fig. 2). Indeed as pointed out in Lovejoy et al. (2007) on the basis of state-of-the-art dropsonde data, the original Kolmogorov law is isotropic and doesn't appear to hold anywhere in the atmosphere (at least at scales above $\approx$5 m)! However, the recognition that an anisotropic generalization of the Kolmogorov law could account for the horizontal statistics (with the vertical being controlled by buoyancy force variance fluxes and Bolgiano–Obhukhov statistics) explains how it is possible for the horizontal Kolmogorov law to hold up to planetary scales (see Fig. 1a, for the space-time scaling up to planetary scales, see also Fig. 3 for IR radiances). The classical lifetime–size (*L*) relation is then obtained by using dimensional analysis on ε: $\tau \approx \varepsilon^{-1/3}L^{2/3}$ where *L* is the horizontal extent of a structure (no longer an isotropic 3D estimate of its size). This law has been validated in both Lagrangian and Eulerian frames, see Radkevitch et al. (2008) (Pinel et al. 2014, Fig. 3).

If one estimates ε by dividing the total tropospheric mass by the total solar power that is transformed into mechanical energy (about 4% of the total this is the thermodynamic efficiency of the atmospheric heat engine; see e.g. Pauluis (2011)), then one finds $\varepsilon \approx 1$ mW/Kg which is close to the directly estimated empirical value (it even explains regional variations, see Fig. 2). Using $\varepsilon \approx 1$ mW/Kg, $L = 20{,}000$ km (the largest great circle distance) this value implies that the lifetime of planetary structures and hence the weather–macroweather transition is $\tau_w \approx 5$–10 days. When the theory is applied to the ocean (which is similarly turbulent

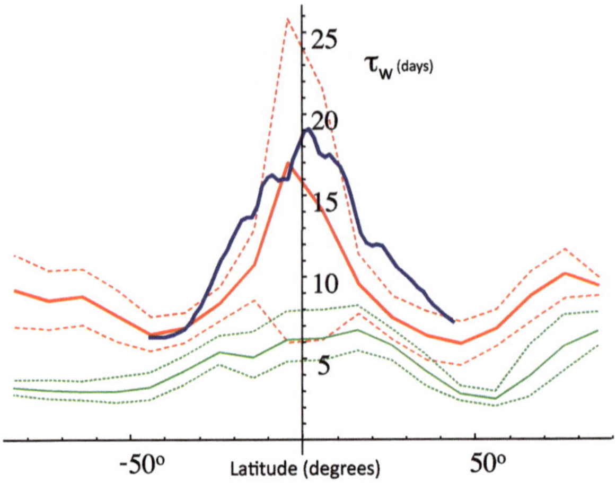

Fig. 2 The weather–macroweather transition scale τ_w estimated directly from break points in the spectra for the temperature (red) and precipitation (green) as a function of latitude with the longitudinal variations determining the dashed one standard deviation limits. The data are from the 138-year long Twentieth Century reanalyses (20CR (Compo et al. 2011)), the τ_w estimates were made by performing bilinear log–log regressions on spectra from 180-day long segments averaged over 280 segments per grid point. The blue curve is the theoretical τ_w obtained by estimating the distribution of ε from the ECMWF reanalyses for the year 2006 (using $\tau_w = \varepsilon^{-1/3}L^{2/3}$ where $L =$ half earth circumference), it agrees very well with the temperature τ_w. τ_w is particularly high near the equator since the winds tend to be lower, hence lower ε. Similarly, τ_w is particularly low for precipitation since it is usually associated with high turbulence (high ε). Reproduced from Lovejoy and Schertzer (2013)

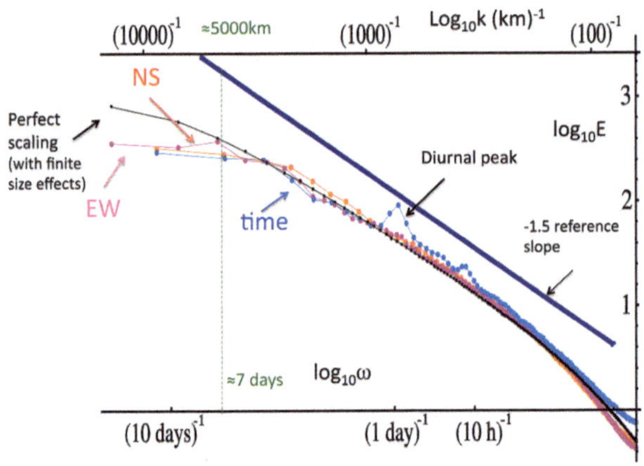

Fig. 3 The zonal, meridional and temporal spectra of 1386 images (2 months of data, September and October 2007) of radiances fields measured by a thermal infrared channel (10.3–11.3 μm) on the geostationary satellite MTSAT over south-west Pacific at resolutions 30 km and 1 h. over latitudes 40°S—30°N and longitudes 80°E—200°E. With the exception of the (small) diurnal peak (and harmonics), the rescaled spectra are nearly identical and are also nearly perfectly scaling (the *black line* shows exact power law scaling after taking into account the finite image geometry. Reproduced from Pinel et al. (2014)

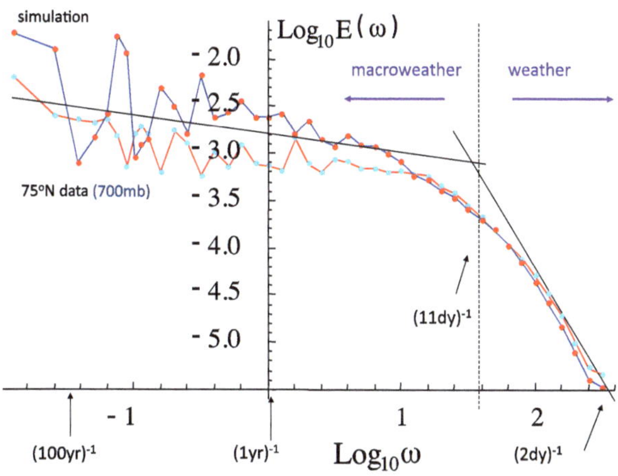

Fig. 4 A comparison of temperature spectra from a grid point of the 20CR data (*bottom, orange line*) and from a turbulence cascade model (*top, blue line*) showing that it well reproduces the weather–macroweather transition. Reproduced from Lovejoy and Schertzer (2013)

with $\varepsilon \approx 10^{-8}$ W/Kg), one obtains a transition at about 1–2 years (also observed, Lovejoy and Schertzer (2010), Fig. 1b). Finally, it can be used to accurately estimate $\varepsilon \approx 40$ mW/Kg on Mars and hence the corresponding Martian transition scale at about 1.8 sols (Fig. 1b, Lovejoy et al. 2014).

From the point of view of turbulent laws, the transition from weather to macroweather is a "dimensional transition" since at time scales longer than τ_w, the spatial degrees of freedom are essentially "quenched" so that the system's dimension is effectively reduced from 1 + 3 to 1 (Lovejoy and Schertzer 2010). Using spectral analysis Fig. 4 shows that simple multifractal turbulence models reproduce the transition. GCM control runs, i.e. with constant external forcings (see Sect. 2.2 and Fig. 5c)—also reproduce realistic macroweather variability, justifying the term "macroweather". However in forced GCMs—as with instrumental and multiproxy data beyond a critical time scale τ_c, the variability starts to increase again (as in the weather regime) and the true climate regime begins; $\tau_c \approx 10$ years in the anthropocene, and $\tau_c \gtrsim 100$ years in the pre-industrial epoch, (see Sect. 2.2, Fig. 5).

In order to understand the key difference between weather, macroweather and the climate, rather than spectra, it is useful to consider typical fluctuations. Classically—for example, in the Kolmogorov law—fluctuations were taken to be differences, i.e. $\Delta T(\Delta t)$:

$$(\Delta T\,(\Delta t))_{\text{diff}} = T(t) - T\,(t - \Delta t) \tag{1}$$

While this is fine for weather fluctuations—these typically increase with scale Δt—it is not adequate for those that typically decrease with Δt, and as we shall see this includes macroweather fluctuations. For these, we often consider "anomalies"; for example, for the temperature anomaly $T(t)$ is the temperature with both the

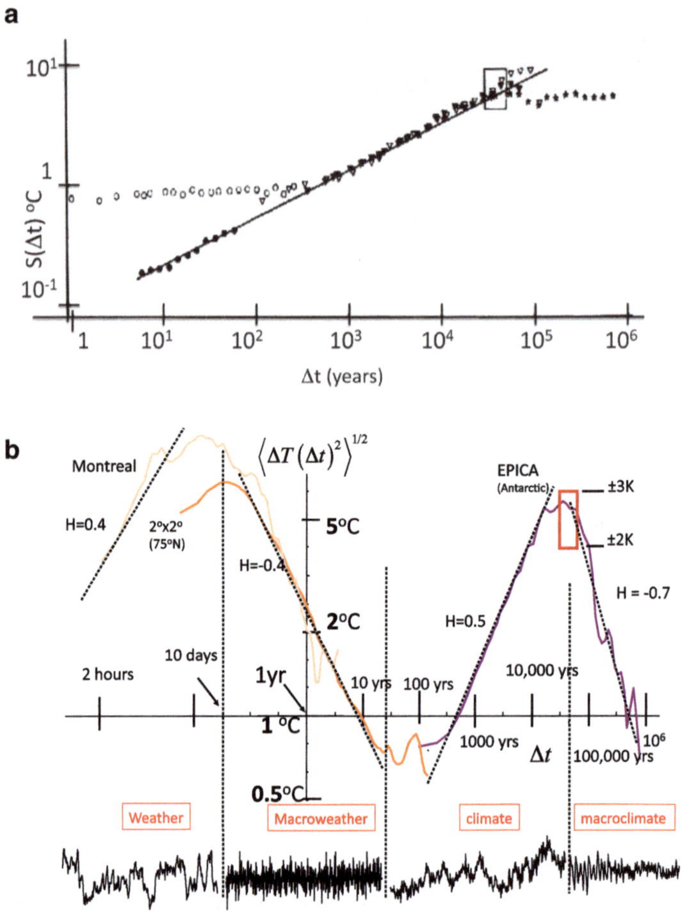

Fig. 5 (**a**) The RMS difference structure function estimated from local (Central England) temperatures since 1659 (*open circles*, *upper left*), northern hemisphere temperature (*black circles*), and from paleo-temperatures from Vostok (Antarctic, *solid triangles*), Camp Century (Greenland, *open triangles*) and from an ocean core (*asterixes*). For the northern hemisphere temperatures, the (power law, linear on this plot) climate regime starts at about 10 years. The rectangle (*upper right*) is the "glacial-interglacial window" through which the structure function must pass in order to account for typical variations of ±2 to ±3 K for cycles with half periods ≈50 kyrs. Reproduced from Lovejoy and Schertzer 1986). (**b**) A composite RMS Haar structure function from (daily and annually detrended) hourly station temperatures (*left*), 20CR temperatures (1871–2008 averaged over 2° pixels at 75°N) and paleo-temperatures from EPICA ice cores (right) over the last 800 kyrs. The glacial–interglacial window is shown upper right rectangle. Adapted from Lovejoy (2015a). (**c**) Haar fluctuation analysis of globally, annually averaged outputs of past Millenium simulations over the pre-industrial period (1500–1900) using the NASA GISS E2R model with various forcing reconstructions. Also shown (*thick*, *black*) are the fluctuations of the pre-industrial multiproxies showing that they have stronger multi centennial variability. Finally, (*bottom*, *thin black*) are the results of the control run (no forcings), showing that macroweather (slope < 0) continues to millennial scales. Reproduced from Lovejoy et al. (2013).

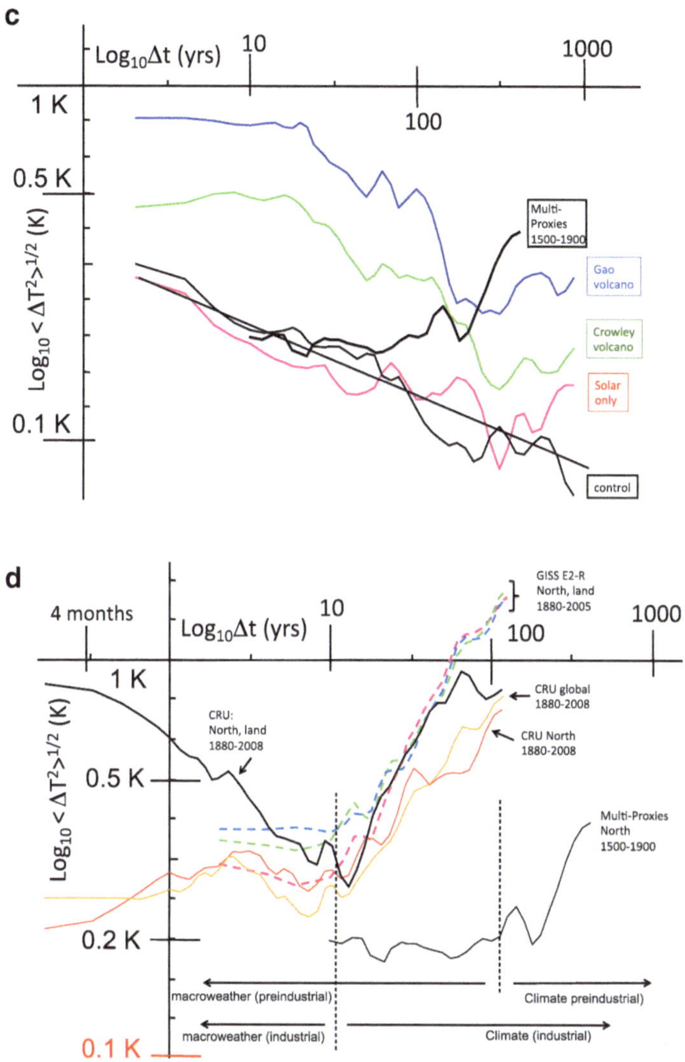

Fig. 5 (continued) (**d**) Haar fluctuation analysis of Climate Research Unit (CRU, HadCRUtemp3 temperature fluctuations), and globally, annually averaged outputs of past Millenium simulations over the same period (1880–2008) using the NASA GISS model with various forcing reconstructions (*dashed*). Also shown are the fluctuations of the pre-industrial multiproxies showing the much smaller centennial and millennial scale variability that holds in the pre-industrial epoch. Reproduced from (Lovejoy et al. 2013)

annual cycle and the overall mean of the series removed so that $\langle T \rangle = 0$ where "$\langle . \rangle$" indicates averaging. For such zero mean anomaly series $T(t)$, define the Δt resolution anomaly fluctuation by:

$$(\Delta T(\Delta t))_{\text{anom}} \approx \frac{1}{\Delta t}\int_{t-\Delta t}^{t} T(t')\,dt' \tag{2}$$

(as for differences, in $\Delta T(\Delta t)$ we suppressed the t dependence since we assume that the fluctuations are statistically stationary). Since $T(t)$ fluctuates around zero, averaging it at larger and larger Δt tends to decrease the fluctuations so that the decreasing classical anomaly fluctuations and the increasing difference fluctuations will each have restricted and incompatible ranges of validity.

In general, average fluctuations may either increase or decrease depending on the range of Δt considered so that we must define fluctuations in a more general way; wavelets provide a fairly general framework for this. A simple expedient combines averaging and differencing while overcoming many of the limitations of each: the Haar fluctuation (from the Haar wavelet). It is simply the difference of the mean over the first and second halves of an interval:

$$(\Delta T(\Delta t))_{\text{Haar}} = \frac{2}{\Delta t}\int_{t-\Delta t/2}^{t} T(t')\,dt' - \frac{2}{\Delta t}\int_{t-\Delta t}^{t-\Delta t/2} T(t')\,dt' \tag{3}$$

(see Lovejoy and Schertzer (2012) for these fluctuations in a wavelet formalism). In words, the Haar fluctuation is the difference fluctuation of the anomaly fluctuation, it is also equal to the anomaly fluctuation of the difference fluctuation. In regions where the fluctuations decrease with scale we have:

$$\begin{aligned} (\Delta T(\Delta t))_{\text{Haar}} &\approx (\Delta T(\Delta t))_{\text{anom}} \qquad (\text{decreasing with } \Delta t) \\ (\Delta T(\Delta t))_{\text{Haar}} &\approx (\Delta T(\Delta t))_{\text{diff}} \qquad (\text{increasing with } \Delta t) \end{aligned} \tag{4}$$

In order for Eq. (4) to be reasonably accurate, the Haar fluctuations in Eq. (3) need to be multiplied by a calibration factor; here, we use the canonical value 2 although a more optimal value could be tailored to individual series.

Over ranges where the dynamics have no characteristic time scale, the statistics of the fluctuations are power laws so that:

$$\langle |\Delta T(\Delta t)|^q \rangle \propto \Delta t^{\xi(q)} \tag{5}$$

the left-hand side is the qth order structure function and $\xi(q)$ is the structure function exponent. "< >" indicates ensemble averaging; for individual series this is estimated by temporal averaging (over the disjoint fluctuations in the series). The first order ($q = 1$) case defines the "fluctuation exponent" $\xi(1) = H$:

$$\langle |\Delta T(\Delta t)| \rangle \propto \Delta t^{H} \tag{6}$$

In the special case where the fluctuations are quasi-Gaussian, $\xi(q) = qH$ and the Gaussian white noise case corresponds to $H = -1/2$. More generally, there will be "intermittency corrections" so that:

$$K(q) = qH - \xi(q) \tag{7}$$

where $K(q)$ is a convex function with $K(1) = 0$. $K(q)$ characterizes the multifractality associated with the intermittency.

Equation (6) shows that the distinction between increasing and decreasing mean fluctuations corresponds to the sign of H. It turns out that the anomaly fluctuations are adequate when $-1 < H < 0$ whereas the difference fluctuations are adequate when $0 < H < 1$. In contrast, the Haar fluctuations are useful over the range $-1 < H < 1$ which encompasses virtually all geoprocesses, hence its more general utility. When H is outside the indicated ranges, then the corresponding statistical behaviour depends spuriously on either the extreme low or extreme high frequency limits of the data.

2.2 *The low Frequency Macroweather Limit and the Transition to the Climate*

We have argued that there is a drastic statistical transition in all the atmospheric fields at time scales of 5–10 days, and that the basic equations have no characteristic time scale. However, it was noted since (Lovejoy and Schertzer 1986) (Fig. 5a) that global temperature differences tend to increase in a scaling manner right up to the ice age scales: the glacial-interglacial "window" at about 50 kyrs (a half cycle) over which fluctuations are typically of the order ± 2 to ± 4 K.

Figure 5a shows the root mean square second order structure function defined by difference fluctuations $\left\langle \Delta T(\Delta t)^2_{\mathrm{diff}} \right\rangle^{1/2}$ for both local and hemispherically averaged temperatures. From the above discussion, we anticipate that it will give spurious results in the regions where the true fluctuations decrease with scale; indeed, the local (central England) series (upper left in Fig. 5a and ocean cores beyond $\approx$100 kyrs, upper right) are spuriously flat (i.e., the differences do not reflect the underlying scaling of the fluctuations that are in fact decreasing over these ranges). This is confirmed using more modern data as well as Haar rather than difference fluctuations, in Fig. 5b that shows a composite of temperature variability over the range of scales of hours to nearly a million years. From Fig. 5b, it can be seen that the drastic weather–macroweather spectral transition corresponds to a change in the sign of H for $H > 0$ to $H < 0$, i.e. from fluctuations increasing to fluctuations decreasing with scale. The bottom of the figure shows extracts of typical data at the corresponding resolutions, when $H > 0$, the signal "wanders" like a drunkard's walk, when $H < 0$, successive fluctuations tend to cancel out.

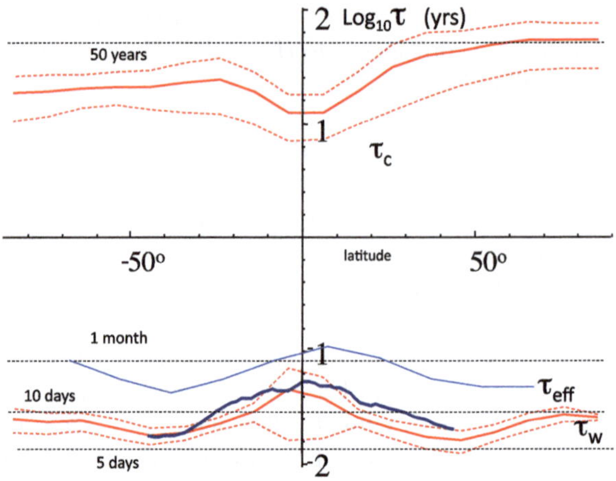

Fig. 6 Variation of τ_w (*bottom*) and τ_c (*top*) as a function of latitude as estimated from the 138-year long 20CR reanalyses, 700 mb temperature field (the τ_c estimates are only valid in the anthropocene). The *bottom red* and *thick blue* curves for τ_w are from Fig. 2; also shown at the *bottom* is the effective external scale (τ_{eff}) of the temperature cascade estimated from the European Centre for Medium-Range Weather Forecasts interim reanalysis for 2006 (*thin blue*). The top τ_c curves were estimated by bilinear log–log fits on the Haar structure functions applied to the same 20CR temperature data. The macroweather regime is the regime between the top and bottom curves

Moving to the longer time scales, one may also note that beyond a decade or two, the fluctuations again increase with scale. In reality, as one averages from weeks to months to years, the temperature fluctuations are indeed averaged out, appearing to converge to a fixed climate. However, starting at decades, this apparent fixed climate actually starts to fluctuate, varying up to ice age scales in much the same way as the weather varies (with nearly the same exponent $H \approx 0.4$, see Fig. 5b). While the adage says "The climate is what you expect, the weather is what you get", the actual data indicate that "Macroweather is what you expect, the climate is what you get".

The annual and decadal scales in Fig. 5a, b are from the anthropocene, it is important to compare this with the pre-industrial variability. This comparison is shown in detail in Fig. 5c, d that includes comparisons with GCM outputs. From the figures we see that in the anthropocene, macroweather ends (scale τ_c) at around a decade or so; Fig. 6 gives estimates of τ_c averaged over fixed latitudes showing that it is a little shorter in the low latitudes. We have seen (Fig. 4) that without external forcing, turbulence models when taken to their low frequency limit reproduce macroweather statistics; the same is true of GCMs in their "control run" mode (Fig. 5c). These results are important for macroweather forecasting since they represent a potential calculable climate perturbation to the otherwise (pure internal variability) macroweather behaviour.

In order to reproduce the low frequency climate regime characterized by increasing fluctuations, we therefore need something new: either a new source of internal variability or external forcings. Figure 5d shows that whereas in the anthropocene, the GCMs with Green House Gas (GHG) forcings do a good job of reproducing the variability, in the pre-industrial period (Fig. 5c), their centennial and millennial scale variability seems to be too weak (at least when using current estimates of "reconstructed" solar and volcanic forcings (Lovejoy et al. 2013)).

The usual way to understand the low frequencies is to consider them as responses to small perturbations, indeed, even the strong anthropogenic forcing is less than 1% of the mean solar flux and may be considered this way. This smallness is the usual justification for making the approximation that the external forcings (whether of natural or anthropogenic origin) yield a roughly linear response, indeed, this is the basis of linearized energy balance models and it can also be supported from a dynamical systems point of view (Ragone et al. 2015).

In order to avoid confusion, it is worth making these notions more precise. For simplicity, consider the atmosphere with fixed external radiative forcing $F(\underline{r})$ at location $\underline{r}$, (e.g. corresponding to GCM control runs). For this fixed forcing, the (stochastic) temperature field is:

$$T_F(\underline{r},t) = \langle T_F(\underline{r})\rangle + T'_F(\underline{r},t) \tag{8}$$

where the ensemble average is independent of time (since the past forcing is fixed) and T' (with $\langle T'\rangle = 0$) is the random deviation. If we identify $\langle T_F(\underline{r})\rangle$ with the climate and $T'_F(\underline{r},t)$ with the internal variability, then:

$$\begin{aligned} &T_{F,\mathrm{internal}}(\underline{r},t) = T_F(\underline{r},t) - T_{F,\mathrm{clim}}(\underline{r})\,; \quad T_{F,\mathrm{clim}}(\underline{r}) = \langle T_F(\underline{r},t)\rangle\,; \\ &T_{F,\mathrm{internal}}(\underline{r},t) = T'_F(\underline{r},t) \end{aligned} \tag{9a}$$

For simplicity, we have ignored the annual cycle, the internal variability is somewhat different than the notion of temperature anomalies discussed in Sect. 4.

Now increase the forcing from $F(\underline{r}) \to F(\underline{r}) + \Delta F(\underline{r},t)$ so that the climate part increases from $\langle T_F(\underline{r})\rangle \to \langle T_{F+\Delta F}(\underline{r},t)\rangle$ i.e. $T_{F,\mathrm{clim}}(\underline{r}) \to T_{F+\Delta F,\mathrm{clim}}(\underline{r},t)$ and:

$$\Delta T_{\Delta F,\mathrm{clim}}(\underline{r},t) = T_{F+\Delta F,\mathrm{clim}}(\underline{r},t) - T_{F,\mathrm{clim}}(\underline{r}) \tag{9b}$$

is the change in the climate response to the changed forcing. The generalized climate sensitivity λ can then be defined as:

$$\lambda(\underline{r},t) = \frac{\partial T_{F,\mathrm{clim}}(\underline{r},t)}{\partial F(\underline{r},t)} \approx \frac{\Delta T_{F,\mathrm{clim}}(\underline{r},t)}{\Delta F(\underline{r},t)} \tag{10}$$

GCMs make many realizations (sometimes from many models—"multimodel ensembles") and this equation may be used to determine the climate response and generalized sensitivity (the more common equilibrium and transient climate sensitivities are discussed momentarily). If t is a future time, then $T_{F+\Delta F}(\underline{r},t)$ is a prediction of the future state of the atmosphere including the internal variability and the changed forcing, whereas $T_{F+\Delta F,\mathrm{clim}}(\underline{r},t)$ is called a climate "projection".

Sometimes climate projections and sensitivities are estimated from single GCM model runs by estimating the ensemble averages by temporal averages over decadal time scales.

We can now state the linear response assumption:

$$\Delta T_{\text{clim}}\left(\underline{r},t\right) = G\left(\underline{r},t\right) * \Delta F\left(\underline{r},t\right) \tag{11}$$

where $G(\underline{r},t)$ is the system Green's function, in this context, it is also known as the Climate Response Function (CRF), "$*$" means convolution. Equation (11) is the most general statement of linearity for systems whose physics is the same at all times and locations (it assumes that only the differences in times and locations between the forcing and the responses are important). To date, applications of CRFs have been limited to globally averaged temperatures and forcings so that the spatial ($\underline{r}$) dependence is averaged out; for simplicity, below we drop the spatial dependence.

The CRF is only meaningful if the system is linear, in which case it is the response of the system to a Dirac function forcing. The simplest CRF is itself a Dirac function possibly with a lag $\Delta t \geq 0$, i.e. $G(t) = \lambda\delta(t - \Delta t)$, (sensitivity λ). Such CRFs have been used with some success by Lean and Rind (2008) and Lovejoy (2014a) to account for both anthropogenic and natural forcings. Rather than characterize the system by a response to Dirac forcing, it is more usual to characterize it by its responses to a step function $F(t)$ (the Equilibrium Climate Sensitivities, ECS) or to a linearly increasing $F(t)$ ("ramps"; Transient Climate Responses, TCR). Since step functions and ramps are simply the first and second integrals of the Dirac function, if the response is linear (Eq. 11), then knowledge of these responses as functions of time is equivalent to the CRF (note that usually the ECS is defined as the response after an infinite time, and TCR after a finite conventional period of 70 years).

Traditionally, Green's functions are deduced from linear differential operators arising from linear differential equations. For example, by treating the ocean as a homogeneous slab, the linearized energy balance equation may be used to determine the CRF, but the latter is an integer ordered ordinary differential equation for the mean global temperature which leads to exponential CRFs (e.g. Schwartz 2012; Zeng and Geil 2017). Such CRFs are unphysical since they break the scaling symmetry of the dynamics; the dynamical ocean is better modelled as a hierarchy of slabs each with its own time constant (rather than a unique slab with a unique constant). To model this in the linear energy balance framework requires introducing differential terms of fractional order; these generally lead to the required scaling CRFs (SCRF) and will be investigated elsewhere.

Rather than determine the CRF from differential operators, they can be determined directly from the symmetries of the problem. In this case (considering only the temporal CRF, $G(t)$), the three relevant symmetries are: (a) that the physics is stationary in time, (b) that the system is causal, (c) that there is no characteristic time scale. From these three symmetries we obtain $G(t) \propto t^{H_R-1}\Theta(t)$ where H_R is the SCRF response exponent and $\Theta(t)$ is the Heaviside function ($=0$ for $t < 0$, $=1$ for $t \geq 0$), necessary to ensure causality of the response.

Before continuing, we must note that such pure power law SCRFs are unusable due to either high or low frequency divergences; in this context, the divergences are aptly called "runaway Green's function effect" (Hébert and Lovejoy 2015) so that truncations are needed. For forcings that have infinite "impulses" (such as step functions or ramps whose temporal integrals diverge), when $H_R > 0$ low frequency temperature divergences will occur, unless $G(t)$ has a low frequency cutoff whereas whenever $H_R < 0$, the cutoff must be at high frequencies. For example, Rypdal (2015) and Rypdal and Rypdal (2014) use an SCRF with exponent $H_R > 0$ (without cutoff) so that low frequency temperature divergences occur unless all the forcings return to zero quickly enough. This is why Hebert et al. (2017) use $H_R < 0$ but introduce a high frequency cutoff τ in order to avoid the divergences: $G(t) = \lambda_H (t/\tau + 1)^{H_R - 1}\Theta(t)$; λ_H is a generalized sensitivity. In this case, the cutoff should correspond to the smallest time scale over which the linear approximation is valid. While the most general (space-time) linear approximation (i.e. with $G(\underline{r},t)$) may be valid at shorter time scales, if we reduce the problem to a "zero dimensional" (globally averaged) series $T(t)$, then clearly a linear response is only possible at scales over which the ocean and atmosphere are strongly coupled. The breakthrough in understanding and quantifying this was to use Haar fluctuations to show that the coupling of air temperature fluctuations over land and SST fluctuations abruptly change from very low to very high at the ocean weather-ocean macroweather transition scale of $\tau = 1$–2 years (see Fig. 7). A truncated SCRF with this τ and with $H_R \approx -0.5$ allows (Hebert et al. 2017) to make future projections based on historical forcings as well as to accurately project the forced response of GCM models.

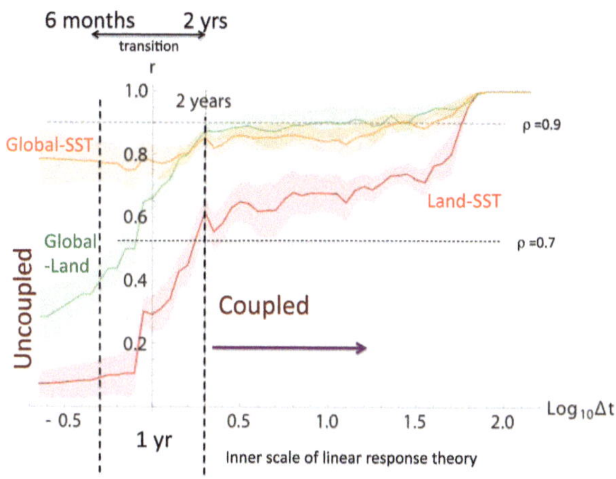

Fig. 7 The correlations quantifying the coupling of global, land and ocean temperature fluctuations. At each scale Δt, the correlation coefficient ρ of the corresponding Haar fluctuations was calculated for each pair of the monthly resolution series. The key curve is the correlation coefficient of globally averaged air over land with globally averaged sea surface temperature (SST, *bottom*, *red*). One can see that there is a sharp transition at $\tau \approx 1$–2 years from very low correlations, to very high correlations corresponding to uncoupled and coupled fluctuations. Reproduced from Hebert et al. (2017)

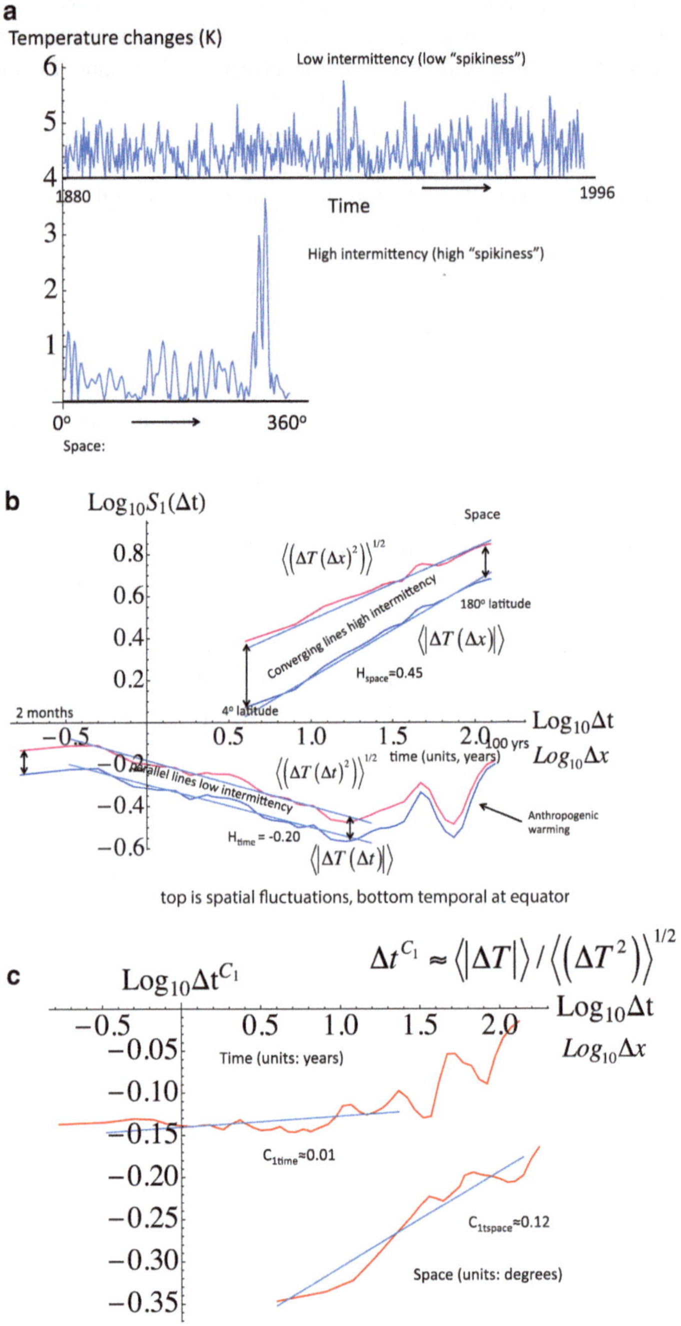

Fig. 8 (**a**) A comparison of temporal and spatial macroweather series at 2° resolution. The top are the absolute first differences of a temperature time series at monthly resolution (from 80°E, 10°N,

2.3 Climate Zones and Intermittency: In Space and Time

We have argued that macroweather is the dynamical regime of fluctuations with time scales between the lifetimes of planetary structures (τ_w) and the climate regime where either new (slow) internal processes or external forcings begin to dominate (τ_c). We have seen that a key characteristic is that mean fluctuations tend to decrease with time scale so that the macroweather fluctuation exponent $H < 0$. However in general, fluctuations require an infinite hierarchy of exponents for their characterization (the entire function $K(q)$ in Eq. (7)). In particular, when $K(q)$ is large, the process is typically "spikey" with the spikes distributed in a hierarchical manner over various fractal sets.

To see this, consider the data shown in Fig. 8a (macroweather time series and spatial transects, top and bottom, respectively). Fig. 8b compares the root mean square (RMS, exponent $\xi(2)/2$) and mean fluctuation (exponent $H = \xi(1)$) of macroweather temperature temporal data (bottom) and for the transect (top). When the system is Gaussian, $\xi(q) = qH$ so that $K(q) = 0$) and we obtain $\xi(2)/2 = \xi(1)$ so that the lines in the figure will be parallel. We see that to a good approximation this is indeed true of the nonspikey temporal series (Fig. 8a, top). However, the spatial transect is highly spikey (Fig. 8a, bottom) and the corresponding statistics (the top lines in Fig. 8b) tend to converge at large Δt. To a first approximation, it turns out that $\xi(2)/2 - \xi(1) \approx K'(1) = C_1$ which characterizes the intermittency near the mean. However, there is a slightly better characterization of C_1 (described in Lovejoy and Schertzer (2013), Chap. 11), using the intermittency function (see Fig. 8c and caption) whose theoretical slope (for ensemble averaged statistics) is exactly $K'(1) = C_1$. As a point of comparison, recall that fully developed turbulence in the weather regime typically has $C_1 \approx 0.09$, (see Lovejoy and Schertzer (2013), Table 4.5). The temporal macroweather intermittency ($C_1 \approx 0.01$) is indeed small whereas the spatial intermittency is large ($C_1 \approx 0.12$).

The strong spatial intermittency is the statistical expression of the existence of climate zones (Lovejoy and Schertzer 2013). However we shall see that due to space-time statistical factorization (next subsection), each region may be forecast separately. In addition, a low intermittency (Gaussian) approximation can be made

Fig. 8 (continued) 1880–1996, displaced by 4 K for clarity), and the bottom is the series of absolute first differences of a spatial latitudinal transect (annually averaged, 1990 from 60°N), as a function of longitude. Both use data from the 20CR. One can see that while the top is noisy, it is not very "spikey". (**b**) The first order and RMS Haar fluctuations of the series and transect from (**a**). One can see that in the spikey transect, the fluctuation statistics converge at large lags (time scale Δt), the rate of the converge is quantified by the intermittency parameter C_1. The series (*bottom*) is less spikey, converges very little and has low C_1 (see (**c**)). (**c**) A comparison of the intermittency function $F = \langle|\Delta T|\rangle(\langle|\Delta T|^{1+\Delta q}\rangle)/(\langle|\Delta T|^{1-\Delta q}\rangle)^{1/\Delta q}$ (more accurate than the approximation indicated in the figure) for the series and transect in the (**a**) and (**b**), quantifying the difference in intermittencies: in time $C_1 \approx 0.01$, in space, $C_1 \approx 0.12$. Since $K'(1) = C_1$, when Δq is small enough (here, $\Delta q = 0.1$ was used), we have $F(\Delta t) = \Delta t^{C_1}$. The break in the temporal scaling at about 20–30 years is due to anthropogenic forcings for the temporal statistics. Note that in spite of this Gaussian approximation for forecasts, there is evidence that the 5th and higher moments of the temperature fluctuations diverge (i.e. power probability distributions) so that the Gaussian approximation fails badly for the extreme 3% or so of the fluctuations (see Lovejoy and Schertzer (1986) and Lovejoy (2014a)).

2.4 Scaling, Space-Time Statistical Factorization and Size-Lifetime Relations

In the previous section we saw that there was evidence for scaling separately both in space and in time with the former being highly intermittent (multifractal) and the latter being nearly Gaussian (Fig. 8). However, in order to make stochastic macroweather forecasts, we need to understand the *joint* space-time macroweather statistics and these turn out to be quite different from those in the weather regime. For the latter, recall that there exist well-defined statistical relations between weather structures ("meso-scale complexes", "storms", "turbulence", etc.) of a given size L and their lifetimes τ. Indeed, the textbook space-time "Stommel" diagrams that adorn introductory meteorology textbooks show log spatial scale versus log temporal scale plots with boxes or circles corresponding to different morphologies and phenomenologies and these typically occupy the diagonals. These diagrams are usually interpreted as implying that each factor of two or so in spatial scale corresponds to fundamentally different dynamical processes, each with its own typical spatial extent and corresponding lifetime. However, as pointed out in Schertzer et al. (1997), the part of the diagram occupied by realistic structures and processes are typically not only on diagonals (implying a scaling space-time relation), but are on the precise diagonal whose slope has the value 2/3, theoretically predicted by the (Lagrangian, co-moving) size-lifetime relation discussed above: $\tau = \varepsilon^{-1/3} L^{2/3}$. The usual interpretation is an example of the "phenomenological fallacy" (Lovejoy and Schertzer 2007): rather than refute the scaling hypothesis, the Stommel diagrams support it.

As usual, the Eulerian (fixed frame) space-time relations are much easier to determine empirically, although theoretically their relation to Lagrangian statistics is not trivial. In a series of papers based on high resolution lidar data (Lilley et al. 2008; Lovejoy et al. 2008; Radkevitch et al. 2008) and then geostationary IR data (Fig. 3, Pinel et al. (2014)), an argument by Tennekes (1975) about the small structures being "swept" by larger ones was extended to the (atmospheric) case assuming that there was no scale separation between small and large horizontal scales. It was concluded that the corresponding Eulerian (i.e. fixed frame) space-time relation generally had space-time spectra of the form:

$$P_{xyt}\left(k_x, k_y, \omega\right) = [\![\left(k_x, k_y, \omega\right)]\!]^{-s} \tag{12}$$

where P_{xyt} is the space-time spectra density:

$$P_{xyt}\left(k_x, k_y, \omega\right) \propto \left\langle \left|\tilde{T}\left(k_x, k_y, \omega\right)\right|^2 \right\rangle \tag{13}$$

and $[\![(k_x, k_y, \omega)]\!]$ is the wavenumber (k_x, k_y)–frequency (ω) scale function nondimensionalized by the large scale turbulent velocities (i.e. using ε and the size of the earth). The analogous (real space) second order joint space-time structure function statistics:

$$S_{xyt}\left(\Delta x, \Delta y, \Delta t\right) = \left\langle \Delta T(\Delta x, \Delta y, \Delta t)^2 \right\rangle \tag{14}$$

were of the form:

$$S_{xyt}\left(\Delta x, \Delta y, \Delta t\right) = [\![(\Delta x, \Delta y, \Delta t)]\!]^{\xi(2)} \tag{15}$$

where $[\![(\Delta x, \Delta y, \Delta t)]\!]$ is the real space (nondimensional) scale function for horizontal lag $(\Delta x, \Delta y)$ and temporal lag Δt. The scale functions relevant here satisfy the isotropic scaling: $[\![\lambda^{-1}(\Delta x, \Delta y, \Delta t)]\!] = \lambda^{-1}[\![(\Delta x, \Delta y, \Delta t)]\!]$ and $[\![\lambda(k_x, k_y, \omega)]\!] = \lambda[\![(k_x, k_y, \omega)]\!]$ where λ is a scale reduction factor. This is directly confirmed in Fig. 3 for IR radiances.

In the simplest cases (with no mean advection and ignoring weak scaling singularities associated with waves (Pinel and Lovejoy 2014)), and retaining only a single spatial lag Δx, and wavenumber k_x, the nondimensional scale functions reduce to the usual vector norms, i.e. they are of the form:

$$[\![(\Delta x, \Delta t)]\!] = \left(\Delta x^2 + \Delta t^2\right)^{1/2} \tag{16}$$

$$[\![(k_x, \omega)]\!] = \left(k_x^2 + \omega^2\right)^{1/2} \tag{17}$$

With $s = d + \xi(2)$ with $d =$ the dimension of space-time, in this example $d = 2$.

In order to define a relationship between a structure of extent L with the lifetime τ, we can use S_{xt}. For example, if we wait at a fixed location ($\Delta x = 0$) for a time τ, we may ask what distance L must we go at a given instant ($\Delta t = 0$) in order to expect the same typical fluctuation? This gives us an implicit relation between L and τ: $S_{xt}(0, \tau) = S_{xt}(L, 0)$; in this simple case (Eqs. 15 and 16) this implies $\tau = L$ for the nondimensional variables so that the dimensional relationship would correspond to a constant speed relating space and time. A similar relation would be obtained by using the same argument in Fourier space on the spectral density P.

What is the space-time relation in macroweather where we consider temporal averages over periods $>\tau_w$, typically months or longer? In this case, we average over many lifetimes of structures of all sizes, so clearly size-lifetime relations valid in the weather regime must break down. Lovejoy and Schertzer (2013) and Lovejoy and de Lima (2015) argued on theoretical, numerical and empirical grounds that—

at least to a good approximation—the result is statistical space-time factorization (SSTF). The application of the SSTF to the second order statistics means:

$$\begin{aligned} P_{xt}(k_x,\omega) &= P_x(k_x)\,P_t(\omega) \\ R_{xt}(\Delta x,\Delta t) &= R_x(\Delta x)\,R_t(\Delta t); \end{aligned} \tag{18}$$

Note that in real space we have used correlation functions $R_{xt}(\Delta x, \Delta t) = \langle T(t,x)\, T(t-\Delta t, x-\Delta x)\rangle$ rather than Haar structure functions S; in macroweather ($H < 0$), they are essentially equivalent. However for small lags in time, one effectively goes outside the macroweather regime and $\Delta t = 0$ is problematic. When both $H_t < 0$ and $H_x < 0$ we can avoid issues that arise at small Δt, Δx by using correlation functions (Fig. 9a) (for the case $H_t < 0$, $H_x > 0$, see Sect. 10.3 of Lovejoy and Schertzer (2013)).

Using the autocorrelations to obtain space-time macroweather relations, we obtain $R_{xt}(0,\tau) = R_{xt}(L,0)$ so that using factorization and the identity $R_t(0) = R_x(0)$ the implicit τ-L relation is:

$$R_t(\tau) = R_x(L) \tag{19}$$

This is valid if both space and time have $H < 0$; if there is scaling, we have $R_t(\tau) \propto \tau^{H_t}$ and $R_x(L) \propto L^{H_x}$ with exponents $H_t < 0$, $H_x < 0$. The lifetime of a macroweather structure of size L is thus:

$$\tau \propto L^{H_x/H_t} \tag{20}$$

which—unless $H_x = H_t$ —is quite different from the lifetime-size relationship in the weather regime; Fig. 9a shows that $\tau \propto L^{0.65}$, for macroweather temperature and precipitation. Fig. 9a, shows that empirically the factorization works well for both temperature and precipitation data, and Fig. 9b shows that it is also (even better) obeyed by the GISS E2R GCM; Del Rio Amador 2017 shows that it holds very accurately for 36 CMIP5 control runs.

It turns out that the SSTF is important for macroweather forecasting. This is because, using means square estimators, it implies that no matter how strong the correlations (teleconnections), if one has long time series at each point, pixel or region, that no further improvement can be made in the forecast by adding co-predictors such as the temperature data at other locations (Del Rio Amador 2017). This effectively means that the original nonlinear initial value PDE problem has been effectively transformed into a linear but fractional ordered ODE "past value" problem, we pursue this in the next sections.

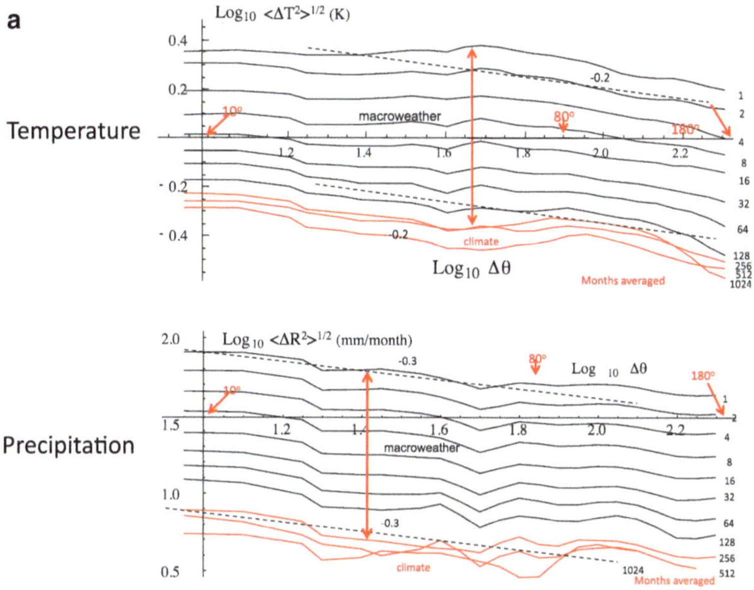

Fig. 9 (**a**) The joint space (Δθ i.e. angle subtended) time (Δt) RMS fluctuations of temperature (*top*, adapted from (Lovejoy 2017)) and precipitation (*bottom*, adapted from (Lovejoy and de Lima 2015)). In both cases, zonal spatial anomaly fluctuations are given for data averaged over 1, 2, 4, ..., 1024 months (since the temporal $H < 0$ this is an anomaly fluctuation). The temperature data are from the HadCRUtemp3 database and the precipitation data from the Global Historical Climate Network, both at 5°, monthly resolutions and spanning the twentieth century. On this log–log plot, SSTF implies $S_{\theta t}(\Delta\theta, \Delta t) = S_\theta(\Delta\theta)S_t(\Delta t)$ so that the curves will be parallel. If in addition they respect spatial scaling, then they will be linear, and if they respect the temporal scaling, then as we double the temporal resolution (*top* to *bottom*), they will be equally spaced (separated by log 2^H). Eventually (*red*), the temporal scaling breaks down (at $\tau_c \approx 256$ months). Over the regimes where both SSTF and scaling hold we have for temperature, $S_{\theta,t}(\Delta\theta, \Delta t) \approx \Delta\theta^{-0.2}\Delta t^{-0.3}$ and for precipitation $S_{\theta,t}(\Delta\theta, \Delta t) \approx \Delta\theta^{-0.3}\Delta t^{-0.4}$. The double headed red arrows show the corresponding total predicted range over macroweather time scales. (**b**) The same as (**a**), but for temperature fluctuations from GISS-E2R historical simulations from 1850. In this case, rather than using anomalies (which were the only data available for (**a**)), we used the difference between two realizations of the same historical simulation (i.e. with identical external boundary conditions) obtained by slightly varying the initial conditions. The temporal behaviour of this plot shows rapidly the model climate is approached under temporal averaging, and how it varies as a function of angular scale. Again we see that the joint fluctuations have nearly exactly the same shapes (confirming SSTF); over the ranges where the scaling holds, the joint structure function is:$S_{\theta,t}(\Delta\theta, \Delta t) \approx \Delta\theta^{0.3}\Delta t^{-0.4}$. This plot shows that GCMs obey the SSTF very accurately, a fact confirmed in Sect. 4 by the success by which they can be predicted by SLIMM

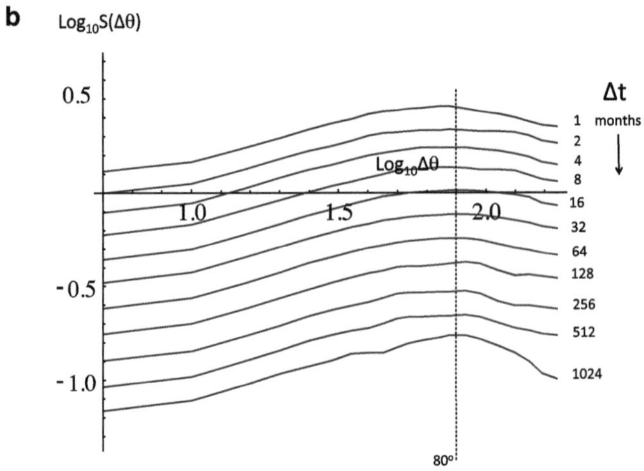

Fig. 9 (continued)

3 Macroweather Forecasting

3.1 The Fractional Gaussian Noise Model and some of Its Characteristics

We have argued that macroweather is scaling but with low intermittency, so that a Gaussian forecasting model may be an acceptable approximation. The simplest such model is fractional Gaussian noise (fGn). We now give a brief summary of some useful properties of fGn; for a longer review, see Lovejoy et al. (2015) and for an extensive mathematical treatment see (Biagini et al. 2008).

Over the parameter range of interest $-1/2 < H < 0$, fGn is essentially a smoothed Gaussian white noise and its mathematical definition raises similar issues. For our purposes, it is most straightforward to use the framework of generalized functions and start with the unit Gaussian white noise $\gamma(t)$ which has $\langle\gamma\rangle = 0$ and is "δ correlated":

$$\left\langle\gamma(t)\gamma\left(t'\right)\right\rangle = \delta\left(t - t'\right) \tag{21}$$

where "δ" is the Dirac function. The H parameter fGn $G_H(t)$ is thus:

$$G_H(t) = \frac{c_H}{\Gamma(1/2+H)}\int_{-\infty}^{t}\left(t - t'\right)^{-(1/2-H)}\gamma\left(t'\right)dt';\ -1 < H < 0 \tag{22}$$

The constant c_H is a constant chosen so as to make the expression for the statistics particularly simple, see below. Mathematically $\gamma(t)$ is thus the density of the Wiener process $W(t)$, often written $\gamma(t)dt = dW$: just as the Dirac function is only meaningful when integrated, the same is true of $\gamma(t)$. For fGn, we shall see below that $G_H(t)dt = dB_{H'}$ where $B_{H'}$ is a generalization of the Wiener process, fractional Brownian motion (fBm, parameter $H' = 1 + H$) and $B_{H'}$ reduces to a Wiener process when $H' = 1/2$. $G_H(t)$ is thus the (singular) density of an fBm measure. In practice, we will always consider $G_H(t)$ smoothed over finite resolutions so that whether we define $G_H(t)$ indirectly via fBm or directly as a smoothing of Eq. (22) the result is equivalent.

We can see by inspection of Eq. (22) that $G_H(t)$ is statistically stationary and by taking ensemble averages of both sides of Eq. (22) we see that the mean vanishes: $\langle G_H(t)\rangle = 0$. When $H = -1/2$, the process $G_{-1/2}(t)$ itself is simply a Gaussian white noise. Although we justified the use of fGn as the simplest scaling process, it could also be introduced as the solution of a stochastic fractional ordered differential equation:

$$\frac{d^{H+1/2}T}{dt^{H+1/2}} = \gamma(t) \tag{23}$$

the solution of which is $T(t) \propto G_H(t)$.

Now, take the average of G_H over τ, the "τ resolution anomaly fluctuation":

$$G_{H,\tau}(t) = \frac{1}{\tau}\int_{t-\tau}^{t} G_H\left(t'\right) dt' \tag{24}$$

If c_H is now chosen such that:

$$c_H = \left(\frac{\pi}{2\cos(\pi H)\,\Gamma(-2H-2)}\right)^{1/2} \tag{25}$$

then we have:

$$\left\langle G_{H,\tau}(t)^2\right\rangle = \tau^{2H};\ -1 < H < 0 \tag{26}$$

This shows that a fundamental property of fGn is that in the small scale limit ($\tau \to 0$), the variance diverges and H is scaling exponent of the root mean square (RMS) value. This singular small scale behaviour is responsible for the strong power law resolution effects in fGn. Since $\langle G_H(t)\rangle = 0$, sample functions $G_{H,\tau}(t)$ fluctuate about zero with successive fluctuations tending to cancel each other out; this is the hallmark of macroweather.

Anomalies

An anomaly is the average deviation from the long-term average and since $\langle G_H(t)\rangle = 0$, the anomaly fluctuation over interval Δt is simply G_H at resolution Δt rather than τ:

$$(\Delta G_{H,\tau}(\Delta t))_{\text{anom}} = \frac{1}{\Delta t}\int\limits_{t-\Delta t}^{t} G_{H,\tau}(t')\,dt' = \frac{1}{\Delta t}\int\limits_{t-\Delta t}^{t} G_{H}(t')\,dt' = G_{H,\Delta t}(t);\ \Delta t > \tau \tag{27}$$

Hence using Eq. (26):

$$\left\langle (\Delta G_{H,\tau}(\Delta t))^2_{\text{anom}} \right\rangle = \Delta t^{2H};\ -1 < H < 0 \tag{28}$$

Differences

In the large Δt limit we have:

$$\left\langle (\Delta G_{H,\tau}(\Delta t))^2_{\text{diff}} \right\rangle \approx 2\tau^{2H}\left(1 - (H+1)(2H+1)\left(\frac{\Delta t}{\tau}\right)^{2H}\right) \tag{29}$$

Since $H < 0$, the differences asymptote to the value $2\tau^{2H}$ (double the variance). Notice that since $H < 0$, the differences are not scaling with Δt.

Haar Fluctuations

For the Haar fluctuation we obtain:

$$\left\langle (\Delta G_{H,\tau}(\Delta t))^2_{\text{Haar}} \right\rangle = 4\Delta t^{2H}\left(2^{-2H} - 1\right);\ \Delta t \geq 2\tau \tag{30}$$

this scales as Δt^{2H} and does not depend on the resolution τ. This relation can be used to estimate the spatial variation of H, Fig. 10 gives the spatial distribution using 20CR data. It can be seen that H is near zero over the oceans and is lower over land, typical values being −0.1 and −0.3, respectively. Below, we see that this corresponds to large memory (and hence forecast skill) over oceans and lower memory and skill over land.

Autocorrelations

$$\begin{aligned}\langle G_{\tau,H}(t)G_{\tau,H}(t-\Delta t)\rangle &= R\left(\widehat{\Delta t}\right)\\ &= \frac{\tau^{2H}}{2}\left[\left(\widehat{\Delta t}+1\right)^{2H+2} + \left(\widehat{\Delta t}-1\right)^{2H+2} - 2\widehat{\Delta t}^{2H+2}\right];\ \widehat{\Delta t} = \frac{\Delta t}{\tau}\end{aligned} \tag{31}$$

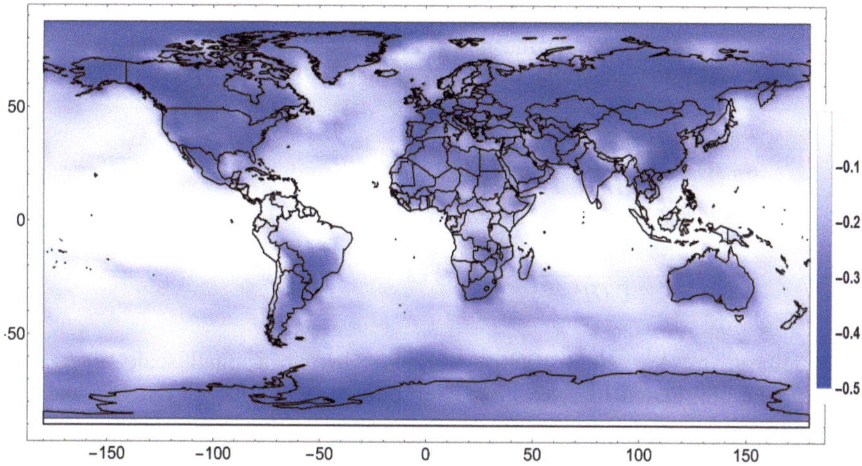

Fig. 10 The spatial distribution of the exponent H estimated at $5^\circ \times 5^\circ$ resolution using monthly resolution data from the NCEP reanalyses (1948–2010) and estimated by a maximum liklihood method. The mean was -0.11 ± 0.09

Spectra

Since fGn is stationary, its spectrum is given by the Fourier transform of the autocorrelation function. Note that in the above, $\Delta t > 0$; since the autocorrelation is symmetric for the Fourier transform with respect to Δt, we use the absolute value of Δt. We obtain:

$$E(\omega) = \frac{\Gamma(3+2H)\sin \pi H}{\sqrt{2\pi}} |\omega|^{-\beta};\ \beta = 1 + 2H \tag{32}$$

Relation to fBm

It is more common to treat fBm whose differential $dB_{H'}(t)$ is given by:

$$dB_{H'} = G_H(t)dt;\ H' = H + 1;\ 0 < H' < 1 \tag{33}$$

so that:

$$\Delta B_{H'}(\tau) = B_{H'}(t) - B_{H'}(t-\tau) = \int_{t-\tau}^{t} G_{H'}(t')\,dt' = \tau G_{H',\tau}(t) \tag{34}$$

with the property:

$$\left\langle \Delta B_{H'}(\Delta t)^2 \right\rangle = \Delta t^{2H'} \tag{35}$$

While this defines the increments of $B_{H'}(t)$ and shows that they are stationary, it does not completely define the process. For this, one conventionally imposes $B_{H'}(0) = 0$, and this leads to the usual definition:

$$B_{H'}(t) = \frac{c_{H'}}{\Gamma(H'+1/2)} \int_{-\infty}^{0} \left((t-s)^{H'-1/2} - (-s)^{H'-1/2} \right) \gamma(s) ds + \frac{c_{H'}}{\Gamma(H'+1/2)} \int_{0}^{t} (t-s)^{H'-1/2} \gamma(s) ds \tag{36}$$

(Mandelbrot and Van Ness 1968). Whereas fGn has a small scale divergence that can be eliminated by averaging over a finite resolution τ, the fGn integral $\int_{-\infty}^{t} G_H(t')\,dt'$ on the contrary has a low frequency divergence. This is the reason for the introduction of the second term in the first integral in Eq. (36): it eliminates this divergence at the price of imposing $B_{H'}(0) = 0$ so that fBm is nonstationary (although its increments are stationary, Eq. (34)).

A comment on the parameter H is now in order. In treatments of fBm, it is usual to use the parameter H confined to the unit interval, i.e. to characterize the scaling of the increments of fBm. However, fBm (and fGn) are very special scaling processes, and even in low intermittency regimes such as macroweather—they are at best approximate models of reality. Therefore, it is better to define H more generally as the fluctuation exponent (Eq. 6); with this definition, H is also useful for more general (multifractal) scaling processes although the interpretation of H as the "Hurst exponent" is only valid for fBm). When $-1 < H < 0$, the mean at resolution τ (Eq. 24) defines the anomaly fluctuation, so that H is equal to the fluctuation exponent for fGn, in contrast, for processes with $0 < H < 1$, the fluctuations scale as the mean differences and Eq. (35) shows that H' is the fluctuation exponent for fBm. In other words, as long as an appropriate definition of fluctuation is used, H and $H' = 1 + H$ are fluctuation exponents of fGn, fBm, respectively. The relation $H' = H + 1$ follows because fBm is an integral order 1 of fGn. Therefore, since the macroweather fields of interest have fluctuations with mean scaling exponent $-1/2 < H < 0$, we use H for the fGn exponent and $½ < H' < 1$ for the corresponding integrated fBm process.

We can therefore define the resolution τ temperature as:

$$T_\tau(t) = \sigma_T G_{H,\tau}(t) = \sigma_T \frac{B_{H'}(t) - B_{H'}(t-\tau)}{\tau} \tag{37}$$

Using Eq. (26), the τ resolution temperature variance is thus:

$$\langle T_\tau^2 \rangle = \sigma_T^2 \tau^{2H} \tag{38}$$

From this and the relation $T_\tau(t) = \sigma_T G_{H,\tau}(t)$, we can trivially obtain the statistics of $T_\tau(t)$ from those of $G_{H,\tau}(t)$.

3.2 Mean Square (MS) Estimators for fGn and the ScaLIng Macroweather Model (SLIMM)

The Mean Square (MS) estimator framework is a general framework for predicting stochastic processes, it determines predictors that minimize the prediction error variance, see, e.g., Papoulis (1965). Since Gaussian processes are completely determined by their second order statistics, the MS framework therefore gives optimum forecasts for fGn.

Our problem is to use data $T_\tau(s)$ at times $s < 0$ (or equivalently, the innovations $\gamma(s)$) to predict the future temperature $T_\tau(t)$ at times $t > 0$. Denoting this predictor by $\widehat{T}_\tau(t)$ MS theory then shows that the latter is given by a linear combination of data, i.e. either the $T_\tau(s)$ or equivalently by a linear combination of past white noise "innovations" $\gamma(s)$:

$$\begin{aligned} \widehat{T}_\tau(t) &= \int_{-\tau_0<s\leq 0} M_T(t,s)\, T_\tau(s) ds \\ \widehat{T}_\tau(t) &= \int_{-\tau_0<s\leq 0} M_\gamma(t,s)\, \gamma(s) ds \end{aligned} \tag{39}$$

where M_T, M_γ are the predictor kernels based on past temperatures and past innovations, respectively, and the range of integration is over all available data, the range $-\tau_0 < s \leq 0$. The simplest problems are those where the range extends to the infinite past ($\tau_0 \to \infty$), but practical predictions require the solution for finite τ_0.

The prediction error is thus:

$$E_T(t) = T_\tau(t) - \widehat{T}_\tau(t) \tag{40}$$

and from MS theory, the basic condition imposed by minimizing the error variance $\langle E_T^2(t) \rangle$ is:

$$\left\langle E_T(t) \widehat{T}_\tau(t) \right\rangle = \langle E_T(t) T_\tau(s) \rangle = \langle E_T(t) \gamma(s) \rangle = 0;\ t > 0;\ s \leq 0 \tag{41}$$

This equation states that the (future) prediction error $E_T(t)$ is statistically independent of the predictor $\widehat{T}_\tau(t)$ or, equivalently, it is independent of the past data $T_\tau(s)$, $\gamma(s)$ upon which the predictor is based. This makes intuitive sense: if there was a nonzero correlation between the available data and the prediction error, then there would still information in the data that could be used to improve the predictor and reduce the error. Since GCM forecasts are not MS, they do not satisfy this orthogonality condition. On the one hand, this explains how they can have negative skill (see below), on the other, it justifies complex GCM post-processing that exploit past data to reduce the errors. Indeed, a condition used to optimize post-processing corrections is actually close to the orthogonality condition.

In Lovejoy et al. (2015), the mathematically simplest predictor was given in the case of infinite past data but using the innovations $\gamma(s)$:

$$\widehat{T}_\tau(t) = \int\limits_{-\infty}^{0} M_\gamma(t,s)\,\gamma(s)ds$$
$$M_\gamma(t,s) = \tfrac{c_H\sigma_T}{\tau\Gamma(H+3/2)}\left[(t-s)^{H+1/2} - (t-\tau-s)^{H+1/2}\right] \tag{42}$$

The error is:

$$E_T = T_\tau(t) - \widehat{T}_\tau(t)$$
$$= \frac{c_H\sigma_T}{\tau\Gamma(H+3/2)}\left[\int\limits_0^t (t-s)^{H+1/2}\gamma(s)ds - \int\limits_0^{t-\tau}(t-\tau-s)^{H+1/2}\gamma(s)ds\right] \tag{43}$$

Since $\widehat{T}(t)$ depends only on $\gamma(s)$ for $s < 0$ and E_T on $\gamma(s)$ for $s > 0$, it can be seen by inspection that the orthogonality condition (Eq. 41) holds. Using this MS predictor, we can define the Mean Square Skill Score (MSSS) or "skill" for short:

$$\mathrm{MSSS} = S_k(t) = 1 - \frac{\left\langle E_T(t)^2\right\rangle}{\left\langle T_\tau(t)^2\right\rangle} \tag{44}$$

For MS forecasts, we can use the orthogonality condition to obtain equivalently;

$$S_k(t) = \frac{\left\langle \widehat{T}_\tau^2(t)\right\rangle}{\left\langle T_\tau^2(t)\right\rangle} \tag{45}$$

which shows that for MS forecasts, the skill is the same as the fraction of the variance explained by the predictor.

Using the predictor (Eq. 42) we can easily obtain the skill for fGn forecasts:

$$S_k(\lambda) = \left[\tfrac{F_H(\infty) - F_H(\lambda)}{F_H(\infty) + \frac{1}{2H+2}}\right];\ \lambda = t/\tau;\ \lambda \geq 1 \tag{46}$$

where the auxiliary function F_H is given by:

$$F_H(\lambda) = \int\limits_0^{\lambda-1}\left((1+u)^{H+1/2} - u^{H+1/2}\right)^2 du;\ \lambda \geq 1 \tag{47}$$

with:

$$F_H(\infty) = \pi^{-1/2}2^{-(2H+2)}\Gamma(-1-H)\,\Gamma(3/2+H) \tag{48}$$

and the asymptotic expression:

$$F_H(\lambda) = F_H(\infty) - \frac{(H+1/2)^2}{-2H}\lambda^{2H} + \ldots \tag{49}$$

(Lovejoy et al. 2015). For any system that has quasi-Gaussian statistics and scaling fluctuations with $-1/2 < H < 0$ the theoretical skill, Eq. (46) represents a stochastic predictability limit, of similar fundamental significance to the usual deterministic predictability limits arising from sensitive dependence on initial conditions. In Sect. 4.2, we show that CMIP5 GCMs can indeed be predicted to nearly this limit using the MS approach outlined here.

Although the MSSS is commonly used for evaluating forecasts, the correlation coefficient between the hindcast and the temperature is occasionally used:

$$\rho_{\widehat{T},T}(t,\tau) = \frac{\left\langle \widehat{T}_\tau(t) T_\tau(t) \right\rangle - \left\langle \widehat{T}_\tau(t) \right\rangle \langle T_\tau(t) \rangle}{\left\langle \widehat{T}_\tau(t)^2 \right\rangle^{1/2} \left\langle T_\tau(t)^2 \right\rangle^{1/2}} \tag{50}$$

Since $\langle T \rangle = 0$, the upper right cross term vanishes and using orthogonality $\left\langle T_\tau(t) \widehat{T}_\tau(t) \right\rangle = \left\langle \widehat{T}_\tau(t)^2 \right\rangle$ we obtain:

$$\rho_{\widehat{T},T}(t,\tau) = S_k(t,\tau)^{1/2} \tag{51}$$

Therefore, MS forecast skill can equivalently be quantified using either correlations or MSSS.

Figure 11a shows the theoretical skill as a function of H for different forecast horizons. To underscore the huge memory implied by the power law kernel M_γ, we can compare the fGn kernel with that of the exponential kernels that arise in auto-regressive (AR) type processes. This is relevant here since the main existing stochastic macroweather forecasts techniques ("Linear Inverse Modelling", LIM, see the next subsection) are vector AR processes that reduce to scalar AR processes in an appropriately (diagonalized) frame. If for simplicity we consider only forecasts one time step into the future (i.e. horizon τ, for a process resolution τ), then the fraction $f(\lambda)$ of the predictor variance that is due to innovations at times $\lambda\tau$ or further in the past can be written in the same form as for fGn:

$$f(\lambda) = \frac{I(\lambda)}{I(0)};\ I(\lambda) = \int_{-\infty}^{-\lambda} (g(s) - g(s-1))^2 ds \tag{52}$$

where $g(s) = (-s)^{1/2+H}$ for fGn (for SLIMM predictions) and $g(s) = e^s$ for AR processes. The comparison is shown in Fig. 11b, it can be seen that almost all the information needed to forecast an AR process is in the most recent three steps, whereas for SLIMM, with $H = -0.1$ (appropriate for forecasting the globally averaged temperature), roughly 20% comes from innovations more than 1000 steps in the past. Significantly, we will see that this does not mean that we need such long series to make good forecast; this is because even relatively short series with $H = -0.1$ have information from the distant past; this is discussed below.

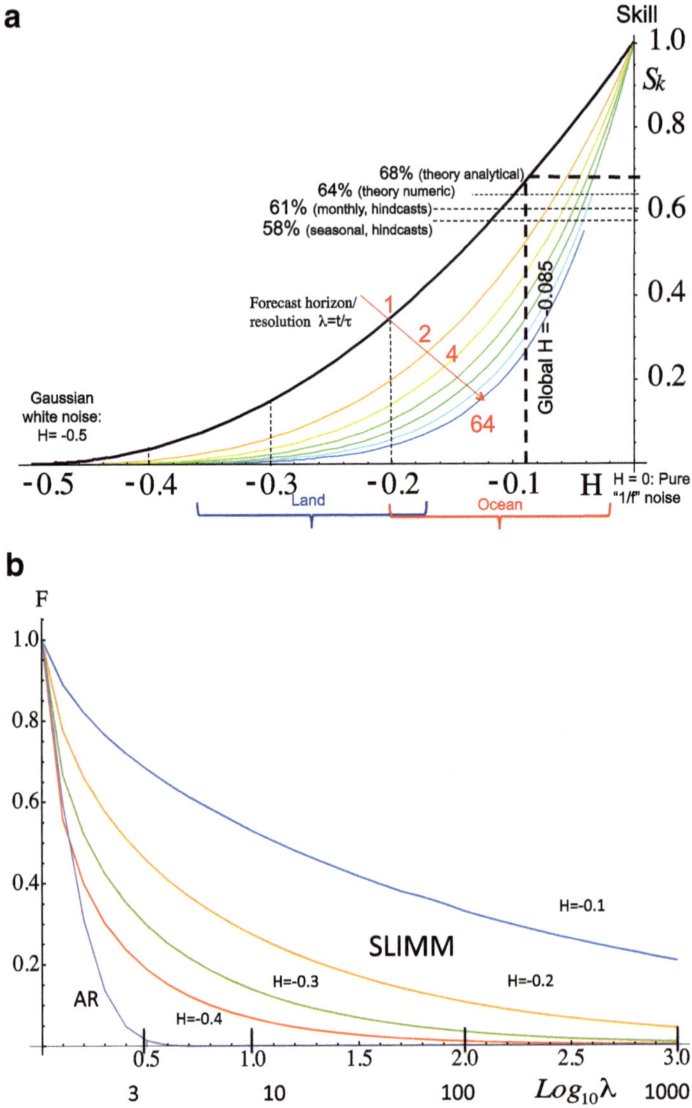

Fig. 11 (**a**) Forecast skill for nondimensional forecast horizons λ = (horizon/resolution) = 1, 2, 4, 8, …, 64 (*left* to *right*) as functions of H. For reference, the rough empirical values for land, ocean and the entire globe (the value used here, see below) are indicated by dashed vertical lines. The horizontal lines show the fraction of the variance explained (the skill, S_k, Eq. (46)) in the case of a forecast of resolution τ data at a forecast horizon $t = \tau$ ($\lambda = 1$; corresponding to forecasting the anomaly fluctuation one time unit ahead). (**b**) The fraction of the prediction variance of a forecast one time step ahead that is due to innovations further in the past than λ time units (one unit = resolution τ). The right four curves are for SLIMM ($H = -0.1, -0.2, -0.3, -0.4$), and the far left curve is for an auto-regressive process $F = f(\lambda)$ = Fraction of total memory used in forecasts one step into the future

3.3 SLIMM Prediction Skill and Alternative Stochastic Macroweather Prediction Systems

Following Hasselmann (1976) who proposed the use of stochastic differential equations to understand low frequency weather (i.e. macroweather), attempts have been made to use this for monthly, Seasonal to Interannual forecasts. The basic idea is to model the atmosphere as an Ornstein-Uhlenbeck process, i.e. the solution of $\frac{dT}{dt} + T/\tau = \gamma(t)$ where τ is the basic time scale and γ is a white noise forcing. The idea is that the weather acts essentially as a random white noise perturbation to the temperature T. Fourier analysis shows that the spectrum is $E(\omega) \propto 1/(\omega^2 + \tau^{-2})$ so that at high frequencies, $E(\omega) \propto \omega^{-2}$ whereas at low frequencies, $E(\omega) \approx$ constant. The process is thus an (unpredictable) white noise; this can be seen directly by taking the low frequency limit $dT/dt \approx 0$ in the equation. From an empirical point of view, there are two scaling regimes (exponents $\beta = 0, 2$), corresponding to $H = (\beta - 1)/2 = -1/2$ and $H = ½$, respectively, but neither is realistic: for example, the true values for the temperature are closer to ≈ -0.1, ≈ 0.4 for macroweather, weather respectively with the former showing significant spatial variations, see Fig. 10. The key point is that models based on integer order differential equations implicitly assume that the low frequencies are unpredictable whereas on the contrary, the temporal scaling implies long range dependencies, a large memory. From the point of view of differential equations, we thus require terms of fractional order (see Eq. (22)).

Over the decades, the Hasselman inspired approach has been significantly developed, in the framework of "Linear Inverse Modelling" (LIM), sometimes also called the "Stochastic Linear Framework" (SLF), although the latter is somewhat a misnomer since it restrictively excludes fractional ordered (but still linear) terms (for LIM, SLF see, e.g., Penland (1996), Penland and Sardeshmuhk (1995), Sardeshmukh et al. (2000), and Newman (2013)). The essential development is the extension of scalar Ornstein-Uhlenbeck processes to vector processes with each component being a significant macroweather variable (e.g. an El Nino index, an ocean temperature at a particular grid point, etc.). Typical implementations such as described in Newman (2013) involve 20 components (implying hundreds of empirical parameters). When diagonalized, the system reduces to decoupled Ornstein-Uhlenbeck processes whose longest characteristic times are about 1 year, and beyond this, the system has little skill, see Fig. 12a.

Because its theoretical basis is weak and it involves a large number of empirical parameters, LIM is an example of what is commonly termed an "empirically based" approach. Other such approaches have been proposed, notably by Suckling et al. (2016) and they have had some success by using carefully chosen climate indices that are linearly related to macroweather variables of interest and using empirically determined time delays. In contrast, SLIMM is based on fundamental space-time scale symmetries that we argue are respected by the dynamical equations.

In order to use SLIMM for forecasts, it is important to first remove the low frequency responses to anthropogenic forcings, failure to do so (Baillie and Chung

2002) leads to poor results. For annually, globally averaged temperatures, it turns out that reasonable results can be obtained using the CO_2 radiative forcing (proportional to logCO_2 concentration) as a linear surrogate for all anthropogenic forcings (Fig. 12b). SLIMM then forecasts the internal variability: the residuals. The reason that this works so well is presumably that all anthropogenic effects are linked through the economy and the economy is well characterized by energy use and hence by CO_2 emissions.

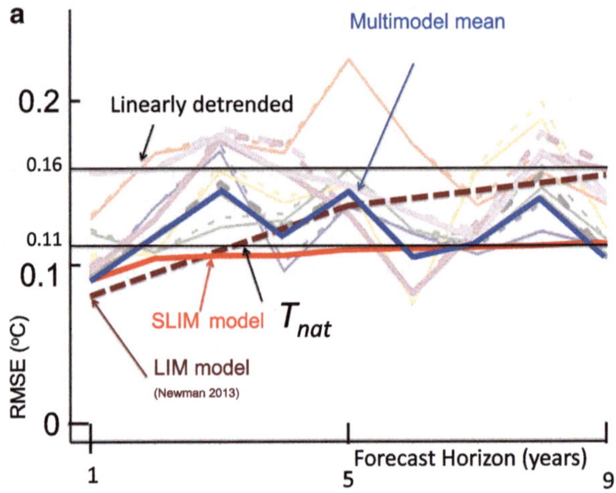

Fig. 12 (**a**) A comparison of Root Mean Square Error (RMSE) of hindcasts of various global annual temperatures for horizons of 1–9 years: the (GCMbased) ENSEMBLES experiment (from (Garcıa-Serrano and Doblas-Reyes 2012), LIM (Newman 2013) and SLIMM (Lovejoy et al. 2015). The light lines are from individual members of the ENSEMBLE experiment; the heavy line is the multimodel ensemble. This shows the RMSE comparisons for the global mean surface temperatures compared to NCEP/NCAR (2 m air temperatures). Horizontal reference lines indicate the standard deviations of T_{nat} (*bottom horizontal line*, the RMS of the residuals after removing the anthropogenic forcing using the CO_2 as a linear surrogate, itself nearly equivalent to the pre-industrial variability (Lovejoy 2014a)) and of the RMS of the residuals of the linearly detrended temperatures (*top horizontal line*). Also shown are the RMSE for the LIM model and the SLIMM. Adapted from Lovejoy et al. (2015). (**b**) The NASA GISS globally, annually averaged temperature series from 1880–2013 plotted as a function of CO_2 radiative forcing. The regression slope indicated corresponds to 2.33 ± 0.22 K/CO_2 doubling. The internal variability forecast by SLIMM are the residuals (see (**c**)). Adapted from Lovejoy (2014b). (**c**) (*Top*): The residuals temperature of (**b**) after the low frequency anthropogenic rise has been removed (*blue*) with the hindcast from 1998 (*red*). (*Bottom left*): The anomaly defined as the average natural temperature (i.e., residual) over the hindcast horizon (*blue*), *red* is the hindcast. (*Bottom right*): The temperature since 1998 (*blue*) with hindcast (*red*), a blowup of the hindcast part of the top right. Adapted from Lovejoy (2015b). (**d**) This shows the kernel $M_T(t,s)$ (Eq. (39), the discrete case) when the data extends to $s_0 = \tau_0$ in the past with parameter $H = -0.1$. Note the strong weighting on both the most recent (*right*) and the most ancient available data (*left*). Reproduced from Del Rio Amador (2017)

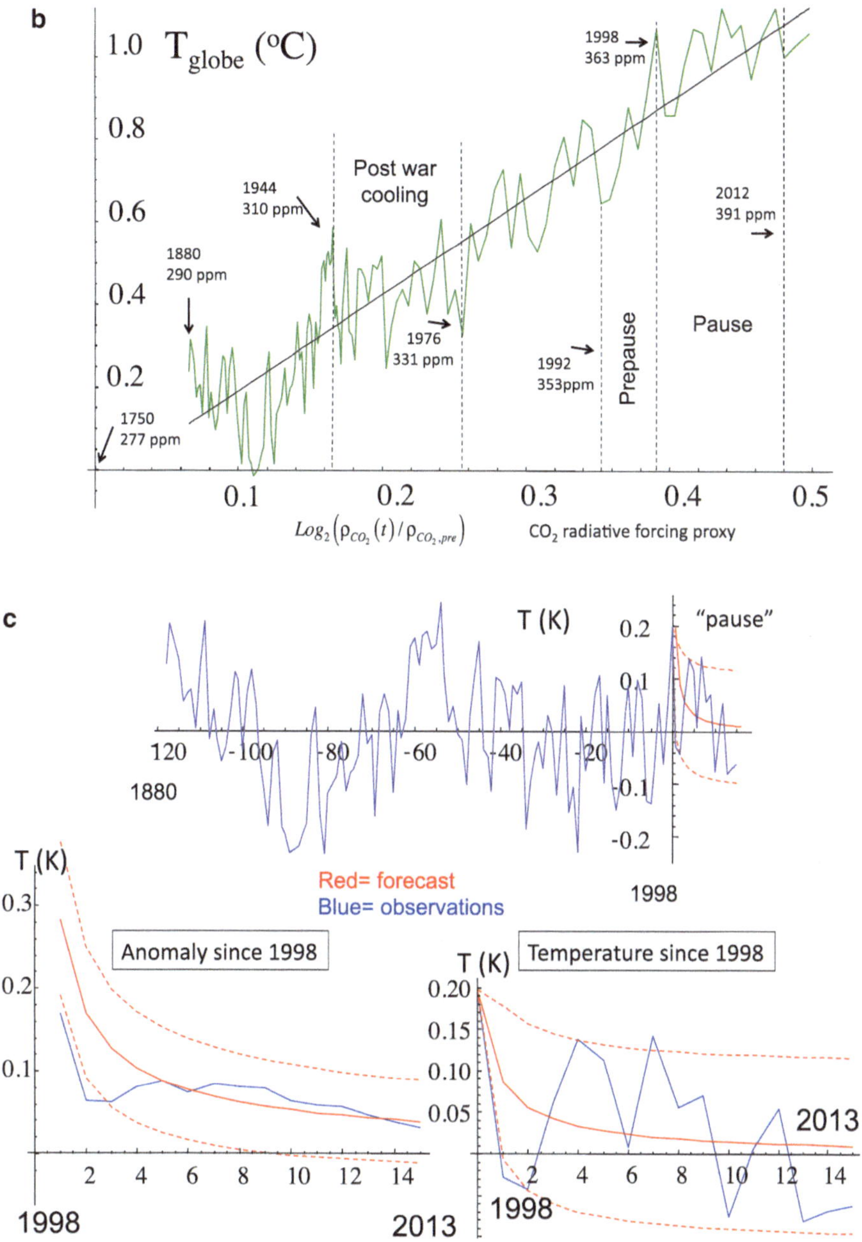

Fig. 12 (continued)

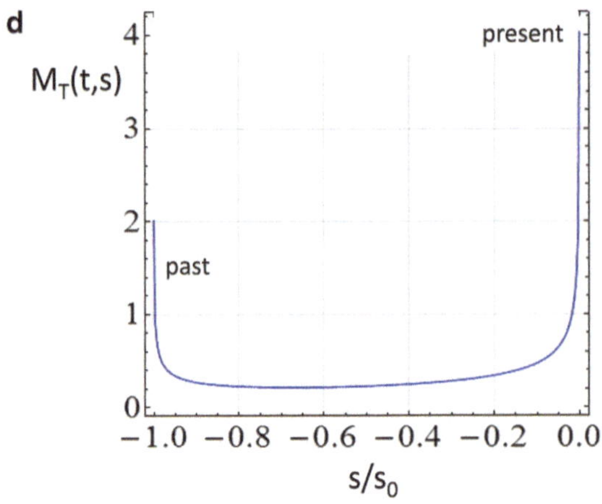

Fig. 12 (continued)

When SLIMM hindcasts are made for hemispheric and global scales (Lovejoy et al. 2015), they are generally better than LIM and GCM forecasts (Fig. 12a). In addition, Lovejoy (2015b) made global scale SLIMM forecasts and showed that they could accurately (to within about ±0.05 °C for three year anomalies) forecast the so-called "pause" in the warming (1998–2015). In comparison, CMIP3 GCM predictions were about 0.2 °C too high. While the cause of the GCM over-prediction is currently debated (e.g., Schmidt et al. 2014; Guemas et al. 2013; Steinman et al. 2015), the SLIMM prediction was successful large because as Fig. 12b shows, the pause was simply a natural cooling event that followed the enormous "pre-pause" 1992–1998 warming, with all of this superposed on a rising anthropogenic warming trend.

The SLIMM forecast technique showed that the fGn model was worth pursuing. However, the original technique was based on M_γ, i.e. finding the optimum predictor using the innovations $\gamma(s)$ directly (obtained by numerically inverting Eq. (22)) and assuming that the available data extended into the infinite past. It is much more convenient to use the past data $T(s)$ and to take into account the fact that the past data are only finite in extent. Since an fGn process at resolution τ is the average of the increments of an fBm, process, it suffices to forecast fBm so that in the operational version of SLIMM described below, we therefore availed ourselves of the mathematical solution of the prediction problem of finding the kernel $M_T(t,s)$ in Eq. (39) for both finite and infinite past data. Gripenberg and Norros (1996) mathematically solved the fBm solution with $1/2 < H' < 1$ and this was numerically investigated by Hirchoren and D'attellis (1998).

We saw that the (infinite past) innovation kernel M_γ (Eq. 42) gave a strong (even singular) weight to the recent past, forecasting AR processes has an analogous strong weighting of the recent data. However, Gripenberg and Norros (1996) found something radically new in the case of finite data: the most ancient available data also had a singular weighting! In their words, this was because "the closest witnesses to the unobserved past have special weight", see Fig. 12d for a graphical example.

4 Stochastic Predictability Limits and Forecast Skill

4.1 Stochastic Predictability Limits: StocSIPS Hindcasting Skill Demonstrated on CMIP5 Control Runs

We are used to the deterministic predictability limits that arise from the "butterfly effect"—sensitive dependence on initial conditions—we argued that this limit (the inverse Lyapunov exponent of the largest structures) was roughly given by the lifetime of planetary structures: $\tau_w = \varepsilon^{-1/3}L^{2/3}$ (Schertzer and Lovejoy 2004). However, we also argued that when taken way beyond this limit, that both the GCMs and the atmosphere should be considered stochastic. More precisely, we argued that fGn provides a good approximation for the temporal variability, and that due to SSTF, attempting to use spatial correlations for co-predictors may not lead to an improvement when compared to direct predictions that exploit the huge memory of the system. However, SSTF does not necessarily extend from temperatures to other series such as climate indices. It is possible that use of the latter as co-predictors may yield larger skills.

Since fGn has stochastic predictability limits that determine its skill, Eq. (46), these should therefore be relevant in both GCMs and in real macroweather. However, in the latter and in externally forced GCMs, as discussed in Sect. 4.2 there are low frequency responses to climate forcings, and these must be forecast separately (using linearity Eq. (11)) from the internal macroweather variability modelled by fGn processes. This means that the best place to test our predictors is on unforced GCMs, i.e. on control runs. For this purpose we used 36 globally and monthly averaged CMIP5 model control runs. For each, we estimated the relevant exponent H by determining the value that made the predictor best satisfy the orthogonality condition (Eq. 41); this was slightly more accurate than using either spectra or Haar fluctuation analysis (Del Rio Amador 2017). While each model had somewhat different exponents, we found a mean $H = -0.11 \pm 0.09$ theoretically implying a huge memory (see, e.g., Fig. 11a, b). We used the discrete M_T kernel (following (Hirchoren and Arantes 1998)) and produced 12-month hindcasts comparing both the theoretical skill and the actual hindcast skill, see Fig. 13a. Figure 13b shows that the control runs were hindcast very nearly to their theoretical limits. It is thus quite plausible that the theoretical stochastic predictability limit Eq. (46) really is an upper bound on the skill of macroweather forecasts.

4.2 Regional Forecasting

In the previous section, we saw that without external forcings, we can make global scale macroweather forecasts that nearly attain their theoretical limits, and in Sect. 3.3 (the pause), we already indicated that by appropriately removing the low

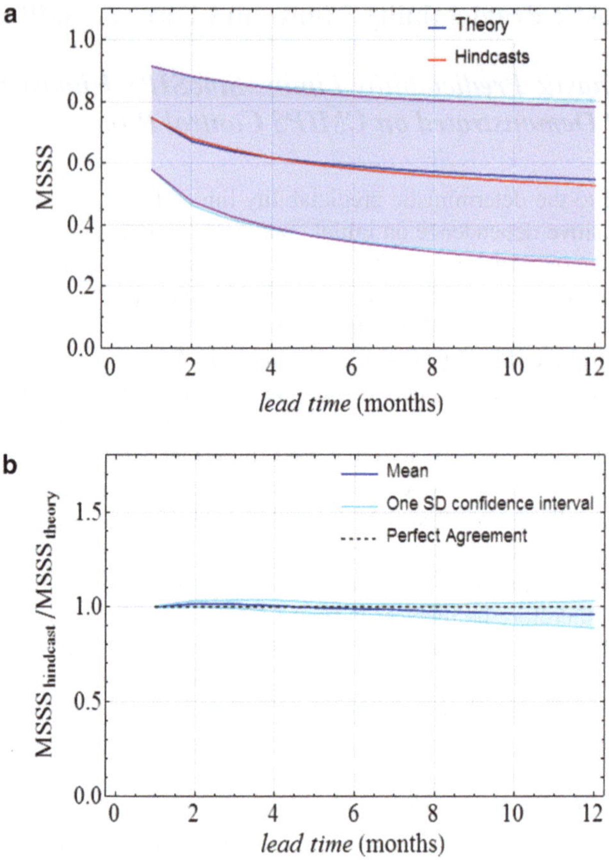

Fig. 13 (**a**) The MSSS for hindcasting 36 CMIP5 GCM control runs, each at least 2400 months long. Each GCM had a slightly different *H* and hence different theoretical predictability. The graph shows that both the means and the spreads of theory and practice (SLIMM hindcasts) agree very well. Reproduced from Del Rio Amador (2017). (**b**) The ratio of the actual MSSS hindcast skill to theortical MSSS skill evaluated for the CMIP5 control runs used in (**a**). Reproduced from Del Rio Amador (2017)

frequencies (in that case, the anthropogenic forcings), we could also make accurate global scale real world forecasts. Due to SSTF, we argued that if at a given location long series were available, they could be forecast directly, that using information at other locations as co-predictors would not increase the overall skill. In this section, we therefore discuss regional forecasts at 5° resolution. This resolution was chosen because it is the smallest that is available from both historical data and reanalysis data sets that we used.

The various steps in the forecast are illustrated in Fig. 14 using the pixel over Montreal as an example. The first step is to remove the low frequencies that are

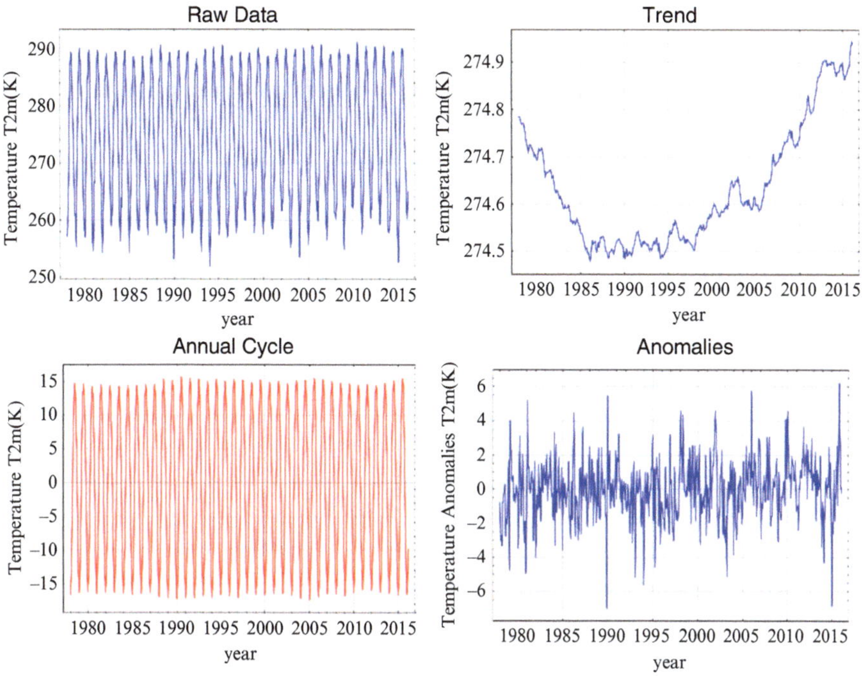

Fig. 14 An example of forecasting the temperature at Montreal using the National Centers for Environmental Prediction (NCEP) reanalysis (at $5^{\circ} \times 5^{\circ}$ resolution). The top left shows the raw monthly data, the bottom left shows the mean annual cycle as deduced using a (causal) 30-year running estimate, the upper right shows the low frequency (a causal 30-year running average) trend and the bottom right shows the resulting anomalies that were forecast by SLIMM. Reproduced from Del Rio Amador (2017)

not due to internal macroweather variability; failure to remove them will lead to serious biases since the SLIMM forecast assumes a long-term mean equal to 0 and the ensemble forecast is always towards this mean. The low frequencies have both a mean component (mostly anthropogenic in origin but also one due to internal variability) and a strong annual cycle that slowly evolves from one year to the next. Using the knowledge (Fig. 5d) that the scaling is broken at decadal scales, we can use a high pass filter to separate out these from the internal variability. Similarly, the annual cycle can be forecast by using the past thirty years of data in order to make running estimates of the relevant Fourier coefficients (only keeping those for the annual cycle and 6, 4 and 3 month harmonics). The various steps are shown in Fig. 14. Finally the anomalies (lower right) were forecast using SLIMM. The regional variation of the skill of the resulting StocSIPS hindcasts is shown in Fig. 15a, we can see that it is close to the theoretical maximum.

4.3 StocSIPS-CanSIPS Comparison

The previous section reminded us that real world forecasts must estimate, remove and separately forecast the nonmacroweather low frequencies, the higher frequency internal fGn-like component. The overall model, including this "pre-processing" is called the Stochastic Seasonal to Interannual Prediction System (StocSIPS, see the website: http://www.physics.mcgill.ca/StocSIPS/), it is comparable in scope to the Canadian Seasonal to Interannual Prediction System (CanSIPS (Merryfield et al. 2011)) and the European Seasonal to Interannual Prediction System (EuroSIPS, http://www.ecmwf.int/en/forecasts/documentation-and-support/long-range/seasonal-forecast-documentation/eurosip-user-guide/multi-model), but of course is based directly on a stochastic rather than a deterministic-stochastic

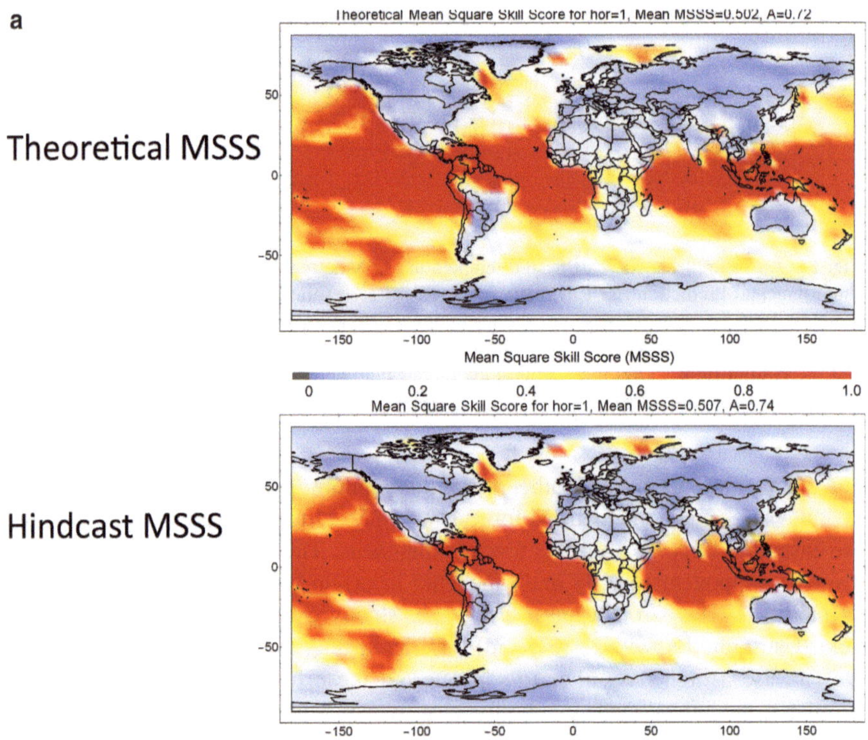

Fig. 15 (**a**) Theoretical (*top*) versus empirical (*bottom*) hindcast skill for 1 month hindcasts using Period Sep, 1980–Dec, 2015. Reference: NCEP Reanalysis. The theory and practice are very close. Reproduced from Del Rio Amador (2017). (**b**) The MSSS, shown for the actuals and estimated from hindcasts from six of the 12 "producing centres", adapted from the WMO web site (accessed in April 2016). To aid in the interpretation, an example is given by the *black arrow*: when the MSSS $= -5$, the Mean Square Error (MSE) is 5 times the amplitude of the anomaly variance. It can be seen that actuals' error variances are typically several times the anomaly variances leading to significant negative skill over most of the earth. Reproduced from Del Rio Amador (2017)

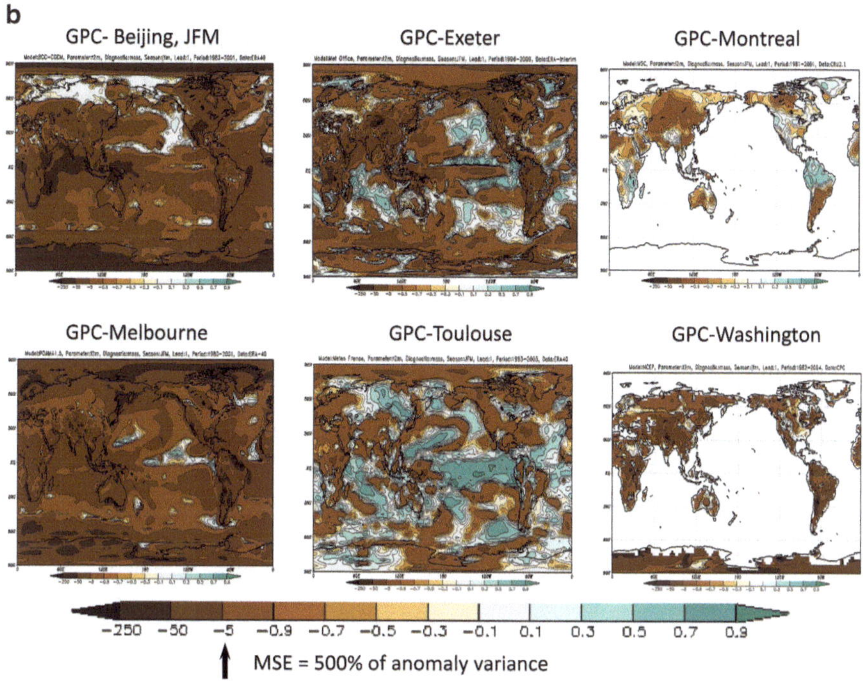

Fig. 15 (continued)

(GCM) model. Indeed, according to the World Meteorological Organization (WMO) site (http://www.wmo.int/pages/prog/wcp/wcasp/gpc/gpc.php), there are 12 international "producing" centres; StocSIPS based at McGill would be the 13th. Although completely unfunded, since April 2016, it has provided operational monthly, seasonal and annual temperature forecasts at 5° resolution.

As the previous section showed, SLIMM can forecast GCM control runs to nearly their theoretical stochastic predictability limits. However, we must evaluate the full StocSIPS system (pre-processing plus SLIMM) and compare it with conventional approaches. We singled out CanSIPS, which since 2010 is the institutional product of Environment Canada, for particularly close comparisons. Every month, CanSIPS makes monthly through annual temperature and precipitation forecasts; the publically available maps are only over Canada, but we accessed the global products and made global hindcast comparisons since 1980. The CanSIPS products are based on "multimodel ensemble" consisting of 10 realizations of the CanCM3 and 10 realizations of the CanCM4 GCM.

Before continuing, recall the method by which GCMs currently produce macroweather forecasts. The first step is the initialization; when CanSIPS is initialized it uses reanalyses from the European Centre for Medium-range Weather Forecasts (ECMWF) and these are data-model "hybrids" obtained by assimilating meteorological observations into the ECMWF GCM. The problem is that both the reanalyses and CanSIPS have their own different climatologies so that the latter cannot directly ingest the ECMWF reanalyses, instead, the ECMWF initial values must be converted into ECMWF anomalies. These anomalies are used to determine the CanSIPS initial values, the "actuals". The process can be symbolically written as:

$$\begin{aligned} T_{\mathrm{CanSIPS}}\left(\underline{r}, t\right) &= \overline{T}_{i(t),\mathrm{CanSIPS}}\left(\underline{r}\right) + T'_{\mathrm{CanSIPS}}\left(\underline{r}, t\right) \\ T_{\mathrm{ECMWF}}\left(\underline{r}, t\right) &= \overline{T}_{i(t),\mathrm{ECMWF}}\left(\underline{r}\right) + T'_{\mathrm{ECMWF}}\left(\underline{r}, t\right) \end{aligned} \tag{53}$$

where the overbar represents the climatological temperature $\overline{T}_i\left(\underline{r}\right)$ at position r, for the month number $i = 1, 2, \ldots, 12$ and the primes indicate the anomalies which are functions of both position and time (i(t) denotes the month number of time t). The conventional way to define $\overline{T}_i\left(\underline{r}\right)$ is to use the averages over the previous 30 i^{th} months (at each location/pixel $\underline{r}$). Aside from the annual cycle (that was deliberately ignored in Sect. 2.2), the anomalies differ from the internal variability because they are based on temporal rather than ensemble averages and they have contributions from external forcings.

CanSIPS is thus initialized $T_{\mathrm{CanSIPS}}\left(\underline{r}, 0\right)$ using the ECMWF anomaly at time $t = 0$:

$$T_{\mathrm{CanSIPS}}\left(\underline{r}, 0\right) = \overline{T}_{i(0),\mathrm{CanSIPS}}\left(\underline{r}\right) + T'_{\mathrm{ECMWF}}\left(\underline{r}, 0\right) \tag{54}$$

The forecasts $\widehat{T}_{\mathrm{CanSIPS}}\left(\underline{r}, t\right)$ (at $t > 0$, indicated with circonflex) are then made using the 20 member CanSIPS ensemble followed by complex (and time consuming) post-processing that primarily correct for the "model drift" and poor climate sensitivity. "Model drift" refers to the tendency of model temperatures (even in control runs) to display low frequency variations that are usually attributed to slow (mostly ocean) processes, artefacts that are not fully "balanced" when the model is initialized. Since the model does not have perfect representation of the sensitivity to anthropogenic effects, the corresponding systematic errors also contribute a further low frequency "drift". Both are removed (to some extent) using hindcasts over the previous 5-year period in an attempt to estimate (and remove) spurious linear trends (Merryfield et al. 2011). Unfortunately, 5 years is too short to properly estimate the trend (the true trends are buried in the macroweather noise until a decade or so in scale, see Fig. 5d) so that the internal 5-year variability is thus spuriously removed in the post-processing.

In spite of these manipulations, the final result $\widehat{T}_{\mathrm{CanSIPS}}\left(\underline{r}, t\right)$—i.e. an "actual"—is seriously in error as can be seen in Fig. 15b: which shows that the actuals' error variance is typically several times larger than the anomaly variance. Due to this, the publically available macroweather forecasts are of the anomalies $\widehat{T'}_{\mathrm{CanSIPS}}\left(\underline{r}, t\right) = \widehat{T}_{\mathrm{CanSIPS}}\left(\underline{r}, t\right) - \overline{T}_{i(t),\mathrm{CanSIPS}}\left(\underline{r}, t\right)$. For these anomalies, the comparison with StocSIPS is much closer, see Fig. 16. The figure shows that even for anomalies over most of the globe, for 2 months and longer, StocSIPS has higher skill. StocSIPS' increased

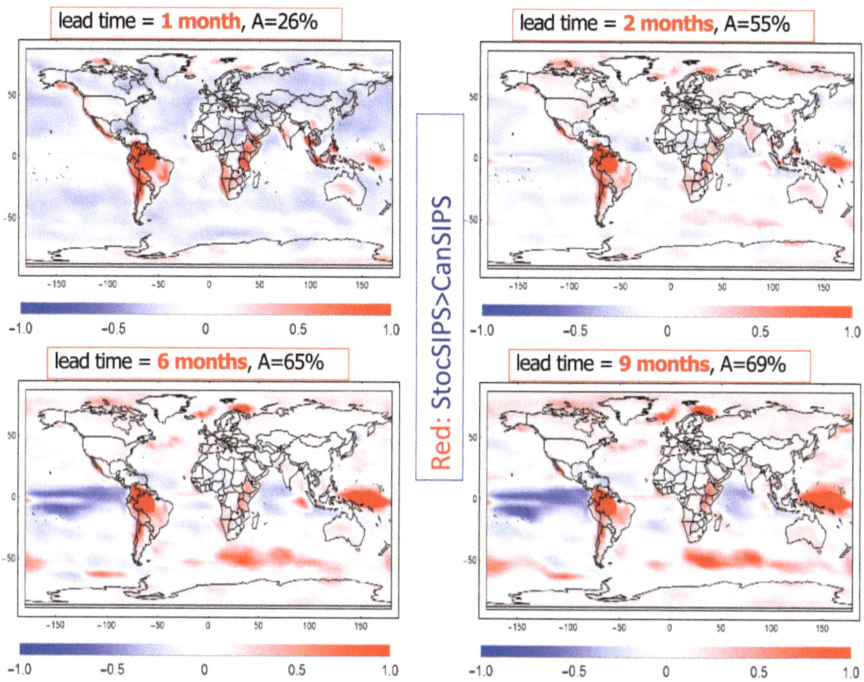

Fig. 16 The differences in MSSS for CanSIPS and StocSIPS anomaly hindcasts over the period 1980–2010 for lead times of 1, 2, 6, 9 months, red indicates regions over which StocSIPS has higher skill. It may be seen that for 2 months and longer, this is over most of the globe. StocSIPS' increased skill is particularly noticeable over land, probably due to the fact that the CanSIPS ocean model is still within its deterministic predictability limit of 1–2 years. Reproduced from Del Rio Amador (2017)

skill is particularly noticeable over land, probably due to the fact that the CanSIPS ocean model is still within its deterministic predictability limit of 1–2 years making its ocean forecast reasonably accurate. This impression is bolstered in Fig. 17 which compares CanSIPS at 6 months and StocSIPS at 2 years (the skill is comparable), and also in Fig. 18 that shows that StocSIPS' relative advantage grows with lead time and is particularly strong over land.

Although we have not discussed it in this review, StocSIPS actually provides forecasts of the probability distributions (both mean, discussed up until now, and the standard deviation about the mean). This can be used for various probabilistic forecasts. For example, Fig. 19a, b shows a typical seasonal forecast and its validation. In Fig. 19a we see that the StocSIPS anomaly forecasts generally follow the data better than CanSIPS. In Fig. 19b, we see that for this location and date, that the StocSIPS forecast was both more accurate and less uncertain than the CanSIPS forecast. This was true for both the actuals and the anomalies. This can be seen since not only is the dashed red StocSIPS mean closer to the NCEP validation

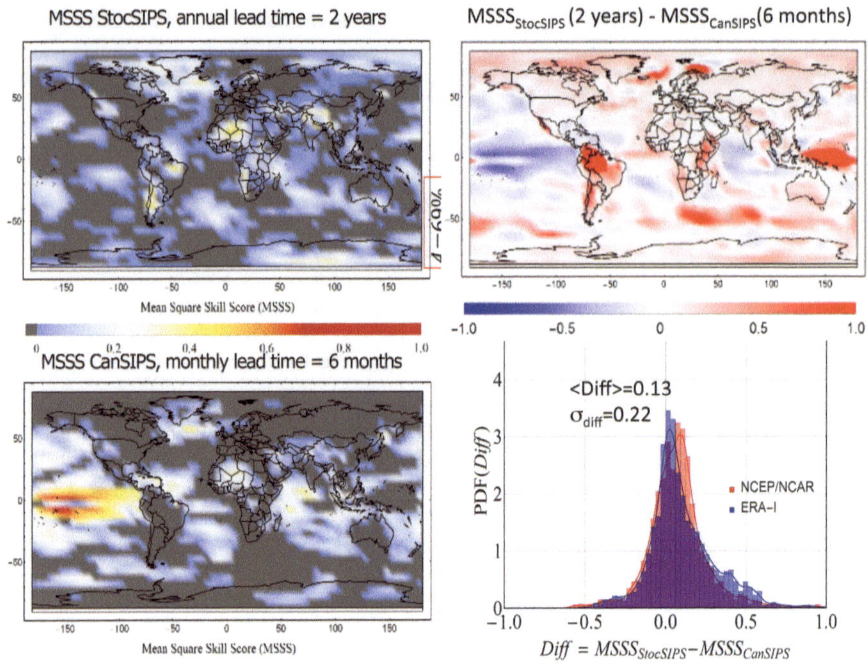

Fig. 17 A comparison of the MSSS for StocSIPS at 2 year lead times (*top left*) and CanSIPS at 6 months (*bottom left*). The map of their differences (top right) and histogram of the differences lower right using both the ECMWF interim reanalyses (ERA-I, *red*) and NCEP reanalyses (*blue*) show that the 2 year StocSIPS forecast is somewhat better than the CanSIPS 6 month forecast. Reproduced from Del Rio Amador (2017)

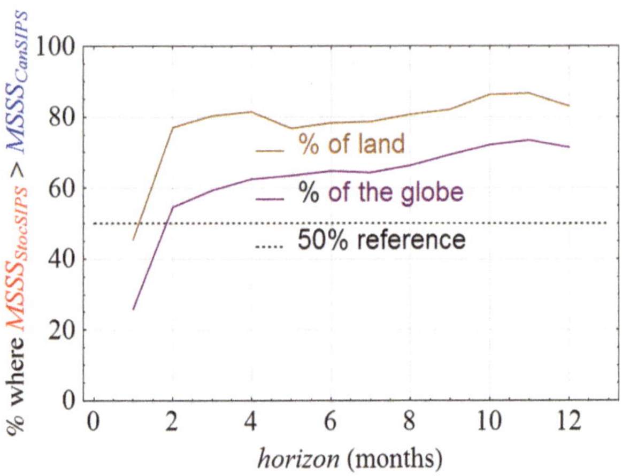

Fig. 18 The relative skill of StocSIPS and CanSIPS anomaly hindcasts (1980–2010) over the globe and over land only showing that StocSIPS' relative advantage increase systematically with lead time and is particularly strong over land. Reproduced from Del Rio Amador (2017)

(dashed black) than the CanSIPS dashed blue, but the uncertainties (the spreads in the probability densities) is narrower for the StochSIPS forecast. Other probabilistic forecasts that can readily be produced by StocSIPS include tercile forecasts: i.e. the probabilities of the forecast temperature being below, above or equal to the local climatology; see the StocSIPS site for examples.

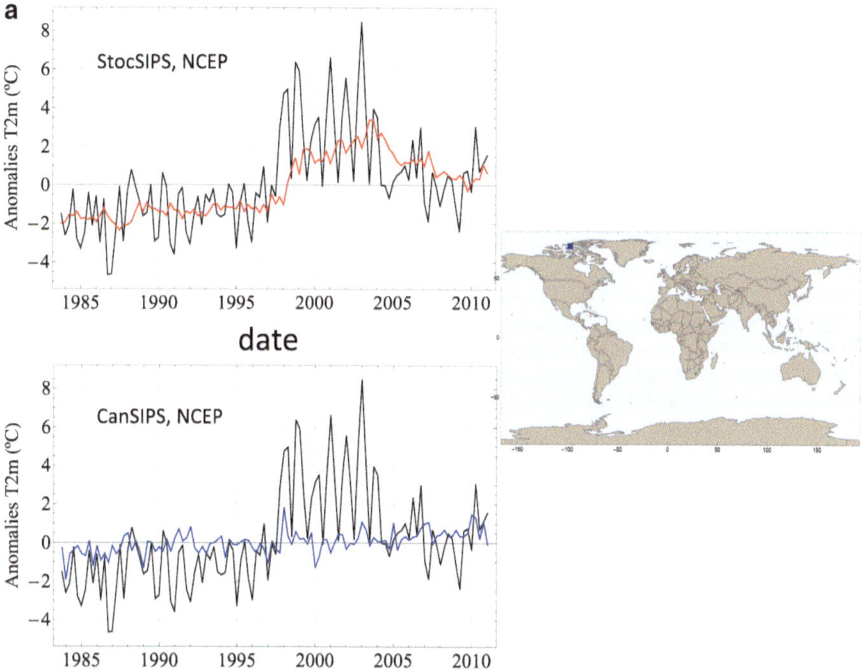

Fig. 19 (**a**) StocSIPS (*top*, *red*) and CanSIPS (*bottom*, *blue*) seasonal forecasts, two seasons ahead for temperature anomalies at 97.5 W, 77.5 N (see *blue point* on the map at *right*). The forecasts are compared with the NCEP reanalysis anomalies (*black*) that are calculated with respect to the period 1980–2010. It can be seen that StocSIPS is much closer to the data (see also (**b**)). (**b**) The histograms of seasonal forecasts, two seasons ahead for DJF (2009–2010) using data up to $t = 0 =$ JJA 2009, location the same as in (**a**) (*top* actuals, *bottom*, anomalies, StocSIPS in *red*, CanSIPS in *blue*, NCEP data in *black*). The *dashed black lines* are the NCEP validation data for DJF, the *black* probability density curves show the spread of the climatological variations based on past NCEP reanalyses (1981–2010), the variability is thus placed around the observed DJF temperature. The StocSIPS and CanSIPS dashed lines (*red* and *blue*) are their respective forecasts for DJF, the curves represent the estimated uncertainties in the forecast. For both actuals and anomalies StocSIPSs forecasts are sharper—their probability density functions (PDFs) are narrower and more peaked; they are also more accurate since the red dashed lines (the StocSIPS forecasts) are closer to validation data (the *black dashed lines*)

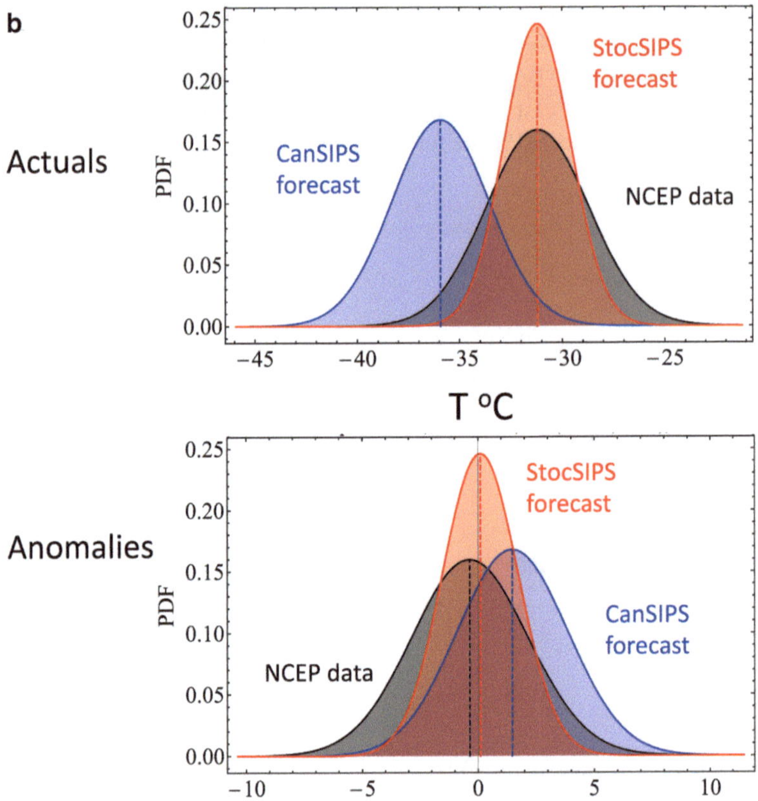

Fig. 19 (continued)

5 Conclusions

Over the last decades, it has become increasingly clear that at weather scales, atmospheric dynamics are governed by both deterministic laws of continuum mechanics and by stochastic turbulence laws. Although the GCM equations do not acknowledge the existence of atoms or molecules, they are nevertheless compatible with statistical mechanics. Similarly they are also believed to be compatible with the turbulence laws and indeed, they obey them quite accurately. Over the same period, the GCM approach has—with the development of ensemble forecasting and stochastic parametrizations—itself evolved into a stochastic one, making it tempting to make weather forecasts directly using the turbulence laws. However the weather regime is highly intermittent, and it involves vector multifractal processes, whose corresponding mathematical prediction problem has yet to be solved. The GCM approach to weather prediction is thus the only one currently available.

The situation is radically different at time scales beyond the GCM deterministic predictability limit—in macroweather. On the one hand, GCMs have large errors associated with unrealistic model climatologies, especially poor representations of the annual cycle, and they also display model drift and unrealistic sensitivities to anthropogenic effects. On the other hand, macroweather "turbulence" (the extension of turbulence models to the macroweather regime) has low intermittency so that Gaussian models are useable approximations (fractional Gaussian noise, fGn). In addition, a new symmetry: statistical space-time factorization essentially decouples space and time so that mean square predictions can conveniently be made for each spatial location independently. Physically this means that even though strong spatial correlations exist (including "teleconnections"), if one has a long enough history at a given point, this spatial information is also implicit in the series so that using data at other spatial locations as co-predictors does not necessarily improve the forecast. The factorization is not exact and does not necessarily apply to other series such as climate indices so that there may be future scope for finding co-predictors and improve the skill.

The ideal testing ground for this approach is in GCM control runs since this is closest to pure fGn. We found that the ScaLIng Macroweather Model (SLIMM) based on an fGn model applied to temperatures from GCM control runs (i.e. pure macroweather processes, no changes in external forcings) is nearly able to attain the maximum theoretical stochastic predictability limit, verifying that GCMs well obey the macroweather laws upon which SLIMM is based and raising the possibility that these stochastic predictability limits are true GCM limits. With respect to usual stochastic forecasts based on exponential correlations (Auto Regressive, or Linear Inverse Modelling), the radically new feature of SLIMM is its exploitation of the huge long range memory. The SLIMM prediction kernel thus has singular weighting to both the most recent data and the most ancient data since the latter contain the maximum information of the distant past.

Applying SLIMM to real data requires pre-processing to remove non-macroweather processes in particular to remove low frequency anthropogenic effects and—for regional forecasts—the annual cycle. The overall resulting system (i.e. pre-processing plus SLIMM) is the STOChastic Seasonal to Interannual Prediction System (StocSIPS). We compared StocSIPS with one of the leading GCM macroweather products: CanSIPS. Even without any co-predictors or other use of spatial correlations, we showed that StocSIPS was much superior to CanSIPS for forecasting "actuals": this was due to StocSIPS' ability to essentially forecast the climatology (especially the annual cycle). However, even for anomaly forecasts, StoSIPS was superior to CanSIPS for lead times of 2 months or longer and its relative advantage grew with the forecast lead time, the advantage was particularly important over land where for 2 months and longer StocSIPS was superior over more than $\approx$80% the earth's land surface.

Aside from its increased skill, StocSIPS has other advantages. For example, at the moment, seasonal forecasts for the city of Montreal (or other localized region) are highly indirect. First data from all atmospheric fields from all over the world must be assimilated. Then the model—on grids typically several hundred kilometres across—is integrated forward in time. Anomalies are calculated, and post-processing is performed to make low frequency corrections for some of the known biases. Finally, the Montreal temperature anomaly is estimated by "downscaling" from the large pixel scale to the local city scale. This can be done either using sophisticated (but complex) nested regional models (of GCM type) or via ad hoc statistical methods based on local climatology. In contrast, if long enough (preferably several decades) of monthly or seasonal data are available, StocSIPS simply removes the low frequencies (including the annual cycle), separately forecasts the anomalies and low frequencies and adds them to produce the forecast. The overall saving in computational speed is estimated to be of the order of 10^7 (about 10^5 to 10^6 for global forecasts on $5^o \times 5^o$ grids). Finally, StocSIPS directly forecasts the conditional ensemble average, i.e. effectively the results of an infinite ensemble whereas CanSIPS uses only 20 members.

StocSIPS can be directly extended to other fields such as wind or precipitation which instead are known to have macroweather statistics roughly satisfying the SLIMM requirements (low intermittency temporal macroweather scaling with $-1/2 < H < 0$ and space-time statistical factorization (SSTF), Lovejoy and de Lima (2015) and Fig. 9a). But StocSIPS' main advantage may be its ability to directly forecast other fields, such as insolation, wind power or degree-days, that can currently only be very indirectly forecast by GCMs. Other future extensions of StocSIPS could include drought indices and the prediction of extremes.

Acknowledgements We thank Lydia Elias, Hannah Wakeling and Weylan Thompson for undergraduate summer contributions in developing StocSIPS. We thank OURANOS for funding Lydia Elias's summer work. We thank Dave Clark, Norberto Majlis and Yosvany Martinez (Environment Canada) for regular discussions. Hydro Quebec is thanked for partial support of L. Del Rio Amador during his PhD. The project itself was unfunded, there were no conflicts of interest.

References

Baillie, R.T., and S.-K. Chung. 2002. Modeling and forecasting from trend-stationary long memory models with applications to climatology. *International Journal of Forecasting* 18: 215–226.

Biagini, F., Y. Hu, B. Øksendal, and T. Zhang. 2008. *Stochastic calculus for fractional Brownian motion and applications*. London: Springer-Verlag.

Chen, W., S. Lovejoy, and J.P. Muller. 2016. Mars' atmosphere: The sister planet, our statistical twin. *Journal of Geophysical Research—Atmospheres* 121: 11968–11988. doi:10.1002/2016JD025211.

Compo, G.P., et al. 2011. The twentieth century reanalysis project. *Quarterly J. Roy. Meteorol. Soc.* 137: 1–28. doi:10.1002/qj.776.

Del Rio Amador, L. 2017. *The stochastic seasonal to interannual prediction system*. Montreal: McGill University.

Garcıa-Serrano, J., and F. J. Doblas-Reyes (2012), On the assessment of near-surface global temperature and North Atlantic multi-decadal variability in the ENSEMBLES decadal hindcast, Climate Dynamics, 39, 2025–2040 doi: 10.1007/s00382–012–1413-1. .

Gripenberg, G., and I. Norros. 1996. On the Prediction of Fractional Brownian Motion. *Journal of Applied Probability* 33: 400–410.

Guemas, V., F.J. Doblas-Reyes, I. Andreu-Burillo, and M. Asif. 2013. Retrospective prediction of the global warming slowdown in the past decade. *Nature Climate Change* 3: 649–653.

Hasselmann, K. 1976. Stochastic climate models, part I: Theory. *Tellus* 28: 473–485.

Hébert, R., and S. Lovejoy. 2015. The runaway Green's function effect: Interactive comment on "Global warming projections derived from an observation-based minimal model" by K. Rypdal. *Earth System Dynamics Discovery* 6: C944–C953.

Hebert, R., S. Lovejoy, and A. de Vernal. 2017. A scaling model for the forced climate variability in the anthropocene. *Climate Dynamics*. (in preparation).

Hirchoren, G.A., and D.S. Arantes. 1998. Predictors for the discrete time fractional Gaussian processes. In *Telecommunications symposium. ITS '98 proceedings*, SBT/IEEE international, 49–53. Sao Paulo: IEEE.

Hirchoren, G.A., and C.E. D'attellis. 1998. Estimation of fractal signals, using wavelets and filter banks. *IEEE Transactions on Signal Processing* 46 (6): 1624–1630.

Kolesnikov, V.N., and A.S. Monin. 1965. Spectra of meteorological field fluctuations. *Izvestiya, Atmospheric and Oceanic Physics* 1: 653–669.

Lean, J.L., and D.H. Rind. 2008. How natural and anthropogenic influences alter global and regional surface temperatures: 1889 to 2006. *Geophysical Research Letters* 35: L18701. doi:10.1029/2008GL034864.

Lilley, M., S. Lovejoy, D. Schertzer, K.B. Strawbridge, and A. Radkevitch. 2008. Scaling turbulent atmospheric stratification. Part II: Empirical study of the the stratification of the intermittency. *Quarterly Journal of the Royal Meteorological Society* 134: 301–315. doi:10.1002/qj.1202.

Lovejoy, S. 2014a. Scaling fluctuation analysis and statistical hypothesis testing of anthropogenic warming. *Climate Dynamics* 42: 2339–2351. doi:10.1007/s00382-014-2128-2.

———. 2014b. Return periods of global climate fluctuations and the pause. *Geophysical Research Letters* 41: 4704–4710. doi:10.1002/2014GL060478.

———. 2015a. A voyage through scales, a missing quadrillion and why the climate is not what you expect. *Climate Dynamics* 44: 3187–3210. doi:10.1007/s00382-014-2324-0.

———. 2015b. Using scaling for macroweather forecasting including the pause. *Geophysical Research Letters* 42: 7148–7155. doi:10.1002/2015GL065665.

———. 2017. How accurately do we know the temperature of the surface of the earth? *Climate Dynamics*. (in press).

Lovejoy, S., and M.I.P. de Lima. 2015. The joint space-time statistics of macroweather precipitation, space-time statistical factorization and macroweather models. *Chaos* 25: 075410. doi:10.1063/1.4927223.

Lovejoy, S., and D. Schertzer. 1986. Scale invariance in climatological temperatures and the local spectral plateau. *Annales Geophysicae* 4B: 401–410.

———. 2007. Scale, scaling and multifractals in geophysics: Twenty years on. In *Nonlinear dynamics in geophysics*, ed. J.E.A.A. Tsonis. New York, NY: Elsevier.

———. 2010. Towards a new synthesis for atmospheric dynamics: Space-time cascades. *Atmospheric Research* 96: 1–52. doi:10.1016/j.atmosres.2010.01.004.

———. 2012. Haar wavelets, fluctuations and structure functions: Convenient choices for geophysics. *Nonlinear Processes in Geophysics* 19: 1–14. doi:10.5194/npg-19-1-2012.

———. 2013. *The weather and climate: Emergent laws and multifractal cascades.*, 496 pp. Cambridge: Cambridge University Press.

Lovejoy, S., A.F. Tuck, S.J. Hovde, and D. Schertzer. 2007. Is isotropic turbulence relevant in the atmosphere? *Geophysical Research Letters* 34: L14802. doi:10.1029/2007GL029359.

Lovejoy, S., D. Schertzer, M. Lilley, K.B. Strawbridge, and A. Radkevitch. 2008. Scaling turbulent atmospheric stratification. Part I: Turbulence and waves. *Quarterly Journal of the Royal Meteorological Society* 134: 277–300. doi:10.1002/qj.201.

Lovejoy, S., D. Schertzer, and D. Varon. 2013. Do GCM's predict the climate.... or macroweather? *Earth System Dynamics* 4: 1–16. doi:10.5194/esd-4-1-2013.

Lovejoy, S., J.P. Muller, and J.P. Boisvert. 2014. On Mars too, expect macroweather. *Geophysical Research Letters* 41: 7694–7700. doi:10.1002/2014GL061861.

Lovejoy, S., L. del Rio Amador, and R. Hébert. 2015. The ScaLIng Macroweather Model (SLIMM): Using scaling to forecast global-scale macroweather from months to decades. *Earth System Dynamics* 6: 1–22. http://www.earth-syst-dynam.net/6/1/2015/. doi:10.5194/esd-6-1-2015.

Mandelbrot, B.B., and J.W. Van Ness. 1968. Fractional Brownian motions, fractional noises and applications. *SIAM Review* 10: 422–450.

Merryfield, W.J., B. Denis, J.-S. Fontecilla, W.-S. Lee, S. Kharin, J. Hodgson, and B. Archambault. 2011. The Canadian Seasonal to Interannual Prediction System (CanSIPS): An overview of its design and operational implementation*Rep.*, 51pp. *Environment Canada.*

Newman, M. 2013. An empirical benchmark for decadal forecasts of global surface temperature anomalies. *Journal of Climate* 26: 5260–5269. doi:10.1175/JCLI-D-12-00590.1.

Panofsky, H.A., and I. Van der Hoven. 1955. Spectra and cross-spectra of velocity components in the mesometeorlogical range. *Quarterly Journal of the Royal Meteorological Society* 81: 603–606.

Papoulis, A. 1965. *Probability, random variables and stochastic processes*. New York, NY: Mc Graw Hill.

Pauluis, O. 2011. Water vapor and mechanical work: a comparison of carnot and steam cycles. *Journal of the Atmospheric Sciences* 68: 91–102. doi:10.1175/2010JAS3530.1.

Penland, C. 1996. A stochastic model of IndoPacific sea surface temperature anomalies. *Physica D* 98: 534–558.

Penland, C., and P.D. Sardeshmuhk. 1995. The optimal growth of tropical sea surface temperature anomalies. *Journal of Climate* 8: 1999–2024.

Pinel, J., and S. Lovejoy. 2014. Atmospheric waves as scaling, turbulent phenomena. *Atmospheric Chemistry and Physics* 14: 3195–3210. doi:10.5194/acp-14-3195-2014.

Pinel, J., S. Lovejoy, and D. Schertzer. 2014. The horizontal space-time scaling and cascade structure of the atmosphere and satellite radiances. *Atmospheric Research* 140–141: 95–114. doi:10.1016/j.atmosres.2013.11.022.

Radkevitch, A., S. Lovejoy, K.B. Strawbridge, D. Schertzer, and M. Lilley. 2008. Scaling turbulent atmospheric stratification. Part III: Empirical study of space-time stratification of passive scalars using lidar data. *Quarterly Journal of the Royal Meteorological Society* 134: 317–335. doi:10.1002/qj.1203.

Ragone, F., V. Lucarini, and F. Lunkeit. 2015. A new framework for climate sensitivity and prediction: A modelling perspective. *Climate Dynamics* 46: 1459–1471. doi:10.1007/s00382–015–2657-3.

Richardson, L.F. 1926. Atmospheric diffusion shown on a distance-neighbour graph. *Proceedings of the Royal Society* A110: 709–737.

Rypdal, K. 2015. Global warming projections derived from an observation-based minimal model. *Earth System Dynamics Discussions* 6: 1789–1813. doi:10.5194/esdd-6-1789-2015.

Rypdal, M., and K. Rypdal. 2014. Long-memory effects in linear response models of Earth's temperature and implications for future global warming. *Journal of Climate* 27 (14): 5240–5258. doi:10.1175/JCLI-D-13-00296.1.

Sardeshmukh, P., G.P. Compo, and C. Penland. 2000. Changes in probability assoicated with El Nino. *Journal of Climate* 13: 4268–4286.

Schertzer, D., and S. Lovejoy. 1985. The dimension and intermittency of atmospheric dynamics. In *Turbulent shear flow*, ed. L.J.S. Bradbury et al., 7–33. Berlin: Springer-Verlag.

———. 1995. From scalar cascades to Lie cascades: Joint multifractal analysis of rain and cloud processes. In *Space/time variability and interdependance for various hydrological processes*, ed. R.A. Feddes, 153–173. New York, NY: Cambridge University Press.

———. 2004. Uncertainty and predictability in geophysics: Chaos and multifractal insights. In *State of the planet, frontiers and challenges in geophysics*, ed. R.S.J. Sparks and C.J. Hawkesworth, 317–334. Washington, DC: American Geophysical Union.

Schertzer, D., S. Lovejoy, F. Schmitt, Y. Chigirinskaya, and D. Marsan. 1997. Multifractal cascade dynamics and turbulent intermittency. *Fractals* 5: 427–471.

Schertzer, D., I. Tchiguirinskaia, S. Lovejoy, and A.F. Tuck. 2012. Quasi-geostrophic turbulence and generalized scale invariance, a theoretical reply. *Atmospheric Chemistry and Physics* 12: 327–336. doi:10.5194/acp-12-327-2012.

Schmidt, G.A., D.T. Shindell, and K. Tsigaridis. 2014. Reconciling warming trends. *Nature Geoscience* 7: 158–160.

Schwartz, S.E. 2012. Determination of Earth's transient and equilibrium climate sensitivities from observations over the twentieth century: Strong dependence on assumed forcing. *Surveys in Geophysics* 33: 745–777.

Steinman, B.A., M.E. Mann, and S.K. Miller. 2015. Atlantic and Pacific multidecadal oscillations and Northern Hemisphere temperatures. *Science* 347: 988–991. doi:10.1126/science.1257856.

Suckling, E.B., E. Hawkins, G. Jan van Oldenborgh, and J.M. Eden. 2016. An empirical model for probabilistic decadal prediction: A global analysis. *Climate Dynamics (submitted).*

Tennekes, H. 1975. Eulerian and Lagrangian time microscales in isotropic turbulence. *Journal of Fluid Mechanics* 67: 561–567.

Vallis, G. 2010. Mechanisms of climate variaiblity from years to decades. In *Stochastic physics and climate modelliing*, ed. P.W.T. Palmer, 1–34. Cambridge: Cambridge University Press.

Van der Hoven, I. 1957. Power spectrum of horizontal wind speed in the frequency range from 0.0007 to 900 cycles per hour. *Journal of Meteorology* 14: 160–164.

Zeng, X., and K. Geil. 2017. Global warming projection in the 21st Century based on an observational data driven model. *Geophysical Research Letters*. (in press).

Regime Change Detection in Irregularly Sampled Time Series

Norbert Marwan, Deniz Eroglu, Ibrahim Ozken, Thomas Stemler, Karl-Heinz Wyrwoll, and Jürgen Kurths

Abstract Irregular sampling is a common problem in palaeoclimate studies. We propose a method that provides regularly sampled time series and at the same time a difference filtering of the data. The differences between successive time instances are derived by a transformation costs procedure. A subsequent recurrence analysis is used to investigate regime transitions. This approach is applied on speleothem-based palaeoclimate proxy data from the Indonesian–Australian monsoon region. We can clearly identify Heinrich events in the palaeoclimate as characteristic changes in dynamics.

Keywords Irregular sampling • Recurrence plot • Recurrence quantification analysis • Regime change • Indonesian–Australian monsoon • Heinrich events

---

N. Marwan (✉)
Potsdam Institute for Climate Impact Research (PIK), 14473 Potsdam, Germany
e-mail: marwan@pik-potsdam.de

D. Eroglu
Potsdam Institute for Climate Impact Research (PIK), 14473 Potsdam, Germany

Institute of Physics, Humboldt-Universität zu Berlin, Robert-Koch-Platz 4, 10099 Berlin, Germany
e-mail: deniz.eroglu@physik.hu-berlin.de

I. Ozken
Department of Physics, Ege University, 35100 Izmir, Turkey

T. Stemler
School of Mathematics and Statistics, The University of Western Australia, Crawley, WA 6009, Australia

K.-H. Wyrwoll
School of Earth and Environment, The University of Western Australia, Crawley, WA 6009, Australia

J. Kurths
Potsdam Institute for Climate Impact Research (PIK), 14473 Potsdam, Germany

Institute of Applied Physics of the Russian Academy of Sciences, 46 Ulyanova St., Nizhny Novgorod 603950, Russia

A.A. Tsonis (ed.), *Advances in Nonlinear Geosciences*,
DOI 10.1007/978-3-319-58895-7_18

1 Introduction

In the last decades, palaeoclimate research has experienced an exciting progress with ever-higher resolution and better age control high-resolution records, innovative technologies and types of proxies, as well as new data series analysis approaches, such as speleothem-based proxies, fluid inclusion analysis and laser ablation techniques, and complex network-based data analysis. (Dennis et al., 2001; Kennett et al., 2012; McDermott, 2001; McRobie et al., 2015; Rehfeld et al., 2013). This progress helps greatly to increase our understanding of past climate variation and the mechanisms behind the climate system, but also to assess future climate-related vulnerability of our society. Of particular interest are critical transitions, such as tipping points or regime shifts, because they can bring the climate system into another mode of operation (Lenton et al., 2008; Scheffer et al., 2012). Identifying tipping points from measurements is no simple task. Several approaches have been proposed, such as testing for slowing down and increase of the autocorrelation (Scheffer et al., 2009), reconstructing potentials of the dynamics by using the modality of the data distribution (Livina et al., 2010), using a modified detrended fluctuation analysis (DFA) (Livina and Lenton, 2007), or the concept of stochastic resonance (Braun et al., 2011). While dynamical transitions are rather obvious when they appear in the first two moments (i.e. in mean or variance), they can be hidden when superimposed by signals of different time scales or by noise, issues frequently observed in palaeoclimate time series. For such problems, the application of methods from nonlinear time series analysis is a well-accepted perspective, e.g., by using the fluctuation of similarity (FLUS) (Malik et al., 2012). Another promising tool for the identification of subtle transitions is the framework of recurrence plots (Marwan et al., 2007). Recurrence plots and their quantification consider the evolution of neighbouring states in a phase space. Besides characterizing different classes of dynamics or testing for synchronization and nonlinear interrelationships and couplings of multiple systems, it allows to test for dynamical regime changes with respect to different properties, such as changes in the geometry of the attractor, in the predictability of states, or in the intermittency behaviour (Donner et al., 2011; Eroglu et al., 2014; Marwan et al., 2007). The recurrence plot framework has been successfully applied to investigate past transitions, e.g., in the Asian monsoon system (Marwan et al., 2013) and in the East African climate (Donges et al., 2011), and to uncover a seesaw effect within the East Asian and Indonesian–Australian summer monsoon system (Eroglu et al., 2016).

However, most palaeoclimate proxy records (independent of the actual archive) come with the challenge of irregular sampling. While sampling in the field or in the lab is often done on a regular depth/length axis, varying sedimentation or growth rates result in variable time–depth relationships and in time series with non-equidistant sampling points in the time-domain (Breitenbach et al., 2012). The most common procedure is data preprocessing using linear interpolation. However, interpolation can lead to a positive bias in autocorrelation estimation (and, thus, an overestimation of the persistence time) and a negative bias in cross correlation analysis (Rehfeld et al., 2011). Therefore, several approaches have been suggested for analysing irregularly sampled time series (Ozken et al., 2015; Rehfeld and Kurths, 2014; Rehfeld et al., 2011; Scargle, 1982; Stoica and Sandgren, 2006).

In the following we will focus on a recently proposed technique that is based on a measure that compares spike trains by quantifying the effort it needs to transform one spike train to the other one (Hirata and Aihara, 2009; Victor and Purpura, 1997). This measure corresponds to a modified difference filter (a common practice to remove low-frequency variation and trends), where we determine the differences by a criterion of how close subsequent short segments of an unevenly sampled time series are by determining the cost needed to transform one segment into the following one (Ozken et al., 2015). Such comparison of successive segments has some similarity with the FLUS method (Malik et al., 2012), but instead uses the transformation cost as the similarity measure, and is thus directly applicable on irregularly sampled time series. We illustrate this approach by analysing a speleothem-based palaeoclimate record with respect to regime transitions.

2 Methods

2.1 Transformation Costs Time Series

Cumulative trends or low-frequency variations are common in palaeoclimate proxy records, but are often undesirable and can cause difficulties in the analysis. One frequently used solution is the difference filter, where the values of the proxy record are replaced by the differences of subsequent values, $y(t-\Delta t/2) = x(t) - x(t-\Delta t)$, with Δt the sampling time of a regularly sampled time series. Another, even more challenging problem is the irregular sampling frequently occurring in palaeoclimate proxy records. The *transformation costs time series* (TACTS) approach tries to overcome both problems by transforming irregularly sampled time series to regular ones and simultaneously using the transformation cost as the difference value. This procedure induces less loss of information compared to traditional interpolation procedures.

The core of the TACTS method is to measure the shortest distance (transformation cost) between two data segments by using two different processes: (1) *shifting points* in time which causes changes in the amplitude for marked data and (2) *adding–deleting* operations. The process starts with dividing the data into small and equally sized segments. These segments can have different number of points, because the points are not equally sampled. The transformation costs between all sequence windows are then calculated by

$$p(S_a, S_b) = \overbrace{\sum_{(\alpha,\beta)\in C} \{\lambda_0 |t_a(\alpha) - t_b(\beta)| + \lambda_k |L_a(\alpha) - L_b(\beta)|\}}^{\text{shifting}} + \underbrace{\lambda_S(|I| + |J| - 2|C|)}_{\text{adding/deleting}}. \tag{1}$$

The equation states two distinct operations for two essential processes. If the operation is *shifting*, then the first part of the equation involves, otherwise the *adding–deleting* operation involves as the second part. In the first part, the summation is over the pairs $(\alpha, \beta) \in C$, where C is the set of points that will be shifted in time and changed in amplitude. α and β are the αth event in the first segment (S_a) and the βth event in the second segment (S_b). The amplitude of points which are αth and βth elements of S_a and S_b are denoted by $L_a(\alpha)$ and $L_b(\beta)$, respectively. The data-adapted constants λ_0 and λ_k are given by

$$\lambda_0 = \frac{M}{\text{total time}} \tag{2a}$$

$$\lambda_k = \frac{M-1}{\sum_i^{M-1} |x_i - x_{i+1}|}. \tag{2b}$$

where M is the total number of events, and x_i is the amplitude of ith element in the time series.

In the second part of Eq. (1), I and J are sets of indices of the events in S_a and S_b, respectively. The parameter λ_S is the cost of deleting or adding processes and is used as an optimization parameter. The selection of optimum λ_S is the following: first we calculate total cost time series for the entire range of $\lambda_S \in [0, 4]$ with step size $\Delta\lambda_S = 0.01$. Then we examine frequency distributions for each cost time series. Since each cost value is independent of the others, we expect to have a normal distributed histogram and choose the optimal λ_S according to the best fit on normal distribution.

Equation (1) is a metric distance function, satisfying the following three conditions:

- $p(S_a, S_b) \geq 0$ (positive)
- $p(S_a, S_b) = p(S_b, S_a)$ (symmetric)
- $p(S_a, S_c) \leq p(S_a, S_b) + p(S_b, S_c)$ (triangle inequality).

Now we illustrate the method for two consecutive segments. Irregularly sampled data is equally spaced into small windows which are given as state a ($S_a = \{a_\alpha\}_{\alpha=1}^4$) and state b ($S_b = \{b_\beta\}_{\beta=1}^3$). The costs computed between the states and all details are given in Fig. 1 step by step.

Note that the decision of which operation process to minimize costs is important. The transformation by shifting costs $\lambda_0|t_a(\alpha) - t_b(\beta)| + \lambda_1|L_a(1) - L_b(1)|$ and deleting and adding a point costs $2\lambda_S$. Here we chose the least cost operation to either shift or delete/add. Therefore, in the algorithm, we consider all these possibilities and chose the operation carefully.

The final appearance of the cost time series is as follows: assume that we have an irregularly sampled time series $\{u_i\}_{i=1}^N$, where N is the number of points. The data is divided into a set of W-sized n segments and each segment has a minimum of a certain number of points, therefore,

$$TACTS = \{p(W_i, W_i + 1)\}_{i=1}^{n-1}$$

for all sequence windows. This leads to an equally sampled and detrended time series. The resulting cost values series can be considered as the difference filtered time series with a regularly sampled time axis and can be further analysed with standard or advanced time series analysis tools, e.g., in order to detect regime shifts (Fig. 1).

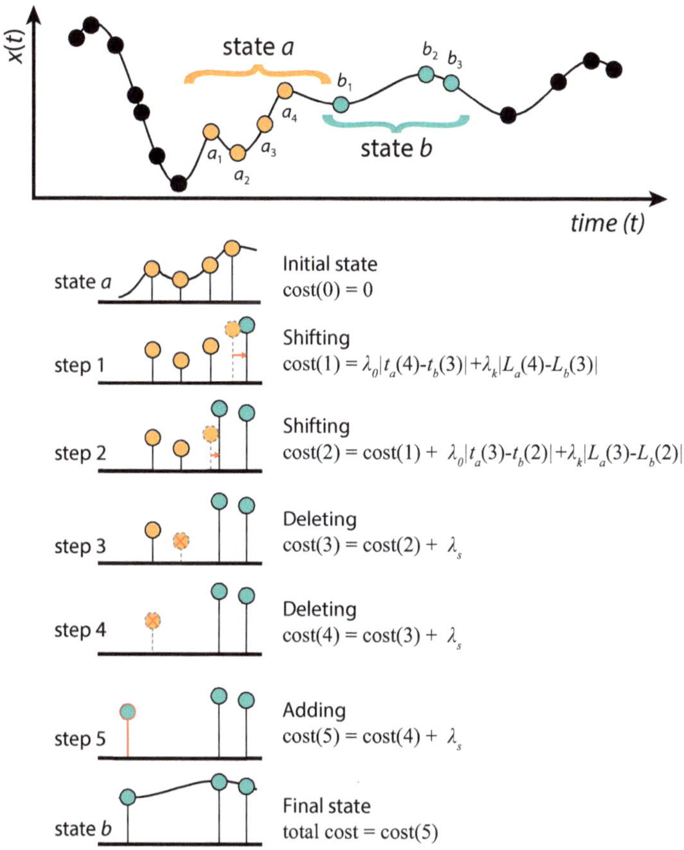

Fig. 1 Illustration of the transformation cost time series method, which finds the minimum transformation cost between two data segments such as state *a* and state *b* in the *top panel*. In five steps state *a* is transformed into state *b*. At steps 1 and 2, we apply *shifting* a point in time and, as a consequence of shifting, changing the amplitude of the point. These operations cost regarding to first part of Eq. (1). Steps 3 and 4 are *deleting* and step 5 is *adding* a point; each of these operations costs a constant λ_s. The costs are written next to the related processes according to Eq. (1)

2.2 Recurrence Analysis

Recurrence is a ubiquitous property of many dynamical systems. Slight changes in observed recurrence behaviour allow to infer changes in the dynamics (Marwan, 2011; Marwan et al., 2007). In order to investigate recurrence properties, recurrence plots and recurrence quantification analysis have been developed (Marwan, 2008; Marwan et al., 2007). A recurrence plot is the graphical representation of those times j at which a system recurs to a previous state $\mathbf{x}_i$:

$$R_{i,j} = \Theta(\varepsilon - \|\mathbf{x}_i - \mathbf{x}_j\|), \qquad i,j = 1,\ldots,N \tag{3}$$

with Θ the Heaviside function, ε a recurrence threshold, $\|\mathbf{x}_i - \mathbf{x}_j\|$ the Euclidean distance between two states $\mathbf{x}_i$ and $\mathbf{x}_j$ in the phase space, and N the number of observations (or time series length). Such a recurrence plot consists of typical large-scale and small-scale features that can be used to interpret the dynamics visually. Important features are diagonal lines: similar evolving epochs of the phase space trajectory cause diagonal structures parallel to the main diagonal in the recurrence plot. The length l of such diagonal line structures of at least length $l_{\min}$ depends on the dynamics of the system (periodic, chaotic, stochastic) (Fig. 2) and can be directly related with dynamically invariant properties, like K_2 entropy (Marwan et al., 2007). Therefore, recurrence quantification analysis (RQA) uses the features

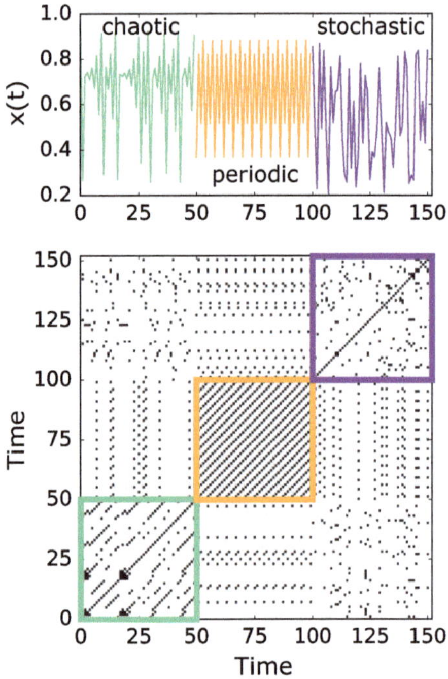

Fig. 2 Example of a recurrence plot for changing dynamics from chaotic via periodic to stochastic dynamics, each lasting 50 time steps. In the periodic region, continuous long diagonal lines are observed, in the chaotic region, shorter diagonals and single points appear, and in the stochastic part, we find almost only single points

within the recurrence plots for defining measures of complexity. For example, the distribution $P(l)$ of line lengths l is used by several measures of complexity in order to characterize the system's dynamics in terms of predictability/determinism or laminarity. The measure *determinism* DET is the fraction of recurrence points (i.e. $R_{i,j} = 1$) that form diagonal lines and can be computed by

$$\mathrm{DET} = \frac{\sum_{l_{\min}}^{N} l \cdot P(l)}{\sum_{i,j=1}^{N} R_{i,j}}. \tag{4}$$

In order to study the time-dependent behaviour of a system or time series, RQA measures can be computed within a moving window, applied on the time series. The window has size w and is moved with a step size s over the data in such a way that succeeding windows overlap with $w - s$. This technique can detect chaos-period and also more subtle chaos–chaos transitions (Marwan et al., 2007), or different kinds of transitions between strange non-chaotic behaviour and period or chaos (Ngamga et al., 2007). Moreover, the reliability of several RQA measures was investigated by their scaling properties with respect to critical points in the dynamics (Afsar et al., 2015).

3 Palaeoclimate Regime Transition

To illustrate the power of the techniques we advocate here, we choose as illustrating example a speleothem δ^{18}O record from the Secret Cave at Gunung Mulu in Borneo/Indonesia (Carolin et al., 2013). This particular record has been interpreted as a time series of the dynamics of the East Asian–Indonesian–northwest Australia monsoon. This monsoon regime provides a circulation regime that strongly links both hemispheres and serves as a major heat source, playing a significant role at planetary scale (Chang et al., 2006; McBride, 1987). Central to its geography is the Maritime Continent which provides a core region of monsoon activity (Chang et al., 2004; Ramage, 1968). A transect in regional precipitation patterns from the northern part of the Maritime Continent to the northern margin of Australia coincides with a change from the dominance of the boreal summer monsoon to the austral summer monsoon (Chang et al., 2004, 2006; Robertson et al., 2011). The transect captures key palaeoproxy monsoon records and has the potential to provide details of the function of the monsoon regime over Quaternary time scales (Ayliffe et al., 2013; Carolin et al., 2013; Denniston et al., 2013; Partin et al., 2007). Imbedded in some of these records are short-lived millennial and centennial scale events, and, more general, relatively short-lived phases of climate instability.

While the full proxy record is around 100,000 years, we only analyse the last 62,000 years of the δ^{18}O record (Fig. 3a). Before the 62,000 years many gaps appear and the data become too sparse to give any useful information about. The record used for the analysis contains about 1200 data points. Time intervals between measurements are irregular and follow a Gamma distribution with a skewness of

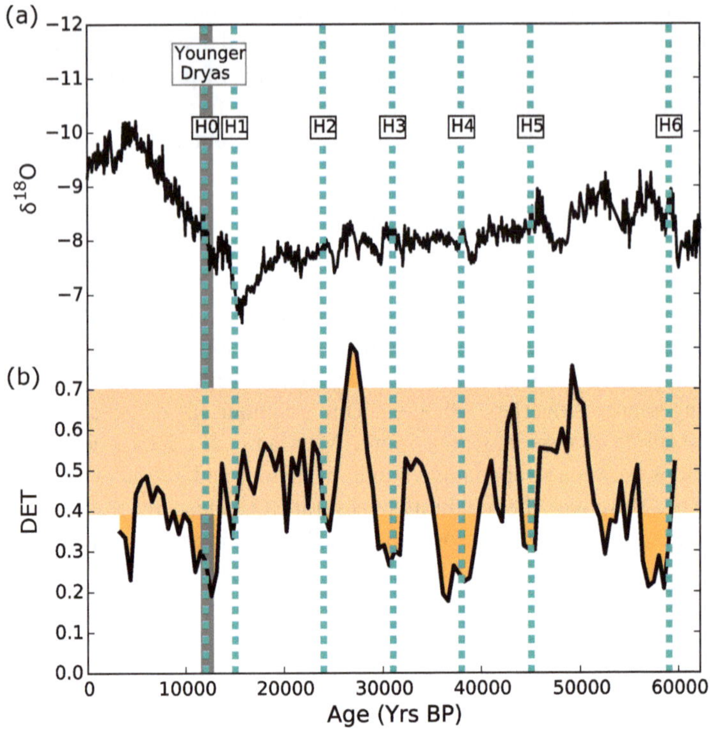

Fig. 3 (**a**) δ^{18}O record of Secrete Cave, Borneo. (**b**) RQA-determinism *DET*, Eq. (4), time series resulting from the transformation cost time series. The *light orange band* of the *DET* indicates the 90% confidence interval. The *vertical lines* H1–H6 give the six Heinrich events as well as H0, the Younger-Dryas

4.9. In our analysis we use a window length of ≈210 years to calculate the TACTS. While the parameters $\lambda_{0,k}$ are determined by Eq. (2), we optimize $\lambda_S = 1.07$.

The next step is to analyse the regularly sampled TACTS with RQA using a sliding window method. We consider 30 data points (or 6200 years) of the TACTS as our window size. Given the average number of points in the proxy record, 30 data points of the TACTS correspond to approximately 100–140 points in the original proxy. Using an overlap of 90% of consecutive windows, we determine the *DET* [Eq. (4)] for each window with length of 6200 years (Fig. 3b). The recurrence threshold is selected to be $\epsilon = 20\%$ of the standard deviation of the data in the particular window. The advantage of this ϵ selection scheme is that it allows us to analyse proxy records with inherent non-stationarity. In addition, we determine the statistical significance of *DET* using the bootstrapping method as outlined in Marwan et al. (2013) (*light red band* in Fig. 3b).

The determinism *DET* indicates several distinct regime changes in the time series from less to more predictable (and vice versa) dynamics (Fig. 3b). Most minima of *DET*, signified as periods of decreased predictability, coincide with the so-called

Heinrich events (H1 to H6). Heinrich events are identified in the North Atlantic sediments as layers of ice-rafted debris, associated with the coldest phase just before the Dansgaard–Oeschger events, and result from episodic discharge of icebergs in the Hudson Bay region (Clement and Peterson, 2008; McNeall et al., 2011).

Heinrich events are well represented in the Chinese speleothem and loess record as periods of weakened summer monsoon and intensified winter monsoon (An, 2014). In their interactions with the Siberian Mongolian High of the East Asian Winter Monsoon they can be expected to trigger cold surges which leave their imprint in the proxy palaeoclimate record (Wyrwoll et al., 2016). During the East-Asian Winter Monsoon (EAWM), the Siberian High with its central pressure reaching in excess of 1035 hPa dominates much of the Eurasian continent. Strong northwesterly flows occur at its eastern margins, where one branch of the flow separates and first is directed eastward into the subtropical western Pacific and then tends southward in the direction of the South China Sea. These cold air 'excursions', also described as 'cold surges', are channeled by the trough southwards and are a characteristic feature of the EAWM (Lau and Chang, 1987). Their path is in part related to relief controls of the Tibetan Plateau. Cold surges transport absolute vorticity and water vapour up-stream of the South China Sea to the Equator (Koseki et al., 2013) and lead to the flare-up of convective activity over the Maritime Continent (Chan and Li, 2004). In the Borneo region, cold surges enhance surface cyclonic circulation triggering the Borneo Vortex, which leads to deep convection giving rise to heavy rainfall events (Koseki et al., 2013; Ooi et al., 2011).

It is noteworthy that in raw δ^{18}O record from the Secret Cave the Heinrich events are almost indistinguishable from other variations in the time series. In the original work by Carolin et al., H1 to H6 were detected by visual comparison of the record to others (e.g. NGRIP), but the Younger Dryas (coinciding with the H0 event) was not detected (Carolin et al., 2013). However, our method clearly extracts these events, including the previously not detected Younger Dryas, and highlights the hidden impact of such distal forcing. Moreover, it allows an objective, quantitative analysis, while Carolin et al. rely on the subjective method of matching extreme proxy values with specific dates. At present, the Borneo Vortex leaves a strong climate signal on the regional precipitation patterns (Ooi et al., 2011). We propose that the prominence of the 'instability climate phases', coincident with the timing of Heinrich events in the Borneo record, is an expression of regional controls that are linked to the operation of the Borneo Vortex. The claim draws attention to the need to give more consideration to specific regional controls in explaining the palaeoclimate proxy record rather than simply appeal to global or hemispheric controls.

4 Conclusion

We have used the Secret Cave δ^{18}O record from Borneo to illustrate the usefulness of the novel TACTS method for analysing palaeoclimate records. TACTS can transform irregularly sampled time series into a regularly sampled cost time series.

This is an important step, since most modern time series analysis methods—like the RQA used here—require a regular sampled time series as an input. Furthermore, the TACTS method is less biased than interpolation methods frequently used to transform irregularly sampled into regularly sampled data sets. This transformation only requires three parameters. The two parameters $\lambda_{0,k}$ are given by the average amplitude and frequency of the record [see Eq. (2)], while λ_S needs to be optimized. Being a difference filter, the TACTS method lends itself naturally for palaeoclimate investigations, where proxy records often have some non-stationarity and usually need to be detrended. As we have shown the detrending is build into the TACTS method, therefore we do not need this additional step in our time series analysis.

Applying the TACTS and RQA approach on palaeoclimate data from the Secret Cave speleothem, we were able to identify regime changes in the monsoon activity during the last 62,000 years. We report on several distinct regime changes coinciding with the Heinrich events H1 to H6 and therefore add quantitative evidence of these impacts to previous, more qualitative studies (Carolin et al., 2013). Moreover, our analysis clearly unveils that also the Younger Dryas had an impact on the monsoon activity over the Maritime Continent.

Given that irregular sampling of proxy records is quite common in Earth science, the TACTS method has large potential in quantitative Earth science without prior modification or preprocessing the data.

Acknowledgements This work was supported by grants from the Leibniz Association, grant SAW-2013- IZW-2 (Gradual environmental change versus single catastrophe—Identifying drivers of mammalian evolution) and the European Union's Horizon 2020 Research and Innovation programme under the Marie Skłodowska-Curie grant agreement No 691037 (RISE project QUantitative palaeoEnvironments from SpeleoThems QUEST). We thank Sebastian Breitenbach for fruitful discussions and support.

References

Afsar, O., D. Eroglu, N. Marwan, J. Kurths. 2015. *Europhysics Letters* 112(1): 10005.

An, Z., ed. 2014. *Late Cenozoic climate change in Asia*. Developments in Paleoenvironmental Research, vol. 16. Dordrecht: Springer. doi:10.1007/978-94-007-7817-7.

Ayliffe, L.K., M.K. Gagan, J.X. Zhao, R.N. Drysdale, J.C. Hellstrom, W.S. Hantoro, M.L. Griffiths, H. Scott-Gagan, E. St Pierre, J.A. Cowley, and B.W. Suwargadi. *Nature Communications* 4 (May): 2908. doi:10.1038/ncomms3908.

Braun, H., P. Ditlevsen, J. Kurths, and M. Mudelsee. 2011. *Paleoceanography* 26(3): PA3214. doi:10.1029/2011PA002140.

Breitenbach, S.F.M., K. Rehfeld, B. Goswami, J.U.L. Baldini, H.E. Ridley, D.J. Kennett, K.M. Prufer, V.V. Aquino, Y. Asmerom, V.J. Polyak, H. Cheng, J. Kurths, and N. Marwan. 2012. *Climate of the Past* 8(5): 1765. doi:10.5194/cp-8-1765-2012.

Carolin, S.A., K.M. Cobb, J.F. Adkins, B. Clark, J.L. Conroy, S. Lejau, J. Malang, and A.A. Tuen. 2013. *Science (New York, N.Y.)* 340(2013): 1564. doi:10.1126/science.1233797.

Chan, J.C.L., and C.Y. Li. 2004. *East Asian monsoon*. World Scientific Series on Asia-Pacific Weather and Climate, ed. C.P. Chang, vol. 2, 54–106. Singapore: World Scientific.

Chang, C.P., P. Harr, J. McBride, and H.H. Hsu. 2004. In *East Asian monsoon*. World Scientific Series on Meteorology of East Asia, ed. C.P. Chang, vol. 2, 107–150. Singapore: World Scientific.

Chang, C.P., Z. Wang, and H. Hendon. 2006. In *The Asian monsoon*, Springer Praxis Books, 89–127. Berlin/Heidelberg: Springer.

Clement, A.C., and L.C. Peterson. 2008. *Reviews of Geophysics* 46(2006): 1. doi:10.1029/2006RG000204.

Dennis, P., P. Rowe, and T. Atkinson. 2001. *Geochimica et Cosmochimica Acta* 65(6): 871. doi:10.1016/S0016-7037(00)00576-7.

Denniston, R.F., K.H. Wyrwoll, V.J. Polyak, J.R. Brown, Y. Asmerom, A.D. Wanamaker Jr., Z. LaPointe, R. Ellerbroek, M. Barthelmes, D. Cleary, J. Cugley, D. Woods, and W.F. Humphreys. 2013. *Quaternary Science Reviews* 78: 155. doi: http://dx.doi.org/10.1016/j.quascirev.2013.08.004.

Donges, J.F., R.V. Donner, M.H. Trauth, N. Marwan, H.J. Schellnhuber, and J. Kurths. 2011. *Proceedings of the National Academy of Sciences* 108(51): 20422. doi:10.1073/pnas.1117052108.

Donner, R.V., J. Heitzig, J.F. Donges, Y. Zou, N. Marwan, and J. Kurths. 2011. *European Physical Journal B* 84: 653. doi:10.1140/epjb/e2011-10899-1.

Eroglu, D., N. Marwan, S. Prasad, and J. Kurths. 2014. *Nonlinear Processes in Geophysics* 21: 1085. doi:10.5194/npg-21-1085-2014.

Eroglu, D., F.H. McRobie, I. Ozken, T. Stemler, K.H. Wyrwoll, S.F.M. Breitenbach, N. Marwan, and J. Kurths. 2016. *Nature Communications* 7: 12929. doi:10.1038/ncomms12929.

Hirata, Y., and Aihara, K. 2009. *Journal of Neuroscience Methods* 183(2): 277. doi:10.1016/j.jneumeth.2009.06.030.

Kennett, D.J., S.F.M. Breitenbach, V.V. Aquino, Y. Asmerom, J. Awe, J.U.L. Baldini, P. Bartlein, B.J. Culleton, C. Ebert, C. Jazwa, M.J. Macri, N. Marwan, V. Polyak, K.M. Prufer, H.E. Ridley, H. Sodemann, B. Winterhalder, and G.H. Haug. 2012. *Science* 338(6108): 788. doi:10.1126/science.1226299.

Koseki, S., T.Y. Koh, and C.K. Teo. 2013. *Quarterly Journal of the Royal Meteorological Society* 139(675): 1566. doi:10.1002/qj.2052.

Lau, K.M., and C.P. Chang. 1987. *Monsoon meteorology*, 161–202. Oxford: Oxford University Press.

Lenton, T.M., H. Held, E. Kriegler, J.W. Hall, W. Lucht, S. Rahmstorf, and H.J. Schellnhuber. 2008. *Proceedings of the National Academy of Sciences* 105(6): 1786. doi:10.1073/pnas.0705414105.

Livina, V.N., and T.M. Lenton. 2007. *Geophysical Research Letters* 34(3): L03712. doi:10.1029/2006GL028672.

Livina, V.N., F. Kwasniok, and T.M. Lenton. 2010. *Climate of the Past* 6(1): 7. doi:10.5194/cp-6-77-2010.

Malik, N., Y. Zou, N. Marwan, and J. Kurths. 2012. *Europhysics Letters* 97(4): 40009. doi:10.1209/0295-5075/97/40009.

Marwan, N. 2008. *European Physical Journal: Special Topics* 164(1): 3. doi:10.1140/epjst/e2008-00829-1.

Marwan, N. 2011. *International Journal of Bifurcation and Chaos* 21(4): 1003. doi:10.1142/S0218127411029008.

Marwan, N., M.C. Romano, M. Thiel, and J. Kurths. 2007. *Physics Reports* 438(5–6): 237. doi:10.1016/j.physrep.2006.11.001.

Marwan, N., S. Schinkel, and J. Kurths. 2013. *Europhysics Letters* 101: 20007. doi:10.1209/0295-5075/101/20007.

McDermott, F. 2001. *Science* 294(5545): 1328. doi:10.1126/science.1063678.

McBride, J.L. 1987. In *Monsoon meteorology*, ed. Chang, C.P., and T.N. Krishnamurti, 203–23. Oxford, UK: Oxford University Press.

McNeall, D., P.R. Halloran, P. Good, and R.A. Betts. 2011. *Wiley Interdisciplinary Reviews: Climate Change*. 2(5): 663. doi:10.1002/wcc.130.

McRobie, F.H., T. Stemler, and K.H. Wyrwoll. 2015. *Quaternary Science Reviews* 121: 120. doi:10.1016/j.quascirev.2015.05.011. http://dx.doi.org/10.1016/j.quascirev.2015.05.011.

Ngamga, E.J., A. Nandi, R. Ramaswamy, M.C. Romano, M. Thiel, and J. Kurths. 2007. *Physical Review E* 75(3), 036222. doi:10.1103/PhysRevE.75.036222.

Ooi, S.H., A.A. Samah, and P. Braesicke. 2011. *Journal of Geophysical Research Atmospheres* 116(21). doi:10.1029/2011JD015991.

Ozken, I., Eroglu, D., Stemler, T., Marwan, N., Bagci, G.B., and Kurths, J. 2015. *Physical Review E* 91(6): 062911. doi:10.1103/PhysRevE.91.062911.

Partin, J.W., K.M. Cobb, J.F. Adkins, B. Clark, and D.P. Fernandez. 2007. *Nature* 449(7161): 452. doi:10.1038/nature08125.

Ramage, C.S. 1968. *Monthly Weather Review* 96(6): 365.

Rehfeld, K., and Kurths, J. 2014. *Climate of the Past* 10(1): 107. doi:10.5194/cp-10-107-2014.

Rehfeld, K., N. Marwan, J. Heitzig, and J. Kurths. 2011. *Nonlinear Processes in Geophysics* 18(3): 389. doi:10.5194/npg-18-389-2011.

Rehfeld, K., N. Marwan, S.F.M. Breitenbach, and J. Kurths. 2013. *Climate Dynamics* 41(1): 3. doi:10.1007/s00382-012-1448-3.

Robertson, A., V. Moron, J.H. Qiam, C.P. Chang, F. Tangan, E. Aldrian, T. Koh, and L. Jueng. 2011. *The global monsoon system: research and forecast*, ed. Chang, C.P., Y. Ding, N.C. Lau, R. Johnson, B. Eang, and T. Yasunari, 85–109. Singapore: World Scientific.

Scargle, J.D. 1982. *The Astrophysical Journal* 263: 835. doi:10.1086/160554.

Scheffer, M., J. Bascompte, W.A. Brock, V. Brovkin, S.R. Carpenter, V. Dakos, H. Held, E.H. van Nes, M. Rietkerk, and G. Sugihara. 2009. *Nature* 461(7260): 53. doi:10.1038/nature08227.

Scheffer, M., S.R. Carpenter, T.M. Lenton, J. Bascompte, W. Brock, V. Dakos, J. van de Koppel, I.A. van de Leemput, S.A. Levin, E.H. van Nes, M. Pascual, and J. Vandermeer. 2012. *Science (New York, N.Y.)* 338(6105): 344. doi:10.1126/science.1225244.

Stoica, P., and Sandgren, N. 2006. *Digital Signal Processing* 16(6): 712. doi:10.1016/j.dsp.2006.08.012.

Victor, J.D., and Purpura, K.P. 1997. *Network: Computation in Neural Systems* 8(2): 127. doi:10.1088/0954-898X_8_2_003.

Wyrwoll, K.H., J. Wei, Z. Lin, Y. Shao, and F. He. 2016. *Quaternary Science Reviews* 149: 102. doi:10.1016/j.quascirev.2016.04.015.

Topological Data Analysis: Developments and Applications

Francis C. Motta

Abstract Topological Data Analysis (TDA) and its mainstay computational device, persistent homology (PH), has established a strong track record of providing researchers across the data-driven sciences with new insights and methodologies by characterizing low-dimensional geometric structures in high-dimensional data. When combined with machine learning (ML) methods, PH is valued as a discriminating-feature extraction tool. This work highlights many of the recent successes at the intersection of TDA and ML, introduces some of the foundational mathematics underpinning TDA, and summarizes the efforts to strengthen the bridge between TDA and ML. Thus, this document is a launching point for experimentalists and theoreticians to consider what can be learned from the shape of their data.

Keywords Topological data analysis • Persistent homology • Machine learning

1 Introduction

When first encountering ideas in topology it can be instructive to view a topological space as a generalization of a metric space. The quantitative dissimilarity between points defined by a metric function, which gives geometric form to the space, is replaced by the set-theoretic condition of mutual membership in the so-called open subsets—the collection of which is only required to contain the whole space, the empty set, and be closed under arbitrary unions and finite intersections. This relaxation allows the geometry of a space to be quite radically deformed without altering the topology. Topologies ignore the gamut of transformations that rotate, stretch, shrink, grow, and twist so long as they don't tear. Among the qualities which cannot be altered, if the topology is to be preserved, is the number of holes in a space.

There was a time when mathematicians jokingly referred to topology as a branch of mathematics so pure it would never be applied—after all, what use is

F.C. Motta (✉)
Mathematics Department, Duke University, Durham, NC, USA
e-mail: motta@math.duke.edu

A.A. Tsonis (ed.), *Advances in Nonlinear Geosciences*,
DOI 10.1007/978-3-319-58895-7_19

a subject that can't distinguish between a coffee cup and a doughnut? Recently, a computational paradigm known as Topological Data Analysis (TDA) has emerged as an applied branch of topology.

TDA represents a set of computational methods aimed at extracting, quantifying, and characterizing latent geometric structure in data. These tools are being widely used by researchers across the scientific disciplines, and are being adapted to a variety of applications. For example, topological methods for quantifying periodicity in time-series data (Perea and Harer, 2015) are valued by systems biologists who collect and study gene expression time series in the hopes of identifying genes participating in periodic processes (Perea et al., 2015). Also, TDA is proving relevant to the study of neuronal activity data (Chung et al., 2009b; Dabaghian et al., 2012; Singh et al., 2008), and characterizing the intrinsic geometry of neuron firing correlation matrices has suggested geometric organization of place neurons in the mouse hippocampus (Giusti et al., 2015). Furthermore, TDA has proven useful for characterizing defects in patterned surfaces and crystal structure (Hiraoka et al., 2016; Pearson et al., 2015), which is of interest to condensed matter physicists and important to manufacturing processes at the nanoscale.

The focus of many researchers is the use of TDA tools to extract robust and discriminating features, useful for data classification and other machine-learning (ML) tasks. Included in the growing number of medical classification tasks, TDA methods have helped reveal a new subgroup of breast cancer (Nicolau et al., 2011), and have revealed shape-based data features associated with brain disorders such as autism (Chung et al., 2009a), epilepsy (Wang et al., 2014), and Alzheimer's disease (Pachauri et al., 2011). Recently a growing interest in developing and applying TDA methods to time-varying data has emerged: Topological features of driver behavior have been shown to enhance multi-target tracking technologies used by government and law enforcement agencies (Rouse et al., 2015), while TDA methods were combined with classic approaches in nonlinear dynamical systems analysis to attack the ML problem of human action recognition (Venkataraman et al., 2016).

The predominant tool used by researchers to extract informative topological features from data is *persistent homology* (PH) (Edelsbrunner and Harer, 2008; Zomorodian and Carlsson, 2005). PH may be regarded as a far-reaching generalization of (single-linkage) hierarchical clustering (Florek et al., 1951), taking a multiscale approach to characterizing topological structure in data and encoding this information in a compact representation. What follows is introduction to *persistent homology* (PH) and the significant results needed to justify its use as a data analysis device. By way of examples, Sect. 2 more precisely defines the types of structures in data that PH can reveal. Section 3 motivates PH's multiscale philosophy, while Sect. 4 formally defines PH and discusses some of the mathematical foundation on which the tool is based. Section 5 describes the output of PH calculations, the *persistence diagram* (PD), and discusses its virtues as a representative of shape-based data features. Finally, Sect. 6 highlights numerous efforts to map persistence diagrams into spaces with additional structure to make the homological features extracted by PH more powerful and flexible when combined with statistical approaches and ML.

The goal of this work is to introduce researchers, especially those working in geosciences, to TDA to encourage greater access to this burgeoning data analysis field. What follows assumes only a familiarity with basic set and vector space operations and softens many formalities, electing for a more intuitive approach. The hope is that a careful reading will provide a foundation for anyone interested in applying PH or in pursuing the references herein, which give more complete treatments of the mathematics.

2 Homological Structures in Data

Collectors and analysts of data, from every branch of science, often ask the same fundamental question: How dissimilar is this data set from that data set? This question takes many forms: In what ways do disease survivors differ from those who succumb? How well does simulated data from a model match real data from an experiment? Can we compare the present state of a system to its past states to predict its future state? PH is a lens through which to view a data set, a stable transformation of data that empowers us to answer the question of data proximity by measuring (dis)similarity in terms of topological structures.

A first type of topological structure, one which is fundamental to classification problems, is the number of clusters, groups, or components into which a data set is divided. However, number of components is a very coarse description of shape, and may not be the most important or defining characteristic of a data set; from a topologist's perspective it is just the first—in a sense made precise in Sect. 4—of an infinite sequence of *topological invariants*[1] which characterize the shape of a space. These invariants are known as the *homology groups* and the properties they capture are, loosely speaking, the number of n-dimensional holes in a topological space. Slightly more precisely, an n-dimensional hole in a space is formed by the absence of an n-dimensional object whose $(n-1)$-dimensional boundary remains. Homology is the algebraic language that defines these notions of boundary and hole, and enables the computation of such objects.

Figures 1, 2, and 3 offer three data sets which exemplify homological structure in two forms of data: scalar fields and data point clouds. Both data types are ubiquitous in experimental and computational sciences and are appropriate and well-represented in the many studies which exploit PH. Consider first Fig. 1 (left) showing a time series of daily maximum temperature in degrees Celsius, computed as an average of the maximum daily temperatures gathered by land surface stations from around the globe and reported in the Daily Global Historical Climatology Network (Menne et al., 2017). The right panel gives the corresponding point cloud generated by a coordinate-delay reconstruction of the time series (Kantz and

[1]Properties preserved under *homeomorphism*: a continuous bijective function with continuous inverse.

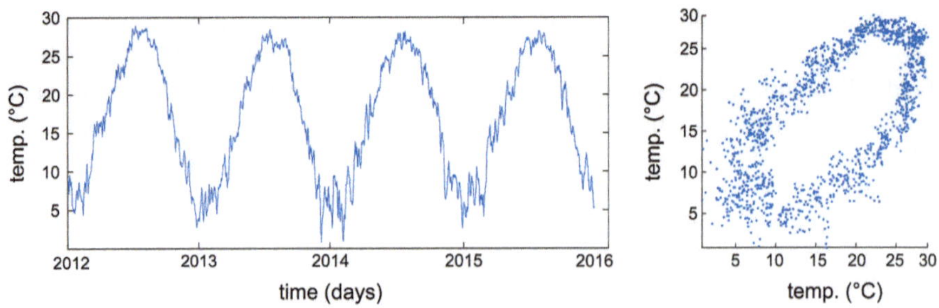

Fig. 1 Average daily maximum temperature in degrees Celsius between January 1, 2012 and December 31, 2015 shown as a time series (*left*) and a two-dimensional point cloud (*right*). The point cloud was derived by a delay-coordinate embedding of the time series, $x(t)$, $t = 1, \ldots, 1461$, where each data point is of the form $(x(t), x(t + 50))$

Schreiber, 1997, Chap. 3.2) into $\mathbb{R}^2$, computed by pairing time series values lagged approximately 7 weeks apart. The point cloud forms an apparent loop and noisily encloses a one-dimensional hole, showcasing the fact that periodic phenomena are inherently circular.

Figure 2 represents a distinct, but equally common form of data: a scalar field defined over an equally spaced grid. Figure 2 (top) shows brightness temperature fields derived from hyperspectral data collected by the GOES-13 satellite imaging radiometer (Munro et al., 2005), as it captured Hurricane Danny on the twenty-first of August, 2015 at two times near the storm's peak intensity. At both times, the coldest cloud tops are seen to be punctured by the eye of the storm, offering a partially obscured view of warmer surfaces below. Danny's eye introduces a hole in some of the *sublevel sets*[2] of the temperature surface (Fig. 2, bottom). At the time Hurricane Danny is nearing its maximum intensity the range of temperatures over which this hole in the sublevel sets persists is significantly larger than at the earlier time, when the storm was less intense and the eye less distinct. In other words, there is a difference in the prominence of the peaks forming the inverted caldera in the temperature surfaces near the eye. The number of connected components in the sublevel sets is also a distinguishing feature.

Finally, Fig. 3 shows examples of higher dimensional holes (two-dimensional voids) in data. Shown are *level sets*[3] of pressure and wind speed at a snapshot of a highly spatially resolved Weather Research and Forecasting (WRF) model simulation of Hurricane Isabel (Kuo et al., 2017) as she made landfall off the Eastern coast of the United States. The level sets of low pressure form the boundaries

[2]The *sublevel sets* of a real-valued function, $f : X \to \mathbb{R}$, are the subsets $f^{-1}((-\infty, a]) = \{x \in X | f(x) \leq a\} \subset X$.

[3]The *level sets* of a real-valued function, $f : X \to \mathbb{R}$, are the subsets $f^{-1}(\{a\}) = \{x \in X | f(x) = a\} \subset X$.

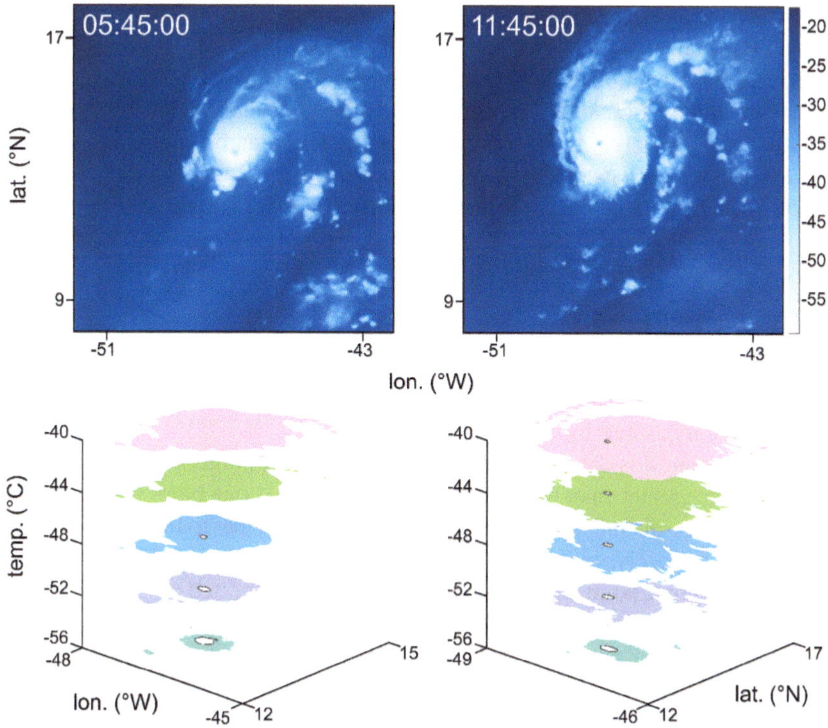

Fig. 2 Brightness temperature surfaces captured by infrared channel 3 of the GOES-13 satellite imager showing Hurricane Danny on August 21, 2015 as the storm approached its peak intensity (*top*) as well as sublevel sets of the corresponding surfaces (*bottom*). Sublevels are indicated by the height of the regions, e.g., the highest region indicates longitude and latitude coordinates where the brightness temperature is less than or equal to −40. The eye of the storm is shown as a one-dimensional hole in the sublevel sets, and persists over a range of levels depending on its prominence

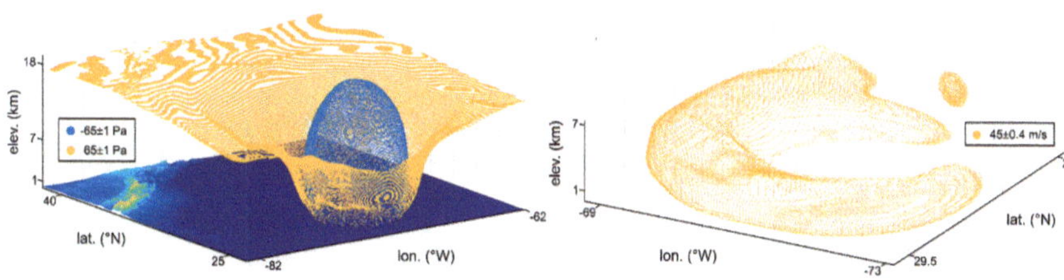

Fig. 3 Approximate level sets of several meteorological variables from a WRF simulation of Hurricane Isabel, showing examples of two-dimensional holes in point cloud data. Level sets of low pressure enclose spherical voids (*left*) while level sets of high wind speed enclose toroidal voids (*right*)

of concentric, spherical voids, while the strongest winds around Isabel's eye are strikingly toroidal.

A careful consideration of these examples might reveal that even readily apparent components, holes, and voids are, strictly speaking, not represented by real data: A point cloud is merely a finite subset of a metric space and contains no holes or voids. PH is a framework that resolves this inconsistency and provides robust and quantified descriptions of the homology of data by adopting a multiscale approach.

3 Multiscale Philosophy

Figure 4 illustrates a planar point regarded as collection of data points whose dissimilarity is measured by the usual Euclidean metric, $d : \mathbb{R}^2 \times \mathbb{R}^2 \to \mathbb{R}$. It is evident that within this data set there are three groups of points whose intragroup similarity is far greater than their intergroup similarity. If one blurs their eyes the points become less distinct and eventually some clump together to form the apparent groups. More precisely, imagine disks, $D(r,x) \equiv \{y \; |d(x,y) \leq r\}$, of radius r centered on each data point, and consider the number of connected groups formed by the union of these disks (Fig. 4b). In a sense, this is a relaxation of the requirement of a metric that the distance between distinct points be strictly greater than 0, as points at small, non-zero distance become indistinguishable members of the same group. This illustrates a hallmark of data science: the process of clustering in the hopes of separating a data set into meaningful groups.

For the cartoon data there is an ostensible choice of connectivity threshold (Fig. 4c), but this is almost never the case with real data which may be high-dimensional and not so easily visualized, or may exist only as a data cloud in an abstract metric space. So, in order to avoid the unpleasantness of choosing a threshold at which to declare two points belonging to the same group, it is common to instead consider the evolution of clusters as the threshold is increased from zero to

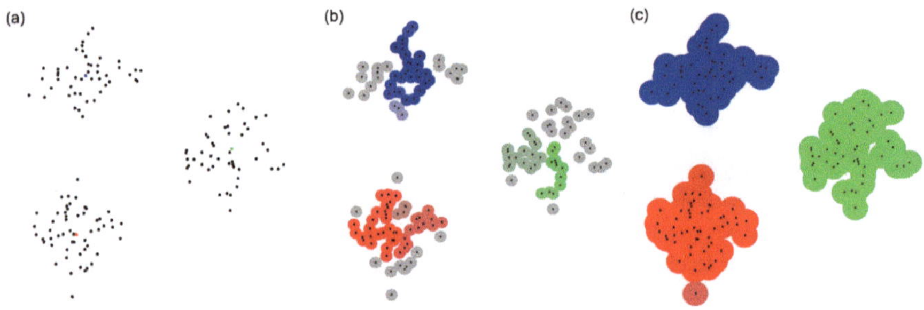

Fig. 4 (**a**) Cartoon data point cloud in the plane with distance measured by the usual Euclidean metric along with disks of radius (**b**) $r_1 > 0$ and (**c**) $r_2 > r_1$ centered on the data points. Disk darkness range of radii of data point membership in a cluster

the maximum pairwise dissimilarity between points, with two groups merging if the minimum distance between any of their members is less than the current threshold. This multiscale approach is known as *single-linkage hierarchical clustering* (Florek et al., 1951). From this perspective, it is the observation that the union of disks of radius r form exactly three groups over a large range of radii that justifies the impression that the data consists of three clusters.

Generalizing the cartoon example, often data is in the form of a finite subset of a metric space, $X \subset (M, d)$, where the metric $d : M \times M \to [0, \infty)$ defines a notion of dissimilarity. Thinking this way, at each threshold radii r, the union of disks

$$B_r(X) \equiv \bigcup_{x \in X} D(r, x) \subset M$$

forms a subspace of M. An alternative but equivalent view is that the family $\mathcal{B}_r(X)$ is defined by sublevel sets of the function which measures the distance in M to X:

$$B_r(X) = d_X^{-1}((-\infty, r]),$$

where $d_X(y) = \inf_{x \in X}\{d(x, y)\}$, for each $y \in M$. Either perspective yields a one-parameter, *nested family* of topological spaces such that if $r_1 < r_2$, then $B_X(r_1) \subset B_X(r_2)$. In this way, agglomerative clustering is equivalent to tracking the evolution and merging of the connected components of the one-parameter family of topological spaces. PH generalizes multiscale clustering by not only tracking the evolution of the components of the one-parameter family of spaces across scales, but also the evolution of the higher-dimensional structures, such as holes and voids.

The two characterizations of $B_r(X)$ are suggestive of the two types of data to which PH is commonly applied: (1) point clouds, given structure by a measure of data point dissimilarity that can be used to parameterize a nested family of spaces, and (2) real-valued functions defined on a topological space whose sublevel sets form the nested family of subspaces of the domain. In both cases it is common to build a sequence of topological spaces known as a *simplicial complexes* that are easily stored and analyzed by a computer and may be viewed as approximations of $B_r(X)$ (de Silva and Ghrist, 2007). Since many forms of image data as well as scalar functions are often represented as a data-cube defined over a grid of equally spaced domain values, an analog of a *simplicial complex* known as a *cubical complex* (Allili et al., 2001; Wood et al., 2011) is also commonly used to approximate a sequence of sublevel sets.

4 Complexes, Homology, and Persistence

Simplicial homology is not only the most theoretically accessible homology theory, since it makes the notion of an n-dimensional hole both precise and transparent, it is also the practical computational framework for PH as a data analysis tool because it makes homology entirely computable. For a more complete treatment of homology theory, consult a text on algebraic topology such as Hatcher's *Algebraic Topology* (Hatcher, 2002).

4.1 Simplicial Complexes

Figure 5 (left) shows a finite collection of data points $X = \{\mathsf{a}, \mathsf{b}, \mathsf{c}, \mathsf{d}, \mathsf{e}, \mathsf{f}, \mathsf{g}, \mathsf{h}, \mathsf{i}\}$ and the corresponding space

$$B_r(X) = \bigcup_{x \in X} D(r, x)$$

for some fixed $r > 0$. Although it is plainly clear that $B_r(X)$ is a space with a two connected component, getting a computer to store, manipulate, and compute with such an indiscrete object is unnecessarily difficult, to say nothing of tracking the entire family $B_r(X)$ as r is varied. For the task of tracking connected components, it is sufficient to instead consider the nested family of graphs $\mathcal{G}_r(X) = (X, E_r)$, where vertices are data points and where an edge $\{x, y\} \in E_r$ exists if and only if $d(x, y) \leq 2r$. The components in $B_r(X)$ are exactly described by the connected subgraphs in $\mathcal{G}_r(X)$. Indeed, $\mathcal{G}_r(X)$—which has been superimposed on $B_r(X)$ in Fig. 5 (center)—consists of two connected subgraphs. Increasing r from 0 to the diameter of the point cloud gives rise to a finite sequence of graphs that change only when edges are introduced at the finitely many pairwise distances between

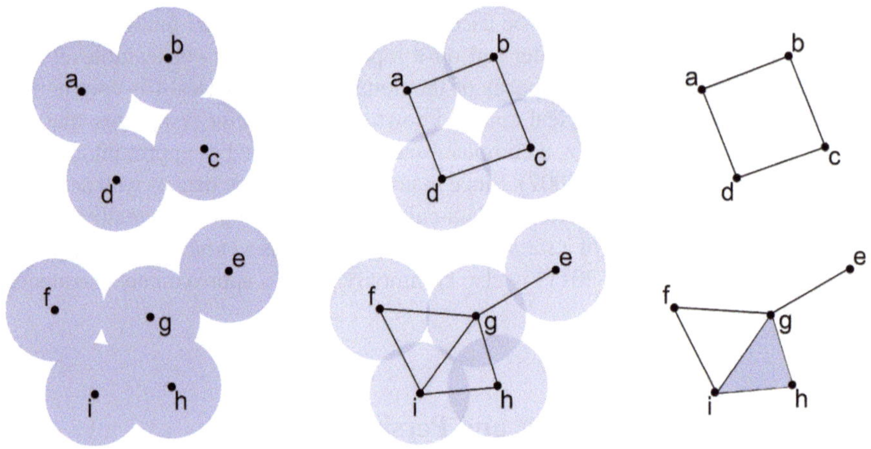

Fig. 5 Union of disks surrounding cartoon data point cloud in the plane with distance measured by the usual Euclidean metric (*left*) along with a geometric realization of the graph determined by pairwise intersections of disks (*center*) and the geometric Čech simplicial complex realized by mutual intersections of disks (*right*)

points in X. By representing a family of topological spaces by a list of combinatorial objects (graphs) the problem of tracking the merging of components is made computable (Sibson, 1973).

In light of our discussion of holes, it is also plainly clear that $B_r(X)$ has some! In particular, the disks centered on a, b, c, d and f, g, i each, respectively, enclose a one-dimensional vacancy in the space. Correspondingly, there are cycles a-b-c-d-a and f-g-h-f in $\mathcal{G}_r(X)$. Realizing that g-h-i-g also forms a cycle, despite the fact that the radius is large enough that there is a point of triple-intersection of the corresponding disks, shows that a graph structure alone is insufficient to capture the higher-dimensional holes in $B_r(X)$. This inadequacy is remedied by generalizing a graph to a *simplicial complex* by allowing "edges" (henceforth called *simplices*) with more than two elements.

Definition An *abstract simplicial complex* S is a collection of sets which is closed under taking subsets. In other words, if S is a simplicial complex and $\sigma \in S$ is a *simplex*, then all subsets of σ, which we'll call its *faces*, are also simplices in S. If the largest set in S has $m + 1$ elements, we say S is a simplicial complex of *dimension* $|S| \equiv m$.

Just as the graph $\mathcal{G}_r(X)$ is represented geometrically in Fig. 5 (center)—with the data points being vertices and edges being line segments between data points—there is a natural *geometric realization* of an abstract simplicial complex as a subspace of a Euclidean space by further taking triples to be faces of triangles, e.g., $\{\mathsf{g}, \mathsf{h}, \mathsf{i}\}$ is the triangle bounded by the three edges $\{\mathsf{g}, \mathsf{h}\}$, $\{\mathsf{g}, \mathsf{i}\}$, and $\{\mathsf{h}, \mathsf{i}\}$. Likewise, 4-element subsets can be realized as filled tetrahedra whose boundary consists of the 4 triangles defined by its 3-elements subsets, and so on. Some care must be taken to ensure that only abstract simplices which have non-empty intersection, i.e., which share a face, intersect in the geometric realization and that the intersection is along the corresponding geometric face. This can be achieved by representing the singleton sets in the abstract simplicial complex as *affinely independent vectors*[4] in some Euclidean space $\mathbb{R}^M$, for sufficiently large M, and mapping abstract simplices to the convex hulls of the corresponding subsets of vectors. Observe that the convex hull of three affinely independent vectors is a triangle, of four such vectors is a tetrahedra, etc.

The disks surrounding f, g, and h only intersect in pairs. This is reflected by inclusion of the edges $\{\mathsf{b}, \mathsf{c}\}$, $\{\mathsf{c}, \mathsf{d}\}$, and $\{\mathsf{b}, \mathsf{d}\}$ in $\mathcal{G}_r(X)$. On the other hand, it is the fact that all three disks, $D(r, \mathsf{g})$, $D(r, \mathsf{h})$, and $D(r, \mathsf{i})$, overlap that ensures they do not form a hole. This observation justifies the inclusion of the triangular face bounded by the cycle e-f-g-e in the geometric simplicial complex representation of $B_r(X)$. The triple-intersection therefore adds the triple $\{\mathsf{e}, \mathsf{f}, \mathsf{g}\}$ to the abstract simplicial complex representation of $B_r(X)$. Figure 5 (right) shows the geometric realization of the abstract simplicial complex containing subsets σ if and only if the intersection of all the disks centered at the elements of σ are non-empty. This is known as the *Čech complex.*

---

[4]A set of vectors $\{v_0, \ldots, v_n\} \subset \mathbb{R}^M$ is *affinely independent* if the set $\{v_i - v_0 | i = 1, \ldots, n\}$ is linearly independent.

Definition Let $X = \{x_1, \ldots, x_n\} \in (M, d)$ be a finite collection of points drawn from the metric space M with dissimilarity function $d : M \times M \to [0, \infty)$. Then for $r \geq 0$, the *Čech complex* of X with connectivity parameter r is the abstract simplicial complex

$$\check{C}_r(X) = \left\{ \sigma \subseteq X \;\middle|\; \bigcap_{x \in \sigma} D(r, x) \neq \emptyset \right\}.$$

Notably, $\check{C}_r(X)$ is a better representation of $B_r(X)$ than $\mathcal{G}_r(X)$ as it properly encodes the number of connected components and the number of holes. More generally and more formally, the so-called Nerve Lemma (Borsuk 1948; Hatcher 2002, Sect. 4.G) guarantees that the geometric realization of $\check{C}_r(X)$ is homotopy equivalent to $B_r(X)$, and thus faithfully represents much of the topology of the union of disks. Visualizing the continuous contraction of the union of disks onto the geometric realization of its Čech complex is suggestive of why this result holds true in general.

While the Nerve Lemma endows the Čech complex with a very attractive property, in practice $\check{C}_r(X)$ is difficult to store and compute as a search through all subsets of size n becomes quickly intractable as n grows. An alternative to the theoretically appealing Čech complex is the more computable *Vietoris–Rips complex*, or simply the *Rips complex* that benefits from only requiring storage of the graph $\mathcal{G}_r(X)$.

Definition Let $X = \{x_1, \ldots, x_n\}$ be a finite collection of data and $d : X \times X \to [0, \infty)$ be a measure of dissimilarity between data points in X. Then for $r \geq 0$, the *Vietoris–Rips complex* of X with connectivity parameter r is the abstract simplicial complex

$$\mathcal{R}_r(X) = \left\{ \sigma \subseteq X \;\middle|\; d(x_i, x_j) \leq r, \text{for all } x_i, x_j \in \sigma \right\}.$$

The distinguishing feature of the Rips complex is that all higher-dimensional simplices are completely determined by the 1-simplices,[5] the graph structure determined by pairwise dissimilarities. This also suggests the reason for the carefully chosen language in the definition of the Rips complex: $\mathcal{R}_r(X)$ is not determined by the data X being a subset of an ambient metric space. In this way, a Rips complex may be built on any set of discrete objects endowed with pairwise dissimilarities, and thus is a reflection of the intrinsic geometry of data.

No matter which simplicial complex construction is chosen, it is a general property that a nested family of simplicial complexes,

$$\mathcal{S}_0 \subseteq \mathcal{S}_1 \subseteq \cdots \subseteq \mathcal{S}_m,$$

[5]Often called the 1-skeleton.

may be regarded as the sublevel sets of a real-valued function, $f : \mathcal{S}_m \to \mathbb{R}$ defined on simplices, with the property that if $\sigma_1 \subset \sigma_2$ then $f(\sigma_1) \leq f(\sigma_2)$. This *monotonicity* condition on f ensures that its sublevel sets are themselves simplicial complexes. Said another way, if we imagine varying the parameter r, the monotonicity of f guarantees that an n-simplex does not appear before any of its faces. To illustrate, let X be a data point cloud, $d : X \times X \to [0, \infty)$ the measure of pairwise dissimilarities, and $\mathcal{S}$ the simplicial complex consisting of all subsets X. Define $f : \mathcal{S} \to \mathbb{R}$ by $f(\sigma) = \max_{x,y \in \sigma}\{d(x,y)\}$. Then the nested family of Rips complexes $\mathcal{R}_r(X)$ is given by sublevel sets of f. In particular,

$$\mathcal{R}_r(X) = f^{-1}((-\infty, r]).$$

4.2 *Persistent Homology*

The preceding section introduced a one-dimensional hole in a simplicial complex as a graph cycle (a loop of 1-simplices) that is not the boundary of two-dimensional simplices in the complex. By formalizing the notions of cycles, boundaries, and holes in simplicial complexes, simplicial homology extends these ideas to all dimensions. The strategy employed is algebraic: one constructs vector spaces that encode collections of n-simplices and then relates them via natural maps to collections of $(n-1)$-simplices representing their boundary. This is done by treating each n-simplex as an independent basis vector of a finite-dimensional vector space over a field $\mathbb{F}$.

Definition Let $\mathcal{S}$ be a simplicial complex of dimension m. For each $n = 0, \ldots, m$ define the set of *n-chains* to be

$$C_n(\mathcal{S}) = \{\alpha_0 x_0 + \cdots + \alpha_n x_n \mid \{x_0, \ldots, x_n\} \in \mathcal{S}, \alpha_i \in \mathbb{F}\}.$$

By choosing $C_n(\mathcal{S}) = \{0\}$[6] for $n < 0$ and $n > m$, further define the boundary maps $\partial_n : C_n(\mathcal{S}) \to C_{n-1}(\mathcal{S})$,

$$\begin{aligned}\partial_n(\{x_0, \ldots, x_n\}) &= \sum_{j=0}^{n} (-1)^j \{x_0, \ldots, \widehat{x_j}, \ldots, x_n\} \\ &= \sum_{j=0}^{n} (-1)^j \{x_0, \ldots, x_{j-1}, x_{j+1}, \ldots, x_n\},\end{aligned}$$

on simplices, and extend to all n-chains in $C_n(\mathcal{S})$ by linearity.

[6]The trivial vector space over $\mathbb{F}$ consisting only of the 0 vector.

The notion of an *oriented simplex* (Hatcher, 2002, Sect. 2.1) provides a proper explanation for the coefficient, ± 1, in the definition of ∂_n. That said, $\mathbb{F}$ is often taken to be the field with two elements, $\mathbb{F}_2 \equiv \mathbb{Z}/2\mathbb{Z} \cong \{0, 1\}$, in which case $-1 \equiv 1$ and an n-chain $\sigma_1 + \cdots + \sigma_k \in C_n(\mathcal{S})$ may be regarded as simply the subset of the n simplices represented in the formal sum, without needing to define orientation on the simplices. Choosing $\mathbb{F} \cong \mathbb{Z}/p\mathbb{Z}$ for small prime p is also a reasonable choice and can yield additional information about the orientability of a complex.[7]

The linear boundary maps are endowed with the desirable property that $\partial_{n-1} \circ \partial_n = 0$ for each n, which intuitively reflects the property that the boundary of a simplex does not itself have a boundary. Said another way, this property ensures that the image of ∂_n is contained in the kernel of ∂_{n-1}, or that those collections of $(n-1)$-simplices which *could* form the boundary of a collection of n-simplices do not themselves have a boundary. With these ingredients we can formalize what is meant by a hole in a simplicial complex: a boundaryless collection of (n-1)-simplices for which the collection of n-simplices it could enclose is absent from the complex.

Definition Given the *chain complex* of vector spaces

$$\cdots \to C_{n+1}(\mathcal{S}) \xrightarrow{\partial_{n+1}} C_n(\mathcal{S}) \xrightarrow{\partial_n} C_{n-1}(\mathcal{S}) \to \cdots,$$

define the subspace of *n-cycles* (chains without boundary) to be $Z_n \equiv \ker(\partial_n)$ and the subspace of *n-boundaries* to be $B_n \equiv \mathrm{im}(\partial_{n+1}) = \partial_{n+1}(C_{n+1}(\mathcal{S}))$. Further define the *n-th order homology group* of $\mathcal{S}$ to be the quotient of vector spaces $H_n(\mathcal{S}) \equiv Z_n(\mathcal{S})/B_n(\mathcal{S})$.[8]

The elements of $H_n(S)$ are equivalence classes of n-cycles that are not boundaries of $(n+1)$-chains. Two n-cycles are in the same class (called *homologous*) if they differ by a boundary. More formally, if $\gamma \in Z_n(\mathcal{S})$ is an n-cycle, then $[\gamma] \equiv \gamma + B_n(\mathcal{S}) \in H_n(\mathcal{S})$. Importantly, boundaries are in the same class as the 0 cycle. So, filling an n-dimensional hole amounts to eliminating a class $[\gamma] \in H_n(S)$ by making the cycle γ into the boundary of an $(n+1)$-chain so that $[\gamma] = [0] \in H_n(S)$ since 0 and γ differ by a boundary, namely γ.

Of course $H_n(S)$ is also a vector space over $\mathbb{F}$ by defining the vector sum $[\gamma] + [\alpha] \equiv [\gamma + \alpha]$. The dimension of $H_n(S)$ is known as the n-th Betti number of S, denoted $\beta_n(S) \equiv \dim(H_n(S))$, and thus represents the number of linearly independent n-dimensional holes. This suggests an important fact about homology as it applies to the computation of PH and the extraction of topological structure from data: a hole may be represented by more than one cycle. For example, consider the subcomplex $\mathcal{S} = \{\{\mathsf{g},\mathsf{h},\mathsf{i}\}, \{\mathsf{f},\mathsf{g}\}, \{\mathsf{g},\mathsf{h}\}, \{\mathsf{h},\mathsf{i}\}, \{\mathsf{f},\mathsf{i}\}, \{\mathsf{g},\mathsf{i}\}, \mathsf{f}, \mathsf{g}, \mathsf{h}, \mathsf{i}\}$ in Fig. 5

[7] More generally, the sets of n-chains may be defined to be the free abelian groups with coefficients taken from a commutative ring. In this setting the boundary maps are homomorphisms (Hatcher, 2002).

[8] If $\mathbb{F}$ is chosen to be a commutative ring, the boundaries and cycles form subgroups which explains the terminology homology *groups*.

(right), consisting of one filled and one empty triangle connected along an edge $\{\mathsf{g},\mathsf{i}\}$. Taking $\mathbb{F} = \{0, 1\}$, the cycle $\gamma \equiv \{\mathsf{g},\mathsf{i}\} + \{\mathsf{f},\mathsf{i}\} + \{\mathsf{f},\mathsf{g}\}$ enclosing the hole is homologous to $\alpha \equiv \{\mathsf{f},\mathsf{i}\}+\{\mathsf{f},\mathsf{g}\}+\{\mathsf{g},\mathsf{h}\}+\{\mathsf{h},\mathsf{i}\}$ since $\gamma-\alpha = \{\mathsf{g},\mathsf{i}\}+\{\mathsf{g},\mathsf{h}\}+\{\mathsf{h},\mathsf{i}\}$, which is a boundary. Geometrically this is realized by the continuous collapse of the filled triangle onto the edge $\{\mathsf{g},\mathsf{i}\}$.

The goal to track the evolution of geometric structure of a data set across scales can now be formalized in the language of homology. Let X be a data point cloud and for each $r \geq 0$ consider the Rips simplicial complex, $\mathcal{R}_r(X)$, built on X. If $s \leq t$, then $\mathcal{R}_s(X) \subseteq \mathcal{R}_t(X)$. Therefore, if $[\gamma] \in H_n(\mathcal{R}_s(X))$, then $[\gamma] \in H_n(\mathcal{R}_t(X))$, although the cycles which are homologous to γ may have changed, i.e., the homology will be altered as components merge or as holes disappear.

Definition Let $\mathcal{S}_x$ be a nested family of simplicial complexes. The *n-th order persistent homology groups* of the family of complexes are the quotients

$$H_n^{x,y} \equiv Z_n(\mathcal{S}_x)/(B_n(\mathcal{S}_y) \cap Z_n(\mathcal{S}_x)),$$

for $x \leq y$. The dimension of $H_n^{x,y}$ is the *n-th persistent Betti number*, denoted $\beta_n^{x,y}$.

In particular, $\beta_n^{x,x} = \beta_n(\mathcal{S}_x)$ accounts for the n-holes in the complex $\mathcal{S}_x$ built at the parameter value $x \in \mathbb{R}$. Furthermore, for each ordered pair (x, y) with $x \leq y$, the number $\beta_n^{x,y}$ counts the number of linearly independent n-th order homology classes which appeared (were *born*) in a complex $\mathcal{S}_r$, with $r \leq x$ and which have not merged with a class that was born in an earlier homology group, i.e., are still present in $\mathcal{S}_y$ (have not *died*) (Edelsbrunner and Harer, 2010, Chap. VII.1).

Practically speaking, if X has N elements, then there are only finitely many different Rips complexes, parameterized by the unique pairwise distances, r_j, between points in the cloud. Between these critical connectivity thresholds the homology remains the same, but increasing the scale from r_j to the scale r_k, the homology will change and so too might the Betti numbers. For example, assuming unique pairwise distances there are 82 simplicial complexes, $\mathcal{S}_0 \subset \cdots \subset \mathcal{S}_{81}$, in the nested family of Rips simplicial complexes built by varying the connectivity radius on data points in Fig. 5. A subset of these are shown in Fig. 6, highlighting the appearance and disappearance of zero- and one-dimensional homological features. The chain complex of $\mathcal{S}_0$ is

$$\to 0 \xrightarrow{\partial_1} C_0(\mathcal{S}_0) \xrightarrow{\partial_0} 0 \to,$$

which implies that $H_0(\mathcal{S}_0) \cong \mathbb{F}^9$ since $\ker(\partial_0) = C_0(\mathcal{S}_0) \cong \mathbb{F}^9$ with basis consisting of the nine 0-simplices, while $\mathrm{im}(\partial_1) \cong \{0\}$. On the other hand, the chain complex of $\mathcal{S}_1$ is

$$\to 0 \to C_1(\mathcal{S}_1) \xrightarrow{\partial_1} C_0(\mathcal{S}_1) \xrightarrow{\partial_0} 0 \to$$

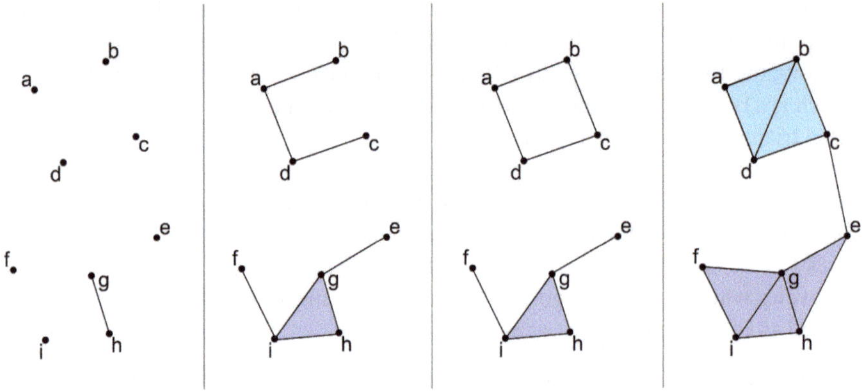

Fig. 6 *From left to right*, geometric realizations of abstract Rips complexes $\mathcal{S}_1, \mathcal{S}_8, \mathcal{S}_9, \mathcal{S}_{13}$

and $C_1(\mathcal{S}_1) \cong \mathbb{F}$ with basis consisting of the single 1-simplex $\{\mathsf{g}, \mathsf{h}\}$. Taking $\mathbb{F} = \mathbb{Z}/2\mathbb{Z}$, $\partial_1(\{\mathsf{g}, \mathsf{h}\}) = \{\mathsf{g}\} + \{\mathsf{h}\}$, and so $\mathrm{im}(\partial_1)$ is a one-dimensional subspace of $C_0(\mathcal{S}_1)$, spanned by the 0-chain $\{\mathsf{g}\} + \{\mathsf{h}\}$. Now $\{\mathsf{g}\}$ and $\{\mathsf{h}\}$ differ by the boundary $\{\mathsf{g}\} + \{\mathsf{h}\}$, since $\{\mathsf{g}\} - \{\mathsf{h}\} = \{\mathsf{g}\} + \{h\}$ over the field with two elements. Thus $[\mathsf{g}] = [\mathsf{h}] \in H_0(\mathcal{S}_1)$ and the dimension of the 0-th order homology group has decreased by one.

Still working with $\mathbb{Z}/2\mathbb{Z}$ coefficients, the appearance of a one-dimensional hole in simplicial complex $\mathcal{S}_9$ will be realized by the fact that the 1-cycle, $\gamma \equiv \{\mathsf{a}, \mathsf{b}\} + \{\mathsf{a}, \mathsf{d}\} + \{\mathsf{b}, \mathsf{c}\} + \{\mathsf{c}, \mathsf{d}\} \in Z(\mathcal{S}_9)$ may be taken as a representative of the nontrivial class, $[\gamma] \neq [0]$, which is contained in $H_1(\mathcal{S}_9)$ but not in $H_1^{8,9} = Z_1(\mathcal{S}_8)/(B_1(\mathcal{S}_9) \cap Z_1(\mathcal{S}_8))$. Thus $[\gamma]$ is born at connectivity parameter r_9. This hole disappears in S_{13} with the introduction of the edge $\{\mathsf{b}, \mathsf{d}\}$ because the Rips construction insists on the addition of the 2-simplices $\{\mathsf{a}, \mathsf{b}, \mathsf{d}\}$ and $\{\mathsf{b}, \mathsf{c}, \mathsf{d}\}$. Explicitly,

$$\partial_2(\{\mathsf{a}, \mathsf{b}, \mathsf{d}\} + \{\mathsf{b}, \mathsf{c}, \mathsf{d}\}) = \{\mathsf{a}, \mathsf{b}\} + \{\mathsf{a}, \mathsf{d}\} + \{\mathsf{b}, \mathsf{c}\} + \{\mathsf{c}, \mathsf{d}\} + 2\{\mathsf{b}, \mathsf{d}\} = \gamma,$$

is now a boundary and so $[\gamma] = [0]$ in the quotient $H_1(\mathcal{S}_{13}) = Z_1(\mathcal{S}_{13})/B_1(\mathcal{S}_{13})$. The class $[\gamma]$ merges with a class born before it, and so it dies at the connectivity parameter r_{13}.

5 Persistence Diagrams and the Shape of Data

The preceding section demonstrated how PH captures the character of the intrinsic geometry of a point cloud by the scales at which homological features are born and the scales at which they die. Moreover, a multiscale homological description of a scalar function may be encoded in the pairings of critical levels between which the homology of sublevel sets remain unchanged. For each dimension $n = 0, 1, \ldots$,

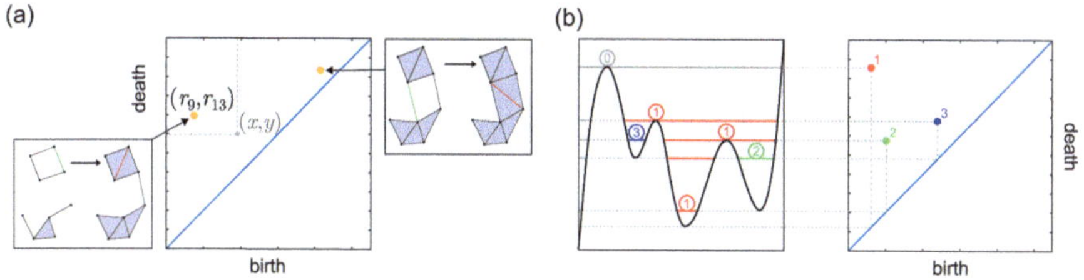

Fig. 7 Example H_1 PD of (**a**) Rips family of simplicial complexes built on cartoon data set from Fig. 5 and (**b**) H_0 PD of nested family of sublevel sets of the given univariate real-valued function. Sublevel sets are drawn as *solid lines* at the corresponding levels. Sublevel set numbers delineate connected components and match the numbers of the persistence pairs computed by PH. Connected components are born at local minima and die when they merge at a local maximum. Component 1 is observed to die when it merges with component 0 born at some lower level not shown

the PH calculation generates a finite collection of ordered pairs, (b_i, d_i), specifying the birth and death parameters of each n-dimensional homological feature. Plotting these ordered pairs as a multiset of points in the plane associates to some data its *persistence diagrams* (PDs). Necessarily $d_i \geq b_i$ and so points in a PD appear above the diagonal. The *persistence* of a feature is taken to be its death value minus its birth value.

Figure 7a shows the H_1 PD of the Rips family of simplicial complexes built on the point cloud given in Fig. 5. The point (x, y) captures only one persistence pair, (r_9, r_{13}), (i.e., $\beta_1^{x,y} = 1$) showing that there is only one hole born at a smaller connectivity parameter than x which persists at connectivity parameter y. As shown, this feature dies at connectivity parameter r_{13}. Figure 7b shows the H_0 PD for the family of sublevel sets of a simple function. New connected components in the sublevel sets are born at local minima and die when the components merge at a local maximum, with the component born earlier persisting.

TDA folklore says that highly persistent pairs in a PD are real topological features, while short-lived pairs may be regarded as "topological noise." For instance, component 3 in Fig. 7b is spawned from a mere wrinkle in the function, compared with the prominent overall dip in the function characterized by component 1. Features near the diagonal may exist only because of the finiteness of the data and the topological approximations made by constructing complexes (Ghrist, 2008). This philosophy is partially rooted in the notion that data represents a finite sampling of a topological space and that given enough data (a dense enough sampling of this space) the topological noise would be eliminated and the true homology of the space would be revealed by the persistence calculation. However, in the view that PH is transformation that can illuminate important differences in data, all regions of a diagram may be relevant, except perhaps those points whose persistence cannot be validated due to finite precision in the acquisition of data. In fact, several studies have found that for classification problems it need not be the regions of highest persistence that are most discriminating (Adams et al., 2016; Bendich et al., 2016; Rouse et al., 2015).

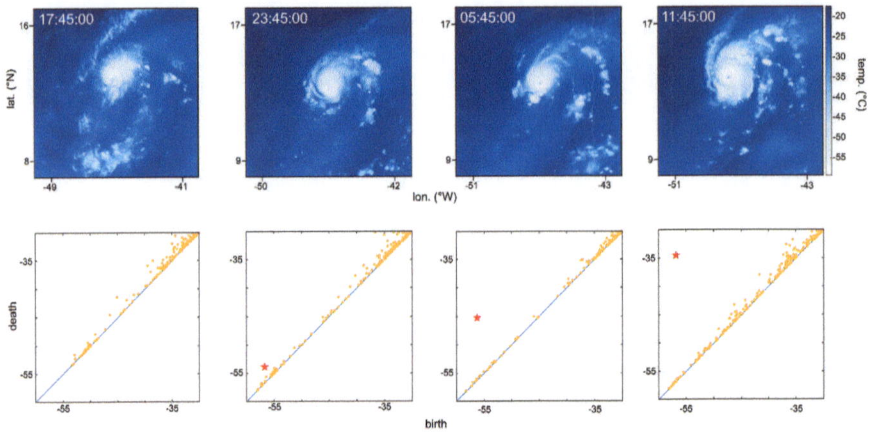

Fig. 8 Brightness temperature surfaces, interpolated to a 400 × 400 grid, captured by infrared channel 3 of the GOES-13 satellite imager showing *from left to right* the temporal evolution of Hurricane Danny between August 20 and August 21, 2015 (*top*) with corresponding persistence diagrams of the families of cubical complexes defined by sublevel sets (*bottom*). The *star* in the PDs corresponds to the hole in the sublevel sets formed by the eye

Consider in Fig. 8 the persistence diagrams encoding the evolution of one-dimensional holes in the sublevel sets of four snapshots of the brightness temperature field of Hurricane Danny over the 18 h preceding its peak intensity. Because the birth and death coordinates in these PDs reflect levels of the brightness temperature surface, they are in the units of degrees Celsius for this example. For clarity, the diagrams have been restricted to low temperature birth and death values. H_1 PH classes with large birth values will correspond to topological circles first formed in sublevel sets near the highest temperatures in the images, i.e., the sea surface, while the features born at low levels reflect structures in the colder cloud tops.

The corresponding images (Fig. 8, top) capture the storm as it rapidly intensified from a category 1 (75 knots) to a category 3 (110 knots). Naturally, as the images change in time, so too do the persistence diagrams. An important question is, how are changes in the underlying data reflected in changes in the PDs? If PH is to be useful as a lens through which to view data, it cannot be too sensitive to perturbations, since data is commonly corrupted by noise.

As Hurricane Danny intensifies its eye becomes more prominent, carving a one-dimensional hole across a larger range of sublevel sets. Correspondingly a feature emerges from the diagonal of the PD and its persistence steadily grows (Fig. 8, bottom). This observation highlights two distinct but related facts: (1) introducing a small perturbation in a function may give rise to a new homological feature but, if so, it will be short-lived and thus appear as a point near the diagonal in the sublevel set PD and (2) as a function is continuously varied, so too are the features in the corresponding PD.

To make the notion of stability of PDs precise, a measure of dissimilarity between diagrams must be defined. Let the space of persistence diagrams, denoted Per, be the collection of all finite multisets of points above the diagonal, each augmented with countably infinitely many copies of the diagonal. Now Per may be given the structure of a metric space in several ways.

Definition Given two diagrams $P, Q \in \text{Per}$, define the *bottleneck distance* between P and Q to be

$$W_\infty(P, Q) \equiv \inf_{\gamma : P \to Q} \sup_{p \in P} \|p - \gamma(p)\|_\infty,$$

where $\gamma : P \to Q$ is a bijection from P to Q. Also, for each $q > 0$ define the *q-Wasserstein distance* between P and Q to be

$$W_q(P, Q) \equiv \inf_{\gamma : P \to Q} \left(\sum_{p \in P} \|p - \gamma(p)\|_\infty^q \right)^{1/q}.$$

Both measures of distance between diagrams depend on an optimal matching between the pairs they contain. This explains the need to add countably infinitely many copies of the diagonal to each multiset of above-diagonal points since two diagrams may represent data sets with distinct numbers of multiscale homological features, and including the diagonals ensures that a matching between features will exist—although it may assign (perhaps short-lived) features to a point of zero persistence on the diagonal. Although this requirement is technical, it is well-motivated by the fact that small perturbations in the data may add homological features, but they will be near the diagonal, as shown in Fig. 8 and alluded to with the wrinkle in Fig. 7b.

Giving a metric structure to Per gives a precise way to relate measures of distance between data to the measures of distance between diagrams and shows, under some mild assumptions, that the association of a function to its sublevel set PD is a continuous transformation. More precisely, given some technical hypotheses about f, g and X, if D_f and D_g are persistence diagrams of the nested family of sublevel sets of functions $f, g : X \to \mathbb{R}$, then

$$W_\infty(D_f, D_g) \leq C\|f - g\|_\infty$$

and

$$W_q(D_f, D_g) \leq C\|f - g\|_\infty^{1-k/q},$$

for some $C > 0$ and $k > 1$ that depends on properties of X, for each $q \geq k, q < \infty$ (Edelsbrunner and Harer 2010, Chap. VIII.2; Cohen-Steiner et al. 2007). The implication is that the persistence transformation can be made into an α-Hölder continuous function ($0 \leq \alpha \leq 1$) between metric spaces if Per is endowed with the metric W_q for sufficiently large q, and is in fact Lipschitz for the bottleneck metric. A similar stability statement holds for data point clouds: the bottleneck distance between the PDs of the families of Rips complexes built on two point clouds—thought of as finite metric spaces—is bounded by twice the (Edwards, 1975) distance between the clouds (Chazal et al., 2009). Thus, a smoothly varying point cloud will have continuously varying persistence diagrams, and PDs will be insensitive to noise. These results establish PH as a stable measurement, capturing shape-based differences in data, and justify the use of certain statistical and ML methods in conjunction with PDs.

Although numerous studies have explored the statistics of persistence diagrams (Mileyko et al., 2011; Munch et al., 2015; Turner et al., 2014), there are some peculiarities of Per as a metric-measure space that should give pause. For instance, the average of two diagrams need not be unique if the optimal matching between points in those diagrams is not unique (Mileyko et al., 2011). Also, the calculation of both the bottleneck and the q-Wasserstein distances relies on finding optimal matchings between points in the two diagrams by solving a bipartite graph matching problem (Edelsbrunner and Harer, 2010, Chap. VIII.4). Classical algorithms to do this (Hopcroft and Karp, 1971; Kuhn, 1955) quickly become impractical as the number of points in the diagrams grows. However, recent work toward faster algorithms is promising to lower this computational hurdle (Kerber et al., 2016).

6 Coordinatizing Diagrams

In addition to the limitations imposed by the computational complexity of computing the bottleneck and Wasserstein distances, many well-established modern ML protocols rely on structure that Per doesn't have. For instance, common implementations of support vector machines (Burges, 1998), neural networks (Zhang, 2000), and decision tree classifiers (Safavian and Landgrebe, 1991) all rely on vector space and/or inner product structure. Thus, much effort has been made to "coordinatize" persistence diagrams and map them into spaces with additional structure (Adams et al., 2016; Adcock et al., 2016; Bubenik, 2015; Carrière et al., 2015; Di Fabio and Ferri, 2015; Reininghaus et al., 2015). Figure 9 shows illustrations of several of these "homological feature vector" transformations of PDs. Some comparisons of the potential advantages and disadvantages between these methods have been made in the literature (Adams et al., 2016) and so this section highlights some of the concerns relevant to any approach that aims to strengthen the connection between ML and PH by adding structure to PDs.

One simple but effective way to generate a vector from a PD is to superimpose a grid over a region above the diagonal and to each grid element assign the number of homological features it contains (Fig. 9a). Despite its simplicity, this binning procedure was successfully exploited to perform regression analysis on collections of PDs derived from brain artery structure data (Bendich et al., 2016),

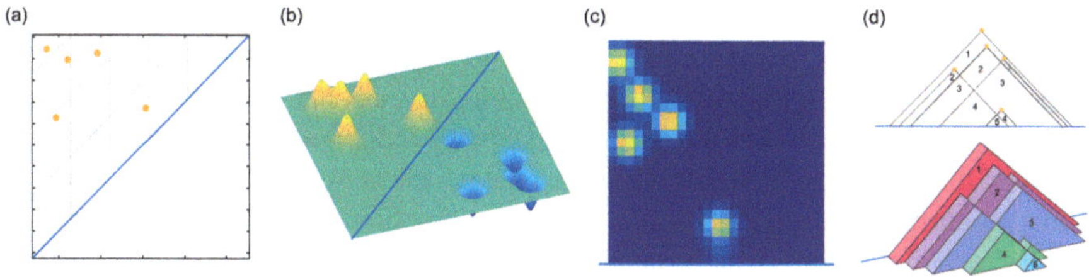

Fig. 9 Illustrations of coordinatizations of PDs. (**a**) Original PD with superimposed grid representing coordinates of an integer valued vector. (**b**) Persistence surface derived by summing Gaussians centered on persistence points and negative Gaussians centered on pairs mirrored over the diagonal. (**c**) Persistence images gotten by integrating a weighted sum of Gaussians centered on persistence pairs over grid elements in the birth-persistence plane. (**d**) Persistence landscapes essentially summarizing for each (x, y) above the diagonal, the number of homological features (b, d) with $b \leq x$ and $d \geq y$, i.e., persistence pairs which are above and the to left of the point (x, y) counted by $\beta_n^{x,y}$

and thereby gave new insights into structural changes associated with arterial aging. That said, the stability enjoyed by PDs is lost by the binning vectorization: An arbitrarily small perturbation of the underlying data may alter the birth or death of a homological feature so that the corresponding point in the PD moves from one bin to another. Moreover, in light of the discussion in Sect. 5 regarding the emergence of points from the diagonal, it is apparent that perturbations creating new homological features will also introduce a discontinuity.

A solution to the discontinuity introduced by finite-persistence points moving between bins, which has been explored by a number of authors (Adams et al., 2016; Donatini et al., 1998; Ferri et al., 1997; Reininghaus et al., 2015), is to replace points in a persistence diagram with continuous functions defined on the plane. By summing these functions over the points of finite-persistence, a PD is transformed into a surface. For example, several studies have proposed replacing each persistence pair, (b, d), with a 2D Gaussian function

$$G(x, y) \equiv \exp((-(b - x)^2 - (d - y)^2)/\sigma),$$

centered at (b, d), with spread controlled by σ. The resulting surface may be viewed as a point in a function space (Reininghaus et al., 2015), or may be further converted into a finite-dimensional vector by assigning to each element of a superimposed grid the integral of the surface over that grid element (Adams et al., 2016).

Without further modification these transformations into vector spaces still lack stability because of the emergence of points from the diagonal. Several approaches to restore stability have been proposed. For example, in Reininghaus et al. (2015) PDs are stably mapped to a kernel, which can then be fed to a number of different ML methods such as kernel SVM (Hofmann et al., 2008), by considering inner products between smooth surfaces which vary continuously with persistence pairs and which vanish along the diagonal. These surfaces are constructed by taking the sum of Gaussians centered on persistence pairs along with negative Gaussians on persistence pairs that have been mirrored across the diagonal (Fig. 9b). Similarly, in Adams et al. (2016), a transformation to a finite-dimensional vector referred to as a *persistence image* was shown to maintain stability for certain measures of distance between diagrams provided that the sum of Guassians is weighted by a function that vanishes along the diagonal (Fig. 9c). Finally, a functional summary of a PD called a *persistence landscape* was proposed, and its stability and statistical properties were analyzed, in Bubenik (2015) (Fig. 9d). This transformation maps a PD into the normed vector space of functions $f : \mathbb{Z}_+ \times \mathbb{R} \to [-\infty, \infty]$, which has many desirable properties favorable to statistical analysis.

Each of the methods discussed in this section has been applied to both toy and real data sets in conjunction with statistical and ML methods requiring more structure than a measure of dissimilarity between diagrams. That said, the value of PH and the effectiveness of the various PD transformations is decidedly problem specific. Because the range of data-driven sciences gaining new insights by using topological analysis methods is already broad and growing, it would be beneficial to have a comparison study of the strengths, weaknesses, and appropriateness of each method.

References

Adams, Henry, Tegan Emerson, Michael Kirby, Rachel Neville, Chris Peterson, Patrick Shipman, Sofya Chepushtanova, Eric Hanson, Francis Motta, and Lori Ziegelmeier. 2017. Persistence images: A stable vector representation of persistent homology. *Journal of Machine Learning Research* 18: 1–35

Adcock, Aaron, Erik Carlsson, and Gunnar Carlsson. 2016. The ring of algebraic functions on persistence bar codes. *Homology, Homotopy and Applications* 18(1): 381–402.

Allili, M., K. Mischaikow, and A. Tannenbaum. Oct 2001. Cubical homology and the topological classification of 2D and 3D imagery. In *Proceedings 2001 international conference on image processing (Cat. No.01CH37205)*, vol. 2, 173–176.

Bendich, Paul, James S. Marron, Ezra Miller, Alex Pieloch, and Sean Skwerer. 2016. Persistent homology analysis of brain artery trees. *The Annals of Applied Statistics* 10(1): 198–218.

Borsuk, Karol. 1948. On the imbedding of systems of compacta in simplicial complexes. *Fundamenta Mathematicae* 35(1): 217–234.

Bubenik, Peter. 2015. Statistical topological data analysis using persistence landscapes. *The Journal of Machine Learning Research* 16(1): 77–102.

Burges, Christopher JC. 1998. A tutorial on support vector machines for pattern recognition. *Data Mining and Knowledge Discovery* 2: 121–167.

Carrière, Mathieu, Steve Y. Oudot, and Maks Ovsjanikov. 2015. Stable topological signatures for points on 3D shapes. *Computer Graphics Forum* 34: 1–12.

Chazal, Frédéric, David Cohen-Steiner, Marc Glisse, Leonidas J. Guibas, and Steve Y. Oudot. 2009. Proximity of persistence modules and their diagrams. In *Proceedings of the twenty-fifth annual symposium on computational geometry*, SCG '09, 237–246. New York, NY: ACM.

Chung Moo K., Peter Bubenik, and Peter T. Kim. 2009a. Persistence diagrams of cortical surface data. In *Information processing in medical imaging*, vol. 21, 386–397. Berlin: Springer.

Chung, Moo K., Vikas Singh, Peter T. Kim, Kim M. Dalton, and Richard J. Davidson. 2009b. Topological characterization of signal in brain images using min-max diagrams. In *Medical image computing and computer-assisted intervention—MICCAI 2009*, ed. Guang-Zhong Yang, David J. Hawkes, Daniel Rueckert, Alison Noble, and Chris Taylor, vol. 5762, 158–166. Berlin/Heidelberg: Springer.

Cohen-Steiner, David, Herbert Edelsbrunner, and John Harer. 2007. Stability of persistence diagrams. *Discrete and Computational Geometry* 37(1): 103–120.

Dabaghian, Yu, Facundo Memoli, Loren Frank, and Gunnar Carlsson. 2012. A topological paradigm for hippocampal spatial map formation using persistent homology. *PLoS Computational Biology* 8(8): e1002581.

de Silva, Vin, and Robert Ghrist. 2007. Coverage in sensor networks via persistent homology. *Algebraic and Geometric Topology* 7: 339–358.

Di Fabio, Barbara, and Massimo Ferri. 2015. Comparing persistence diagrams through complex vectors. In *International conference on image analysis and processing 2015 part I*, ed. Murino, V., and E. Puppo. Lecture Notes in Computer Science, vol. 9279, 294–305. Heidelberg: Springer.

Donatini, Pietro, Patrizio Frosini, and Alberto Lovato. 1998. Size functions for signature recognition. In *SPIE's international symposium on optical science, engineering, and instrumentation*, 178–183.

Edelsbrunner, Herbert, and John Harer. 2008. Persistent homology—a survey. *Contemporary Mathematics* 453: 257–282.

Edelsbrunner, Herbert, and John Harer. 2010. *Computational topology: an introduction*. Providence, RI: American Mathematical Society.

Edwards, David A. 1975. The structure of superspace. In *Studies in topology*, ed. Nick M. Stavrakas and Keith R. Allen, 121–133. New York, NY: Academic.

Ferri, Massimo, Patrizio Frosini, Alberto Lovato, and Chiara Zambelli. 1997. Point selection: a new comparison scheme for size functions (with an application to monogram recognition). In *Computer vision ACCV'98*, 329–337. Berlin/Heidelberg: Springer.

Florek, K., J. Łukaszewicz, J. Perkal, H. Steinhaus, and S. Zubrzycki. 1951. Sur la liaison et la division des points d'un ensemble fini. *Colloquium Mathematicae* 2(3–4): 282–285.

Ghrist, Robert. 2008. Barcodes: The persistent topology of data. *Bulletin of the American Mathematical Society* 45(1): 61–75.

Giusti, Chad, Eva Pastalkova, Carina Curto, and Vladimir Itskov. 2015. Clique topology reveals intrinsic geometric structure in neural correlations. *Proceedings of the National Academy of Sciences* 112(44): 13455–13460.

Hatcher, Allen. 2002. *Algebraic topology*. Cambridge: Cambridge University Press.

Hiraoka, Yasuaki, Takenobu Nakamura, Akihiko Hirata, Emerson G. Escolar, Kaname Matsue, and Yasumasa Nishiura. 2016. Hierarchical structures of amorphous solids characterized by persistent homology. *Proceedings of the National Academy of Sciences* 113(26): 7035–7040.

Hofmann, Thomas, Bernhard Schölkopf, and Alexander J. Smola. 2008. Kernel methods in machine learning. *Annals of Statistics* 36(3): 1171–1220.

Hopcroft, John E., and Richard M. Karp. 1971. A N5/2 algorithm for maximum matchings in bipartite. In *Proceedings of the 12th annual symposium on switching and automata theory (Swat 1971)*, SWAT '71, 122–125. Washington, DC: IEEE Computer Society.

Kantz, Holger, and Thomas Schreiber. 1997. *Nonlinear time series analysis*. Cambridge Nonlinear Science Series. Cambridge/New York: Cambridge University Press. Originally published: 1997.

Kerber, Michael, Dmitriy Morozov, and Arnur Nigmetov. 2016. Geometry helps to compare persistence diagrams. In *2016 Proceedings of the eighteenth workshop on algorithm engineering and experiments (ALENEX)*, 103–112.

Kuhn, Harold W. 1955. The Hungarian method for the assignment problem. *Naval Research Logistics Quarterly* 2: 83–97.

Kuo, Wei, Bill Wang, Cindy Bruyere, Tim Scheitlin, and Don Middleton. 2017. Hurricane Isabel data produced by the weather research and forecast (WRF) model. Courtesy of NCAR, and the U.S. National Science Foundation (NSF). http://www.vets.ucar.edu/vg/isabeldata/readme.html.

Menne, M.J., I. Durre, B. Korzeniewski, S. McNeal, K. Thomas, X. Yin, S. Anthony, Ray R., R.S. Vose, B.E. Gleason, and T.G. Houston. 2017. Global historical climatology network - daily (GHCN-daily), version 3.22. NOAA National Climatic Data Center. ftp://ftp.ncdc.noaa.gov/pub/data/ghcn/daily/readme.txt.

Mileyko, Yuriy, Sayan Mukherjee, and John Harer. 2011. Probability measures on the space of persistence diagrams. *Inverse Problems* 27(12): 124007.

Munch, Elizabeth, Katharine Turner, Paul Bendich, Sayan Mukherjee, Jonathan Mattingly, and John Harer. 2015. Probabilistic fréchet means for time varying persistence diagrams. *Electronic Journal of Statistics* 9(1): 1173–1204.

Munro, John, Peter Landecker, and Martin Gale. Goes n data book section 3. Technical Report 2, National Aeronautics and Space Administration, Feb 2005. Copyright ©2006 Boeing. Unpublished work.

Nicolau, Monica, Arnold J. Levine, and Gunnar Carlsson. 2011. Topology based data analysis identifies a subgroup of breast cancers with a unique mutational profile and excellent survival. *Proceedings of the National Academy of Sciences* 108(17): 7265–7270.

Pachauri, Deepti, Christian Hinrichs, Moo K. Chung, Sterling C. Johnson, and Vikas Singh. 2011. Topology-based kernels with application to inference problems in Alzheimer's disease. *IEEE Transactions on Medical Imaging* 30(10): 1760–1770.

Pearson, Daniel A., R. Mark Bradley, Francis C. Motta, and Patrick D. Shipman. Dec 2015. Producing nanodot arrays with improved hexagonal order by patterning surfaces before ion sputtering. *Physical Review E* 92: 062401.

Perea, Jose A., and John Harer. 2015. Sliding windows and persistence: An application of topological methods to signal analysis. *Foundations of Computational Mathematics* 15(3): 799–838.

Perea, Jose A., Anastasia Deckard, Steve B. Haase, and John Harer. 2015. Sw1pers: Sliding windows and 1-persistence scoring; discovering periodicity in gene expression time series data. *BMC Bioinformatics* 16(1): 257.

Reininghaus, Jan, Stefan Huber, Ulrich Bauer, and Roland Kwitt. 2015. A stable multi-scale kernel for topological machine learning. In *Proceedings of the IEEE Conference on Computer Vision and Pattern Recognition*, 4741–4748.

Rouse, David, Adam Watkins, David Porter, John Harer, Paul Bendich, Nate Strawn, Elizabeth Munch, Jonathan DeSena, Jesse Clarke, Jeffrey Gilbert, Peter Chin, and Andrew Newman. 2015. Feature-aided multiple hypothesis tracking using topological and statistical behavior classifiers. In *Signal processing, sensor/information fusion, and target recognition XXIV*, vol. 9474, 94740L–94740L–12.

Safavian, S.R., and D. Landgrebe. May 1991. A survey of decision tree classifier methodology. *IEEE Transactions on Systems, Man, and Cybernetics* 21(3): 660–674.

Sibson, R. 1973. Slink: An optimally efficient algorithm for the single-link cluster method. *The Computer Journal* 16(1): 30–34.

Singh, Gurjeet, Facundo Memoli, Tigran Ishkhanov, Guillermo Sapiro, Gunnar Carlsson, and Dario L. Ringach. 2008. Topological analysis of population activity in visual cortex. *Journal of Vision* 8(8): 11.

Turner, Katharine, Yuriy Mileyko, Sayan Mukherjee, and John Harer. 2014. Fréchet means for distributions of persistence diagrams. *Discrete and Computational Geometry* 52(1): 44–70.

Venkataraman, Vinay, Karthikeyan Natesan Ramamurthy, and Pavan Turaga. Sept 2016. Persistent homology of attractors for action recognition. In *2016 IEEE international conference on image processing (ICIP)*, 4150–4154. Washington, DC: IEEE.

Wang, Yuan, Hernando Ombao, and Moo K. Chung. 2014. Persistence landscape of functional signal and its application to epileptic electroencaphalogram data (unpublished).

Wood, Peter John, Adrian P. Sheppard, and Vanessa Robins. 2011. Theory and algorithms for constructing discrete morse complexes from grayscale digital images. *IEEE Transactions on Pattern Analysis and Machine Intelligence* 33 (undefined): 1646–1658.

Zhang, Guoqiang Peter. 2000. Neural networks for classification: A survey. *IEEE Transactions on Systems, Man, and Cybernetics, Part C (Applications and Reviews)* 30(4): 451–462.

Zomorodian, Afra, and Gunnar Carlsson. 2005. Computing persistent homology. *Discrete and Computational Geometry* 33(2): 249–274.

Nonlinear Dynamical Approach to Atmospheric Predictability

C. Nicolis

Abstract The principal properties of initial condition and of model errors along with their repercussions on atmospheric predictability are reviewed. A general nonlinear dynamics-inspired approach is developed, from which generic trends are derived. The main ideas are illustrated on selected low-order models capturing the principal qualitative aspects of the phenomena of interest.

Keywords Nonlinear dynamics • Stochastic processes • Predictability • Error growth

1 Introduction

The variability of atmospheric and climate dynamics over a wide range of time and space scales are well-established facts (Lorenz, 1984; Nicolis and Nicolis, 1987). A typical example is provided by the daily evolution of air temperature at a particular location (Fig. 1). One observes small scale irregular fluctuations that are never reproduced in an identical fashion, superimposed on the large-scale regular seasonal cycle of solar radiation. A second illustration of variability pertains to the much larger scale of global climate. All elements at our disposal show indeed that the earth's climate has undergone spectacular changes in the past, like the succession of glacial–interglacial periods. Figure 2 represents the variation of the volume of continental ice over the last million years as inferred from the evolution of the composition of marine sediments in oxygen 16 and 18 isotopes. Again, one is struck by the intermittent character of the evolution, as witnessed by a marked aperiodic component masking to a great extent an average time scale of 100,000 years that is sometimes qualified as the Quaternary glaciation "cycle." An unexpected corollary is that the earth's climate can switch between quite different modes over a short time (in the geological scale), of the order of a few thousand years.

A fundamental consequence of the aperiodicity of the atmospheric and climate dynamics is the well-known difficulty to make reliable predictions. Contrary to

C. Nicolis (✉)
Institut Royal Météorologique de Belgique, 3 av. Circulaire, 1180 Brussels, Belgium
e-mail: cnicolis@oma.be

A.A. Tsonis (ed.), *Advances in Nonlinear Geosciences*,
DOI 10.1007/978-3-319-58895-7_20

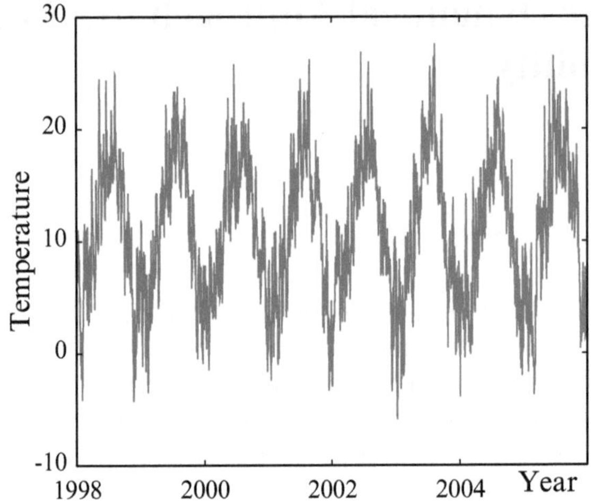

Fig. 1 Mean daily temperature at Uccle (Brussels) between January 1st 1998 and December 31, 2006

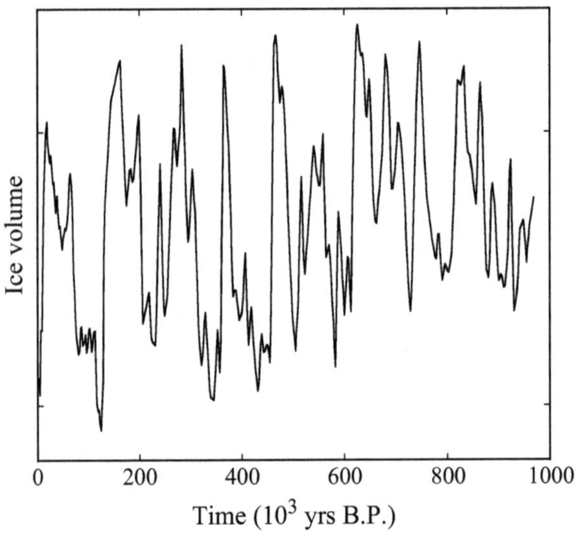

Fig. 2 Evolution of the global ice volume on earth during the last million years as inferred from oxygen isotope data

simple periodic or multiperiodic phenomena for which a long-term prediction is possible, predictions in meteorology are limited in time. The most plausible (and currently admitted) explanation is based on the realization that the atmosphere displays sensitivity to the initial conditions: a small uncertainty in the initial data used in a prediction model (usually referred as "error") is amplified in the course of the evolution (Lorenz, 1969). Such uncertainties are inherent in the very process of experimental measurement. A great deal of effort is devoted in atmospheric sciences in the development of *data assimilation* techniques aiming to reduce them as much as possible (Kalnay, 2003), but it is part of the laws of nature that they will never be fully eliminated. Now, sensitivity to the initial conditions happens to be the principal signature of deterministic chaos. The chaotic character of atmospheric dynamics is by now a widely accepted fact, compatible both with the analysis of the available data and the modeling of atmospheric phenomena. It is often referred as the *butterfly effect* and is epitomized in the provocative and by now famous Lorenz's statement "Predictability: Does the flap of a butterfly's wing in Brazil set a tornado in Texas?" (Lorenz, 1993; Nicolis and Nicolis, 2009; Tsonis, 1992).

It is important to realize that much like experiment, modeling is also limited in practice by a finite resolution (of the order of several kilometers) and the concomitant omission of "subgrid" processes, e.g., local turbulence. Furthermore, many of the parameters present are not known to a great precision. In addition to initial errors prediction must thus cope with *model errors*, reflecting the fact that a model is only an approximate representation of nature (Schubert and Schang, 1996; Tribbia and Baumhefner, 1988). This raises the problem of sensitivity of atmospheric dynamics to the parameters present in the description of the different processes.

If the dynamics were simple neither of these two types of errors would matter. But this is manifestly not the case in the atmosphere, where nonlinear couplings, bifurcations, and abrupt transitions are part of everyday reality. Initial and model errors can thus be regarded as probes revealing the fundamental instability and complexity underlying the atmosphere.

In this chapter the principal properties of initial condition and of model errors are analyzed from a nonlinear dynamics perspective and their repercussions on predictability are assessed. We set up a general formulation applicable to wide classes of situations, from which some generic trends can be derived. We subsequently illustrate the main ideas on selected low-order models, i.e., models involving a limited number of key variables aimed to capture the principal qualitative aspects of the phenomena of interest.

2 Formulation: Minimal Case

As stated in the Introduction, modeling captures only a part of reality. We may thus expect (Fig. 3) that if a model "lives" in a certain phase space spanned by a set of variables which we will denote as x-variables and involves a certain set of parameters μ, then the full system we want to describe (to which we may refer as "nature") will

(a) "live" in an extended phase space spanned not only by x-type variables but also by additional, y-variables not expressible straightforwardly in terms of the x-variables;
(b) involve parameters μ_N whose values are different from those of the model parameters μ.

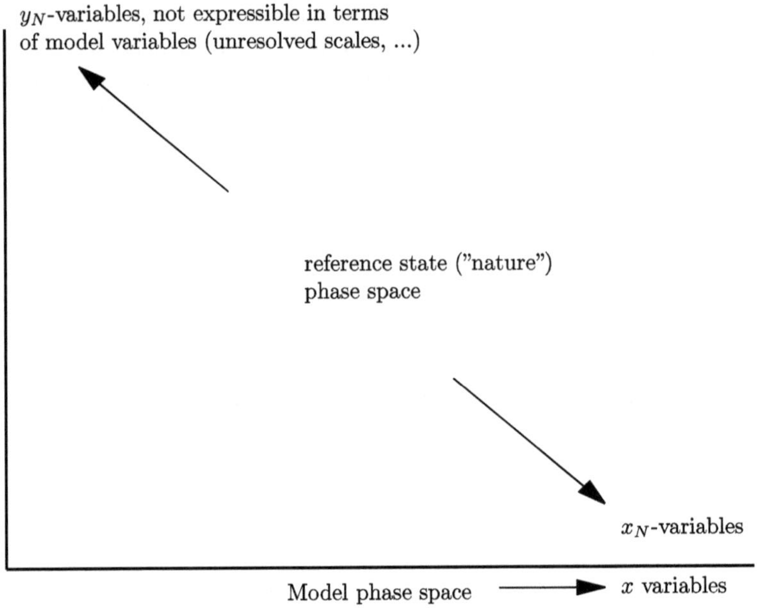

Fig. 3 Schematic representation of phase spaces of the model and of the reference system ("nature")

Let us write the evolution laws of the model variables $\mathbf{x} = (x_1, \ldots, x_n)$ in the form

$$\frac{d\mathbf{x}}{dt} = \mathbf{f}(\mathbf{x}, \mu) \tag{1}$$

where $\mathbf{f} = (f_1, \ldots, f_n)$ are, typically, nonlinear functions of $x_1, \ldots, x_n$. "Nature" will then be described by an amended form of (1), in which some extra terms associated with physical processes not accounted for by the model are incorporated. We first limit ourselves for clarity to the case where model and "nature" span the same phase space (i.e., they involve the same number of variables). We arrive then at evolution equations for nature's variables $\mathbf{x}_N = (x_{N_1}, \ldots, x_{N_n})$ in the form (Nicolis, 2003)

$$\begin{aligned}\frac{d\mathbf{x}_N}{dt} &= \mathbf{f}_N(\mathbf{x}_N, \mu_N) \\ &= \mathbf{f}(\mathbf{x}_N, \mu_N) + \eta\mathbf{G}(\mathbf{x}_N, \mu_N)\end{aligned} \tag{2a}$$

where $\eta\mathbf{G}$ stands for the difference between the form of the full ($\mathbf{f}_N$) and of the model ($\mathbf{f}$) evolution laws.

The task of prediction consists in inferring from Eqs. (1)–(2a) the behavior of the error $\mathbf{u}$,

$$\mathbf{u}(t) = \mathbf{x}(t) - \mathbf{x}_N(t) \tag{2b}$$

Clearly, a full answer to this problem for arbitrary values of the parameters μ and μ_N and for the most general forms of $\mathbf{f}$ and $\mathbf{f}_N$ constitutes an impossible task. We therefore focus on a more restricted version of the problem for which generic results can be obtained, in which:

- The magnitude of initial error $|\mathbf{u}_0|$ is small.
- The values of model and nature parameters are close,

$$\mu = \mu_N + \delta\mu, \qquad |\delta\mu/\mu_N| \ll 1$$

- The evolution laws $\mathbf{f}_N$ and $\mathbf{f}$ are close, in the sense that

$$\eta = \gamma\delta\mu, \qquad \gamma \text{ being finite}$$

Subtracting Eq. (2a) from (1), expanding $\mathbf{f}_N$, μ_N around $\mathbf{f}$, μ and keeping only linear terms on the grounds of the above assumptions one arrives then at a closed equation for the linearized evolution of the error in the form

$$\frac{d\mathbf{u}}{dt} = J \cdot \mathbf{u} + \boldsymbol{\Phi}\delta\mu \tag{3}$$

where J is the Jacobian matrix

$$J = (\partial\mathbf{f}/\partial\mathbf{x})_N \tag{4a}$$

and $\boldsymbol{\Phi}$ is the model error source term,

$$\boldsymbol{\Phi} = (\partial\mathbf{f}/\partial\mu)_N - \gamma\mathbf{G}_N \tag{4b}$$

To solve Eq. (3) we first write the formal solution of Eq. (1) as

$$\mathbf{x}(t) = \mathbf{F}^t(\mathbf{x}_0, \mu) \tag{5}$$

where $\mathbf{x}_0$ is the initial state and $\mathbf{F}^t$ a smooth function such that for finite t and for each given $\mathbf{x}_0$ (and μ) there exists only one $\mathbf{x}(t)$. Decomposing $\mathbf{x}_0$ and $\mathbf{x}(t)$ as in Eq. (2b), expanding $\mathbf{F}^t$ around $\mathbf{x}_0$ and neglecting terms beyond the linear ones in $|\mathbf{u}_0|$ one is led to

$$\begin{aligned}\mathbf{u}(t) &= \frac{\partial\mathbf{F}^t(\mathbf{x}_0, \mu)}{\partial\mathbf{x}_0} \cdot \mathbf{u}_0 \\ &= M(t, \mathbf{x}_0) \cdot \mathbf{u}_0\end{aligned} \tag{6}$$

Here M has the structure of an $n \times n$ matrix and is referred to as the *fundamental matrix* (Nicolis, 2003; Nicolis and Nicolis, 2012). An analysis of this equation in systems giving rise to chaotic dynamics shows that in the limit of long times

$|\mathbf{u}(t)|$ increases exponentially along certain phase space directions, and decreases exponentially or follows a power law in t along the remaining ones. To express the privileged status of this exponential dependence it is natural to consider the logarithm of $|\mathbf{u}(t)|/|\mathbf{u}_0|$ divided by the time t (Eckmann and Ruelle, 1985),

$$\sigma(\mathbf{x}_0) = \frac{1}{t} \ln \frac{|\mathbf{u}(t)|}{|\mathbf{u}_0|} \tag{7}$$

in the double limit where $|\mathbf{u}_0|$ tends to zero and t tends to infinity. A more detailed description consists in considering perturbations along the different phase space directions and evaluating the quantities $\sigma_j(\mathbf{x}_0)$ $i,j = 1 \dots n$ corresponding to them. We refer to these quantities as the *Lyapunov exponents* . They can be ordered in size, $\sigma_1 \geq \cdots \geq \sigma_n$ and for a generic initial perturbation $\sigma(\mathbf{x}_0)$ in (7) coincides with σ_1. It can be shown that the σ_j's are intrinsic properties of the dynamical system at hand, in the sense that they are independent of the way one measures distances in phase space.

One can now check straightforwardly that the formal solution of Eq. (3) is given by

$$\mathbf{u}(t) = M(t,0) \cdot \mathbf{u}_0 + \delta\mu \int_0^t dt' M(t,t') \boldsymbol{\Phi}(t') \tag{8}$$

The first part of this expression features the fundamental matrix M, introduced in Eq. (6), which governs the propagation of the initial error up to the running time t. The second part arises entirely from model deficiencies. It is the sum total of contributions in which deficiencies arising at a time t' between 0 and t (which may be thought of as "effective" errors) are propagated from t' to the running time t by, once again, the fundamental matrix M.

The principal quantity to be evaluated using the above formalism is the mean error $|\mathbf{u}(t)|$ or, more conveniently the mean quadratic error $\mathbf{u}^2(t)$, averaged over both the attractor of the reference system and the initial errors. Different cases may be envisaged as discussed below.

3 Growth of Initial Errors in Absence of Model Error

In absence of model error the right-hand side of Eq. (8) reduces to its first term. Figure 4 depicts the resulting evolution of the mean instantaneous error for the logistic map

$$x_{n+1} = 4\mu x_n(1 - x_n) \tag{9}$$

a prototypical discrete-time dynamical system giving rise to chaotic dynamics, to which large classes of continuous-time dynamical systems can be mapped under an appropriate transformation (Schuster, 1988).

Fig. 4 Time dependence of the mean error for the logistic map at $\mu = 1$ starting from 100,000 initial conditions scattered on the attractor. The mean value of the initial error is $\epsilon \approx 10^{-4}$

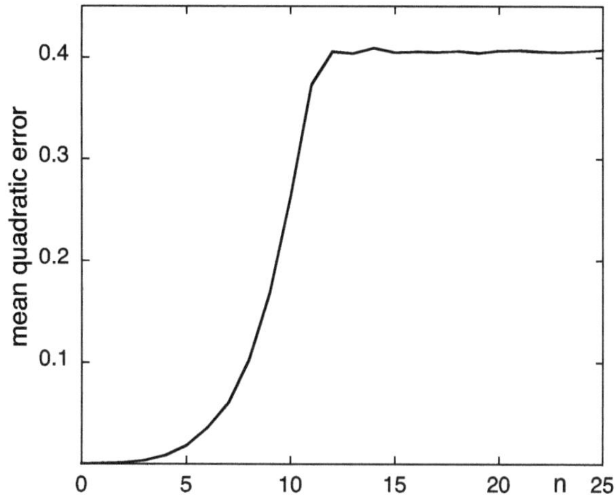

This type of evolution actually turns out to be universal, shared by all systems giving rise to deterministic chaos. Three different stages may be distinguished (Nicolis, 1992): an initial (short time) "induction" stage during which errors grow exponentially while remaining small; an intermediate "explosive" stage displaying an inflexion point situated at a value t^* of t depending logarithmically on the norm ϵ of the initial error, $t^* \approx \ln(1/\varepsilon)$ where errors suddenly attain appreciable values; and a final stage, where the mean error reaches a saturation level of the order of the size of the attractor and remains constant thereafter. The mechanism ensuring this saturation is the reinjection of the trajectories that would first tend to escape owing to the instability of motion, back to a subset of phase space that is part of the attractor.

The first stage reflects local properties driven by the largest Lyapunov exponent $\sigma_{\max}$ [cf. Eq. (7)] and is fully accounted for by the linearized approach [first term in Eq. (8)]. In contrast, the remaining two stages depend on global properties. In particular, in the second stage, the linear dependence of mean quadratic error in time indicates diffusive propagation of the error on the attractor. Finally, in the saturation stage errors scan the structure of the attractor as a whole. Clearly, beyond a time horizon of the order of $\sigma_{\max}^{-1}$ and, a fortiori, beyond the time t^* of the inflexion point in Fig. 4 errors attain a macroscopic level and predictions become random.

In actual fact, when confronted with the problem of predicting the evolution of a concrete system, the observer is led to follow the growth of a (at best) small but *finite* error over a *transient*, usually limited period of time. In this context the quantity of interest is a finite time version of Eq. (7) which now depends on both t and $\mathbf{x}_0$, and the averaged error becomes

$$\langle u_t^2(\epsilon)\rangle = \epsilon^2 \int d\mathbf{x}_0 \; \rho_s(\mathbf{x}_0) \; \exp\{2t\sigma_{\max}(t, \mathbf{x}_0)\} \tag{10}$$

where $\rho_s(\mathbf{x}_0)$ is the invariant probability distribution on the attractor, showing that error growth amounts to studying, for finite times, the average over the attractor of an exponential function $\langle \exp\{2t\sigma_{\max}(t,\mathbf{x}_0)\}\rangle$. To recover for such t's the picture of a Lyapunov exponent-driven exponential amplification of the error one needs to identify (10) with the exponential of the average of $\sigma_{\max}(t,\mathbf{x}_0)$, which is the conventional Lyapunov exponent $\sigma_{\max}$. In a typical attractor this is not legitimate since the expansion rates are position-dependent, in which case the average of a nonlinear function like the exponential in Eq. (10) cannot be reduced to the exponential of an averaged argument. This property reflects the fluctuations to which the local Lyapunov exponents are subjected.

Writing, in analogy with (10),

$$\langle u_t^2(\epsilon)\rangle = \epsilon^2 \exp\{2t\sigma_{\text{eff}}\} \tag{11}$$

one shows that σ_{eff} is t-dependent, starting at $t = 0$ with a value significantly larger than $\sigma_{\max}$. This entails that error growth is neither driven by the Lyapunov exponent nor follows an exponential law but behaves, actually, in a *superexponential* fashion. This property further complicates the problem of prediction of complex systems (Nicolis et al., 1995).

In a multivariate system, in addition to the norm $|\mathbf{u}(t)|$ of the error vector it is important to have information on the directions along which error is likely to grow most rapidly. In general the directions corresponding to the different expansion and contraction rates are not orthogonal to each other. As it turns out this non-orthogonality provides an additional mechanism of superexponential error growth beyond the one due to the variability of the local Lyapunov exponents, related to the fact that certain linear combinations of perturbations or errors may grow more rapidly than perturbations or errors along a particular direction. In a different vein, a multivariate dynamical system possesses several Lyapunov exponents, some of which are negative. For short times all these exponents are expected to take part in the error dynamics. Since a typical attractor associated with a chaotic system is fractal, a small error displacing the system from an initial state on the attractor may well place it outside the attractor. Error dynamics might then involve a transient prior to the re-establishment of the attractor, during which errors would decay in time.

An important class of multivariate systems are spatially extended systems. Here it is often convenient to expand the quantities of interest in series of appropriate basis functions the members of which represent the different spatial scales along which the phenomenon of interest can develop and, in particular, the different scales along which an initial error can occur. The ideas outlined above imply, then, that the predictability properties of a phenomenon depend in general on its spatial scale.

In summary, error growth dynamics is itself subjected to strong variability since not all initial errors grow at the same rate. As a result the different predictability indexes such as $\sigma_{\max}$ or σ_{eff}, the saturation level, and the time t^* to reach the inflexion point provide only a partial picture, since in reality the detailed evolution depends upon the way the different possible error locations and directions are weighted. This variability is illustrated in Fig. 5 depicting the transient evolution of the probability distribution of the error in a model system.

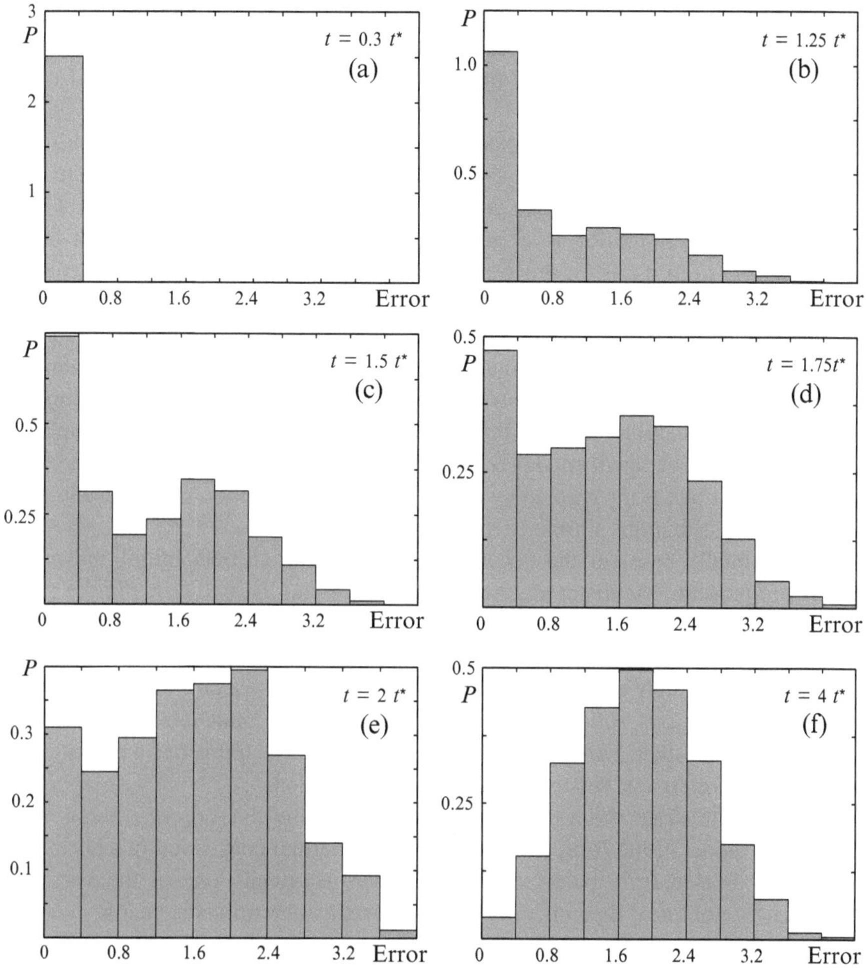

Fig. 5 Snapshots of the probability density of the error in a three-variable model system for six different stages of the evolution, (**a**) to (**f**). Time is normalized to the value corresponding to the inflexion point t^* of the mean error curve ($t^* \approx 20$ time units)

4 Growth of Model Errors

We start with the case where initial errors are absent. The right-hand side of Eq. (8) reduces then to its second term. For smooth functions $\boldsymbol{\Phi}(t)$ as encountered in typical situations the integral over time should behave proportionally to t for short times, entailing that the mean square error should vary as t^2 (Nicolis, 2003). The proportionality factor multiplying this dependence is just the average of $\boldsymbol{\Phi}$ at $t = 0$ over the invariant distribution of the attractor of the reference system (the "nature"), multiplied by the error in the parameter $\delta\mu$. Since the action of the fundamental matrix $M(t,t')$ is not manifested in this time limit, instability of motion and the largest Lyapunov exponent in particular do not play here a crucial role. This is to be contrasted with what happens in the growth of initial errors considered in the previous section.

To analyze the later stages of model error growth one needs to augment Eq. (2a) by nonlinear terms in $\mathbf{u}$ and $\delta\mu$. Simulations on model systems show that error growth follows then, much like initial error, a curve similar to that of Fig. 4. Interestingly, the saturation level attained is finite, practically independent of the smallness of $\delta\mu$, as it reflects the average of typical quadratic distances between any two points of the reference attractor: as time grows the representative points of the reference and approximate systems become increasingly phase shifted, even though the attractors on which they lie may be quite close. We have here a signature of the zero Lyapunov exponent, associated with the borderline between asymptotic stability and instability. Notice that similarly to initial error, the dynamics of individual (non-averaged) model errors are subject to high variability in the form of intermittent bursts interrupted by periods of low error values, giving rise to error probability distributions similar to those of Fig. 5.

Let us finally consider the behavior of the error when both initial and model errors are present (Nicolis et al., 2009). We notice that the two terms in the right-hand side of Eq. (8) vary according to different time dependencies (essentially, exponential and linear) and start with different initial values ($\mathbf{u}_0$ and zero). One may thus legitimately expect that there should typically be a crossover time where the two contributions, reflecting the role of initial and of model error, respectively, will match each other. Furthermore, since neither of these terms has a definite sign an extremum on $|\mathbf{u}^2(t)|$ versus time is not to be excluded.

We now illustrate the validity of these conjectures on Lorenz's three-mode truncation model of the Boussinesq equations of thermal convection in a horizontal fluid layer heated from below (Lorenz, 1963), historically one of the very first examples of nonlinear dynamical systems generating deterministic chaos:

$$\frac{dx}{dt} = \sigma(-x+y)$$

$$\frac{dy}{dt} = rx - y - xz$$

$$\frac{dz}{dt} = xy - bz$$

Here x, y, z are normalized Fourier mode amplitudes of temperature and bulk velocity. Model error is accounted here by slight variations of parameter r, while an initial error of 10^{-4} sampled from a uniform distribution is applied to each of the three variables. Figure 6 summarizes the main results. Since model errors are initially zero, the initial stage of the dynamics of global (initial plus model) error is bound to be dominated by the growth of initial condition errors. For long times both initial and model errors attain a finite level, depending, as mentioned

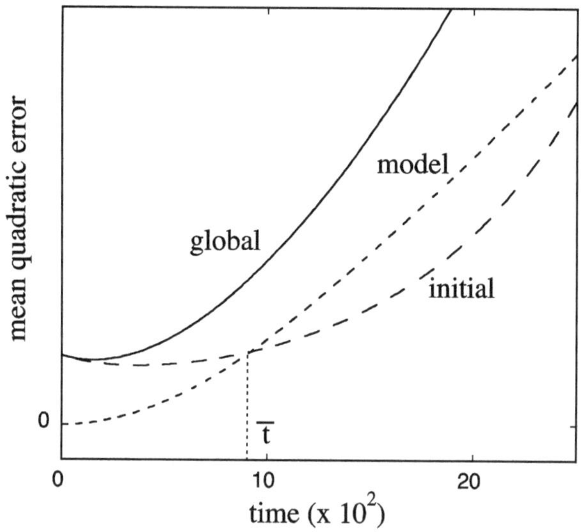

Fig. 6 Time dependence of the mean quadratic error of the Lorenz model (Lorenz, 1963) for parameter values giving rise to deterministic chaos, starting from 10^5 initial conditions scattered on the attractor in the presence of initial condition errors (*dashed line*), model errors (*dotted line*), and both initial condition and model errors (*full line*). $\bar{t}$ denotes the crossover time whereby both sources of errors attain equal magnitudes. The initial error $\epsilon = 10^{-4}$, sampled from a uniform probability distribution, is applied to each of the three variables. Parameter values are $b = 8/3$, $\sigma = 10$, $r = 28$, and the model error in r is 1.5×10^{-4}

earlier, on the characteristics of the attractor of the reference system. Between these two extremes one witnesses a crossover between the growth of the two types of error occurring at some intermediate time $\bar{t}$. Beyond this time the classical butterfly effect is then superseded by an effect reflecting the sensitivity of the evolution laws themselves toward small errors. This constitutes an additional irreducible limitation in the prediction of complex systems.

5 The Role of Unresolved Scales

We next outline an extension of the formulation developed in the preceding sections accounting for the role of the unresolved scales. We suppose that a more comprehensive and satisfactory description of the processes to be modeled, to which we already referred in Sect. 2 as "nature," is afforded by an enlarged form of Eq. (1) displaying two types of variables: a set $\mathbf{x}_N = \{x_{N_1} \dots x_{N_n}\}$ spanning the same phase space Γ_n as the model variables $\mathbf{x}$, as well as an extra set $\mathbf{y}_N = \{y_{N_1} \dots y_{N_m}\}$ spanning an m-dimensional phase space Γ_m. Furthermore, in addition to the parameters μ, which now take (generally unknown) values μ_N, there exists an extra set of parameters ϵ. Nature's phase space velocity is thus a vector in the $n+m$ dimensional space $\Gamma_n \otimes \Gamma_m$ consisting of a part $\mathbf{v}_N \in \Gamma_n$ and a part $\mathbf{w}_N \in \Gamma_m$, such that (Nicolis, 2004)

$$\frac{d\mathbf{x}_N}{dt} = \mathbf{v}_N(\mathbf{x}_N, \mathbf{y}_N, \mu_N, \epsilon) \tag{12a}$$

and

$$\frac{d\mathbf{y}_N}{dt} = \mathbf{w}_N(\mathbf{x}_N, \mathbf{y}_N, \mu_N, \epsilon) \tag{12b}$$

It will be assumed that the solutions of Eqs. (12a) and (12b) satisfy sufficiently strong ergodic properties entailing, in particular, that in the limit of long times they remain confined in certain attracting invariant sets of the corresponding phase spaces.

We first derive from the above equations an expression for the behavior of the mean quadratic error in Γ_n subspace, defined by the Euclidean norm

$$\langle \mathbf{u}^2 \rangle_t = \langle (\mathbf{x} - \mathbf{x}_N)^2 \rangle \tag{13}$$

Here the brackets denote the average over an ensemble of initial conditions, taken to be identical for both $\mathbf{x}$ and $\mathbf{x}_N$ spanning nature's attractor. To this end we write the formal solution of (1) and (12a) as

$$\mathbf{x}(t) = \mathbf{x}(0) + \int_0^t d\tau \mathbf{f}[\mathbf{x}(\tau), \mu]$$
$$\mathbf{x}_N(t) = \mathbf{x}_N(0) + \int_0^t d\tau \mathbf{v}_N[\mathbf{x}_N(\tau), \mathbf{y}_N(\tau), \mu_N, \epsilon] \tag{14}$$

Subtracting these relations and assuming that there is no error arising from uncertainties in the initial conditions, $\mathbf{x}(0) = \mathbf{x}_N(0)$, we get from (14),

$$\langle u^2 \rangle_t = \int_0^t d\tau \int_0^t d\tau' \langle [\mathbf{v}_N(\tau) - \mathbf{f}(\tau)] \rangle \langle [\mathbf{v}_N(\tau') - \mathbf{f}(\tau')] \rangle \tag{15}$$

This equation features the time correlation function of the excess phase velocity of the system compared to nature in the Γ_n subspace. If this quantity is not proportional to a delta function in $\tau - \tau'$—and this will be so as long as one keeps track of the deterministic origin of the quantities concerned—a Taylor series expansion in t can be performed straightforwardly in Eq. (15) in the regime of short times. To the dominant order this leads to

$$\langle u^2 \rangle_t = t^2 \int d\mathbf{x}_N(0) d\mathbf{y}_N(0) \rho_{N_s}[\mathbf{x}_N(0), \mathbf{y}_N(0)]$$
$$\times \{\mathbf{v}_N[\mathbf{x}_N(0), \mathbf{y}_N(0), \mu_N, \epsilon] - \mathbf{f}[\mathbf{x}_N(0), \mu]\}^2 \tag{16}$$

where ρ_{N_s} is the invariant density on nature's attractor. Since the factor multiplying t^2 in (16) is nonvanishing we conclude that the mean quadratic error exhibits a universal t^2 behavior, found already in Sect. 4 in the case where the model and nature span the same phase space. This behavior would be transformed to a proportionality in t if the $\mathbf{v}_N - \mathbf{f}$'s are delta correlated as it is often assumed in dealing with certain types of model error. This point is taken up in more detail in Sect. 7.

We emphasize that our formulation holds beyond the t^2 regime. In particular, Eq. (15) is an exact expression valid for all times. Since the projection of nature's attractor in Γ_n is different from the model attractor, the behavior described in Eq. (15) corresponds actually to a transient evolution prior to reaching the model attractor. On the other hand, $\mathbf{v}_N$ is the Γ_n projection of the tangential velocity on nature's attractor. When both model and nature span the same phase space, $\mathbf{x}$, μ, and $\mathbf{f}$ can be developed around $\mathbf{x}_N$, μ_N, and $\mathbf{v}_N$. Then the difference $\mathbf{f} - \mathbf{v}_N$ gives rise to the model's Jacobian matrix evaluated on nature's attractor acting on the error vector $\mathbf{u}$, plus an inhomogeneous term. This introduces a coupling between error dynamics and the Lyapunov exponents as in Sects. 2 and 3. Now, as seen clearly from Eq. (16), $\langle u^2 \rangle$ depends for short times on the structure of the support of the invariant density ρ_{N_s}. This depends, in turn, explicitly on the attractor dimensionality, but only weakly on the Lyapunov exponents. To identify a more direct connection with Lyapunov exponents one needs to specify the way the $\mathbf{y}_N$ variables are coupled to the $\mathbf{x}_N$s. This problem will be addressed below. Specifically, we analyze a form of Eqs. (12) where $\mathbf{y}_N$ represents a set of variables evolving on a fast time scale compared to $\mathbf{x}_N$. As the model equations are limited solely to the slow variables, the model error is expected to depend on the way the elimination of the fast variables is carried out. Examples of the situation just described include the parameterization of radiative properties of clouds or of the pressure field by suitable diagnostic relations.

To formulate the above idea quantitatively we write Eqs. (12a)–(12b) in the form

$$\frac{d\mathbf{x}_N}{dt} = \mathbf{v}_N(\mathbf{x}_N, \mathbf{y}_N, \mu_N) \tag{17a}$$

and

$$\epsilon \frac{d\mathbf{y}_N}{dt} = \mathbf{w}_N(\mathbf{x}_N, \mathbf{y}_N, \mu_N, \epsilon) \qquad \epsilon << 1 \tag{17b}$$

Under certain conditions of smoothness of $\mathbf{v}_N$ and $\mathbf{w}_N$, a classical theorem of nonlinear analysis due to Tikhonov (Wasow, 1965) asserts that to the dominant order in ϵ one can set the left-hand side of (17b) to zero. Then the right-hand side reduces to an algebraic equation (the "diagnostic" relation):

$$\mathbf{w}_N(\mathbf{x}_N, \mathbf{y}_N, \mu_N, \epsilon) = 0 \tag{18a}$$

which defines the "slow manifold" S in $\Gamma_n \otimes \Gamma_m$. Using this relation one can express $\mathbf{y}_N$ as a function of $\mathbf{x}_N$ provided that certain suitable invertibility conditions are satisfied:

$$\mathbf{y}_N = \mathbf{h}(\mathbf{x}_N, \mu_N) \tag{18b}$$

Substituting into Eq. (17a) one obtains

$$\left(\frac{d\mathbf{x}_N}{dt}\right)_S = \mathbf{v}_N[\mathbf{x}_N, \mathbf{h}(\mathbf{x}_N, \mu_N), \mu_N] \equiv \mathbf{f}(\mathbf{x}_N, \mu_N) \tag{19}$$

where the subscript S stands for the projection on the slow manifold and $\mathbf{f}$ will play the role of the phase space velocity of the model equations:

$$\frac{d\mathbf{x}}{dt} = \mathbf{f}(\mathbf{x}, \mu) \tag{20}$$

We are interested in the error generated in the Γ_n subspace. Setting

$$\mathbf{x} = \mathbf{x}_N + \mathbf{u}, \qquad \mathbf{h}(\mathbf{x}_N) = \mathbf{y}_N + \mathbf{e}, \qquad \mu = \mu_N + \delta\mu$$

we write

$$\frac{d\mathbf{u}}{dt} = -\mathbf{v}_N(\mathbf{x}_N, \mathbf{y}_N, \mu_N) + \mathbf{f}(\mathbf{x}_N + \mathbf{u}, \mu_N + \delta\mu) \tag{21}$$

The next step is to expand $\mathbf{f}$ in Taylor series in $\mathbf{u}$ and $\delta\mu$, $\mathbf{v}_N$ in $\mathbf{y}_N - \mathbf{h}_N(\mathbf{x}_N, \mu_N)$, and keep first order terms. This is legitimate as long as these excess quantities remain small. While this is so for the $\mathbf{u}$ and $\delta\mu$ expansions—at least in the short time regime—it may break down for the $\mathbf{y}_N$ expansion if the inversion of the diagnostic relation (18a) produces turning points in the function $\mathbf{h}$ of Eq. (18b) (Wasow, 1965). The excess of $\mathbf{y}_N$ over $\mathbf{h}$ can then become large during a short time interval whose duration is determined by ϵ, owing to the discontinuous jumps that the phase space trajectory is bound to undergo once it reaches these turning points. It can be shown (Andronov et al., 1966) that this phenomenon is a generic mechanism giving rise to relaxation oscillations. Barring for the time being such deviations we obtain from Eq. (21)

$$\begin{aligned}\frac{d\mathbf{u}}{dt} = \left(\frac{\partial \mathbf{f}}{\partial \mathbf{x}}\right)_N \mathbf{u} + \left(\frac{\partial \mathbf{f}}{\partial \mu}\right)_N \delta\mu \\ - \frac{\partial \mathbf{v}_N}{\partial \mathbf{y}_N}[\mathbf{y}_N - \mathbf{h}(\mathbf{x}_N, \mu_N)]\end{aligned} \tag{22}$$

where the last term in the right-hand side accounts for the deviations of the full trajectories from the slow manifold. If the evolution is started on nature's attractor, the sum of the last two terms in Eq. (22) can be regarded as a well-defined function of time,

$$\boldsymbol{\Phi}(t) = \left\{ \frac{\partial}{\partial \mu} \mathbf{f}[\mathbf{x}_N(t), \mu] \right\}_{\mu_N} \delta\mu \\ - \frac{\partial \mathbf{v}_N[\mathbf{x}_N(t), \mathbf{y}_N(t)]}{\partial \mathbf{y}_N} \{\mathbf{y}_N(t) - \mathbf{h}[\mathbf{x}_N(t), \mu_N]\} \tag{23}$$

Only the first term of this function is accessible on the basis of information available from the model equations. Nevertheless, on inspecting the second term of Eq. (23) one recognizes one of the off-diagonal blocks of nature's Jacobian matrix. This suggests that, for the class of systems considered, there is a non-trivial interference between error dynamics in $\mathbf{x}$ space and the extra variables not retained in the model equations.

Equation (22) can be formally solved by using the fundamental matrix $\mathbf{M}(t, t')$, introduced in Sect. 2, associated here with the model Jacobian matrix $\partial \mathbf{f}/\partial \mathbf{x}$ evaluated on nature's attractor. Remembering that $\mathbf{u}(0) = 0$ we obtain,

$$\langle \mathbf{u}^2 \rangle_t = \int_0^t d\tau' \int_0^t d\tau'' \mathbf{M}(t, \tau')\mathbf{M}(t, \tau'') \cdot \langle \boldsymbol{\Phi}(\tau')\boldsymbol{\Phi}(\tau'') \rangle \tag{24}$$

This expression is reminiscent of the formulation of model error when the model and nature span the same phase space (see Sect. 4). To determine the short time behavior the right-hand side can be expanded in powers of t. If the $\boldsymbol{\Phi}$'s are not delta correlated, this will yield

$$\langle \mathbf{u}^2 \rangle \approx t^2 \langle \boldsymbol{\Phi}^2(0) \rangle \tag{25}$$

where the average is taken over the invariant probability density on nature's attractor. To this order the model's Lyapunov exponents are not intervening, since $\mathbf{M}$ is set equal to unity. They will start playing at order t^3. We also notice that if $\boldsymbol{\Phi}(t)$ is modeled as a Gaussian white noise process, the presence of a delta function in Eq. (24) will remove one t factor and one will obtain instead of Eq. (25) a mean quadratic error proportional to time. From the standpoint of our approach, this limit appears legitimate if the y variables induce a very weakly correlated chaos in nature's evolution. There is evidence that this happens in high-dimensional phase spaces descriptive of spatially extended systems.

As an illustration of the general setting summarized above we consider a generic atmospheric model of a scalar meteorological variable z around a latitude circle (Lorenz and Emmanuel, 1998),

$$\frac{dz_i}{dt} = (z_{i+1} - z_{i-2})z_{i-1} - z_i + F \qquad (i = 1, \ldots, 2k) \tag{26a}$$

where i are equidistant grid points along the circle and F is a forcing parameter. We define coarse variables x_j as averages of z_i's over two adjacent grid points, and fine scale variables as differences of such z_i's. Introducing a new, coarser grid by lumping two successive initial grid points into a single new one and neglecting variability within this grid leads then to the model equations for the x (coarse) variables

$$\frac{dx_i}{dt} = \left(\frac{x_{i+1}}{2} - 1\right) x_i - \frac{x_{i-1}^2}{2} + F \qquad (i = 1, \ldots, k) \qquad (26b)$$

We take $k = 8$ and $F = 10$ or $F = 12$. In both cases the full, as well as the model, systems possess chaotic solutions. Furthermore, the spectrum of Lyapunov exponents of the reference system ("nature") for $F = 10$ is bounded, both from below and from above, by the one for $F = 12$. Finally, the model's largest and smallest Lyapunov exponents are less (in absolute value) that the corresponding exponents of nature by more than a factor three.

Figure 7a depicts the global behavior of the mean quadratic error, Eq. (24), as evaluated numerically using the reference and model equations (26). Following an initial increase, an inflexion point is rapidly reached and subsequently the error saturates after about one time unit at a plateau whose value is higher the larger the parameter F.

A more detailed view of the early stages of the increase is provided in Fig. 7b. The full and dashed lines stand, respectively, for the results of the full numerical evaluation and for those of the evaluation of the initial t^2 regime given by the approximate expression of Eq. (25). We see that the lifetime of this latter regime becomes shorter in the case of $F = 12$, for which the most negative Lyapunov exponent of nature exceeds in absolute value the corresponding exponent obtained in the case $F = 10$. This confirms the existence of connections between the behavior of the model error and the indicators of the underlying dynamics. Finally, the dotted lines represent the contribution combining the t^2 and t^3 terms.

6 Error Dynamics in Extended-Range Forecasts

The possibility to produce reliable atmospheric forecasts not only for short lead times but also for time periods up to a season or a year is of obvious fundamental interest and practical concern. An early attempt at addressing the problem and exploring the physics behind was reported in an important paper by Shukla (1981) in which extended-range forecasts were formulated in terms of time averages. This author raised then the issue of the predictability of time averages and stressed its differences with classical predictability involved in ordinary weather forecasts, noticing that they are determined by different physical processes. He carried out 60-day integrations of a general circulation model starting with different initial conditions and evaluated the variance of the errors among the first 30-day averages,

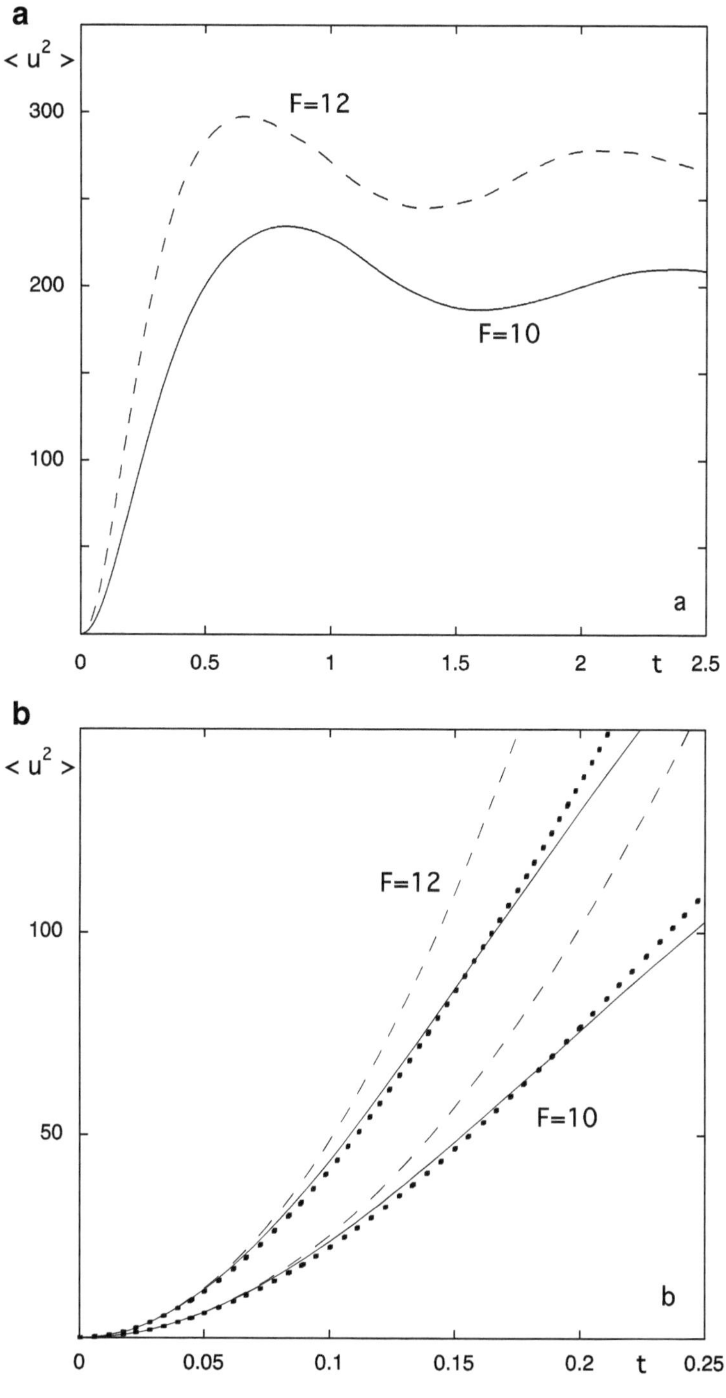

Fig. 7 (**a**) Global behavior of the mean quadratic error between the solutions of Eqs. (26a) and (26b) with $k = 8$, $F = 10$ (*full line*), and $F = 12$ (*dashed line*). (**b**) Short time behavior of $\langle u^2 \rangle$ (*full lines*), the initial t^2 regime (*dashed line*), and the combination of both t^2 and t^3 (*dotted line*). The number of realizations used for the averaging is 10^4

concluding that predictability was sufficiently secured up to such periods. An open question left in these investigations was, what determines the most appropriate averaging periods for time-averaged predictions.

More recently, thanks to advances in data observing and processing and to increasing computer power long-lead forecasts of certain properties such as temperature and precipitation are routinely issued (Molteni et al., 2011). In this context the relative roles of time averaging and ensemble forecasting in the quality of a seasonal forecast have been studied (Smith et al., 2014). Intrinsic limitations arising from widely varying local predictability and from the occurrence of transitions have also been pointed out and analyzed using a low-order model (Palmer, 1993).

In the present section a nonlinear dynamics perspective of time averaging associated with extended-range forecasts is proposed (Nicolis, 2016). We start with the laws governing the evolution of small errors arising from incomplete specification of the initial conditions or from imperfect modeling as outlined in the preceding sections. We derive general expressions for the corresponding errors at the level of the time averages and analyze them in a number of representative situations, with emphasis on the role of the complexity of the underlying dynamics. The results will reveal some unexpected connections between the averaging period and the magnitude of the associated error, suggesting optimal strategies for the choice of this period depending on the intrinsic properties of the system of interest and the values of the parameters involved.

Let $\mathbf{x} = (x_1, \ldots, x_n)$ be a set of variables describing the state of the atmosphere at a certain level of resolution. Their instantaneous values $\mathbf{x}(t)$ will depend on the structure of the evolution laws, on the initial state $\mathbf{x}_0$ and a set of parameters $\mu_1 \ldots, \mu_m$ which, depending on the case, may account for environmental forcings or for the effect of unresolved scales. In a deterministic setting one expects a dependence of the form of Eq. (5), where $\mathbf{F}^t$ is a smooth, one-valued mapping of $\mathbf{x}_0$ on $\mathbf{x}(t)$.

As stressed throughout the preceding sections, the growth of small errors arising from the finite precision in the initial state or from imperfections inherent in modeling is responsible for intrinsic limitations in the predictability of the future states of a system. To formulate this problem for the purposes of the present section at the level of description afforded by Eq. (5) we introduce the perturbed trajectory

$$\mathbf{y}(t) = \mathbf{F}^t(\mathbf{x}_0 + \boldsymbol{\epsilon}, \mu + \delta\mu) \tag{27}$$

assuming for the time being that both the reference and the perturbed systems span the same phase space. The task of "ordinary" prediction consists then in inferring from Eqs. (5) and (27) the behavior of the error $\mathbf{u}(t)$,

$$\mathbf{u}(t) = \mathbf{y}(t) - \mathbf{x}(t) \qquad (\mathbf{u}_0 = \boldsymbol{\epsilon}) \tag{28}$$

under the choice of an appropriate norm as, e.g., the Euclidean norm $|\mathbf{u}(t)|$ of the vector $\mathbf{u}(t)$.

As stated in the beginning of this section, in many instances one is led to inquire about the predictability of the average value of an observable over a time interval T,

$$\overline{\mathbf{x}}_T = \frac{1}{T}\int_0^T dt\mathbf{x}(t) \tag{29}$$

Clearly, the predictability properties of $\overline{\mathbf{x}}_T$ will depend on the behavior of the quantity

$$|\overline{\mathbf{u}}|_T = \frac{1}{T}\left|\int_0^T dt(\mathbf{y}(t) - \mathbf{x}(t))\right| \tag{30}$$

An expression for the short time behavior of $\mathbf{u}(t)$ and $|\overline{\mathbf{u}}|_T$ in the limit of small initial errors and small deviations $\delta\mu$ of parameter values can be obtained by expanding in $\boldsymbol{\epsilon}$ and $\delta\mu$ and retaining the first non-trivial terms:

$$\mathbf{u}(t) = \frac{\partial \mathbf{F}^t(\mathbf{x}_0,\mu)}{\partial \mathbf{x}_0}\cdot\boldsymbol{\epsilon} + \frac{\partial \mathbf{F}^t(\mathbf{x}_0,\mu)}{\partial \mu}\delta\mu \tag{31}$$

where the quantity

$$M(t,\mathbf{x}_0) = \frac{\partial \mathbf{F}^t(\mathbf{x}_0,\mu)}{\partial \mathbf{x}_0}$$

is the fundamental matrix of the system introduced in Sect. 2. As seen in Sects. 2 and 3, in the limit of long times $|\mathbf{u}(t)|$ increases exponentially along the unstable directions of the tangent manifold of the reference trajectory $\mathbf{x}_t$ on $\mathbf{x}_0$ and decreases exponentially or follows a power law along the stable directions. These directions are related, in turn, to the eigenvalues and eigenfunctions of time-ordered products of M over the interval $(0,t)$.

Combining (30) and (31) we obtain the short time behavior of the time average error,

$$|\overline{\mathbf{u}}|_T = \frac{1}{T}\left|\int_0^T dt\{(M(t,\mathbf{x}_0)\cdot\boldsymbol{\epsilon} + \frac{\partial \mathbf{F}^t(\mathbf{x}_0,\mu)}{\partial \mu}\delta\mu\}\right| \tag{32}$$

Expressions (29)–(32) invite the following comments:

A. In Absence of Parametric Model Error ($\delta\mu = 0$)

(i) If the dynamics displays sensitivity to the initial conditions, then in the short time regime [Eq. (32)] $|\overline{\mathbf{u}}|_T$ averaged over an ensemble of trajectories on the attractor is expected to increase. Its growth is, however, slower than the one of the instantaneous error (Buizza and Leutbecher, 2015; Nicolis, 2016).

(ii) In the limit $T \to \infty$ the linearized approach fails. On the other hand, Eq. (30), which remains valid, shows that the long time average error is just the difference of the statistical averages of two trajectories of a dynamical system emanating from two different initial conditions. In an ergodic system where all trajectories span the full phase space available—as expected to be the case for the atmospheric dynamics—this difference is bound to tend to zero,

$$|\overline{\mathbf{u}}|_T \to 0 \qquad \text{as} \qquad T \to \infty \tag{33}$$

Combining with point (i) above, we conclude that there is bound to exist then an averaging time T^* at which $|\overline{\mathbf{u}}|_T$ becomes maximum. This is at first sight counter-intuitive, as one would tend to believe that increasing T's lead to smoother and hence more predictable records.

(iii) For stable dynamical systems $|\overline{\mathbf{u}}|_T$ is expected to decay in a basically monotonic fashion, possibly with a slight modulation around a mean envelope.

B. In Presence of Parametric Model Error ($\delta\mu \neq 0$)

Owing to the contribution in $\delta\mu$ in Eq. (32), $|\overline{\mathbf{u}}|_T$ is expected to grow for short T's. On the other hand, in the limit $T \to \infty$ $|\overline{\mathbf{u}}|_\infty$, as deduced from Eq. (30), will typically settle at a value of $O(|\delta\mu|)$, as the reference and the perturbed trajectories will span two different attractors. The existence or not of a maximum of $|\overline{\mathbf{u}}|_T$ at some value T^* will depend then on the initial slope of the $\delta\mu$-part of Eq. (32), the value of initial error $|\boldsymbol{\epsilon}|$ and the value of $|\delta\mu|$. As we saw in Sect. 4, in a continuous-time system model error due to parametric uncertainty grows linearly with time t in the short time regime (Nicolis, 2003). At the level of the averages this linear t-dependence which we write as αt will give rise to a T-dependence in the form $(1/T)(\alpha T^2/2) = \alpha T/2$. In other words, averaging reduces the initial growth rate of model errors by a factor of 2.

We now illustrate the approach to the predictability of time averages summarized above on two representative low-order models:

– Lorenz's 3-mode truncation of the Boussinesq equations of thermal convection (Lorenz, 1963) introduced already in Sect. 4,

$$\frac{dx}{dt} = \sigma(-x + y)$$
$$\frac{dy}{dt} = rx - y - xz$$
$$\frac{dz}{dt} = xy - bz \qquad \text{(model I)}$$

– Lorenz's low-order atmospheric model (Lorenz, 1984).

$$\frac{dx}{dt} = -ax - y^2 - z^2 + aF$$
$$\frac{dy}{dt} = -y + xy - bxz + G$$
$$\frac{dz}{dt} = bxy + xz - z \qquad \text{(model II)}$$

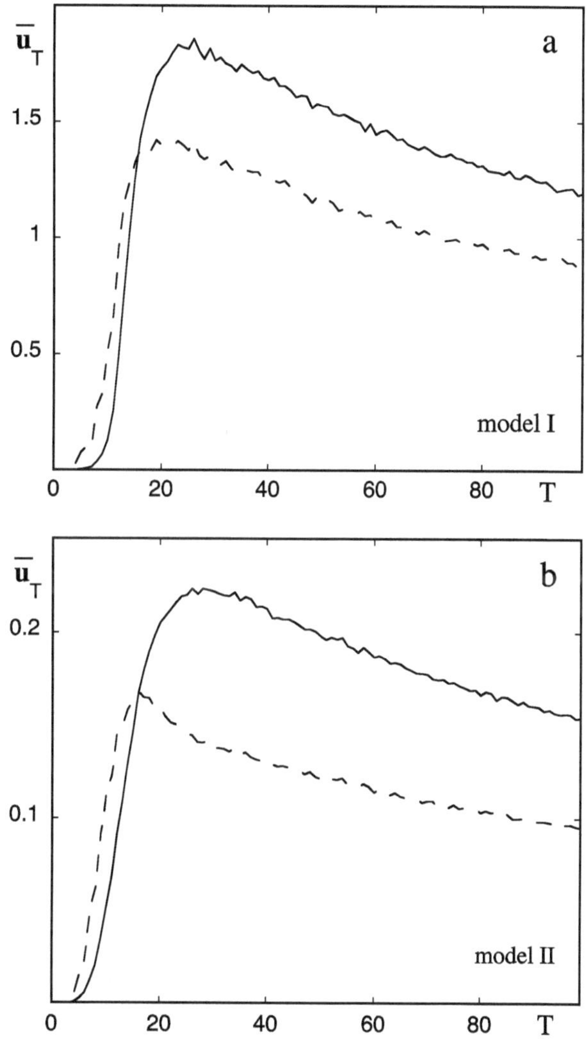

Fig. 8 Dependence of the Euclidean norm of the mean error (*full lines*) and its standard deviation (*dashed lines*) on the averaging period T for the Lorenz models (**a**) I and (**b**) II. An initial error $\epsilon = 0.001$ sampled from a uniform probability distribution varying between -0.5 and 0.5 is applied to the x component. Parameters are $b = 8/3, \sigma = 10, r = 28$ (model I) and $a = 0.25$, $b = 6, F = 16, G = 3$ (model II). The number of ensembles is 10^4

where x, y, z refer now to the average wind velocity and its spatial variability and F, G are forcing parameters.

Figure 8a, b depicts the dependence of the Euclidean norm of the mean error $\bar{\mathbf{u}}_T$ (full lines) and its standard deviation (dashed lines) on the averaging time T for models I(a) and II(b), averaged over 10,000 ensembles, for an initial error in the x component of $\epsilon = 0.001$ multiplied by a random number sampled from a uniform distribution in the interval $[-0.5, +0.5]$. We use standard parameter values (see caption) for which the values of the positive Lyapunov exponent $\sigma_{\max}$ are 0.92 for model I and 0.56 for model II. In both cases a maximum is observed at some value of averaging time, illustrating the ubiquity of this property in unstable systems giving rise to deterministic chaos as argued on very general grounds in this section. The high variability around the mean already encountered in connection with Fig. 5 is also confirmed. Notice that the value of T at maximum is larger for model II compared to model I, owing presumably to the relative magnitudes of the positive Lyapunov exponents as a result of which in model II the takeoff time of the error is longer than in model I.

7 Can Prediction Errors Be Controlled?

Prediction is one of the main objectives of scientific endeavor. As seen in the preceding sections, the possibility to accomplish this task properly may be compromised by the presence of irreducible sources of errors. A natural question to be raised is, then, to what extent a predictive model can be augmented by an appropriate control algorithm allowing one to keep in check, to the extent of the possible, the development of errors that would tend to reach an unacceptable level.

There exists as yet no comprehensive answer to this question. A growing trend is to model error source terms by stochastic forcings of different kinds, to be added to the model equations. This procedure is especially tempting when error source terms arise from the generally poor accounting of processes not directly expressible in terms of the model variables, as is the case of phenomena evolving on short time and space scales that are not resolved by the model at hand.

As pointed out earlier there exist actually two kinds of predictability indices, pertaining to the short time behavior and to the saturation level of the error. Furthermore, in a complex system one should not limit the predictability analysis to the mean error but should address the variability around the mean as well. Typically, a control in the form of a stochastic forcing tends to enhance the variability of the processes involved as compared to that predicted by the model and thus to bring it closer to the natural variability. On the other hand, in the short time regime it enhances mean error and hence deteriorates the model performance. As regards the saturation level, we will see in this section that its action is system dependent. There exists a range of parameters where both mean error and variability can be corrected in the desired sense, but this is only one out of many possibilities. In short, the trends are not only non-universal but are also in many cases conflicting about the desired goals (Nicolis, 2005).

Our stating point are Eqs. (12a)–(12b). The forecasting model is a projected version of these equations whereby the y_N variables are expressed in terms of the x_N ones by the diagnostic relations (18b),

$$\mathbf{y}_N^0 = \mathbf{h}(\mathbf{x}_N, \mu_N) \tag{34}$$

The evolution of the remaining x variables is given by

$$\frac{d\mathbf{x}^0}{dt} = \mathbf{f}(\mathbf{x}^0, \mu) \tag{35a}$$

where the phase space velocity vector **f**

$$\mathbf{f} = \mathbf{v}_N(\mathbf{x}, \mathbf{h}(\mathbf{x}), \mu) \tag{35b}$$

and, in addition, the parameter μ is given a value generally different from μ_N. This implies, in particular, that the model variables are not necessary in a one-to-one correspondence with the $\mathbf{x}_N$ "nature" variables but may be complex combinations of them.

In what follows it will be assumed that $\mathbf{v}_N$ and **f** differ (in norm) by a small quantity:

$$\mathbf{v}_N(\mathbf{x}_N, \mathbf{y}_N, \mu_N) = \mathbf{f}(\mathbf{x}, \mu) + \eta \mathbf{G}(\mathbf{x}_N, \mathbf{y}_N, \mu_N) \qquad |\eta| << 1 \tag{36}$$

where vector **G** is a certain function of the full set of nature's variables. The smallness of η reflects the proximity of μ to μ_N and of the actual $\mathbf{y}_N$, generally a complex function of time and parameters, to the function featured in Eq. (34).

Throughout this section we will be interested in the model and nature's climatologies. These are associated with the ensemble averages of $\mathbf{x}_N$ and **x** type variables over the invariant probability densities p_N and p, respectively, attained by the system in the limit of long times,

$$\mathbf{m}_N \equiv \langle \mathbf{x}_N \rangle = \int d\mathbf{x}_N d\mathbf{y}_N \mathbf{x}_N p_N(\mathbf{x}_N, \mathbf{y}_N) \tag{37}$$

$$\mathbf{m} \equiv \langle \mathbf{x} \rangle = \int d\mathbf{x} \mathbf{x} p(\mathbf{x}) \tag{38a}$$

Angle brackets denote ensemble averages over the realizations of the process. In a similar vein one may define higher order moments associated with the variability around m_N and m:

$$m_{kl} = \int d\mathbf{x} x_k x_l p(\mathbf{x}) \tag{38b}$$

etc. The error associated with a climatological forecast will be given by a suitably normalized difference $\mathbf{m} - \mathbf{m}_N$. The question we address here is whether this error can be minimized by replacing the model evolution laws (35) by the augmented set of equations

$$\frac{d\mathbf{x}}{dt} = \mathbf{f}(\mathbf{x}, \mu) + \mathbf{g}(\mathbf{x}) \cdot \mathbf{R}(t) \tag{39a}$$

where **g** is an $n \times n$ matrix and **R** a vector whose components are uncorrelated white noises

$$< R_i(t) R_j(t') = q_i^2 \delta_{ij}^{\mathrm{kr}} \delta(t - t') \tag{39b}$$

where q_i^2 are the variances of the noise. In Eq. (39a) the term $\mathbf{g}(\mathbf{x}) \cdot \mathbf{R}(t)$ accounts for the variability associated with the extra term (in $\eta\mathbf{G}$) present in Eq. (36) in which as mentioned already the excess of $\mathbf{y}_N$ over $\mathbf{h}$ exhibits a complex time dependence, here assimilated to an uncorrelated Markov noise. It is understood that $\mathbf{g}$ is of order 1 and q_j^2 are of the order η^2 in order to match the strength of the term in η in Eq. (36).

As well known from the theory of stochastic differential equations (Gardiner, 1983) when white noise is coupled multiplicatively to the system's state variables (i.e., when $\mathbf{g}$ in (39) depends non-trivially on $\mathbf{x}$) the evolution can be mapped into a Fokker–Planck type equation. This equation can be written in two different ways according to whether the Itô or the Stratonovich interpretation is adopted. We here choose the first alternative, known to be the most appropriate one when a continuous-time stochastic process is obtained as the limit of an underlying discrete-time process. Under these conditions one has (Gardiner, 1983):

$$\frac{\partial p(\mathbf{x},t)}{\partial t} = -\sum_{i=1}^{n} \frac{\partial}{\partial x_i} f_i(\mathbf{x},\mu)p + \frac{1}{2}\sum_{ijk}^{n} q_k^2 \frac{\partial}{\partial x_i}\frac{\partial}{\partial x_j} g_{ik}(\mathbf{x})g_{jk}(\mathbf{x})p \tag{40}$$

7.1 Moment Equations

Multiplying both sides of Eq. (40) by $\mathbf{x}$ and integrating over $\mathbf{x}$ we obtain the evolution equation of model's climatology $\mathbf{m}$ [Eq. (38a)],

$$\frac{dm_i}{dt} = \langle f_i(\mathbf{x},\mu)\rangle \qquad i = 1,\ldots,n \tag{41a}$$

Notice the absence of an explicit contribution of noise at that level. Equation (41a) is not closed, as the average of a nonlinear function like $\mathbf{f}$ differs from the function itself evaluated at the average value of the state variable. One may, however, expand $\mathbf{x}$ around $\mathbf{m}$,

$$\mathbf{x} = \mathbf{m} + \delta\mathbf{x} \tag{41b}$$

The first order terms of (41a) in $\delta\mathbf{x}$ vanish identically by definition. The first non-trivial term accounting for the variability is therefore provided by the second order terms of the expansion which involve the matrix of second order derivatives of f_i. If $\mathbf{f}$ is a quadratic function of its variables as in typical problems of interest in atmospheric dynamics, expansion of $\mathbf{f}$ around $\mathbf{m}$ will yield the exact result:

$$\frac{dm_i}{dt} = f_i(\{m_j\},\mu) + \frac{1}{2}\sum_{jk} \frac{\partial^2 f_i}{\partial x_j \partial x_k}\mathbf{V}_{jk} \tag{42a}$$

where $\mathbf{V}_{jk}$ is the covariance matrix,

$$\mathbf{V}_{jk} = \langle \delta x_j \delta x_k \rangle \tag{42b}$$

and the second derivative factors are x-independent. In the presence of higher order nonlinearities (42a) would still make sense as the dominant part of an expansion around the mean state, assuming (as is indeed the case at least in the long time limit) a small variability around the climatological mean.

Equation (42a) shows that in the steady-state regime the climatological mean is related to the variability around it. The connection is mediated by the matrix of second derivatives of the successive components, f_i, of the phase space velocity **f**, referred as the Hessian matrix. This quantity is both system and parameter dependent. One may therefore anticipate that noise corrections need not act in the same way for the mean state and for the variability.

To evaluate m_i and $\mathbf{V}_{jk}$ we now multiply both sides of (40) by $x_j x_k$ and integrate over $\mathbf{x}$. We obtain

$$\frac{d}{dt} m_{jk} = \langle x_j f_k + x_k f_j \rangle + \sum q_l^2 \langle g_{jl} g_{kl} \rangle \tag{43}$$

To proceed further we introduce the decomposition (41b) and expand, as before, **f** around **m**. We also perform a similar expansion for g_{jl}. For a quadratic function **f**, a generic case in atmospheric dynamics at least at the level of the primitive equations, **g** will be at most linear in **x**, otherwise the stochastic evolution equations will exhibit divergent behavior. Furthermore we switch from the higher order moments m_{jk} etc. to the associated variances $\mathbf{V}_{jk}$ etc. [Eq. (42b)] by subtracting from Eq. (43) Eq. (42a) applied to k and l, after multiplying them by m_l and m_k respectively and summing over. One obtains in this way after some algebra

$$\begin{aligned}\frac{d}{dt}\mathbf{V}_{jk} &= \sum_l (J_{kl} V_{lj} + V_{kl} J_{lj}^{+}) + \sum_{lpq} q_l^2 \frac{\partial g_{jl}}{\partial x_q} \frac{\partial g_{kl}}{\partial x_q} V_{pq} + \sum_l q_l^2 g_{jl} g_{kl} \\ &+ \frac{1}{2} \sum_{pq} \left(\frac{\partial^2 f_k}{\partial x_p \partial x_q} V_{jpq} + \frac{\partial^2 f_j}{\partial x_p \partial x_q} V_{kpq} \right)\end{aligned} \tag{44a}$$

The first sum in the right-hand side of this equation displays the Jacobian matrix J associated with the deterministic part of the evolution, and its adjoint J^{+}. The last term contains the contributions from the third order variances

$$\mathbf{V}_{jpq} = \langle \delta x_j \delta x_p \delta x_q \rangle \tag{44b}$$

All coefficients multiplying variances are either to be evaluated at the average state **m** (this is the case of J, J^{+} and $g_{jl} g_{kl}$) or are **x**-independent (this is the case of the first **g** derivatives and the second **f** derivatives).

Equations (42a) and (44a) are the first main result of our formulation. In this first form, however, they contain information on both the model's intrinsic variability through the terms obtained by setting $q_l^2 = 0$, and on the extra variability introduced by the noise through the q_l^2 terms. It is this latter variability that is our main concern here. Let θ_k, θ_{kl}, etc., be the corrections to the moments when $q_l^2 \neq 0$. In order to disentangle them from the intrinsic variability we set

$$m_k = m_k^{(0)} + \theta_k$$
$$V_{kl} = V_{kl}^{(0)} + \theta_{kl}$$

etc., and expand Eqs. (42a) and (44a) to the first order in θ, assumed to be of the order of q^2 and of the smallness parameter η in Eq. (36). The steady-state version of the resulting equations reads

$$\sum_j J_{lj}^{(0)}\theta_j = -\frac{1}{2}\sum_{jk}\left(\frac{\partial^2 f_i}{\partial x_j \partial x_k}\right)^{(0)}\theta_{jk} \tag{45}$$

$$\sum_l (J_{kl}^{(0)}\theta_{lj} + \theta_{kl}J_{lj}^{+(0)}) = -\sum_l q_l^2 g_{jl}^{(0)} g_{kl}^{(0)} - \sum_{lpq} q_l^2 \frac{\partial g_{jl}}{\partial x_p}\frac{\partial g_{kl}}{\partial x_p} V_{pq}^{(0)}$$
$$-\sum_{lp}\left\{\left(\frac{\partial J_{kl}}{\partial m_p}\right)^{(0)} V_{jl}^{(0)} + \left(\frac{\partial J_{lj}^{+}}{\partial m_p}\right)^{(0)} V_{kl}^{(0)}\right\}\theta_p \tag{46}$$

The superscript zero in these equations denotes evaluation of the corresponding quantity at mean and variance values corresponding to the model's intrinsic variability.

Equations (45)–(46) are exact first order versions of Eqs. (42a) and (44a), with the sole exception that the noise correction to the third order variance in (44a) has been considered to be a higher order effect. They constitute our second main result. Their interest is that for any given model they allow for an explicit evaluation of θ_j and θ_{kl} in terms of the noise variances $\{q_l^2\}$, as they constitute a system of linear inhomogeneous equations with respect to these variables. The evaluation procedure may be summarized as follows.

(i) Express the noise correction to the climatological mean θ_j from (45) by inverting the Jacobian matrix evaluated at $m^{(0)}$

$$\theta_j = -\frac{1}{2}\{(J^{(0)})^{-1}\}_{ji}\sum_{kl}\left(\frac{\partial^2 f_i}{\partial x_k \partial x_l}\right)^{(0)}\theta_{kl} \tag{47}$$

(ii) Substitute θ_j from this expression into the last term of Eq. (46), combine the resulting terms in $\theta_{kl}$with those of the left-hand side in the general form

$$(\mathbf{A}\theta + \theta\mathbf{A}^{+}) = \mathbf{D} \tag{48}$$

where $\theta = \{\theta_{kl}\}$ and $\mathbf{A}, \mathbf{D}$ are $n \times n$ matrices.

(iii) Invert Eq. (48) to obtain θ_{kl} in the form

$$\theta_{kl} = \sum_m a_{klm} q_m^2 \tag{49}$$

and finally substitute into Eq. (47) to obtain θ_j in terms of the variances of the noise and the indicators of the intrinsic dynamics of the model.

(iv) Evaluate nature's means $\mathbf{m}_N$ from Eq. (37) and the associated variances. Express the (suitably normalized) differences $\mathbf{m} - \mathbf{m}_N$, etc., in terms of the model structure and the properties of the correcting noise using Eqs. (47)–(49). Determine the extent to which these differences can be minimized by suitable tuning of the variance of the noise and/or the way it is coupled to the model variables.

7.2 *Illustration on a Simple Example*

We illustrate the procedure outlined in the preceding section on the simple case where both the model and nature span a one-dimensional phase space. We emulate intrinsic variability by an additive white noise, both for the model and for nature, and introduce in the model equation an extra correcting noise term aiming to counteract the model error. The principal interest of this example is that calculations can be carried out systematically and in all detail. We also notice that in some situations this setting may have some elements of reality as, for instance, in the analysis of the effect of sea surface temperature anomalies in global energy balance (Frankignoul and Hasselmann, 1977). The model equation is written as [cf. Eq. (39a)]

$$\frac{dx}{dt} = f(x) + R_m(t) + \gamma x R_c(t) \tag{50}$$

where the intrinsic, R_m, and correcting, R_c, noises are uncorrelated white noises of variance q_m^2 and q_c^2, respectively. The correcting noise was taken to be multiplicative. As for nature's equation, it contains a first part in the form of Eqs. (12a) and (36), augmented by a white noise $R_N(t)$ of variance q_N^2 emulating natural variability

$$\frac{dx_N}{dt} = f(x_N) + \eta G(x_N) + R_N(t) \tag{51}$$

The Fokker–Planck equations associated with (50) and (51) are

$$\frac{\partial p}{\partial t} = -\frac{\partial}{\partial x} f(x)p + \frac{q_m^2}{2}\frac{\partial^2 p}{\partial x^2} + \frac{q_c^2}{2}\gamma^2 \frac{\partial^2}{\partial x^2} x^2 p \tag{52}$$

$$\frac{\partial p_N}{\partial t} = -\frac{\partial}{\partial x_N}\{f(x_N)p + \eta G(x_N)\}p_N + \frac{q_N^2}{2}\frac{\partial^2 p_N}{\partial x_N^2} \tag{53}$$

Since there is no intrinsic variability in the absence of noise the natural reference value is here the steady-state solution, $\bar{x}$ of the noise-free model equation

$$f(\bar{x}) = 0 \tag{54}$$

The Jacobian $J^{(0)}$ reduces to $f'(\bar{x})$ and the model climatology as deduced from the procedure outlined earlier reads

$$m = \bar{x} + \left(\frac{q_m^2}{4} + \frac{q_c^2\gamma^2}{4}\bar{x}^2\right)\frac{f''(\bar{x})}{(f'(\bar{x}))^2} \tag{55}$$

where the prime denotes derivation with respect to x. Nature's climatology is likewise given by

$$m_N = \bar{x} - \frac{\eta G(\bar{x})}{f'(\bar{x})} + \frac{q_N^2}{4}\frac{f''(\bar{x})}{(f'(\bar{x}))^2} \tag{56}$$

where the corrections to $\bar{x}$ are again limited to the first order in η and q^2. Comparing (55) and (56) we see that the climatological model error can be counteracted provided the variance of the correcting noise satisfies

$$q_c^2 = \frac{1}{\gamma^2\bar{x}^2}\left\{-4\eta\frac{f'(\bar{x})}{f''(\bar{x})}G(\bar{x}) + (q_N^2 - q_m^2)\right\} \tag{57a}$$

This condition is to be fulfilled as long as q_c^2 remains positive. Now the stability of the reference state $\bar{x}$ entails that $f'(\bar{x}) < 0$. The positivity of the right-hand side of Eq. (57a) imposes, therefore, the condition

$$4\eta|f'(\bar{x})|\frac{G(\bar{x})}{f''(\bar{x})} > q_m^2 - q_N^2 \tag{57b}$$

For completeness we also compile the expressions of the model and nature's variabilities around their climatological means under the same conditions as above:

$$V = \langle\delta x_m^2\rangle = -\frac{q_m^2 + q_c^2\gamma^2\bar{x}^2}{2f'(\bar{x})} \tag{58a}$$

$$V_N = \langle\delta x_N^2\rangle = -\frac{q_N^2}{2f'(\bar{x})} \tag{58b}$$

To fix ideas, consider the specific example

$$v_N = kx_N\left(1 - \frac{x_N}{N}\right) + \eta G(x_N); \qquad \eta G(x_N) = -\frac{k}{N}\eta x_N^2 \tag{59a}$$

$$f = kx\left(1 - \frac{x}{N}\right); \qquad \gamma = k \tag{59b}$$

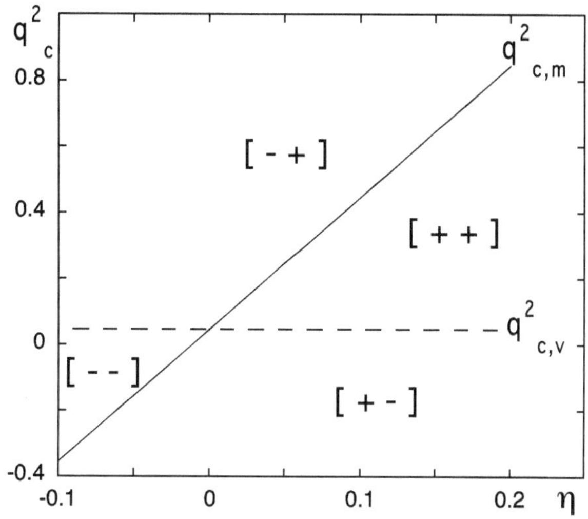

Fig. 9 State diagram of the variances of correcting noise q_c^2, versus the model error amplitude, η, between nature Eq. (59a) and the model Eq. (59b) as obtained from the theoretical expression, Eq. (60). Pluses and minuses in the sectors delimited by the two lines $q_{c,m}^2$ and $q_{c,V}^2$ refer to the signs of the resulting error committed in the mean and variance of the model, respectively. Parameter values are $q_N^2 = 0.05$, $q_m^2 = 0.005$, $N = 2$, and $k = 0.5$

where N is a parameter. We have

$$\bar{x} = N, \qquad m = N - \frac{q_m^2}{2Nk} - \frac{q_c^2}{2}Nk, \qquad V = \frac{q_m^2}{2k} + \frac{q_c^2}{2}N^2 k \tag{59c}$$

$$m_N = N(1-\eta) - \frac{q_N^2}{2Nk}, \qquad V_N = \frac{q_N^2}{2k} \tag{59d}$$

The condition given by Eq. (57a) and the analogous condition for the variances, expressing that the correcting noise counteracts the model error for the climatological means and the variability around them, read, respectively,

$$\begin{aligned} q_{c,m}^2 &= \frac{1}{k^2N^2}(q_N^2 - q_m^2) + \frac{2\eta}{k} \\ q_{c,V}^2 &= \frac{1}{k^2N^2}(q_N^2 - q_m^2) \end{aligned} \tag{60}$$

Figure 9 summarizes the information contained in these relations in the form of a "state diagram" where q_c^2 is plotted against η . The pluses and minuses in the sectors delimited by the resulting two lines refer, from left to right, to m being larger (+) or smaller (−) than m_N and to V being larger (+) or smaller (−) than V_N. As can be seen for any nonvanishing model error $\eta \neq 0$ it is impossible to counteract simultaneously the error in both the means and in the variability, since the two lines $q_{c,m}^2$ and $q_{c,V}^2$ cross only at $\eta = 0$. This is further illustrated in Fig. 10, where the relative errors for the means, $(m - m_N)/m_N$ and the variances, $(V - V_N)/V_N$ are plotted against q_c^2 for a range of values traversing regions [+−], [++] and [−+] of Fig. 9, keeping other parameters fixed. The dashed lines stand for

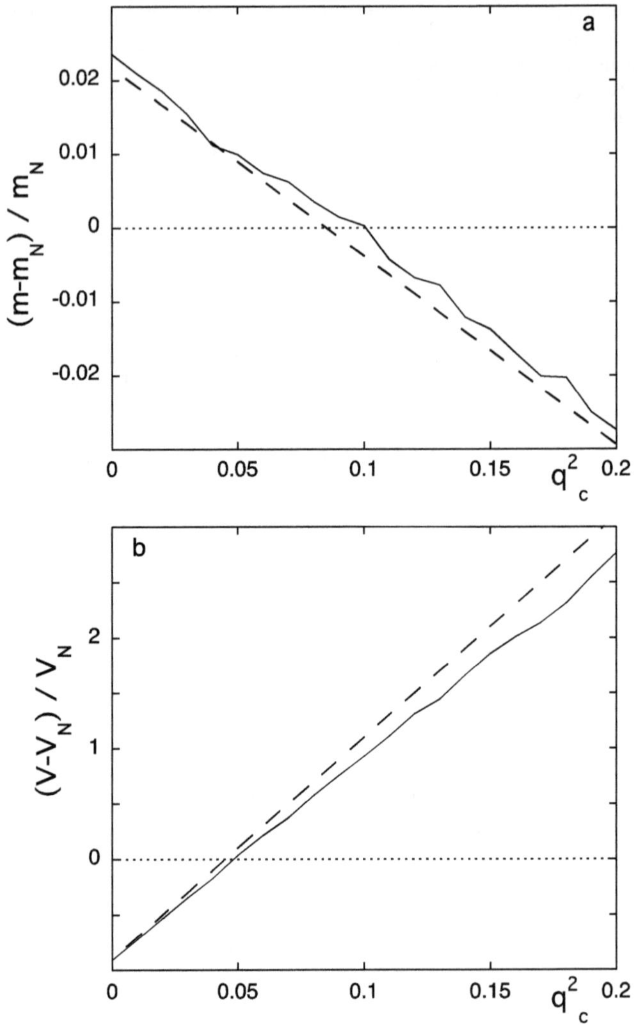

Fig. 10 Relative error in: (**a**) the mean, and (**b**) the variance obtained from the model, Eqs. (59b), versus the amplitude of a correcting multiplicative white noise forcing. In each case the *dashed line* refers to the theoretical expressions of Eqs. [(59c), (59d)], and the *full line* to the numerical stochastic simulation of the reference, Eq. (59a), and model, Eq. (59b), after averaging over 50,000 time units. Parameter values as in Fig. 9 with smallness parameter $\eta = 0.01$

the analytic relations (59c)–(59d), whereas the full lines refer to the results obtained by solving the Langevin equations (50) and (51) corresponding to the particular model considered. The agreement is quite satisfactory.

8 Conclusions

Sensitivity to the initial conditions as symbolized by the butterfly effect and sensitivity to the parameters as well as to the representation of subgrid processes are deeply rooted into the physics of the atmosphere. They impose irreducible limitations to prediction in that the measured or computed values of the different observables are contaminated by errors that tend to grow, from the regime of short lead times to the asymptotic one of the long-term predictions.

In this chapter we outlined a systematic approach to the dynamics of prediction errors. We addressed, successively, short time behavior (Sects. 2–5), extended-range forecasts (Sect. 6) and, finally, asymptotic behavior in connection with climatological properties (Sect. 7). In each case emphasis was placed on the fundamental mechanisms governing error growth and on the possibility to sort out generic features and trends, thanks to systematic analytical evaluations based on the presence of well-defined smallness parameters. This allowed us to disentangle processes governed essentially by linearized laws such as the early stages of initial condition and model errors from those in which nonlinear effects were playing an essential role, such as the existence of a maximum value of the error in extended-range forecasts as a function of the averaging time (Sect. 6), or the compromise in achieving a correct variability versus correct averages (Sect. 7). These approaches were illustrated on prototypical model systems capturing salient features of atmospheric and climate dynamics. On the grounds of their generality they also provide insights on detailed numerical prediction models and on problems of practical concern, such as optimal choices of averaging periods in extended-range forecasts or of the characteristics of the correcting noises in the representation of subscale processes.

A distinctive feature of our formulation is the intertwining of deterministic and probabilistic concepts and tools. The rationale for this is that, owing to the growth of errors, a single deterministic trajectory loses rapidly its operational significance. One is led then to consider ensembles of trajectories and to evaluate averages over the individual realizations. In the context of atmospheric dynamics this procedure is referred to as *ensemble forecasts* (Wilks, 2011). Its merit is to sort out systematic quantitative trends in relation with the indicators of the intrinsic dynamics, that would remain masked in a purely deterministic setting based on individual trajectories. This view was especially crucial in the analysis of Sect. 7, where the problem of error control was mapped into a probabilistic problem governed by the Fokker–Planck equation. The nonlinearity and instability inherent in the deterministic description were substituted here by a description based on a linear evolution law (the Fokker–Planck equation) possessing strong stability properties and leading to a unique steady-state solution in the long-time regime.

Despite spectacular recent improvements in operational forecasting, our understanding of the fundamentals of predictability and error growth remains incomplete. A field in which this limitation is especially apparent is the prediction of extreme values. Contrary to traditional prediction averaging is here to a large extent irrelevant, as the fine structure of both the trajectories and of the probability distributions begins to matter. In a different vein, in many instances of interest some of the parameters present in a problem are subjected to variations in space and time in connection, for instance, with anthropogenic effects or the well-known variability of solar influx. Systematic, dynamics-driven approaches like the one outlined in this chapter are likely to be at the origin of progress in addressing such challenging problems from a new angle.

Acknowledgements This work is supported, in part, by the Science Policy Office of the Belgian Federal Government.

References

Andronov, A.A., A. Vitt, and C. Khaikin. 1966. *Theory of oscillators*. Oxford: Pergamon.

Buizza, R., and M. Leutbecher. 2015. The forecast skill horizon. *Quarterly Journal of the Royal Meteorological Society* 141: 3366–3382.

Eckmann, J.P., and D. Ruelle. 1985. Ergodic theory of chaos and strange attractors. *Reviews of Modern Physics* 57: 617–656.

Frankignoul, C., and K. Hasselmann. 1977. Stochastic climate models II: application to sea-surface temperature anomalies and thermocline circulation. *Tellus* 29: 289–305.

Gardiner, C. 1983. *Handbook of stochastic methods*. Berlin: Springer.

Kalnay, E. 2003. *Atmospheric modeling, data assimilation and predictability*. Cambridge: Cambridge University Press.

Lorenz, E.N. 1963. Deterministic non-periodic flow. *Journal of the Atmospheric Sciences* 20, 130–141.

Lorenz, E.N. 1969. Atmospheric predictability as revealed by naturally occurring analogs. *Journal of the Atmospheric Sciences* 26: 636–646.

Lorenz, E.N. 1984. Irregularity: a fundamental property of the atmosphere. *Tellus* 36: 98–110.

Lorenz, E.N. 1993. *The essence of chaos*. Seattle: University of Washington Press.

Lorenz, E.N., and K.A. Emmanuel. 1998. Optimal sites for supplementary weather observations: simulation with a small model. *Journal of the Atmospheric Sciences* 55: 399–414.

Molteni, F., T. Stockdale, M.A. Balmaseda, G. Balsamo , R. Buizza, L. Ferranti, L. Magnusson, K. Mogensen, T.N. Palmer, and F. Vitart. 2011. The new ECMWF seasonal forecast system (System 4). Technical Memorandum 656, ECMWF, Reading, UK.

Nicolis, C. 1992. Probabilistic aspects of error growth in atmospheric dynamics. *Quarterly Journal of the Royal Meteorological Society* 118: 553–568.

Nicolis, C. 2003. Dynamics of model error: some generic features. *Journal of the Atmospheric Sciences* 60: 2208–2218.

Nicolis, C. 2004. Dynamics of model error: the role of unresolved scales revisited. *Journal of the Atmospheric Sciences* 61: 1740–1759.

Nicolis, C. 2005. Can error source terms in forecasting models be represented as Gaussian Markov noises? *Quarterly Journal of the Royal Meteorological Society* 131: 2151–2170.

Nicolis, C. 2016. Error dynamics in extended-range forecasts. *Quarterly Journal of the Royal Meteorological Society* 142: 1222–1231.

Nicolis, C., and G. Nicolis. 1987. *Irreversible phenomena and dynamical systems analysis in geosciences*. Dordrecht: Reidel.

Nicolis, C., and G. Nicolis. 2009. The butterfly effect. *Scholarpedia* 4(5): 1720.

Nicolis, G., and C. Nicolis. 2012. *Foundations of complex systems*, 2nd ed. Singapore: World Scientific.

Nicolis, C., S. Vannitsem, and J.F. Royer. 1995. Short-range predictability of the atmosphere: mechanisms for superexponential error growth. *Quarterly Journal of the Royal Meteorological Society* 121:705–722.
Nicolis, C., L. Perdigao, and S. Vannitsem. 2009. Dynamics of prediction errors under the combined effect of initial condition and model errors. *Journal of the Atmospheric Sciences* 66: 766–778.
Palmer T.N. 1993. Extended-range atmospheric prediction and the Lorenz model. *Bulletin of the American Meteorological Society* 74: 49–65.
Schubert, S., and Y. Schang. 1996. An objective method for inferring sources of model error. *Monthly Weather Review* 124: 325–340.
Schuster, H.G. 1988. *Deterministic chaos*. Weinheim: VCH Verlag.
Shukla, J. 1981. Dynamical predictability of monthly means. *Journal of the Atmospheric Sciences* 38:2547–2572.
Smith, L.A., H. Du, E. Suckling, F. Niehörster. 2014. Probabilistic skill in ensemble seasonal forecasts. *Quarterly Journal of the Royal Meteorological Society* 141: 1085–1100.
Tribbia, J.J., D.P. Baumhefner. 1988. The reliability of improvements in deterministic short-range forecasts in the presence of initial state and modeling deficiencies. *Monthly Weather Review* 116: 2276–2288.
Tsonis, A. 1992. *Chaos from theory to applications*. London: Springer.
Wasow, W. 1965. *Asymptotic expansions for ordinary differential equations*. New York: Interscience.
Wilks D.S. 2011. *Statistical methods in the atmospheric science*. New York: Academic.

Linked by Dynamics: Wavelet-Based Mutual Information Rate as a Connectivity Measure and Scale-Specific Networks

Milan Paluš

Abstract Experimentally observed networks of interacting dynamical systems are inferred from recorded multivariate time series by evaluating a statistical measure of dependence, usually the cross-correlation coefficient, or mutual information. These measures reflect dependence in static probability distributions, generated by systems' evolution, rather than coherence of systems' dynamics. Moreover, these "static" measures of dependence can be biased due to properties of dynamics underlying the analyzed time series. Consequently, properties of local dynamics can be misinterpreted as properties of connectivity or long-range interactions. We propose the mutual information rate as a measure reflecting coherence or synchronization of dynamics of two systems and not suffering by the bias typical for the "static" measures. We demonstrate that a computationally accessible estimation method, derived for Gaussian processes and adapted by using the wavelet transform, can be effective for nonlinear, nonstationary, and multiscale processes. The discussed problem and the proposed method are illustrated using numerically generated data of coupled dynamical systems as well as gridded reanalysis data of surface air temperature as the source for the construction of climate networks. In particular, scale-specific climate networks are introduced.

Keywords Complex networks • Dynamical systems • Entropy rate • Mutual information rate • Wavelet transform • Climate networks • Scale-specific networks

1 Introduction

"More is different," the simple sentence of the most creative (Soler, 2017) physicist P. W. Anderson (1972) reflects the complex reality in which the behavior of complex systems, consisting of many interacting elements, cannot be explained by a simple

M. Paluš (✉)
Department of Nonlinear Dynamics and Complex Systems, Institute of Computer Science, Czech Academy of Sciences, Pod Vodárenskou věží 2, 182 07 Prague 8, Czech Republic
e-mail: mp@cs.cas.cz

A.A. Tsonis (ed.), *Advances in Nonlinear Geosciences*,
DOI 10.1007/978-3-319-58895-7_21

extrapolation of the laws describing the behavior of a few elements. Studying systems of many interacting elements as complex networks (Albert and Barabási, 2002; Boccaletti et al., 2006; Havlin et al., 2012; Newman et al., 2006) is an intensively developing paradigm in which statistical physics embraced the graph theory. In the graph-theoretical characterization of complex networks, a network is considered as a graph $G = (V, E)$, where V is a set of nodes (or vertices) and E is a set of edges (or links) where each edge represents a connection between two nodes. In the case of weighted graphs a weight $w_{i,j}$ is assigned to each edge $e_{i,j}$, connecting the vertices v_i and v_j, by the weight function $W : E \to \mathbf{R}$. The graph $G = (V, E)$ is characterized by the adjacency matrix A whose elements $a_{i,j} = w_{i,j}$; and $a_{i,i} = 0$ by definition. We will consider undirected graphs, i.e. $a_{i,j} = a_{j,i}$. A special case of graphs are unweighted graphs, also known as binary graphs, since $a_{i,j}$ can attain either the value 1 if $e_{i,j} \in E$, or the value 0 otherwise.

In this study we will consider networks of interacting, possibly stochastic, dynamical systems. In the network paradigm, each system represents a node of the network. Consider that the interactions among the nodes (dynamical systems) are not known. However, we can observe and record evolution of each dynamical system. A series of measurements done on such a system in consecutive instants of time $t = 1, 2, \ldots$ is usually called a time series $\{x(t)\}$. In order to infer a network from a multivariate time series $\{x_i(t)\}$ usually some measure of statistical dependence between components $\{x_i(t)\}$ and $\{x_j(t)\}$, recorded from the nodes v_i and v_j, respectively, is estimated. This measure, or a transformation thereof, is considered as a weight $w_{i,j}$ assigned to the edge $e_{i,j}$. The networks of this type are known as interaction networks (Bialonski et al., 2010) or functional networks. The latter term have been spread from neurophysiology where the statistical association of neural activities in two distinct parts of the brain is called the functional connectivity (Friston, 1994), as opposed to a structural, anatomical connectivity given by an existence of a physical link (Bullmore and Sporns, 2009). Neurophysiology is probably the most active and influential scientific field where the functional networks are constructed and studied; making use of a huge amount of multivariate data recording various modes of brain activity (Achard et al., 2006; Bullmore and Sporns, 2009; Reijneveld et al., 2007). The interaction networks, however, are studied also in different areas such as climatology (Donges et al., 2009a,b; Steinhaeuser et al., 2012; Tsonis and Roebber, 2004; Tsonis et al., 2006; Yamasaki et al., 2008, 2009) or economy and finance (Onnela et al., 2004; Schweitzer et al., 2009). Since the existence of a link in an interaction network is inferred from an estimate of a statistical dependence measure, the strength and even the existence of a link bear some level of uncertainty. Kramer et al. (2009) propose a systematic statistical procedure for the inference of functional connectivity networks from multivariate time series yielding as the output both the inferred network and a quantification of uncertainty of the number of edges. Paluš et al. (2011) present differences in the topology of interaction networks with edges derived either from the largest absolute correlations or from the statistically most significant absolute correlations. Bialonski et al. (2010) demonstrate that a spatial sampling can lead to an occurrence of spurious structures in interaction networks constructed from time series sampled in spatially extended systems and propose tailored random networks as a suitable null hypothesis to be tested (Bialonski et al., 2011). Hlinka et al. (2012) observed that a spurious small-world topology emerged in interaction networks constructed using correlations of time series generated by randomly connected dynamical systems. While Bialonski et al. (2010) attribute spurious topologies to sampling problems and finite-precision, finite-length time series, Hlinka et al. (2012) see the problem in partial transitivity—an inherent property of the correlation coefficient. Also Zalesky et al. (2012) observed that the networks in which connectivity was measured using the correlation coefficient were inherently more clustered than random networks, while partial correlation networks were inherently less clustered than random networks. Therefore, in a similar line with Bialonski et al. (2011), also Zalesky et al. (2012) propose to use a sort of null networks in order to explicitly normalize for the inherent topological structure found in the correlation networks.

In this study we will focus on the dynamics underlying time series used for the construction of interaction networks. We will demonstrate how "dynamical memory" influences the bias in estimations of "static" dependence measures such as the absolute correlation coefficient or the mutual information. We will propose the mutual information rate as a measure reflecting dependence of dynamics of two systems or processes. We will introduce a computationally accessible algorithm that can be effective for quantification of the coherence or synchronization of nonlinear, nonstationary, and multiscale processes and thus can be used for the construction of interaction networks from experimental time series recorded in natural complex systems.

2 Dependence

Consider two discrete random variables X and Y with sets of values Ξ and Υ, respectively. The probability distribution function (PDF) $p_X(x)$ for the variable X, for simplicity denoted as $p(x)$, is $p(x) = \Pr\{X = x\}$, $x \in \Xi$. The probability distribution function $p(y)$ for the variable Y is defined in the full analogy; and the joint PDF $p(x, y)$ is $\Pr\{(X, Y) = (x, y)\}$, $x \in \Xi$, $y \in \Upsilon$. Uncertainty in a random variable, say X, is characterized by its entropy

$$H(X) = -\sum_{x\in\Xi} p(x) \log p(x). \tag{1}$$

The joint entropy $H(X, Y)$ of X and Y is

$$H(X, Y) = -\sum_{x\in\Xi}\sum_{y\in\Upsilon} p(x, y) \log p(x, y). \tag{2}$$

The two variables X and Y are independent if and only if $p(x, y) = p(x)p(y)$, i.e.

$$\log \frac{p(x,y)}{p(x)p(y)} = 0.$$

The average digression from independence, i.e., the averaged value of $\log \frac{p(x,y)}{p(x)p(y)}$ is known as mutual information

$$I(X;Y) = \sum_{x\in\Xi} \sum_{y\in\Upsilon} p(x,y) \log \frac{p(x,y)}{p(x)p(y)}. \tag{3}$$

The mutual information can be expressed using the entropies (1), (2) as

$$I(X;Y) = H(X) + H(Y) - H(X,Y). \tag{4}$$

Thus the mutual information $I(X;Y)$ quantifies the decrease of uncertainty in $H(X,Y)$ due to the dependence between X and Y, i.e., it measures the average amount of common information, contained in the variables X and Y. The mutual information is a measure of general statistical dependence for which the following statements hold:

- $I(X;Y) \geq 0$,
- $I(X;Y) = 0$ iff X and Y are independent.

In practice, however, the PDF's are not known and we only have a set of measurements $\{x_1, x_2, \ldots, x_N\}$ for the variable X and $\{y_1, y_2, \ldots, y_N\}$ for the variable Y. Estimation of the entropies (1), (2) and the mutual information (3) can be done using some of suitable estimators, for review see Hlaváčková-Schindler et al. (2007).

A common measure of linear dependence is the (Pearson's) correlation coefficient. First, we compute the mean of all measurements $\{x_1, x_2, \ldots, x_N\}$ as

$$\bar{x} = \frac{1}{N} \sum_{i=1}^{N} x_i$$

and the variance

$$\sigma^2 = \frac{1}{N-1} \sum_{i=1}^{N} (x_i - \bar{x})^2$$

and transform the measurements into a data with a zero mean and a unit variance

$$\widetilde{x_i} = \frac{x_i - \bar{x}}{\sigma}. \tag{5}$$

After the same procedure with the measurements of the variable Y, the correlation coefficient of X and Y is

$$C(X,Y) = \frac{1}{N}\sum_{i=1}^{N} \widetilde{x_i}\widetilde{y_i}. \tag{6}$$

Without loss of generality, in the following we will suppose that considered data or time series have (or have been transformed in order to have) a zero mean and a unit variance.

Suppose that the variables X and Y have a bivariate Gaussian distribution. Then their mutual information $I(X;Y)$ can be expressed using their correlation coefficient $C(X,Y)$ (see, e.g. Paluš et al. (1993) and references therein)

$$I(X;Y) = -\frac{1}{2}\log\left(1 - C^2(X,Y)\right). \tag{7}$$

The correlation coefficient (6) and the mutual information (3) are the measures of dependence which reflect the digression of the "static" bivariate distribution $p(x,y)$ from the product $p(x)p(y)$. We use the term "static" in order to stress that both the correlation coefficient (6) and the bivariate PDF $p(x,y)$ which determines the mutual information (3) are given by the set of pairs $\{(x_1,y_1),(x_2,y_2),\ldots,(x_N,y_N)\}$ irrespectively of the order of the pairs. Any permutation of the pairs (x_i,y_i) yields the same result.

3 Dynamics

Let us consider n discrete random variables $X_1,\ldots,X_n$ with values $(x_1,\ldots,x_n) \in \Xi_1 \times \cdots \times \Xi_n$. The PDF for an individual X_i is $p(x_i) = \Pr\{X_i = x_i\}$, $x_i \in \Xi_i$, the joint PDF for the n variables $X_1,\ldots,X_n$ is $p(x_1,\ldots,x_n) = \Pr\{(X_1,\ldots,X_n) = (x_1,\ldots,x_n)\}$. The joint entropy of the n variables $X_1,\ldots, X_n$ with the joint PDF $p(x_1,\ldots,x_n)$ is

$$H(X_1,\ldots,X_n) \\ = -\sum_{x_1\in\Xi_1}\cdots\sum_{x_n\in\Xi_n} p(x_1,\ldots,x_n)\log p(x_1,\ldots,x_n). \tag{8}$$

A stochastic process $\{X_i\}$ is an indexed sequence of random variables $X_1,\ldots,X_n$, characterized by the joint PDF $p(x_1,\ldots,x_n)$. Uncertainty in a variable X_i is characterized by its entropy $H(X_i)$. The rate at which a stochastic process "produces" uncertainty is measured by its entropy rate

$$h = \lim_{n\to\infty}\frac{1}{n}H(X_1,\ldots,X_n). \tag{9}$$

In practice we will deal with a time series $\{x(t)\}$, $t = 1, 2, \ldots, N$. While considering measurements $\{x_1, x_2, \ldots, x_N\}$ of a random variable X, its values x_i are typically considered mutually independent, i.e., obtained by independent, random draws from a PDF $p(x)$. On the other hand, a time series $\{x(t)\}$ reflects a temporal evolution of a process or a system, and typically the values $x(t)$ and $x(t+\tau)$, where τ is a time lag, are not independent. The level of dependence between $x(t)$ and $x(t+\tau)$ reflects a "dynamical memory" of the temporal evolution of an underlying process or system. The decrease of the dependence between $x(t)$ and $x(t+\tau)$, with increasing τ, i.e., the rate at which a process "forgets" its history depends on complexity of the temporal evolution of a process or a system and we will refer to this complexity as "temporal dynamics," or shortly as "dynamics."

Since a time series $\{x(t)\}$ reflects the dynamics of an underlying process or system, a stochastic process $\{X_i\}$ characterized by the joint PDF $p(x_1, x_2 \ldots, x_n)$ which typically differs from the product $p(x_1)p(x_2)\ldots p(x_n)$, is an appropriate theoretical concept for the study of time series. Thus a time series is considered as a realization of a stochastic process $\{X_i\}$ and should not be equated with a set of measurements of a single variable X with a PDF $p(x)$. The entropy rate (9) is a useful characterization of the dynamics of a system or a process underlying the time series $\{x(t)\}$. In information theory the entropy rate (9) is considered as a measure of production of information of an information source (Cover and Thomas, 1991).

Alternatively, a time series $\{x(t)\}$ can be considered as a projection of a trajectory of a dynamical system, evolving in a measurable state space. Kolmogorov, who introduced the theoretical concept of classification of dynamical systems by information rates, was inspired by information theory and generalized the notion of the entropy of an information source. The Kolmogorov–Sinai entropy (KSE thereafter) or metric entropy (Petersen, 1989) is a topological invariant, suitable for the classification of dynamical systems or their states, and is related to the sum of the system's positive Lyapunov exponents (Pesin, 1977). The concept of entropy rates is common to theories based on philosophically opposite assumptions (randomness vs. determinism) and is ideally applicable for the characterization of complex processes, where possibly deterministic rules are always accompanied by random influences.

As a potentially useful quantitative characterization of the dynamics, the entropy rate has become a target of many numerical algorithms using experimental time series as their input. Particularly intensive development, focused on the estimation of the metric entropy, has started with the advent of the methods for the reconstruction of chaotic dynamics in the 1980s. Grassberger and Procaccia (1983a) used the concept of Rényi entropy (Cover and Thomas, 1991) to redefine the KSE in terms of the Rényi entropy of order two and proposed an estimator of the metric entropy K_2 using their celebrated correlation integral (Grassberger and Procaccia, 1983b). The method has been extended into numerous version, e.g. by Cohen and Procaccia (1985). Schouten et al. (1994) treated the correlation integral as a probability distribution and derived a maximum-likelihood estimator of the KSE. Pawelzik and Schuster (1987) consider the full spectrum of generalized metric entropies K_q. Fraser (1989) pointed to an interesting relation between an n-dimensional version of the mutual information and the KSE of a dynamical system underlying studied time series. Paluš (1997a) studied this relation in detail and confirmed its validity by comparing the KSE estimates with the values of the positive Lyapunov exponents of the studied chaotic systems. Reliable KSE estimates, however, require large amounts of data. Therefore Paluš (1996a) proposed "coarse-grained entropy rates" which relate the KSE to the rate of the decrease of a finite-precision mutual information of a time series and its time-lagged twins. Also bounded by a finite precision and a limited amount of real data, Pincus (1991) introduced an approximate entropy based on a difference of the correlation integrals.

The entropy rate reflects how quickly a system "forgets" its history. In the case of chaotic dynamical systems the metric entropy is related to a time interval which a dynamical system takes to return to a close vicinity of some of its previous states. Baptista et al. (2010) propose two formulas to estimate the KSE and its lower bound from the recurrence times of chaotic systems. The recurrence plots (Marwan et al., 2007) give a number of useful dynamical quantities including the KSE.

A time series of measurements of a finite precision can be conveniently converted into a sequence of symbols from a finite set of values. Bandt and Pompe (2002) introduced the concept of permutation entropy for symbolic sequences and demonstrate its relations to the KSE. Lesne et al. (2009) studied entropy rate estimators for short symbolic sequences based on block entropies and Lempel–Ziv complexity (Ziv and Lempel, 1978). Kennel et al. (2005) developed an algorithm for estimating the entropy rate of Markov models using weighted context trees. The entropy rates can also be computed using the causal state machine-based estimator (Crutchfield and Young, 1989; Haslinger et al., 2010; Shalizi et al., 2001).

Let us return from symbolic sequences to continuous stochastic processes. Let a stochastic process $\{X_i\}$ is a zero mean, stationary, Gaussian process with power spectral density $\Phi(\omega)$, where ω is a normalized frequency. Then its entropy rate h_G, apart from a constant term, is (Paluš, 1997b; Pinsker, 1964)

$$h_G = \frac{1}{2\pi} \int_0^{2\pi} \log \Phi(\omega) d\omega. \tag{10}$$

4 Dynamics and Connectivity

In order to understand the notion of temporal dynamics of a process and its characterization using the entropy rate, let us consider the autoregressive process (ARP)

$$x(t) = c \sum_{k=1}^{10} a_k x(t-k) + \sigma e(t), \tag{11}$$

where $a_{k=1,\ldots,10} = 0, 0, 0, 0, 0, 0.19, 0.2, 0.2, 0.2, 0.2$, $\sigma = 0.01$ and $e(t)$ is a Gaussian noise with a zero mean and a unit variance. The parameter c modulates

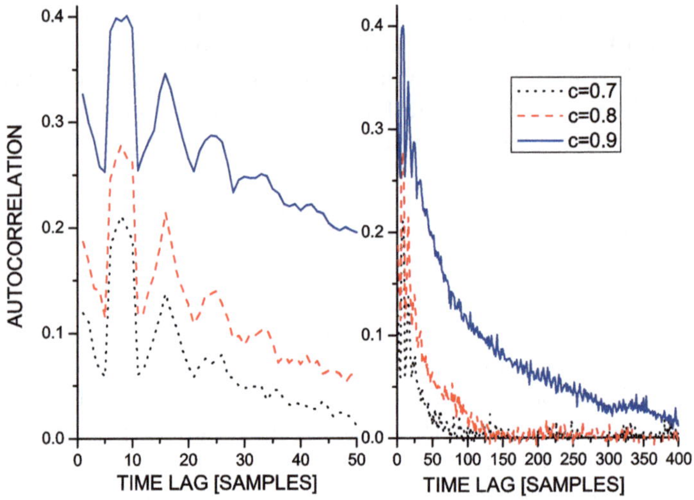

Fig. 1 Autocorrelation function for the autoregressive process (11) for different values of the coefficient c: $c = 0.7$ (the *dotted black line*), $c = 0.8$ (the *dashed red line*), and $c = 0.9$ (the *solid blue line*). Time lags 1–400 samples (*right panel*), the detail for time lags 1–50 samples (*left panel*)

the proportion of the deterministic part of the process which is a function of the history of the process, to the noise part of the process. The greater the coefficient c, the stronger the memory, i.e., the dependence between $x(t)$ and $x(t + \tau)$. This effect is demonstrated in Fig. 1, where the autocorrelation function $C(x(t), x(t + \tau))$ as a function of the time lag τ is plotted for different values of the coefficient c. For $c = 0.7$ (the dotted black line) the autocorrelation function (ACF) has the lowest values and vanishes (fluctuates with values close to zero) for time lags around 100 samples; for $c = 0.8$ (the dashed red line) the ACF has higher values and vanishes about the time lag equal to 150 samples, while for $c = 0.9$ (the solid blue line) the ACF has the largest values and requires more than 400 samples of the time lag to vanish. The ACF reflects the fact that increasing c the dynamical memory of the process (11) is stronger and longer lasting.

How these differences in the dynamical memory or in the dynamics are reflected in the entropy rate? We generate realizations of the ARP (11) with different c and compute the entropy rates h_G according to Eq. (10). Figure 2a presents the entropy rate h_G for 100 realizations of the ARP (11) with c increasing from 0.5 to 0.9. The entropy rate of such ARP's monotonically decreases with increasing c. A higher entropy rate means that the process generates uncertainty at a higher rate so that it forgets its history more quickly. Predictability of a process with a higher entropy rate is worse and possible for a shorter prediction horizon than predictability of a process with a lower entropy rate.

Time series $\{x_i(t)\}$ recording temporal evolution of different systems or subsystems of a complex system might reflect different dynamics yielding different

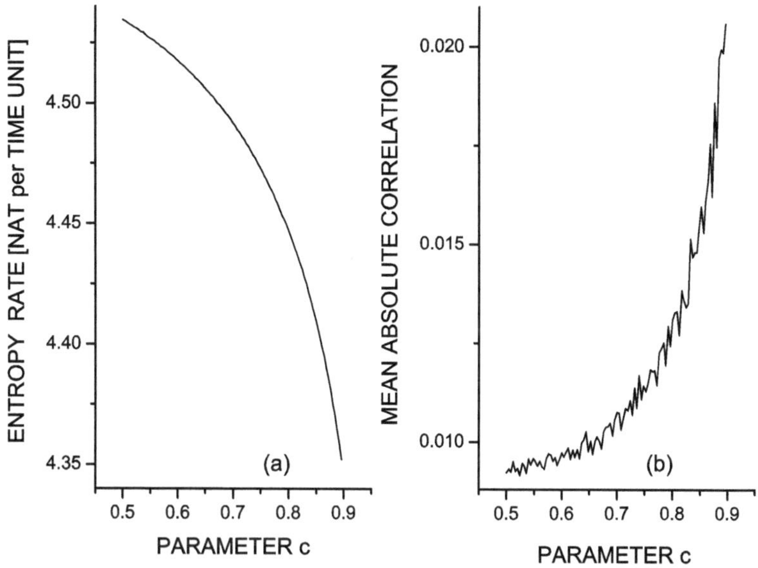

Fig. 2 (**a**) Entropy rate h_G for the autoregressive process (11) as a function of the parameter c. (**b**) Dependence of the mean absolute cross-correlation between independent realizations of the autoregressive process (11) on the parameter c

entropy rates. As we have noted in the Introduction, the connectivity in complex networks constructed from multivariate time series, i.e., the existence and the strength of links between nodes are inferred using dependence measures such as the mutual information (3) and the correlation (6). The absolute value of the latter is typically used, while the mutual information is always non-negative. Applying the definitions (3) and (6) to time series $\{x(t)\}$ and $\{y(t)\}$, they are treated as sets $\{x_i\}$ and $\{y_i\}$ of measurements of random variables X and Y. The computed $C(X, Y)$ or $I(X; Y)$ do not reflect the dynamics of $\{x(t)\}$ and $\{y(t)\}$. Indeed, the pairs $(x(t), y(t))$ would yield the same values of $C(X, Y)$ or $I(X; Y)$ independently of their temporal order. The computed values of $C(X, Y)$ or $I(X; Y)$ are, however, only estimates of the true dependence between processes generating the datasets $\{x(t)\}$ and $\{y(t)\}$. The estimates have some bias, giving a mean digression from the true value, and a variance giving the range of fluctuations of the estimates around their mean value.

Using the above defined ARP (11) we can study the behavior of the correlation estimates for time series with different dynamics. In particular, we can generate realizations of the ARP (11) with different c's and thus with different entropy rates. Now, let us study the distribution of the cross-correlations between *independent* realizations of the process (11) for different values of the parameter c. For each c we generate 8192 process realizations, each realization consisting of 16,384 samples. Figure 3 presents histograms of cross-correlations between independent realizations of ARP (11) for two different c's. The mean value is always correctly equal to zero; however, the variance increases with increasing c, i.e., with decreasing the entropy

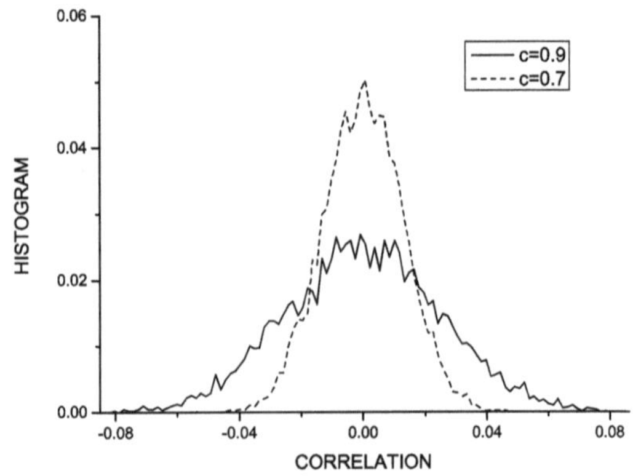

Fig. 3 Histograms of cross-correlations between independent realizations of the autoregressive process (11) for two different values of the parameter c

rate. As a consequence, when considering the *absolute* correlations, or a non-negative dependence measure such as the mutual information, its mean value has an increasing upward bias with the decreasing entropy rate. This effect is illustrated in Fig. 2b. In this example the bias in the absolute correlations reaches relatively small values 0.01–0.02. These values, however, are obtained for time series of 16,384 samples. In Sect. 8 we will show that in real time series of 512 samples the bias can reach such values as 0.4. For even shorter and/or more regular (lower entropy rate) time series the bias can be even higher (Paluš, 2007).

5 Mutual Information Rate

Instead of treating time series $\{x(t)\}$ and $\{y(t)\}$ as sets $\{x_i\}$ and $\{y_i\}$ of measurements of random variables X and Y, now let us consider the time series $\{x(t)\}$ and $\{y(t)\}$ as realizations of stochastic processes $\{X_i\}$ and $\{Y_i\}$, characterized by PDF's $p(x_1, \ldots, x_n)$ and $p(y_1, \ldots, y_n)$, respectively. In the analogy of generalization of the entropy (1) to the entropy rate (9) in order to characterize dynamics of a process, now we generalize the mutual information (3) to the mutual information rate (MIR) (Cover and Thomas, 1991) as

$$i(X_i; Y_i) = \lim_{n\to\infty} \frac{1}{n} I(X_1, \ldots, X_n; Y_1, \ldots, Y_n). \tag{12}$$

While the mutual information $I(X; Y)$ evaluates the difference between the bivariate PDF $p(x, y)$ and the product of the univariate PDF's $p(x)p(y)$, the MIR (12) is the limit value of the mutual information $I(X_1, \ldots, X_n; Y_1, \ldots, Y_n)$ evaluating the difference between the $2n$-variate PDF $p(x_1, \ldots, x_n, y_1, \ldots, y_n)$ and the product of the two n-variate PDF's $p(x_1, \ldots, x_n)p(y_1, \ldots, y_n)$. The MIR quantifies the dependence between the sequences of states $X_1, \ldots, X_n$ of the process $\{X_i\}$ and states $Y_1, \ldots, Y_n$ of the process $\{Y_i\}$. In the case of dynamical systems the MIR reflects coherent dynamics or a common evolution of two systems whose trajectories are projected onto the time series $\{x(t)\}$ and $\{y(t)\}$.

For pairs of dynamical systems that are either mixing, or exhibit fast decay of correlations, or have sensitivity to initial conditions, Baptista et al. (2012) have proposed a way how to calculate MIR and its upper and lower bounds in terms of Lyapunov exponents, expansion rates, and capacity dimension. In general, estimators of MIR are well elaborated for symbolic dynamics, extending the estimators of the entropy rates. Shlens et al. (2007) further develop the estimator of Kennel et al. (2005) and applied it in order to estimate the information transfer between a stimulus and neural spike trains. Blanc et al. (2011) extended the entropy rate estimator for symbolic sequences (Lesne et al., 2009) and compared several estimators adapted for the estimation of the MIR between coupled dynamical systems in a symbolic representation, including the Lempel–Ziv (Ziv and Lempel, 1978) and the causal state machine-based estimator (Crutchfield and Young, 1989; Haslinger et al., 2010; Shalizi et al., 2001).

Considering continuous stochastic processes, for zero mean, Gaussian stochastic processes $\{X_i\}$, $\{Y_i\}$, characterized by power spectral densities (PSD) $\Phi_X(\omega)$, $\Phi_Y(\omega)$ and cross-PSD $\Phi_{X,Y}(\omega)$, the MIR can be expressed (see Pinsker 1964) as

$$i_G(X_i; Y_i) = -\frac{1}{4\pi}\int_0^{2\pi} \log(1 - |\Gamma_{X,Y}(\omega)|^2)d\omega, \tag{13}$$

using the magnitude-squared coherence

$$|\Gamma_{X,Y}(\omega)|^2 = \frac{|\Phi_{X,Y}(\omega)|^2}{\Phi_X(\omega)\Phi_Y(\omega)}. \tag{14}$$

Now we can return to the ARP (11) and use its independent realizations generated with different values of the parameter c and thus characterized by different entropy rates, in order to study the bias of dependence measures in relation to dynamics (entropy rate). In Fig. 4a we study again the absolute cross-correlations of independent realizations the ARP (11) as a function of the parameter c. The mean values are the same as in Fig. 2b; however, here we illustrate also the variance as the bars mean $\pm\,\sigma$. We can see that with increasing c (decreasing the entropy rate) both the mean and variance of the absolute cross-correlations increase. Using the computationally feasible formula (13) for the mutual information rate, in Fig. 4b we present means and variances for the MIR estimates for independent realizations the ARP (11) as a function of the parameter c. There is some positive bias, represented by the mean MIR, which is low, randomly fluctuating and independent of the dynamics of the evaluated time series, i.e., independent of the parameter c. Also the variances of MIR are independent of c, they are practically the same for all values of c—the positions of bars $\pm\sigma$ in Fig. 4b are given by the fluctuations in the mean value. The mutual information rate is a measure of dependence between dynamics of systems or processes, and unlike the static measures, its bias does not depend on the complexity of dynamics.

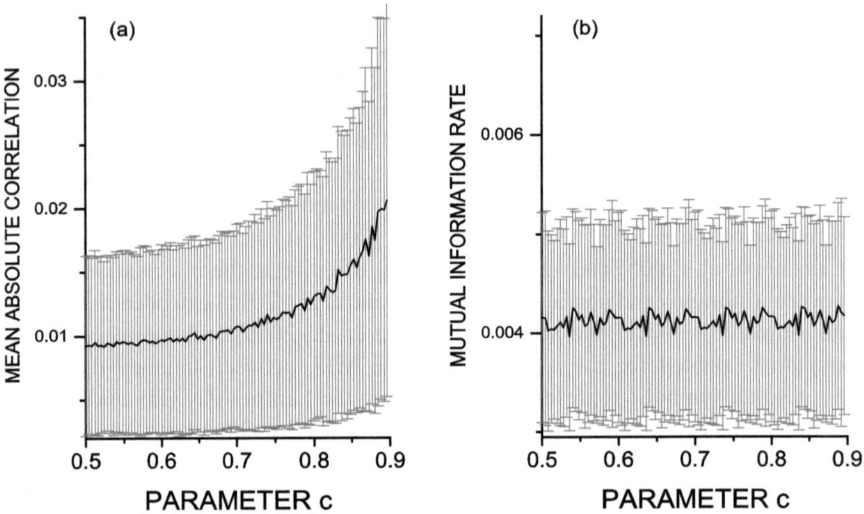

Fig. 4 (**a**) Mean (*solid line*) and variance (bars $\pm\sigma$ above and below the mean value) of the absolute cross-correlation between independent realizations of the autoregressive process (11) as a function of the parameter c. (**b**) The same as (**a**) but for the mutual information rate (13). Note that scales in (**a**) and (**b**) are different

6 Information Rates of Gaussian Processes and Dynamical Systems

The formulas (10) for the entropy rate and (13) for the mutual information rate of Gaussian processes can be efficiently evaluated using the fast Fourier transform (FFT). The question is, however, how applicable are these formulas for real-world time series recorded from complex, possibly nonlinear systems. Using a number of paradigmatic chaotic dynamical systems, Paluš (1997b) inquired a relation between the Kolmogorov–Sinai entropy of a dynamical system and the entropy rate of a Gaussian process with the same spectrum as the sample spectrum of the time series generated by the dynamical system. An extensive numerical study suggests that such a relation as a nonlinear one-to-one function exists when the Kolmogorov–Sinai entropy varies smoothly with variations of system's parameters, but is broken near bifurcation points. Although the formula (10) does not give values numerically close to the true values of the Kolmogorov–Sinai entropy of studied dynamical systems, it allows a relative quantification and distinction of different states of nonlinear systems. In a practical application, the formula (10) was used in order to characterize changing complexity of dynamics of neuronal oscillations on route to an epileptic seizure (Jiruska et al., 2010). A strongly nonlinear character of the neuronal activity of epileptogenic brain regions has been confirmed, e.g., by Casdagli et al. (1996).

In order to demonstrate how the formula (13) for the mutual information rate of Gaussian processes reflects changes in the dependence of dynamics of two coupled

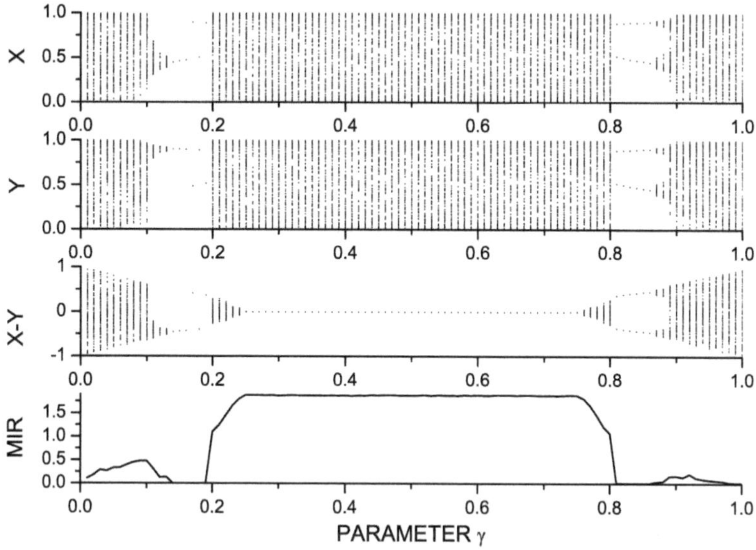

Fig. 5 *Top three panels*: bifurcation diagrams of two coupled logistic maps (*from the top*: x, y, and $x{-}y$), for the control parameter value $a = 4$, corresponding to fully chaotic maps when uncoupled, as a function of the coupling coefficient γ. *Bottom panel*: the mutual information rate (13) between $\{X\}$ and $\{Y\}$, computed using the FFT, as a function of the coupling coefficient γ

nonlinear dynamical systems on their route to synchronization we will use two well-known dynamical systems with chaotic behavior. As an example of a discrete-time system let us borrow the symmetrically coupled logistic maps from Blanc et al. (2011) where the system $\{X\}$ is represented by the time series $\{x_n\}$ and the system $\{Y\}$ by the time series $\{y_n\}$:

$$x_{n+1} = \gamma f_a(x_n) + (1 - \gamma)f_a(y_n) \tag{15}$$

$$y_{n+1} = (1 - \gamma)f_a(x_n) + \gamma f_a(y_n) \tag{16}$$

where γ is the coupling coefficient and varies between 0 and 1. The function $f_a(x_n) \equiv ax_n(1 - x_n)$. It is known that, in the uncoupled case, $a = 4$ gives a chaotic behavior. The latter is demonstrated in the bifurcation diagrams in Fig. 5 where for small γ both the system $\{X\}$ and $\{Y\}$ are chaotic and not synchronized. For $0.13 < \gamma < 0.2$ a zone of periodic behavior appears, followed by the fully chaotic regime from γ approaching 0.2. The two systems become fully synchronized from $\gamma \approx 0.25$—in the bifurcation diagram the difference $x - y$ stays on the zero value, i.e., the trajectories of the systems $\{X\}$ and $\{Y\}$ are identical. Then we observe a quasi-symmetry about $\gamma = 0.5$, i.e., the synchronized behavior ends for $\gamma > 0.75$ and we observe the chaotic, periodic, and again chaotic behavior of the unsynchronized systems. This development is reflected in the mutual information rate (13), depicted in the bottom of Fig. 5. With γ increasing from zero also the

MIR gradually increases; however, it falls down to zero for the interval of periodic dynamics. Thus the MIR is not simply a measure of dependence of dynamics, it rather quantifies an information transfer between systems and processes. In the case of periodic systems with the zero entropy rate (KSE), also the MIR is zero. In the subsequent chaotic regimes the MIR quickly increases with γ approaching the synchronization threshold. During the fully synchronized regime the MIR stays on its maximum value. It is interesting to compare Fig. 5 with Fig. 4 in Blanc et al. (2011) where the authors present results of their four MIR estimators, stating that the Lempel–Ziv estimator (Ziv and Lempel, 1978) and the causal state machine-based estimator (Crutchfield and Young, 1989; Haslinger et al., 2010; Shalizi et al., 2001) gave the most faithful results. The latter are qualitatively equivalent to the results obtained using the formula (13) for the MIR of Gaussian processes, estimated using the FFT (the bottom graph of Fig. 5). The qualitative equivalence means that although the values of the MIR estimates are different, the shapes of the MIR dependence on the coupling parameter γ are very similar.

As an example of a continuous-time system we will consider the unidirectionally coupled Rössler systems, studied also by Paluš and Vejmelka (2007), given by the equations

$$\begin{aligned}\dot{x}_1 &= -\omega_1 x_2 - x_3\\ \dot{x}_2 &= \omega_1 x_1 + a_1\, x_2\\ \dot{x}_3 &= b_1 + x_3(x_1 - c_1)\end{aligned} \tag{17}$$

for the autonomous system $\{X\}$, and

$$\begin{aligned}\dot{y}_1 &= -\omega_2 y_2 - y_3 + \varepsilon(x_1 - y_1)\\ \dot{y}_2 &= \omega_2 y_1 + a_2\, y_2\\ \dot{y}_3 &= b_2 + y_3(y_1 - c_2)\end{aligned} \tag{18}$$

for the response system $\{Y\}$. We will use the parameters $a_1 = a_2 = 0.15$, $b_1 = b_2 = 0.2$, $c_1 = c_2 = 10.0$, and frequencies $\omega_1 = 1.015$ and $\omega_2 = 0.985$, i.e., the two systems are similar, but not identical.

Figure 6a presents four Lyapunov exponents (LE) of the coupled systems (the two negative LE's are not shown) as functions of the coupling strength ε. One positive and one zero LE of the driving system $\{X\}$ are constant, while the LE's of the driven system $\{Y\}$ which are positive and zero without a coupling or with a weak coupling decrease with increasing ε. The two systems can enter a synchronized regime when the originally positive LE of the response system becomes negative. After a transient negativity and a return to zero, the originally positive LE of the driven system $\{Y\}$ becomes decreasing and negative for $\varepsilon > 0.15$ (Fig. 6a). The mutual information rate (13) between $\{X\}$ and $\{Y\}$ computed using the FFT (the solid blue line in Fig. 6b) gradually increases with the increasing coupling strength $0 < \varepsilon < 0.15$, however, shows a steep increase at or after the synchronization

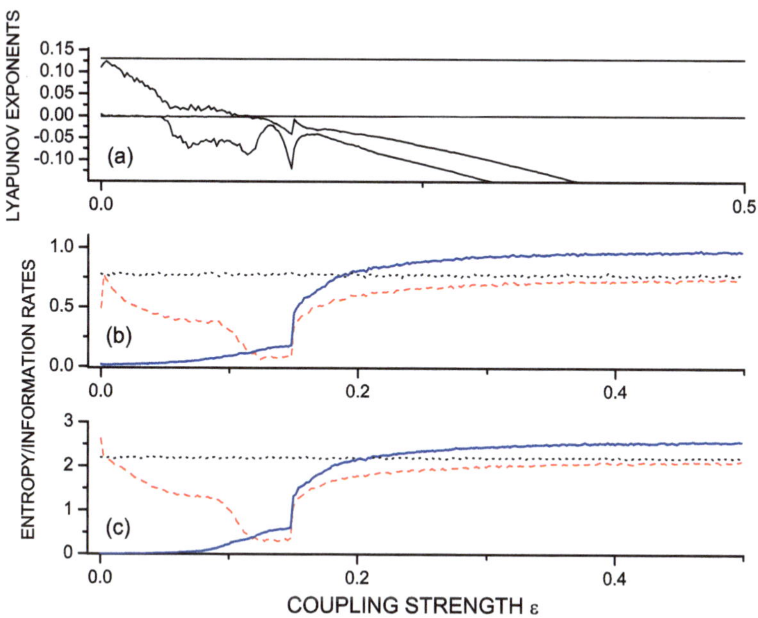

Fig. 6 (**a**) Two largest Lyapunov exponents of the drive $\{X\}$ (the *constant lines*) and the response $\{Y\}$ (the *decreasing lines*), (**b**) the entropy rates (10) for the drive $\{X\}$ (the *dotted black line*) and the response $\{Y\}$ (the *dashed red line*) and the mutual information rate (13) between $\{X\}$ and $\{Y\}$ (the *solid blue line*) computed using the FFT; (**c**) the same as in (**b**), but computed using the CCWT; for the unidirectionally coupled Rössler systems (17), (18), as functions of the coupling strength ε. The Lyapunov exponents are measured in nats per a time unit; the entropy and information rates are measured in units of nats per sample

threshold at $\varepsilon \approx 0.15$. Then it again increases slowly to its asymptotic value in the synchronized state. At the coupling strength $\varepsilon \approx 0.15$ also the entropy rate (10) of the driven system $\{Y\}$ (the dashed red line in Fig. 6b) steeply increases and then continues in a gradual increase and asymptotically approaches the entropy rate (10) of the autonomous system $\{X\}$ (the dotted black line in Fig. 6b). Due to this behavior Paluš et al. (2001) described the route to synchronization as an adjustment of information rates. It is important that even the mutual information rate (13) of Gaussian processes, computed using the FFT of time series generated by the studied systems, reflects both the gradual increase of coupling as well as the sudden transient into synchronization.

7 MIR and Networks of Dynamical Systems

In Sect. 5 we have introduced the mutual information rate as a quantity measuring the dependence between dynamics of two systems or processes. Unlike the static measures such as the correlation coefficient or the mutual information of random variables, the estimates of the MIR do not suffer by a bias dependent on the character of dynamics underlying analyzed time series. Blanc et al. (2011) also show that the MIR is independent of time lag between time evolutions of studied systems. Together with Blanc et al. (2011) and Baptista et al. (2012) we propose the MIR as an association measure suitable for inferring interaction networks from multivariate time series generated by coupled dynamical systems. Specifically in this paper we propose to use the formula (13) for the mutual information rate of Gaussian processes. Although Gaussian processes are inherently linear, in Sect. 6 we have demonstrated that the MIR (13) computed using the FFT of time series generated by the studied nonlinear dynamical system was able to distinguish not only synchronized from unsynchronized states, but also different levels of dependence between dynamics of the studied systems due to different strengths of their coupling. These observations, however, cannot assure a general applicability of the MIR (13) for natural nonlinear systems. Before constructing networks from experimental multivariate time series it is necessary to test for a presence of nonlinearity in studied time series and assess its actual effect on the inference and quantification of dependence relations present in the data. It is not surprising that such studies have been done in the same areas where the research based on the complex networks paradigm is very active.

Functional brain networks are frequently constructed using time series from sequences of functional magnetic resonance imaging (fMRI) (Achard et al., 2006; Bullmore and Sporns, 2009). Hlinka et al. (2011) demonstrate that the linear correlation coefficient is a sufficient measure of functional connectivity in resting-state fMRI data. Potential new information brought by nonlinear measures such as the mutual information is relatively minor and negligible in comparison with natural intra- and inter-subject variability. Hartman et al. (2011) confirm this finding in specific computations of graph-theoretical measures from fMRI brain networks. Also spatio-temporal dependence structures in electrophysiological data such as the electroencephalogram (EEG) are characterized within the complex networks paradigm (Bullmore and Sporns, 2009; Reijneveld et al., 2007). Nonlinear character of the EEG in epilepsy is known (Casdagli et al., 1996), some level of nonlinearity can be detected also in normal human EEG recordings (Paluš, 1996b). Distinction of different physiological and/or pathological brain states observed using nonlinear measures can successfully be reproduced by a proper application of standard tools derived from the theory of linear stochastic processes (Theiler and Rapp, 1996). While the latter findings characterized single-channel EEG signals, the character of dependence between EEG signals from different parts of the scalp are relevant for the construction of the EEG brain networks. Nonlinear measures have been applied in order to distinguish different consciousness states using the so-called multichannel attractor embedding (Matousek et al., 1995). Changes in dependence structures in multichannel EEG data which have been described by a nonlinear measure such as the correlation dimension from the multichannel embedding (Matousek et al., 1995), however, can be equivalently captured by a linear measure extracted from a correlation matrix (Paluš et al., 1992).

Using an equivalent approach, climate networks (Donges et al., 2009a,b; Steinhaeuser et al., 2012; Tsonis and Roebber, 2004; Tsonis et al., 2006; Yamasaki et al., 2008, 2009) are constructed using multivariate time series of long-term records of meteorological variables such as the air temperature or pressure. Already in the 1980s a number of researchers attempted to infer nonlinear dynamical mechanisms from meteorological data and claimed detections of a weather or climate attractor of a low dimension (Fraedrich, 1986; Nicolis and Nicolis, 1984; Tsonis and Elsner, 1988). Other authors pointed to a limited reliability of chaos-identification algorithms and considered the observed low-dimensional weather/climate attractors as spurious (Grassberger, 1986; Lorenz, 1991). Paluš and Novotná (1994) even found the air temperature data well-explained by a linear stochastic process, when the dependence between a temperature time series $\{x(t)\}$ and its lagged twin $\{x(t+\tau)\}$ was considered. Hlinka et al. (2014) extended the later result to the dependence between the monthly time series of the gridded whole-Earth air temperature reanalysis data. These results do not mean that the dynamics underlying records of meteorological data is linear. For instance, a search for repetitive patterns on specific temporal scales in the air temperature and other meteorological data has led to an identification of oscillatory phenomena possibly possessing a nonlinear origin and exhibiting phase synchronization between oscillatory modes extracted either from different types of climate-related data or data recorded at different locations on the Earth (Feliks et al., 2010; Paluš and Novotná, 2004, 2006, 2009, 2011). The studies of Hlinka et al. (2013, 2014) merely state that for inferring general dependence and causal relations, the approaches derived for Gaussian processes perform very well and nonlinear approaches do not bring substantial new information.

These arguments and the fact that the mutual information rate estimator, computed using the FFT and the formula (13) for the MIR of Gaussian processes is computationally less demanding that estimators for general nonlinear processes, form the basis for our recommendation of the MIR (13) as a measure suitable for inference of networks from experimental multivariate time series recorded from complex systems of various origins. There is still a serious demand for the amount and stationary character of the analyzed data, since the computation of the magnitude-squared coherence (14) is based on dividing the time series into a number of segments over which the complex cross-spectrum (the numerator in Eq. (14) right-hand side) is averaged. In many cases time series from natural complex systems are relatively short and nonstationary. Nonstationarity in the sense of changing relationships between time series with time leads to changes in the strength and even the existence of links in interaction networks during some time intervals. The complex network paradigm copes with this phenomenon using the concept of temporal networks (Holme and J. Saramäki, 2012) or evolving networks. The latter approach assumes approximate step-wise stationarity of the analyzed time series, and a standard "static" network is inferred in a relatively short time window which is "sliding" over the whole time interval spanned by the available experimental time series. The time evolution of graph-theoretical characteristics is then studied with respect to a time evolution and/or an occurrence of marked events in the studied complex system. This approach has been successfully applied in the EEG brain networks (Bialonski et al., 2013; Kuhnert et al., 2010; Lehnertz et al., 2014), as well as in the climate networks (Radebach et al., 2013). An alternative approach, applied in the field of climate networks, is "picking-up" a number of unequal-length subsets of the whole time series, tight to an occurrence of some phenomenon (e.g. El Niño) and performing the summation in the formula (6) for the correlation coefficient only using the selected subsets of the data (Tsonis and Swanson, 2008). Neither the latter approach, nor the evolving network strategy can be applied when using the standard FFT-based evaluation of the MIR (13).

In order to cope with nonstationarity we propose to use a wavelet transform instead of the Fourier transform. In particular, the complex continuous wavelet transform (CCWT) is applied in order to convert a time series $x(t)$ into a set of complex wavelet coefficients $W(t,f)$:

$$W(t,f) = \int_{-\infty}^{\infty} \psi(t')x(t-t')dt' \tag{19}$$

using the complex Morlet wavelet (Torrence and Compo, 1998):

$$\psi(t) = \frac{1}{\sqrt{2\pi\sigma_t^2}} \exp\left(-\frac{t^2}{2\sigma_t^2}\right) \exp(2\pi i f_0), \tag{20}$$

where σ_t is the bandwidth parameter, and f_0 is the central frequency of the wavelet. σ_t determines the rate of the decay of the Gauss function, its reciprocal value $\sigma_f = 1/\pi\sigma_t$ determines the spectral bandwidth. In order to keep the wavelet representation close to the original MIR (13) evaluation based on the FFT, we use a set of equidistantly spaced central wavelet frequencies in the relevant frequency range given by the time series length and its sampling frequency, instead of the power-law pyramidal scheme, usually used in the wavelet context. Then the product of the complex wavelet coefficients $W_X(t,f)W_Y^*(t,f)$, as well as the norms $|W_X(t,f)|$, $|W_Y(t,f)|$ are averaged over time t. Finally, the wavelet magnitude-squared coherence

$$|\Gamma_{X,Y}^W(f)|^2 = \frac{|W_{X,Y}(f)|^2}{|W_X(f)||W_Y(f)|} \tag{21}$$

is used in the summation over the set of the central wavelet frequencies according to Eq. (13). Here $W_{X,Y}(f)$, $|W_X(f)|$, and $|W_Y(f)|$ stand for the time averages of $W_X(t,f)W_Y^*(t,f)$, $|W_X(t,f)|$, and $|W_Y(t,f)|$, respectively.

Let us return to the unidirectionally coupled Rössler systems (17), (18). Using the wavelet representation we can recompute both the mutual information rate (13) and the entropy rate (10) as functions of the coupling strength ε (Fig. 6c). While the CCWT-based estimators give different values than the FFT-based estimators, they agree in the qualitative sense that the curves of the ε-dependence of the MIR in Fig. 6b, c (solid/blue curves) are the same. The two estimators also give a good agreement in the ε-dependence of the entropy rates (dashed/red curves in Fig. 6b, c), there is just a small difference in the entropy rates of the driven system for very small values of ε.

It is important that the wavelet-based estimator of the MIR (13) gives a relative distinction of coupling regimes with different coupling strengths. This is a property which we expect from an association measure suitable for the inference of interaction networks from multivariate time series. It will assign proper weights to network edges, an edge of two more strongly coupled nodes (dynamical systems) will obtain a greater weight than edges connecting nodes with a weaker coupling. For the construction of the binary networks, a greater value of MIR for strongly coupled nodes assure an existence of an edge by exceeding a chosen threshold or a critical value given by a statistical test. For establishing statistically significant links we propose to use the surrogate data strategy as described in Paluš (2007) and Paluš and Vejmelka (2007). The temporal averaging of the product of the complex wavelet coefficients $W_X(t,f)W_Y^*(t,f)$ might evoke a temptation to randomize the phases φ of the complex wavelet coefficients $W(t,f) \equiv A(t,f)\exp(i\varphi(t,f))$. Generating the FFT-based surrogate data (Theiler et al., 1992), the set of the original phases of the Fourier coefficient is substituted by a set of independent, identically distributed (IID) phases randomly sampled from a uniform distribution on the interval $(0, 2\pi)$. However, the phase differences of the wavelet coefficients of two signals are not IID, even if the underlying processes are independent. Using random IID phases in the summation of $W_X(t,f)W_Y^*(t,f)$ would underestimate the critical values in the test for independence and false edges would be inferred. The character of the (long-range) dependence of the phase differences in $W_X(t,f)W_Y^*(t,f)$ of independent processes depends on the central wavelet frequency. Therefore, it is more convenient to generate surrogate data and estimate the MIR from them as in the usual surrogate data test strategy (Paluš, 2007; Paluš and Vejmelka, 2007).

Unlike in the FFT-based MIR estimation, we apply the CCWT on the whole time interval of available data. Then we either average the product of the complex wavelet coefficients $W_X(t,f)W_Y^*(t,f)$, as well as the norms $|W_X(t,f)|$, $|W_Y(t,f)|$, over the whole time interval or apply the sliding-window strategy of the evolving networks (Radebach et al., 2013) or the strategy of Tsonis and Swanson (2008) of the averaging over time intervals selected according to an occurrence of a specific phenomenon. Using the strategies that cope with nonstationarity, however, one should consider a smoothing effect of the wavelet coefficients for low frequencies (large time scales). Since time series from natural complex systems frequently reflect processes with a $1/f$ spectrum, the wavelet coefficients for low frequencies have much greater weights than the coefficients for high frequencies and effects of short-living phenomena can be masked in the resulted MIR estimates. Therefore we recommend to limit the final summation in the MIR formula (13) to higher frequencies or shorter time scales in which the effect of short-living phenomena is not attenuated. The latter idea can be generalized and even for stationary time series one can restrict the MIR evaluation to a specific range of time scales, i.e. to a specific spectral band. Then a scale-specific or frequency-specific connectivity is evaluated and *scale-specific* or *frequency-specific interaction networks* can be studied.

Until now we have considered the MIR $i(X_i; Y_i)$ of two stochastic processes $\{X_i\}$, $\{Y_i\}$. Constructing a network of n nodes, i.e., n dynamical systems, we will consider n time series as realizations of n stochastic processes $\{X_i^k\}$, $k = 1, \ldots, n$. (For simplicity we consider n univariate time series/stochastic processes, an equivalent of a multivariate stochastic process with n components. The considerations here can be generalized to n multivariate stochastic processes with various numbers n_i of components.) Then we can evaluate the standard bivariate MIR $i(X_i^k; X_i^l)$ for each pair of components. In order to distinguish direct from indirect interactions we can also consider conditional (partial) MIR $i(X_i^k; X_i^l | X_i^j; j = 1, \ldots, n, j \neq k, j \neq l)$ which quantifies the "net" dependence between the two processes without an influence of the remaining $n - 2$ processes.

For the evaluation of the conditional MIR, in the framework of Gaussian processes, we will follow the work of Schelter et al. (2006) who extended the notion of partial correlations to the partial mean phase coherence.

For each pair of processes $\{X_i^k\}$, $\{X_i^l\}$ and each central wavelet frequency $f \in \{f_1, f_2, \ldots, f_{N_f}\}$ we evaluate the time-averaged complex wavelet coherence

$$\Gamma_{k,l}^W(f) = \frac{W_{k,l}(f)}{\sqrt{|W_k(f)||W_l(f)|}}. \tag{22}$$

Thus for each f we obtain a complex $n \times n$ matrix $\Gamma^W(f)$. This complex matrix is inverted, $\Omega(f) = (\Gamma^W(f))^{-1}$. Using the entries of the inverted complex matrix $\Omega(f)$ we evaluate the conditional wavelet coherence as

$$\Upsilon_{k,l}(f) = \frac{\Omega_{k,l}(f)}{\sqrt{|\Omega_{k,k}(f)||\Omega_{l,l}(f)|}}. \tag{23}$$

Finally, the magnitude-squared conditional wavelet coherence is used in the summation according to Eq. (13) in order to obtain the conditional MIR

$$i_G(X_i^k; X_i^l | X_i^j; j = 1, \ldots, n, j \neq k, j \neq l)$$
$$= -\frac{1}{2N_f} \sum_{f=f_1}^{f_{N_f}} \log(1 - |\Upsilon_{k,l}(f)|^2). \tag{24}$$

8 Climate Networks

Understanding the complex dynamics of the Earth atmosphere and climate is a great scientific challenge with a potentially high societal impact. In their seminal paper, Tsonis and Roebber (2004) have proposed to study the climate system as a complex network. Since then the field of climate networks is rapidly developing and expanding in the scope of methodology as well as applications. The spatio-temporal dynamics of the atmosphere is captured by multivariate time series of long-term recordings of meteorological variables. Typically, such instrumental data are preprocessed and interpolated in order to assign a time series of a variable to each node of a regular angular grid covering the Earth surface, as well as slices of the atmosphere at various altitude or air pressure levels. Such gridded time series of meteorological variables, available due to, e.g., the NCEP/NCAR reanalysis project (Kalnay et al., 1996) are usually, although not exclusively, used for the construction of climate networks. Monthly (Donges et al., 2009b; Tsonis and Swanson, 2008) or daily (Gozolchiani et al., 2008; Yamasaki et al., 2008) surface air temperature data are frequently used, however, equipotential heights (Donges et al., 2011; Tsonis et al., 2008), sea surface temperature, humidity, precipitation and related data (Malik et al., 2012; Steinhaeuser et al., 2012) and other meteorological data are also analyzed. Individual grid-points, characterized by time series of a chosen meteorological variable, are considered as nodes (vertices) of a climate network, while links (edges) are inferred from some, mostly statistical association between the time series related to the two nodes at the edge's end-points. The most common association measure is the Pearson's correlation coefficient (Tsonis and Roebber, 2004; Tsonis and Swanson, 2008), however, also the Spearman's rank correlation coefficient is used (Carpi et al., 2012), and more general, nonlinear measures are tested, e.g., the bivariate mutual information (Donges et al., 2009a,b), and the mutual information of ordinal time series (Barreiro et al., 2011; Deza et al., 2013) or measures from the phase synchronization analysis (Yamasaki et al., 2009) and the event synchronization analysis (Malik et al., 2012).

In the following study we use monthly mean values of the near-Surface Air Temperature (SAT) from the NCEP/NCAR reanalysis (Kalnay et al., 1996). We include the data up to the latitudes 87.5° in the grid of 2.5° × 2.5° which leads to 10,224 grid-points or network nodes. The temporal interval of 624 months starting in January 1958 and ending in December 2009 is used for the inference of the network using the correlation coefficient and the CCWT-based MIR estimator. For the FFT-based estimator the temporal interval is extended backward by 16 months (starting in September 1956) in order to have 5 segments of 128 monthly samples.

In order to avoid trivial correlations due to seasonal temperature variability, the annual cycle has been removed from each SAT time series. The SAT anomalies (SATA thereafter) have been computed by subtracting the averages for each month from related samples, e.g., the average January temperature was subtracted from all January samples, etc.

As the first step of the network analysis we compute the correlation coefficients $c_{i,j}$ for each pair of nodes $i,j = 1,\dots,N_N = 10{,}224$. We use the matrix of the absolute correlations $C_{i,j} = |c_{i,j}|$ in order to obtain the adjacency matrix $A_{i,j}$ of the binary network, defined as: $A_{i,j} = 1$ iff $C_{i,j} > c_T$, otherwise $A_{i,j} = 0$. $A_{i,i} = 0$ by definition. The total number of existing edges divided by the number of all possible edges in known as the network density (or edge density) ϱ. Following Donges et al. (2009a,b) we choose the threshold c_T such that the resulting network density is $\varrho = 0.005$.

The basic characterization of connectivity of a node i is its degree, or degree centrality k_i

$$k_i = \sum_{j=1}^{N_N} A_{i,j}, \tag{25}$$

giving the number of nodes to which the node i is connected. Since the reanalysis data are defined on a grid which is regular in the angular coordinates, the geographic distances of the grid-points depend on the latitude λ_i. In order to correct for this dependence, for the climate networks defined on the regular angular grid the area weighted connectivity (Tsonis et al., 2006) is defined as

$$\mathrm{AWC}_i = \frac{\sum_{j=1}^{N_N} A_{i,j} \cos(\lambda_j)}{\sum_{j=1}^{N_N} \cos(\lambda_j)}. \tag{26}$$

The AWC can be interpreted as the fraction of the Earth's surface area a vertex is connected to.

The AWC computed for each node of the SATA network based on the absolute correlations with $\varrho = 0.005$ (AC-network in the following) is mapped in Fig. 7a. According to this analysis the most connected nodes ("hubs") of the climate network lie in the tropical areas of the Pacific and Indian Oceans. The hub in the tropical Pacific include the so-called El Niño areas. The El Niño/Southern Oscillation (ENSO) is a dominant mode of the global atmospheric circulation variability which quasiperiodically causes shift in winds and ocean currents centered in the Tropical Pacific region and is linked to anomalous weather/climate patterns worldwide (Sarachik et al., 2010). The global influence of the El Niño phenomenon is used to explain the observation that the El Niño area constitutes the principal hub of the climate network.

Let us characterize the dynamics of the SATA time series using the entropy rate (10). The FFT-based estimator assign an entropy rate value to each grid-point, i.e. to each node of the network, so they can be mapped in the same way as the AWC. The entropy rate map is presented in Fig. 7b. The correspondence between the lowest entropy rates and the highest AWC of the AC-network is indisputable. In order to assess a bias in the correlation estimator we need time series which are independent, but have the same dynamics (the same entropy rate) as the original SATA time series. Such time series can be generated using the FFT-based surrogate data algorithm (Paluš, 2007; Theiler et al., 1992). The fast Fourier transform is applied to a time series, the magnitudes of the complex Fourier coefficients are preserved, but their phases are randomized. Using different sets of random phases the inverse FFT generates a number of independent realizations of a Gaussian process with the same spectrum [and thus with the same entropy rate (10)] as that of the original time series. A potential digression from the Gaussian distribution is solved by a histogram transformation known as the amplitude adjustment. We use

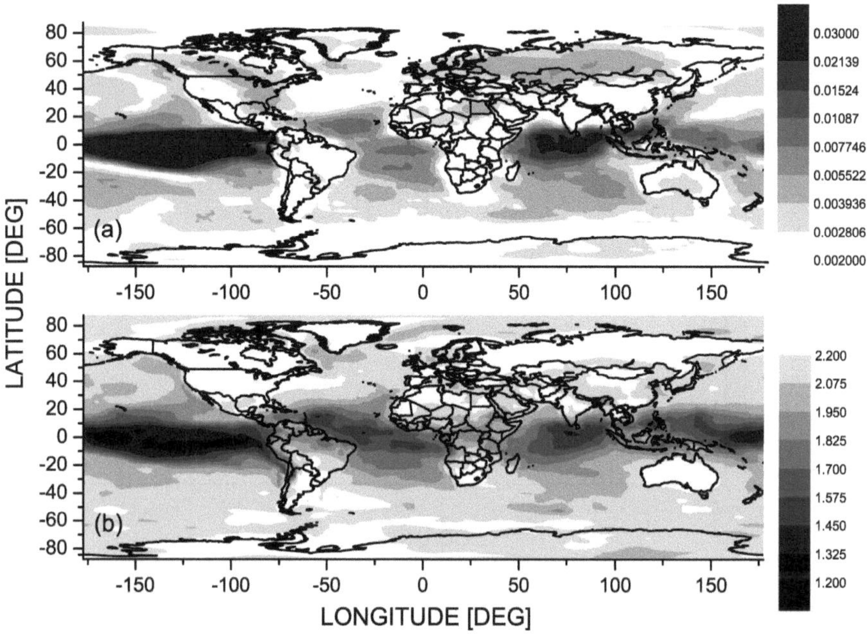

Fig. 7 (**a**) Area weighted connectivity for the SATA climate network with the density $\varrho = 0.005$ obtained by the uniform thresholding of the absolute correlations. (**b**) Complexity of each node SATA time series measured by the Gaussian process entropy rate h_G. Note the reversed gray scales, the *black color* corresponds to the largest AWC in (**a**), while in (**b**) it corresponds to the lowest entropy rates

both the FT surrogate data and the amplitude-adjusted FT (AAFT) surrogate data; however, they give equivalent results. Generating a large number of realizations of the FT (AAFT) surrogate data for the SATA time series we can estimate distributions of the correlations of independent surrogates of the SATA series from various grid-points. Figure 8 compares such histogram for SATA-surrogates for a pair of grid-points from a low entropy rate area (the El Niño area) and from a high entropy rate area (an Euro-Asian area on 60°N). While in the Northern hemisphere high entropy rate area the correlation bias (the cross-correlation of realizations of independent processes) scarcely reaches over ± 0.1, in the tropical Pacific areas the cross-correlation bias can reach values close to ± 0.4.

As an alternative we construct a climate network using the MIR (13) and again we threshold the MIR values in order to obtain the network density $\varrho = 0.005$. The area weighted connectivity for the MIR-networks is mapped in Fig. 9, where we can compare the results for both the FFT- and CCWT-based estimators. The results are quite similar. A few small differences can be caused by the fact that the FFT-based estimator used segments of 128 samples and thus cannot include the connectivity on large time scales as the CCWT estimator which utilizes 624 samples in one whole segment. The differences between the MIR-network (Fig. 9) and the

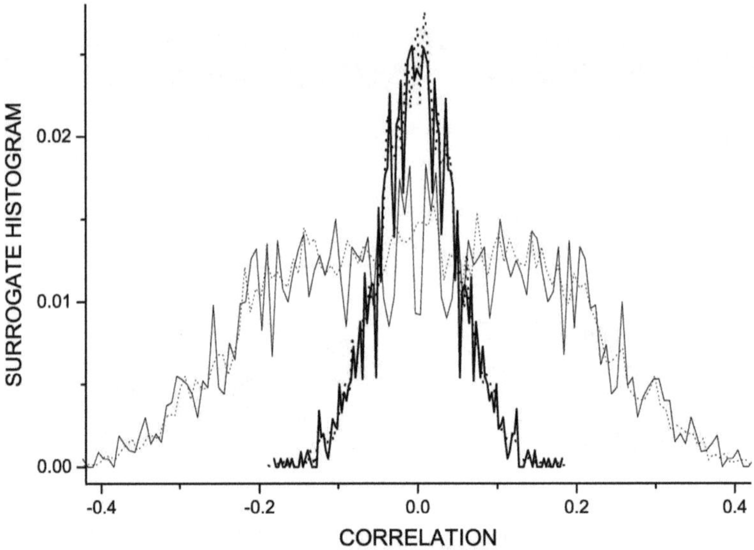

Fig. 8 Histograms of cross-correlations of independent FT (*solid lines*) and amplitude-adjusted FT (*dotted lines*) surrogate data for the SATA of a pair of nodes from the low entropy rate area (the *thin lines*, the nodes with the latitude 0°, the longitude 90°W and 10°S, 120°W) and a pair from the high entropy rate area (the *thick lines*, the nodes 60°N, 25°E and 60°N, 75°E)

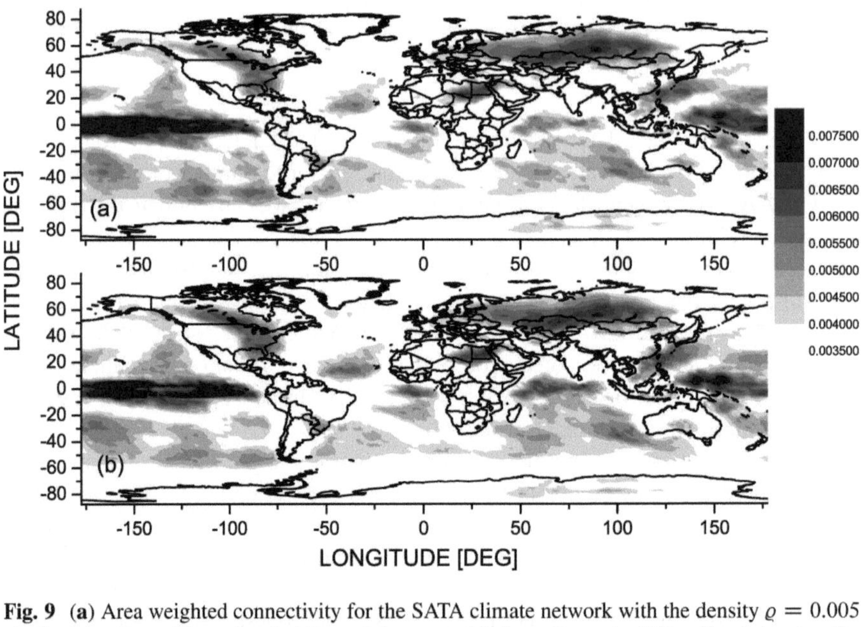

Fig. 9 (**a**) Area weighted connectivity for the SATA climate network with the density $\varrho = 0.005$ obtained by the uniform thresholding of the mutual information rate (13) estimated using the Fourier transform (**a**) and the continuous complex wavelet transform (**b**). Note that the scale is different from that in Fig. 7a

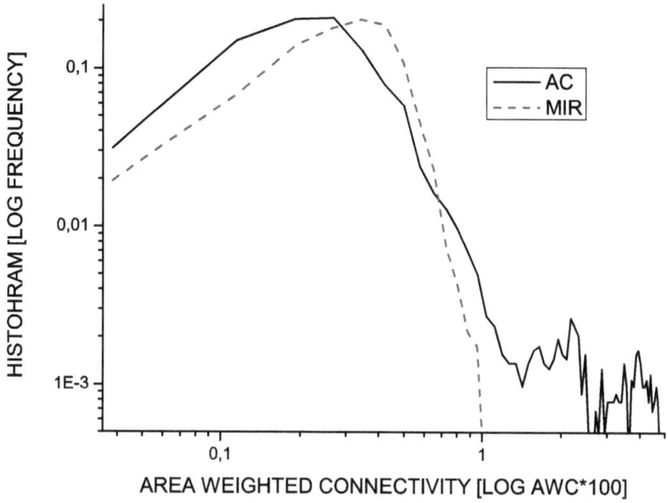

Fig. 10 Histograms (64 bins) of the area weighted connectivity (AWC*100) for the SATA climate network with the density $\varrho = 0.005$ obtained by the uniform thresholding of the absolute correlations (*solid black line*), and of the mutual information rate (13) (*dashed red line*)

AC-network (Fig. 7a) are much larger and more important. In the comparison with the AC-network, the very connected hub in the Indian Ocean almost disappears in the MIR network. The hub in the El Niño area survives; however, it is weaker and confined to a smaller area. The connectivity in the continental areas of the Northern hemisphere increases in the MIR-network. This comparison, however, cannot give an answer which network representation is closer to the physical reality.

In the analogy with degree distributions, studied in the complex network theory, in Fig. 10 we present histograms estimating the distributions of the area weighted connectivity. The AWC distribution for the AC-network (the solid black line) shows a heavy irregular tail of extreme AWC values, while the AWC distribution for the MIR-network (the dashed red line) shows a distribution bounded by a fast probability decay for large AWC values, well captured by a Poisson distribution. Scholz (2010) obtained such distributions using a node-similarity network model. Each node has a set of features, quantified as coordinates in an Euclidean space. Based on a random data set, two nodes are defined as connected (similar) when their Euclidean distance is below a certain threshold. Using a small threshold only very similar (close) nodes are connected. This represents a sparsely connected network showing typically scale-free power-law like distributions. Increasing the threshold, more densely connected networks are modelled with node degree distributions very similar to that of the MIR-network (the dashed red line in Fig. 10).

Bivariate histograms estimating the joint probability distribution of the SATA entropy rate and the AWC for the studied climate networks are presented in Fig. 11. In the AC-network the extremely high AWC values are tight to the nodes with the low entropy rate of the SATA time series (Fig. 11a). Together with the histograms

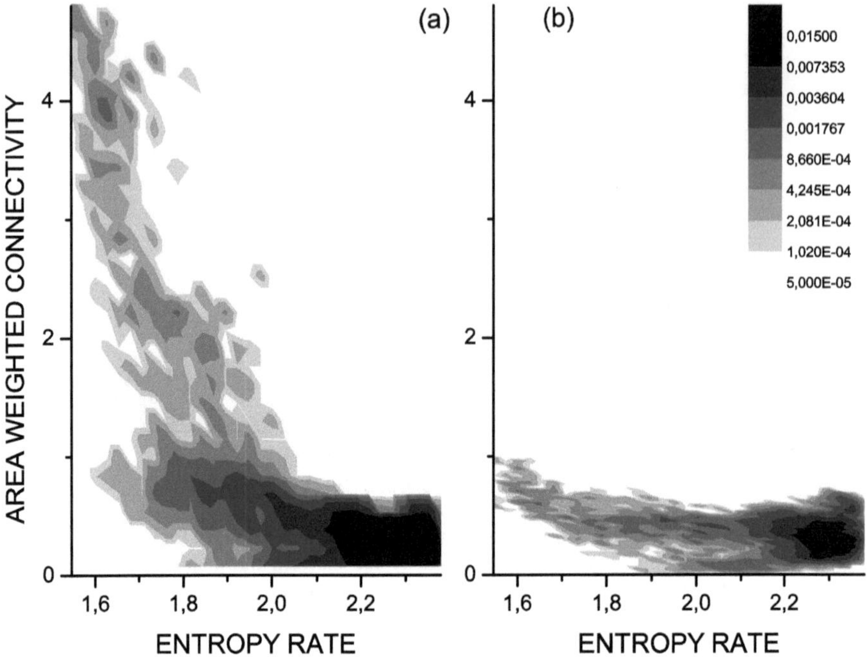

Fig. 11 (**a**) Gray-coded bivariate histograms (32 × 32 bins) reflecting the joint probability distribution of the SATA entropy rate (10), cf. Fig. 7b, and the area weighted connectivity (AWC*100) for the SATA climate network with the density $\varrho = 0.005$ obtained by the uniform thresholding of the absolute correlations, cf. Fig. 7a. (**b**) The same as in (**a**), but considering the area weighted connectivity for the SATA climate network with the density $\varrho = 0.005$ obtained by the uniform thresholding of the mutual information rate (13), cf. Fig. 9b

in Fig. 10 this picture supports the conclusion that the very high AWC values of the nodes characterized by low entropy rates are probably consequences of the bias in the absolute correlation estimations. The MIR-network lacks extreme AWC values; however, some tendency for the preference of higher AWC values in the nodes with low entropy rates remains (Fig. 11b). Since the MIR should not be biased upward by the low entropy rate, this dependence reflects the physical reality: The nodes in the El Niño area are the hub of the climate networks. Their increased connectivity reflects distant influences of the ENSO phenomenon. On the other hand, the quasiperiodic ENSO behavior in certain frequency ranges increases the dynamical memory/decreases the entropy rate of the SATA time series in the El Niño area. We will demonstrate these phenomena using the scale-specific connectivity in the next Section.

9 Scale-Specific Climate Networks

Using the idea of the scale-specific connectivity reflected by the CCWT-based MIR estimates in which the summation over the wavelet scales (central wavelet frequencies) is restricted to a chosen scale range (Sect. 7) we will study scale-specific SATA climate networks. Starting with the MIR estimate restricted to the wavelet time scales corresponding to the periods 4–6 years, in Fig. 12a we map the AWC for the scale-specific SATA climate network for the time scales 4–6 years (SSCN(4–6yr) thereafter). As in the previous cases we consider the binary network with $\varrho = 0.005$. The hub of this network, i.e., the highest scale-specific connectivity in the time scales 4–6 years is located in the tropical Pacific area. It is not surprising since the oscillatory modes in the range of quasi-biennial oscillations (QBO, periods 2–3 years) and quasi-quadrennial oscillations (QQO, the periods fluctuating between 3 and 7 years) have been detected in the quasi-periodic ENSO dynamics (Jiang et al., 1995; Kondrashov et al., 2005). The scale-specific MIR-network for the QBO scale 2–3 years (not presented) has practically the same AWC geographical distribution as the MIR-network for the used QQO range 4–6 years (Fig. 12a). It is interesting to note that the SATA oscillatory mode with the periods 2–3 years is not simply a higher harmonic (Sheppard et al., 2011) of the mode

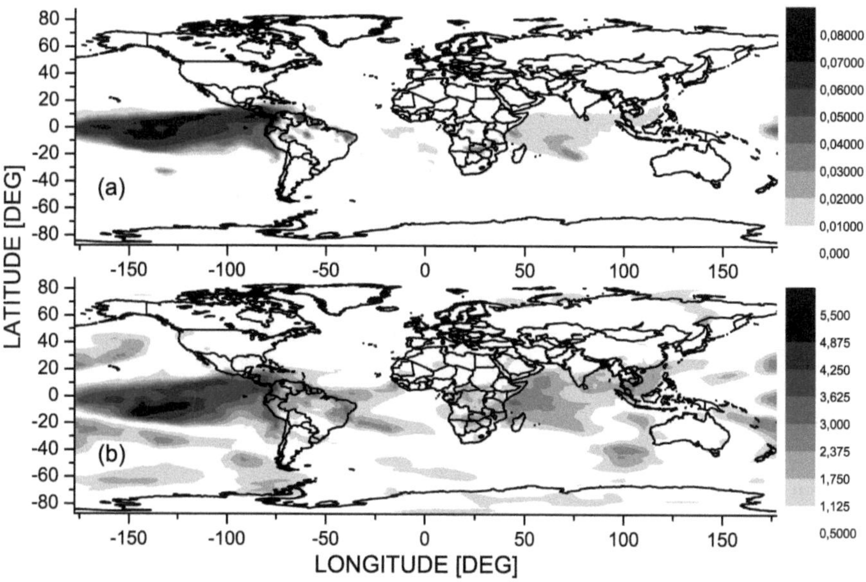

Fig. 12 (**a**) Area weighted connectivity for the scale-specific SATA climate network with the density $\varrho = 0.005$ obtained by the uniform thresholding of the mutual information rate (13) estimated using the continuous complex wavelet transform within the scales related to the periods 4–6 years. (**b**) Dependence of the SATA time series on the Southern Oscillation index measured by MIR (13) estimated using the CCWT within the scales related to the periods 4–6 years

with the periods 4–6 years, since the test for the 1:2 phase coherence between these modes did not reject the null hypothesis of phase independence of these oscillatory modes.

The ENSO is characterized by several indices derived from the sea surface temperature and the Southern Oscillation index (SOI, see http://www.cru.uea.ac.uk/cru/data/soi/ for the data and their description) which is defined as the normalized air pressure difference between Tahiti and Darwin. The MIR quantifying the dependence within the time scales 4–6 years between the SOI and the SATA time series in each grid-point is illustrated in Fig. 12b. The hubs of the SSCN(4–6yr) in the tropical Pacific and Indian Oceans (Fig. 12a) are parts of the areas connected to the ENSO within this time scale (Fig. 12b). However, the areas connected to the ENSO are quite more extended in the Pacific Ocean, tropical Atlantic Ocean and in the Indian and Southern Ocean. Also large continental areas in the Central and Southern America, areas in Africa and some areas in Asia and Northern America have the SATA variability in the time scale 4–6 years connected to the ENSO. This extended ENSO scale-specific connectivity is apparently reflected also in the broad-band connectivity and confirms the role of the hub of the global climate networks for the ENSO tropical Pacific area, as we have observed in the previous Section. The quasi-periodic dynamics plays an important role in the ENSO area temperature variability, e.g. the QQO mode explains almost 40% of the variability of the sea surface temperature anomalies (Jiang et al., 1995). This fact explains the low entropy rate of the SATA time series in this area and the dependence between the AWC and the entropy rate in Fig. 11b.

Oscillatory phenomena with the period around 7–8 years have been observed in the air temperature and other meteorological data by many authors (see Paluš and Novotná (2009, 2011) and references therein). Therefore, in the following we will focus on the scale-specific climate network with the connectivity given by the CCWT-based MIR estimate with the wavelet coherence summation restricted to the wavelet scales corresponding to the periods 7–8 years (SSCN(7–8yr) thereafter). Again we consider the binary network with $\varrho = 0.005$. The AWC of the SSCN(7–8yr) is mapped in Fig. 13a. Consistently with the observation of the 7–8 year cycle in a number of European locations (Paluš and Novotná, 2004, 2007), the SSCN(7–8yr) has a hub in a large area in Europe, but also in Western Asia and Greenland. A strong hub of the SSCN(7–8yr) lies in the tropical Atlantic and also in the Pacific areas different from the ENSO area. The 7–8 year cycle in the European SAT is connected with the North Atlantic Oscillation (Paluš and Novotná, 2009, 2011).

The North Atlantic Oscillation (NAO) is a dominant pattern of the atmospheric circulation variability in the extratropical Northern Hemisphere. On the global scale, the NAO has a climate significance that rivals the Pacific ENSO (Marshall et al., 2001) since it influences the air temperature, precipitation, occurrence of storms, wind strength and direction in the Atlantic sector and surrounding continents. The NAO is characterized by the NAO index (NAOI, see http://www.cru.uea.ac.uk/cru/data/nao/ for the data and their description). While the quasi-periodic dynamics of the ENSO is apparent in the SOI, the NAOI has rather a red-noise-like character

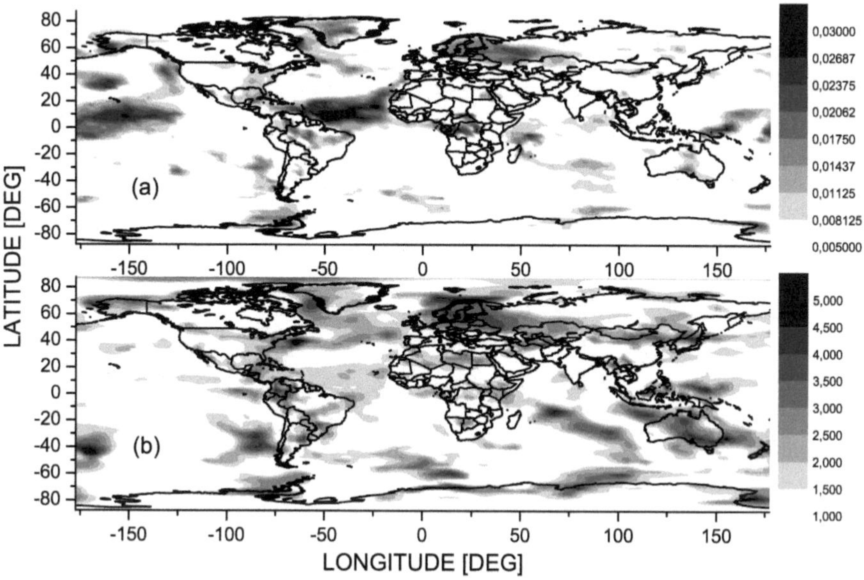

Fig. 13 (**a**) Area weighted connectivity for the scale-specific SATA climate network with the density $\varrho = 0.005$ obtained by the uniform thresholding of the MIR (13) estimated using the CCWT within the scales related to the periods 7–8 years. (**b**) Dependence of the SATA time series on the North Atlantic Oscillation index measured by MIR (13) estimated using the CCWT within the scales related to the periods 7–8 years

(Fernandez et al., 2003). However, sensitive detection methods such as the Monte-Carlo singular system analysis (Paluš and Novotná, 2004) uncovered in the NAO dynamics several oscillatory components from which the cycle with the period around 7–8 years is the most prominent (Feliks et al., 2010; Paluš and Novotná, 2007, 2009, 2011). The NAO 7- to 8-year oscillatory mode is phase synchronized with related modes in the SAT in large areas of Europe (Paluš and Novotná, 2011) as well as with other weather- and climate-related variables in various areas through the Earth (Feliks et al., 2010, 2013).

The MIR quantifying the dependence within the time scales 7–8 years between the NAOI and the SATA time series in each grid-point is illustrated in Fig. 13b. We can see that all the hubs in Fig. 13a lie in the areas where the SATA time series are dependent on the NAOI in this scale, with the exception of the Pacific tropical area between 125° and 180°W (Fig. 13b). For the better understanding of the topology of the SSCN(7–8yr) we quantify the dependence between the SATA time series in the node 150°W, 5°N, using the MIR estimated within the wavelet scales related to the periods 7–8 years, see Fig. 14a. Apparently this hub (the Pacific tropical area between 125° and 180°W) is connected just with other areas in the Pacific Ocean and disconnected from the rest of the SSCN(7–8yr).

Using the same scale-specific connectivity, we can see that the node at 50°W, 7.5°N in the tropical Atlantic hub is connected to other hubs of the SSCN(7–8yr),

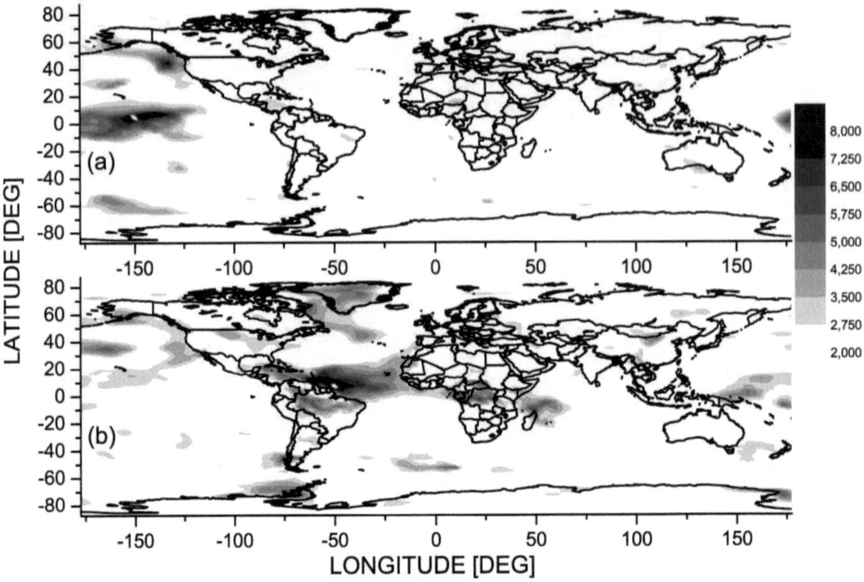

Fig. 14 (**a**) Dependence of the SATA time series from each node with the SATA time series in the node with the longitude 150°W and the latitude 5°N, measured by MIR (13) estimated using the CCWT within the scales related to periods 7–8 years. (**b**) The same as in (**a**), but for the node 50°W, 7.5°N. The reference node can be seen as a white pixel in the black background

but not to the hub in Europe, see Fig. 14b. (And, of course, the Pacific tropical hub is not connected to any other hub.) The map of the scale-specific connectivity of a node in the European hub (the node 22.5°E, 60°N lying close to the SW Finland Baltic coast, see Fig. 15a) confirms the disconnection between the tropical Atlantic and the European hubs; however, the latter is connected to many areas all over the world.

The above-mentioned phase synchrony between the 7- to 8-year oscillatory mode in the NAO and in the European SAT time series evokes the hypothesis that, at least a part of, the connectivity in the SSCN(7–8yr) is induced by the NAO and its worldwide influence. In order to test this hypothesis we construct a version of the scale-specific SSCN(7–8yr) in which, however, the connectivity is given by the scale-specific MIR conditioned on the NAO index. In particular, for each pair of nodes the two SATA time series and the NAOI time series are used to construct the 3×3 wavelet coherence matrix $\Gamma^W(f)$ which is then inverted and the MIR conditioned on the NAOI is evaluated according to Eqs. (23) and (24). This measure quantifies the scale-specific dependence between the two SATA series without a possible influence of the NAO 7- to 8-year oscillatory mode. The AWC map for this conditional SSCN(7–8yr) in Fig. 15b shows that the hub in Europe and W Asia disappears. It means that the scale 7–8 year-specific mutual connectivity of the SATA time series in the areas in Europe and W Asia and their connections to

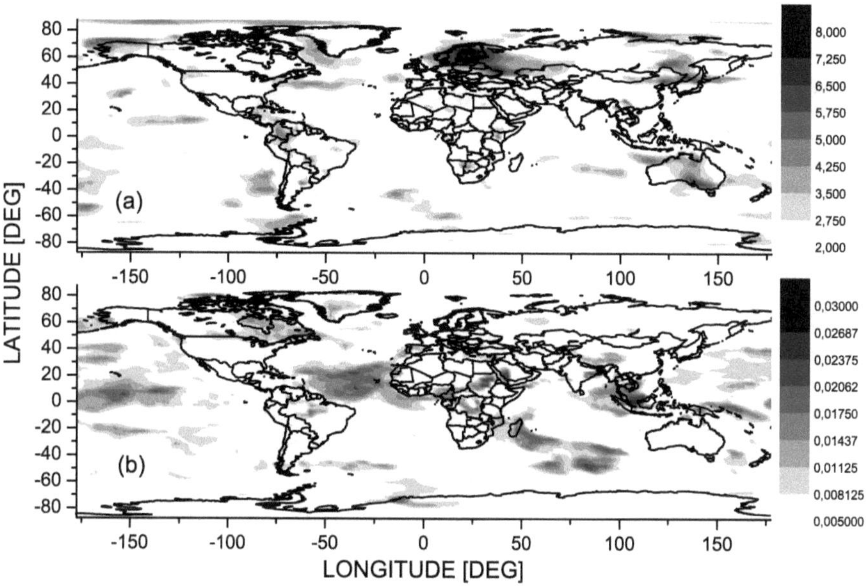

Fig. 15 (**a**) Dependence of the SATA time series from each node with the SATA time series in the node 22.5°E, 60°N, measured by MIR (13) estimated using the CCWT within the scales related to periods 7–8 years. (**b**) Area weighted connectivity for the conditional scale-specific SATA climate network with the density $\varrho = 0.005$ obtained by the uniform thresholding of the MIR (13) estimated using the CCWT within the scales related to the periods 7–8 years. The MIR between each two nodes is taken conditionally on the NAO index

other areas in the world (see Fig. 15a) is induced by the NAO. It is interesting that the hub in the tropical Atlantic survived the conditioning on the NAO, since Feliks et al. (2010, 2013) track the NAO 7- to 8-year oscillatory mode to an oscillation of a similar period in the position and strength of the Gulf Stream's sea surface temperature front in the North Atlantic. The position of the tropical Atlantic hub coincides with the sink region and a warm loop of the Gulf stream. So it seems that the tropical Atlantic area plays a role in the emergence of the 7–8 year oscillations in nonlinear atmosphere–ocean interactions in the Northern Atlantic, in particular in the dynamics of the NAO. Then the NAO induces this oscillatory mode in temperature variability in large areas in Europe, Asia as well as in other regions in the world (Fig. 13b). These observations concur with some findings of Feliks et al. (2013) and need further study and understanding. Detailed insight into the related atmospheric circulation phenomena is, however, out of the scope of this paper. Here we wanted to demonstrate the potential of the MIR estimated using the CCWT which gives the possibility to study either the total, or scale-specific or conditional connectivity in networks of interacting dynamical systems or spatio-temporal phenomena in a discrete approximation within the complex networks paradigm.

10 Conclusion

Using a simple example of the autoregressive process we have demonstrated how increasing dynamical memory, reflected, e.g., in stronger autocorrelations, leads to increasing bias in estimating dependence measures such as the absolute value of the correlation coefficient. Similar behavior can be observed also in estimates of the mutual information (Paluš and Vejmelka, 2007), or the mean phase coherence (Xu et al., 2006). We have observed how this phenomenon can bias the connectivity in climate networks since the time evolution of the air temperature anomalies, recorded in different geographical areas, has different dynamics. Also in other research fields where interaction/functional networks are inferred from experimental time series this problem can influence the results and skew their interpretation. For instance, many studies of EEG functional networks reported a changed network connectivity in different conscious states, however, changed EEG dynamics had been reported earlier in similar experimental conditions. The mutual information rate can be the dependence measure which can help to distinguish changes in connectivity and long-range synchrony from changes in the dynamics of network nodes. Also other authors (Baptista et al., 2012; Blanc et al., 2011) propose the MIR as an association measure suitable for inferring interaction networks from multivariate time series generated by coupled dynamical systems. Blanc et al. (2011) stress the independence of the MIR of the time lag which can occur between the time evolutions of two interacting systems or processes. This property might be particularly important considering the observation of Martin et al. (2013) regarding the construction of the climate networks from daily air temperature and geopotential height data. Inference of time lags in which the maximum cross-correlation occurs is unreliable and can lead to physically unrealistic large lags and even to the inclusion of non-existing links to the network.

In this paper we have proposed a computationally accessible algorithm based on the MIR of Gaussian processes, adapted by using the wavelet transform. We have demonstrated that this algorithm can be effective for nonlinear, nonstationary, and multiscale processes. Using the examples of the climate networks we have presented the ability of the scale-specific and conditional MIR to attribute different hubs of the climate network to different atmospheric circulation phenomena. We believe that the introduced approach can help in further understanding of complex systems and their dynamics which can be observed and recorded in the form of multivariate time series.

Acknowledgements The author would like to thank Professor A.A. Tsonis for many inspiring discussions and the kind invitation to the workshop on nonlinear dynamics in geosciences.

This study was supported by the Ministry of Education, Youth and Sports of the Czech Republic within the Program KONTAKT II, Project No. LH14001, and in its initial stage by the Czech Science Foundation, Project No. P103/11/J068.

References

Achard, S., R. Salvador, B. Whitcher, J. Suckling, and E. Bullmore. 2006. A resilient, low-frequency, small-world human brain functional network with highly connected association cortical hubs. *The Journal of Neuroscience* 26(1): 63–72.

Albert, R., and A.-L. Barabási. 2002. Statistical mechanics of complex networks. *Reviews of Modern Physics* 74: 47–97.

Anderson, P.W. 1972. More is different. *Science* 177(4047): 393–396.

Bandt, C., and B. Pompe. 2002. Permutation entropy: a natural complexity measure for time series. *Physical Review Letters* 88(17): 174102.

Baptista, M., E. Ngamga, P.R. Pinto, M. Brito, and J. Kurths. 2010. Kolmogorov-Sinai entropy from recurrence times. *Physics Letters A* 374(9): 1135–1140.

Baptista, M.S., R.M. Rubinger, E.R. Viana, J.C. Sartorelli, U. Parlitz, and C. Grebogi. 2012. Mutual information rate and bounds for it. *PLoS One* 7(10): e46745.

Barreiro, M., A.C. Marti, and C. Masoller. 2011. Inferring long memory processes in the climate network via ordinal pattern analysis. *Chaos: An Interdisciplinary Journal of Nonlinear Science* 21(1): 013101.

Bialonski, S., and K. Lehnertz. 2013. Assortative mixing in functional brain networks during epileptic seizures. *Chaos: An Interdisciplinary Journal of Nonlinear Science* 23(3): 033139.

Bialonski, S., M.-T. Horstmann, and K. Lehnertz. 2010. From brain to earth and climate systems: small-world interaction networks or not? *Chaos: An Interdisciplinary Journal of Nonlinear Science* 20(1): 013134.

Bialonski, S., M. Wendler, and K. Lehnertz. 2011. Unraveling spurious properties of interaction networks with tailored random networks. *PLoS One* 6(8): e22826.

Blanc, J.-L., L. Pezard, and A. Lesne. 2011. Delay independence of mutual-information rate of two symbolic sequences. *Physical Review E* 84: 036214.

Boccaletti, S., V. Latora, Y. Moreno, M. Chavez, and D.-U. Hwang. 2006. Complex networks: structure and dynamics. *Physics Reports* 424(4–5): 175–308.

Bullmore, E., and O. Sporns. 2009. Complex brain networks: graph theoretical analysis of structural and functional systems. *Nature Reviews Neuroscience* 10(3): 186–198.

Carpi, L., P. Saco, O. Rosso, and M. Ravetti. 2012. Structural evolution of the tropical pacific climate network. *The European Physical Journal B* 85(11): 1–7.

Casdagli, M., L. Iasemidis, J. Sackellares, S. Roper, R. Gilmore, and R. Savit. 1996. Characterizing nonlinearity in invasive {EEG} recordings from temporal lobe Epilepsy. *Physica D: Nonlinear Phenomena* 99(23): 381–399.

Cohen, A., and I. Procaccia. 1985. Computing the Kolmogorov entropy from time signals of dissipative and conservative dynamical systems. *Physical Review A* 31: 1872–1882.

Cover, T., and J. Thomas. 1991. *Elements of information theory*. New York: Wiley.

Crutchfield, J.P., and K. Young. 1989. Inferring statistical complexity. *Physical Review Letters* 63: 105–108.

Deza, J., M. Barreiro, and C. Masoller. 2013. Inferring interdependencies in climate networks constructed at inter-annual, intra-season and longer time scales. *The European Physical Journal Special Topics* 222(2): 511–523.

Donges, J., Y. Zou, N. Marwan, and J. Kurths. 2009a. The backbone of the climate network. *Europhysics Letters* 87: 48007.

Donges, J., Y. Zou, N. Marwan, and J. Kurths. 2009b. Complex networks in climate dynamics. *The European Physical Journal Special Topics* 174(1): 157–179.

Donges, J.F., H.C. Schultz, N. Marwan, Y. Zou, and J. Kurths. 2011. Investigating the topology of interacting networks. *The European Physical Journal B* 84: 635–651.

Feliks, Y., M. Ghil, and A.W. Robertson. 2010. Oscillatory climate modes in the Eastern Mediterranean and their synchronization with the North Atlantic oscillation. *Journal of Climate* 23: 4060–4079.

Feliks, Y., A. Groth, A.W. Robertson, and M. Ghil. 2013. Oscillatory climate modes in the Indian monsoon, North Atlantic, and tropical pacific. *Journal of Climate* 26(23): 9528–9544.

Fernandez, I., C.N. Hernandez, and J.M. Pacheco. 2003. Is the North Atlantic oscillation just a pink noise? *Physica A: Statistical Mechanics and Its Applications* 323: 705–714.

Fraedrich, K. 1986. Estimating the dimensions of weather and climate attractors. *Journal of the Atmospheric Sciences* 43(5): 419–432.

Fraser, A.M. 1989. Information and entropy in strange attractors. *IEEE Transactions on Information Theory* 35(2): 245–262.

Friston, K.J. 1994. Functional and effective connectivity in neuroimaging: a synthesis. *Human Brain Mapping* 2(1–2): 56–78.

Gozolchiani, A., K. Yamasaki, O. Gazit, and S. Havlin. 2008. Pattern of climate network blinking links follows El Niño events. *Europhysics Letters* 83(2): 28005.

Grassberger, P. 1986. Do climatic attractors exist? *Nature* 323: 609–612.

Grassberger, P., and I. Procaccia. 1983a. Estimation of the Kolmogorov entropy from a chaotic signal. *Physical Review A* 28(4): 2591–2593.

Grassberger, P., and I. Procaccia. 1983b. Measuring the strangeness of strange attractors. *Physica D: Nonlinear Phenomena* 9(12): 189–208.

Hartman, D., J. Hlinka, M. Paluš, D. Mantini, and M. Corbetta. 2011. The role of nonlinearity in computing graph-theoretical properties of resting-state functional magnetic resonance imaging brain networks. *Chaos: An Interdisciplinary Journal of Nonlinear Science* 21(1): 013119.

Haslinger, R., K.L. Klinkner, and C.R. Shalizi. 2010. The computational structure of spike trains. *Neural Computation* 22(1): 121–157.

Havlin, S., D. Kenett, E. Ben-Jacob, A. Bunde, R. Cohen, H. Hermann, J. Kantelhardt, J. Kertész, S. Kirkpatrick, J. Kurths, J. Portugali, and S. Solomon. 2012. Challenges in network science: applications to infrastructures, climate, social systems and economics. *The European Physical Journal Special Topics* 214(1): 273–293.

Hlaváčková-Schindler, K., M. Paluš, M. Vejmelka, and J. Bhattacharya. 2007. Causality detection based on information-theoretic approaches in time series Analysis. *Physics Reports* 441(1): 1–46.

Hlinka, J., M. Paluš, M. Vejmelka, D. Mantini, and M. Corbetta. 2011. Functional connectivity in resting-state FMRI: is linear correlation sufficient? *NeuroImage* 54(3): 2218–2225.

Hlinka, J., D. Hartman, and M. Paluš. 2012. Small-world topology of functional connectivity in randomly connected dynamical systems. *Chaos: An Interdisciplinary Journal of Nonlinear Science* 22(3): 033107.

Hlinka, J., D. Hartman, M. Vejmelka, J. Runge, N. Marwan, J. Kurths, and M. Paluš. 2013. Reliability of inference of directed climate networks using conditional mutual information. *Entropy* 15(6): 2023–2045.

Hlinka, J., D. Hartman, M. Vejmelka, D. Novotná, and M. Paluš. 2014. Non-linear dependence and teleconnections in climate data: sources, relevance, nonstationarity. *Climate Dynamics* 42(7–8): 1873–1886.

Holme, P., and J. Saramäki. 2012. Temporal networks. *Physics Reports* 519(3): 97–125.

Jiang, N., J. Neelin, and M. Ghil. 1995. Quasi-quadrennial and quasi-biennial variability in the equatorial pacific. *Climate Dynamics* 12(2): 101–112.

Jiruska, P., J. Csicsvari, A.D. Powell, J.E. Fox, W.-C. Chang, M. Vreugdenhil, X. Li, M. Palus, A.F. Bujan, R.W. Dearden, et al. 2010. High-frequency network activity, global increase in neuronal activity, and synchrony expansion precede epileptic seizures in vitro. *The Journal of Neuroscience* 30(16): 5690–5701.

Kalnay, E., M. Kanamitsu, R. Kistler, W. Collins, D. Deaven, L. Gandin, M. Iredell, S. Saha, G. White, J. Woollen, et al. 1996. The ncep/ncar 40-year reanalysis project. *Bulletin of the American Meteorological Society* 77(3): 437–471.

Kennel, M.B., J. Shlens, H.D. Abarbanel, and E. Chichilnisky. 2005. Estimating entropy rates with Bayesian confidence intervals. *Neural Computation* 17(7): 1531–1576.

Kondrashov, D., S. Kravtsov, A.W. Robertson, and M. Ghil. 2005. A hierarchy of data-based ENSO models. *Journal of Climate* 18(21): 4425–4444.

Kramer, M.A., U.T. Eden, S.S. Cash, and E.D. Kolaczyk. 2009. Network inference with confidence from multivariate time series. *Physical Review E* 79: 061916.

Kuhnert, M.-T., C.E. Elger, and K. Lehnertz. 2010. Long-term variability of global statistical properties of epileptic brain networks. *Chaos: An Interdisciplinary Journal of Nonlinear Science* 20(4): 043126.

Lehnertz, K., G. Ansmann, S. Bialonski, H. Dickten, C. Geier, and S. Porz. 2014. Evolving networks in the human epileptic brain. *Physica D: Nonlinear Phenomena* 267(0): 7–15.

Lesne, A., J.-L. Blanc, and L. Pezard. 2009. Entropy estimation of very short symbolic sequences. *Physical Review E* 79: 046208.

Lorenz, E.N. 1991. Dimension of weather and climate attractors. *Nature* 353: 241–244.

Malik, N., B. Bookhagen, N. Marwan, and J. Kurths. 2012. Analysis of spatial and temporal extreme monsoonal rainfall over South Asia using complex networks. *Climate Dynamics* 39(3–4): 971–987.

Marshall, J., Y. Kushnir, D. Battisti, P. Chang, A. Czaja, R. Dickson, J. Hurrell, M. McCartney, R. Saravanan, and M. Visbeck. 2001. North Atlantic climate variability: phenomena, impacts and mechanisms. *International Journal of Climatology* 21(15): 1863–1898.

Martin, E.A., M. Paczuski, and J. Davidsen. 2013. Interpretation of link fluctuations in climate networks during El Niño periods. *Europhysics Letters* 102(4): 48003.

Marwan, N., M. C. Romano, M. Thiel, and J. Kurths. 2007. Recurrence plots for the analysis of complex systems. *Physics Reports* 438(56): 237–329.

Matousek, M., J. Wackermann, M. Palus, A. Berankova, V. Albrecht, and I. Dvorak. 1995. Global dimensional complexity of the EEG in healthy volunteers. *Neuropsychobiology* 31(1): 47–52.

Newman, M., A.-L. Barabási, and D.J. Watts. 2006. *The structure and dynamics of networks*. Princeton: Princeton University Press.

Nicolis, C., and G. Nicolis. 1984. Is there a climatic attractor? *Nature* 311: 529–532.

Onnela, J., K. Kaski, and J. Kertész. 2004. Clustering and information in correlation based financial networks. *The European Physical Journal B* 38(2): 353–362.

Paluš, M. 1996a. Coarse-grained entropy rates for characterization of complex time series. *Physica D: Nonlinear Phenomena* 93(1–2): 64–77.

Paluš, M. 1996b. Nonlinearity in normal human EEG: cycles, temporal asymmetry, nonstationarity and randomness, not chaos. *Biological Cybernetics* 75(5): 389–396.

Paluš, M. 1997a. Kolmogorov entropy from time series using information-theoretic functionals. *Neural Network World* 7: 269–292.

Paluš, M. 1997b. On entropy rates of dynamical systems and gaussian Processes. *Physics Letters* A 227(5–6): 301–308.

Paluš, M. 2007. From nonlinearity to causality: statistical testing and inference of physical mechanisms underlying complex dynamics. *Contemporary Physics* 48(6): 307–348.

Paluš, M., and D. Novotná. 1994. Testing for nonlinearity in weather Records. *Physics Letters A* 193(1): 67–74.

Paluš, M., and D. Novotná. 2004. Enhanced Monte Carlo singular system analysis and detection of period 7.8 years oscillatory modes in the monthly NAO index and temperature records. *Nonlinear Processes in Geophysics* 11(5–6): 721–729.

Paluš, M., and D. Novotná. 2006. Quasi-biennial oscillations extracted from the monthly NAO index and temperature records are phase-synchronized. *Nonlinear Processes in Geophysics* 13(3): 287–296.

Paluš, M., and D. Novotná. 2007. Common oscillatory modes in geomagnetic activity, NAO index and surface air temperature records. *Journal of Atmospheric and Solar - Terrestrial Physics* 69(17–18): 2405–2415.

Paluš, M., and D. Novotná. 2009. Phase-coherent oscillatory modes in solar and geomagnetic activity and climate variability. *Journal of Atmospheric and Solar - Terrestrial Physics* 71(8–9): 923–930.

Paluš, M., and D. Novotná. 2011. Northern hemisphere patterns of phase coherence between solar/geomagnetic activity and ncep/ncar and era40 near-surface air temperature in period 78 years oscillatory modes. *Nonlinear Processes in Geophysics* 18(2): 251–260.

Paluš, M., and M. Vejmelka. 2007. Directionality of coupling from bivariate time series: how to avoid false causalities and missed connections. *Physical Review E* 75: 056211.

Paluš, M., I. Dvorak, and I. David. 1992. Spatiotemporal dynamics of human EEG. *Physica A: Statistical Mechanics and Its Applications* 185(1–4): 433–438.

Paluš, M., V. Albrecht, and I. Dvořák. 1993. Information theoretic test for nonlinearity in time-series. *Physics Letters A* 175(3–4): 203–209.

Paluš, M., V. Komarek, Z. Hrncir, and K. Sterbova. 2001. Synchronization as adjustment of information rates: detection from bivariate time series. *Physical Review E* 63(4): 046211.

Paluš, M., D. Hartman, J. Hlinka, and M. Vejmelka. 2011. Discerning connectivity from dynamics in climate networks. *Nonlinear Processes in Geophysics* 18(5): 751–763.

Pawelzik, K., and H.G. Schuster. 1987. Generalized dimensions and entropies from a measured time series. *Physical Review A* 35: 481–484.

Pesin, Y.B. 1977. Characteristic Lyapunov exponents and smooth ergodic theory. *Russian Mathematical Surveys* 32(4): 55.

Petersen, K.E.. 1989. *Ergodic theory*. Cambridge: Cambridge University Press.

Pincus, S.M. 1991. Approximate entropy as a measure of system complexity. *Proceedings of the National Academy of Science* 88(6): 2297–2301.

Pinsker, M.S. 1964. *Information and information stability of random variables and processes*. San Francisco: Holden-Day.

Radebach, A., R.V. Donner, J. Runge, J.F. Donges, and J. Kurths. 2013. Disentangling different types of El Niño episodes by evolving climate network analysis. *Physical Review E* 88: 052807.

Reijneveld, J.C., S.C. Ponten, H.W. Berendse, and C.J. Stam. 2007. The application of graph theoretical analysis to complex networks in the brain. *Clinical Neurophysiology* 118(11): 2317–2331.

Sarachik, E.S., and M.A. Cane. 2010. *The El Niño-Southern Oscillation phenomenon*. Cambridge: Cambridge University Press.

Schelter, B., M. Winterhalder, R. Dahlhaus, J. Kurths, and J. Timmer. 2006. Partial phase synchronization for multivariate synchronizing systems. *Physical Review Letters* 96: 208103.

Scholz, M. 2010. Node similarity as a basic principle behind connectivity in complex networks. ArXiv 1010.0803.

Schouten, J.C., F. Takens, and C.M. van den Bleek. 1994. Maximum-likelihood estimation of the entropy of an attractor. *Physical Review E* 49: 126–129.

Schweitzer, F., G. Fagiolo, D. Sornette, F. Vega-Redondo, A. Vespignani, and D. White. 2009. Economic networks: the new challenges. *Science* 325(5939): 422.

Shalizi, C., and J. Crutchfield. 2001. Computational mechanics: pattern and prediction, structure and simplicity. *Journal of Statistical Physics* 104(3–4): 817–879.

Sheppard, L.W., A. Stefanovska, and P.V.E. McClintock. 2011. Detecting the harmonics of oscillations with time-variable frequencies. *Physical Review E* 83: 016206.

Shlens, J., M.B. Kennel, H.D. Abarbanel, and E. Chichilnisky. 2007. Estimating information rates with confidence intervals in neural spike trains. *Neural Computation* 19(7): 1683–1719.

Soler, J.M. 2007. A rational indicator of scientific creativity, *Journal of Informetrics* 1(2): 123–130.

Steinhaeuser, K., A. Ganguly, and N. Chawla. 2012. Multivariate and multiscale dependence in the global climate system revealed through complex networks. *Climate Dynamics* 39(3–4): 889–895.

Theiler, J., and P.E. Rapp. 1996. Re-examination of the evidence for low-dimensional, nonlinear structure in the human electroencephalogram. *Electroencephalography and Clinical Neurophysiology* 98(3): 213–222.

Theiler, J., S. Eubank, A. Longtin, B. Galdrikian, and J.D. Farmer. 1992. Testing for nonlinearity in time series: the method of surrogate data. *Physica D: Nonlinear Phenomena* 58: 77–94.

Torrence, C., and G.P. Compo. 1998. A practical guide to wavelet analysis. *Bulletin of the American Meteorological Society* 79(1): 61–78.

Tsonis, A.A., and J.B. Elsner. 1988. The weather attractor over very short timescales. *Nature* 333: 545–547.

Tsonis, A., and P. Roebber. 2004. The architecture of the climate network. *Physica A: Statistical Mechanics and Its Applications* 333: 497–504.
Tsonis, A., and K. Swanson. 2008. Topology and predictability of El Niño and La Niña networks. *Physical Review Letters* 100(22): 228502.
Tsonis, A., K. Swanson, and P. Roebber. 2006. What do networks have to do with climate? *Bulletin of the American Meteorological Society* 87(5): 585–596.
Tsonis, A., K. Swanson, and G. Wang. 2008. On the role of atmospheric teleconnections in climate. *Journal of Climate* 21(12): 2990–3001.
Xu, L., Z. Chen, K. Hu, H.E. Stanley, and P.C. Ivanov. 2006. Spurious detection of phase synchronization in coupled nonlinear oscillators. *Physical Review E* 73: 065201.
Yamasaki, K., A. Gozolchiani, and S. Havlin. 2008. Climate networks around the globe are significantly affected by El Niño. *Physical Review Letters* 100(22): 228501.
Yamasaki, K., A. Gozolchiani, and S. Havlin. 2009. Climate networks based on phase synchronization analysis track El-Niño. *Progress of Theoretical Physics Supplement* 179(179): 178–188.
Zalesky, A., A. Fornito, and E. Bullmore. 2012. On the use of correlation as a measure of network connectivity. *NeuroImage* 60(4): 2096–2106.
Ziv, J., and A. Lempel. 1978. Compression of individual sequences via variable-rate coding. *IEEE Transactions on Information Theory* 24(5): 530–536.

Non-Extensive Statistical Mechanics: Overview of Theory and Applications in Seismogenesis, Climate, and Space Plasma

G.P. Pavlos, L.P. Karakatsanis, A.C. Iliopoulos, E.G. Pavlos, and A.A. Tsonis

Abstract In this small review, the theoretical framework of non-extensive statistical theory, introduced by Constantino Tsallis in 1988, is presented in relation with the q-triplet estimation concerning experimental time series from climate, seismogenesis, and space plasmas systems. These physical systems reveal common dynamical, geometrical, or statistical characteristics. Such characteristics are low dimensionality, typical intermittent turbulence multifractality, temporal or spatial multiscale correlations, power law scale invariance, non-Gaussian statistics, and others. The aforementioned phenomenology has been attributed in the past to chaotic or self-organized critical (SOC) universal dynamics. However, after two or three decades of theoretical development of the complexity theory, a more compact theoretical description can be given for the underlying universal physical processes which produce the experimental time series complexity. In this picture, the old reductionist view of universality of particles and forces is extended to the modern universality of multiscale complex processes from the microscopic to the macroscopic level of different physical systems. In addition, it can be stated that a basic and universal organizing principle exists creating complex spatio-temporal and multiscale different physical structures or different dynamical scenarios at every physical scale level. The best physical representation of the underline universal organizing principle is the well-known entropy principle. Tsallis introduced a q-entropy (S_q) as a non-extensive (q-extension) of the Boltzmann–Gibbs (BG) entropy (for $q = 1$, the BG entropy is restored) and statistics in order to describe efficiently the rich phenomenology that complex systems exhibit. Tsallis q-entropy could be a strong candidate for entropy principle according to which nature creates complex structures everywhere, from the microscopic to the macroscopic level, trying to succeed the extremization of the Tsallis entropy. In addition, this Sq

G.P. Pavlos • L.P. Karakatsanis (✉) • A.C. Iliopoulos • E.G. Pavlos
Department of Electrical and Computer Engineering, Research Team of Chaos and Complexity, Democritus University of Thrace, Xanthi 67100, Greece
e-mail: lkaraka@gmail.com

A.A. Tsonis
Department of Mathematical Sciences, Atmospheric Sciences Group, University of Wisconsin - Milwaukee, Milwaukee, WI 53201, USA

Hydrologic Research Center, San Diego, CA, USA

A.A. Tsonis (ed.), *Advances in Nonlinear Geosciences*,
DOI 10.1007/978-3-319-58895-7_22

entropy principle is harmonized with the q extension of the classic and Gaussian central limit theorem (q-CLT). The q-extension of CLT corresponds to the Levy a-stable extension of the Gaussian attractor of the classic statistical theory. The q-CLT is related to the Tsallis q-triplet theory of random time series with non-Gaussian statistical profile. Moreover Tsallis q-extended entropy principle can be used as the theoretical framework for the unification of some new dynamical characteristics of complex systems such as the spatio-temporal fractional dynamics, the anomalous diffusion processes and the strange dynamics of Hamiltonian and dissipative dynamical systems, the intermittent turbulence theory, the fractional topological and percolation phase transition processes according to Zelenyi and Milovanov non-equilibrium and non-stationary states (NESS) theory, as well as the non-equilibrium renormalization group theory(RNGT) of distributed dynamics and the reduction of dynamical degrees of freedom.

Keywords Non-extensive statistics • Tsallis q-triplet • Complexity • Climate • Space plasmas • Magnetosphere • Solar wind • Sunspot • Solar flares

1 Introduction

The last 20 years, complexity theory advances rapidly giving emphasis to nonlinear dynamics, fractal theory, and the fractional calculus and statistical physics. The basic tool for the comparison of complexity theory with the physical reality is the nonlinear time series analysis. Near thermodynamic equilibrium statistics and dynamics are two separated but fundamental elements of the physical theory. Also, at thermodynamical equilibrium, nature reveals itself as a Gaussian and macroscopically uncorrelated process simultaneously with unavoidable or inevitable and objective deterministic character. However, modern evolution of the scientific knowledge reveals the equilibrium characteristics of the physical theory as an approximation or the limit of a more synthetic physical theory which is characterized as complexity theory. The new physical characteristics of complexity theory can be manifested as a physical system is driven in far from equilibrium states.

At 1988, Constantino Tsallis for describing complex systems introduced the q-generalization of Boltzmann–Gibbs (BG) entropy and statistical mechanics (Tsallis 1988). In particular, Tsallis non-extensive statistical mechanics (Tsallis 2004a) includes the generalization of BG statistical mechanics, inspired by multifractal theory, extending BG entropy principle to the Tsallis q-entropy principle where the BG statistical theory is a special case corresponding to the value $q = 1$.

Tsallis theory indicates that Nature works at every level creating physical states corresponding to q-entropy extremes, or more general types of entropy in contrast with the historical Gaussian thermodynamical equilibrium states. Non-equilibrium critical phenomena and phase transition processes exhibit a rich phenomenology which includes, among others, heavy tailed power laws, scale invariance and multiscale correlations, long-range spatio-temporal correlations, multifractal and hierarchical structures, anomalous diffusion, anomalous transports, self-organized percolations states, coherent states with memory, topological phase transition and fractal topology-fracton states, etc. All these characteristics, which can be observed in almost every complex system, can be understood as a general local–non-local ordering principle of Nature included in the q-entropy principle of Tsallis non-extensive statistical mechanics.

Moreover, as usually happens with any novel physical theory, new mathematical concepts are used. Examples are non-Euclidian geometry and Riemannian smooth manifolds for relativity theory, Hilbert spaces and operators for quantum theory, etc. Similarly, q-mathematics and fractional calculus are used for the mathematical formulation of Tsallis statistical mechanics theory and Tsallis q-entropy principle. The importance of the generalized principle of Tsallis q-entropy is proportional to that of the famous time arrow. Time being irreversible means that every natural process is associated with entropy production. Thus, Nature generates novelty, information, ordered structures, and long-range correlations. Mathematically, based on the fractional calculus, integration–derivation, the well-known Langevin and Fokker–Planck (FP) equations are generalized to fractional ones, the solutions of which correspond to multifractal spatio-temporal structures. These structures are described by singular and fractional functions and correspond to q-entropy extremes. Thus, the q-entropy principle is related to the holistic, multiscale, and globally correlated spatio-temporal structures.

The Tsallis q-entropy is also related to strange kinetics and fractal–multifractal profile of the phase space, which causes long-range spatio-temporal correlations and non-Gaussian probability distribution functions of the dynamical fluctuations as well as non-Poisson temporal distributions (Zaslavsky 2002). These characteristics, along with critical exponents, power laws or heavy tails of probability distribution functions, as well as singularities and spatial–temporal fractality of non-differentiable distribution of the physical functions and physical properties–quantities, are related to the strange topology of phase space caused by the q-entropy principles of Tsallis theory as the dynamical system tries to achieve extremization of q-entropy states (Alemany and Zanette 1994).

Therefore, far from equilibrium statistics and dynamics can be unified through the Tsallis non-extensive statistics included in Tsallis q-entropy theory (Tsallis 2009) and the fractal generalization of dynamics included in theories developed by Ord (1983), Nottale (2006), Castro (2005), Zaslavsky (2002), Shlesinger et al. (1993), Tarasov (2005), El-Nabulsi (2005), Goldfain (2007)), and others.

For the last 6 years and in a series of papers, we applied tools from nonlinear time series analysis and new concepts included in Tsallis theory in various complex systems, such as space plasmas, earthquakes, climate, brain dynamics, and recently in DNA and material dynamics. The results verified the presence of non-extensive statistical mechanics characteristics in all the aforementioned complex systems (Karakatsanis and Pavlos 2008; Karakatsanis et al. 2012; Karakatsanis et al. 2013;

Pavlos et al. 2011, 2012a, b, 2014, 2015, 2016; Iliopoulos et al. 2012, 2015a, b, 2016a; Iliopoulos 2016b). The paper is organized as follows: In Sect. 2 we present the basic theoretical framework of Tsallis non-extensive statistics in Sect. 3 a brief summary of results concerning the estimation of Tsallis q-triplet for various geophysical systems is given, and finally in Sects. 4 and 5 we present the theoretical interpretations of the q-triplet results and the closing remarks of this study.

2 The General Framework of Tsallis Statistics

Non-extensive Tsallis statistical theory is connected to the q-extension of exponential and logarithmic functions and the q-extension of a Fourier transform (FT), (Tsallis 2009). The q-extension of mathematics underlying the q-extension of statistics is presented under the solution of the nonlinear equation

$$\frac{dy}{dx} = y^q, (y(0) = 1, q \in R) \tag{1}$$

Its solution is the q-exponential function e_q^x

$$e_q^x \equiv [1 + (1 - q)\, x]^{1/(1-q)} \tag{2}$$

The q-extension of logarithmic function is the reverse of e_q^x

$$\ln_q x \equiv \frac{x^{1-q} - 1}{1 - q} \tag{3}$$

The q-logarithm satisfies the property

$$\ln_q (x_A x_B) = \ln_q x_A + \ln_q x_B + (1 - q) \left(\ln_q x_A\right) \left(\ln_q x_B\right) \tag{4}$$

According to the pseudo-additive property of the q-logarithm, a generalization of the product and sum as the q-product and q-sum can be introduced in (1)

$$x \otimes_q y \equiv e_q^{\ln_q x + \ln_q y} \tag{5}$$

$$x \oplus_q y \equiv x + y + (1 - q)\, xy \tag{6}$$

Moreover, in the context of the q-generalization of the Central Limit Theorem (CLT) the q-extension of FT can be introduced in Eq. (1)

$$F_q [p] (\xi) \equiv \int_{-\infty}^{+\infty} dx e_q^{ix\xi [p(x)]^{q-1}} p(x), (q \leq 1) \tag{7}$$

Tsallis, inspired by multifractal analysis (Tsallis 2009), proposed that the BG entropy

$$S_{\mathrm{BG}} = -k \sum p_i \ln p_i = k < \ln (1/p_i) > \tag{8}$$

cannot describe all the rich phenomenology of nonlinear dynamic systems, since BG statistical theory presupposes ergodicity of the underlying dynamics in the system phase space. However, the complexity of dynamics is far beyond from simple ergodic, therefore Tsallis introduced a generalization of BG entropy based on the extended concept of q-entropy:

$$S_q = k \left(1 - \sum_{i=1}^{N} p_i^q \right) / (q-1) = k < \ln_q (1/p_i) > \tag{9}$$

For a continuous state space, we have

$$S_q = k \left[1 - \int [p(x)]^q dx \right] / (q-1) \tag{10}$$

For a system of particles and fields with short-range correlations in their immediate neighborhoods, the Tsallis q-entropy S_q asymptotically leads to BG entropy (S_{BG}) corresponding to $q = 1$. For probabilistically dependent or correlated system A and B, it can be proven that

$$\begin{aligned} S_q (A+B) &= S_q(A) + S_q (B/A) + (1-q) S_q(A) S_q (B/A) \\ &= S_q(B) + S_q (A/B) + (1-q) S_q(B) S_q (A/B) \end{aligned} \tag{11}$$

where $S_q(A) \equiv S_q\left(\{p_i^A\}\right)$,$S_q(B) \equiv S_q\left(\{p_i^B\}\right)$, $S_q(B/A)$, and $S_q(A/B)$ are the conditional entropies of systems A, B. When the systems are probabilistically independent, then relation (11) changes to

$$S_q (A+B) = S_q(A) + S_q(B) + (1-q) S_q(A) S_q(B) \tag{12}$$

The first part of $S_q(A+B)$ is additive $(S_q(A) + S_q(B))$ while the second part is multiplicative including long-range correlations supporting the macroscopic ordering phenomena.

2.1 *Intermittent Turbulence and Multifractality Via Non-Extensive Statistics*

The fractal–multifractal structuring of phase space, caused by the nonlinear dynamics includes islands, cantori, and stickiness and is related to singular measures, singular (irregular) functions of space and time (fractal functions), as well as to scale invariance properties and multiscale interaction causing long-range correlations and hierarchical structures (Shlesinger et al. 1987; Shlesinger 1988; Shlesinger et al. 1993; Zaslavsky 2002).

The q-extended statistical mechanics and the Tsallis q-distributions correspond to general power law probability functions in phase space with local singularities (α) related to the singularity spectrum functions $f(\alpha)$ and the generalized fractal dimension spectrum functions $D_{\overline{q}}$ (Theiler 1990; Arneodo et al. 1995). The multifractal structure of the phase space can be described by the generalized Rényi fractal dimensions

$$D_{\overline{q}} = \frac{1}{\overline{q}-1}\lim_{\lambda\to 0}\frac{\log\sum_{i=1}^{N} p_i^{\overline{q}}}{\log\lambda}, \tag{13}$$

where $p_i \sim \lambda^{\alpha(i)}$ is the local probability at location (i) in the phase space, λ is the local size of phase space, and $a(i)$ is the local singularity (r) point wise dimension of the dynamics. The Rényi $\overline{q}$ numbers (different from the q-index of Tsallis statistics) take values in the entire region $(-\infty\ , +\infty)$ of real numbers. The spectrum of distinct local point wise dimensions $\alpha(i)$ is given by the estimation of the function $f(\alpha)$ defined by the scaling of the density $n(a,\lambda)$ $\lambda^{-f(a)}$, where $n(a,\lambda)da$ is the number of local regions that have a scaling index between a and $a+da$. This reveals $f(a)$ as the fractal dimension of points with scaling index a. The fractal dimension $f(a)$ which varies with a shows the multifractal character of the phase space dynamics which includes interwoven sets of singularity of strength a, by their own fractal measure $f(a)$ of dimension (Halsey et al. 1986; Theiler 1990).

The multifractal spectrum $D_{\overline{q}}$ of the Renyi dimensions can be related to the spectrum $f(a)$ of local singularities using the following:

$$\sum p_i^{\overline{q}} = \int d\alpha' p\left(\alpha'\right)\lambda^{-f(\alpha\prime)}d\alpha' \tag{14}$$

$$\tau\left(\overline{q}\right) \equiv \left(\overline{q}-1\right)D\overline{q} \overset{\min_{a}}{=} \overline{q}\alpha - f\left(\alpha\right) \tag{15}$$

$$a\left(\overline{q}\right) = \frac{d\left[\tau\left(\overline{q}\right)\right]}{d\overline{q}} \tag{16}$$

$$f\left(\alpha\right) = \overline{q}\alpha - \tau\left(\overline{q}\right) \tag{17}$$

2.2 Tsallis Central Limit Theorem Extension and q-Triplet

According to the Tsallis q-extension of the entropy principle, any stationary random variable can be described as the stationary solution of a generalized fractional diffusion equation. For metastable stationary solutions of a stochastic process, the maximum entropy principle of BG statistical theory can faithfully be described by the maximum (extreme) of the Tsallis q-entropy function. Extremization of Tsallis q-entropy corresponds to the q-generalized form of the normal distribution function

$$p_q(x) = A_q \sqrt{\beta} e_q^{-\beta\left(x<x>_q\right)^2} \tag{18}$$

where $A_q = \sqrt{(q-1)/\pi}\Gamma\left(1/(q-1)\right)/\Gamma\left((3-q)/[2/(q-1)]\right)$ for $q>1$,
and $A_q = \sqrt{(1-q)/\pi}\Gamma\left((5-3q)/[2(1-q)]\right)/\Gamma\left((2-q)/(1-q)\right)$ for $q<1$, $\Gamma(z)$ being the Riemann function.

The q-extension of statistics also includes q-extension of the CLT, which can faithfully describe non-equilibrium long-range correlations in a complex system. The normal CLT concerns Gaussian random variables (x_i) for which the sum $Z = \sum_{i=1}^{N} x_i$ gradually tends to a Gaussian process as $N \to \infty$, while its fluctuations tend to zero, in contrast to the possibility of non-equilibrium fluctuations with long-range correlations. Using the FT q-extension, we can prove that q-independence means independence for $q = 1$ (normal CLT), but for $q \neq 1$ it means strong correlation (q-extended CLT). In this case ($q \neq 1$), the number of allowed states $W_{A_1+A_2+\cdots+A_N}$ in a system composed of $(A_1, A_2, \ldots, A_N)$ subsystems is expected to be less than $W_{A_1+A_2+\cdots+A_N} = \Pi_{i=1}^{N} W_{A_i}$ where $W_{A_1}, W_{A_2}, \ldots, W_{A_N}$ are the possible states of the subsystems. This means self-organization of dynamics for $q \neq 1$ and development of long-range correlations in space and time.

The deeper theoretical foundation of Tsallis q-triplet is included in the extended q-extended CLT (Umarov et al. 2008). In particular, the extended q-CLT states that an appropriately scaled limit of sums of q_k correlated random variables is a q_{k-1}-Gaussian, which is the q_k^*-Fourier image of a q_k^*-Gaussian. The q_k, q_k^* are sequences

$$q_k = \frac{2q + k(1-q)}{2 + k(1-q)} \quad \text{and} \quad q_k^* = q_{k-1} \quad \text{for } k = 0, \pm 1, \pm 2, \ldots \tag{19}$$

including the triplet (P_{att}, P_{cor}, P_{scl}), where P_{att}, P_{cor}, and P_{scl} are parameters of attractor, correlation, and scaling rate, respectively, and corresponds to the q-triplet (q_{sens}, q_{rel}, q_{stat}) according to the relations.

$$\left(P_{\text{att}}, P_{\text{cor}}, P_{\text{scl}}\right) \equiv \left(q_{k-1}, q_k, q_{k+1}\right) \equiv \left(q_{\text{sens}}, q_{\text{rel}}, q_{\text{stat}}\right) \tag{20}$$

The parameter $P_{\text{att}} \equiv q_{\text{sens}} \equiv q_{k-1}$ describes the non-ergodic q-entropy production of the multiscale correlated process as the system shifts to the state of the q_{att}-Gaussian, where the q-entropy is extremized in accordance with the generalization of the Pesin's theorem (Tsallis 2004b)

$$K_{\text{qsen}} \equiv \lim_{t\to\infty} \lim_{W\to\infty} \lim_{M\to\infty} \frac{S_q\left(P_i(t)\right)/k}{t} = \lambda_{\text{qsen}} \tag{21}$$

The parameter $P_{\mathrm{cor}} \equiv q_{\mathrm{rel}} \equiv q_k$ describes the q-correlated random variables participating to the dynamical process of the q-entropy production and the relaxation process towards the stationary state. The parameter $P_{\mathrm{slc}} \equiv q_{\mathrm{stat}} \equiv q_{k+1}$ describes the scale invariance profile of the stationary state corresponding to the scale invariant q-Gaussian attractor as well as to an anomalous diffusion process mirrored at the variance scaling according to general, asymptotically scaling, from

$$N^D P_x(x) \sim G\left(\frac{x}{N^D}\right) \tag{22}$$

where $P_x(x)$ is the probability function of the self-similar statistical attractor G and D is the scaling exponent characterizing the anomalous diffusion process (Baldovin and Stella 2007)?

$$\left\langle x^2 \right\rangle \sim t^{2D} \tag{23}$$

The non-Gaussian multiscale correlation can create the intermittent multifractal structure of the phase space mirrored also in the physical space multifractal distribution of the turbulent dissipation field. The multiscale interaction at non-equilibrium critical NESS creates the heavy tail and power law probability distribution function obeying the q-entropy principle. The singularity spectrum of a critical NESS corresponds to extremized Tsallis q-entropy.

The q_{stat} Index and Non-Extensive Physical States

A long-range-correlated meta-equilibrium non-extensive physical process can be described by the nonlinear differential equation (Tsallis 2009)

$$\frac{d\left(p_i Z_{q_{\mathrm{stat}}}\right)}{dE_i} = -\beta q_{\mathrm{stat}}\left(p_i Z_{q_{\mathrm{stat}}}\right)^{q_{\mathrm{stat}}} \tag{24}$$

The solution of this equation corresponds to the probability distribution

$$p_i = e_{q_{\mathrm{stat}}}^{-\beta_{\mathrm{stat}} E_i} / Z_{q_{\mathrm{stat}}} \tag{25}$$

where $\beta_{q_{\mathrm{stat}}} = \frac{1}{KT_{\mathrm{stat}}}$,and $Z_{q_{\mathrm{stat}}} = \sum_j e_{q_{\mathrm{stat}}}^{-\beta q_{\mathrm{stat}} E_j}$. Then the probability distribution is given by

$$p_i \propto \left[1 - (1-q)\, \beta_{q_{\mathrm{stat}}} E_i\right]^{1/1-q_{\mathrm{stat}}} \tag{26}$$

for discrete energy states $\{E_i\}$ and by

$$p(x) \propto \left[1 - (1-q)\, \beta_{q_{\mathrm{stat}}} x^2\right]^{1/1-q_{\mathrm{stat}}} \tag{27}$$

for continuous x states of $\{X\}$, where the values of the magnitude X correspond to the state points of the phase space. Distributions functions (26) and (27) correspond to the attracting stationary solution of the extended (anomalous) diffusion equation related to the nonlinear dynamics of the system. The stationary solutions $p(x)$ describe the probabilistic character of the dynamics on the attractor set of the phase space. The non-equilibrium dynamics can evolve on distinct attractor sets, depending upon the control parameters, while the q_{stat} exponent can change as the attractor set of the dynamics changes.

The q_{sen} Index and the Entropy Production Process

Entropy production is related to the general profile of the attractor set of the dynamics. The profile of the attractor can be described by its multifractality as well as by its sensitivity to initial conditions. The sensitivity to initial conditions can be expressed as

$$\frac{d\xi}{dt} = \lambda_1 \xi + \left(\lambda_{q_{\text{sen}}} - \lambda_1\right) \xi^{q_{\text{sen}}} \tag{28}$$

where ξ is the trajectory deviation in the phase space: $\xi \equiv \lim_{\Delta(x) \to 0}\{\Delta x(t)/\Delta x(0)\}$ and $\Delta x(t)$ is the distance between neighboring trajectories (Tsallis 2002). The solution of Eq. (28) is given by

$$\xi = \left[1 - \frac{\lambda q_{\text{sen}}}{\lambda_1} + \frac{\lambda q_{\text{sen}}}{\lambda_1} e^{(1-q_{\text{sen}})\lambda_1 t}\right]^{\frac{1}{1-q_{\text{sen}}}} \tag{29}$$

The q_{sen} exponent is related to the multifractal profile of the attractor set according to

$$\frac{1}{q_{\text{sen}}} = \frac{1}{a_{\min}} - \frac{1}{a_{\max}} \tag{30}$$

where $a_{\min}(a_{\max})$ corresponds to zero points of the multifractal exponent spectrum $f(a)$, that is $f(a_{\min}) = f(a_{\max}) = 0$.

The deviations of neighboring trajectories and the multifractality of the dynamic attractor set are related to the chaotic phenomenon of entropy production according to Kolmogorov–Sinai entropy production theory and the Pesin's theorem. The q-entropy production can be expressed as

$$K_q \equiv \lim_{t\to\infty} \lim_{W\to\infty} \lim_{N\to\infty} \frac{< S_q > (t)}{t} \tag{31}$$

where W is the number of non-overlapping windows in the phase space and N the state points in the windows according to $\sum_{i=1}^{W} N_i = N$. S_q is estimated using the probability $P_i(t) \equiv N_i(t)/N$. According to Tsallis, K_q is finite only for $q = q_{\text{sen}}$

The q_{rel} Index and the Relaxation Process

Thermodynamic fluctuation–dissipation theory (Chame and De Mello 1994) is based on the Einstein original diffusion theory (Brownian motion theory). Diffusion is a physical mechanism for extremization of entropy. If ΔS denote the deviation of entropy from its equilibrium value S_0, then the probability of a proposed fluctuation is given by

$$P \sim \exp\left(\Delta S/k\right) \tag{32}$$

The Einstein–Smoluchowski theory of Brownian motion was extended to the general FP diffusion theory of non-equilibrium processes. The potential of FP equation may include many meta-equilibrium stationary states near or far away from thermodynamical equilibrium. Macroscopically, relaxation to the equilibrium stationary state of some dynamical observable $O(t)$ related to system evolution in the phase space can be described by the form of general form

$$\frac{d\Omega}{dt} \simeq -\frac{1}{\tau}\Omega \tag{33}$$

where $\Omega(t) \equiv [O(t) - O(\infty)]/[O(0) - O(\infty)]$ describes relaxation of the macroscopic observable $O(t)$ towards its stationary state value. The non-extensive generalization of fluctuation–dissipation theory is related to the general correlated anomalous diffusion processes (Tsallis 2009). The equilibrium relaxation process is transformed to the meta-equilibrium non-extensive relaxation process according to

$$\frac{d\Omega}{dt} = -\frac{1}{T_{q_{\text{rel}}}}\Omega^{q_{\text{rel}}} \tag{34}$$

the solution of this equation is given by:

$$\Omega(t) \simeq e_{q_{\text{rel}}}^{-t/\tau_{\text{rel}}} \tag{35}$$

The autocorrelation function $C(\tau)$ or the mutual information $I(\tau)$ can be used as candidate observables $\Omega(t)$ for estimation of q_{rel}. However, in contrast to the linear profile of the correlation function, the mutual information includes the nonlinearity of the underlying dynamics and it is proposed as a more faithful index of the relaxation process and the estimation of the Tsallis exponent q_{rel}.

3 Applications of Tsallis Statistics in Various Geophysical Systems

In this section, we presented the q-triplet estimation of Tsallis statistics concerning experimental time series from climate, seismogenesis, and space plasmas systems.

3.1 Seismogenesis

In Iliopoulos et al. (2012) and Pavlos et al. (2014) seismogenesis of Greece and adjacent areas was studied using Tsallis statistics. Time series considered were interevent times and moment magnitude. The results revealed the efficiency of Tsallis statistics in describing the non-Gaussian statistics of the time series. In the following we present results (as an example) concerning the estimation of Tsallis q-triplet for Hellenic seismogenesis (Pavlos et al. 2014). Figure 1a–f presents the time series of magnitude data (Fig. 1a), the Tsallis q-Gaussian distribution (Fig. 1b) along with the fitting (Fig. 1c), the multifractal spectrum (Fig. 1d), the generalized dimension function (Fig. 1e), and finally the auto-mutual information function in log–log

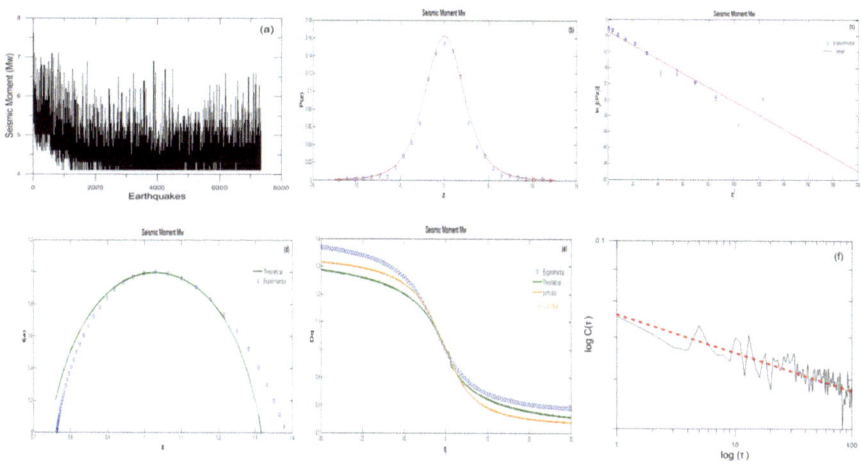

Fig. 1 (**a**) Time series of Seismic Moment (Mw) time series (**b**) PDF $P(z_i)$ (*blue dots*) vs. $z_i q$ Gaussian function (*red line*) that fits $P(z_i)$ for the seismic moment (**c**) Linear Correlation between $\ln_q P(z_i)$ and $(z_i)^2$ where $q_{\text{stat}} = 1.43 \pm 0.123$ for the seismic moment (**d**) Multifractal spectrum of seismic moment time series (*blue dots*) with solid line a sixth-degree polynomial. We calculate the $q_{\text{sen}} = -0.6957 \pm 0.0052$. (**e**) $D(q)$ vs. q of the seismic moment time series (*blue dots*) (**f**) Log–log plot of the self-correlation coefficient $C(\tau)$ vs. time delay τ for the seismic moment time series. We obtain the best fit with $q_{\text{rel}} = 17.418 \pm 1.514$

plot correspondingly. The Tsallis q-triplet values were found to be different from unity and to satisfy the relation: $q_{\mathrm{sen}} < 1 < q_{\mathrm{stat}} < q_{\mathrm{rel}} : -0.6957 < 1 < 1.43 < 17.418$ (for more details concerning the specific results, see Pavlos et al. 2014).

3.2 Climate

Geopotential Height Spatial Series

In this paragraph, we present new results concerning spatial series corresponding to Geopotential Height index. The data are taken from the website of National Oceanic and Atmospheric Administration (NOAA) (https://www.esrl.noaa.gov/psd/data/gridded/data.ncep.reanalysis.pressure.html). They include the global distribution of the Geopotential Height (GH) (Kalnay et al. 1996) index of month average values in 1000 mb pressure level, during the period (1948–2012) measured in meters. The spatial coverage of the measurement consists by 2.5 × 2.5 degree global grids (144 × 73) from 0.0E to 357.5E, 90.0N to 90.0S. We constructed the spatial distribution of the GH index for the month January of specific years 1948 (shown as an example in Fig. 2a), 1960, 1970, 1980, 1990, 2000, and 2012 and for different regions, namely: (a) all planet, (b) north hemisphere, and (c) south hemisphere, respectively.

In order to proceed statistical analysis, we apply the first difference filter (an example is shown in Fig. 2c) at the spatial series of GH index in order to exclude the low periodicity component. In this way, we focus on the rapid fluctuations of the raw data.

In Fig. 3a–f we presented the PDF $P(z_i)$ vs. z_i q Gaussian function that fits $P(z_i)$, the multifractal spectrum, and the log–log plot of the self-correlation coefficient $I(\tau)$ vs. time delay τ for the GH spatial series (JAN, 1960) for north and south hemisphere. In both cases we observed clearly non-Gaussian statistics. The results showed that the q-triplet values were found to satisfy the relation $q_{\mathrm{sen}} < 1 < q_{\mathrm{stat}} < q_{\mathrm{rel}}$.

Furthermore, in Fig. 4a–c, we show the estimation of the q-triplet indices (q_{stat}, q_{sen}, q_{rel}) for the spatial distribution of the GH index for the areas: globally, north hemisphere and south hemisphere over decades. Moreover, in Fig. 4a, we present the values of Tsallis q_{stat} index estimated for the spatial distribution of the GH comparing in three different areas and in almost decennial cycle for the month January. In all cases, the value $q_{stat} > 1$ reveals the presence of long-range correlations with underlying dynamics characterized by non-Gaussian (q-Gaussian) distributions and a strong non-extensive character, attaining values between 1.38 ± 0.01 and 1.75 ± 0.02 globally, 1.26 ± 0.02 and 1.48 ± 0.02 for north hemisphere, and 1.41 ± 0.02 and 1.80 ± 0.02 for south hemisphere. The comparison of Tsallis q_{stat} values revealed that the spatial series of GH in these areas can be described

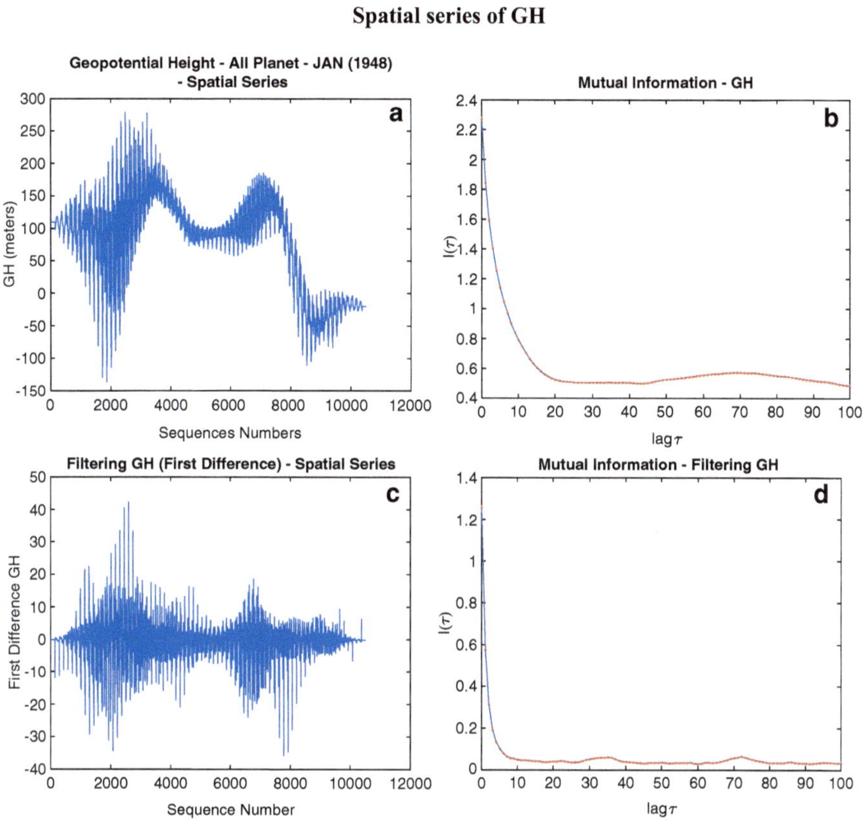

Fig. 2 (**a**) The GH spatial series for all planet by grid 2.5 × 2.5° the time of January of 1948. (**b**) The mutual information of the corresponding GH index. (**c**) The filtering signal of GH index (filter of first differences) (**d**) The mutual information of the filtering GH index

by Tsallis q-Gaussian distributions with similar q_{stat} indices, indicating similar statistical complexity.

The most interesting finding reported in Fig. 4a is the spatial distribution of the q_{stat} values which showed a downward slope, especially in globally and south hemisphere and in January of 2012 the q_{stat} values are almost identical for all regions. Similar, in Fig. 4b we showed the q_{sen} index estimated for the same spatial series and regions. In all cases estimated the $q_{sen} < 1$ for all signals and for all regions, a result which indicates a power law character for sensitivity of initial conditions, attaining values between -0.39 ± 0.08 and 0.89 ± 0.13 globally, -0.36 ± 0.01 and -0.70 ± 0.05 for north hemisphere, and -0.52 ± 0.03 and 1.30 ± 0.10 for south hemisphere. It seems that the profile of the q_{sen} values of the planet and the north hemisphere is the same, but the profile of the south hemisphere showed wide fluctuations between the decades.

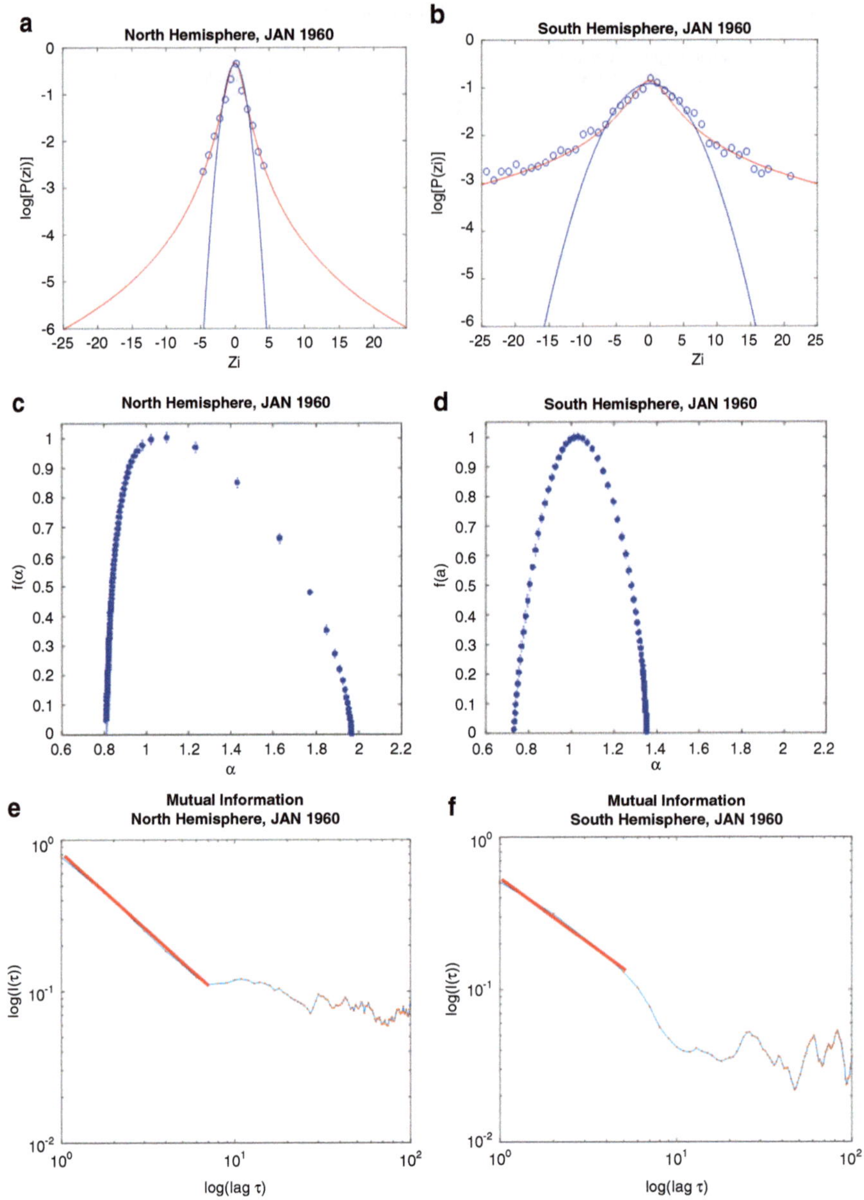

Fig. 3 (**a, b**) PDF $P(z_i)$ vs. $z_i q$ Guassian function that fits $P(z_i)$ for the GH spatial series (JAN, 1960) for north and south hemisphere, (**c, d**) Multifractal spectrum of GH spatial series (JAN, 1960) for north and south hemisphere, (**e, f**) Log–log plot of the self-correlation coefficient I(τ) vs. time delay τ for the GH spatial series of GH time series (JAN, 1960) for north and south hemisphere

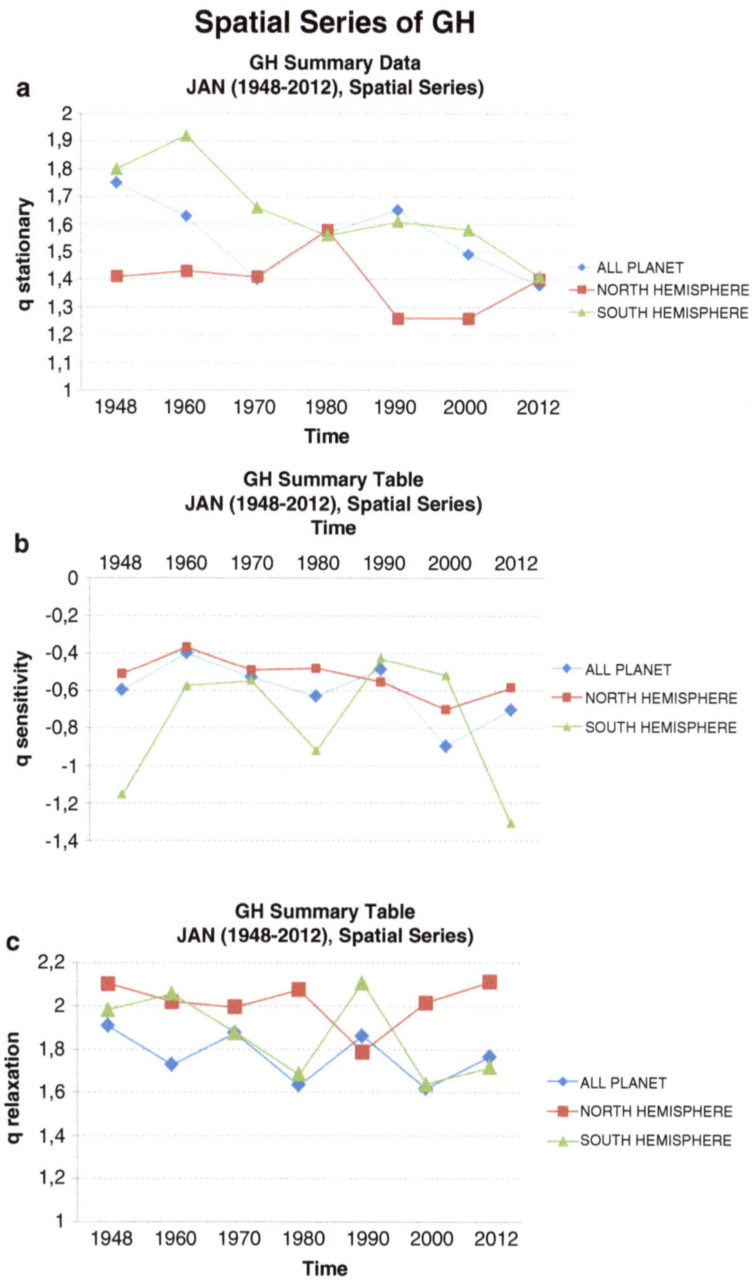

Fig. 4 (**a–c**) The Tsallis q-triplet indices (q_{stat}, q_{sen}, q_{rel}) for the spatial series of GH for specific regions: all planet, north hemisphere, and south hemisphere

Finally, in Fig. 4c we present the results for the q_{rel} index. It provides a description on how fast the system relaxes on the metastable state. In all cases, the results showed that $q_{rel} > 1$ with a similar profile, attaining values between 1.62 ± 0.01 and 1.91 ± 0.01 globally, 1.79 ± 0.02 and 2.11 ± 0.01 for north hemisphere, and 1.64 ± 0.01 and 2.11 ± 0.01 for south hemisphere. In particular, the profile of q_{rel} values is similar at the regions of globally and south hemisphere with a downward slope.

Moreover, noticeable differences of the q-triplet estimated at distinct local or temporal regions were found. The analysis is giving significant information identifying and characterizing the dynamical characteristics of the earth's climate and the ability of complexity and self-organization. Specifically, the results of the q stationary reveals that the GH sequences are characterized by long-range correlations and "memory character" or "persistent behavior" or patterns of GH raw data. The evolution of q stationary over decades and specific areas showed specific patterns with analogous trends. Similar behavior showed the results from the values variation of q sensitivity and q relaxation at the spatial of GH index. Clearly, although all temporal or spatial regions include information and exhibit a complex character, there are differences in the degree of complexity and therefore in the time needed for the transition to a new state of equilibrium, after being disturbed.

Temperature and Rainfall

In Pavlos et al. (2014) we presented results concerning the q-triplet Tsallis statistics for the air temperature and rainfall experimental data sets from the weather station 20046 Polar GMO in E.T. Krenkelja for the period 1/1/1960–31/12/1960, shown in Fig. 5a–l correspondingly. In both cases, we observed clearly non-Gaussian statistics. The q-triplet values were found to satisfy the relation $q_{sen} < 1 < q_{stat} < q_{rel}$ for the two data sets from atmospheric dynamics.

3.3 Space Plasma

Magnetosphere

In this paragraph, we present results concerning data "following" a shock event from near Earth IP plasma at L1 to the Earth's magnetosphere and magnetotail. We used the data of four spacecraft, ACE, Cluster 4 (Tango), Themis-E, and Themis-C. The results are summarized in Fig. 6a–c (for extended information concerning the results, see Pavlos et al. 2016). In Fig. 6 we show the Tsallis q-triplet results, namely Tsallis q_{stat} (Fig. 6a), Tsallis q_{sen} (Fig. 6b), and Tsallis q_{rel} (Fig. 6c) indices. In all cases, the results showed $q_{stat} > 1$, $q_{sen} < 1$, and $q_{rel} > 1$ suggesting that the underlying dynamics characterized by non-Gaussian (q-Gaussian) distributions, with a power law behavior for sensitivity of initial conditions and a q_{rel}-exponential

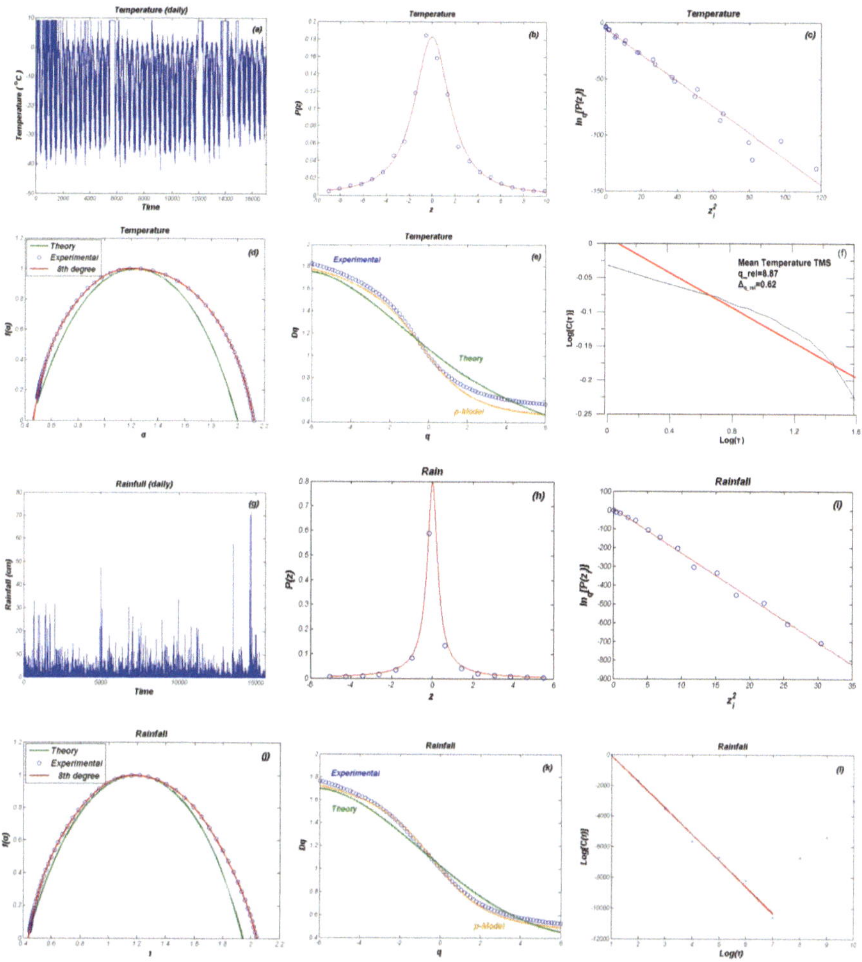

Fig. 5 (**a**) Time series of Temperature. (**b**) PDF $P(z_i)$ vs. $z_i q$ Guassian function that fits $P(\mathrm{z_i})$ for the Temperature. (**c**) Linear Correlation between $\ln_q P(z_i)$ and $(z_i)^2$ where $q_{\mathrm{stat}} = 1.89 \pm 0.08$ for the Temperature. (**d**) Multifractal spectrum of Temperature time series with solid line a 8° polynomial. We calculate the $q_{\mathrm{sen}} = 0.407 \pm 0.013$. (**e**)$D(q)$ vs. q of the Temperature time series. (**f**) Log–log plot of the self-correlation coefficient $C(\tau)$ vs. time delay τ for the Temperature time series. We obtain the best fit with $q_{\mathrm{rel}} = 8.87 \pm 0.62$. (**g**) Time series of Rainfall. (**h**) PDF $P(z_i)$ vs. $z_i q$ Gaussian function that fits $P(z_i)$ for the Rainfall. (**i**) Linear Correlation between $\ln_q P(z_i)$ and $(z_i)^2$ where $q_{\mathrm{stat}} = 2.21 \pm 0.06$ for the Rainfall. (**j**) Multifractal spectrum of Rainfall time series with solid line a 8° polynomial. We calculate the $q_{\mathrm{sen}} = 0.444 \pm 0.007$. (**k**) $D(q)$ vs. q of the Rainfall time series. (**l**) Log–log plot of the self-correlation coefficient $C(\tau)$ vs. time delay τ for the Rainfall time series. We obtain the best fit with $q_{\mathrm{rel}} = 6.04 \pm 0.47$

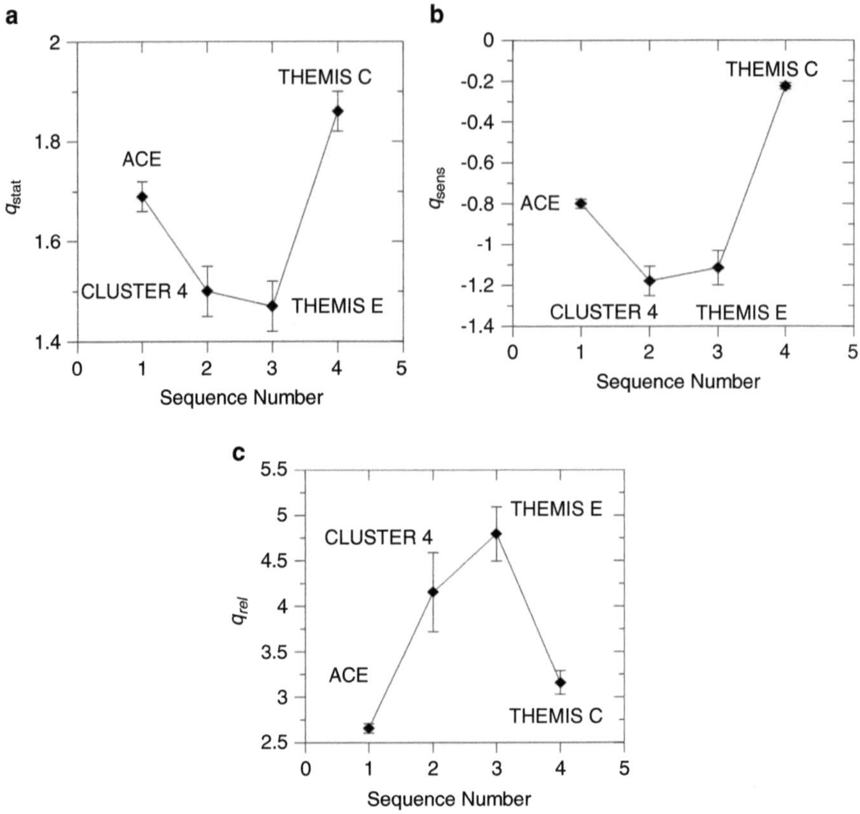

Fig. 6 Tsallis q-triplet during shock period for all spacecraft: (**a**) $q_{\rm stat}$ index, (**b**) $q_{\rm ses}$ index, (**c**) $q_{\rm rel}$ index

decay relaxation of the system to meta-equilibrium non-extensive stationary states, for all regions. In addition, the comparison of the results presented in Fig. 6 reveals similar physical character of the plasma at the distant magnetotail (THEMIS-C) and at the interplanetary medium (ACE). Both spacecraft observe strong non-extensive and intermittent (multifractal) profile of the plasma system in comparison with the other spacecraft (CLUSTER-4, THEMIS-E) which are located near the front side of the bowshock and the Earth magnetopause. This observational result indicates weak self-organization of the space plasma near the Earth (CLUSTER-4, THEMIS-E) in comparison with the distant plasma in the Earth magnetotail (THEMIS-C) or the interplanetary medium (ACE). This difference in plasma behavior could be due to the strong interaction of space plasma with the near Earth environment which destroys or doesn't permit the development of strong extensivity and self-organization.

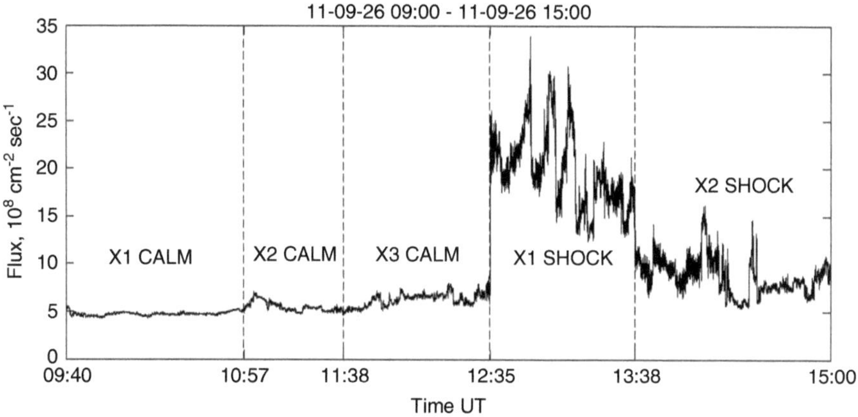

Fig. 7 Time series of the ion flux divided into five segments, namely x1calm, x2calm, x3calm, x1shock, x2shock

Solar Wind

In this paragraph, we present results (for more details, see Pavlos et al. 2015) concerning the evolution and the gradual phase transition of the dynamics of solar wind from calm to shock. In order to achieve this, we divided an ion flux time series, consisting of 604.510 counts, into five segments, as shown in Fig. 7. The first three segments (x1calm, x2calm, x3calm) correspond to the calm period time series (previous to shock), while the other two to the main shock period (x1shock) and its relaxation (x2shock).

The results concerning Tsallis q-triplet as the solar wind dynamics evolves towards the shock event and its relaxation are summarized in Fig. 8a–c which presents the index q_{stat}, along with the bar errors (Fig. 8a), the valuation of q_{sen} index values, along with the bar errors (Fig. 8b), and q_{rel} indexes values, along with error estimation (Fig. 8c). As it can be seen all indices are different from unity, while differences in triplet values concerning calm and shock period show a gradual development of non-Gaussian, non-extensive solar wind dynamics, which reach its peak in the main shock event (x1shock), characterized with of low entropy production and a fastest speed approach to metastable stationary state(s).

Sunspot Time Series

In this study Pavlos et al. (2014) present the q-triplet of the sunspot index by using data of Wolf number. Especially, we use the Wolf number, known as the international sunspot number measures the number of sunspots and group of sunspots on the surface of the sun computed by the formula $R=k^*(10g+s)$ where s is the number of individual spots, g is the number of sunspot groups, and k is a

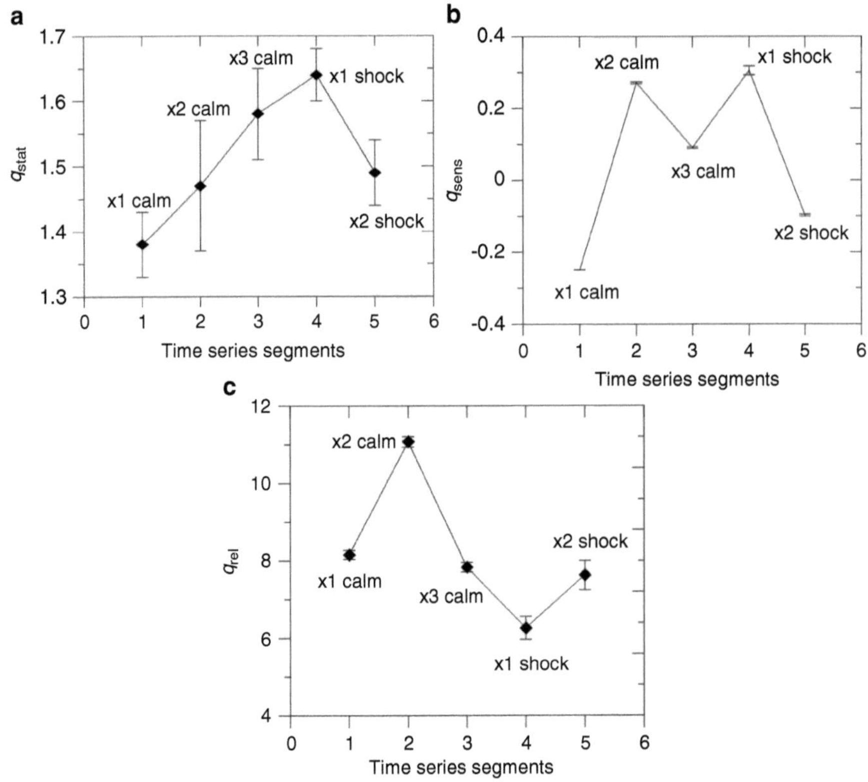

Fig. 8 (**a**) The index q_{stat} for five ion flux time series segments of the solar wind. (**b**) The q_{sen} values for the five ion flux time series segments of the solar wind. (**c**) The index q_{rel} for the five ion flux time series segments of the solar wind

factor that varies with location known as the observatory factor. We analyze a period of 184 years. We clearly observe non-Gaussian statistics to the system of sunspot index. The *q*-triplet values satisfy the relation $q_{sen} < 1 < q_{stat} < q_{rel}$ (Fig. 9) for the sunspot time series (see Table 1).

Solar Flares Time Series

Similarly, in this study Pavlos et al. (2014) present the *q*-triplet of the daily solar flares index. Moreover, we analyze the daily Flare Index of the solar activity that was determined using the final grouped solar flares obtained by National Geophysical Data Center (NGDC). It is calculated for each flare using the formula: $Q = (i{*}\,t)$, where "*i*" is the importance coefficient of the flare and "*t*" is the duration of the flare in minutes. To obtain final daily values, the daily sums of the index for the total surface are divided by the total time of observation of that day. The data

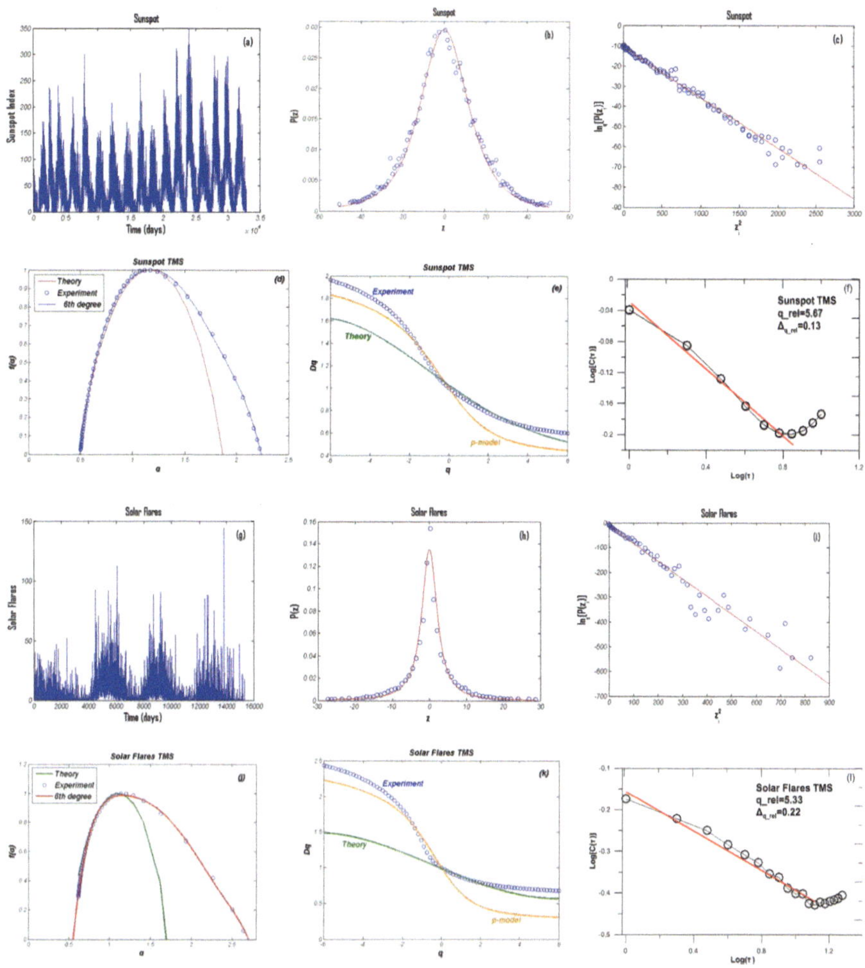

Fig. 9 (**a**) Time series of Sunspot Index concerning the period of 184 years. (**b**) PDF $P(z_i)$ vs. $z_i q$ Guassian function that fits $P(z_i)$ for the Sunspot Index. (**c**) Linear Correlation between $\ln_q P(z_i)$ and $(z_i)^2$ where $q_{\text{stat}} = 1.53 \pm 0.04$ for the Sunspot Index. (**d**) Multifractal spectrum of Sunspot Index time series with solid line a sixth-degree polynomial. We calculate the $q_{\text{sen}} = 0.368 \pm 0.005$. (**e**) $D(q)$ vs. q of the Sunspot Index time series. (**f**) Log–log plot of the self-correlation coefficient $C(\tau)$ vs. time delay τ for the Sunspot Index time series. We obtain the best fit with $q_{\text{rel}} = 5.67 \pm 0.13$. (**g**) Time series of Solar Flares concerning the period of 184 years. (**h**) PDF $P(z_i)$ vs. $z_i q$ Guassian function that fits $P(z_i)$ for the Solar Flares. (**i**) Linear Correlation between $\ln_q P(z_i)$ and $(z_i)^2$ where $q_{\text{stat}} = 1.90 \pm 0.05$ for the Solar Flares. (**j**) Multifractal spectrum of Solar Flares time series with *solid line* a sixth-degree polynomial. We calculate the $q_{\text{sen}} = 0.308 \pm 0.005$. (**k**) $D(q)$ vs. q of the Solar Flares time series. (**l**) Log–log plot of the self-correlation coefficient $C(\tau)$ vs. time delay τ for the Solar Flares time series. We obtain the best fit with $q_{\text{rel}} = 5.33 \pm 0.22$

Table 1 Summarize parameter values of solar dynamics including the sunspot and the solar flare dynamics: The q-triplet $(q_{\text{sen}}, q_{\text{stat}}, q_{\text{rel}})$ of Tsallis

System	q_sen	q_stat	q_rel $(C(\tau))$
Solar (sunspot index)	0.368 ± 0.005	1.53 ± 0.04	5.672 ± 0.127
Solar (flares index)	0.308 ± 0.005	1.870 ± 0.005	5.33 ± 0.22

covers time period from 1/1/1996 to 31/12/2007. We clearly observe non-Gaussian statistics for solar flares time series, but the non-Gaussianity of solar flares was found much stronger than the sunspot index. The q-triplet values satisfy the relation $q_{\text{sen}} < 1 < q_{\text{stat}} < q_{\text{rel}}$ (Fig. 9) for the solar flares time series (see Table 1).

4 Theoretical Interpretations of q-Triplet Results

The Tsallis q-triplet of the non-extensive statistical theory is related to the non-Gaussian dynamics of the system when $q_{\text{sen}} \neq q_{\text{stat}} \neq q_{\text{rel}} \neq 1$. In this direction, we present some interested theoretical concepts which help to understand the physical meaning of the observational results of this study.

Earthquakes, climate, and space plasma systems are typical cases of stochastic spatio-temporal distribution of physical magnitudes such as force–fluids fields and matter fields. However, the classical Hydrodynamical (HD) and Magnetohydrodynamical (MHD) description for climate and space plasmas or the Boltzmann–Maxwell statistical mechanics are inefficient to describe the non-equilibrium state of the complex system as they include smooth and differentiable spatial–temporal functions (HD–MHD theory) or Gaussian statistical processes (Boltzmann–Maxwell statistical mechanics), correspondingly.

The results of this study are related also to modern theoretical concepts such as fractal topology (Zelenyi and Milovanov 2004), turbulence theory (Frisch 1996), strange dynamics (Zaslavsky 2002), percolation theory (Milovanov 1997), anomalous diffusion theory and anomalous transport theory (Shlesinger et al. 1993; Milovanov 2001), fractional dynamics (Zaslavsky 2002; Tarasov 2005, 2006, 2013), and non-equilibrium RG theory (Chang 1992).

4.1 *Fractional Calculus*

The differentiable nature of magnitudes with smooth distributions of the macroscopic picture of physical processes is a natural consequence of the Gaussian microscopic randomness which, through the classical CLT, is transformed to the macroscopic, smooth, and differentiable processes. The classical CLT is related to the condition of microscopic and macroscopic time-scale separation, where at the long-time limit the memory of the microscopic non-differentiable character is lost. On the other hand, the q-extension of CLT induces the nonexistence of time-scale separation between microscopic and macroscopic scales as the result of multiscale global correlations which produce fractional dynamics and singular functions of spatio-temporal dynamical physical variables.

The non-local character is evident in both cases of fractional derivative and integral on a fractal set. The non-local character of fractional calculus is related to multiscale and self-similar character of the fractal structure. The fractional extension of integral and differential calculus can be used for the description of the non-local multiscale phenomena described by fractional Maxwell's Equations (fME), or the fractional Magnetohydrodynamics (fMHD) of fractal plasma states, or the fractional Fokker–Planck Equation (fFPE) of fractal media (Tarasov 2005, 2013). The solution of the fractional equations corresponds to fractional non-differentiable singular self-similar functions as we can observe at the experimental data. Generally, fractional differential integral equations have as solutions non-differentiable (singular) spatio-temporal distribution functions of physical magnitudes.

4.2 *Anomalous Diffusion and Strange Dynamics*

Nonlinear dynamics can create fractal structuring of the phase space and global correlations in the nonlinear system. For non-extensive systems the entire phase space is dynamically not entirely occupied (the system is not ergodic), but only a scale-free-like part of it is visited yielding a long-standing (multi)-fractal-like occupation. According to Milovanov and Zelenyi (2000), Tsallis entropy can be rigorously obtained as the solution of a nonlinear functional equation referred to the spatial entropies of the subsystems involved including two principal parts. The first part is linear (additive) and leads to the extensive Boltzmann–Gibbs entropy. The second part is multiplicative corresponding to the non-extensive Tsallis entropy referred to the long-range correlations. The fractal–multifractal structuring of the phase space makes the effective number W_{eff} of possible states, namely those whose probability is non-zero, to be smaller ($W_{\text{eff}} < W$) than the total number of states. This is the statistical manifestation of self-organization process.

The dynamics in the topologically anomalous phase space corresponds to a random walk process which is scale invariant in spatial and temporal self-similarity transform

$$\widehat{R} : t' \rightarrow \lambda_t t, \xi' = \lambda_\xi \xi \tag{36}$$

The spatial–temporal scale invariance causes strong spatial and temporal correlations mirrored in singular self-similar temporal and spatial distribution functions which satisfy the fractional generalization of classical Fokker–Planck–Kolmogorov equation (FFPK-equation) (Zaslavsky 2002):

$$\frac{\partial^{\beta} P}{\partial t^{\beta}} = \frac{\partial^{\alpha}}{\partial(-\xi)^{\alpha}}(\mathrm{AP}) + \frac{1}{2}\frac{\partial^{2\alpha}}{\partial(-\xi)^{2\alpha}}(\mathrm{BP}) \tag{37}$$

where $P \equiv P(\xi, t)$ is the probability density of the state (ξ) at the time (t). The critical components (α, β) correspond to the fractal dimensions of the spatial–temporal non-Gaussian distributions of the spatial–temporal functions–processes or probability distributions. The quantities A, B are given by

$$A = \lim_{\Delta t \to 0} \frac{\langle\langle|\Delta\xi^{\alpha}|\rangle\rangle}{(\Delta t)^{\beta}}, \mathrm{B} = \lim_{\Delta t \to 0} \frac{\langle\langle|\Delta\xi^{2\alpha}|\rangle\rangle}{(\Delta t)^{\beta}} \tag{38}$$

where $\ll \ldots \gg$ denotes a generalized convolution operator (Zaslavsky 2002).

The FFPK equation is an archetype fractional equation of fractional stochastic dynamics in a (multi)-fractal phase space with fractal temporal evolution caused by the self-similar and multiscale structure of islands around islands, responsible for the flights and trappings of the dynamics. The "spatial" random variable can be any physical variable, such as position in physical space, velocity in the velocity space or a dynamic field space (magnetic or electric) at a certain position in physical space, etc., underlying to the nonlinear chaotic dynamics. The fractional dynamics of plasma includes fractal distribution of field and currents, as well as fractal distribution of energy dissipation field.

The fractional temporal derivative $\partial^{\beta}/\partial t^{\beta}$ in kinetic equations allows one to take fractal-time random walks into account, as the temporal component of the strange dynamics in fractal-turbulent media. The waiting times follow the power law distribution $P(\tau)\ \ \tau^{-(1+\beta)}$ since the "Levy flights" of the dynamics also follow the power law of distribution.

The asymptotic (root mean square of the displacement) of the transport process is given by $\langle|\xi|^2\rangle = 2Dt^{\mu}$, while the generalized transport coefficient μ depends on the values of the fractal coefficients (α, β), according to the relation $\mu = \frac{\beta}{\alpha}$ (Shlesinger et al. 1993). The parameter (β) has the meaning of the fractal dimension of an "active" time while the parameter (α) is related to the spatial fractal dimension in the percolating fractal plasma system.

The solution of the fractal kinetic equation corresponds to Levy distributions and asymptotically to Tsallis q-Gaussians. According to Alemany and Zanette (1994), the set of points visited by the random walker can reveal a self-similar fractal structure produced by the extremization of Tsallis q-entropy. The q-Gaussian distribution of the fractal structure created by the strange dynamics and the extremized q-entropy asymptotically corresponds to the Levy distribution $P(\xi)\ \ \xi^{-1-\gamma}$ where the q-exponent is related to the Levy exponent γ by $q = \frac{3+\gamma}{1+\gamma}$. The Levy exponent γ corresponds to the fractal structure of the points visited by the random walker. According to Alemany and Zanette (1994) and Tsallis (2009), the fractal extension of dynamics includes simultaneously the q-extension of statistics as well as the fractal extension RNG theory in the fractional Fokker–Planck–Kolmogorov Equation (FFPK).

The q-statistics of Tsallis corresponds to the meta-equilibrium solutions of the FFPK equation (Tarasov 2005; Tsallis 2009). Also, the meta-equilibrium states of FFPK equation correspond to the fixed points of Chang non-equilibrium RNG theory for space plasmas (Chang 1992; Zaslavsky 2002). The anomalous topology of phase space dynamics includes inherently the statistics as a consequence of its multiscale and multifractal character. From this point of view the non-extensive character of thermodynamics constitutes a kind of unification between statistics and dynamics. From a wider point of view the FFPK equation is a partial manifestation of a general fractal extension of dynamics. According to Tarasov (2005), the Zaslavsky's equation can be derived from a fractional generalization of the Liouville and BBGKI equations. According also to Tarasov (2005, 2006), the fractal extension of dynamics including the dynamics of particles or fields is based on the fact that the fractal structure of the spatially distributed matter (particles, fluids, and fields) can be replaced by a fractional continuous model. In this generalization the fractional integrals can be considered as approximations of integrals on fractals. Also, the fractional derivatives are related with the development of long-range correlations and localized fractal structures.

4.3 Fractal Topology, Critical Percolation, and Stochastic Dynamics

In this section, we follow Milovanov (2012) and present some basic concepts concerning topological aspects of percolating random fields, which can explain the complex and non-extensive character of various complex systems.

For any random field distribution $\psi\left(\vec{x}\right)$ in the n-dimensional space (E^n) there exists a critical percolation threshold which divides the space E^n into two topological distinct parts: Regions where $\psi\left(\vec{x}\right) < h_c$ marked as "empty" and regions where $\psi\left(\vec{x}\right) > h_c$, marked as "filled." When $\psi\left(\vec{x}\right) \neq h_c$, one of these parts will include an infinite connected set which is said to percolate. As the threshold h changes, we can find the critical threshold h_c where the topological phase transition occurs, namely the non-percolating part starts to percolate. The random field may be a spatial distribution of physical random magnitudes or it can correspond to the random distribution of physical properties in the phase space of the underlying dynamics.

The geometry of the percolating set at the critical state $(h \to h_c)$ is a typical fractal set for length scales between microscopic distances and percolation correlation length which diverges. The statistically self-similar geometry includes power law behavior of the "mass" density of the fractal set such as "fractal mass density" $x^{D\text{-}n}$, where x is the length scale, D is the Hausdorff fractal dimension which must be smaller than the dimensionality (n) of the embedding Euclidean space. In addition to the parameter D of the fractal dimension, there is the index of connectivity θ which describes the "shape" of the fractal set and may be different for fractals even with equal values of the fractal dimension D. The index of connectivity θ is defined as characterizing the shortest (geodesic) line connecting two different points on the fractal set by the relation $d_\theta = (2+\theta)/2$, where d_θ is the minimal Hausdorff dimension of the minimal (geodesic) line for all possible homeomorphisms that transform the fractal F into a fractal F′. The geodesic line on a self-similar fractal set (F) is a self-affine fractal curve whose own Hausdorff fractal dimension is equal to $(2+\theta)/2$. The index of connectivity plays an essential role in many dynamical phenomena on fractals, while it is a topological invariant of the fractal set F.

From the fractal dimension D and the connectivity index θ we can define a hybrid parameter $d_s = \frac{2D}{2+\theta}$ which is known as the spectral or the fracton dimension which represents the density of states for vibrational excitations in fractal network termed as fractons (Milovanov 2012). The root mean square displacement of the random walker on the fractal set is given by

$$\left\langle |\xi|^2 \right\rangle \sim t^{2/2+d_\theta} = t^{1/d_\theta} \tag{39}$$

where d_θ is the fractal dimension of the self-affine trajectory on the fractal set. Also, the spectral dimension which measures the probability of the random walker to return to the origin is given by

$$P(t) \sim t^{-\frac{d_s}{2}} \tag{40}$$

while the Hausdorff fractal dimension D is a structural characteristic of the fractal structure F, the spectral dimension d_s mirrors the dynamical properties such as wave excitation and diffusion. The fractal dimension d_f of the fractal structure F of a percolating random field distributed in the E^n Euclidian space is given by $d_f = n - \beta/\nu$, where β, ν are the universal critical exponents of the critical percolation state (Milovanov 2012).

4.4 Renormalization Group (RNG) Theory and Phase Space Transition

The multifractal and multiscale intermittent turbulent character of the complex dynamics in the various physical systems justifies the application of RNG theory for the description of the scale invariance and the development of long-range correlation of the complex systems' intermittent turbulence state. Generally, and according to Chang (1992) a complex system can be described by generalized Langevin stochastic equations of the general type:

$$\frac{\partial \varphi_i}{\partial t} = f_i\left(\vec{\varphi}, \vec{x}, t\right) + n_i\left(\vec{x}, t\right) \quad i = 1, 2, \ldots \tag{41}$$

where f_i corresponds to the deterministic process as concerns the dynamical variables $\phi\left(\overrightarrow{x},t\right)$ and n_i to the stochastic components (fluctuations). Generally, f_i are nonrandom forces corresponding to the functional derivative of the free energy functional of the system. According to Chang (1992) the behavior of a nonlinear stochastic system far from equilibrium can be described by the density functional P, defined by path integration of the system's stochastic Lagrangian:

$$P\left(\overrightarrow{\varphi}\left(\overrightarrow{x},t\right)\right) = \int D\left(\overrightarrow{x}\right)\exp\left\{-i\cdot\int L\left(\overrightarrow{\dot{\varphi}},\overrightarrow{\varphi},\overrightarrow{x}\right)d\overrightarrow{x}\right\}dt \qquad (42)$$

where $L\left(\overrightarrow{\dot{\varphi}},\overrightarrow{\varphi},x\right)$ is the stochastic Lagrangian of the system, which describes the full dynamics of the stochastic system. Moreover, the far from equilibrium renormalization group theory applied to the stochastic Lagrangian L generates the singular points (fixed points) in the affine space of the stochastic distributed system. At fixed points the system reveals the character of criticality, as near criticality the correlations among the fluctuations of the random dynamic field are extremely long-ranged and there exist many correlation scales. Also, close to dynamic criticality certain linear combinations of the parameters, characterizing the stochastic Lagrangian of the system, correlate with each other in the form of power laws and the stochastic system can be described by a small number of relevant parameters characterizing the truncated system of equations with low or high dimensionality and strong self-organization ordering process.

According to these theoretical results of Chang's theory, the stochastic distributed system can exhibit low dimensional chaotic or high dimensional SOC like behavior, including fractal or multifractal structures with power law profiles. The power laws are connected to the near criticality phase transition process which creates spatial and temporal correlations as well as strong or weak reduction (self-organization) of the infinite dimensionality corresponding to a spatially distributed system. First and second phase transition processes can be related to discrete fixed points in the affine dynamical (Lagrangian) space of the stochastic dynamics. The SOC like behavior of plasma dynamics corresponds to the second phase transition process as a high dimensional process at the edge of chaos. The process of strong and low dimensional chaos can be related to a first order phase transition process. The probabilistic solution (42) of the Eq. (41) of the generalized Langevin equations may include Gaussian or non-Gaussian processes as well as normal or anomalous diffusion processes depending upon the critical state of the system.

From this point of view, a SOC or low dimensional intermittent chaos or distinct non-extensive q-statistical states with different values of the Tsallis q-triplet depends upon the type of the critical fixed (singular) point in the functional solution space of the system. When the stochastic system is externally driven or perturbed, it can be moved from a particular state of criticality to another characterized by a different fixed point and different dimensionality or scaling laws. Thus, a SOC state could be a special kind of critical dynamics of an externally driven stochastic system, while SOC and low dimensional chaos can coexist in the same dynamical system as processes manifested by different kinds of fixed (critical) points in its solution space. Due to this fact, a complex systems' dynamics may include high dimensional SOC process or low dimensional chaos or other more general dynamical process corresponding to various q-statistical states.

5 Closing Remarks

In this review, we presented results concerning non-extensive statistics in distributed systems complex dynamics corresponding to earthquakes, climate, and space plasma systems. The results of this study show clearly the non-Gaussian character of the above systems and the existence of multiscale strong correlations from the microscopic to the macroscopic level. In particular, the estimation of Tsallis q-triplet statistics revealed the possibility of dynamical non-equilibrium phase transition processes and percolation topological phase transition related to the space plasma and climate systems (Zelenyi and Milovanov 2004).

The aforementioned results indicate the inefficiency of classical HD—MHD or classical statistical theories based on the classical central limit theorem to explain the complexity of the distributed systems dynamics, since these theories include smooth and differentiable spatial–temporal functions (HD-MHD theories) or Gaussian statistics (Boltzmann–Maxwell statistical mechanics). The differentiable nature of smooth distribution of the macroscopic picture of physical processes is a natural consequence of the Gaussian microscopic randomness which, through the classical CLT, is transformed to the macroscopic, smooth, and differentiable processes. The classical CLT is related to the condition of time-scale separation, where at the long-time limit the memory of the microscopic non-differentiable character is lost. On the contrary, the results of this study indicate the presence of non-Gaussian non-extensive statistics with heavy tails probability distribution functions, which are related to the q-extension of central limit theorem. The q-extension of CLT induces the non-existence of time-scale separation between microscopic and macroscopic scales as the result of multiscale global correlations.

These multiscale global correlations are the basis for the fractal, multifractal structure of the distributed systems, producing fractional dynamics which can be described by the singular character of the spatio-temporal dynamical physical variables. Thus, a generalization from the classical field-particle dynamics of classical continues mechanical systems or flow dynamical systems towards the fractional dynamics. The fractional extension of integral and differential calculus can be used for the description of the non-local multiscale phenomena described by the corresponding nonlinear equations of fractal media. Therefore, according to the experimental data analysis of this study and the theoretical framework of fractional dynamics, we can conclude that the nonlinear distributed dynamical systems in study are globally hierarchical, self-similar, and scale invariant physical systems which are characterized by nonlinear and non-local internal fractional dynamics, maintaining the hierarchical structure of the intermittent turbulence. In this direction, the nonlinear distributed dynamical systems can include fracton excitations and fracton dynamics where fracton formations are waves on fractal structures. Fracton dynamics can cause the oscillations of statistical parameters observed during phase transition events.

In addition, the physical interpretation of our results indicates the possibility for the existence of phase transitions events from a weak non-equilibrium (quasi)-stationary state (NESS) to a strong NESS as the outcome of cluster interaction in the distributed nonlinear dynamical systems. These states (NESS) can have the topology of a percolating fractal set, including multiscale interactions of fields and particles and can be related to the simultaneous development of numerous instabilities interfering with each other. The structural stability of the NESS as a symmetric turbulent phase is maintained due to multiscale correlations creating the existence of local extremes of the free energy.

Summarizing, Tsallis q-entropy principle can reliably explain the self-similar hierarchical turbulent structuring and phase transition processes presented in this study for different types of distributed nonlinear dynamical systems. These systems that live far from equilibrium can reveal meta-equilibrium stationary states (NESS) as critical percolation states. These non-equilibrium states, similar to Boltzmann–Gibbs thermodynamical meta-equilibrium states, can be produced as the system tends to obtain extremization of Tsallis q-entropy (S_q). The quantitative change of the non-extensive Tsallis statistics can be related to the renormalization group theory (RGT) change of the fixed points (NESS) in the dynamical parameter space of the dynamics. The internal mechanism for this is the anomalous diffusion process in the physical space or the anomalous random walk in a hierarchical and multifractal structured phase space. The dynamics in the multifractal phase space or physical space is described by the fractional equations (e.g., Langevin and the corresponding FFPK equations). Moreover, we conjecture that the meta-equilibrium stationary states can be obtained also as the fixed points of a fractional renormalization flow equation in a fractal parameter space. Also, the hierarchical, self-similar, multiscale, and multifractal structure of the distributed system at critical percolation and intermittent turbulent states can be described by the solution of the fractional Langevin equations, as the N-point correlation functions related to the functional derivative of the q-partition function Z_q defined in the framework of non-extensive Tsallis statistical mechanics-thermodynamics.

References

Alemany, P.A., and D.H. Zanette. 1994. Fractal random walks from a variational formalism for Tsallies entropies. *Physical Review E* 49 (2): R956–R958.

Baldovin, F., and A.L. Stella. 2007. Central limit theorem for anomalous scaling due to correlations. *Physical Review E* 75 (02): 020101(R).

Castro, C. 2005. On non-extensive statistics, chaos and fractal strings. *Physica A* 347: 184.

Chame, A., and E.V.L. De Mello. 1994. The fluctuation-dissipation theorem in the framework of the Tsallis statistics. *Journal of Physics A: Mathematical and General* 27 (11): 3663.

Chang, T. 1992. Low-dimensional behavior and symmetry braking of stochastic systems near criticality can these effects be observed in space and in the laboratory. *IEEE* 20 (6): 691–694.

El-Nabulsi, A.R. 2005. A fractional approach to nonconservative Lagrangian dynamical systems. *FIZIKA A* 14: 289–298.

Frisch, U. 1996. *Turbulence*, 310. Cambridge: Cambridge University Press. ISBN 0521457130.

Goldfain, E. 2007. Chaotic dynamics of the renormalization group flow and standard model parameters. *International Journal of Nonlinear Science* 3: 170–180.

Halsey, T.C., et al. 1986. Fractal measures and their singularities: The characterization of strange sets. *Physical Review A* 33 (2): 1141.

Iliopoulos, A.C., G.P. Pavlos, E.E. Papadimitriou, and D.S. Sfiris. 2012. Chaos, self organized criticality, intermittent turbulence and non-extensivity revealed from seismogenesis in North Aegean area. *International Journal of Bifurcation and Chaos* 22 (9): 1250224.

Iliopoulos, A.C., N.S. Nikolaidis, and E.C. Aifantis. 2015a. Portevin–Le Chatelier effect and Tsallis nonextensive statistics. *Physica A: Statistical Mechanics and its Applications* 438: 509–518.

Iliopoulos, A.C., G.P. Pavlos, L. Magafas, L. Karakatsanis, M. Xenakis, and E. Pavlos. 2015b. Tsallis *q*-triplet and stock market indices: the cases of S & P 500 and TVIX. *Journal of Engineering Science and Technology Review* 8 (1): 34–40.

Iliopoulos, A.C., M. Tsolaki, and E.C. Aifantis. 2016a. Tsallis statistics and neurodegenerative disorders. *Journal of the Mechanical Behavior of Materials* 25 (3–4): 129–139.

Iliopoulos, A.C. 2016b. Complex systems: Phenomenology, modeling, analysis. *International Journal of Applied & Experimental Mathematics* 1: 105.

Kalnay, E., et al. 1996. The NCEP/NCAR 40-year reanalysis project. *Bulletin of the American Meteorological Society* 77: 437–470.

Karakatsanis, L.P., and G.P. Pavlos. 2008. SOC and chaos into the solar activity. *Nonlinear Phenomena in Complex Systems* 11 (2): 280–284.

Karakatsanis, L.P., G.P. Pavlos, and D.S. Sfiris. 2012. Universality of first and second order phase transition in solar activity. Evidence for non-extensive Tsallis statistics. *International Journal of Bifurcation and Chaos* 22 (9): 1250209.

Karakatsanis, L.P., G.P. Pavlos, and M.N. Xenakis. 2013. Tsallis non-extensive statistics, intermittence turbulence, SOC and chaos in the solar plasma, part two: Solar flare dynamics. *Physica A* 392 (18): 3920–3944.

Milovanov, A.V. 1997. Topological proof for the Alexander-Orbach conjecture. *Physical Review E* 56 (3): 2437–2446.

———. 2001. Stochastic dynamics from the fractional Fokker-Planck-Kolmogorov equation: Large-scale behavior of the turbulent transport coefficient. *Physical Review E* 63 (4): 047301.

———. 2012. Percolation models of self-organized critical phenomena. arXiv: 207.5389.

Milovanov, A.V., and L.M. Zelenyi. 2000. Functional background of the Tsallis entropy: "coarse-grained" systems and "kappa" distribution functions. *Nonlinear Processes in Geophysics* 7: 211–221.

Nottale, L. 2006. Fractal space-time, non-differentiable and scale relativity. *Invited contribution for the Jubilee of Benoit mandelbrot.*

Ord, G.N. 1983. Fractal space-time: a geometric analogue of relativistic quantum mechanics. *Journal of Physics A: Mathematical and General* 16: 1869.

Pavlos, G.P., A.C. Iliopoulos, V.G. Tsoutsouras, D.V. Sarafopoulos, D.S. Sfiris, L.P. Karakatsanis, and E.G. Pavlos. 2011. First and second order non-equilibrium phase transition and evidence for non-extensive Tsallis statistics in Earth's magnetosphere. *Physica A* 390 (15): 2819–2839.

Pavlos, G.P., L.P. Karakatsanis, M.N. Xenakis, D. Sarafopoulos, and E.G. Pavlos. 2012a. Tsallis statistics and magnetospheric self-organization. *Physica A* 391 (11): 3069–3080.

Pavlos, G.P., L.P. Karakatsanis, and M.N. Xenakis. 2012b. Tsallis non-extensive statistics, intermittent turbulence, SOC and chaos in the solar plasma. Part one: Sunspot dynamics. *Physica A* 391 (24): 6287–6319.

Pavlos, G.P., et al. 2014. Universality of Tsallis non-extensive statistics and time series analysis: Theory and applications. *Physica A* 395 (1): 58–95.

Pavlos, G.P., L.P. Karakatsanis, A.C. Iliopoulos, E.G. Pavlos, M.N. Xenakis, P. Clark, et al. 2015. Measuring complexity, nonextensivity and chaos in the DNA sequence of the major histocompatibility complex. *Physica A: Statistical Mechanics and its Applications* 438: 188–209.

Pavlos, G.P., O.E. Malandraki, E.G. Pavlos, A.C. Iliopoulos, and L.P. Karakatsanis. 2016. Non-extensive statistical analysis of magnetic field during the March 2012 ICME event using a multi-spacecraft approach. *Physica A: Statistical Mechanics and its Applications* 464: 149–181.

Shlesinger, M.F., B.J. West, and J. Klafter. 1987. Levy dynamics of enhanced diffusion: Application to turbulence. *Physical Review Letters* 58: 1100–1103.

Shlesinger, M.F. 1988. Fractal time in condensed matter. *Reviews in Physical Chemistry* 39: 269–290.

Shlesinger, M.F., G.M. Zaslavsky, and J. Klafter. 1993. Strange kinetics. *Nature* 363: 31.

Tarasov, V.E. 2005. Fractional Liouville and BBGKI equations. *Journal of Physics: Conferences Series* 7: 17–33.

———. 2006. Magnetohydrodynamics for fractal media. *Physics of Plasmas* 13: 052107.

———. 2013. Review of some promising fractional physical models. *International Journal of Modern Physics B* 27 (9): 1330005.

Theiler, J. 1990. Estimating fractal dimension. *JOSA A* 7 (6): 1055–1073.

Tsallis, C. 1988. Possible generalization of Boltzmann-Gibbs statistics. *Journal of Statistical Physics* 52 (1–2): 479–487.

———. 2002. Entropic non-extensivity a possible measure of complexity. *Chaos, Solitons and Fractals* 13: 371–391.

Tsallis C. 2004a. Non-extensive statistical mechanics: construction and physical interpretation. In *Non-extensive entropy – interdisciplinary applications*, ed. G.M. Murray & C. Tsallis, 1–53. Oxford: Oxford University Press.

Tsallis, C. 2004b. What should a statistical mechanics satisfy to reflect nature? *Physica D* 193: 3–34.

———. 2009. *Introduction to non-extensive statistical mechanics*. New York: Springer.

Umarov, S., et al. 2008. On a q-central limit theorem consistent with non-extensive statistical mechanics. *Milan Journal of Mathematics* 76: 307–328.

Zaslavsky, G.M. 2002. Chaos, fractional kinetics, and anomalous transport. *Physics Reports* 371: 461–580.

Zelenyi, L.M., and A.V. Milovanov. 2004. Fractal topology and strange kinetics: From percolation theory to problems in cosmic electrodynamics. *Pysics-Uspekhi* 47 (8): 749–788.

Spatial Patterns of Peak Flow Quantiles Based on Power-Law Scaling in the Mississippi River Basin

Gabriel Perez, Ricardo Mantilla, and Witold F. Krajewski

Abstract This study explores the spatial variability of peak flows for different drainage area sizes in the Mississippi River Basin (MRB) based on the power-law relation between flood quantiles (Q_p) and drainage areas (A) expressed as $Q_p = \alpha_p A^{\theta_p}$. The aim is to reveal consistent regional flood patterns within the MRB. The authors use 5137 streamflow gauges with peak flow records and the USGS Hydrologic Unit Code (HUC) catchment organization framework to estimate the scaling parameters (α_p and θ_p) at multiple spatial disaggregation levels, including the complete Mississippi River Basin (MRB), six major MRB sub-regions (HUC-2), and finally 84 medium-scale catchments (HUC-4). The analysis at the HUC-4 level exposes remarkable regional flood patterns in θ_p and α_p, which are used to estimate peak flows at 2.33 and 100 years of return periods at multiple spatial scales including 1, 100, 1000, and 10,000 km^2 drainage areas. The results expose a peak flow quantile relation that varies as a function of region and drainage area, demonstrating that the regions with the higher peak flows quantiles are varying with respect to the watershed size along the MRB. Mainly, we found that the cluster of higher floods extends from the center to the eastern MRB for drainage areas from 1 to 10,000 km^2. Conversely, the clusters of lower 2.33-year floods are preserved in the western MRB for the same range of drainage areas. The results presented in this study demonstrate that the flood-producing mechanisms are varying with respect to the drainage area size and regions, providing a starting point for a quantitative description of physical processes that dominate the variability of flood-producing mechanisms, a critical step in the design of parsimonious continental scale hydrological models.

Keywords Scaling of floods • Peak flow quantiles • Mississippi River Basin • Spatial patterns

---

G. Perez • R. Mantilla • W.F. Krajewski (✉)
IIHR-Hydroscience and Engineering, Department of Civil and Environmental Engineering, The University of Iowa, Iowa City, IA, 52242, USA
e-mail: witold-krajewski@uiowa.edu

A.A. Tsonis (ed.), *Advances in Nonlinear Geosciences*,
DOI 10.1007/978-3-319-58895-7_23

1 Introduction

Engineering design and various aspects of water resources management rely on the empirical methodology of Regional Flood Frequency Analysis (RFFA) to estimate flood quantiles at ungauged sites. This methodology depends on observations of annual maximum flows over homogeneous regions (e.g., Smith et al. 2015; Srinivas et al. 2008; Haddad et al. 2012; Wan Jaafar and Han 2012). However, many regions in the world remain poorly gauged or have experienced dramatic changes in land use or climate that make past observations less useful. To remedy this situation, we need methodologies for the estimation of flood frequencies that are based on physical principles of water movement and general knowledge of the geographic and geomorphologic setting of the upstream catchment at the location of interest.

An important step in taking the leap from RFFA to physics-based estimations of flood frequencies is the identification of scaling patterns revealed by data in the physical system in which floods occur (i.e., watersheds and river networks). Fuller (1914) was the first to connect the power scaling structure to a statistical framework for peak flow data from the United States; however, significant questions have emerged and many remain unanswered in terms of the physical controls and hydrologic variables that are governing the power-law scaling structure in peak flows. More recently, several studies (Ogden and Dawdy 2003; Gupta 2004; Gupta et al. 2010; Ayalew et al. 2015; Gupta et al. 2015) have presented evidence that the power-law relation between flood quantiles and drainage area is not a regional feature but instead emerges in nested basins. The power-law structure in peak flows see Eq. (1) represents the systematic increase in the maximum discharge (Q) for a specific quantile (p) as a function of the drainage area (A) as,

$$Q_p = \alpha_p A^{\theta_p} \tag{1}$$

The rate of increase is controlled by two scaling parameters: the intercept (α_p) and the scaling exponent (θ_p). A detailed explanation of the origins and early developments of the flood scaling methods have been summarized by Dawdy et al. (2012).

A diversity of studies have explored different approaches to quantifying the variables that control the value of α and θ. A number of researchers have quantified the role of rainfall properties such as intensity, duration, and spatial coverage as key players in determining the scaling parameter values (e.g., Gupta et al. 1996; Jothityangkoon and Sivapalan 2001; Mandapaka et al. 2009; Robinson and Sivapalan 1997). Mantilla et al. (2006) studied the flood scaling in real river networks, generalizing results from previous studies (Gupta and Waymire 1998; Menabde and Sivapalan 2001; Morrison and Smith 2001). Furey and Gupta (2007) evaluated the flood scaling dynamic for 148 rainfall-runoff events, demonstrating the strong influence of depth, duration, and spatial variability of excess rainfall on the scaling parameters. These results encouraged new studies to more deeply explore the scaling structure vis-à-vis rainfall properties. Ayalew et al. (2014a, b, 2015) demonstrated clear connections between rainfall properties and scaling parameters at different spatial scales. Ayalew et al. (2014a) used rainfall-runoff model simulation results to study how the rainfall intensity, duration, hillslope overland velocity, and channel flow velocity affect the scaling parameters in three small basins of 252, 520, and 1082 km^2 in a spatial scale study of the Cedar River Basin with drainage area of 17,000 km^2. In a subsequent study, Ayalew et al. (2015) analyzed actual data and showed the interplay between duration and depth of excess runoff with the scaling parameters for 51 rainfall-runoff events at the mesoscale Iowa River basin with a drainage area of 32,400 km^2, demonstrating that even at this large scale, flood scaling still dominates. In a more extensive recognition of the scaling parameters structure, Kroll (2014) shows the scaling exponent structure in the United States, defining 18 water regions. In this same direction Medhi and Tripathi (2015) explain the connections between basin attributes and scaling exponents, defining homogenous regions based on the region-of-influence method, showing evidence of simple-scaling for regions in which snowfall dominates the total precipitation. In addition, their results suggest small flood scaling exponents for regions with large soil moisture storages and high evapotranspiration losses, and large fractions of overland flow compared to base flow. These studies represent an outstanding advance in the understanding of the flood scaling structure for several spatial domains, range of basin sizes, and their connection with rainfall and catchment properties. However, none of these studies have demonstrated how the differences of scaling parameters are controlling the flood magnitude for different drainage size areas in a specific large spatial domain such as the Mississippi River Basin (MRB).

We organize this research in three specific aspects: (1) characterizing the spatial structure in α and θ for different scales within the MRB; (2) evaluating changes in α and θ for different quantiles and spatial regions; and (3) unmasking regional differences in flood magnitudes and flood frequency signatures for specific drainage areas.

Regarding (1) researchers have explored in depth the existence of flood scaling for flood quantiles and flood events in different basin sizes (Furey and Gupta 2007; Ayalew et al. 2014a, b, 2015; Medhi and Tripathi 2015), but the power-law structure for flood quantiles in a large domain such as the MRB is still unknown. Therefore, research is needed to determine the upper bound, if one exists, in the spatial limit over the watershed domain.

Regarding (2) we need to improve our understanding of the characterization of α and θ across space. Although we know that scaling parameters change across different hydrologic conditions (Medhi and Tripathi 2015), we want to determine if these changes exhibit gradual or abrupt shifts in space thus enabling possible connections between scaling parameters and spatial patterns in hydrologic signatures.

Regarding (3), the flood scaling framework allows us to analyze floods in specific drainage areas; therefore, we will use the different values of α and θ across the spatial domain to compare the flood changes among spatial locations, drainage areas, and flood quantiles, representing at the same time the different hydrologic conditions behind flood processes. These results should help to identify the spatial locations and drainage area magnitudes in which flood quantiles are high.

This document is organized as follows. In Sect. 2 we describe the study area and peak flow data, including the hydrologic variability in the region, watershed boundaries and spatial units of analysis, number of peak flow gauges, and the type of regression analysis to estimate the scaling parameters. In Sect. 3, we present procedures used to reveal the different flood patterns for diverse drainage areas and flood quantiles. Subsequently, we report and discuss results in Sect. 4, emphasizing characteristics of flood scaling across scales for different sub-regions, and provide insights into regional homogeneity based on flood scaling, and flood patterns for different watershed sizes. In Sect. 5, we conclude by addressing the importance and consequences of the main findings of this research, proposing future work around the connections between scaling parameters and mechanistic processes behind floods. Finally, we include two appendices describing the procedure to test the regional homogeneity (Appendix 1), and the identification of simple-scaling or multi-scaling in the different analysis units in the MRB (Appendix 2).

2 Peak Flow Data in the Mississippi River Basin

One of the largest continental basins in the world, the Mississippi River Basin is characterized by diverse hydrologic, climatic, and geomorphologic settings. The MRB drains an area of almost 3 million km^2 and is significantly impacted by human activity due to industrial and agricultural practices. From a hydroclimatological perspective, there are strong gradients in rainfall, snowfall, evapotranspiration, and temperature across the watershed at multiple temporal scales. These geophysical properties make the MRB a good candidate to understand the spatial variability in the scaling structure of floods. Our study includes only those stream gauge locations that drain watersheds smaller than 10,000 km^2 to guarantee that the information represents the flood diversity inside a particular HUC partition, avoiding biases introduced by gauges in large rivers that flow through a HUC (e.g., the Mississippi River) that may be influenced by large-scale regulation and with flow regimes that result from integrating multiple climate regimes.

Streamflow in the MRB is routinely estimated at 7587 gauged locations that record peak flows, managed by the US Geological Survey (USGS), facilitating our analyses. The USGS peak flow data is easily accessible by web services (http://nwis.waterdata.usgs.gov/usa/nwis/peak). The USGS records maximum annual floods at specific gauge sites, which are the inputs for the quantile estimation related to different probabilities of exceedance (also expressed as the "return period"). We used the guidelines for determining peak flow frequency outlined in the USGS Bulletin 17B. This procedure uses the probability distribution Log Pearson Type 3 with the incorporation of outlier treatments, flows affected by regulation in dams, estimation of the regional skew, and historical flood information. The complete method is incorporated in the software PeakFQ (Flynn et al. 2006). We used the

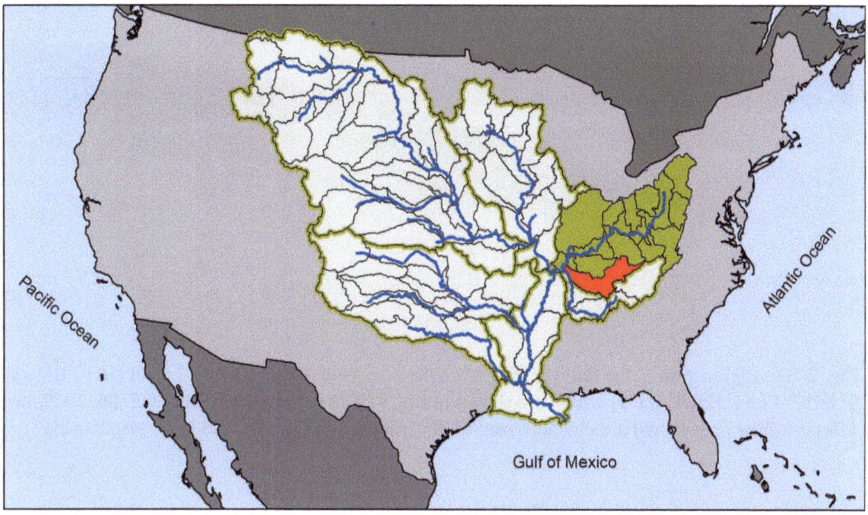

Fig. 1 Division of the complete Mississippi River Basin into six HUC2 regions and 84 HUC4 sub-regions. HUC2 region 05 and HUC4 region 0531 are highlighted in solid *colors*

PeakFQ software to estimate the peak flow quantiles for each location. We did not include peak flow data affected by dams or gauges with annual peak flow records reported as zero. This last condition arose because some gauges present long records of annual floods, but some have values equal to zero. These could be a consequence of long dry periods in small basins or instrument errors. For the regression analysis section we only considered the gauges with more than 10 years of record period, a total of 5137 gauges.

In order to analyze the variability of scaling parameters α and θ over different spatial scales and quantiles, we segmented the Mississippi River Basin. For this purpose, we used three levels of spatial discretization. We use the spatial hydrological units (HUC), defined by the USGS (Seaber et al. 1987). The largest spatial unit is the complete Mississippi River Basin. The second is the HUC-2 level with six sub-regions, and finally the HUC-4 level which partitions the MRB into 84 sub-regions. Figure 1 shows the spatial definition of the three levels of analysis and Fig. 2 illustrates an example of the spatial segmentation in the flood scaling from Level 1 to Level 3.

3 Scaling Patterns of Flood Data

We selected two methods to describe spatial patterns of scaling in peak flows in the MRB. First, we calculated a Flood Severity Index (SI), defined as the ratio between the peak flows with a return period of 100 years and the mean annual flood. In this

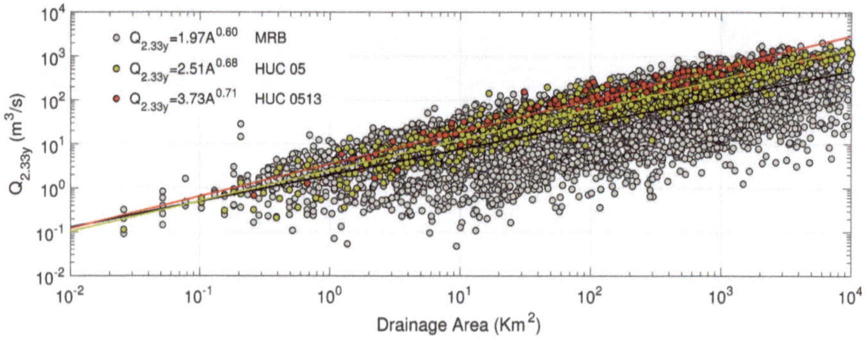

Fig. 2 Disaggregation in the flood scaling from the complete MRB to HUC-2 level (HUC 05) and to HUC-4 level (HUC 0531) which are shown in Fig. 1. The models are based on 5136, 1029, and 110 peak flow gauges and are characterized by R^2 values of 0.27, 0.88, and 0.93, respectively

calculation, the mean annual flood is represented as the peak flow with a return period of 2.33 years. We used these results as the starting point in creating the relationship between the spatial pattern of floods, physical controls, and hydrologic conditions across space.

Second, we fitted a power-law function between peak flows and drainage using a Weighted Least Square (WLS) regression in each of the regions defined by the HUC partitions of the MRB. The WLS reduced the uncertainty in the estimation of the scaling parameters, because peak flow gauges with few records will have larger uncertainty in the estimation of peak flow quantiles in comparison with peak flow gauges with larger records. Therefore, the length of the peak flow record is used as weight in the WLS regression. To illustrate the effect of considering the WLR rather than a standard Ordinary Least Square (OLS) regression, Fig. 3 shows the comparison of scaling parameters (α and θ) for the 84 sub-regions at HUC-4 level for the 2.33 and 100 years of return periods. Although Fig. 3 shows small changes in α and θ between WLS and OLS, these could be translated as important differences in the estimation of the peak flow quantiles along the MRB. For this reason, it is important to use the WLS regression to reduce the uncertainty introduced by peak flow gauges with few records. The regression for the entire MRB included all 5137 gauges. The number of gauges in the HUC-2 decomposition, which defines six sub-regions of the MRB, ranges from 338 to 1601. Finally, the number of gauges defined by the HUC-4 decomposition with 84 sub-regions ranges from 6 to 195, with 90% of the sub-regions containing more than 20 gauges.

We conducted a separate analysis to explore the variability of peak flows and regional homogeneity across space. We calculated the residuals from the power-law function for the three analysis levels and displayed them spatially. These residuals were organized according to their signs (positive or negative residuals). This approach helped us recognize the existence of spatial clusters in the distribution of peak flows, caused by regions with higher or lower floods. This procedure provides a qualitative method to assess the regional homogeneity of floods in which the flood scaling is described only by drainage area (Gupta and Dawdy 1995).

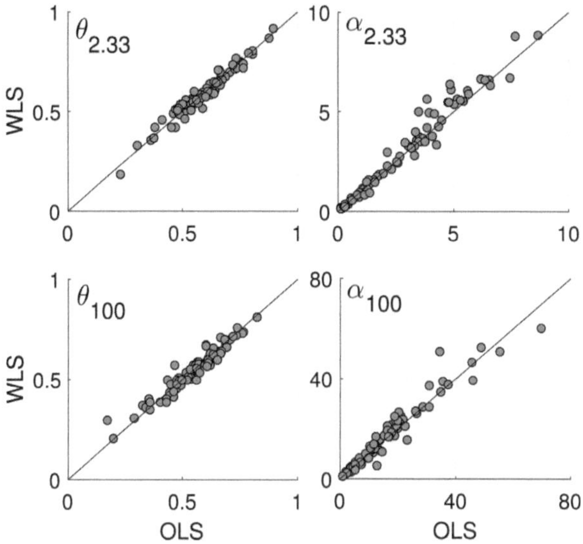

Fig. 3 Comparison of scaling parameters between WLS and OLS regressions

We mapped the values of θ_p and α_p for the 84 sub-regions to reveal the spatial patterns behind the scaling parameters for the 2.33- and 100-year return periods. These plots represent the spatial signatures of floods synthetized in two parameters (θ and α) across the drainage areas and return periods. In particular, we used the estimated power-law formulas for each HUC-4 partition to estimate peak flow quantiles of 2.33- and 100-year return periods for drainage areas of 1, 100, 1000, and 10,000 km^2. Finally, we estimated the flood SI for the same range of drainage area based on the power-law formulas. These results allow us to analyze the spatial shifts of floods for different flood quantiles and spatial scales.

4 Discussion and Analysis of Results

4.1 First Insights into Spatial Patterns of Peak Flows

Figure 4 shows the SI for all 5137 peak flow gauges. The map shows a spatial pattern over the MRB that reveals strong differences between the eastern and western parts of the basin. This variability is attributed to the differences in precipitation and runoff generation mechanisms across the large spatial domain. Western areas present SIs between 6 and 10, with a cluster in the northwest border with values around 2.

Although the current study focuses on identifying flood patterns across the spatial scales and flood quantiles rather than explaining the processes that govern these

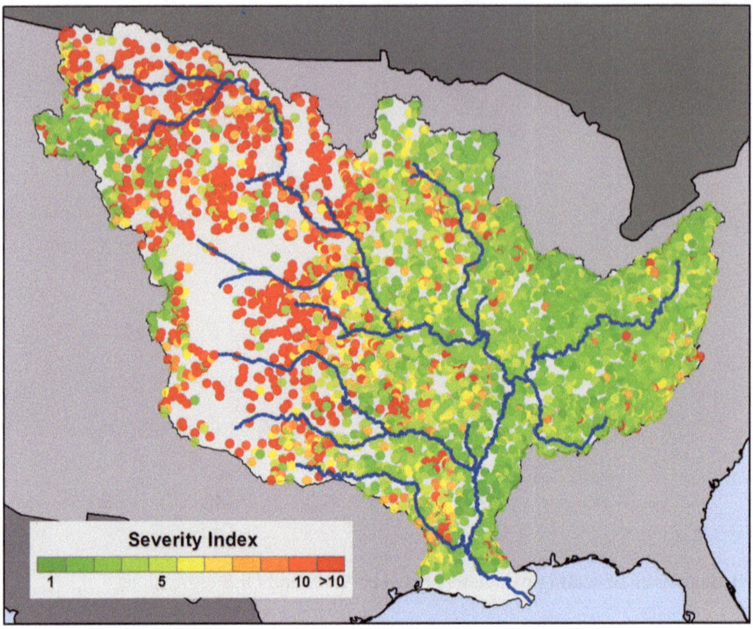

Fig. 4 Flood Severity Index for the 5,137 peak flow gauges

patterns, we discuss some insights behind the spatial structure of the SI. Higher SI values in the west are connected to differences in flood mechanisms behind the frequent floods (i.e., the 2.33-year flood) and the infrequent ones (i.e., 100-year flood). In general, floods in this region are described with a combination of snowmelt in the winter season and higher precipitation in the spring season. We can assume that the occurrence of periodic floods is more connected to one of these processes rather than both simultaneously. However, the 100-year flood (low probability of occurrence) could be connected to combinations of extreme conditions of these processes. An example of this dynamic are the floods in Montana, in which the periodic floods are related to only heavy rainfall in the spring; the higher floods (with a return period greater than 50 years) are a result of a long period of snow accumulation without intermediate melting time, in conjunction with a high soil moisture content and a high rainfall in the region (Parrett et al. 1984). We could link the western cluster with SI values of 2 by the strong orographic controls in the southwest region of Montana. This control incentivizes convection of moisture, generating more rainfall in the area; becoming the dominant flood generation process. Consequently, the magnitude in the mean annual flood and the 100-year flood in this region is not very different (a Flood Severity Index from 1 to 2). Quantification of the relative role of these mechanisms would require the implementation of physics-based models to confirm or reject the hypothesis.

In contrast, eastern Mississippi presents index values around 2 and 3; moving toward the center of the basin the dispersion increases, showing values between 2 and 5. This increase in SI allows us to speculate about a mix of processes behind the mean annual flood and the 100-year flood in the eastern region. Some findings reported by Lavers and Villarini (2013) and Villarini et al. (2014) have shown that climatological signatures from tropical cyclones and atmospheric rivers play an important role in the flood structure in the east and central part of the basin. Note that the central region of the MRB presents a high variance in SI values, with a transition between low values in the eastern region to high values in the western region.

The results from examining the SI show a certain degree of spatial structure. Nevertheless, the analysis mixes peak flow gauges draining different watershed drainage areas, which can conceal scale-dependent differences in a region. To reveal those differences, we analyze the power-law structure between drainage area and flood quantiles described by power-law scaling (see Sect. 4.2) to expose the spatial structure of floods across watershed scales (see Sect. 4.3).

4.2 Representations of Patterns in Flood Scaling

The results for the 2.33-year flood at the MRB level show $\alpha_{2.33}$ equal to 1.97 and $\theta_{2.33}$ equal to 0.60. On the other hand, at the HUC-2 level, the $\alpha_{2.33}$ ranges from 0.85 to 5.13, and the $\theta_{2.33}$ ranges from 0.53 to 0.71 (see Fig. 5). At the HUC-4 level, the $\alpha_{2.33}$ and $\theta_{2.33}$ values vary in the ranges of 0.11–8.83, and 0.18–0.91, respectively.

Figure 6 shows the distribution of the scaling parameters for the 2.33- and 100-year return periods, which leads us to use simple-scaling as our null hypothesis for

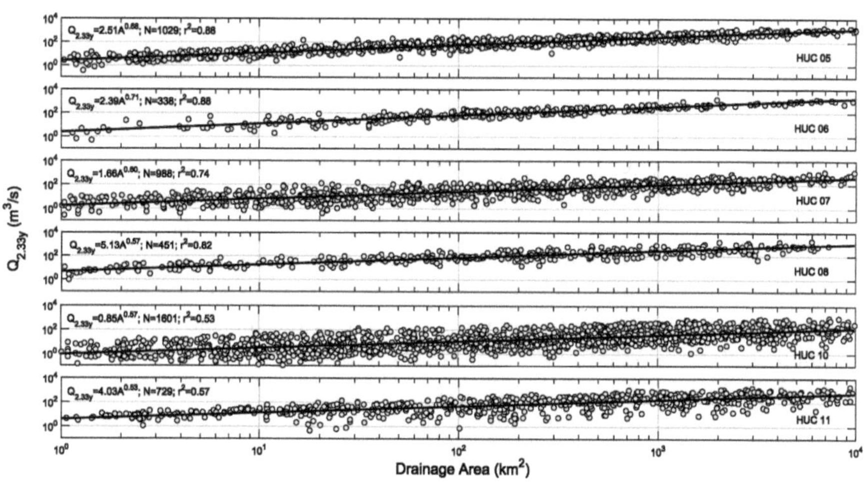

Fig. 5 Power-law regression at spatial domain of HUC-2 level (six sub-regions)

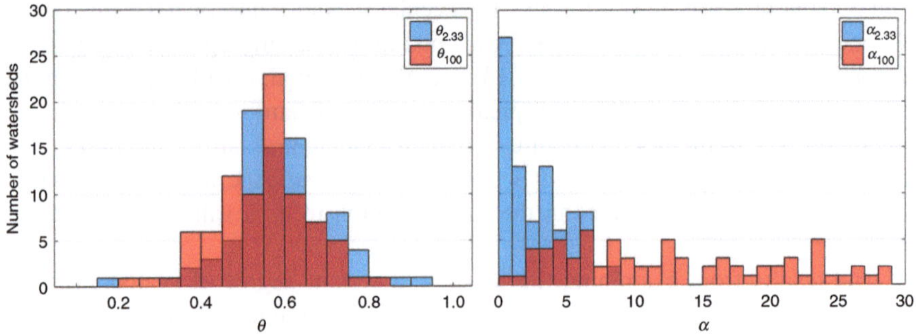

Fig. 6 Histogram of scaling exponents and intercepts for the peak flow of 2.33-year and 100-year return period at the HUC-4 level

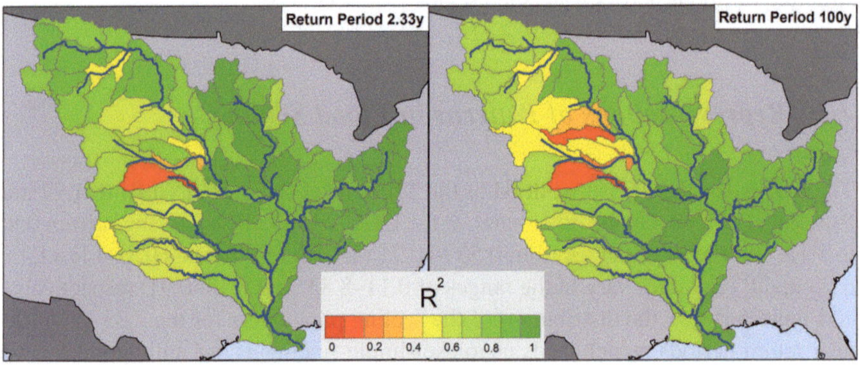

Fig. 7 Adjusted coefficient of determination (R^2) for the power law with 2.33-year floods (*left*) and 100-year floods (*right*) at the HUC-4 level

scaling in peak flow quantiles in the MRB for the HUC-4 level. Gupta and Dawdy (1995) defines simple-scaling when the scaling exponent remains constant through the quantiles, and multi-scaling when the scaling exponents change for different quantiles. In general, we could presume that the simple-scaling dominates in the MRB, by the similarity of distribution of the scaling exponents presented in Fig. 6. A rigorous statistic test must be performed to evaluate the statistically significant difference between the scaling exponents of the 2.33- and 100-year return periods. For this purpose the *Appendix 1* describes the use of the statistic test known as the Potthoff analysis (Potthoff 1966) in order to identify the type of scaling in each sub-region at HUC-4 level in the MRB testing the null hypothesis $H_0 : \theta_{2.33} = \theta_{100}$

We assessed the performance of the power-law function for the three levels of analysis using the adjusted coefficient of determination (R^2) (see Fig. 2 for the MRB, Fig. 5 for the HUC-2 level, and Fig. 7 for the HUC-4 level). For the 2.33-year floods in the MRB we obtained an R^2 of 0.58; at the HUC-2 level (six sub-regions), the R^2 varies from 0.53 to 0.88; and at the HUC-4 level (84 watersheds), the R^2 shows a range from 0.15 to 0.98, with a mean and standard deviation of 0.79 and 0.16,

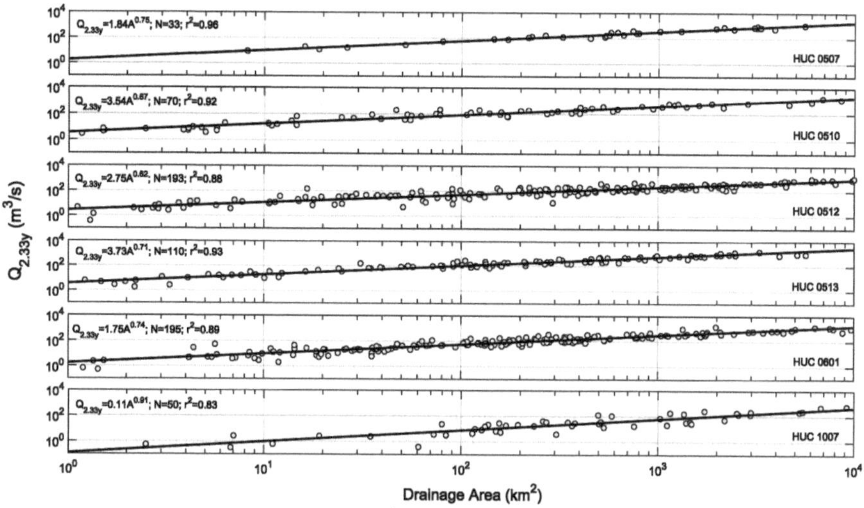

Fig. 8 Power law in six HUC-4 sub-regions (0512, 0513, 0601, 0507, 1007, and 0510)

respectively. On the other hand, for the 100-year floods in the MRB the model has an R^2 of 0.61; at the HUC-2 level the R^2 varies from 0.50 to 0.84; and at the HUC-4 level (84 watersheds), the R^2 shows a range from 0.17 to 0.95, with a mean and standard deviation of 0.74 and 0.17, respectively. In general, the HUC-4 decomposition provides a better performance of the power laws in explaining the scaling structure of floods. Figure 8 shows six of the 84 regressions obtained for the 84 sub-regions at the HUC-4 level for the 2.33-year flood. We found that only 11 of the 84 sub-regions show an R^2 less than 0.3. We hypothesize that the poor values of R^2 (less than 0.3) presented at the HUC-4 level were caused by different hydrologic conditions generating flood in the sub-regions. We explore this further in the following sub-section.

Spatial Clustering of Residuals from Power-Law Functions for Different Decomposition Levels

The determination (or lack) of regional homogeneity is essential to characterize flood-producing mechanisms. In this study, we propose a strategy to determine homogeneity that follows the fundamental idea given by Over and Gupta (1994), in which catchment size is the only variable needed to describe the flood scaling. We use the power-law function between flood quantiles and drainage area to group the peak flow residuals as HOT (peak flow above the linear regression) and COLD (peak flow below the linear regression). This notation is associated with overestimation or underestimation of flood values by the regression. We hypothesize that, if the watershed is homogeneous with respect to the flood-producing mechanism, the HOT

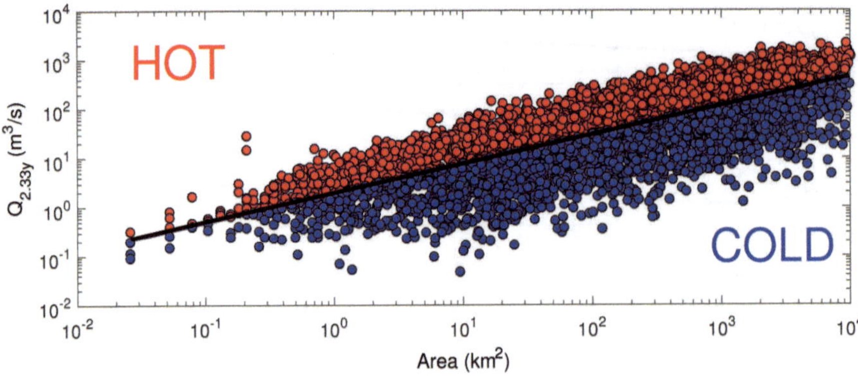

Fig. 9 HOT-COLD plot of the power law with the 5137 peak flow gauges located in the complete MRB. *Red* points identify the peak flow values above the regression and *blue* points identify the peak flow values below the regression

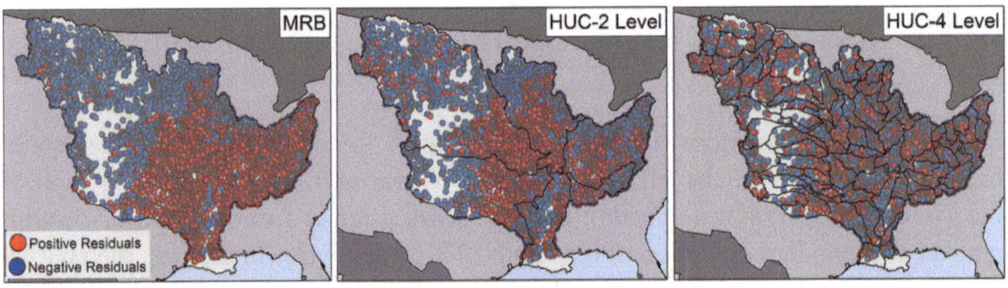

Fig. 10 Spatial pattern for the HOT-COLD analysis in the complete MRB, HUC-2 level, and HUC-4 level. *Red points* identify the peak flow values above the regression and *blue* points identify the peak flow values below the regression

and COLD gauges should be randomly distributed in the watershed, showing that all floods are correctly represented by the parameter of drainage area in the power-law function. By contrast, if there is a strong difference in flood responses inside the watershed, HOT and COLD clusters will be apparent, representing different hydrologic conditions inside the watershed. Figure 9 shows the scaling plot where gauges are classified as having HOT or COLD residuals for the regression analysis at the full MRB level.

We mapped the HOT-COLD classification in the three levels of analysis in the Fig. 10. In the spatial structure at the MRB level, essentially the eastern region has higher floods than the western regions in any range of drainage area. Clearly, this pattern exists because of the strong difference in hydrologic conditions in the two regions. The HOT-COLD results at HUC-2 level display more mixed HOT and COLD patterns than at the full MRB; however, there are still strong clusters of HOT and COLD floods in each of the six sub-regions. Finally, the HUC-4 level, with 84 sub-regions, shows a more evenly spread pattern in the spatial distribution of HOT and COLD sites; however, a close inspection of patterns in the HUC-4 units reveals some units with significant HOT-COLD clusters. Even if at HUC-4 level basins are still presenting a degree of non-homogeneity, the number of gauges inside the basins is limited, therefore it is not possible to go to a more refined level of analysis. To make more evident the existence (or non-existence) of HOT-COLD clusters, we present in the *Appendix 2* the testing of the homogeneity assumption based on the spatial autocorrelation Moran in the randomness evaluation of the power-law residuals in the space for each of the sub-regions HUC-4 level at the MRB.

4.3 Spatial Flood Patterns Based on Power-Law Formulas

In this section, we use the scaling parameters to explore regional flood differences and similarities as these values synthesize the flood processes in a scale-dependent quantity. The spatial structure in the α and θ values for the return periods of 2.33 and 100 years for each HUC-4 level is shown in Fig. 11. Assuming a drainage area equal to 1 km^2, Eq. (1) gives Q_p values equal to α_p. Therefore, the α_p value can be interpreted as the flood quantile for the unitary drainage area (1 km^2). Consequently,

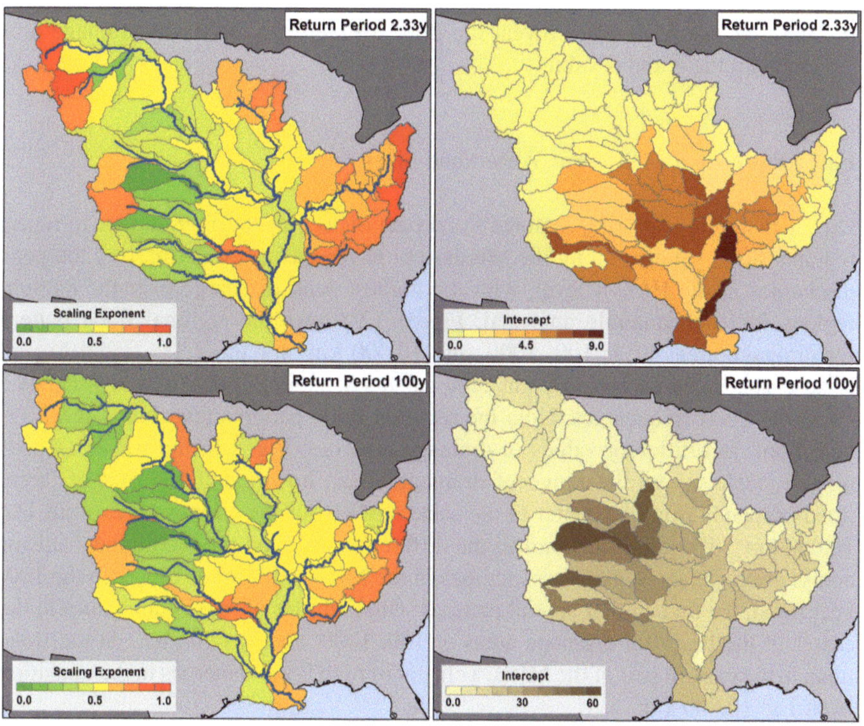

Fig. 11 Scaling exponent and intercept at HUC-4 level for peak flows with return periods of 2.33- and 100-year

these values describe the pattern of floods for small watersheds with 1 km^2 of drainage area along the Mississippi watershed. The α patterns in the west and central part of the Mississippi watershed show strong differences, with $\alpha_{2.33}$ values closer to 1 and 5, respectively, and α_{100} values closer to 10 and 55 in the west and central part, respectively. These results suggest that the 2.33-year floods for a watershed with a size of 1 km^2 in the central region of the Mississippi River Basin are five times stronger in magnitude than the floods in the western part of the basin. The question that remains is if this spatial pattern is preserved for different watershed sizes; this will be addressed in section "Flood Patterns for Different Watershed Sizes".

The θ values represent the slope between drainage areas and floods in the log space, connecting the power rate in the flood increments across drainage areas. As we mentioned earlier, α is controlling floods with a unitary drainage area, however, changes in the α values are also impacting the other spatial scales. This means that a displacement in the intercept will modify the flood magnitude in direct proportion to the A^θ value. By contrast, the θ magnitude affects the flood magnitude differently across the scales, with a potential relation. The spatial patterns in α are completely different from the spatial structure of θ. An example are clusters of high θ values found in the west, east, and north of the basin. These values show a transition to lower values toward the center of the basin, locating finally the lowest cluster of θ in the Midwest with θ values close to 0.2. With the spatial structures in θ and α identified, we proceed to reveal the flood structure in ranges of drainage area governed by the θ and α values.

Flood Patterns for Different Watershed Sizes

Our final step in exploring regional flood patterns is combining θ and α. To reveal regional flood patterns along the Mississippi River Basin, we normalized the peak discharges at the HUC-4 level, with the highest peak discharge over the 84 sub-regions for each quantile analyzed. Figure 12 shows the regional flood pattern for drainage areas of 1, 100, 1000, and 10,000 km^2 for the 2.33-, and 100-year floods. The maps reveal different regional flood patterns for different watershed sizes and flood frequencies. Note the regional flood patterns revealed at the 2.33-year flood and different values of drainage area (see Fig. 12): the western floods remain stable across the range of drainage areas; however, the cluster of larger floods expands from the center to the eastern MRB for the drainage area from 1 to 10,000 km^2. These results confirm the differences in flood-producing mechanisms across drainage areas and across regions in the Mississippi River Basin. In the 100-year flood, we find similar spatial patterns with a cluster of maximum values in the center of the basin for drainage areas of 100, 1000, and 10,000 km^2. At the same time, in the eastern part of the MRB, a cluster of high floods emerges in the transition from 1 to 10,000 km^2.

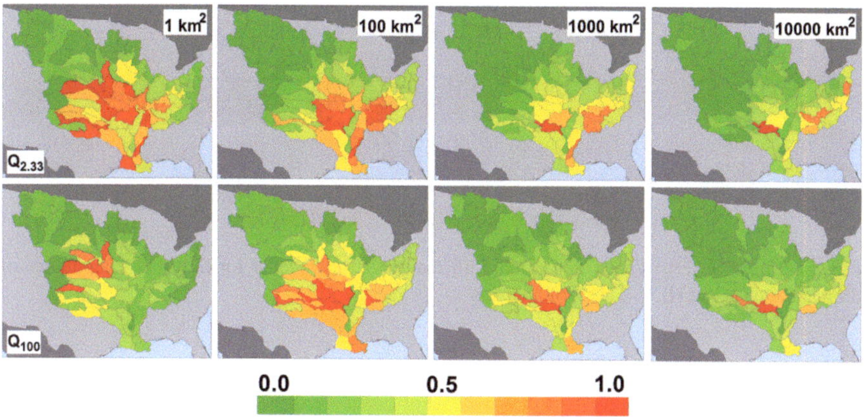

Fig. 12 Normalized peak flows estimated with the power-law model for different return periods (*rows*) and magnitudes of drainage area (*columns*) at the HUC-4 level of analysis

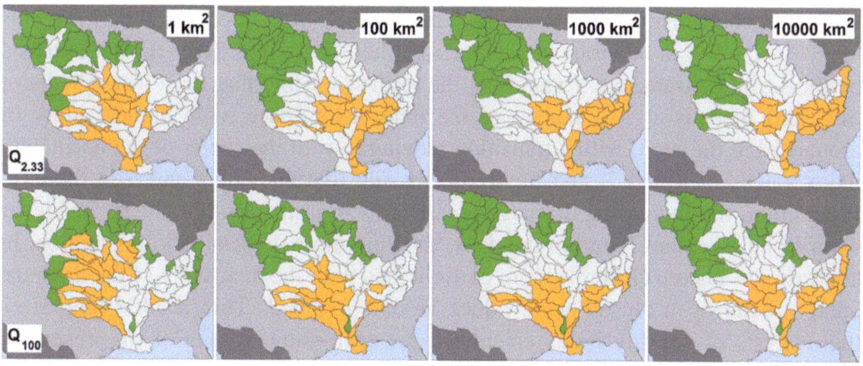

Fig. 13 Top 20 sub-regions with higher (*orange*) and lower (*green*) peak flows for specific drainage area sizes and return period of 2.33 and 100 years at the HUC-4 level of analysis

To highlight the variability of these regional flood patterns with respect to the drainage area, we grouped the HUC-4 level in the Top 20 sub-regions with higher floods and the Top 20 sub-regions with lower floods. Figure 13 shows the transition in space of the Top 20 sub-regions with higher 2.33- and 100-year floods respect to the watershed size, with a displacement from the center to the eastern Mississippi from 1 to 10,000 km^2. Looking at the Top 20 of the lower 2.33-year floods the cluster in the northwestern Mississippi is preserved in the different ranges of watershed sizes. However, the spatial pattern exhibited by the Top 20 of lower 100-year floods is more dispersed with respect to the 2.33-year flood pattern. These results demonstrate that the flood-producing mechanisms change not only in region, but also in drainage area magnitudes.

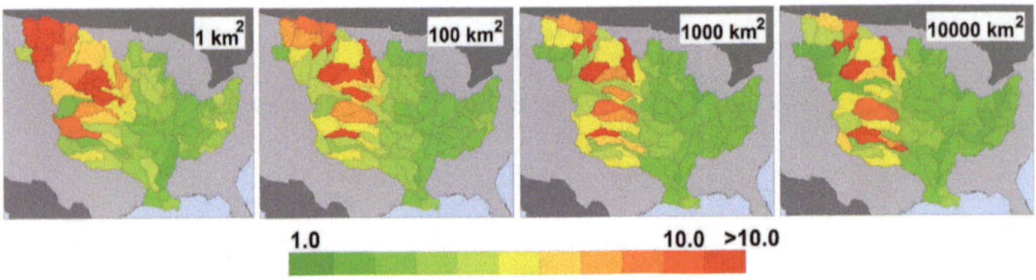

Fig. 14 Flood Severity Index generated with the power-law model for different magnitudes of drainage area at the HUC-4 level of analysis

Severity Index with Flood Scaling.

To improve our description of regional flood patterns for different frequencies and drainage areas, we calculated the severity index using the power-law formulas calculated for each one of the 84 HUC-4 sub-regions. The severity index is defined as the ratio between the 100-year flood and the 2.33-year flood. Figure 14 presents the SI variability in the drainage area across the Mississippi River Basin described by the scaling parameters. Examining the results, we find that for a watershed size of 1 km^2 there is a clear pattern of the index over space, with higher values of around 10 in the west, with a transition moving toward the center of the basin with index values of 4 and 7, and finally decreasing to values of 2 and 5 in the east. In addition, the increasing drainage area begins to transform the cluster of higher severity index values found in the west, showing more dispersion in this region. In contrast, the eastern cluster is more consolidated with the increasing of the drainage area, structuring a cluster with values around 2 and 3. This result summarizes the analysis of severity index calculated for each gauge in Fig. 4 by presenting patterns in flood ratios across sizes of drainage area. These results highlight the importance of discerning the watershed size from the smallest (1 km^2), to medium-sized ($100 < A < 1000$ km^2), to largest (>10,000 km^2) watersheds in flood estimation.

5 Conclusions

The flood scaling analysis performed in this study reveals a diversity of regional flood patterns using scaling parameters (θ and α) of the 2.33- and 100-year floods for different drainage area values. We show that at HUC-4 level of decomposition the power laws represent a satisfactory representation of peak; although the HOT-COLD pattern suggests that in some sub-regions the analysis can be improved with a refined level of analysis (e.g., HUC-6 Level). The number of gauges inside the basins is a limitation to evaluate a more refined scale.

A remarkable result is the shift of regional flood patterns for different drainage sizes and for the 2.33- and 100- year floods at HUC-4 level, in which the relative flood magnitude depends of the catchment size, showing a dynamic dependence of floods related to drainage area and spatial location. This result is especially strong in the spatial transition of the Top 20 regions with higher 2.33- and 100-year from the center to the eastern Mississippi. The results of this analysis provide clear signatures in flood-producing mechanisms that should be explained from physical considerations.

We recognize several caveats in our study. We presented spatial patterns only up to watersheds smaller than 10,000 km^2. This threshold was defined to guarantee the flood diversity inside of a HUC-4, eliminating the influence of gauges over large rivers that integrate different climate regimes and are more likely to be affected by regulation. In addition, we are aware that using a different probability distribution (e.g., heavy tail distributions) could change the outcome of the flood quantile estimations. We decided on using the standard methodology proposed by the USGS in the Bulletin 17B based on the probability distribution Log Pearson Type 3 as the best option for the flood quantile estimation because it is easily replicable, thanks to the USGS PeakFQ software.

Our research represents an effort to quantify the structure of flood scaling in a range of drainage areas and flood quantiles. We report notable regional flood patterns that should relate to physical variables to explain the underlying mechanisms behind flood dynamics across scales.

Acknowledgements The authors thank the Iowa Flood Center for supporting this study. This study builds on the framework presented at the conference Twenty Years of Nonlinear Geophysics and discussed in Gupta et al. (2007).

Appendix 1: Identifying the Scaling Type

Gupta and Dawdy (1995) defines simple-scaling as when the scaling exponent in the power-law regression between flood quantiles and upstream area remains constant for all flood quantiles, and multi-scaling as when the scaling exponents change. We use the Potthoff analysis (Potthoff 1966) in order to test the null hypothesis $H_0: \theta_{2.33} = \theta_{100}$. The Potthoff analysis identifies if there is a significant difference in the regression exponents when data are separated in different groups. This analysis requires performing a multiple linear regression to compare the regression coefficients for different peak flow quantiles.

With a significance level of 5%, we can conclude that if the p-value is less than 0.05 the null hypothesis of simple-scaling should be rejected. However, if the p-value is greater than 0.05 we cannot reject the null hypothesis, leaving open the possibility of simple-scaling in the data. The test relies on three equations,

$$Y^{[i]} = \ln(Q_{2.33}) = \ln(\alpha_{2.33}) + \theta_{2.33}\ln(A) \tag{2}$$

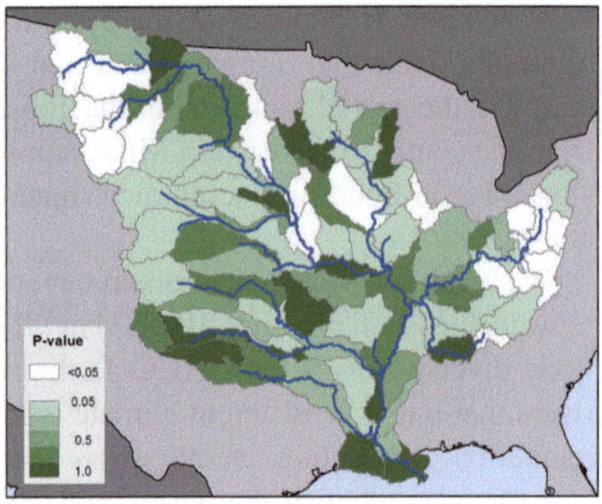

Fig. 15 Spatial pattern of the p-value to test the null hypothesis $H_0 : \theta_{2.33} = \theta_{100}$ in the 84 sub-watershed

$$Y^{[j]} = \ln(Q_{100}) = \ln(\alpha_{100}) + \theta_{100}\ln(A) \tag{3}$$

$$Y^{[i,j]} = a + bX^{[i,j]} + cG^{[i]} + dG^{[i]}X^{[i,j]} \tag{4}$$

X is the vector of drainage area repeated twice, since the drainage area is the same for Eqs. (2) and (3). G is the dichotomous grouping variable (dummy variable) coding one for the region i and zero for the region j. The coefficients in Eq. (4) evaluate the difference in coefficients for Eqs. (2) and (3). For our purposes the coefficient d determines differences between $\theta_{2.33}$ and θ_{100}, therefore we estimate the p-value for the coefficient d to test the null hypothesis $H_0 : \theta_{2.33} = \theta_{100}$.

Figure 15 presents the estimated p-value for the 84 sub-regions at HUC-4 level. Based on this result we report 17 sub-watersheds with multi-scaling with clusters on the western and eastern Mississippi; and 67 sub-watershed with possible simple-scaling with a predominance in the center of the Mississippi River Basin.

Appendix 2: Testing Regional Homogeneity

We assessed regional homogeneity using the Moran spatial autocorrelation (Moran 1950) for the residuals of the scaling power law in space. We group the residuals of the power-law regression as HOT: Positive residuals, and COLD: Negative residuals. The Moran spatial autocorrelation evaluates if a pattern of a spatial variable is clustered, dispersed, or random based on the null hypothesis that the variable (residuals in our case) is randomly distributed in space.

We hypothesize that, if there is regional homogeneity over the scaling of peak flows the groups HOT-COLD should be randomly distributed in the watershed, showing that the floods in the region have a similar hydrologic response. However, if there are strong differences of flood responses along the watershed, HOT and COLD clusters will start to arise, breaking the regional homogeneity assumption in the flood scaling theory. The significance of the test can be evaluated with a

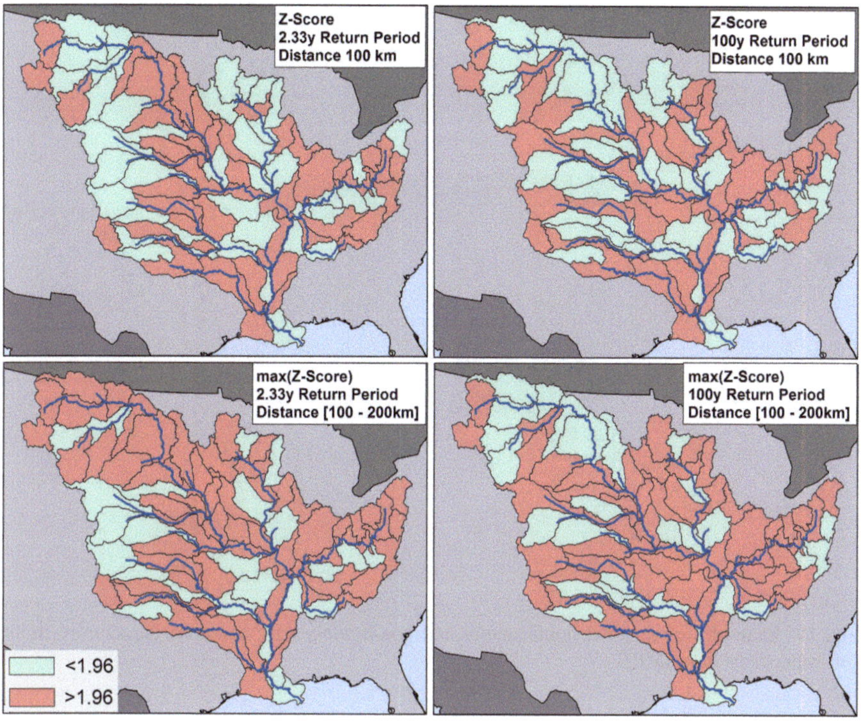

Fig. 16 Map of the *Z*-score in the 84 sub-watersheds for the 2.33- and 100-year return period

Z-score in which a *Z*-score of 1.96 rejects the null hypothesis with a significance level of 5%, in such case, we infer the existence of a clustering pattern of residuals in the space and therefore the regional homogeneity is not obtained. The Moran statistic is based on neighboring elements that are defined with a specific buffer distance. We evaluate the *Z*-score for a search distance between 100 and 200 km as this range seems sufficient to consider the inclusion of different peak flow gauges at the watershed level of HUC-4. The spatial structure of the *Z*-score presented in Fig. 16 demonstrates that the regional homogeneity is independent of frequency in some regions, showing watersheds with *Z*-scores higher than 1.96 for the 2.33-year floods, but lower than 1.96 for the 100-year floods. Also, the results show a dominant pattern of non-homogeneity in the MRB at the HUC-4 level, suggesting that a more refined spatial scale is necessary to obtain a more accurate representation of peak flows through scaling of floods. Figure 17 shows 6 examples of the classification of regional homogeneity based on the *Z*-score. Notice that the HUCs 1010, 1029, and 510 have *Z*-scores less than 1.96 which is related to a regional homogeneity feature (randomness in the HOT-COLD residuals), leading to conclusion that the flood scaling on these watersheds is well represented. However, the HUCs 1027,

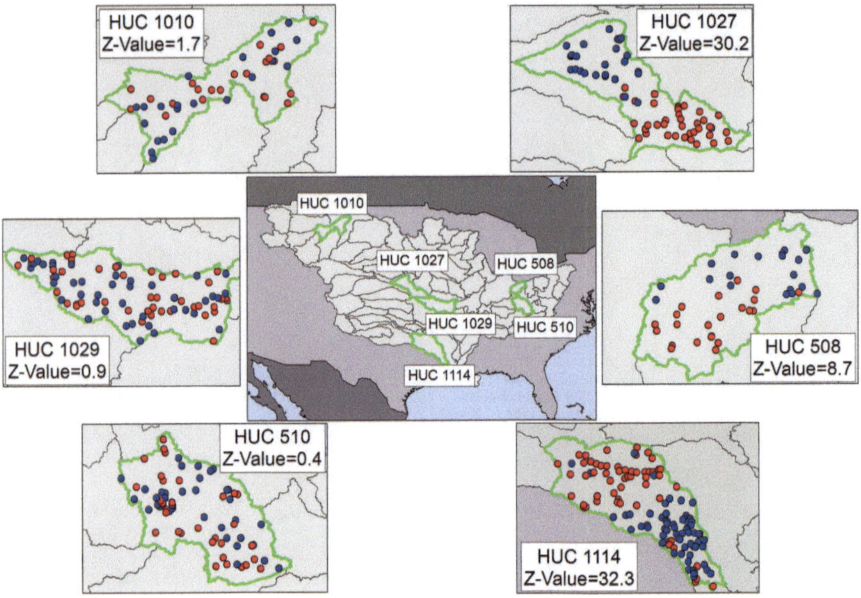

Fig. 17 Examples of regional homogeneity and non-homogeneity based on the *Z*-score in six sub-watersheds in the MRB

508, and 1114 have *Z*-scores higher than 1.96 in which the clustering of the HOT-COLD residuals in the space is obvious. Therefore, these watersheds must be refined based on the HOT-COLD clusters to properly capture the regional signature of floods across the scales.

References

Ayalew, T.B., W.F. Krajewski, and R. Mantilla. 2014a. Connecting the power-law scaling structure of peak-discharges to spatially variable rainfall and catchment physical properties. *Advances in Water Resources* 71: 32–43. doi:10.1016/j.advwatres.2014.05.009.

Ayalew, T.B., W.F. Krajewski, R. Mantilla, and S.J. Small. 2014b. Exploring the effects of hillslope-channel link dynamics and excess rainfall properties on the scaling structure of peak-discharge. *Advances in Water Resources* 64: 9–20. doi:10.1016/j.advwatres.2013.11.010.

Ayalew, T.B., W.F. Krajewski, and R. Mantilla. 2015. Analyzing the effects of excess rainfall properties on the scaling structure of peak discharges: Insights from a mesoscale river basin. *Water Resources Research* 51: 3900–3921. doi:10.1002/2014WR016258.

Dawdy, D.R., V.W. Griffis, and V.K. Gupta. 2012. Regional flood-frequency analysis: How we got here and where we are going. *Journal of Hydrologic Engineering* 17 (9): 953–959. doi:10.1061/(ASCE)HE.1943-5584.0000584.

Flynn, K.M., W.H. Kirby, and P.R. Hummel. 2006. *User's manual for Program PeakFQ, annual flood-frequency analysis using Bulletin 17B guidelines*. Retrieved from http://purl.access.gpo.gov/GPO/LPS97012.

Fuller, W.E. 1914. Flood flows. *Transaction ASCE* 77: 564–617.

Furey, P.R., and V.K. Gupta. 2007. Diagnosing peak-discharge power laws observed in rainfall-runoff events in Goodwin Creek experimental watershed. *Advances in Water Resources* 30 (11): 2387–2399. doi:10.1016/j.advwatres.2007.05.014.
Gupta, V.K., and D.R. Dawdy. 1995. Physical interpretations of regional variations in the scaling exponents of flood quantiles. *Hydrological Processes* 9: 347–361. http://doi.org/10.1002/hyp.3360090309.
Gupta, V.K., and E. Waymire. 1998. Spatial variability and scale invariance in hydrologic regionalization. In *Scale dependece and scale invariance in hydrology*, ed. G. Sposito. Cambridge: Cambridge University Press.
Gupta, V.K., S.L. Castro, and T.M. Over. 1996. On scaling exponents of spatial peak flows from rainfall and river network geometry. *Journal of Hydrology* 187 (1–2): 81–104. doi:10.1016/S0022-1694(96)03088-0.
Gupta, V.K. 2004. Emergence of statistical scaling in floods on channel networks from complex runoff dynamics. *Chaos, Solitons & Fractals* 19(2): 357–365. http://doi.org/10.1016/S0960-0779(03)00048-1.
Gupta, V.K., B.M. Troutman, and D.R. Dawdy 2007. Towards a non- linear geophysical theory of floods in river networks: An overview of 20 years of progress. In *Nonlinear dynamics in geosciences*, ed. A.A. Tsonis, and J.B. Elsner, 121–151. New York: Springer.
Gupta, V.K., R. Mantilla, B.M. Troutman, D. Dawdy, and W.F. Krajewski. 2010. Generalizing a nonlinear geophysical flood theory to medium-sized river networks. *Geophysical Research Letters* 37(11): 1–6. http://doi.org/10.1029/2009GL041540.
Gupta, V.K., T.B. Ayalew, R. Mantilla, and W.F. Krajewski. 2015. Classical and generalized Horton laws for peak flows in rainfall-runoff events. *Chaos (Woodbury, N.Y.)* 25(7): 75408. http://doi.org/10.1063/1.4922177.
Haddad, K., A. Rahman, and J.R. Stedinger. 2012. Regional flood frequency analysis using Bayesian generalized least squares: A comparison between quantile and parameter regression techniques. *Hydrological Processes* 26 (7): 1008–1021. doi:10.1002/hyp.8189.
Jothityangkoon, C., and M. Sivapalan. 2001. Temporal scales of rainfall - runoff processes and spatial scaling of flood peaks: space - time connection through catchment water balance. *Advances in Water Resources* 24 (9–10): 1015–1036. doi:10.1016/S0309-1708(01)00044-6.
Kroll, C. 2014. The prediction of hydrologic statistics in nested watersheds across the United States. *World Environmental and Water Resources Congress*, 2326–2335 Retrieved from http://earthjustice.org/features/campaigns/fracking-across-the-united-states.
Lavers, D.A., and G. Villarini. 2013. Atmospheric rivers and flooding over the central United States. *Journal of Climate* 26 (20): 7829–7836. doi:10.1175/JCLI-D-13-00212.1.
Mandapaka, P.V., W.F. Krajewski, R. Mantilla, and V.K. Gupta. 2009. Dissecting the effect of rainfall variability on the statistical structure of peak flows. *Advances in Water Resources* 32 (10): 1508–1525. doi:10.1016/j.advwatres.2009.07.005.
Mantilla, R., V.K. Gupta, J. Mesa, and O. 2006. Role of coupled flow dynamics and real network structures on Hortonian scaling of peak flows. *Journal of Hydrology* 322: 155–167. doi:10.1016/j.jhydrol.2005.03.022.
Medhi, H., and S. Tripathi. 2015. On identifying relationships between the flood scaling exponent and basin attributes. *Chaos* 25 (7): 075405. doi:10.1063/1.4916378.
Menabde, M., and M. Sivapalan. 2001. Linking space-time variability of river runoff and rainfall fields: A dynamic approach. *Advances in Water Resources* 24 (9–10): 1001–1014. doi:10.1016/S0309-1708(01)00038-0.
Moran, P.A. 1950. Notes on continuous stochastic phenomena. *Biometrika* 17: 37.
Morrison, J.E., and J.A. Smith. 2001. Scaling properties of flood peaks. *Extremes* 4: 5–22. doi:10.1023/A:1012268216138.
Ogden, F.L., and D.R. Dawdy. 2003. Peak discharge scaling in small Hortonian watershed. *Journal of Hydrologic Engineering* 8(2): 64–73. http://doi.org/10.1061/(ASCE)1084-0699(2003)8:2(64).

Over, T.M., and V.K. Gupta. 1994. Statistical analysis of mesoscale rainfall dependence of a random cascade generator on large scale forcing. *Journal of Applied Meteorology*. http://doi.org/10.1175/1520-0450(1994)033<1526:saomrd>2.0.co;2.

Parrett, C., D.D. Carlson, S. Craig, and J. Grodon. 1984. Floods of May 1978 in southeastern montana and northeastern Wyoming. *U.S Geological Survey Professional Paper*, (May).

Potthoff, R.F. 1966. *Statistical aspects of the problem of biases in psychological tests*. North Carolina: Department of Statistics Chapel Hill: University of North Carolina. Institute of Statistics Mimeo Series.

Robinson, J.S., and M. Sivapalan. 1997. Temporal scales and hydrological regimes: Implications for flood frequency scaling. *Water Resources Research* 33 (12): 2981. doi:10.1029/97WR01964.

Seaber, P.R., F.P. Kapinos, and G.L. Knapp. 1987. Hydrologic unit maps: U.S. Geological survey. *Water-Supply Paper* 2294, 63.

Smith, A., C. Sampson, and P. Bates. 2015. Regional flood frequency analysis at the global scale. *Water Resources Research* 51: 539–553. doi:10.1002/2014WR015829.

Srinivas, V.V., S. Tripathi, A.R. Rao, and R.S. Govindaraju. 2008. Regional flood frequency analysis by combining self-organizing feature map and fuzzy clustering. *Journal of Hydrology* 348 (1–2): 148–166. doi:10.1016/j.jhydrol.2007.09.046.

Villarini, G., R. Goska, J.a. Smith, and G.A. Vecchi. 2014. North Atlantic tropical cyclones and U.S. flooding. *Bulletin of the American Meteorological Society* 95 (9): 1381–1388. doi:10.1175/BAMS-D-13-00060.1.

Wan Jaafar, W.Z., and D. Han. 2012. Calibration catchment selection for flood regionalization modeling. *Journal of the American Water Resources Association* 48 (4): 698–706. doi:10.1111/j.1752-1688.2012.00648.x.

Studying the Complexity of Rainfall Within California Via a Fractal Geometric Method

Carlos E. Puente, Mahesh L. Maskey, and Bellie Sivakumar

Abstract A deterministic geometric approach, the fractal–multifractal (FM) method, useful in modeling storm events and recently adapted in order to encode highly intermittent daily rainfall records, is employed to study the complexity of rainfall sets within California. Specifically, sets—from south to north—at Cherry Valley, Merced, Sacramento and Shasta Dam and containing, respectively 59, 116, 115, and 72 years, all ending at water year 2015, are studied. The analysis reveals that: (a) the FM approach provides faithful encodings of all records, by years, with mean square and maximum errors in accumulated rain that are less than a mere 2 and 10%, respectively; (b) the evolution of the corresponding "best" FM parameters, allowing visualization of the inter-annual rainfall dynamics from a reduced vantage point, exhibit a highly entropic variation that prevents discriminating between sites and extrapolating to the future; and (c) the rain signals at all sites may be termed "equally complex," as usage of k-means clustering and conventional phase-space analysis of FM parameters yields comparable results for all sites.

Keywords Fractal-multifractal • Rainfall • Dynamics • Encoding • Climate change • Complexity • Chaos • Phase-space • Classification • Geometry

---

Electronic supplementary material The online version of this chapter (doi:10.1007/978-3-319-58895-7_24) contains supplementary material, which is available to authorized users.

C.E. Puente (✉) • M.L. Maskey
Department of Land, Air and Water Resources, University of California, Davis, Davis, CA 95616, USA
e-mail: cepuente@ucdavis.edu

B. Sivakumar
School of Civil and Environmental Engineering, The University of New South Wales, Sydney, NSW 2052, Australia

Department of Land, Air and Water Resources, University of California, Davis, CA 95616, USA

A.A. Tsonis (ed.), *Advances in Nonlinear Geosciences*,
DOI 10.1007/978-3-319-58895-7_24

Notation

ε_{ac}	Root mean square error in accumulated set
ε_{mac}	Maximum error in accumulated set
ν_j	Nash–Sutcliffe efficiency on records at *j*-day scale
η_j	Nash–Sutcliffe efficiency on accumulated sets at *j*-day scale
ζ_j	Number of zeros present in the sets at *j*-day scale
π_j	Percent of zeros matched in the FM set at *j*-day scale
H_c	Entropy of class distribution
$\overline{\text{OL}}_c$	Average orbit length by classes

1 Introduction

As rainfall is a fundamental input to the hydrologic system, quantifying its temporal and spatial complexity is paramount for the proper planning, design, and implementation of water resource infrastructure. As the process often exhibits complex nonlinear behavior and high-intermittency, it is desirable to develop improved techniques that may allow further understanding of its structure.

With the development of stochastic and fractal notions and the advent of modern computation, a substantial effort has been made in the past few decades at conceptualizing numerous rainfall models. Attempting to capture the erratic, intermittent, random, and, in short, complex nature of rainfall, various frameworks have been proposed. Such include attempts to quantify complexity based on: (a) chaotic features of the records (e.g., Rodríguez-Iturbe et al. 1989; Sharifi et al. 1990; Ghilardi and Rosso 1990; Rodríguez-Iturbe 1991; Jayawardena and Lai 1994; Koutsoyiannis and Pachakis 1996; Sivakumar et al. 1999, 2001a; Peters et al. 2001; Men et al. 2004; Dhanya and Kumar 2010; Jothiprakash and Fathima 2013), (b) nonlinear time series models (e.g., French et al. 1992; Luk et al. 2000; Jin et al. 2005; Ramirez et al. 2005; Nasseri et al. 2008; Kim et al. 2009; Sivakumar 2009; Sivakumar and Singh 2012; Sivakumar et al. 2014), and (c) representations aiming at preserving statistical and fractal and multifractal rainfall properties (e.g., Rodríguez-Iturbe 1986; Gupta and Waymire 1990; Tessier et al. 1993; Lovejoy and Schertzer 2013; Puente and Obregón 1996; Sivakumar 2000, 2004; Sivakumar et al. 2001b; Maskey et al. 2015).

Relevant to this research, Puente (1996) developed a simple geometric procedure, the so-called fractal–multifractal (FM) method, which generates "seemingly random" sets as fractal transformation of multifractal measures without requiring any statistical assumptions. This method, which fits within the modern notion of a fourth paradigm in data-intensive scientific discovery (Hey et al. 2009), produces a vast class of patterns defined over one and higher dimensions that not only preserves key statistical indicators, viz. moments, autocorrelation function, power spectrum, multifractal spectrum, but also captures intricate details and the textures present in the data sets, something which is quite difficult to accomplish using (physical) stochastic models (Puente 2004; Cortis et al. 2009).

While the FM approach has already been used to model: (a) rainfall events (e.g., Puente 1996; Obregón et al. 2002a, b; Huang et al. 2012a, b), (b) daily rainfall sets gathered over a year (Maskey et al. 2015, 2016b; Puente et al. 2017), (c) daily streamflow records over a year (Puente et al. 2017; Maskey et al. 2016a), (d) daily temperature measurements (Puente et al. 2017), and (e) even spatial contaminant plumes (Puente et al. 2001a, b), this article represents the first effort in using the FM method as a tool to quantify rainfall complexity, using for the purpose data sets collected within California.

The organization of the paper is as follows. Given first is an introduction to the FM notions and, in particular, the specific adaptation used to model intermittent rainfall records. This is followed by the methodology employed in fitting specific data sets via a numerical optimization exercise and an explanation of how FM geometric parameters will be used to quantify complexity. Then, the analysis of the records' complexity at four stations, from south to north, Cherry Valley, Merced, Sacramento and Shasta Dam, is advanced, including a study of the inter-annual dynamics and data-mining classifications at each site and a comparison of complexity features in space among the sites. The article concludes with its conclusions and recommendations.

2 The Fractal–Multifractal Method

The transformation of multifractal measures via fractal interpolating functions, leading to the fractal–multifractal (FM) method (Puente 1996), is reviewed here.

A *fractal interpolating function* $f: x \to y$, passing through $N+1$ ordered points in plane $\{(x_n, y_n)|\ x_0 < x_1 < \ldots < x_N\}$ and having a graph $G = \{(x, f(x)|x \in [x_0, x_N] = [0, 1]\}$, is defined as the unique fixed point of N affine maps (Barnsley 1988):

$$w_n \begin{pmatrix} x \\ y \end{pmatrix} = \begin{pmatrix} a_n & 0 \\ c_n & d_n \end{pmatrix} \begin{pmatrix} x \\ y \end{pmatrix} + \begin{pmatrix} e_n \\ f_n \end{pmatrix}, \qquad n = 1, \ldots, N, \tag{1}$$

such that, $G = w_1(G) \cup w_2(G) \cup \ldots \cup w_N(G)$. While the vertical scalings d_n are free parameters satisfying $|d_n| < 1$, the other coefficients in Eq. (1), a_n, c_n, e_n,and f_n are evaluated via the contracting initial conditions:

$$w_n \begin{pmatrix} x_0 \\ y_0 \end{pmatrix} = \begin{pmatrix} x_{n-1} \\ y_{n-1} \end{pmatrix}, \qquad w_n \begin{pmatrix} x_N \\ y_N \end{pmatrix} = \begin{pmatrix} x_n \\ y_n \end{pmatrix}, \quad n = 1, \ldots, N, \tag{2}$$

which guarantee the existence of a stable attractor and yield N systems of linear equations that may be easily solved in terms of the interpolating points and the vertical scalings. Upon successive iterations of the maps, a convoluted "wire" function f, whose graph has a fractal dimension $1 \leq D < 2$, is found.

The notions may be generalized so that a more general attractor, other than a function, is obtained. Such is easily accomplished replacing the contractile initial conditions (Eq. (2)) by more general contractions:

$$w_n\begin{pmatrix}x_0\\y_0\end{pmatrix}=\begin{pmatrix}x_{2n}\\y_{2n}\end{pmatrix},\qquad w_n\begin{pmatrix}x_{2N-1}\\y_{2N-1}\end{pmatrix}=\begin{pmatrix}x_{2n+1}\\y_{2n+1}\end{pmatrix},\quad n=1,\ldots,N, \tag{3}$$

such that the range in x of map w_n becomes the interval $[x_{2n}, x_{2n+1}]$. Notwithstanding the need of additional end-point parameters y_{2n}, y_{2n+1}, a disperse attractor, over a Cantor set (Mandelbrot 1982), is defined whenever the domain of the attractor contains gaps (Huang et al. 2013; Maskey et al. 2015).

Figure 1 illustrates how a disperse attractor is constructed iterating two affine maps whose end-points are $\{(0,0),(0.41,1.08)\}$ and $\{(0.80,4.02),(1,-0.35)\}$, namely:

$$w_1\begin{pmatrix}x\\y\end{pmatrix}=\begin{pmatrix}0.41 & 0\\0.97 & -0.32\end{pmatrix}\begin{pmatrix}x\\y\end{pmatrix} \tag{4}$$

$$w_2\begin{pmatrix}x\\y\end{pmatrix}=\begin{pmatrix}0.19 & 0\\-4.53 & -0.43\end{pmatrix}\begin{pmatrix}x\\y\end{pmatrix}+\begin{pmatrix}0.80\\4.02\end{pmatrix}, \tag{5}$$

and when such are iterated following independent outcomes of a 33–67% biased coin.

As seen, the Monte Carlo procedure, known as the "chaos game" (Barnsley 1988), defines (say after 2^{14} iterations) a Cantorian function from x to y, and also ultimately induces stable projections (histograms) dx and dy, which are hence functionally related and deterministic. While the former is clearly defined over a Cantor set (as there is a gap of 0.39 in end-points over x) and exhibits a *multifractal* structure containing noticeable repetition, the latter, which exhibits ample intermittency, is the derived measure over y found transforming the input measure dx via the fractal function from x to y, hence explaining the notation *fractal–multifractal* approach. For the sake of rainfall modeling, Fig. 1 also includes an adaptation of the notions via set dy_v so that additional zero values may be defined. Such a pattern is simply found trimming dy below a threshold ϕ, in a manner that evokes removing "traces" of rain.

By varying the parameters associated with the construction, the ideas herein yield indeed a host of rainfall-looking sets that shall be used later on to encode rainfall measurements via an inverse problem that uses recorded information (duly normalized) as the target of an FM optimization exercise that depend on the following parameters: (a) the end-points that define where the attractor would pass, (b) the scalings d_n, (c) the frequencies used to carry the iterations p, and (d) the threshold ϕ used to trim the original output. Without a lack of generality, the first member of the first end-point and the second member of the second end-point have been set, respectively, to (0,0) and $(1, y_{2N+1})$. This implies a total of nine geometric FM parameters when iterating two maps.

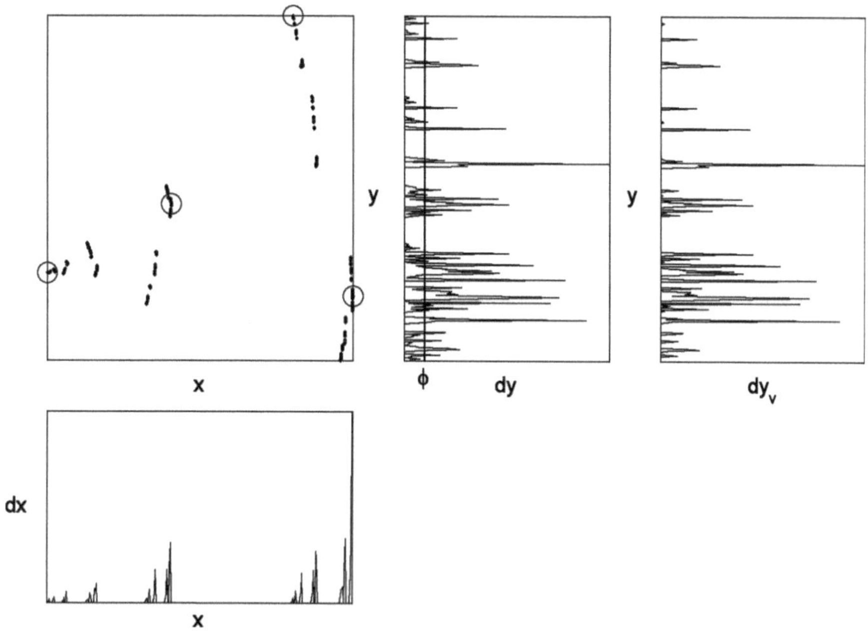

Fig. 1 A generalized FM approach: from a Cantorian texture *dx*, to a projection *dy*, via a disperse attractor from *x* to *y*. The set dy_v is found pruning *dy* below a threshold ϕ

Although, as mentioned earlier, the FM approach do fit within the modern notion of data-intensive scientific discovery (Hey et al. 2009), it is worth remarking that the notions have been assigned a physical interpretation, as the FM sets produced are deterministic realizations of non-trivial stochastic conservative multiplicative cascades, ultimately belonging to the class of "universal multifractals" (Cortis et al. 2013). Although no physical meaning may be assigned to the specific FM parameters nor such may be measured in any obvious way given a pattern (hence requiring an optimization exercise), the collective representations provided by the FM notions are part of a family of sensible physical entities and members of a collection of pertinent geometries with which to attempt to represent rainfall patterns and others (Puente et al. 2017). Certainly, it is not easy to capture the overall geometries of rainfall sets parsimoniously, but the FM approach does hold promise for such a desirable task that also aims at finding a suitable language for describing complexity (Maskey et al. 2015).

3 Methodology

This section explains how the FM approach is used to study the inter-annual and spatial variability of rainfall sets. First, the methodology used to find suitable encodings, for a year of data at a time, is introduced. Second, the validation statistics employed to quantify performance are given. Then, the analysis carried out to classify and quantify rainfall complexity is advanced.

3.1 FM Encodings

Even though the FM methodology is ultimately rather simple and computationally efficient—once a set of parameters is known—the finding of an appropriate representation for a given set is challenging. As there are neither analytical formulas for the attractors nor for the derived measures dy, only a numerical solution is possible. Also, alternative parameter sets may exist, i.e., equifinality (Beven 2006; Huang et al. 2013).

Following previous efforts (e.g., Maskey et al. 2015; Puente et al. 2017), a generalized particle swarm optimization (GPSO) algorithm, with swarm members having dynamic capabilities, is used in the study. Specifically, the GPSO procedure is run 200 times to find that many plausible solutions using swarms made of 500 randomly defined elements—FM parameter values defined uniformly between bounds—and allowing them to evolve following 100 successive iterations. The best parameter values for the 200 runs, even if local optima, are recorded for further study.

In trying to account for the inherent complexity in daily rainfall (Obregón et al. 2002a, b; Huang et al. 2013; Maskey et al. 2015), the objective function to minimize is defined adding three L^2 norms, i.e., root mean square errors, of accumulated rainfall vs. accumulated FM fitted values, at the daily, ε_1, three-day, ε_3, and seven-day, ε_7, scales, over the period of consideration (i.e., a year) plus a few penalties aimed at discarding unacceptable renderings. This gives for the objective function:

$$\varepsilon = \varepsilon_1 + \varepsilon_3 + \varepsilon_7 + \varepsilon_p, \tag{6}$$

$$\varepsilon_j = \sqrt{\frac{1}{M_j}\sum_{i=1}^{M_j}\left(c_{i,j} - \widehat{c}_{i,j}\right)^2}, \qquad j = 1, 3, 7, \tag{7}$$

where M_j is the number of data points at scale j, i.e., $M_1 = 365(366)$, $M_3 = 123$, and $M_7 = 53(54)$, $c_{i,j}$ is the accumulated measured rainfall up to period i for scale j, $\widehat{c}_{i,j}$ is the corresponding value obtained via an FM representation, and ε_p are penalties dealing with the maximum allowable deviations between $c_{i,j}$ and $\widehat{c}_{i,j}$ and information regarding their distribution of zero values. Although GPSO calculations may not always keep the penalty restrictions all the time, the results reported herein do fulfill such constraints.

3.2 Model Performance

To assess the quality of individual FM approximations, various qualifiers are computed at the aforementioned scales. Such include:

(a) Nash–Sutcliffe efficiencies for, accumulated, rainfall vs. FM sets:

$$\eta_j = 1 - \frac{\sum_{i=1}^{M_j} \left(c_{i,j} - \widehat{c}_{i,j}\right)^2}{\sum_{i=1}^{M_j} \left(c_{i,j} - \overline{c}_j\right)^2}, \quad j = 1, 3, 7, \tag{8}$$

where the notation is as in Eq. (7) and $\overline{c}_j$ is the accumulated rainfall mean for scale j;

(b) Nash–Sutcliffe efficiencies of rainfall vs. FM sets (all duly normalized):

$$\nu_j = 1 - \frac{\sum_{i=1}^{M_j} \left(r_{i,j} - \widehat{r}_{i,j}\right)^2}{\sum_{i=1}^{M_j} \left(r_{i,j} - \overline{r}_j\right)^2}, \quad j = 1, 3, 7, \tag{9}$$

where $r_{i,j}$ is the measured rainfall at period i for scale j, $\widehat{r}_{i,j}$ is the corresponding FM value, and $\overline{r}_j$ is the rainfall mean for scale j;

(c) The number of zero values at the three scales, ζ_j in real vs. FM representations; and

(d) The percent of zero values matched by an FM encoding, π_j.

3.3 Complexity Analysis

The encoding of a host of rainfall sets, by year and at various sites, allows visualizing, beyond the annual depth, the dynamics of the process from the vantage point of geometry, i.e., from the perspective of the parameters of the FM representations employed. As shall be illustrated, such FM parameters permit performing more complete inter-annual and spatial comparisons, which enable quantifying the complexity of the records, as follows: (a) yearly patterns may be classified by parameters via data-mining techniques, e.g., k-means clustering (Arthur and Sergi 2007), such that rainfall dynamics may be studied by classes, and (b) parameter values may be studied following a "classical" complexity study, computing correlations of them all and building phase-space diagrams to identify the presence or lack of geometric trends.

4 Rainfall Encodings in California

The FM notions are tested next for daily rainfall sets gathered over water years (October 1st–September 30th) in four sites in California, from south to north: Cheery Valley, Merced, Sacramento, and Shasta Dam, as summarized in Table A1 in the Online Appendix http://puente.lawr.ucdavis.edu/pdf/nag_puente_appendix.pdf. As seen, all sites contain at least 59 years of contiguous sets gathered by NOAA's National Climate Data Center (NCDC) and the average annual rains (from south to north) are 46.7, 12.9, 17.6, and 63.7 in. Prior to FM encoding, all data sets are normalized so that the accumulated depth (over a given year) becomes unity.

Given the geometric intricacies of the records that contain substantial number of zero values, the FM representations used are associated with the Cantorian construction based on the iteration of two maps as highlighted in Fig. 1. As they rely on nine FM geometric parameters, the computed sets have associated compression ratios of about 40:1 (365/9). Since encoding over 360 years of records (see Table A1) takes a substantial amount of time for solving the associated inverse problems (over 12 h per year on a personal computer) and in order to study distinct possible solutions for the optimization exercises, the results reported here not only correspond to the "best" objective functions (over all 200 cases, as explained in Sect. 3) but also include, for sensitivity reasons, up to the best twenty solutions. In what follows and for clarity purposes, the quantity ε_1 is renamed as ε_{ac} (see Eq. (7)), the root mean square error in daily accumulated sets over a year, and the results also report on the maximum daily deviation in accumulated rain over a year, ε_{mac}.

4.1 Examples of FM Encodings

To demonstrate the capability of the FM notions, encodings of four distinct rainfall sets, with varied geometries and for each location, are reported here together with an extensive statistical evaluation of their performance. As an example, Fig. 2 contains rainfall sets (black), best FM encodings (gray), and comparisons of accumulated sets at Cherry Valley for water years ending at 1970, 1980, 1990, and 2000. As readily seen, all the FM sets do capture very well the overall "rough" distribution of rain and, although the rainfall intensities themselves are not perfectly captured—for the locations of all peaks do not necessarily match—the accumulated profiles of FM sets are indeed close to their targets.

The goodness of the best FM representations is further illustrated in Table 1 that contains, by blocks, a host of statistical information at various scales, as in the objective function Eq. (6). As seen in the top block, average and maximum errors in accumulated sets at the daily scale, ε_{ac} and ε_{mac}, are rather small, with magnitudes that do not exceed 2.2 and 9.8%, respectively. As noticed on the second and third blocks, while the Nash–Sutcliffe indices on the accumulated sets, η_j, for a day, 3 days, and 7 days are all close to perfection, the Nash–Sutcliffe indices on the

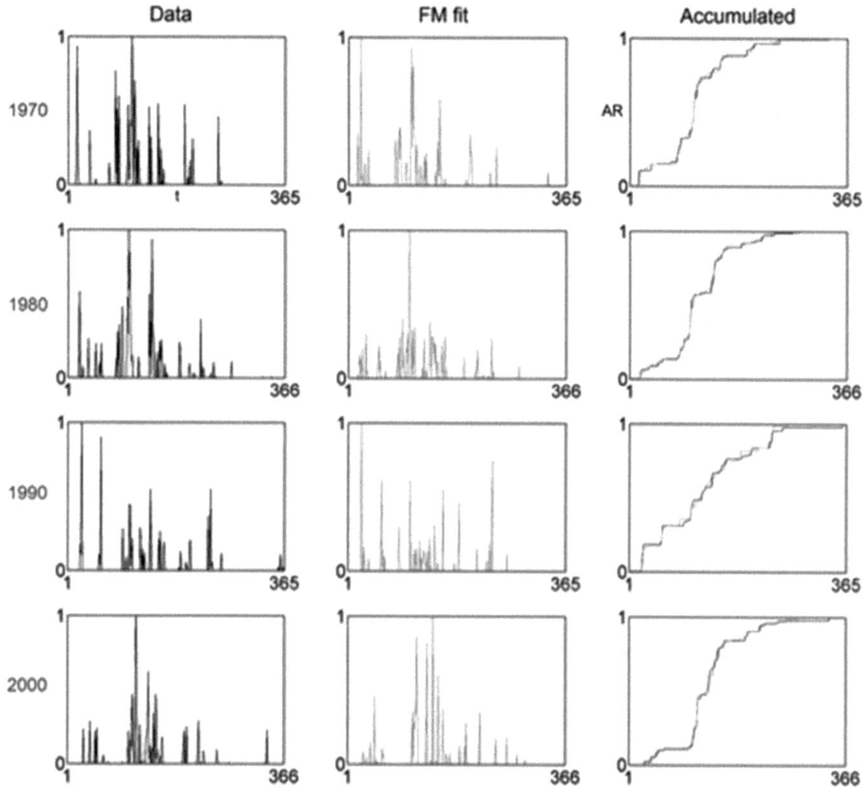

Fig. 2 Examples of daily rainfall records in Cherry Valley for years 1970, 1980, 1990, and 2000 (*black*) and best FM representations (*gray*), followed by their accumulated sets

records themselves, ν_j, exhibit low values at the daily and 3 day scales (due to non-matching rainfall intensities) but reasonable values—above 60%—at the weekly scale. As reported on the last two blocks, the numbers of zeros, ζ_j, in data and FM sets (in parenthesis) are close for all scales and the actual percent of zeros matched by the FM representations, π_j, is consistently higher than 75%, for all scales.

Similar analyses for the other three sites reveal comparably good performance, which, due to space limitations, have been included in the aforementioned Online Appendix. There, the interested reader shall find, for the Merced, Sacramento, and Shasta Dam locations, four examples each of FM fits similar to those in Fig. 2 (Figs. A1, A2 and A3) followed by corresponding statistical information as in Table 1 (Tables A2, A3, and A4). As may be verified, such information supports the usage of the FM method to encode highly intermittent rainfall sets, as follows: (a) all FM representations, for the 12 examples, do resemble the real sets both in texture and overall locations of peaks and cannot be taken apart from data sets by the naked eye, as they do all look reasonable and "real," (b) all FM fits are excellent in accumulated sets as the encoding errors ε_{ac} and ε_{mac} are rather low, always less than merely 2.1

Table 1 Performance of the best FM models for Cherry Valley in Fig. 2 (ε_{ac}, ε_{mac}, η's, ν's, and π's are in percent, ζ's are for data followed by FM fit in parenthesis)

	Period			
Statistics	1970	1980	1990	2000
ε_{ac}	1.8	1.4	2.2	1.8
ε_{mac}	8.0	7.0	8.4	9.8
η_1	99.8	99.9	99.5	99.8
η_3	99.8	99.9	99.5	99.8
η_7	99.8	99.9	99.6	99.8
ν_1	1.0	17.9	−40.0	−40.0
ν_3	36.4	67.9	14.5	29.9
ν_7	65.2	87.1	61.9	71.4
ζ_1	295(300)	276(288)	300(319)	291(320)
ζ_3	85(85)	73(80)	84(91)	81(89)
ζ_7	27(30)	18(26)	26(32)	23(29)
π_1	86.1	85.2	90.7	91.8
π_3	78.8	78.1	80.6	84.0
π_7	77.8	88.9	76.9	78.3

and 9.0%, respectively, (c) all Nash–Sutcliffe indices for accumulated sets, at the scales of 1, 3 and 7 days, η's, remain close to 100%, (d) Nash–Sutcliffe values for the records, ν's, increase with aggregation scale and the numbers at the 7 day scale are typically larger than 65%, as with Cherry Valley, (e) the number of zeros in the records and those in the FM representations, ζ's, are close to each other, and (f) the FM fits do preserve well the location of zeroes in data, with π's that always exceed 54% but that could be, sometimes, as high as 96%.

4.2 Overall Performance

Having studied in detail a few examples of rainfall patterns, this section includes the best FM representations over the whole records available: 59 years for Cherry Valley, 116 for Merced, 115 for Sacramento, and 72 for Shasta Dam. In such spirit, Fig. 3 for Cherry Valley and Figs. A4, A5, and A6 on the Online Appendix for the other sites include the observed rainfall set (top), the corresponding FM fit (middle) obtained by upgrading annual volumes (depths), and their implied accumulated sets over the whole period (bottom).

As readily seen, the textures of FM sets associated with the best solutions (year by year) are indistinguishable from measured sets, not only by the naked eye, but also statistically as illustrated in the previous section. As such, the FM fitted accumulated sets turn out to be excellent renderings of the "real" sets, now over much longer periods of time. The goodness of such overall representations is further

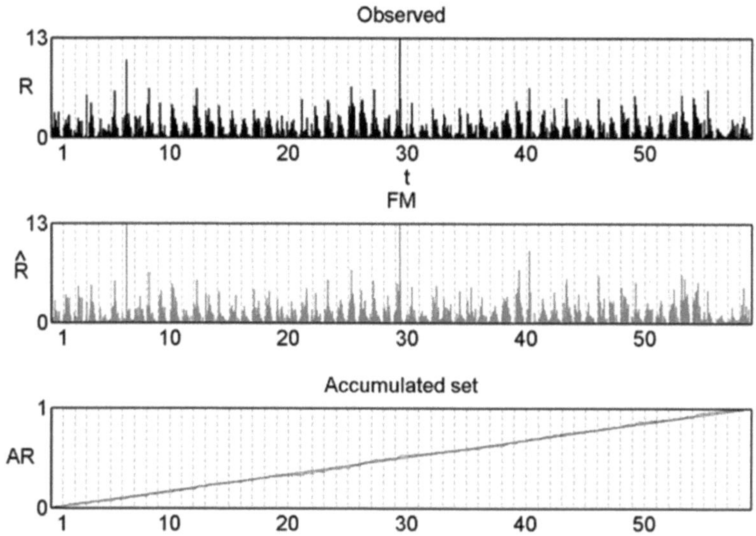

Fig. 3 Rainfall records in Cherry Valley for water years 1956–57 to 2014–15 (*top—black*) and best FM representations (*year by year—gray*) followed by their accumulated sets. The scales of the rain sets are in inches/day

reflected in Table 2, which contains, for all sites, encoding errors and some of the statistical information used earlier at three aggregation scales, but averaged over all years and together with their plus and minus standard deviations. As seen, encoding errors are consistently low, for all sites, with ε_{ac} values that are on the average less than 1.8% and with standard deviations less than 0.3%, and with ε_{mac} values lower than 8% and standard deviations less than 2.1%. Such behavior translates into almost perfect Nash–Sutcliffe values for accumulated sets at all sites, reasonable Nash–Sutcliffe indices for the records at the 7 day aggregation scale, ν_7, with averages above 65% and standard deviations less than 22%, and a large percent of zeroes matched by the FM representations for all scales, as mean values of π's are greater than 65% on the average and as the values for 1 day, π_1, are all greater than 83% with a standard deviation less than 8%.

These results illustrate that the Cantorian-based FM notions are useful to model highly intermittent rainfall sets containing a notorious amount of no-rain activity, with small errors that are within the accuracy of rainfall measurements (Lanza and Vuerich 2009). These results, even if only at four sites, do suggest that the FM geometric parameters of successive sets, year by year, may be used to study the evolution and complexity of rainfall patterns.

Table 2 Overall performance of best FM encoding in all locations (ε_{ac}, ε_{mac}, η's, ν's, and π's are in percent)

	Site			
Statistics	Cherry Valley	Merced	Sacramento	Shasta Dam
ε_{ac}	1.6 ± 0.2	1.6 ± 0.2	1.6 ± 0.2	1.8 ± 0.3
ε_{mac}	7.1 ± 1.5	7.0 ± 1.3	7.1 ± 1.2	7.9 ± 2.1
η_1	99.8 ± 0.1	99.8 ± 0.1	99.8 ± 0.1	99.7 ± 0.1
η_3	99.8 ± 0.1	99.8 ± 0.1	99.8 ± 0.1	99.7 ± 0.1
η_7	99.8 ± 0.1	99.8 ± 0.1	99.8 ± 0.1	99.7 ± 0.1
ν_1	-11 ± 38	-7 ± 32	-9 ± 28	-18 ± 29
ν_3	34 ± 31	34 ± 27	43 ± 23	32 ± 20
ν_7	65 ± 22	68 ± 16	74 ± 13	66 ± 17
π_1	85 ± 8	88 ± 7	91 ± 5	83 ± 8
π_3	76 ± 11	81 ± 10	85 ± 8	75 ± 12
π_7	68 ± 16	76 ± 14	80 ± 10	65 ± 16

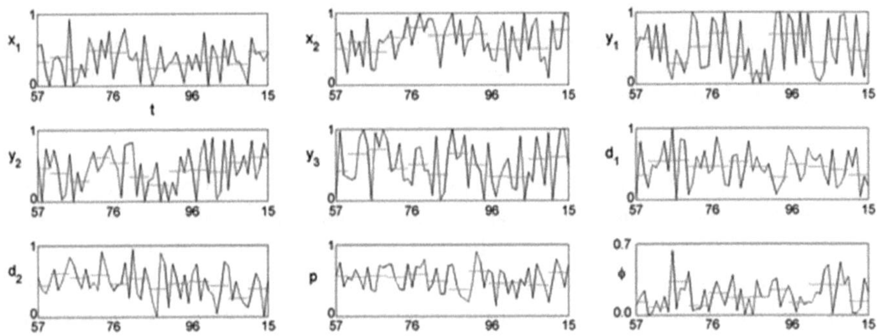

Fig. 4 Evolution of best FM parameters for Cherry Valley (*black*) and averages over 5 years (*gray*)

4.3 *Rainfall Dynamics*

Having sensible approximations of rainfall sets at all sites leads us to hypothesize that the time evolution of the best FM parameters, as defined in Sect. 3, may help elucidate the inter-annual dynamics of rainfall. As such, Fig. 4 and Figs. A7, A8, and A9 (the latter on the Online Appendix) include the time evolution of the best FM parameters for Cherry Valley and the three other sites: the coordinates of the right end-point of the first map, (x_1, y_1), the coordinates of the left end-point of the second map, (x_2, y_2), the y-value of the right end-point of the second map, y_3, the vertical scalings of the two maps, d_1 and d_2, the frequency used to carry the iterations, p, and a rain-trace threshold ϕ. While the x_i values are bounded from the first and last end-points, i.e., from 0 to 1, the y_i values ranged from -5 to 5, and the d_i's between -1 and 1. All of such values are shown normalized in the figures between 0 and 1, superimposing on them local parameter averages computed over 5 years.

As seen for all sites, the FM geometric parameters vary wildly and often swing from high to low values and vice-versa. Such variability is also seen in the averages every 5 years as reported in the graphs. Consistently with the early figures year by year (e.g., Fig. 2), the geometries of successive years do vary significantly, a feature that has already been reported not only for yearly rainfall sets but also for streamflow records (Maskey et al. 2015, 2016a, b). This is corroborated in Table A5 on the Online Appendix, which includes entropy calculations for all parameters at all sites based on histograms containing eight bins resulting in maximum entropies of 3 (based on *log* 2 calculations for a purely uniform case). As seen and as hinted from the evolutions, all entropy values (except for the iteration frequency p and the threshold ϕ that a bit more orderly) reflect near uniformity, as all are between 2.75 and 2.98.

At the end, there are no noticeable trends in the best FM parameters (not even when averaged every 5 years) and such a fact clearly precludes the possibility of readily finding rainfall forecasts from such geometric information or discerning effects due to climate change. As seen, all sites, irrespective of their variable average annual volumes, exhibit a high degree of complexity, a trait that shall be further elaborated later on.

4.4 Sensitivity

In order to further understand the overall effectiveness of the FM method at all sites and in an attempt to find "sub-optimal" solutions that may exhibit parameter trends, Fig. 5 shows the evolution of the encoding errors $\varepsilon_{\mathrm{ac}}$ and $\varepsilon_{\mathrm{mac}}$, but not only for the best solution every year, but also for the best three. As seen, while the three root mean square errors (left) remain close to each other and at values that do not exceed small quantities of 2.4, 2.5, 2.8, and 2.6% from top to bottom (for sites from south to north), the three maximum errors—not explicitly optimized—(right) exhibit an increased variability that is bounded, from top to bottom, by 9.9, 10.0, 10.0, and 14.0%. Noticeably, few years at Shasta Dam exhibit large maximum errors and such may be interpreted as saying that from the point of view of the FM approach such a site is a bit more complex than the others. Although not shown here, a similar analysis of the best 20 solutions confirms the nature of the results. In terms of the employed objective function, the four sites within California may be considered comparable, but in terms of the maximum error in accumulated records (not explicitly accounted for), rainfall at Shasta Dam remains a bit more difficult to encode than in the other sites.

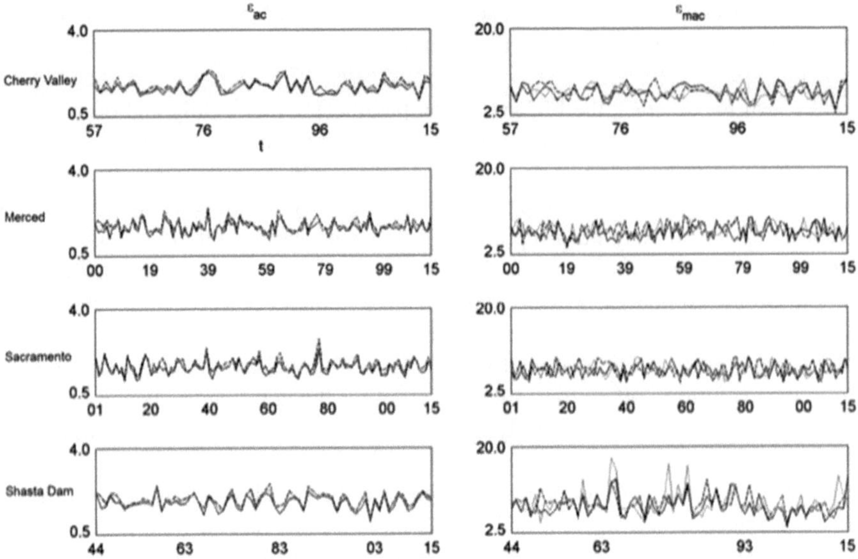

Fig. 5 Evolution of encoding errors ε_{ac} and ε_{mac} at all sites for best three FM solutions: first (*black*), second (*gray*), and third (*dashed*)

Figure 5 suggests that the FM encodings do not possess a single best solution but that there are other close solutions. As such, it becomes natural to inspect how the parameter evolution of such alternative FM representations may look. Figure 6, for Cherry Valley, and Figs. A10, A11, and A12 on the Online Appendix, for the other sites, present such information for the three best FM parameter values. As seen, those solutions that have close encoding errors as reported in Fig. 5 evolve wildly in time, implying solutions in various regions of FM parameter space that reveal the presence of equifinality (Beven 2006). There are indeed distinct parameter configurations yielding close solutions (Hill and Tiedeman 2007), a common feature regarding the solution of inverse problems dealing with complex processes. This fact, and their implied evolutions of parameters, further elaborates the intrinsic complexity of rainfall, as comparisons of this block of figures exhibit rather similar degrees of variability. Certainly, a similar comparison of various solutions in the inherently complex and nonlinear rainfall process may allow further quantifying climate complexity around the globe.

As there is no perceptible trend in the best parameters and as there are several close solutions, a set of alternative parameter values was defined for each site based on fixed lower bounds over time, as included in Table A6 (for un-normalized parameters) on the Online Appendix. Such "filtered" representations resulted in slightly worse performance as reflected in Table 3 when compared to Table 2. Still, however, such new FM parameters led to ample variation in parameters (albeit a bit less), as reflected in Fig. 7 for Cherry Valley and Figs. A13, A14, and A15 (on the Online Appendix) for Merced, Sacramento, and Shasta Dam and as corroborated in

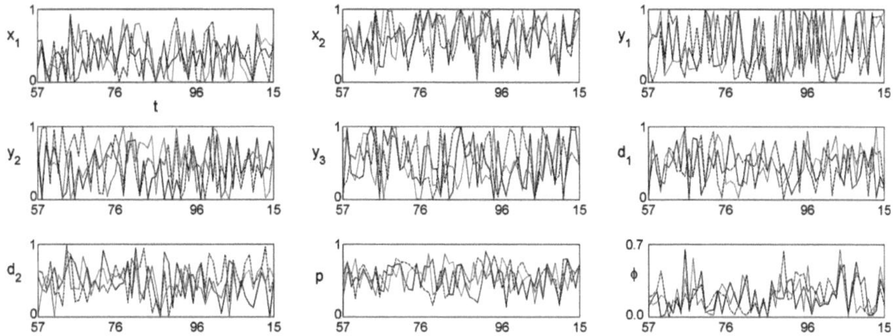

Fig. 6 Evolution of three best FM parameters for Cherry Valley: first (*black*), second (*gray*), and third (*dashed*)

Table 3 Overall performance of filtered FM encoding in all locations (ε_{ac}, ε_{mac}, η's, ν's, and π's are in percent)

	Site			
Statistics	Cherry Valley	Merced	Sacramento	Shasta Dam
ε_{ac}	1.9 ± 0.3	1.8 ± 0.3	1.8 ± 0.3	2.0 ± 0.3
ε_{mac}	7.4 ± 1.4	7.4 ± 1.4	7.5 ± 1.2	8.5 ± 2.6
η_1	99.7 ± 0.1	99.7 ± 0.1	99.7 ± 0.1	99.7 ± 0.2
η_3	99.7 ± 0.1	99.8 ± 0.1	99.8 ± 0.1	99.7 ± 0.2
η_7	99.7 ± 0.1	99.8 ± 0.1	99.8 ± 0.1	99.7 ± 0.2
ν_1	-23 ± 44	-23 ± 40	-13 ± 32	-27 ± 35
ν_3	24 ± 36	24 ± 33	38 ± 25	27 ± 23
ν_7	59 ± 26	60 ± 22	69 ± 16	62 ± 20
π_1	85 ± 7	89 ± 6	90 ± 5	85 ± 7
π_3	76 ± 10	82 ± 9	83 ± 7	76 ± 10
π_7	68 ± 13	77 ± 13	78 ± 9	68 ± 14

the entropy analysis reported in Table A7 on the Online Appendix. Such filtered representations shall be used next together with the best parameters in order to classify rainfall patterns and further assess the inherent complexity of the rainfall records.

5 Classification and Complexity Analysis

5.1 Geometric and Data Classification

As there is ample variation in FM dynamics at all sites, both for best and filtered parameters, this section presents a discretized analysis of the rainfall records via

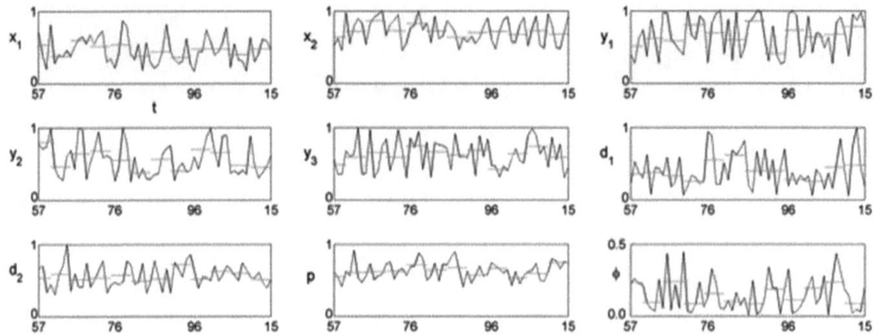

Fig. 7 Evolution of filtered FM parameters for Cherry Valley (*black*) and averages over 5 years (*gray*)

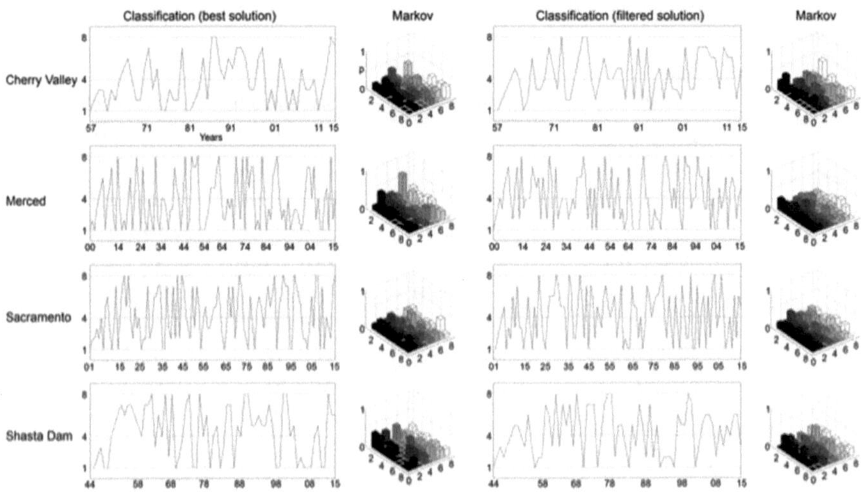

Fig. 8 Evolution of best (*left*) and filtered (*right*) FM rainfall classes obtained via *k*-means clustering of FM parameters for all sites, and their corresponding Markov matrices

FM parameter classifications (for both best and filtered solutions) yielding eight distinct classes using *k*-means clustering, as described in Sect. 3.3.

Figure 8 shows the time evolution of the aforesaid eight classes for all sites (all starting at "class 1" for the initial year), based on best solutions (on the left) and filtered solutions (on the right), together with their implied Markovian matrices summarizing transitions from time to time (to be read from right to left). As seen and as expected, the classification based on filtered parameters differs from that obtained from best parameters and all evolutions, at all sites, exhibit notable swings from low to high classes and vice-versa and broad Markovian matrices.

In order to further qualify the results just discussed, Fig. 9 includes a similar classification into eight classes obtained by using the deciles of the yearly records,

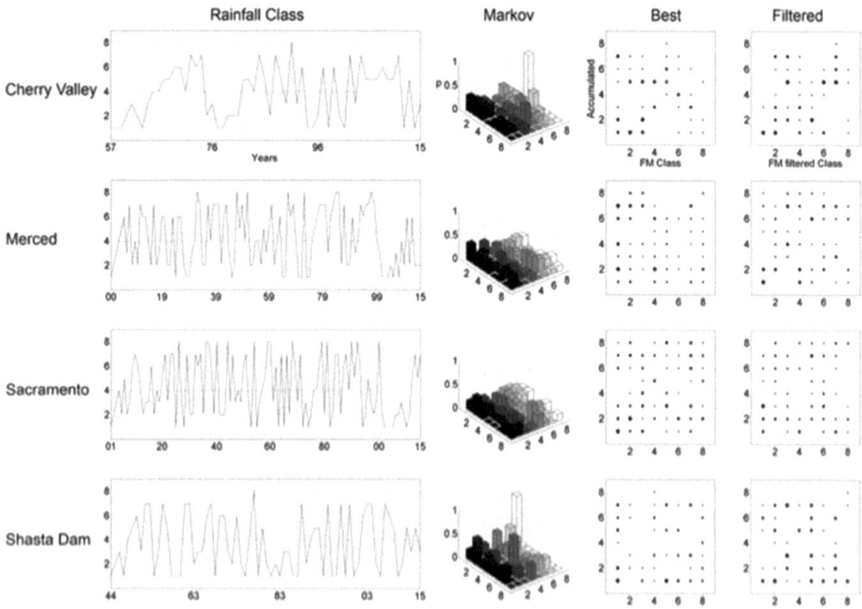

Fig. 9 Evolution of rainfall classes obtained via *k*-means clustering of rainfall deciles and their corresponding Markov matrices, followed by scatterplots comparing decile classes with best and filtered FM classes. Sizes of *circles* are proportional to class repetitions

that is, the same number of parameters as the FM representations in Fig. 8. As seen, the rainfall class evolutions based on deciles exhibit yet similar swings as those based on FM parameters and the corresponding Markov matrices remain fairly broad, except for two noticeable transitions for Cherry Valley and Shasta Dam. As reported in Table A8 on the Online Appendix, the class evolutions for all sites turn out to be, at the end, rather similar in terms of their class entropies, H_c (ranging from 2.71 to 2.96 and hence close to uniformity) and average class orbit lengths, $\overline{OL}_c$, measured, in an absolute sense, over the evolving classes (spanning from 1.98 to 2.77 classes). Although the records for Sacramento may be termed a bit more complex based on these two attributes, there is not enough separation to conclude that sets are not similarly complex nor that any effects due to climate change may be identified.

Figure 9 also includes an inter-comparison between the classes implied by the decile classification and those obtained via best and filtered FM parameters, in the form of scatterplots having a larger circle depending on multiple occurrences of a given combination of classes. As seen, both for best and filtered FM classifications, there is no salient visible patterns that emerge from the analysis but rather very broad diagrams, which give credence to the notion that the three distinct classifications represent different ("orthogonal") views of the records (as other "equifinal" FM solutions would likely produce), which altogether exhibit similar degrees of complexity.

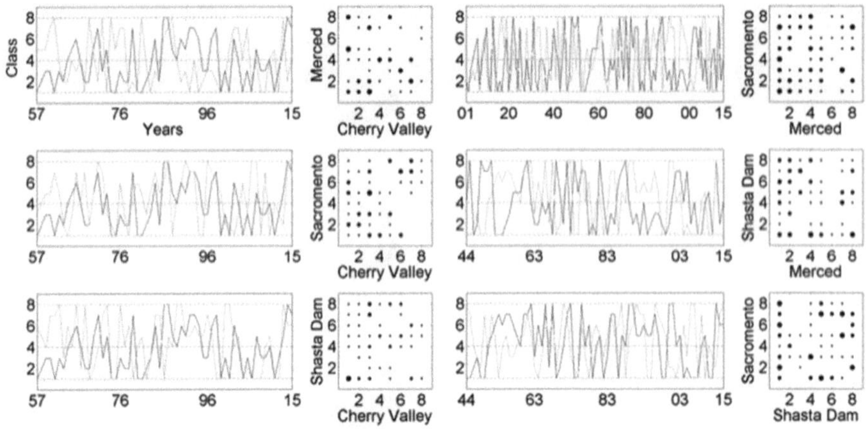

Fig. 10 Pairwise site-comparison of best FM rainfall class evolutions and their scatterplots. The evolutions use *black* for the set in the *x*-axis and *gray* for the one in the *y*-axis. Sizes of *circles* are proportional to class repetitions

5.2 Comparative Analysis in Space

Having explained that rainfall may be considered equally complex from a geometric point of view at the four sites under study, it becomes sensible to compare the class evolutions in space, one site against another. For this purpose, Fig. 10 includes pairwise comparisons associated with the best FM solutions, and Figs. A16 and A17 (on the Online Appendix) do so for the filtered FM and decile classifications, respectively. As exemplified in Fig. 10, such graphs include a visual comparison of the class evolutions for concurrent years and the corresponding scatterplots.

As seen, the rather erratic class evolutions result in fairly uncorrelated scatterplots, which, although having instances where class combinations do not exist (as high classes for Cherry Valley vs. low classes for Sacramento in Fig. 10), do not exhibit noticeable trends. There are no clear correlations among the classes for all pairs of stations for any of the three classifications employed, hence emphasizing the complexity of the rainfall records, not only in time, but also in space, at least for a distance of about 550 miles within California.

5.3 Additional Complexity Analysis

Having obtained "best" FM solutions for every site allows computing autocorrelation functions and phase diagrams for such individual parameters, as usually performed when trying to identify chaotic properties of records, e.g., Sivakumar and Berndtsson (2010). In that spirit, Fig. 11 includes such an analysis for the nine FM parameters at Merced together with the total rainfall depth over the years (in inches) and Figs. A18, A19, and A20 on the Online Appendix do so, in order, for Cherry Valley, Sacramento, and Shasta Dam.

As seen, Fig. 11 is divided into two parts, with the first one including the end-point FM parameters x_1, x_2, y_1, y_2, y_3, and the second comprising the other FM parameters d_1, d_2, p, and ϕ, and the aforementioned total rainfall depth. As observed, for each attribute there are shown five features: (a) the time series itself, (b) the autocorrelation function, (c) the two-dimensional phase diagram with a lag equal to 1 year, (d) the two-dimensional phase diagram with a lag equal to half the total number of years in the records (58 for Merced), and (e) the three-dimensional phase diagram obtained using lags 0, 1, and 2.

As observed in Fig. 11, the annual rainfall depths for Merced, as well as the best FM parameters there, exhibit noticeable swings from high to low and vice-versa and,

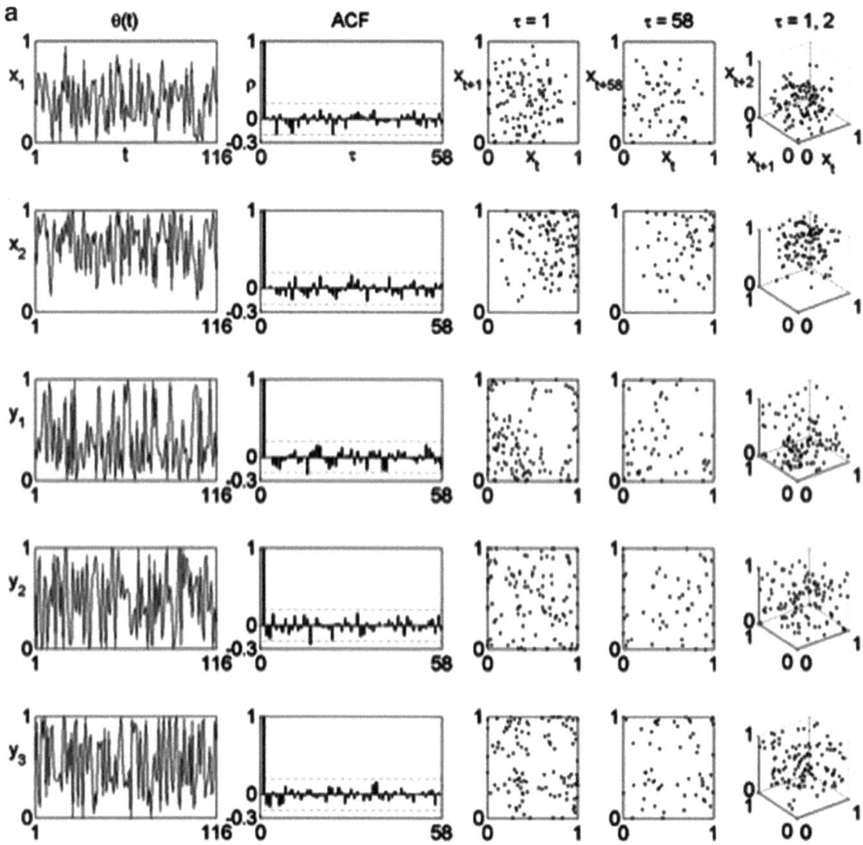

Fig. 11 Complexity analysis of FM parameters for rainfall records in Merced, in order, parameter evolution, $\theta(t)$, parameter autocorrelation, ACF, and three phase-space diagrams, focusing on: (**a**) end-points of affine maps, and (**b**) vertical scalings, iterations proportion, vertical thresholds, and total rainfall depth (in inches)

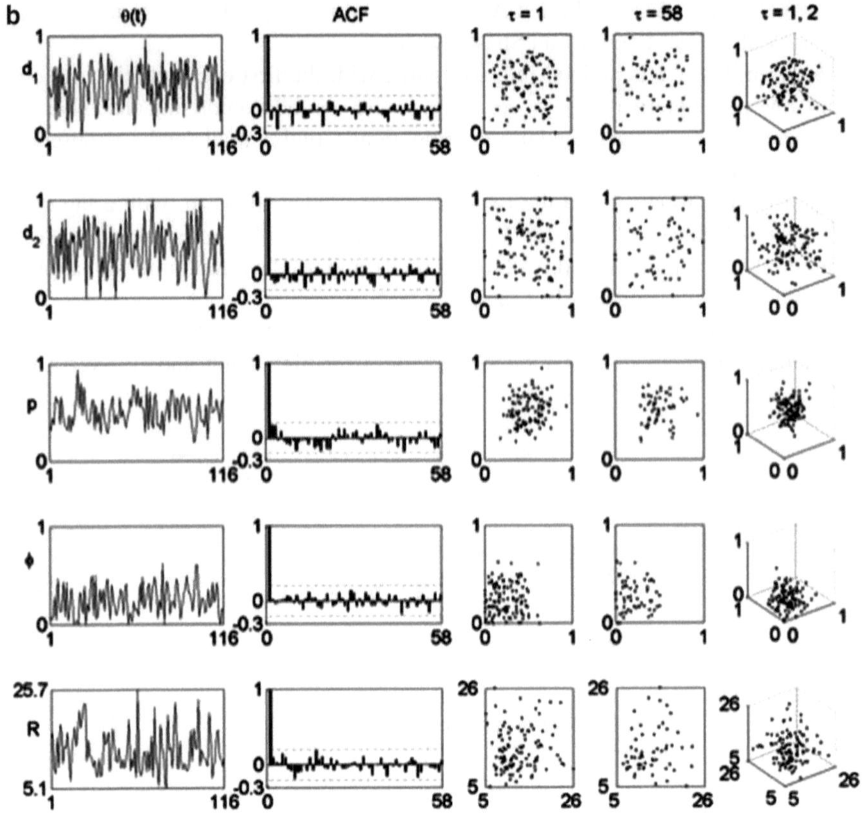

Fig. 11 (continued)

as expected, their corresponding autocorrelations decay rather quickly and remain almost always, within the shown $\pm 1.96/\sqrt{n}$ (where n is the length of records) bands for up to 58 lags. As seen, all phase diagrams for all FM parameters and total rainfall depth exhibit noticeable scattering at the scales considered. The diagrams shown, and others for additional scales not included here, confirm the complexity of the rainfall records, as there is no obvious attractor that may be discerned for any of the geometric attributes of the daily rainfall patterns. As the analysis for the other sites, shown in the corresponding figures on the Online Appendix, reveals rather similar results, it may indeed be surmised that the studied sets lack a low-dimensional structure that may define different degrees of (geometric) complexity for rainfall in space.

6 Conclusions and Further Research

This research illustrates how a Cantorian variant of the fractal–multifractal, FM, approach, which relies on the iteration of two simple maps and uses eight geometric parameters, may be adapted adding a rain-trace threshold parameter to closely encode highly intermittent daily rainfall sets in California. By studying over 360 combined years at (from south to north) Cherry Valley, Merced, Sacramento, and Shasta Dam, it is shown that it is possible to nicely approximate the geometry of individual years at the daily scale (i.e., optimizing their mass functions) with root mean square errors that are less than a mere $1.8 \pm 0.3\%$ and maximum errors in accumulated sets that are less than $7.9 \pm 2.1\%$, both well within measurement errors reported for the rainfall process, Lanza and Vuerich (2009). As the FM encodings also reasonably preserve information pertaining to the distribution of zero rainfall values and Nash–Sutcliffe attributes for rainfall at the weekly scale, the results support the notion that hidden determinism may lie at the root of natural complexity (Puente 1996; Puente and Sivakumar 2007).

Once FM representations are established, the dynamics of the aforementioned nine parameters are used in an attempt to study the inter-annual dynamics of rainfall at the four sites, aiming also at a spatial comparison. The analysis revealed, however, that the evolutions of "best" FM parameters, obtained via an optimization exercise, fail to exhibit any noticeable trends but rather ample variations, for all sites, in a manner that does not reflect any variations due to climate change. As the optimization process revealed the presence of other FM parameter combinations having close objective functions, i.e., equifinality, "filtered" FM sets, narrowing the range of parameter variations, were also defined, but such evolutions also ultimately resulted in non-trivial and highly entropic behaviors, for all sites.

As the evolutions of FM parameters were not useful in discerning differences in the complexity of rainfall among the chosen locations, a more complete analysis of such geometric parameters was carried via classifications (using k-means clustering) and conventional phase-space diagrams. This investigation resulted in further understanding of the rather erratic signals and confirmed that the geometries of the rainfall sets at the four sites, and irrespective of distinct annual rainfall averages, cannot be distinguished from one another, as they all may be termed as "equally complex."

The results of this work emphasize the "deterministic complexity" of the rainfall process, i.e., deterministic, as the individual sets may be represented by the FM method, but complex, as there are no obvious trends in FM parameters over time. The fact that there are equifinal FM solutions certainly suggests further investigating within the space of parameters, using explicit bounds in the numerical search aiming at defining trends. Although the results herein suggest that such may be unlikely, there may still be solutions that could allow discriminating rainfall complexity between sites. Certainly, the analysis should be extended further north to include rainfall sets with less numbers of zero values.

The quantification of rainfall complexity in time and space, as attempted in this work, is certainly a scientifically relevant problem, especially in relation to climate change studies. If trends in dynamics may be discerned, such would have obvious benefits, and if such trends do not exist, that would also be relevant information. Although several questions remain regarding the FM approach, for instance, finding a physical explanation for each of its parameters, it is envisioned that similar analysis may also be carried to assess the complexity of other (less complex) hydro-meteorological attributes such as streamflow and temperature records. Such an avenue of research, with less variable geometric patterns, is also being investigated and will be reported in the future.

Acknowledgements This article is dedicated to Panayiotis Tsonis, whom the first author hugged in Rhodes, as if he was his brother.

References

Arthur, D., and V. Sergi. 2007. K-means++: The advantages of careful seeding. In *SODA '07 proceedings of the eighteenth annual ACM-SIAM symposium on discrete algorithms*, 1027–1035.

Barnsley, M.F. 1988. *Fractals everywhere*. San Diego: Academic Press.

Beven, K. 2006. A manifesto for the equifinality thesis. *Journal of Hydrology* 320 (1): 18–36.

Cortis, A., C.E. Puente, and B. Sivakumar. 2009. Nonlinear extensions of a fractal-multifractal approach for environmental modeling. *Stochastic Environmental Research and Risk Assessment* 23 (7): 897–906.

Cortis, A., C.E. Puente, H.H. Huang, M.L. Maskey, B. Sivakumar, and N. Obregón. 2013. A physical interpretation of the deterministic fractal-multifractal method as a realization of a generalized multiplicative cascade. *Stochastic Environmental Research and Risk Assessment* 28 (6): 1421–1429.

Dhanya, C.T., and D.N. Kumar. 2010. Nonlinear ensemble prediction of chaotic daily rainfall. *Advances in Water Resources* 33 (3): 327–347.

French, M.N., W.F. Krajewski, and R.R. Cuykendall. 1992. Rainfall forecasting in space and time using a neural network. *Journal of Hydrology* 137 (1-4): 1–31.

Ghilardi, P., and R. Rosso. 1990. Comment on "Chaos in rainfall" by I. Rodríguez -Iturbe et al. *Water Resource Research* 26 (8): 1837–1839.

Gupta, V.K., and E. Waymire. 1990. Multiscaling properties of spatial rainfall and river flow distributions. *Journal of Geophysical Research* 95 (D3): 1999–2009.

Hey, T., T. Stewart, and M.T. Kristin. 2009. *The fourth paradigm: Data-intensive scientific discovery*. Vol 1., Redmond, WA: Microsoft Research.

Hill, M.C., and C.R. Tiedeman. 2007. *Effective groundwater model calibration: With analysis of data, sensitivities, predictions and uncertainty*. Hoboken, NJ: John Wiley & Sons, Inc..

Huang, H.H., A. Cortis, and C.E. Puente. 2012a. Geometric harnessing of precipitation records: reexamining four storms from Iowa City. *Stochastic Environmental Research and Risk Assessment* 27 (4): 955–968.

Huang, H.H., C.E. Puente, A. Cortis, and B. Sivakumar. 2012b. Closing the loop with fractal interpolating functions for geophysical encoding. *Fractals* 20 (3–4): 261–270.

Huang, H.H., C.E. Puente, A. Cortis, and J.L. Fernández Martínez. 2013. An effective inversion strategy for fractal–multifractal encoding of a storm in Boston. *Journal of Hydrology* 496: 205–216.

Jayawardena, A.W., and F. Lai. 1994. Analysis and prediction of chaos in rainfall and stream flow time series. *Journal of Hydrology* 153: 23–52.

Jin, Y.H., A. Kawamura, K. Jinno, and R. Berndtsson. 2005. Nonlinear multivariate analysis of SOI and local precipitation and temperature. *Nonlinear Processes Geophys* 12: 67–74.

Jothiprakash, V., and T.A. Fathima. 2013. Chaotic analysis of daily rainfall series in Koyna reservoir catchment area, India. *Stochastic Environmental Research and Risk Assessment* 27 (6): 1371–1381.

Kim, H.S., K.H. Lee, M.S. Kyoung, B. Sivakumar, and E.T. Lee. 2009. Measuring nonlinear dependence in hydrologic time series. *Stochastic Environmental Research and Risk Assessment* 23: 907–916.

Koutsoyiannis, D., and D. Pachakis. 1996. Deterministic chaos versus stochasticity in analysis and modeling of point rainfall series. *Journal of Geophysical Research: Atmospheres* 101 (D21): 26441–26451.

Lanza, L.G., and E. Vuerich. 2009. The WMO field intercomparison of rain intensity gauges. *Atmospheric Research* 94 (4): 534–543.

Lovejoy, S., and D. Schertzer. 2013. *The weather and climate: Emergent laws and multifractal cascades*. Cambridge: Cambridge University Press.

Luk, K.C., J.E. Ball, and A. Sharma. 2000. A study of optimal model lag and spatial inputs to artificial neural network for rainfall forecasting. *Journal of Hydrology* 227 (1): 56–65.

Mandelbrot, B.B. 1982. *The fractal geometry of nature*. Henry Holt and Company.

Maskey, M.L., C.E. Puente, B. Sivakumar, and A. Cortis. 2015. Encoding daily rainfall records via adaptations of the fractal multifractal method. *Stochastic Environmental Research and Risk Assessment* 30: 1917. doi:10.1007/s00477-015-1201-7.

Maskey, M.L., C.E. Puente, and B. Sivakumar. 2016a. A comparison of fractal-multifractal techniques for encoding streamflow records. *Journal of Hydrology* 542: 564–580.

Maskey, M.L., C.E. Puente, B. Sivakumar, and A. Cortis. 2016b. Deterministic Simulation of Highly Intermittent Hydrologic Time Series. *Stochastic Environmental Research and Risk Assessment*. doi:10.1007/s00477-016-1343-2.

Men, B., Z. Xiejing, and C. Liang. 2004. Chaotic analysis on monthly precipitation on Hills Region in Middle Sichuan of China. *Nature and Science* 2 (2): 45–51.

Nasseri, M., K. Asghari, and M.J. Abedini. 2008. Optimized scenario for rainfall forecasting using genetic algorithm coupled with artificial neural network. *Expert Systems with Applications* 35 (3): 1415–1421.

Obregón, N., C.E. Puente, and B. Sivakumar. 2002a. Modeling high resolution rain rates via a deterministic fractal–multifractal approach. *Fractals* 10 (3): 387–394.

———. 2002b. A deterministic geometric representation of temporal rainfall. Sensitivity analysis for a storm in Boston. *Journal of Hydrology* 269 (3–4): 224–235.

Peters, O., C. Hertlein, and K. Christensen. 2001. A complexity view of rainfall. *Physical Review Letters* 88 (1): 018701.

Puente, C.E. 1996. A new approach to hydrologic modelling: derived distribution revisited. *Journal of Hydrology* 187: 65–80.

———. 2004. A universe of projections: May Plato be right? Chaos. *Solitons and Fractals* 19 (2): 241–253.

Puente, C.E., and N. Obregón. 1996. A deterministic representation of temporal rainfall: Result for a storm in Boston. *Water Resources Research* 32 (9): 2825–2839.

Puente, C.E., and B. Sivakumar. 2007. Modeling hydrologic complexity: A case for geometric determinism. *Hydrology and Earth System Sciences* 11: 721–724.

Puente, C.E., O. Robayo, M.C. Díaz, and B. Sivakumar. 2001a. A fractal–multifractal approach to groundwater contamination. 1. Modeling conservative tracers at the Borden site. *Stochastic Environmental Research and Risk Assessment* 15 (5): 357–371.

Puente, C.E., O. Robayo, and B. Sivakumar. 2001b. A fractal–multifractal approach to groundwater contamination. 2. Predicting conservative tracers at the Borden site. *Stochastic Environmental Research and Risk Assessment* 5 (5): 372–383.

Puente, C.E., M.L. Maskey, and B. Sivakumar. 2017. Combining fractals and multifractals to model geoscience records. In *Fractals: Concepts and applications in geosciences*, ed. B. Ghanbarian and A. Hunt. Boca Raton: FL: CRC Press, in press

Ramirez, M.C.V., H.F. de Campos Velho, and N.J. Ferreira. 2005. Artificial neural network technique for rainfall forecasting applied to the Sao Paulo region. *Journal of Hydrology* 301 (1): 146–162.

Rodríguez-Iturbe, I. 1986. Scale of fluctuation of rainfall models. *Water Resources Research* 22 (9): 15S–37S.

———. 1991. Exploring complexity in the structure of rainfall. *Advances in Water Resources* 14 (4): 162–167.

Rodríguez-Iturbe, I., F.B. De Power, M.B. Sharifi, and K.P. Georgakakos. 1989. Chaos in rainfall. *Water Resources Research* 25 (7): 1667–1675.

Sharifi, M.B., K.P. Georgakakos, and I. Rodríguez-Iturbe. 1990. Evidence of deterministic chaos in the pulse of storm rainfall. *Journal of the Atmospheric Sciences* 47 (7): 888–893.

Sivakumar, B. 2000. Fractal analysis of rainfall observed in two different climatic regions. *Hydrological Sciences Journal* 45 (5): 727–738.

———. 2004. Chaos theory in geophysics: past, present and future. *Chaos, Solitons and Fractals* 19 (2): 441–462.

———. 2009. Nonlinear dynamics and chaos in hydrologic systems: Latest developments and a look forward. *Stochastic Environmental Research and Risk Assessment* 23: 1027–1036.

Sivakumar, B., and R. Berndtsson. 2010. *Advances in data-based approaches for hydrologic modeling and forecasting*. Singapore: World Scientific.

Sivakumar, B., and V.P. Singh. 2012. Hydrologic system complexity and nonlinear dynamic concepts for a catchment classification framework. *Hydrology and Earth System Sciences* 16: 4119–4131.

Sivakumar, B., S.Y. Liong, C.Y. Liaw, and K.K. Phoon. 1999. Singapore rainfall behavior: Chaotic? *Journal of Hydrologic Engineering* 4 (1): 38–48.

Sivakumar, B., R. Berndtsson, and M. Persson. 2001a. Monthly runoff prediction using phase space reconstruction. *Hydrological Sciences Journal* 46 (3): 377–387.

Sivakumar, B., S. Sorooshian, H.V. Gupta, and X. Gao. 2001b. A chaotic approach to rainfall disaggregation. *Water Resources Research* 37 (1): 61–72.

Sivakumar, B., F.M. Woldemeskel, and C.E. Puente. 2014. Nonlinear analysis of rainfall variability in Australia. *Stochastic Environmental Research and Risk Assessment* 28 (1): 17–27.

Tessier, Y., S. Lovejoy, and D. Schertzer. 1993. Universal multifractals: Theory and observations for rain and clouds. *Journal of Applied Meteorology* 32: 223–250.

Pandora Box of Multifractals: Barely Open?

Daniel Schertzer and Ioulia Tchiguirinskaia

Abstract Three decades ago, multifractals were a major breakthrough in nonlinear geophysics by providing a general framework to understand, analyze, and simulate fields that are extremely inhomogeneous over a wide range of space-time scales. They have remained on the forefront of nonlinear methodologies, but they are still far from being used or even developed to their full extent. Indeed, they have been too often limited to scalar-valued fields, whereas the relevant geophysical fields are vector fields. This chapter therefore gives new insights on current developments to overcome this limitation. This is done in an inductive manner. For instance, it takes hold on simple considerations on "spherical" and "hyperbolic" rotations to introduce step by step the Clifford algebra of Lévy stable generators of multifractal vectors that have both universal statistical and robust algebraic properties.

Keywords Multifractals • Intermittency • Spatial chaos Symmetry groups • Clifford algebra • Stable Lévy laws • Hyperbolic geometry • Mandelbrot set

1 Introduction

There have been many attempts to analyze and simulate the fluctuations of chaotic systems whose spatial extension is of prime importance, such as turbulence, weather, and climate, therefore to go beyond the dynamical systems with only few degrees of freedom, which were so useful to initiate the "chaos revolution," but cannot help to explore the "spatial chaos" (Lorenz 1991; Tsonis 1992; Schertzer et al. 2002).

This was done at first with the help of mono/uni-scaling approaches, e.g., with the help of spectral analyses or structure functions, and with corresponding simulations of fractional Gaussian noises and motions (Mandelbrot and Van Ness 1968; Mandelbrot 1983). However, multifractal concepts and techniques were needed and developed to grasp the fundamental features of intermittency, which can

D. Schertzer (✉) • I. Tchiguirinskaia
Hydrology Meteorology and Complexity (HM&Co), Ecole des Ponts ParisTech, U. Paris-Est, Champs-sur-Marne, France
e-mail: Daniel.Schertzer@enpc.fr; Ioulia.Tchiguirinskaia@enpc.fr

A.A. Tsonis (ed.), *Advances in Nonlinear Geosciences*,
DOI 10.1007/978-3-319-58895-7_25

be loosely defined as the property that more and more "active" regions of the field are concentrated on smaller and smaller regions of the space-time domain (Benzi et al. 1984; Schertzer and Lovejoy 1984; Parisi and Frisch 1985; Halsey et al. 1986). The level of activity is usually easy to define for scalar fields, but already more involved for vector fields, e.g., a given norm of its gradient (or other type of vector derivatives, e.g., its curl). This loose definition already points out that the definition of a multifractal field is rather independent of the domain dimension, whereas it can be very sensitive to the dimension of the codomain, i.e., the set into which the field values are constrained to fall (Bourbaki 2004). As discussed below, this is even worse for simulations, so that multifractals have been rather limited to scalar-valued fields, therefore to 1D codomains. This unbalance between the dimensions of the domain and codomain has had many unfortunate consequences. At first, this has prevented to deal with key question of complex component interactions of vector fields whereas this was done for dynamical systems (with 1D domain, but with a larger codomain dimension, although rather low). Second, this was achieved by assuming many unrealistic symmetries (isotropy and mirror invariance) instead of studying the nontrivial symmetries corresponding to these interactions. More fundamentally, not only the vector nature of the field is ignored, but also the same is done for the scale change operator.

This vector nature is unfortunately indispensable to answer to challenging questions such as the climatology of (exo-) planets based on first principles (Pierrehumbert 2013) or to fully address the question of the relevance of quasi-geostrophic turbulence and to define an effective, fractal dimension of the atmospheric motions (Schertzer et al. 2012).

This is not only unfortunate, but also more fundamentally unreasonable and illogical to first restrict multifractals to scalar-valued fields then to use the later to try to analyze and simulate vector-valued fields.

In this chapter we present in an inductive manner the neat example of multifractal vector fields generated by a stochastic Clifford algebra, which was on the contrary deduced by Schertzer and Tchiguirinskaia (2015) from the much more general case of Lie cascades (Schertzer and Lovejoy 1995; Chigirinskaya and Schertzer 1996; Chigirinskaya et al. 1998). Here we start from simple and powerful properties of orthogonal rotations and mirror symmetries, such as the spherical and hyperbolic geometries they, respectively, define. Both geometries have been in fact often invoked in fluid dynamics (e.g., elliptic points vs. hyperbolic points) in relation with the long lasting question which of the rotation and the strain is dominant in a given flow (Okubo 1970; Weiss 1991; Haller 2005). It has a much more general scope in this chapter: not only because it concerns a much larger class of processes, but also because it concerns across scale properties. These geometrical features are used in this chapter to introduce almost intuitively the fundamentals of a Clifford algebra, in particular its quadratic form that is in general indefinite, i.e., having a nonunique sign.

Overall, we hope that this chapter will help to open much more widely the Pandora box of multifractals, which seems to us barely open until now.

2 Symmetries and Geometries

2.1 Orthogonal Rotations vs. Mirror Symmetries

Geometric transforms can be added and composed, generating therefore algebras, where the multiplication corresponds to the composition. Figure 1 presents three simple, linear, plane symmetries (*I*, *J*, *K*) and their iterated applications (I^2, J^2, K^2): *I* is the orthogonal rotation (with respect to the axes origin), *J* and *K* are the axial/mirror symmetries, respectively, with respect to the first bissectrix and abscissa axis. The iterated applications of these symmetries show that (1 denoting the identity application):

$$I^2 = -J^2 = -K^2 = IJK = -1 \tag{1}$$

In other words, *I* squares to minus identity (−1), whereas *J* and *K* square to plus identity (1). Obviously, the action of *i* corresponds in the complex plane to a multiplication by the imaginary number *i*, and that of *K* corresponds to the complex conjugation. Both preserve angles, but only the former preserves their orientation, a property required by the strict definition of conformal transforms. Indeed, a mirror symmetry does inverse the angle orientation. This is related to the fact that the former transform is holomorphic, contrary to the latter.

Figure 1 also shows that (*I*, *J*, *K*) are not fully independent in the sense that the composition of two of them yield the third one, preceded by a sign that depends on the order of the composition. This change of sign means that (*I*, *J*, *K*) are anti-commutating:

$$\{I, J\} = \{J, K\} = \{K, I\} = 0 \tag{2}$$

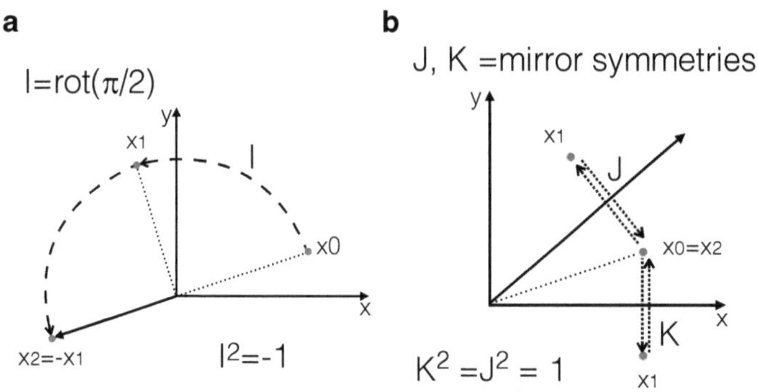

Fig. 1 (**a**) *I* is the orthogonal rotation (with respect to the origin of axes), (**b**) *J* and *K* are the axial/mirror symmetries with respect to the first bissectrix and abscissa axis, respectively. It is easy to check that $I^2 = -J^2 = -K^2 = -1$, where 1 denotes the identity, as well as $J = IK = -KI$ and $IJK = -1$, therefore Eq. (1)

where {.,.} denotes the anti-commutator

$$\{X, Y\} = XY + YX \tag{3}$$

Whereas [.,.] denotes the commutator,

$$[X, Y] = XY - YX \tag{4}$$

which is zero for commutating operators, but in the present case:

$$2I = [J, K]\,; 2J = [I, K]\,; 2K = [J, I] \tag{5}$$

The commutator is a particular case of the celebrated Lie bracket of a Lie algebra. The previously highlighted properties of (I, J, K) show that:

- Two of them are sufficient to generate the basis (1, I, J, K) of a larger algebra often called the algebra of "quasi-quaternions" for reasons discussed below[1]
- Repeated compositions of these operators do not yield any new operator
- The anti-commutator defines a kind of scalar product for which (I, J, K) is orthogonal (with the notation (I, J, K) = (e_1, e_2, e_3)

$$< e_i, e_j > \equiv \{e_i, e_j\}/2 = \delta_{i,j} 1 \tag{6}$$

- And therefore a quadratic form $Q(v) = <v, v>$

Although not needed until now, the matrices corresponding to the symmetries (1, I, J, K) are

$$1 = \begin{bmatrix} 1 & 0 \\ 0 & 1 \end{bmatrix}; I = \begin{bmatrix} 0 & -1 \\ 1 & 0 \end{bmatrix}; J = \begin{bmatrix} 0 & 1 \\ 1 & 0 \end{bmatrix}; K = \begin{bmatrix} 1 & 0 \\ 0 & -1 \end{bmatrix} \tag{7}$$

2.2 *Quaternions*

Equation (1), which summarizes the pseudo-quaternion properties, is very similar to the celebrated "quaternion equation" (Hamilton 1844)[2]:

$$I_2^2 = J_2^2 = K_2^2 = I_2 J_2 K_2 = -1 \tag{8}$$

[1]One may note that they were originally called "coquaternions" (Cookie, 1849) and more recently "split quaternions" (Rosenfeld 1988) or "pseudo-quateriions" (Yaglom, 1968), whreas the name quasi-quaternions was used by Okubo (1978) for a more involved non associative algebra.

[2]The choice of sub-index 2 to distinguish the quaternions from the pseudo-quaternions is due to the fact that their matrix representation corresponds to 2 × 2 block matrices of the latter (see Eq. 9).

The main difference is that all three I_2, J_2, and K_2 are square roots of minus unity (-1), therefore behaves like the orthogonal rotation I. This cannot be achieved neither in the 2D (real) plane, nor in the 3D (real) space. In fact, it requires a 4D (real) hyper-space so that one can conjugate two 2D rotations or two 2D mirror symmetries to obtain new squares of minus unity. This can be easily seen on the classical matrix representation of the quaternions:

$$1_2 = \begin{bmatrix} 1 & 0 \\ 0 & 1 \end{bmatrix}; I_2 = \begin{bmatrix} 0 & -1 \\ 1 & 0 \end{bmatrix}; J_2 = \begin{bmatrix} 0 & -K \\ -K & 0 \end{bmatrix}; K_2 = \begin{bmatrix} -I & 0 \\ 0 & I \end{bmatrix} \tag{9}$$

2.3 Spherical vs. Hyperbolic Geometries

An operator u is said to be a (unitary) direction if it squares to positive or negative identity. Figure 2 illustrates the strong dependence on the sign of u^2 of the exponential transform of a usual geodesic, which is a straight line along a (unitary) direction u. Although this transform is always a geodesic, this occurs for completely different geometries. This difference is easily shown with the help of a broad generalization of the Euler–Moivre identity[3] in the complex plane that states that for any (spherical) angle θ and any real exponent α:

$$(\exp(u\theta))^{\alpha} \equiv \cos(\alpha\theta) + i \sin(\alpha\theta) \tag{10}$$

which merely results from the identity $i^2 = -1$ in the series expansion of the exponential. Similarly, if the unitary direction u is a square root of plus unity ($u^2 = 1$), it is straightforward to obtain:

$$(\exp(u\theta))^{\alpha} \equiv \cosh(\alpha\theta)\, 1 + \sinh(\alpha\theta)\, u \tag{11}$$

where θ and $\alpha\theta$ are now "hyperbolic" angles. The angle $\alpha\theta$ remains a curvilinear coordinate along a geodesic, but in a hyperbolic geometry framework (Milnor 1982), instead of the spherical geometry corresponding to Eq. (10): spherical geodesics are replaced by hyperbolic geodesics, as illustrated in Fig. 2. When the direction u is a square root of minus unity ($u^2 = -1$), we are back to spherical geometry, with an equation similar to Eq. (10):

$$(\exp(u\theta))^{\alpha} \equiv \cos(\alpha\theta)\, 1 + \sin(\alpha\theta)\, u \tag{12}$$

[3]Euler is known for introducing the complex exponential notation ($\alpha = 1$) and Moivre for the identity corresponding to integer α's. Equation (10) summarizes both contributions.

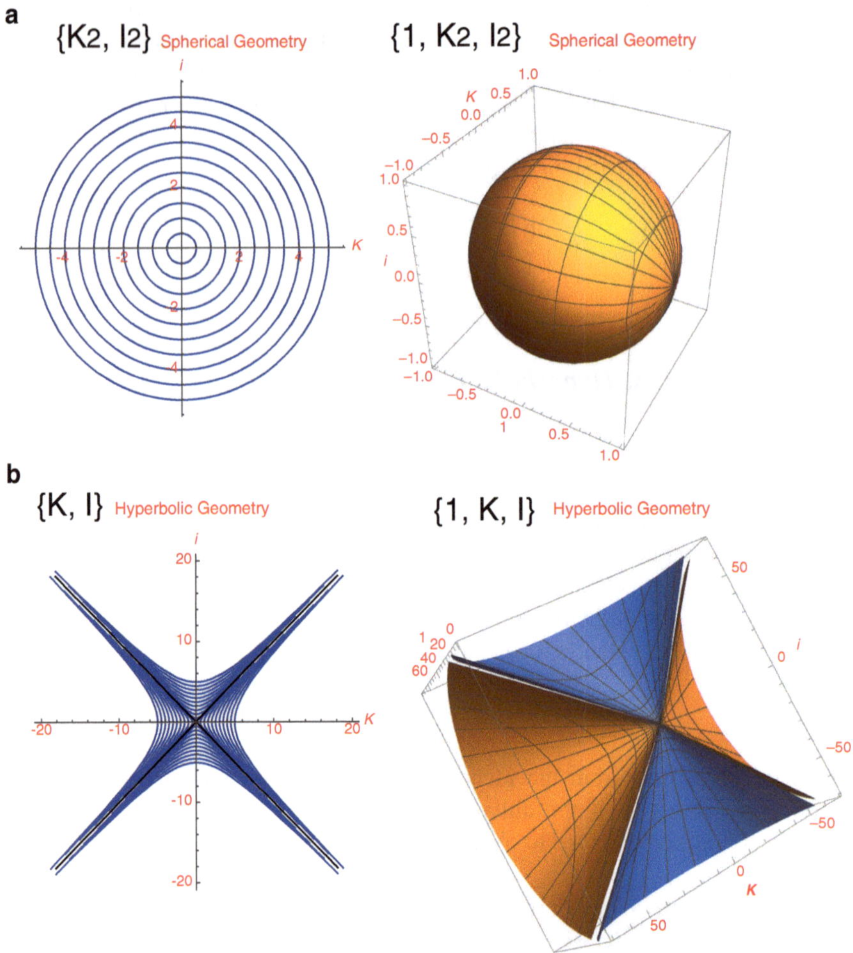

Fig. 2 (**a**) Spherical geodesics, respectively, in the linear spans $\{K_2, I_2\}$ and $\{1_2, K_2, I_2\}$, (**b**) hyperbolic geodesics, respectively, in the linear spans $\{K, I\}$ and $\{1, I, K_2\}$

The two geometries are separated by the light cone[4] defined by the isotropic (or nipotent) operators that square to zero ($u^2 = 0$). The exponential series expansion is then limited to its first two terms:

$$(\exp(u\theta))^{\alpha} \equiv 1 + \alpha\theta\ u \tag{13}$$

[4]Similar to that defined in special and general relativity, i.e. the path of a light flash emanating from a single event.

3 Mandelbrot Set in Spherical and Hyperbolic Geometries

3.1 The Classical M-set

The Mandelbrot set, originally called the M-set (Mandelbrot 1979), has been celebrated for many reasons, including the fact that it is an emblematic link between precursor works of Julia and Fatou on dynamical systems and the more recent concepts of fractals and multifractals. Therefore, it can be used to illustrate the drastic consequences of change from spherical to hyperbolic geometry. The original Mandelbrot set M, called hereafter "the classical M-set" corresponds to the set of complex numbers c's that generate bounded orbits $O(c) = \{c, f_c(c), f_c^2(c), \ldots, f_c^n(c), \ldots\}$ of the simple mapping f_c of the complex plane C with the exponent $\alpha = 2$:

$$f_c : z \to z^{\alpha} + c \tag{14}$$

M is therefore defined by:

$$M = \left\{ c \in C \middle| \sup_{n \in N} \left| f_c^n(c) \right| < \infty \right\} \tag{15}$$

and is consequently the repeller of infinity for the map f_c. Numerically, the M-set is approached with the help of the classical escape algorithm, i.e., with the help of the sets:

$$M_{m,R} = \left\{ c \in C \middle| \sup_{n \leq m} \left| f_c^n(c) \right| < R \right\} \quad M_{m,R} = \left\{ c \in C \middle| \sup_{n \leq m} \left| f^n(c) \right| < R \right\} \tag{16}$$

It is easily shown (Mandelbrot 1979) that $R = 2$ is sufficient for $M_{m,R}$ to converge to M ($\equiv M_{\infty,\infty}$) for $m \to \infty$.

3.2 M-set on Quaternions

With the help of the generalization of the Euler–Moivre identity, there is no difficulty to extend the M-set definition on quaternions, as already done by several authors (Peitgen and Saupe 1988; Gomatam et al. 1995). Figure 3a, b displays two 3D sections of the 4D M-set obtained on quaternions with the help of the classical escape algorithm, more precisely $M_{50,2}$.

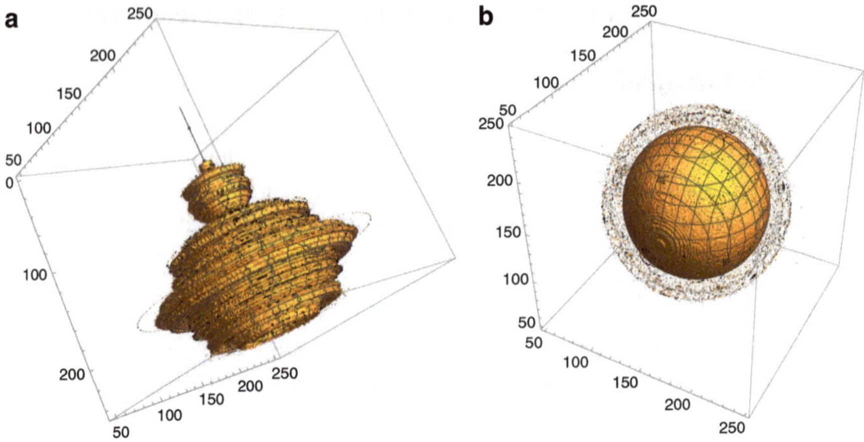

Fig. 3 Intersections of the set M defined on quaternions (**a**) with the 3D linear span $\{1_2, I_2, J_2\}$, (**b**) with $\{I_2, J_2, K_2\}$, which is a sphere (surrounded by a kind of haze, see text). Both figures show that M corresponds to (*spherical*) rotation of the classical M-set around the real/scalar axis $\{1_2\}$

Figure 3a corresponds to the intersection of $M_{50,2}$ with the 3D linear span[5] $\{1_2, I_2, J_2\}$ and therefore displays its real/scalar component (along 1_2) and two imaginary/vector components (along I_2 and J_2). This figure confirms that $M \cap \{1_2, I_2, J_2\}$ is invariant by rotation along the real axis $\{1_2\}$. Therefore, without any surprise, any 2D cut $M \cap \{1_2, aI_2 + bJ_2 + cK_2\}$ is identical to the classical M-set for any real a, b, and c, therefore the intersection with the subspace generated by the real axis and any imaginary axis $(aI_2 + bJ_2 + cK_2)$. This is fundamentally due to the fact that I_2, J_2, and K_2 have an identical role in the quaternion equation [Eq. (8)] and define a spherical geometry. The latter yields, and explains in fact, Fig. 3b that is astonishing at a first glance, but not at the second one: the intersection $M \cap \{I_2, J_2, K_2\}$ of a set as complex as M with the quaternion imaginary space $\{I_2, J_2, K_2\}$ is as simple as a sphere! One may note that this sphere is surrounded by a kind of haze, presumably due to a limited numerical representation of the M-set filaments and their rotation.

3.3 M-set on Pseudo-Quaternions

Now, we consider M-sets on pseudo-quaternions still with the help of the generalized Euler–Moivre identity [Eqs. (11) and (12)]. Figure 4a, b is obtained with the help of the same classical escape algorithm [Eq. (16)] and they are analogs of Fig. 3a, b, but with striking, qualitative differences. More precisely, Fig. 4a

[5] Due to common usage, we are compelled to use curly brackets both for anti-commutators and linear spans, but this should not introduce any confusion.

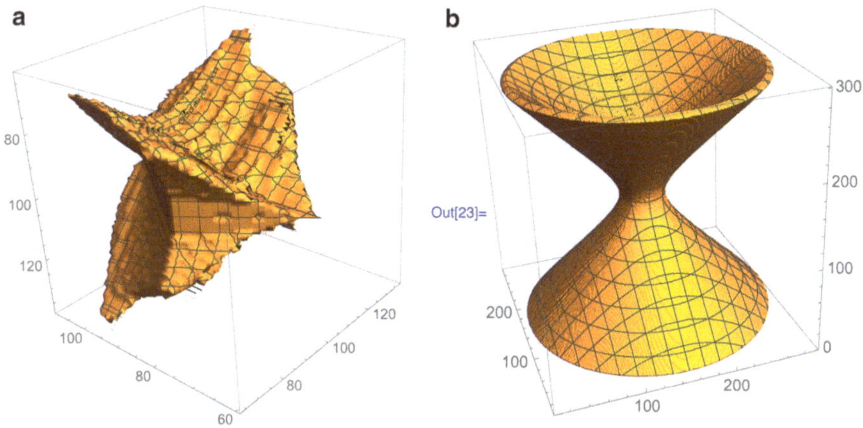

Fig. 4 Intersections of the set M defined on quasi-quaternions (**a**) with the 3D linear span $\{1, I, J\}$, (**b**) with $\{I, J, K\}$ that is a hyperboloid of revolution. Both figures show that M corresponds to hyperbolic rotation of the classical M-set around the real/scalar axis $\{1\}$

corresponds to $M_{50,2} \cap \{1, I, J\}$ and Fig. 4b to $M_{50,2} \cap \{I, J, K\}$. Contrary to Figs. 3a and 4a does not point out a rotational symmetry with respect to the real axis $\{1\}$. Indeed, J and K generate hyperbolic rotations contrary to I, 1_2, I_2, J_2, and K_2 that all generate spherical rotations. This explains that Fig. 4b is no longer a sphere like Fig. 3b, but a hyperboloid of revolution. M seems therefore strongly different from the classical M-set, contrary to the quaternion case. In particular, M is no longer compact (contrary to complex and quaternion case) and the limitation to the distance $R = 2$ in fact truncates M to this distance. Although $M \cap \{1, I\}$ remains identical to the classical M-set, it is spread over hyperbolic geodesics by hyperbolic rotations generated by elements of the pseudo-imaginary space $\{J, K\}$, which again explain that M is no longer bounded. Because of this, the classical definition of the M-set is no longer fully satisfying.

4 From (Pseudo-) Quaternions to Clifford Algebra

Previous sections have pointed out common and distinct properties of the algebra quaternions H (classical notation) and pseudo-quaternions H'. They are in fact both Clifford algebra (Hagen and Scheuermann 2001; Baylis 2004; Trautman and Warszawski 2006), their only difference, but with important consequences, is that their quadratic form, a basic feature of Clifford algebra, has a different signature. A Clifford algebra $Cl(V,Q)$ is indeed defined as being generated by a given vector space V on a field K (in what follows: $K = R$) of operators that can be composed (therefore "upgraded" to operators of higher levels), but are also square roots of the identity times a given quadratic form Q (therefore "downgraded" to scalar operators):

$$\forall v \in V : v^2 = Q(v)1 \tag{17}$$

For $V = Rn$ the quadratic form Q can be put under the canonical diagonal form:

$$Q(v) = v_1^2 + v_2^2 + \cdots + v_p^2 - v_{p+1}^2 - \cdots - v_{p+q}^2 \tag{18}$$

like for every non-degenerate quadratic form on a real vector space, where v_i are the coordinates of v with respect to a given orthogonal basis $\{e_i\}$. The pair (p,q) is the signature of the quadratic form, with $p + q = n$. The corresponding Clifford algebra is classically denoted $Cl_{p,q}(R)$ and is generated by p vectors that square to the positive identity $+1$ and q that square to its negative counterpart -1. The algebra of the pseudo-quaternions H', which is isomorphic to the linear algebra $l(2,R)$ of 2×2 real matrices spanned by 1, I, J, and K [Eq. (7)], can be generated by $V = \{I, J\}$ or $\{K, I\}$ and therefore can be denoted by $Cl_{2,0}(R)$, as well as by $Cl_{1,1}(R)$ because it can also be generated by $\{J, K\}$. This shows that distinct vector spaces V's can generate the same Clifford algebra, i.e., the notation $Cl(V,Q)$ is not one-to-one. On the contrary, the algebra of the quaternions H, spanned by 1_2, I_2, J_2, and K_2 [Eq. (9)], univocally corresponds to $Cl_{0,2}(R)$, although it can be generated by $V = \{I_2, J_2\}$, $\{J_2, K_2\}$or $\{K_2, I_2\}$, but all these spaces have the same signature (0,2). One may note the simpler examples: $Cl_{0,0}(R)$ is isomorphic to R ($V = \varnothing$, no vector, only scalars), $Cl_{0,1}(R)$ to C (a unique generating vector I, which squares to -1, $V = \{I\}$), $Cl_{0,1}(R)$ seems to be nonclassical (with $V = \{J\}$ or $\{K\}$).

Let us mention that two properties are extended in a straightforward manner from V to $Cl(V,Q)$:

- Let $\{e_1, e_2, \ldots, e_n\}$ be an orthogonal basis of V, then $Cl(V, Q)$ admits the basis

$$\{e_{i_1} e_{i_2} \ldots e_{i_{2k}} | 1 \le i_1 < i_2 < \cdots < i_k \le n \quad \text{and} \quad 0 \le k \le n\} \tag{19}$$

where the empty product ($k = 0$) corresponds to the identity. The dimension of $Cl(V, Q)$ is therefore:

$$\dim\left[Cl\left(V, Q\right)\right] = \sum_{k=0,n} \binom{n}{k} = 2^n \tag{20}$$

- The quadratic form Q, initially defined only over V, is in fact defined on the aforementioned basis of $Cl(V, Q)$ [Eq.(19)], and therefore over the whole algebra $Cl(V, Q)$. Indeed, the anticommutation of the e_{i_j} together with the fact they square to the scalar $Q(e_{i_j})$ [Eq.(17)] yields

$$\mathrm{Q}\left(e_{i_1} e_{i_2} \ldots e_{i_{2k}}\right) = \left(1 - 2\delta_{k,2}\right) \mathrm{Q}\left(e_{i_1}\right) \mathrm{Q}\left(e_{i_2}\right) \ldots \mathrm{Q}\left(e_{i_{2k}}\right) \tag{21}$$

where the prefactor corresponds to $(-1)^{(k-1)!}$ that merely results from the $(k-1)!$ permutations to be done to obtain a product of squares e_{ij}^2.

In what follows, Q_v denotes the extension of the initial Q over the whole vector part *Vect(Cl(V, Q))* of Cl(*V*, *Q*), whose signature is denoted by (p_v, q_v), with, as before, p_v and q_v vectors that square to the positive and negative unity, respectively. This signature is not only univocal, contrary to the initial signature (*p*, *q*) of the generating vector space *V*, but also it defines the relevant geometries and geodesics, i.e., spherical or hyperbolic, for the different eigensubspaces of Q_v.

5 A Short Recapitulation of Scalar-Valued Multifractals

In this section we briefly recapitulate the main features of scalar-valued multifractals with the help of a 4-step procedure to simulate continuous in scale, universal multifractals u_λ, where $\lambda = L/l$ is the scale ratio of the outer *L* and inner *l* scales, and is therefore the resolution of the process. These continuous in scale processes are obtained with the help of exponentials of additive processes (Schertzer and Lovejoy 1987), whereas the pioneering works on multiplicative cascades (Yaglom 1966; Mandelbrot 1974), whose generic outcome was later on recognized as multifractal, were obtained by products of identically independently distributed variables along a dyadic (more generally a *p*-adic) tree, i.e., these cascades were discrete in scale and were limited to generate a conservative flux. This procedure corresponds to (see Fig. 5 for illustration):

a. Create a stable sub-generator $\gamma_0^{(\alpha)}$, i.e., an extremely asymmetric Lévy white noise of Lévy stability index α
b. Perform a fractional integration on the sub-generator $\gamma_0^{(\alpha)}$ to obtain a stable generator Γ_λ whose exponential is scaling
c. Take the exponential of the generator Γ_λ to obtain the (conservative) flux ε_λ of given universal multifractal parameters C_1 and α
d. Perform a fractional integration of the forcing $f_\lambda = \varepsilon_\lambda^a$ to obtain a smoother, but non-conservative field u_λ that responds to this forcing.

Combining these features together with those of Clifford algebra (Sect. 4) will enable us to define similar vector-valued multifractals. Some comments are in order and are displayed for each stage in the following sub-sections (more details are available in Schertzer and Tchiguirinskaia (2015)), however, they can be skipped in a first reading.

5.1 Creating a Sub-Generator

Let us recall that a random variable *X* is said to be a Lévy stable variable (Lévy 1937, 1965; Gnedenko and Kolmogorov 1954; Feller 1971; Kahane 1974, 1985; Zolotarev

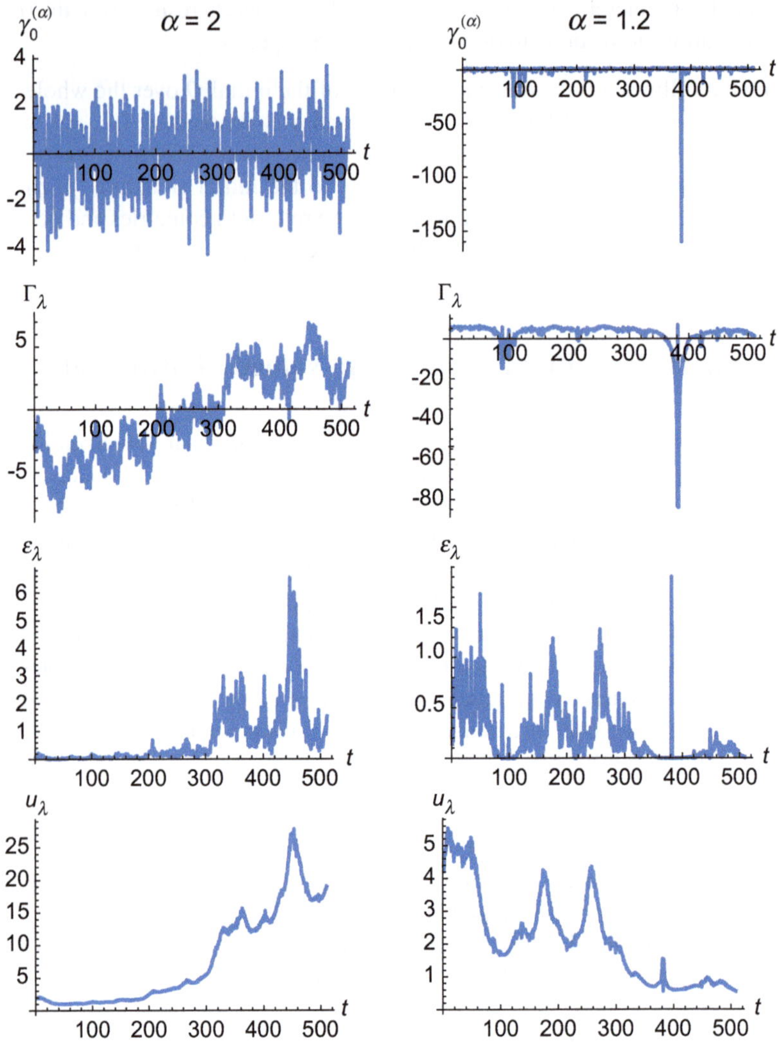

Fig. 5 Illustration of the four steps to generate a scalar-valued multifractal field (see text). The left column corresponds to the Gaussian case ($\alpha = 2$), the right for a Lévy case with $\alpha = 1.2$, both for $C_1 = 0.2$, $a = 1$, $H = 1/9$, and $\lambda = 512$. The horizontal axis is the time $t \in [0, \lambda]$. From top to bottom, (**a**) sub-generators $\gamma_0^{(\alpha)}$, respectively, symmetric ($\alpha = 2$) and extremely asymmetric ($\alpha = 1.2$) with huge negative fluctuations, (**b**) generator Γ_λ obtained by a fractional integration of the sub-generator, so that it is Log(λ) divergent, (**c**) (conservative) flux ε_λ obtained by exponentiation of the generator Γ_λ, (**d**) multifractal field u_λ by fractional integration of order H of ε_λ (adapted from Schertzer and Tchiguirinskaia 2015)

1986) if and only if it is stable under renormalized sums, i.e., it is a fixed point, with the rescaling factor $a(n)$ and centering term $b(n)$, of any n of its independent realizations X_i ($i = 1, n$). This corresponds to ($\overset{d}{=}$ denotes equality in distribution):

$$\forall n \in N, \exists a(n), b(n) \in R : \sum_{i=1,n} X_i \overset{d}{=} a(n)X + b(n) \tag{22}$$

Furthermore, any Lévy stable variable X is attractive for renormalized sum of independent realizations Y_i ($i = 1, n$) of a random variable Y having similar distribution tails:

$$\lim_{n\to\infty} \frac{\sum_{i=1,n} Y_i - b(n)}{a(n)} \overset{d}{=} X \tag{23}$$

i.e., it has a power-law tail whose exponent is the Lévy stability index $\alpha \in]0, 2]$:

$$\forall s >> 1 : \Pr(|X| > s) \approx s^{-\alpha} \tag{24}$$

which is also the critical order[6] of divergence of statistical moments (E[.] denotes the mathematical expectation):

$$\forall q \geq \alpha : \mathrm{E}\left[|X|^q\right] = \infty \tag{25}$$

The Gaussian case ($\alpha = 2$) is the exceptional case whose all statistical moments converge.

5.2 Creating a Stable Generator

A “coloration” of the sub-generator $\gamma_0^{(\alpha)}$ is required to obtain a generator Γ_λ such that its exponential ($\exp(\Gamma_\lambda)$) is scaling, i.e.:

$$\mathrm{E}\left[(\exp(\Gamma_\lambda))^q\right] = \mathrm{E}\left[\exp(q\Gamma_\lambda)\right] \approx \lambda^{K(q)} \tag{26}$$

which means that the generator Γ_λ is log-divergent with the resolution λ. This can be achieved with the help of a fractional integration of order D/α', where D is the domain dimension and $1/\alpha + 1/\alpha' = 1$ (Schertzer and Lovejoy (1991) for details). The resulting generator Γ_λ is also stable, because the linear stability and attractivity of Lévy sub-generators are preserved by fractional integration that is linear as are the definition of stability and attractivity. These properties are transformed by

[6]The moment order q and the index q of a Clifford algebra have nothing else in common, except to be the same alphabetical letter due to respective usages.

exponentiation into a multiplicative stability and attractivity (Schertzer and Lovejoy 1987; Fan 1989; Brax and Pechanski 1991; Schertzer and Lovejoy 1997a, b), if this exponentiation preserves some finite statistics. This is not always granted due to the following inequality:

$$\forall n \in N, \forall X \in R^{+}, q > 0 : \exp(qX) \geq (qX)^{n}/n! \tag{27}$$

which shows that the $\mathrm{E}[\exp(qX)] = \infty$ for any positive variable X with a finite power-law tail exponent α [Eq. (24)] and any order $q > 0$. It is therefore required to only use Lévy stable white noise $\gamma_0^{(\alpha)}$ that are fully asymmetrical, in the sense that they have a power-law tail only for negative values.

5.3 *Creating a Conservative Flux*

The exponentiation of the generator Γ_λ is usually called a flux:

$$\varepsilon_\lambda = \exp\left(\Gamma_\lambda\right) \tag{28}$$

which can be normalized so that it is "conservative," i.e., its average is strictly scale invariant, i.e., conserved for any resolution λ:

$$\forall \lambda \geq 1 : \mathrm{E}\left[\varepsilon_\lambda\right] = \mathrm{E}\left[\varepsilon_1\right] \tag{29}$$

More generally [in agreement with Eq. (26)]:

$$\forall q \in R : \mathrm{E}\left[\varepsilon_\lambda^q\right] = \exp\left(\mathrm{Log}\left(\lambda\right) K(q)\right) \approx \lambda^{K(q)} \tag{30}$$

i.e., the statistical moment of order q of the flux is the (Laplace) characteristic function of the generator and the corresponding cumulant generating function (or second Laplace characteristic function) is $K_\lambda(q) = K(q)\mathrm{Log}(\lambda)$. $K(q)$ is called the "scaling moment function," it satisfies $K(1) = 0$ and has for general expression (Schertzer and Lovejoy 1987):

$$q \geq 0 : K(q) = \frac{C_1}{\alpha - 1}\left(q^\alpha - q\right); q < 0 : K(q) = \infty \tag{31}$$

where C_1 is the codimension of the mean field, which satisfies with the index α:

$$C_1 = \left.\frac{dK(q)}{dq}\right|_{q=1}; C_1\alpha = \left.\frac{d^2K(q)}{dq^2}\right|_{q=1} \tag{32}$$

both relations are useful for determining these parameters. Because of the (multiplicative) attractivity for other multifractals, these multifractals are often called "universal multifractals" (Schertzer and Lovejoy 1997a, b).

5.4 *Creating a Non-Conservative Field*

The last stage corresponds to a fractional integration of the flux raised to a given power a, which can be thus seen as a forcing $f_\lambda = \varepsilon_\lambda^a$ of a fractional differential equation for the non-conservative field u_λ that is smoother (i.e., with stronger auto-correlations) than the forcing f_λ. It has been argued that the overall procedure, often called Fractionally Integrated Flux model (Schertzer et al. 1997), has strong links with various attempts to "renomalize" nonlinear differential equations, particularly the Navier–Stokes equations: the forcing f_λ being the analogue of the renormalized forcing term and the fractional operator that of the renormalized Green function or propagator (Kraichnan 1959a, b; Wilson 1971; Schertzer et al. 1998).

6 Vector-Valued Multifractals

6.1 *Lie/Clifford Algebra of Generators*

The substitution of the products of identically independently distributed variables (Yaglom 1966; Mandelbrot 1974) by exponentials of additive processes (Schertzer and Lovejoy 1987) opens the road to broad generalizations needed to obtain multifractals that are not only continuous in scale (Sect. 5), but also to vector or manifold valued multifractals (Schertzer and Lovejoy 1995; Schertzer and Tchiguirinskaia 2015). In fact the case where the domain and codomain are both scalar already points out in this direction: the (usual) exponential maps the additive group R into the multiplicative group R^+ of positive real numbers and it corresponds to the simplest case of a mapping of a Lie algebra into an associated Lie group (Fig. 6 for illustration). The latter defines in fact the (generalized) exponential transform. The main interest of this transform is that its domain (the Lie algebra) is a vector space, whereas its codomain (the Lie group) has in general a more complex structure being a manifold. More precisely, the Lie algebra is the tangent vector

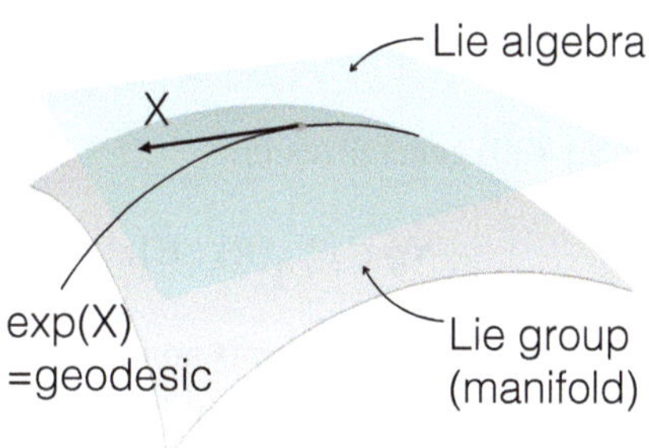

Fig. 6 Schematic of the exponential mapping geodesics of a Lie algebra into geodesics of an associated Lie group

space to the Lie group at its unity (Gilmore 1941; Sattinger and Weaver 1986), is thus the vector space of the generators of the group and maps geodesics (e.g., straight lines of the vector space) into geodesics of the manifold. Following Sect. 4, we proceed like in Sect. 5, but with sub-generators and generators belonging to a real Clifford algebra $Cl_{p,q}(R)$. However, to obtain a similar statistical universality (i.e., stability and attractivity of the generators) with finite statistics requires some further analysis. Indeed, as pointed out in Sects. 2.3 and 3.3, large hyperbolic angles yield very extreme values by exponentiation, whereas this is not the case for spherical angles, and therefore possible divergence of all statistical moments, similar to that of scalar generators that are not fully asymmetric.

Before addressing the general case, it is worthwhile to note that two cases obviously escape from this type of problem:

- Spherical geometry: the signature of the quadratic form Q_v over the vector part of the Clifford algebra is purely negative, i.e., it is $(0, q_v)$
- Gaussian sub-generators and generators ($\alpha = 2$): the signature (p_v, q_v) of Q_v is no longer relevant.

Figure 7 displays 3D snapshots of simulation of a multifractal quaternion velocity field, i.e., the 3D arrows represent the three first components of a field whose space-time domain is 3D + 1 and its codomain is $H = Cl_{0,2}$. This simulation was obtained by generalizing the 4-step procedure for scalar-valued fields (Sect. 5) to (spherical) vector-valued fields.

6.2 Clifford–Laplace Transform and Finite Statistics

To obtain a more systematic assessment on the finiteness of the statistics of the flux, we need to generalize the property for scalar multifractals that the moments of the flux are the Laplace characteristic function of the generators [Eq. (30)]. We therefore need to define a Laplace transform over a Clifford algebra, and at first a scalar product over it. As for any quadratic space, this scalar product is conveniently defined with the help of a polarization identity:

$$\langle X, Y\rangle = \frac{1}{2}\left(Q_v\left(X+Y\right) - Q_v(X) - Q_v(Y)\right) \tag{33}$$

and Eq. (30) generalizes into:

$$\forall q \in C_{p,q}(R) : \mathrm{E}\left[\varepsilon_\lambda^q\right] = \mathrm{E}\left[\exp\left(< q, \Gamma_\lambda >\right)\right] \approx \lambda^{K(q)} \tag{34}$$

where the moment order q is no longer a scalar, but a vector of the Clifford algebra.

Let us mention that behind the similarity between Eqs. (30) and (34), there are important differences due to the involved definitions of Lévy stable vectors, which are furthermore not univocal. These definitions are discussed with some details

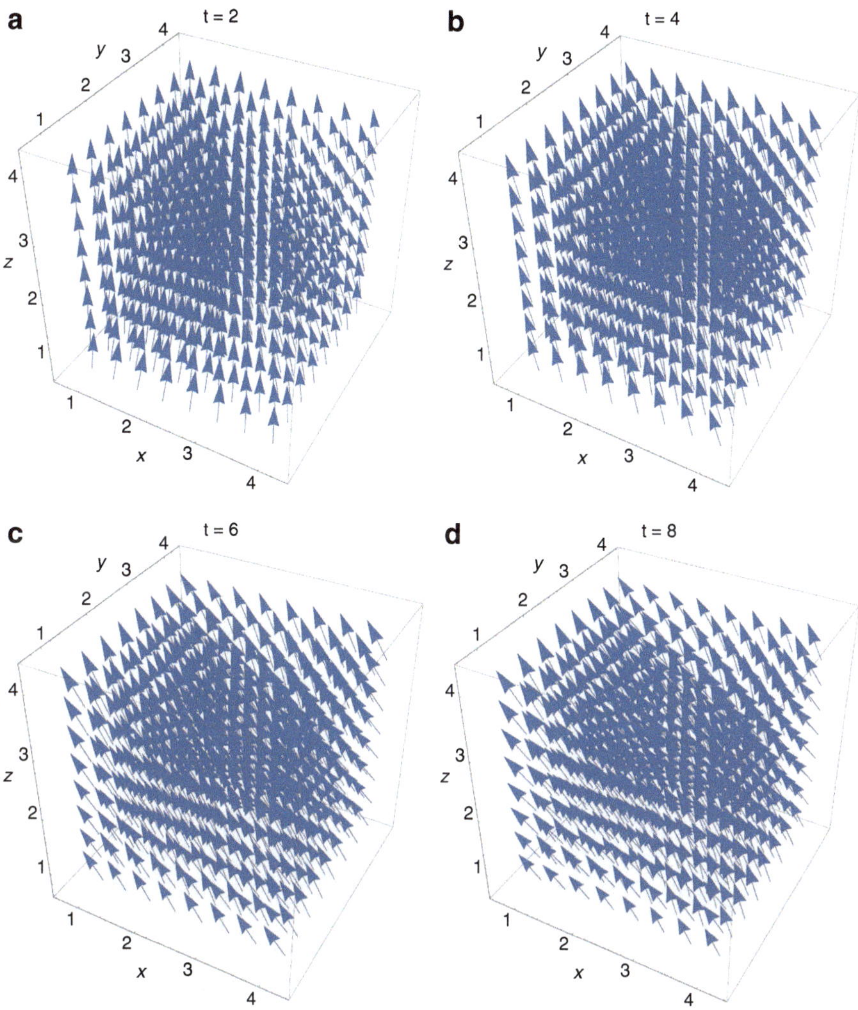

Fig. 7 (**a–h**) Snapshots (time steps t_{2i}, $i = 1{:}8$) of a multifractal simulation of a 3D + 1 intermittent vector field obtained by a quaternion cascade, i.e., with values on $Cl_{0,2}$, see text for details on the numerical simulation (adapted from Schertzer and Tchiguirinskaia 2015)

by Schertzer and Tchiguirinskaia (2015), but it was shown that, in spite of these important technical difficulties, the definition of fully asymmetrical stable Levy variables can be extended to stable Levy vectors to obtain finite statistics (i.e., $K(q) < \infty$) over a hyperbolic subspace of a Clifford algebra $C_{p,q}(R)$. This can be achieved for all "positive" vectors q's, i.e., all their coordinates q_i with respect to the algebra orthogonal basis $\{e_i\}$ are positive (see Fig. 8 for illustration). This is basically due to the fact that, for each i and any positive q_i, the vector $q = q_i e_i$ (no summation over i) is the normal vector of the (hyper-) plane $<q, \Gamma_\lambda> = 0$ that

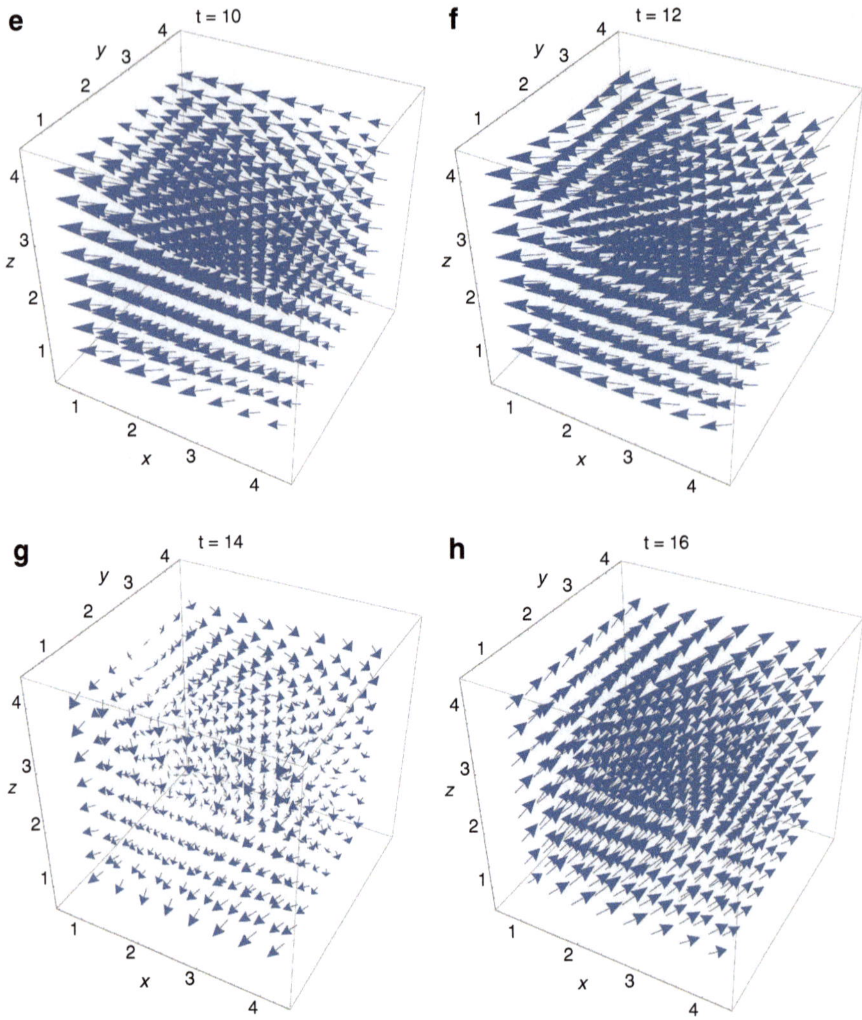

Fig. 7 (continued)

splits into two halves the considered hyperbolic subspace. The fluctuations of the generator should be moderate to avoid divergences on the half subspace defined by $<q, \Gamma_\lambda> > 0$, which contains the vector q. On the contrary, there is no such restriction for the other half subspace defined by $<q, \Gamma_\lambda> \leq 0$.

We can therefore complete the previous list of multifractal vectors with finite statistics (Sect. 6.1) by:

- Hyperbolic geometry: the stable Levy generators should be extremely asymmetrical on hyperbolic eigensubspaces of the quadratic form Q_v (over which the signature of Q_v is positive), whereas this is not required for spherical eigensubspaces.

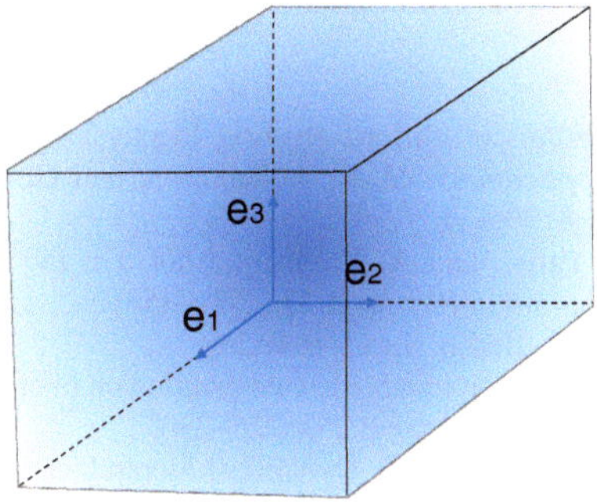

Fig. 8 Schematic of the subspace corresponding to non-negative coordinates (e.g., $q_i \geq 0$) with respect to the orthogonal basis $\{e_i\}$, where statistics are necessarily finite ($|K(q)| < \infty$) and the fluctuations of the generators are moderate (i.e., the probability tail of the extremes falling faster than a power-law)

7 Conclusions and Prospects

The goal of this chapter was to give in an inductive manner new insights on current developments to define stochastic, multifractal vector fields having both universal statistical and robust algebraic properties. It began with simple considerations on orthogonal rotations and mirror symmetries leading to the concept of Clifford algebra. Based on the examples of quaternions and pseudo-quaternions, it emphasized the roles of spherical and hyperbolic geometries, their respective geodesics and the rotations they generate with the help of the exponential transform from the Clifford algebra of group generators to the group itself. This was first illustrated with the help of the generalization of the Mandelbrot set over a Clifford algebra. This demonstrates that a special attention should be paid to hyperbolic rotations. This is particularly useful to define multifractal vector fields having extreme fluctuations, but nevertheless finite statistics. This can be understood by the fact that fields can be seen as flows of particles.

Overall this enables to define vector multifractals with given universality and robustness, based on a stochastic algebra of generators that can be tentatively called a Lévy–Clifford algebra. Due to its generality and its relative simplicity, this algebra should help to resolve many problems encountered on complex systems having nontrivial symmetries and multiscale behavior, and to deeply change our perception of nonlinear processes in geophysics.

Acknowledgements This research was partially supported by the Chair "Hydrology for Resilient Cities," endowed by Veolia (http://www.enpc.fr/node/1073). The authors greatly acknowledge stimulating discussions with participants to the conference "30 Years of Nonlinear Dynamics in Geosciences," in particular with Anastasios Tsonis. They thank him for the quality of this conference and for his generous invitation. Part of this work was done during a summer workshop in the Aspen Center for Physics (supported by National Science Foundation grant PHY-1607761).

References

Baylis, W.E. 2004. Applications of clifford algebras in physics. In *Lectures on clifford (geometric) algebras and applications*, ed. R. Ablamowicz and G. Sobczyk, 91–133. Springer. doi:10.1007/978-0-8176-8190-6_4.

Benzi, R., G. Paladin, G. Parisi, and A. Vulpiani. 1984. On the multifractal nature of fully developed turbulence. *Journal of Physics A* 17: 3521–3531.

Bourbaki, N. 2004. *Integration*. Berlin: Springer Verlag.

Brax, P., and R. Pechanski. 1991. Levy stable law description on intermittent behaviour and quark-gluon phase transitions. *Physics Letter B* 253: 225–230.

Chigirinskaya, Y., and D. Schertzer. 1996. Dynamical hierarchical cascade models, multifractal space-time intermittency and lie structure in turbulence. In *Stochastic models in geosystems*, ed. W.A. Woyczynski and S.S. Molchanov, 57–81. New York: Springer-Verlag.

Chigirinskaya, Y., D. Schertzer, and S. Lovejoy. 1998. An alternative to shell-models: More complete and yet simple model of intermittency. In *Advances in turbulence VII*, ed. U. Frisch, 263–266. Dordrecht: Kluwer Academic Publishers.

Cockle, J. 1849. On systems of algebra involving more than one imaginary. *Philosophical Magazine Series 3* 35: 434–435.

Fan, A.H. 1989. Chaos additif et multiplicatif de Levy. In *Comptes Rendus de l'Académie des Sciences de Paris*, I(308), 151–154.

Feller, W. 1971. *An introduction to probability theory and its applications*, Vol. 2. New York: Wiley.

Gilmore. 1941. *Lie groups*. New York: Wiley.

Gnedenko, B.V., and A.N. Kolmogorov. 1954. *Limit distribution for sums of independent random variables*. Cambridge: Addision-Wesley.

Gomatam, J., J. Doyle, B. Steves, and I. McFarlane. 1995. Generalization of the Mandelbrot set: Quaternionic quadratic maps. *Chaos, Solitons and Fractals* 5 (6): 971–986. doi:10.1016/0960-0779(94)00163-K.

Hagen, H., and G. Scheuermann. 2001. Clifford algebra and flows. Mathematical Methods in CAGD. T. Lyche and L. L. Schumaker (eds.), Vanderbilt University Press, Nashville, TN. 1–9.

Haller, G. 2005. An objective definition of a vortex. *Journal of Fluid Mechanics* 525: 1–26. doi:10.1017/S0022112004002526.

Halsey, T.C., M.H. Jensen, L.P. Kadanoff, I. Procaccia, and B. Shraiman. 1986. Fractal measures and their singularities: The characterization of strange sets. *Physical Review A* 33: 1141–1151.

Hamilton, W.R. 1844. II. On quaternions; or on a new system of imaginaries in algebra. *Philosophical Magazine Series 3* 25 (163): 10–13. doi:10.1080/14786444408644923.

Kahane, J.P. 1974. Sur le modèle de turbulence de Benoit Mandelbrot. *Comptes Rendus (Paris)* 278A: 621–623.

———. 1985. Definition of stable laws, infinitely divisible laws, and Lévy processes. In *Lévy flights and related phenomena in physics*, ed. M. Shlesinger, G. Zaslavsky, and U. Frish, 99–109. Berlin: Springer-Verlag.

Kraichnan, R.H. 1959a. Classical fluctuation-relaxation theorem. *Physical Review* 113: 1181–1182.

———. 1959b. The structure of isotropic turbulence at very high Reynolds numbers. *Journal of Fluid Mechanics* 5: 497–543.

Lévy, P. 1937. *Théorie de l'addition des variables aléatoires*. Paris: Gauthiers Villars.

———. 1965. *Processus stochastiques et mouvement Brownien*. Paris: Gauthiers-Villars.

Lorenz, E.N. 1991. Dimension of weather and climate attractors. *Nature* 353: 241–244.

Mandelbrot, B.B. 1974. Intermittent turbulence in self-similar cascades: Divergence of high moments and dimension of the carrier. *Journal of Fluid Mechanics* 62: 331–350.

———. 1979. Fractal aspects of the iteration of the z-> λz(1-z) for complex λ and z. In *Nonlinear dynamics*, ed. R.H.G. Hellleman, 249–259. New York, NY: Annals of the New York Academy of Sciences.

———. 1983. *The fractal geometry of nature*. San Francisco: Freeman.

Mandelbrot, B.B., and J.W. Van Ness. 1968. Fractional Brownian motions, fractional noises and applications. *SIAM Review* 10: 422–450.

Milnor, J.W. 1982. Hyperbolic geometry: The first 150 years. *Bulletin of the American Mathematical Society* 6 (1): 9–25. doi:10.1090/S0273-0979-1982-14958-8.

Okubo, A. 1970. Horizontal dispersion of floatable trajectories in the vicinity of velocity singularities such as convergencies. *Deep Sea Research* 17: 445–454.

Okubo, S. 1978. Pseudo-quaternion and pseudo-octonion algebras. *Hadronic Journal* 1 (4): 1250–1278.

Parisi, G., and U. Frisch. 1985. On the singularity structure of fully developed turbulence. In *Turbulence and predictability in geophysical fluid dynamics and climate dynamics*, ed. M. Ghil, R. Benzi, and G. Parisi, 84–88. Amsterdam: North Holland.

Peitgen, H.O., and D. Saupe. 1988. *The science of fractal images*. NewYork: Springer-Verlag.

Rosenfeld, B.A. 1988. *A history of non-Euclidean geometry*. New York: Springer-Verlag.

Sattinger, D.H., and O.L. Weaver. 1986. *Lie groups and algebras with applications to physics, geometry and mechanics*. New-York: Springer-Verlag.

Pierrehumbert, R.T. 2013. Strange news from other stars. *Nature Geoscience* 6 (2): 81–83. doi:10.1038/ngeo1711.

Schertzer, D., and S. Lovejoy. 1984. On the dimension of atmospheric motions. In *Turbulence and chaotic phenomena in fluids*, ed. T. Tatsumi, 505–508. Amsterdam: North Holland.

———. 1987. Physical modeling and Analysis of Rain and Clouds by Anisotropic Scaling Multiplicative Processes. *Journal of Geophysical Research*. American Geophysical Union, D 8 (8): 9693–9714. doi:10.1029/JD092iD08p09693.

———, eds. 1991. *Non-linear variability in geophysics*. Kluwer Academic Publishers.

———. 1995. From scalar cascades to Lie cascades: Joint multifractal analysis of rain and cloud processes. In *Space/time variability and interdependance for various hydrological processes*, ed. R.A. Feddes, 153–173. Cambridge: Cambridge University Press.

———. 1997a. Universal multifractals do exist! *Journal of Applied Meteorology* 36: 1296–1303.

———. 1997b. Universal multifractals do exist!: Comments on "A statistical analysis of mesoscale rainfall as a random cascade". *Journal of Applied Meteorology* 36 (9): 1296–1303. doi:10.1175/1520-0450(1997)036<1296:UMDECO>2.0.CO;2.

Schertzer, D., I. Tchiguirinskaia, S. Lovejoy, and A.F. Tuck. 2012. Quasi-geostrophic turbulence and generalized scale invariance, a theoretical reply. *Atmospheric Chemistry and Physics* 12: 327–336.

Schertzer, D., and I. Tchiguirinskaia. 2015. Multifractal vector fields and stochastic Clifford algebra. *Chaos: An Interdisciplinary Journal of Nonlinear Science* 25 (12): 123127. doi:10.1063/1.4937364.

Schertzer, D., S. Lovejoy, F. Schmitt, I. Tchiguirinskaia, and D. Marsan. 1997. Multifractal cascade dynamics and turbulent intermittency. *Fractals* 5 (3): 427–471.

Schertzer, D., M. Larchevêque, and S. Lovejoy. 1998. Beyond multifractal phenomenology of intermittency: Nonlinear dynamics and multifractal renormalization. In *Chaos, fractals and models 96*, ed. G. Iuculano, 53–64. Genova: Italian University Press.

Schertzer, D., I. Tchiguirinskaia, S. Lovejoy, P. Hubert, and H. Bendjoudi. 2002. Which chaos in the rain-runoff process? *J. Hydrological Sciences* 47 (1): 139–148.

Trautman, A., and U. Warszawski. 2006. Clifford algebras and their representations. *Encyclopedia of Mathematical Physics* 1 (2): 518–530. Available at: http://www.fuw.edu.pl/~amt/amt2.pdf.

Tsonis, A.A. 1992. *Chaos: From theory to application*. New York: Plenum.

Weiss, J. 1991. The dynamic of enstrophy transfer in two-dimensional hydrodynamics. *Physica D* 48: 273–294.

Wilson, K.G. 1971. Renormalization group and critical phenomena. I. Renormalization group and the kadanoff scaling picture. *Physical Review B* 4 (9): 3174–3183.

Yaglom, A.M. 1966. The influence on the fluctuation in energy dissipation on the shape of turbulent characteristics in the inertial interval. *Soviet Physics – Doklady* 2: 26–30.

Yaglom, I. 1968. *Complex numbers in geometry*. New York: Academic.

Zolotarev, V.M. 1986. *One-dimensional stable distributions*. Providence, RI: American Mathematical Society.

Complex Networks and Hydrologic Applications

Bellie Sivakumar, Carlos E. Puente, and Mahesh L. Maskey

Abstract Connections are ubiquitous in hydrology. However, understanding the nature and extent of connections in hydrologic systems has and continues to be a tremendous challenge. In recent years, applications of the concepts of complex networks to study connections in hydrologic systems have started to emerge. This chapter aims to offer an overview of the science of complex networks and its applications in hydrology. First, the basic concept of a network, the history of development of network theory, and some important measures of network properties are presented. Next, applications of complex networks in hydrology are reviewed, including studies on spatial connections, temporal connections, and catchment classification. Finally, some remarks on future directions are made.

Keywords Hydrologic systems • Connections • Complex networks • Clustering coefficient • Degree distribution • Rainfall • Streamflow • River networks

1 Introduction

Connections are everywhere in hydrologic systems, and geophysical systems at large. Arguably, the hydrologic cycle is the best example of connections, as every component of this cycle is connected to every other component, directly or indirectly and strongly or weakly. Unraveling the nature and extent of connections in hydrologic systems, as well as their interactions with other geophysical systems, has and continues to be a tremendous challenge.

The last century witnessed the development and application of numerous scientific concepts and methods for studying the nature and extent of connections in

B. Sivakumar (✉)
School of Civil and Environmental Engineering, The University of New South Wales, Sydney, NSW 2052, Australia

Department of Land, Air and Water Resources, University of California, Davis, CA 95616, USA
e-mail: s.bellie@unsw.edu.au; sbellie@ucdavis.edu

C.E. Puente • M.L. Maskey
Department of Land, Air and Water Resources, University of California, Davis, CA 95616, USA

A.A. Tsonis (ed.), *Advances in Nonlinear Geosciences*,
DOI 10.1007/978-3-319-58895-7_26

hydrologic systems; see, for example, Gupta et al. (1986), Salas et al. (1995), Mishra and Coulibaly (2009), and Sivakumar (2017) for some accounts. Such studies have certainly resulted in notable progress in identifying and modeling connections in hydrologic systems. Nevertheless, accurate representations of such connections continue to be elusive. While many factors contribute to this situation (e.g., inherent complexity of hydrologic systems, natural and anthropogenic influences, data and computational constraints), a key reason is the absence of a strong scientific theory that is suitable for representing all types of connections encountered in hydrology. This has also led to calls for a general framework in hydrology (e.g., Dooge 1986; Paola et al. 2006; Sivakumar 2008; Young and Ratto 2009), especially in the face of new challenges, including climate change impacts, water conflicts, and interactions between hydrology and society (e.g., Sivakumar 2011a, b; Montanari et al. 2013).

In the context of connections, network theory or graph theory can offer useful ideas. Although the concept of networks originated in the mid-eighteenth century (Euler 1741) and advanced over the subsequent centuries (Listing 1848; Cayley 1857; Erdös and Rényi 1959, 1960), developments since the late 1990s (Watts and Strogatz 1998; Barabási and Albert 1999; Girvan and Newman 2002) have offered a whole new dimension for studying connections in large, complex, and dynamically evolving systems. These recent developments are put under the broad umbrella of "complex networks."

Applications of the concepts of complex networks in hydrology are just starting to emerge. Thus far, such applications include studies on rainfall monitoring networks (e.g., Malik et al. 2012; Boers et al. 2013; Scarsoglio et al. 2013; Sivakumar and Woldemeskel 2015; Jha et al. 2015), streamflow monitoring networks (e.g., Sivakumar and Woldemeskel 2014; Halverson and Fleming 2015; Braga et al. 2016; Serinaldi and Kilsby 2016; Fang et al. 2017), and river networks (Rinaldo et al. 2006; Zaliapin et al. 2010; Czuba and Foufoula-Georgiou 2014, 2015), among others. Such studies have addressed spatial connections, temporal connections, and catchment classification, among others. In light of recent calls for a general framework in hydrology, the suitability of network theory for such has also been highlighted (Sivakumar 2015).

This chapter aims to offer an overview of the science of complex networks and its applications in hydrology. Section 2 reviews the basic concept of a network and the history of development of network theory. Section 3 describes some of the popular measures for identifying the properties of complex networks. Section 4 reviews the applications of complex networks in hydrology. Section 5 offers some remarks towards the future.

2 Network: Concept and Development

A network or a *graph* is a set of points connected together by a set of lines, as shown in Fig. 1. The points are referred to as *vertices* or *nodes* and the lines are referred to as *edges* or *links*. Therefore, mathematically, a network can be represented as $G = \{V, E\}$, where V is a set of N nodes ($V_1, V_2, \ldots, V_N$) and E is a set of n links. The

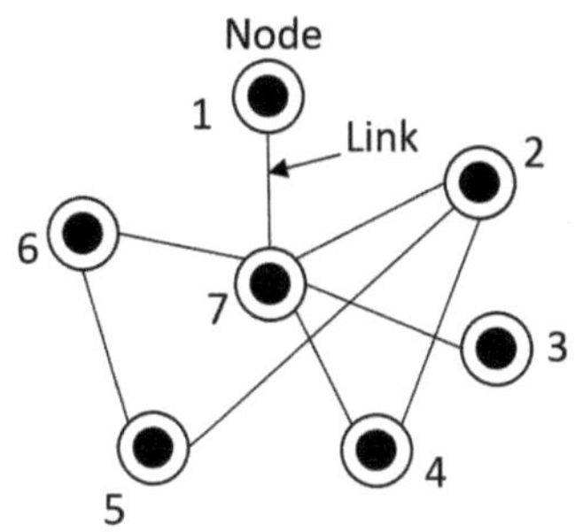

Fig. 1 Concept of a network

network shown in Fig. 1, consisting of a set of identical nodes connected by identical links, is perhaps the simplest form of network. This kind of network, however, is rarely seen in nature and society, since natural and social networks are often far more complex.

The origin of the concept of networks goes back to the works of Leonhard Euler, during the first half of the eighteenth century, on one of the most famous problems in mathematics, the Seven Bridges of Königsberg (Euler 1741), which laid the foundations of the now-popular *graph theory*. Since then, graph theory has seen many key theoretical developments, including topology (Listing 1848), trees (Cayley 1857), and random graph theory (Erdös and Rényi 1959), until the new science of complex networks emerged in the 1990s (Watts and Strogatz 1998).

While the concepts of topology, trees, and random graph theory have and continue to be applied for a wide range of networks, including those encountered in hydrology (especially river/channel networks), they have some important limitations. For example, the random graph theory, the most recent among the above three, assumes that all networks are wired randomly together. Such an assumption, however, is questionable for real networks, since order and determinism are inherent in real systems and networks. Advances in some other areas of complex systems science, which had revealed nonlinear deterministic dynamics, self-organization, and scale-invariance as inherent properties of complex systems (e.g., Lorenz 1963; Mandelbrot 1982; Bak 1996), also led to reconsideration of the assumption of random connections in complex networks. In addition, such concepts are also often not suitable to represent highly irregular, complex, large, and dynamically evolving networks, which are commonplace in reality.

All these led to a renewed and fresh perspective of the study of complex networks in the late 1990s (Watts and Strogatz 1998; Barabási and Albert 1999), under the *new science of networks*. Such studies also led to new discoveries about complex networks, such as small-world networks (Watts and Strogatz 1998), scale-free networks (Barabási and Albert 1999), network motifs (Milo et al. 2002), and community structure (Girvan and Newman 2002). Since then, the science of complex networks has found applications in many different fields. Further details about the science of complex networks and its applications can be found in, for example, Watts (1999), Barabási (2002), and Estrada (2012), among others.

3 Measures of Complex Networks

Within the context of complex networks, a large number of measures have been developed to study the network properties. Such measures include centrality, clustering, adjacency, distance, community structure, bipartivity, fragments, communicability, and global invariants, among others. There are also different sub-measures and methods. A brief description of some of these measures, especially those that have found applications in hydrology, is presented below.

3.1 Degree Centrality

Centrality is one of the most basic measures of a network. While the concept of centrality goes back to the studies of Bavelas (1948) and Leavitt (1951), Jeong et al. (2001) and Newman (2001a) were among the first to use it in the context of complex networks. Several centrality-based measures exist in the literature, including degree centrality, Katz centrality, eigenvector centrality, subgraph centrality, pagerank centrality, vibrational centrality, closeness centrality, betweenness centrality, and information centrality. However, the degree centrality has been one of the most widely used measures.

Degree centrality identifies whether a given node, say i in a network, is more central or more influential than another node in the network. The degree centrality of node i in a network of N nodes is defined as the number of first neighbors (or simply *neighbors*) of node i divided by the total number of possible neighbors (N–1) in the network. Let us consider a selected node i in a network of N nodes, having k_i links which connect it to k_i other nodes. In the network shown in Fig. 2 (left), for example, there are nine nodes (i.e., $N = 9$), with the node i having four links. Therefore, the four nodes corresponding to the four links are the *neighbors* of node

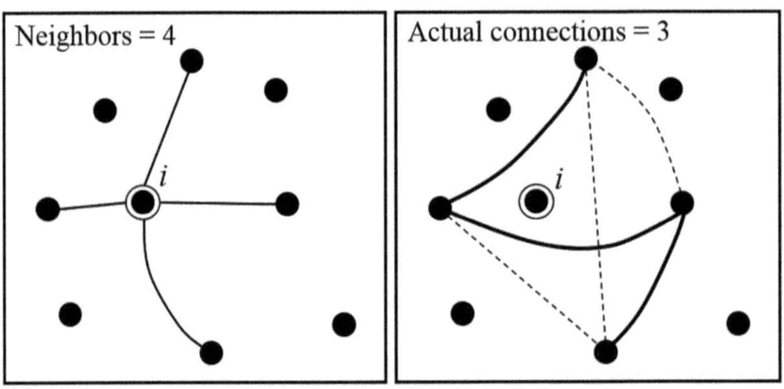

Fig. 2 Network connections and clustering coefficient calculation

i, while the total number of possible neighbors for node i is eight (i.e., N–1). The procedure is repeated for each and every node of the network.

3.2 Clustering Coefficient

One of the most fundamental properties of networks is their tendency to cluster. The concept of clustering has its origin in sociology (Wasserman and Faust 1994), but Watts and Strogatz (1998) were the first to use the concept in the context of complex networks. The tendency of a network to cluster is quantified by the clustering coefficient. There exist several definitions of clustering coefficient (e.g., Watts and Strogatz 1998; Barrat and Weigt 2000; Newman 2001b). However, the method proposed by Watts and Strogatz (1998), which measures the local density, is very widely used.

Let us consider first a selected node i in the network, having k_i links which connect it to k_i other nodes (Fig. 2, left). If the neighbors of the original node i were part of a cluster, there would be $k_i(k_i–1)/2$ links between them. Therefore, with four neighbors of the node i part of the cluster, there are $4(4–1)/2 = 6$ links in the *cluster* of node i (Fig. 2, right). The clustering coefficient of node i is then given by the ratio between the number E_i of links that actually exist between these k_i nodes (solid lines) and the total number $k_i(k_i–1)/2$ (all lines),

$$C_i = \frac{2E_i}{k_i\,(k_i - 1)} \tag{1}$$

The procedure is repeated for each and every node. The average of the clustering coefficients C_i's of all the individual nodes is the clustering coefficient of the whole network C.

The clustering coefficient of the individual nodes and of the entire network can be used to obtain important information about the type of network. For instance, a high clustering coefficient (close to 1.0) indicates a regular network, while a very low clustering coefficient (close to zero), with $C = p$ (where p is the probability of a pair of nodes being connected), indicates a random network. The clustering coefficient of a small-world network and a scale-free network is not only generally smaller than that of the regular network but also considerably larger than that of a comparable random network (i.e., having the same number of nodes and links).

3.3 Degree Distribution

In a network, several structural properties are related to adjacency relationships between nodes. There are many ways to measure the adjacency relations in a network. Some of these are node adjacency, degree distribution, degree–degree correlation, and link adjacency. Among these, the degree distribution is a particularly useful measure (e.g., Barabási and Albert 1999), especially for the identification of the type of network. A brief description is below.

Different nodes in a network may have different number of links. The number of links (k) of a node is called as *node degree*. The degree is an important characteristic of a node, as it allows one to derive many measurements for the network. The spread in the node degrees is characterized by a distribution function $p(k)$, which expresses the fraction of nodes in a network with degree k. This distribution, called *degree distribution*, is often a reliable indicator of the type of network. In a random graph, since the links are placed randomly, a majority of nodes have approximately the same degree, and close to the average degree $\langle k \rangle$ of the network. Therefore, the degree distribution of a completely random graph is a Poisson distribution with a peak at $P(\langle k \rangle)$. Similarly, depending upon the properties of networks, degree distribution can be Gaussian, exponential, or power-law (scale-free) or other.

Among these distributions, the power-law or scale-free distribution has attracted the most attention, as it has been found in many natural and social networks (e.g., Barabási and Albert 1999; Kim et al. 2004; Keller 2005; Clauset et al. 2010). The fractal or scale-free nature of natural and social systems and their ability to also self-organize themselves (e.g., Mandelbrot 1982; Bak 1996; Barnsley 2012) provide further credence to scale-free networks.

3.4 Average Shortest Path Length

A number of distance-based metrics have been developed for studying the topology of networks. These include the average shortest path length, resistance length, and generalized network length, among others. The average path length is considered as one of the most robust measures of network topology. The shortest path length of a node pair i and j is the number of links on the shortest path connecting the node pair. If the node pair is unconnected, then the value of the shortest path length is set to infinite. The average path length (L) of a network with N nodes is the average over all nodes of the shortest path between every combination of node pairs, and is given by:

$$L = \frac{1}{N\,(N-1)} \sum d_{ij} \tag{2}$$

where d_{ij} is the distance between pair i and j.

This definition for average shortest path length, however, diverges if there are unconnected nodes in the network, since the distance between such nodes is set to infinite (Costa et al. 2007). Consideration of only the connected node pairs avoids this divergence problem, but such also introduces a distortion for networks with many unconnected node pairs. The consequence of this is a small value of the average path length, which is expected only for networks with a high number of connections. A closely related measurement is the global efficiency (E) (Latora and Marchiori 2001):

$$E = \frac{1}{N(N-1)} \sum \frac{1}{d_{ij}} \tag{3}$$

where the sum takes all pairs of nodes into account.

The average shortest path length provides important information about the type of network. For example, regular networks, with their high clustering (i.e., stable), have long average path lengths (i.e., inefficient). On the other hand, random networks have short average path lengths (i.e., efficient) but have low clustering (i.e., unstable). Small-world networks have short path lengths and have high clustering and, therefore, are both stable and efficient.

3.5 Community Structure

In many complex networks, nodes cluster together into distinct groups. The properties of these groups are more or less independent of the properties of individual nodes and of the network as a whole. These groups are known as *communities*, and this kind of network structure is known as *community structure*. Identification of communities in a network is particularly useful, since nodes belonging to the same community are more likely to share properties and dynamics.

Many methods have been developed for community detection in networks. These methods include edge betweenness centrality (e.g., Newman and Girvan 2004), greedy optimization (Clauset et al. 2004), leading eigenvector (Newman 2006), walktrap (Pons and Latapy 2006), label propagation (Raghavan et al. 2007), and multilevel modularity optimization (Blondel et al. 2008), among others. Some of these methods rely on the modularity, Q, which quantifies the quality or strength of a community, defined as:

$$Q = \sum_{i} \left(e_{ii} - a_i^2\right) \tag{5}$$

where a_i is the fraction of links in the network which connect to community i and e_{ii} is the fraction of links that exist within the community. A high modularity community will have dense inter-community connections while the intra-community connections are sparse.

4 Hydrologic Applications

Applications of network-based concepts in hydrology started many decades ago, with studies dealing with the earlier concepts of topology, trees, and random graph theory, especially for river/channel networks; see Horton (1945), Strahler (1957), Shreve (1966, 1967, 1969), Scheidegger (1967), Smart (1970), Coffman and Turner (1971), Kirkby (1976), Smart and Werner (1976), Tokunaga (1978), Moon (1980), and Werner (1982) for some earlier studies. However, applications of the new concepts of complex networks have started only recently, and have thus far been limited to rainfall monitoring networks, streamflow monitoring networks, and river networks (e.g., Rinaldo et al. 2006; Zaliapin et al. 2010; Scarsoglio et al. 2013; Sivakumar and Woldemeskel 2014, 2015). Sivakumar (2015) has discussed, with some examples, the general relevance of the concepts of complex networks in hydrology, and also argued for the suitability of complex networks to serve as a generic theory for hydrology.

In what follows, a brief overview of the applications of network theory in hydrology is presented. Details on the applications of complex networks in closely related fields are available elsewhere and, therefore, are not reported here. For instance, there have been a number of applications to climate networks (e.g., Tsonis and Roebber 2004; Tsonis et al. 2006, 2008, 2011; Tsonis and Swanson 2008; Yamasaki et al. 2008; Donges et al. 2009; Donner et al. 2010, 2011; Paluš et al. 2011; Steinhaeuser et al. 2011, 2012; Donner and Donges 2012; Steinhaeuser and Tsonis 2014) and to virtual water networks (e.g., Konar et al. 2011, 2013; Suweis et al. 2011; Carr et al. 2012; Dalin et al. 2012, 2014; D'Odorico et al. 2012; Konar and Caylor 2013; Tamea et al. 2013, 2014; O'Bannon et al. 2014). A much broader overview of the applications of complex networks in geosciences is available in Phillips et al. (2015).

4.1 Connections in Rainfall Data

Malik et al. (2012) examined the spatial and temporal characteristics of extreme (summer) monsoonal rainfall over South Asia. They analyzed the daily gridded rainfall data (1951–2007) from the APHRODITE (Asian Rainfall Highly Resolved Observational Data Integration Towards the Evaluation of Water Resources) project. They employed several complex networks methods, including degree centrality, degree distribution, clustering coefficient, and closeness centrality. Subsequently, Boers et al. (2013) applied network concepts to investigate the spatial characteristics of extreme rainfall synchronicity of the South American Monsoon System (SAMS). They analyzed gridded daily rainfall (1998–2012) obtained from the Tropical Rainfall Measuring Mission (TRMM) 3B42 V7 satellite product. Scarsoglio et al. (2013) applied the complex networks-based methods to study the spatial dynamics of annual precipitation around the globe. They analyzed a 70-year long (January

1941–December 2010) gridded precipitation data from the Global Precipitation Climatology Center (GPCC) Database using several methods, including degree centrality, clustering coefficient, degree distribution, and shortest path length.

Sivakumar and Woldemeskel (2015) examined the spatial connections in rainfall in Australia using concepts of complex networks. They employed the clustering coefficient method and the degree distribution method to monthly rainfall observed over a period of 68 years (1940–2007) at 230 raingage stations across Australia. Their study was the first study that employed complex networks concepts to ground-measured rainfall data, as opposed to the above studies that used gridded rainfall data. They also considered the influence of rainfall correlation threshold (T) on network properties, by carrying out the analysis for seven different thresholds: 0.2, 0.3, 0.4, 0.5, 0.6, 0.7, and 0.8.

Figure 3a, for instance, shows the clustering coefficient results for $T = 0.5$. In these results, a clustering coefficient of 0.0 indicates that there are no actual connections (i.e., no link among the neighbors has correlation exceeding T), while NA indicates that there are no neighbors at all. The results indicate that even nearest stations have very different connectivity properties as part of a network and even distant stations have very similar connectivity properties. A comparison of the results obtained for different threshold values indicates significant changes in connectivity properties with respect to thresholds (figures not shown here; see Sivakumar and Woldemeskel (2015)). Figure 3b shows the actual connections for four selected stations (red circles) from four different regions in Australia, for $T = 0.5$. In these plots, for the station of interest (red circle), a blue circle indicates a station that has a correlation value exceeding the threshold, and a black circle indicates a station that has a correlation value smaller than the threshold. The lines

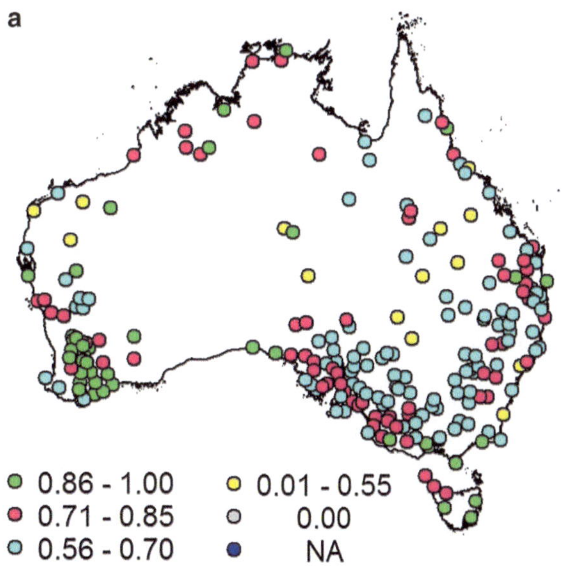

Fig. 3 Network analysis of monthly rainfall data from 230 stations in Australia: (**a**) clustering coefficient values; and (**b**) actual network connections for four selected rainfall stations. The results are for rainfall correlation threshold $T = 0.5$ (Adapted from Sivakumar and Woldemeskel 2015)

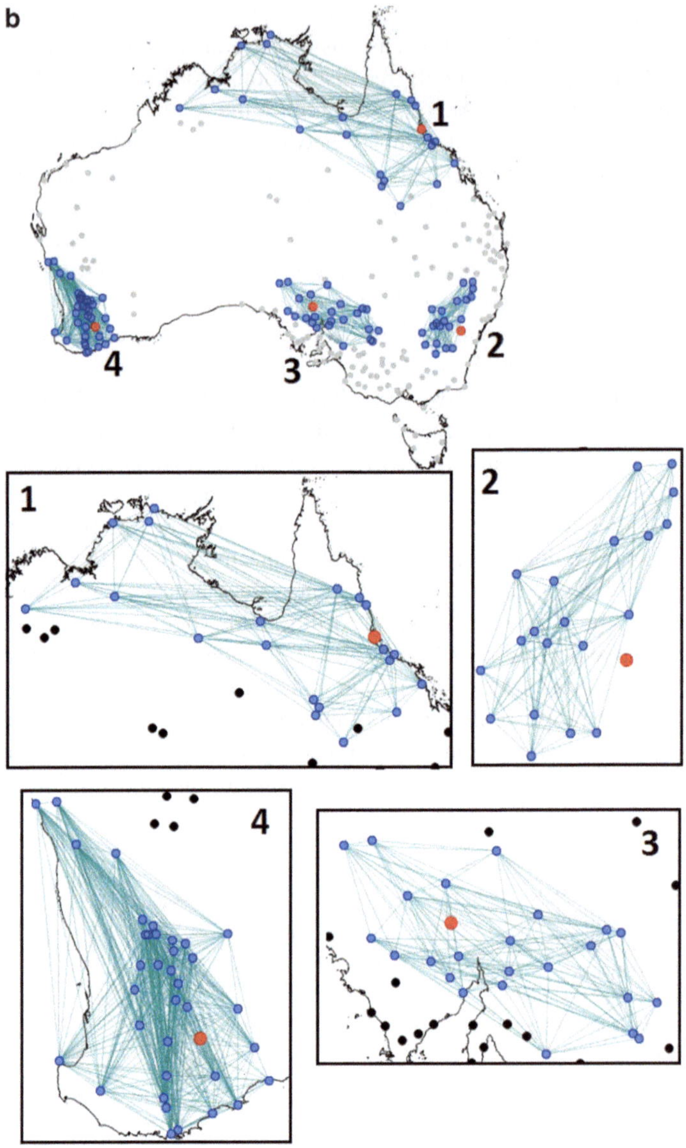

Fig. 3 (continued)

are the actual links among all the links available for the cluster of neighbors (blue circles only). The plots make it abundantly clear that geographic proximity alone does not always result in greater connectivity and that the actual connections can go for very large distances. The connections also reflect the distance and direction of the influencing factors, such as wind. These results are useful in identifying actual neighbors for the purpose of interpolation and other problems associated with spatial rainfall variability, and also have important implications for identifying optimal raingage density and locations. The results from the degree distribution method (see Sivakumar and Woldemeskel (2015)) suggest that the rainfall monitoring network is not a classical random network but more likely an exponentially truncated power-law network.

Jha et al. (2015) attempted to offer hydrologic explanation for the outcomes of network-based methods. They applied the clustering coefficient method to two different raingage networks in Australia: (1) monthly rainfall data over a period of 67 years (1937–2003) from 57 stations in Western Australia; and (2) daily rainfall over a period of 114 years (1890–2003) from 47 stations in the Sydney region. They interpreted the results in terms of topographic properties of raingage stations (latitude, longitude, and elevation) and statistical characteristics of rainfall data (mean, standard deviation, and coefficient of variation).

4.2 Connections in Streamflow Data

Sivakumar and Woldemeskel (2014) employed the concepts of complex networks to study the spatial connections in a streamflow monitoring network in the United States. They applied the degree centrality method and the clustering coefficient method to examine spatial connections in monthly streamflow observed over a period of 52 years (1951–2002) at 639 streamflow gaging stations. The study also investigated the influence of streamflow correlation threshold (T) on these network properties, by considering eight different thresholds: 0.3, 0.4, 0.5, 0.6, 0.7, 0.75, 0.8, and 0.85.

Figure 4a, for instance, shows the degree centrality results for $T = 0.75$. The results indicate that all stations have connections with less than 10% of the stations in the network, and more than one-fourth of the stations have connections to less than just 1% of the other stations. This suggests that only a small proportion of stations has considerable influence in the network, while a large proportion of stations has only very little or almost no influence. The threshold value has significant influence on degree centrality (results not shown here). Figure 4b shows the clustering coefficient results for $T = 0.75$. The results suggest that even nearest stations have significantly different connections and even distant stations have significantly similar connections. The results for the different thresholds (not shown here) suggest that the threshold value has significant influence on clustering coefficient. Figure 4c shows the actual connections for four selected stations (red circles), for $T = 0.75$. The plots clearly indicate that geographic proximity alone does not always result in greater connectivity and that the actual connections can go for very large distances. They also offer some other interesting observations. For instance, despite being in the same region, the two stations in the northwest exhibit significantly different connectivity characteristics, with one showing actual connections mainly within its geographic neighborhood, while the

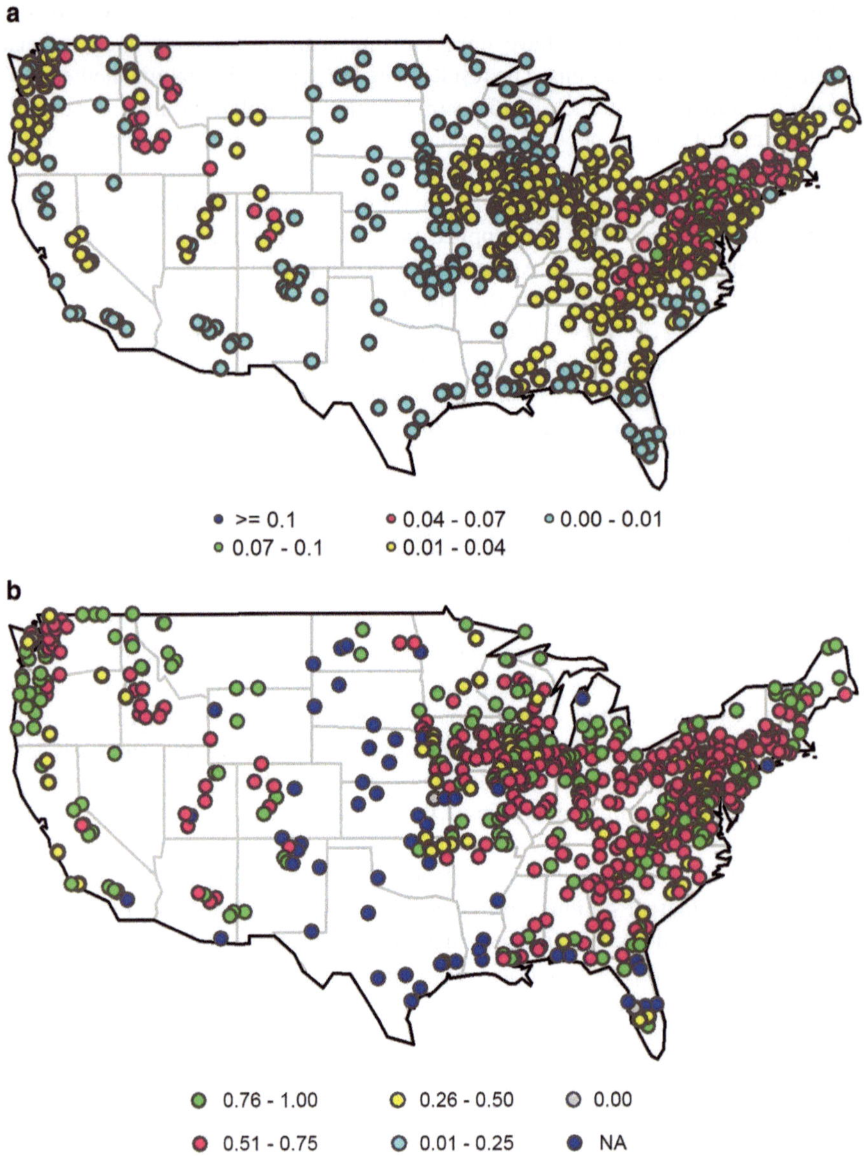

Fig. 4 Network analysis of monthly streamflow data from 639 stations in the United States: (**a**) degree centrality values; (**b**) clustering coefficient values; and (**c**) actual network connections for four selected streamflow stations. The results are for streamflow correlation threshold $T = 0.75$ (Adapted from Sivakumar and Woldemeskel 2014)

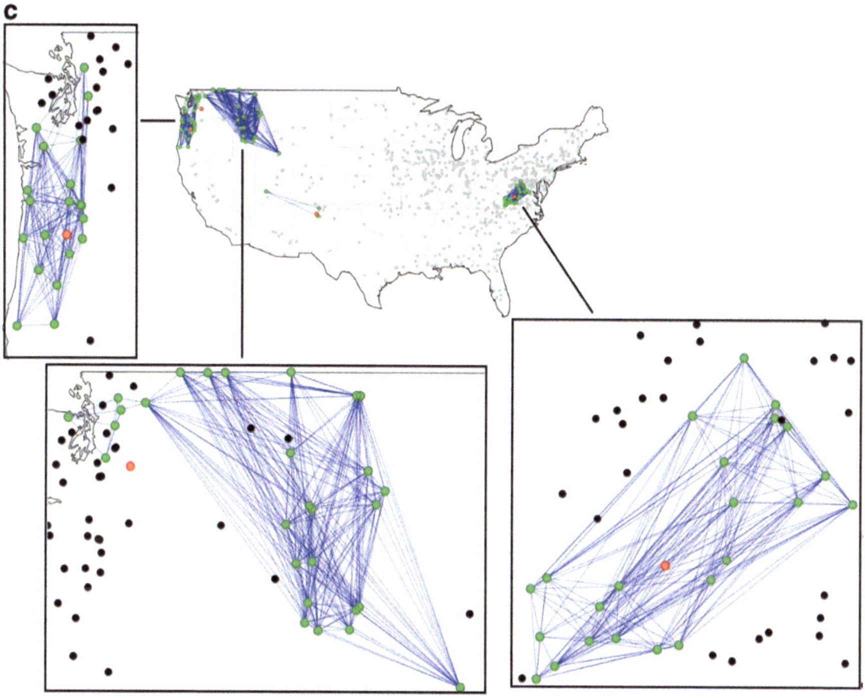

Fig. 4 (continued)

actual connections for the other station exist even, and indeed mainly only, well beyond its geographic neighborhood. These results have important implications for streamflow modeling, including interpolation and extrapolation of data and for predictions in ungaged basins.

Halverson and Fleming (2015) applied the concepts of complex networks to a network of 127 streamflow monitoring stations along the west coast of Canada. In addition to the investigation of whether regional streamflow hydrology might be quantitatively represented as a formal network, their study aimed at assessing whether the results from the network-based methods might offer important information as to the optimal design of streamflow monitoring systems. They employed a host of network-based methods, including clustering coefficient, degree distribution, average shortest path length, and betweenness.

Braga et al. (2016) employed the concepts of complex networks to study temporal dynamics of river flows. Their study involved mapping of the river flow time series into networks using the horizontal visibility graph (HVG). They analyzed daily river flow series over the period 1931–2012 from 141 different stations covering 53 Brazilian rivers. They then employed the degree distribution and clustering coefficient methods as well as their evolutive features to characterize the nature of the networks. They reported that the river discharges in several stations had evolved to become more or less correlated (displaying more or less complex internal network structures) over the years and attributed that behavior to changes in the climate system and other man-made phenomena. Serinaldi and Kilsby (2016) used the directed horizontal visibility graph (DHVG) to study the dynamics of streamflow fluctuations. The DHVG allows a clear visualization and quantification of persistence and irreversibility in flow series. They analyzed daily streamflow time series from 699 streamflow stations in the continental United States. They explored irreversibility by mapping the time series into ingoing, outgoing, and undirected graphs and comparing the corresponding degree distributions. They showed that the degree distributions do not decay exponentially, but tend to follow a subexponential behavior. They reported that the complexity of streamflow dynamics goes beyond a linear representation involving, for instance, the combination of linear processes with short and long range dependence.

4.3 River Networks and Processes

River networks were the first to be studied in hydrology using the concepts of complex networks. Rinaldo et al. (2006) introduced the concepts through a review of theoretical and observational developments on the form and function of natural networks in different contexts in different fields and their relevance in hydrology. They discussed the properties and dynamic origin of the scale-invariant structure of river patterns and its relation to optimal network selection. They argued that purely random or deterministic constructs are unsuitable for a proper description of river networks and other natural network forms. Applying degree distribution, clustering coefficient, and average path length methods, they reported the emergence of nontrivial phase transitions with increasing links-to-nodes ratios with different features like scale-free or small-world networks found for particular cases (Colizza et al. 2004).

Zaliapin et al. (2010) applied the concepts of network theory to study environmental transport problems in river networks, in particular the dynamic topology of directed trees. They described the static geometric structure of a drainage network by a tree (i.e., static tree) and introduced an associated dynamic tree that describes the transport along the static tree. Through application of connectivity concepts (e.g., hierarchical aggregation, clustering), they showed that dynamic trees are also self-similar just as their corresponding static trees, but that their properties differ systematically from those of the corresponding static trees. They also reported an unexpected phase transition in the dynamics of river networks (one from California and two from Italy) and demonstrated universal features of this transition.

Czuba and Foufoula-Georgiou (2014) proposed a simplified network-based predictive framework of sedimentological response in a basin. This framework incorporated network topology, channel characteristics, and transport-process dynamics to perform a nonlinear process-based scaling of the river-network width function to a time-response function. They developed the process-scaling formulation for transport of mud, sand, and gravel, using simplifying assumptions including neglecting long-term storage. They applied the methodology to the Minnesota River basin in the USA. They reported that the network topology and sediment-transport dynamics combined to produce a double-peaked response function for sand, suggesting that there exists a resonant frequency of sediment supply that could lead to an unexpected downstream amplification of sedimentological response. Czuba and Foufoula-Georgiou (2015) extended the above framework to study the internal dynamics of the basin for sediment transport. In particular, they examined how sediment is organized and where sediment accumulates due to the combined effects of river-network structure (topology and associated geometry) and transport dynamics (accounting for slopes, channel morphology, bed shear stress, grain size, etc.). They developed a dynamic connectivity framework and applied it to understand sand transport in the Greater Blue Earth River Network in the Minnesota River basin.

Other complex networks-based studies on river/delta networks include those by Rinaldo et al. (2014), Tejedor et al. (2015a, b, 2016), Masselink et al. (2017), and Passalacqua (2017), among others.

4.4 *Catchment Classification Studies*

Halverson and Fleming (2015) used the concept of community structure for classification of catchments along the west coast of Canada. They applied eight community structure methods (walk trap, fast greedy, leading eigenvector, edge betweenness, multilevel, label propagation, info map, and optimal) to daily streamflow data from a network of 127 monitoring stations. They also used the normalized mutual information (NMI) index to identify the consistency of these methods in classifying catchments. Their study yielded ten communities, each of which was defined by the combination of its median seasonal flow regime and geographic proximity to other communities. They found three of these communities holding 90% of the 127 stations considered. They also presented the representative unit hydrographs for the ten groups, and discussed the classification of stations in terms of elevation and drainage area. They proposed that an idealized sampling network should sample high-betweenness stations as well as small-membership communities which are, by definition, rare or undersampled relative to other communities, while at the same time retaining some degree of redundancy to maintain network robustness.

Fang et al. (2017) introduced the concept of complex networks and community structure to classify catchments in large-scale river basins. Considering the Mississippi River basin as a representative basin, they applied six community detection methods (edge betweenness, greedy algorithm, multilevel modularity optimization, leading eigenvector, label propagation, and walktrap) to daily streamflow data from a network of 1663 stations for catchment classification. They also examined the

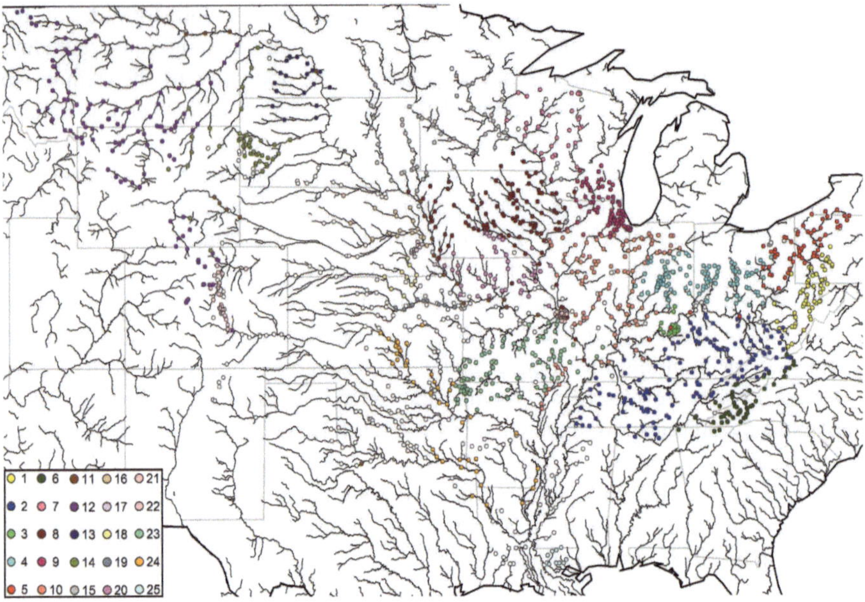

Fig. 5 Community structure concept for classification of catchments in the Mississippi River Basin: Results from multilevel optimization method for streamflow correlation threshold $T = 0.75$. A colored circle represents any particular community with at least ten stations, while an open circle represents any community with less than ten stations (From Fang et al. 2017, with permission)

influence of correlation threshold on classification. They also assessed the consistency among the methods in classifying catchments using the NMI index. They also attempted to explain the community formation in terms of river network/branching and some important catchment/flow properties (drainage area, elevation, mean flow, and flow coefficient of variation).

Figure 5 shows, for instance, the communities identified from the multilevel optimization method when $T = 0.75$, for all the 1663 stations. For better visualization, communities with at least ten stations are shown in colors, while communities with less than ten stations are presented as open circles. Among the important observations are: (1) there is a total of 245 communities among these 1663 stations; (2) 25 communities have at least ten stations (shown in colors and numbered in Fig. 5)—five of these have at least 100 stations and 11 have at least 50 stations; and (3) 172 communities have only one station. As seen, there is a great level of correspondence between the organization of the river network (both in terms of main stems of rivers and in terms of their further branchings) and the catchment communities across almost the entire region of the MRB studied. Similar results are also observed in the case of the other five methods.

The threshold value also has some notable influence in the formation of communities, i.e., size and number. The NMI index values for the six methods for different thresholds indicate a high degree of consistency in the performance among the methods, except for the leading eigenvector method at lower thresholds. Overall, the multilevel optimization method provides the greatest similarity in classification with the rest of the methods, while the leading eigenvector provides the greatest difference against the others. The results also reveal that only a few communities combine to represent a majority of the catchments, with the ten largest communities (roughly 4% of the total number of communities) representing almost two-thirds of the catchments.

5 Final Remarks

The new science of complex networks, a modern development in network theory, offers a new dimension for studying the structure, connections, and dynamics of large, complex, and dynamically evolving systems. The relevance and potential of the concepts of complex networks in hydrologic systems (see Sivakumar 2015) have resulted in some key early applications, including for studying connections in rainfall monitoring networks, streamflow monitoring networks, and river networks. With our improving knowledge, applications of the concepts of complex networks for catchment classification, and some other issues that are currently dominating hydrologic research, have also started to emerge. Despite their preliminary nature, these studies and their outcomes are certainly encouraging, both in advancing the science of complex systems and in studying hydrologic systems.

Looking at the relevance and potential of the concepts of complex networks in hydrology, there is no question that their applications will go a long way, both in breadth and depth. For instance, such concepts are highly useful for prediction of hydrologic systems, interpolation and extrapolation of hydrologic data, identification of optimal (density and locations of) hydrologic monitoring networks, downscaling outputs from global climate models for catchment-scale hydrologic analysis, study of water-energy-food-climate nexus, describing connections in human–water interactions, and formulation of an integrated framework for water planning and management. Therefore, as has been rightly argued (Sivakumar 2015), the science of complex networks has the potential to serve as a generic theory for hydrology. Indeed, it can go even further, to more accurately explain the interactions among all sub-systems of our Earth system. Such will be an immense contribution to the field of geosciences.

Acknowledgments This work is supported by the Australian Research Council (ARC) Future Fellowship grant (FT110100328). Bellie Sivakumar acknowledges the financial support from ARC through this Future Fellowship grant.

References

Bak, P. 1996. *How nature works: the science of self-organized criticality*, 212 pp. New York: Springer-Verlag.

Barabási, A.-L. 2002. *Linked: the new science of networks*. Cambridge, MA: Perseus.

Barabási, A.-L., and R. Albert. 1999. Emergence of scaling in random networks. *Science* 286: 509–512.

Barnsley, F.M. 2012. *Fractals everywhere*. Mineola, New York: Dover Publications.

Barrat, A., and M. Weigt. 2000. On the properties of small-world networks. *The European Physical Journal B* 13: 547–560.

Bavelas, A. 1948. A mathematical model for group structure. *Human Organization* 7: 16–30.

Blondel, V.D., J.-L. Guillaume, R. Lambiotte, and E. Lefebvre. 2008. Fast unfolding of communities in large networks. *Journal of Statistical Mechanics* 2008 (10): P10008.

Boers, N., B. Bookhagen, N. Marwan, J. Kurths, and J. Marengo. 2013. Complex networks identify spatial patterns of extreme rainfall events of the South American Monsoon System. *Geophysical Research Letters* 40: 1–7. doi:10.1002/grl.50681.

Braga, A.C., L.G.A. Alves, L.S. Costa, A.A. Ribeiro, M.M.A. de Jesus, A.A. Tateishi, and H.V. Ribeiro. 2016. Characterization of river flow fluctuations via horizontal visibility graphs. *Physica A* 444: 1003–1011.

Carr, J., P. D'Odorico, F. Laio, and L. Ridolfi. 2012. On the temporal variability of the virtual water network. *Geophysical Research Letters* 39: L06404. doi:10.1029/2012GL051247.

Cayley, A. 1857. On the theory of the analytical forms called trees. *Philosophical Magazine, Ser IV* 13 (85): 172–176.

Clauset, A., M.E.J. Newman, and C. Moore. 2004. Finding community structure in very large networks. *Physical Review E* 70 (6): P066111.

Clauset, A., C. Rohilla Shalizi, and M.E.J. Newman. 2010. Power-law distribution in empirical data. *SIAM Review* 51: 661–703.

Coffman, D.M., and A.K. Turner. 1971. Computer determination of the geometry and topology of stream networks. *Water Resources Research* 7 (2): 419–423.

Colizza, V., V.R. Banavar, A. Maritan, and A. Rinaldo. 2004. Network structures from selection principles. *Physical Review Letters* 92 (19): 198701.

Czuba, J.A., and E. Foufoula-Georgiou. 2014. A network-based framework for identifying potential synchronizations and amplifications of sediment delivery in river basins. *Water Resources Research* 50: 3826–3851.

———. 2015. Dynamic connectivity in a fluvial network for identifying hotspots of geomorphic change. *Water Resources Research* 51: 1401–1421.

Costa, L.F., F.A. Rodriguez, G. Traviesco, and P.R. Villas Boas. 2007. Characterization of complex networks: a survey of measurements. *Advances in Physics* 56 (1): 167–242.

Dalin, C., S. Suweis, M. Konar, N. Hanasaki, and I. Rodriguez-Iturbe. 2012. Modeling past and future structure of the global virtual water trade network. *Geophysical Research Letters* 39: L24402. doi:10.1029/2012GL053871.

Dalin, C., N. Hanasaki, H. Qui, D.L. Mauzerall, and I. Rodriguez-Iturbe. 2014. Water resources transfers through Chinese interprovincial and foreign food trade. *Proceedings of the National Academy of Sciences* 111 (27): 9774–9779.

Donges, J.F., Y. Zou, N. Marwan, and J. Kurths. 2009. Complex networks in climate dynamics. *European Physics Journal* 174: 157–179.

Donner, R.V., and J.F. Donges. 2012. Visibility graph analysis of geophysical time series: potentials and possible pitfalls. *Acta Geophysica* 60 (3): 589–623.

Donner, R.V., Y. Zou, J.F. Donges, N. Marwan, and J. Kurths. 2010. Recurrence networks—a novel paradigm for nonlinear time series analysis. *New Journal of Physics* 12 (3): 033025.

Donner, R.V., M. Small, J.F. Donges, N. Marwan, Y. Zou, R. Xiang, and R. Kurths. 2011. Recurrence-based time series analysis by means of complex network methods. *International Journal of Bifurcation and Chaos* 21 (4): 1019–1046.

Dooge, J.C.I. 1986. Looking for hydrologic laws. *Water Resources Research* 22 (9): 46S–58S.
D'Odorico, P., J. Carr, F. Laio, and L. Ridolfi. 2012. Spatial organization and drivers of the virtual water trade: A community-structure analysis. *Environmental Research Letters* 7: 034007. doi:10.1088/1748–9326/7/3/034007.
Erdös, P., and A. Rényi. 1959. On random graphs, I. *Publicationes Mathematicae Debrecen* 6: 290–297.
———. 1960. On the evolution of random graphs. *Publication of Institute of Hungarian Academy of Sciences* 5: 17–61.
Estrada, E. 2012. *The structure of complex networks: theory and applications*. Oxford University Press, New York, NY, USA.
Euler, L. 1741. Solutio problematis ad geometriam situs pertinentis. *Comment Academic Science Petropolitanae* 8: 128–140.
Fang, F., B. Sivakumar, and F.M. Woldemeskel. 2017. Complex networks, community structure, and catchment classification in a large-scale river basin. *Journal of Hydrology* 545: 478–493.
Girvan, M., and M.E. Newman. 2002. Community structure in social and biological networks. *Proceedings of the National Academy of Sciences* 99 (12): 7821–7826.
Gupta, V.K., I. Rodriguez-Iturbe, and E.F. Wood. 1986. *Scale problems in hydrology: runoff generation and basin response. Water science and technology library series*. Dordrecht, The Netherlands: Springer.
Halverson, M., and S. Fleming. 2015. Complex networks, streamflow, and hydrometric monitoring system design. *Hydrology and Earth System Sciences* 19: 3301–3318.
Horton, R.E. 1945. Erosional development of streams and their drainage basins: Hydrophysical approach to quantitative morphology. *Geological Society of America Bulletin* 56: 275–370.
Jeong, H., S. Mason, A.-L. Barabási, and Z.N. Oltvai. 2001. Lethality and centrality in protein networks. *Nature* 411: 41–42.
Jha, S.K., H. Zhao, F.M. Woldemeskel, and B. Sivakumar. 2015. Network theory and spatial rainfall connections: an interpretation. *Journal of Hydrology* 527: 13–19.
Keller, E.F. 2005. Revisiting 'scale-free' networks. *BioEssay* 27: 1060–1068.
Kim, D.-H., J.D. Noh, and H. Jeong. 2004. Scale-free trees: the skeletons of complex networks. *Physical Review E* 70: 046126.
Kirkby, M.J. 1976. Tests of the random network model, and its application to basin hydrology. *Earth Surface Processes and Landforms* 1 (3): 197–212.
Konar, M., and K.K. Caylor. 2013. Virtual water trade and development in Africa. *Hydrology and Earth System Sciences* 17: 3969–3982.
Konar, M., C. Dalin, S. Suweis, N. Hanasaki, A. Rinaldo, and I. Rodriguez-Iturbe. 2011. Water for food: the global virtual water trade network. *Water Resources Research* 47: W05520. doi:10.1029/2010WR010307.
Konar, M., Z. Hussein, N. Hanasaki, D.L. Mauzerall, and I. Rodriguez-Iturbe. 2013. Virtual water trade flows and savings under climate change. *Hydrology and Earth System Sciences* 17: 3219–3234.
Latora, V., and M. Marchiori. 2001. Efficient behavior of small-world networks. *Physical Review Letters* 87 (19): 198701.
Leavitt, H.J. 1951. Some effects of certain communication patterns on group performance. *Journal of Abnormal and Social Psychology* 46: 38–50.
Listing, J.B. 1848. *Vorstudien zur Topologie*, 811–875. Göttingen: Vandenhoeck und Ruprecht.
Lorenz, E.N. 1963. Deterministic nonperiodic flow. *Journal of the Atmospheric Sciences* 20 (2): 130–141.
Malik, N., B. Bookhagen, N. Marwan, and J. Kurths. 2012. Analysis of spatial and temporal extreme monsoonal rainfall over South Asia using complex networks. *Climate Dynamics* 39: 971–987.
Mandelbrot, B.B. 1982. *The fractal geometry of nature*. New York: W. H. Freeman and Company.
Masselink, R.J.H., T. Heckmann, A.J.A.M. Temme, N.S. Anders, H.P.A. Gooren, and S.D. Deesstra. 2017. A network theory approach for a better understanding of overland flow connectivity. *Hydrological Processes* 31: 207–220.

Milo, R., S. Shen-Orr, S. Itzkovitz, N. Kashtan, D. Chklovskii, and U. Alon. 2002. Network motifs: simple building blocks of complex networks. *Science* 298: 824–827.

Mishra, A.K., and P. Coulibaly. 2009. Developments in hydrometric network design: a review. *Reviews of Geophysics* 47: RG2001. doi:10.1029/2007RG000243.

Montanari, A., G. Young, H.H.G. Savenije, D. Hughes, T. Wagner, L.L. Ren, D. Koutsoyiannis, C. Cudennec, E. Toth, S. Grimaldi, G. Blöschl, M. Sivapalan, K. Beven, H. Gupta, M. Hipsey, B. Schaefli, B. Arheimer, E. Boegh, S.J. Schymanski, G. Di Baldassarre, B. Yu, P. Hubert, Y. Huang, A. Schumann, D.A. Post, V. Srinivasan, C. Harman, S. Thomson, M. Rogger, A. Viglione, H. McMillan, G. Characklis, G. Pang, and V. Belyaev. 2013. "Panta Rhei—Everything Flows": change in hydrology and society—The IAHS Scientific Decade 2013–2022. *Hydrological Sciences Journal* 58 (6): 1256–1275.

Moon, J.W. 1980. On the expected diameter of random channel networks. *Water Resources Research* 16 (6): 1119–1120.

Newman, M.E.J. 2001a. Scientific collaboration networks: II. Shortest paths, weighted networks, and centrality. *Physical Review E* 64: 016132.

———. 2001b. The structure of scientific collaboration networks. *Proceedings of the National Academy of Sciences USA* 98: 404–409.

———. 2006. Finding community structure in networks using the eigenvectors of matrices. *Physical Review E* 74: 036104.

Newman, M.E.J., and M. Girvan. 2004. Finding and evaluating community structure in networks. *Physical Review E* 69: 026113.

O'Bannon, C., J. Carr, D.A. Seekell, and P. D'Odorico. 2014. Globalization of agricultural pollution due to international trade. *Hydrology and Earth System Sciences* 18: 503–510.

Paluš, M., D. Hartman, J. Hlinka, and M. Vejmelka. 2011. Discerning connectivity from dynamics in climate networks. *Nonlinear Processes in Geophysics* 18 (5): 751–763.

Paola, C., E. Foufoula-Georgiou, W.E. Dietrich, M. Hondzo, D. Mohrig, G. Parker, M.E. Power, I. Rodriguez-Iturbe, V. Voller, and P. Wilcock. 2006. Toward a unified science of the Earth's surface: opportunities for synthesis among hydrology, geomorphology, geochemistry, and ecology. *Water Resources Research* 42: W03S10. doi:10.1029/2005WR004336.

Passalacqua, P. 2017. The Delta Connectome: A network-based framework for studying connectivity in river deltas. *Geomorphology* 277: 50–62.

Phillips, J.D., W. Schwanghart, and T. Heckmann. 2015. Graph theory in geosciences. *Earth-Science Reviews* 143: 147–160.

Pons, P., and M. Latapy. 2006. Computing communities in large networks using random walks. *Journal of Graph Algorithms and Applications* 10 (2): 191–218.

Raghavan, U.N., R. Albert, and S. Kumara. 2007. Near linear time algorithm to detect community structures in large-scale networks. *Physical Review E* 76: 036106.

Rinaldo, A., J.R. Banavar, and A. Maritan. 2006. Trees, networks, and hydrology. *Water Resources Research* 42: W06D07. doi:10.1029/2005WR004108.

Rinaldo, A., R. Rigon, J.R. Banavar, A. Maritan, and I. Rodriguez-Iturbe. 2014. Evolution and selection of river networks: Statics, dynamics, and complexity. *Proceedings of the National Academy of Sciences USA* 111 (7): 2417–2424.

Salas, J.D., J.W. Delleur, V. Yevjevich, and W.L. Lane. 1995. *Applied modeling of hydrologic time series*. Littleton, Colorado: Water Resources Publications.

Scarsoglio, S., F. Laio, and L. Ridolfi. 2013. Climate dynamics: a network-based approach for the analysis of global precipitation. *PLoS One* 8 (8): e71129. doi:10.1371/journal.pone.0071129.

Scheidegger, A.E. 1967. On the topology of river nets. *Water Resources Research* 3 (1): 103–106.

Serinaldi, F., and C.G. Kilsby. 2016. Irreversibility and complex network behavior of stream flow fluctuations. *Physica A* 450: 585–600.

Shreve, R.L. 1966. Statistical law of stream numbers. *Journal of Geology* 74: 17–37.

———. 1967. Infinite topologically random channel networks. *Journal of Geology* 75: 178–186.

———. 1969. Stream lengths and basin areas in topologically random channel networks. *Journal of Geology* 77: 397–414.

Sivakumar, B. 2008. Dominant processes concept, model simplification and classification framework in catchment hydrology. *Stochastic Environmental Research and Risk Assessment* 22 (6): 737–748.

———. 2011a. Global climate change and its impacts on water resources planning and management: assessment and challenges. *Stochastic Environmental Research and Risk Assessment* 25 (4): 583–600.

———. 2011b. Water crisis: from conflict to cooperation–an overview. *Hydrological Sciences Journal* 56 (4): 531–552.

———. 2015. Networks: a generic theory for hydrology? *Stochastic Environmental Research and Risk Assessment* 29: 761–771.

———. 2017. *Chaos in hydrology: bridging determinism and stochasticity*, 394 pp. Dordrecht: Springer Science+Business Media.

Sivakumar, B., and F.M. Woldemeskel. 2014. Complex networks for streamflow dynamics. *Hydrology and Earth System Sciences* 18: 4565–4578.

———. 2015. A network-based analysis of spatial rainfall connections. *Environmental Modelling and Software* 69: 55–62.

Smart, J.S. 1970. Use of topologic information in processing data for channel networks. *Water Resources Research* 6 (3): 932–936.

Smart, J.S., and C. Werner. 1976. Applications of the random model of drainage composition. *Earth Surface Processes and Landforms* 1: 219–233.

Steinhaeuser, K., and A.A. Tsonis. 2014. A climate model intercomparison at the dynamics level. *Climate Dynamics* 42: 1665–1670.

Steinhaeuser, K., N.V. Chawla, and A.R. Ganguly. 2011. Complex networks as a unified framework for descriptive analysis and predictive modeling in climate science. *Statistical Analysis and Data Mining* 4: 497–511.

Steinhaeuser, K., A.R. Ganguly, and N.V. Chawla. 2012. Multivariate and multiscale dependence in the global climate system revealed through complex networks. *Climate Dynamics* 39: 889–895.

Strahler, A.N. 1957. Quantitative analysis of watershed geomorphology. *Eos, Transactions American Geophysical Union* 38: 913–920.

Suweis, S., M. Konar, C. Dalin, N. Hanasaki, A. Rinaldo, and I. Rodriguez-Iturbe. 2011. Structure and controls of the global virtual water trade network. *Geophysical Research Letters* 38: L10403. doi:10.1029/2011GL046837.

Tamea, S., P. Allamano, J. Carr, P. Claps, F. Laio, and L. Ridolfi. 2013. Local and global perspectives on the virtual water trade. *Hydrology and Earth System Sciences* 17: 1205–1215.

Tamea, S., J.A. Carr, F. Laio, and L. Ridolfi. 2014. Drivers of the virtual water trade. *Water Resources Research* 50: 17–28.

Tejedor, A., A. Longjas, I. Zaliapin, and E. Foufoula-Georgiou. 2015a. Delta channel networks: 1. A graph-theoretic approach for studying connectivity and steady state transport on deltaic surfaces. *Water Resources Research* 51: 4019–4045.

———. 2015b. Delta channel networks: 2. Metrics of topologic and dynamic complexity for delta comparison, physical inference, and vulnerability assessment. *Water Resources Research* 51: 3998–4018.

Tejedor, A., A. Longjas, R. Caldwell, D.A. Edmonds, I. Zaliapin, and E. Foufoula-Georgiou. 2016. Quantifying the signature of sediment composition on the topologic and dynamic complexity of river delta networks and inferences toward delta classification. *Geophysical Research Letters* 43: 3280–3287.

Tokunaga, E. 1978. Consideration on the composition of drainage networks and their evolution. *Department of Geography/Tokyo Metropolitan University* 13: 1–27.

Tsonis, A.A., and P.J. Roebber. 2004. The architecture of the climate network. *Physica A* 333: 497–504.

Tsonis, A.A., and K.L. Swanson. 2008. Topology and predictability of El Niño and La Niña networks. *Physical Review Letters* 100: 228502.

Tsonis, A.A., K.L. Swanson, and P.J. Roebber. 2006. What do networks have to do with climate? *Bulletin of the American Meteorological Society* 87 (5): 585–595.

Tsonis, A.A., K.L. Swanson, and G. Wang. 2008. Estimating the clustering coefficient in scale-free networks on lattices with local spatial correlation structure. *Physica A* 387: 5287–5294.

Tsonis, A.A., G. Wang, K.L. Swanson, F.A. Rodrigues, and L.F. Costa. 2011. Community structure and dynamics in climate networks. *Climate Dynamics* 37: 933–940.

Wasserman, S., and K. Faust. 1994. *Social network analysis*. Cambridge: Cambridge University Press.

Watts, D.J. 1999. *Small worlds: the dynamics of networks between order and randomness*. Princeton: Princeton University Press.

Watts, D.J., and S.H. Strogatz. 1998. Collective dynamics of 'small-world' networks. *Nature* 393: 440–442.

Werner, C. 1982. Analysis of length distribution of drainage basin parameter. *Water Resources Research* 18 (4): 997–1005.

Yamasaki, K., A. Gozolchiani, and S. Havlin. 2008. Climate networks around the globe are significantly affected by El Niño. *Physical Review Letters* 100: 228501.

Young, P.C., and M. Ratto. 2009. A unified approach to environmental systems modeling. *Stochastic Environmental Research and Risk Assessment* 23: 1037–1057.

Zaliapin, I., F. Foufoula-Georgiou, and M. Ghil. 2010. Transport on river networks: a dynamic tree approach. *Journal of Geophysical Research* 115: F00A15. doi:10.1029/2009JF001281.

Convergent Cross Mapping: Theory and an Example

Anastasios A. Tsonis, Ethan R. Deyle, Hao Ye, and George Sugihara

Abstract In this review paper we present the basic principles behind convergent cross mapping, a new causality detection method, as well as an example to demonstrate it.

Keywords Causality • Nonlinearity • Dynamical systems

1 Convergent Cross Mapping

Convergent cross mapping (CCM) is a powerful new methodological approach that can help distinguish causality from spurious correlation in time series from dynamical systems (Sugihara et al. 2012). The technique is based on the idea that causation can be established if states of the causal variable can be recovered from time series of the affected variable. For example, if past sea surface temperatures can be estimated from time series of sardine abundance, temperature had a measurable and recoverable influence on the population dynamics of sardines (Sugihara et al. 2012). The idea is based on empirical dynamics (EDM) (Sugihara et al. 2012; Sugihara and May 1990; Sugihara 1994) and a theorem proven by Takens (1981) for manifolds and generalized by Sauer et al. (1991) for general non-euclidean attractors and extended to accommodate stochasticity by Stark et al. (2003). The result is generic and broadly states that the essential information of a multidimensional dynamical system is retained in the times series of any single variable of that system.

Thus, CCM uses Takens' idea to detect if two variables belong to the same dynamical system. A brief video introduction is available at http://tinyurl.com/EDM-intro.

A.A. Tsonis (✉)
Department of Mathematical Sciences, Atmospheric Sciences Group, University of Wisconsin - Milwaukee, Milwaukee, WI 53201, USA

Hydrologic Research Center, San Diego, CA, USA
e-mail: aatsonis@uwm.edu

E.R. Deyle • H. Ye • G. Sugihara
Scripps Institution of Oceanography, University of California San Diego, La Jolla, CA 92093, USA

A.A. Tsonis (ed.), *Advances in Nonlinear Geosciences*,
DOI 10.1007/978-3-319-58895-7_27

Consider two time series of length L, $\{X\} = \{X(1), X(2), \ldots, X(L)\}$ and $\{Y\} = \{Y(1), Y(2), \ldots, Y(L)\}$. We begin by forming the lagged-coordinate vectors $\underline{x}(t) = <X(t), X(t-\tau), X(t-2\tau), \ldots, X(t-(E-1)\tau)>$ for $t = 1+(E-1)\tau$ to $t = L$. This set of vectors is the "reconstructed manifold" or "shadow manifold" $\boldsymbol{M_X}$. Note that the term "shadow manifold" includes attractors defined on fractal sets. To generate a cross-mapped estimate of $Y(t)$, denoted by $\hat{Y}(t)|\boldsymbol{M_X}$, we begin by locating the contemporaneous lagged-coordinate vector on $\boldsymbol{M_X}$, $\underline{x}(t)$, and find its $E + 1$ nearest neighbors. Note that $E + 1$ is the minimum number of points needed for a bounding simplex in an E-dimensional space (see note on simplex projection below). Next, denote the time indices (from closest to farthest) of the $E + 1$ nearest neighbors of $\underline{x}(t)$ by $t_1, \ldots t_{E+1}$. These time indices corresponding to nearest neighbors to $\underline{x}(t)$ on $\boldsymbol{M_X}$ are used to identify points (neighbors) in Y (a putative local neighborhood on $\boldsymbol{M_Y}$) to estimate $Y(t)$ from a locally weighted mean of the $E + 1$ $Y(t_i)$ values.

$$\widehat{Y}(t)\Big|M_X = \sum w_i Y(t_i) \quad i = 1 \ldots E + 1$$

where w_i is a weighting based on the distance between $x(t)$ and its ith nearest neighbor on $\boldsymbol{M_X}$ and $Y(t_i)$ are the contemporaneous values of Y. The weights are determined by

$$w_i = u_i / \sum u_j \quad j = 1 \ldots E + 1$$

where

$$u_i = \exp\left\{-d\left[\underline{x}(t), \underline{x}(t_i)\right] / d\left[\underline{x}(t), \underline{x}(t_1)\right)\right\}$$

and $d[\underline{x}(s), \underline{x}(t)]$ is the Euclidean distance between two vectors. Cross mapping from Y to X is defined analogously.

Effectively, if variable X is influencing Y, then causality is established if states of the causal variable X can be recovered from the time series history of Y. Simply put, CCM measures the extent to which the historical record of the affected variable Y (or its proxies) reliably estimates states of a causal variable X (or its proxies), which is quantified by calculating the correlation coefficient ρ between predicted and observed X. If the skill of cross mapping increases with the length of the time series, a direct or indirect causal effect of X on Y can be inferred. The relative level to which predictive skill converges ("CCM skill" hereafter) can be viewed as an estimator of the strength of the causal link. Convergence occurs with additional data as the underlying attractor manifold becomes denser, and nearest neighbors get closer. Figure 1 shows schematically the above procedure.

Significance of CCM is most easily evaluated by looking at the CCM skill with the largest possible library. Although CCM skill is often quantified using Pearson's correlation coefficient, there are a number of problems that can arise from interpreting the significance of CCM predictions using the standard (linearly derived) critical values of r. Notably, the traditional confidence intervals make

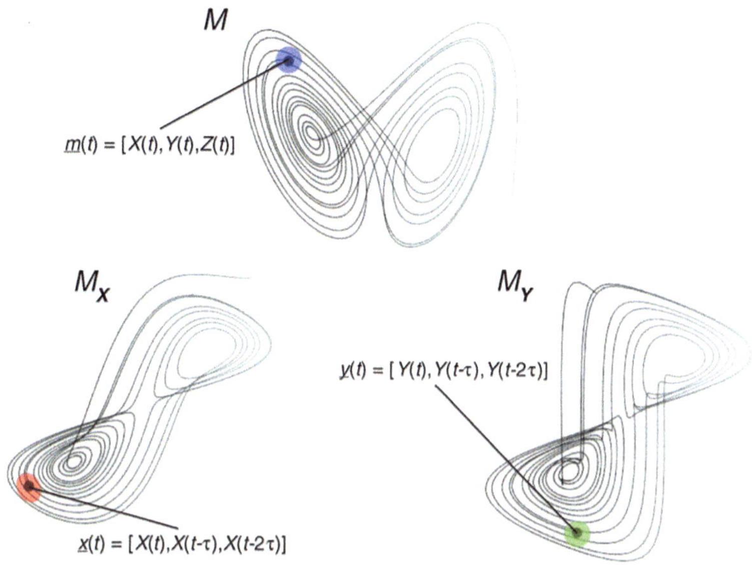

Fig. 1 Convergent cross mapping (CCM) tests for correspondence between shadow manifolds. This example based on the canonical Lorenz system (a coupled system in X, Y, and Z; Eq. S7 without V) shows the attractor manifold for the original system (M) and two shadow manifolds, M_X and M_Y, constructed using lagged-coordinate embeddings of X and Y, respectively (lag $= \tau$). Because X and Y are dynamically coupled, points that are nearby on M_X (e.g., within the *red ellipse*) will correspond temporally to points that are nearby on M_Y (e.g., within the *green circle*). That is, the points inside the *red ellipse* and *green circle* will have corresponding time indices (values for t). This enables us to estimate states across manifolds using Y to estimate the state of X and vice versa using nearest neighbors. With longer time series, the shadow manifolds become denser and the neighborhoods (ellipses of nearest neighbors) shrink, allowing more precise cross map estimates

assumptions about independence of observations and normally distributed values that are unlikely to be met when studying time series with nonlinear dynamics. The more rigorous approach to determining significance of CCM is to use surrogate time series to simulate null distributions. The simplest approach is to create null time series by randomly permuting the time indices of predictee time series X. The distribution generated encapsulates the likelihood that a random variable (with the same distribution of values as X) would produce a CCM skill of a given amount.

Different surrogate approaches can be useful in testing null hypotheses of specific relevance to the application at hand. In the case below, phase randomized surrogates (Ebisuzaki 1997) are used due to the relatively strong spectral character of the time series. If there is strong linear correlation between variables, then the surrogate time series can be randomized together so that the pairwise linear correlation is preserved but the dynamics of the time series are destroyed. Finally, if there is a strong periodicity, e.g., due to the annual cycle in climate, then surrogates can be designed to distinguish true CCM predictability from shared periodic forcing.

CCM can identify bidirectional causality when variables are mutually coupled (the primary case covered by Takens (1981)), as well as unidirectional causality when X influences Y but Y has no effect on X–as occurs when X is an external forcing variable. As explained in reference (Sugihara et al. 2012), CCM applies in dynamic systems, in contrast to the celebrated Granger causality (Granger 1969) framework (see Appendix 1) which is aimed at purely stochastic systems that exhibit linear "separability" (independence between variables) in which case Taken's theorem does not apply. Specifically, CCM addresses cases not covered by Granger involving interdependent (nonlinear) dynamic systems–i.e., cases where Granger's assumption of separable piece-wise independence is explicitly violated.

2 The S-Map Test for Nonlinear Dynamics

It is a good practice to establish presence of nonlinear dynamics in the time series that are tested for causality. To determine whether a time series reflects linear or nonlinear processes we compare the out-of-sample forecast skill of a linear model versus an equivalent nonlinear model. To do this, we apply a two-step procedure: (1) we use simplex projection (Sugihara and May 1990) to identify the best embedding dimension, and (2) we use this embedding in the S-map procedure (Sugihara 1994) to assess the nonlinearity of the time series.

S-maps are an extension of standard linear autoregressive models, however, with S-maps the jacobian coefficients depend on the location of the predictee $\boldsymbol{y}_t$ in an E-dimensional embedding. Here, new linear autoregressive coefficients (the jacobian elements) are recalculated (from the library of a predictant set $\boldsymbol{X}$) by singular value decomposition (SVD) for each new prediction. Thus, "S"-maps involve "sequentially" recalculated jacobians (linear approximations) as the system travels along its attractor. In this calculation, the weight given to each vector in the library depends on how close that vector $\boldsymbol{x}_t$ is to the predictee $\boldsymbol{y}_t$. The extent of this weighting is determined by the parameter θ.

As above, we generate an E-dimensional embedding from points in the library using lagged coordinates to obtain an embedded time series with vectors $\boldsymbol{x}_t \in \mathbb{R}^{E+1}$, where $x_t(1) = 1$ is the constant term in the solution of Eq. (2) below. Let the time series observation in the prediction set T_p time steps forward be $Y_{t+Tp}(1) = Y(t)$.

Then the forecast for Y_t is

$$\widehat{Y}_t = \sum_{j=0}^{E} C_t(j) X_t(j) \tag{1}$$

For our analysis, we chose $T_P = 1$. For each E-dimensional predictee vector $\boldsymbol{y}_t$, $\boldsymbol{C}$ is the jacobian matrix solved by SVD using the library set as follows:

$$\boldsymbol{B} = \boldsymbol{A}\boldsymbol{C}, \tag{2}$$

where $B_i = w(\|\boldsymbol{x}_i - \boldsymbol{y}_t\|)y_i$, $A_{ij} = w(\|\boldsymbol{x}_i - \boldsymbol{y}_t\|)x_i(j)$, and $w(d) = e^{-\theta d_{it}/\bar{d}}$, $\theta \geq 0$, d_{it} is the distance between $\boldsymbol{y}_t$ and the ith neighbor vector $\boldsymbol{x}_i$ in the library embedding, and the scale vector, $\bar{d}$, is the average distance between neighbors in the library. Note that $\boldsymbol{A}$ has dimension $n \times (E + 1)$, where n = size of the library. Again, a different map is generated for each forecast, with the weightings in each map depending on the location of the predictee in the E-dimensional state-space. This weighting procedure is governed by the tuning parameter θ, where $\theta = 0$ gives a global linear map, and increasing values of θ give increasingly local or nonlinear mappings. When $\theta = 0$, all vectors are more or less weighted equally so a single (global) linear map can be used for all predictions. In the case where $\theta > 0$, vectors closest to the predictee in state-space are weighted more heavily in the SVD solution. Such forecasts emphasize local information in the library set, and are therefore nonlinear.

A note on simplex projection and on determining embedding dimension: Simplex projection is a nearest-neighbor forecasting algorithm (Stark et al. 2003) that involves tracking the forward trajectory of nearby points in a lag coordinate embedding. To determine E for computing $\boldsymbol{M}_X$ and for the S-map analysis, an exploratory series of embedding dimensions (E) are used to discover the value of E that best unfolds the attractor and minimizes the trajectory crossings or singularities. Thus, the best E is the dimension that gives the best prediction skill, and is the value used in the S-map procedure as well as in cross mapping.

3 An Example

The example deals with causality between galactic cosmic rays (CR) and global temperature. The basic principles behind a possible connection between galactic cosmic rays and global temperature are as follows: It has been known since the invention of the cloud chamber in 1911 by Charles Thomson Rees Wilson that ionizing radiation leads to atmospheric cloud nucleation. While the prime source of ionizing radiation in the global troposphere is CR, the flux of CR reaching the troposphere depends on the solar wind. The solar wind is a stream of ionized gases that blows outward from the Sun, and its intensity varies strongly with the level of surface activity on the Sun. The Earth's magnetic field shields the planet from much of the solar wind, deflecting that wind like water around the bow of a ship. When the solar activity is great, solar wind is strong, swiping away cosmic rays arriving at the top of the atmosphere. These cosmic rays are hypothesized to impact cloud formation, cloudiness, and therefore global temperature. The net radiative effect of cloudiness depends on the difference between incoming solar radiation and outgoing longwave radiation (OLR). Increased cloudiness in the upper troposphere reduces OLR thereby resulting in warming of the planet. Increased cloudiness in the lower troposphere causes less incoming radiation and therefore cooling of the planet. Data suggest (Stark et al. 2003) that the amount of CR is positively correlated with the amount of low-level clouds, but has no effect on middle- or high-level clouds.

While this is still an open question (see also references Granger 1969; Marsh and Svensmark 2000; Rawal et al. 2013), the reduction in CR flux in times of high solar activity is hypothesized to result in less cloud nucleation, fewer cloud condensation nuclei (CCN), and consequently reduced low-level cloud amounts. This, in turn, leads to a higher solar radiation flux at the Earth's surface, and warmer temperatures. Conversely, weaker solar wind results in cooler temperatures. The actual chemical processes and reactions involved in this problem are complex, but a growing body of experimental and theoretical work has uncovered a chemical pathway by which CR ionization may increase nucleation rates to levels appropriate for CCN (see references Harrison et al. 2011; Kikby et al. 2011; Svensmark et al. 2007; Enghoff et al. 2011; Svesmark et al. 2013; Zhang et al. 2012; Yu 2005; Duplissy et al. 2010, and the references therein). This suggests a superficially simple network linking the Sun, CR, and global climate, with the interaction between the Sun and CR having a potential influence on the climate system. However, reasonable this may be, as described in a 2006 review (Foukal et al. 2006), "The suggested mechanisms are, however, too complex to evaluate meaningfully at present."

To date, attempts at finding observational evidence for the link between solar activity/CR and climate have relied on simple linear cross-correlation or spectral coherence analysis (Stark et al. 2003; Neff et al. 2001; Bond et al. 2001; Svensmark 1998; Usoskin et al. 2004; Shaviv 2003). Although suggestive, it is well known that such statistical analyses cannot actually establish causation and indeed can be highly misleading in moderate to weakly coupled nonlinear dynamic systems, where "mirage correlations" (spurious correlations that come and go and even change sign) are common (Sugihara et al. 2012). Indeed, the singular case where correlation is valid in nonlinear dynamic systems would require strong coupling (the so-called synchrony) between the solar-mediated CR forcing and the climate system, which is unlikely on decadal time scales due to embedded nonlinearities in the climate system. Thus, in order to test for causal linkage between cosmic rays and global temperature, we apply CCM.

For this analysis we use the aa index (Nayaud 1972) as the cosmic rays proxy. This index is a well-documented proxy that characterizes magnetic activity resulting from the interaction between solar wind and Earth's magnetic field (stronger solar wind ➔ stronger magnetic disturbances ➔ higher aa index). More discussion on this issue follows later. The global temperature (GT) record we used is HadCRUT3 set of UK's Met Office. There is increasing uncertainty prior to 1900 in both data series, so we confine our analysis to the post-1900 period and use yearly averages. The chosen interval represents a compromise between noise in the data and sample size. Both time series exhibit a positive trend, however, the GT warming trend is more distinct and dominates the much smaller superimposed interannual fluctuations (Fig. 2a). This is in contrast to the CR record where large interannual variation dominates the twentieth century signal.

To verify that we are dealing with a nonlinear dynamic system rather than a purely stochastic one, we analyze each time series separately using S-maps presented above. Evidence for nonlinear dynamics is demonstrated if forecast performance improves as the S-map model is tuned toward nonlinear solutions

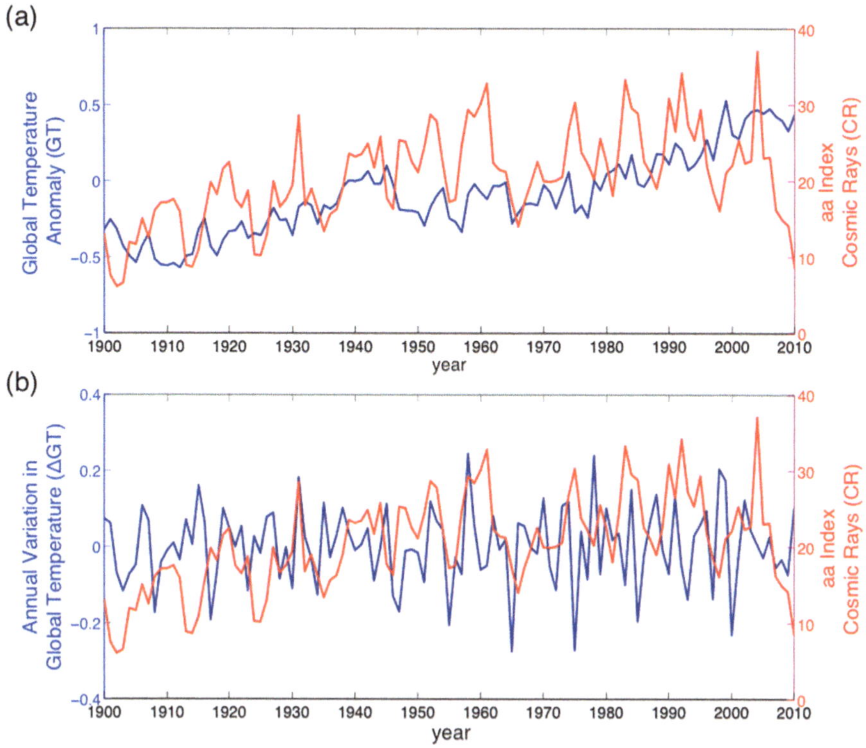

Fig. 2 (**a**) Annual values for the aa index (CR proxy) and normalized GT, and (**b**) annual variations (first-differences) in GT (ΔGT) and CR. Despite a correlation between the CR and raw GT time series ($\rho = 0.38$), there is no measureable dynamic causality on the century-long time scale (Fig. 7). However, on the annual time scale, even though CR and ΔGT are not correlated ($\rho = 0.02$), evidence suggests that are dynamically coupled (Figs. 4 and 5)

($\theta > 0$, where θ is the nonlinear tuning parameter). The results presented in Fig. 3 show that while CR and ΔGT both exhibit evidence for nonlinear dynamics, the raw GT time series does not. It is likely that evidence for nonlinearity is masked by the strong linear trend dominating the raw GT record over the twentieth century. However, bottom Fig. 3 shows that the nonlinear dynamics in temperature variability for this period can be unmasked by taking first differences of GT. That is, while the strong overall twentieth century warming trend (linear trend $\rho = 0.8$) lacks the signature of nonlinear dynamics, year-to-year temperature variability (ΔGT) shows evidence of nonlinear dynamics operating on the annual time scale. The S-map test also demonstrates nonlinear dynamics in the CR (aa) record where the relatively rapid non-trend fluctuations are large compared to the secular increase. Thus the best result with CCM (tests for nonlinear dynamic coupling between variables) should be expected when testing for causality between first differenced GT (year-to-year temperature variability or ΔGT) and the raw CR time series.

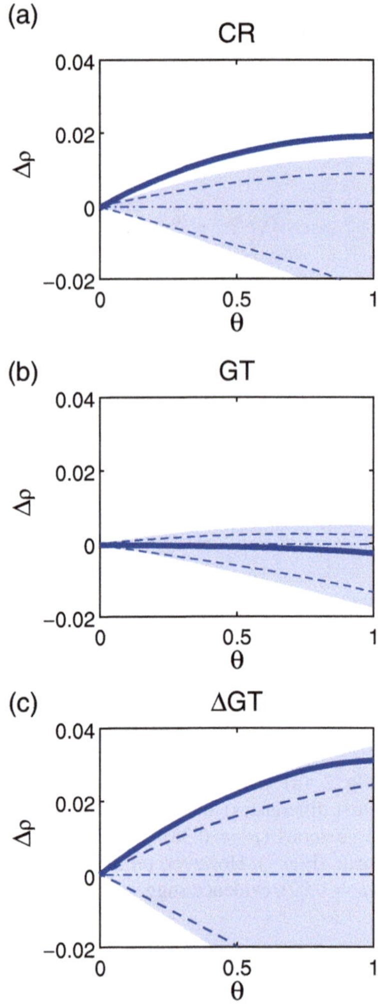

Fig. 3 The S-map analysis of (**a**) the CR time series (aa proxy), (**b**) the GT time series, and (**c**) the first-differenced ΔGT time series. $\Delta\rho$ is the difference in the correlation between actual and predicted values between a linear model (global linear map) and an equivalent nonlinear model (local or nonlinear mappings). In a sense, it is a measure of the curvature of the manifold. Evidence for nonlinear dynamics is demonstrated if predictability improves as the S-map model parameter θ is tuned toward nonlinear solutions ($\theta > 0$). The shaded area is the 5th/95th and the dashed blue line the 10th/90th percentile confidence intervals using surrogate data (see text for details on surrogate data). The figure shows that while CR and ΔGT both show statistical nonlinear state dependent dynamics, the raw GT time series does not

Figure 4 shows the CCM results. More specifically, it shows the correlation coefficient between actual and predicted values (ρ) as a function of sample size L when CR cross maps ΔGT (red) and when ΔGT cross maps CR (blue). Here the optimum embedding dimension is $E = 5$, and the optimal time lags used for CR cross mapping ΔGT and for ΔGT cross mapping CR are 3 and −2, respectively. These lags are based on maximizing cross map signal strength (i.e., maximizing cross map correlations). Clearly, there is no evidence for a causal effect of ΔGT on CR, as witnessed by the lack of convergence (no cross map improvement as the sample size increases) when cross mapping from CR to ΔGT. This indicates (as one should expect!) that information about global temperature is not present in the cosmic rays time series. However, cross mapping from ΔGT to CR succeeds. We

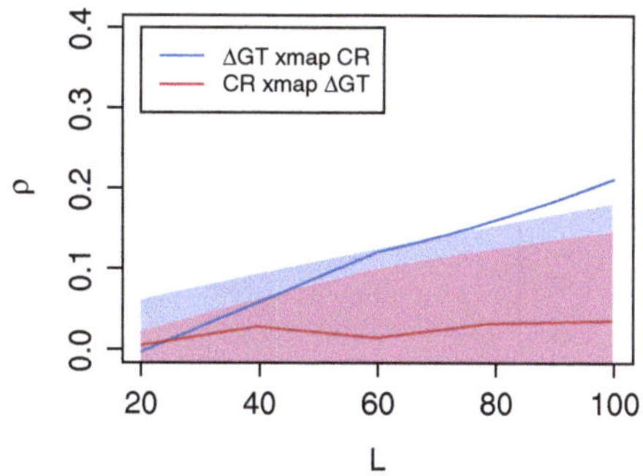

Fig. 4 Results of CCM analysis between the CR time series and the annual variations in GT (ΔGT). Although there is no correlation between these variables ($\rho = 0.02$), convergence (increasing and significant ρ with longer time series; *blue line*) suggests that year-to-year changes in GT are causally forced by galactic CR (i.e., ΔGT cross maps CR, information about CR is encoded in ΔGT). As explained in the text, a comparison with phase randomized surrogate data (*shaded areas*) shows that this result is significant at the 5% level. Lack of convergence (*red line*) confirms, as expected, that ΔGT has no causal influence on CR

observe convergence as L increases indicating that information about cosmic rays is recoverable in the ΔGT record. Thus CCM shows that there is a modest causal effect of cosmic rays on annual global temperature fluctuations.

These results are qualitatively robust to choice of embedding dimension and are statistically significant ($p < 0.05$). The first surrogate data analysis shown in Fig. 4 involves standard phase randomized surrogates ($n = 1000$) generated by inverting the spectra for ΔGT and CR and randomizing the phases (Ebisuzaki 1997). The blue shaded area depicts the 0.05 and 0.95 intervals of CCM results for observed ΔGT cross mapping surrogate CR, and the red shaded area shows the actual CR cross mapping surrogate ΔGT. Again this result is robust, providing independent verification when surrogates are generated as best-fit AR1 time series (Fig. 5). Finally, as a null check, we applied CCM analysis to the CR time series and model generated annual variations in global temperature (ΔGT) generated by the CCSM4 NCAR model—an IPCC AR5 model lacking any mechanism for cosmic rays to affect temperature. As expected (Fig. 6) there is no significant cross mapping between the historical cosmic rays time series and ΔGT from the model.

As might be expected from the S-map analysis there is no detectable convergence with the raw GT data containing the twentieth century warming trend (Fig. 7). The non-convergent cross map signal is consistent with a statistical association that is non-causal in terms of dynamic coupling (Sugihara et al. 2012). Indeed, the cross map estimates contain less information than is contained in the linear trend of GT ($\rho = 0.8$), reflecting little beyond the incidental cross-correlation between CR and GT ($\rho = 0.38$). Lack of convergence combined with a failure to manifest

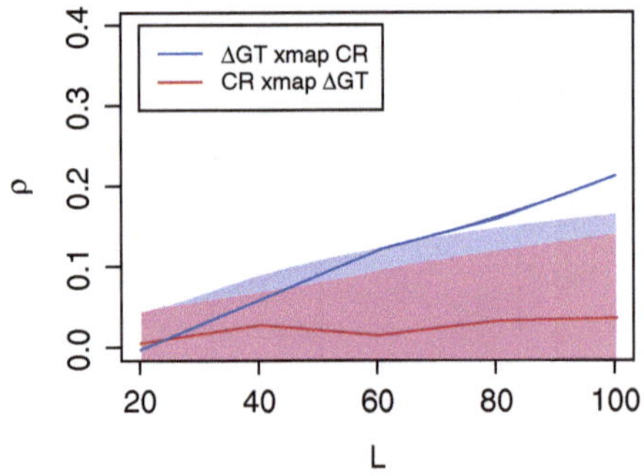

Fig. 5 As in Fig. 3 but with AR1 surrogates. Results of CCM analysis between the CR time series and the annual variations in global temperature (ΔGT). Convergence (increasing and significant ρ with longer time series) (*blue line*) shows that year-to-year changes in global temperature are causally forced by galactic cosmic rays. Lack of convergence (*red line*) shows, as expected, that ΔGT has no causal influence on CR. Results of the AR1 surrogates are nearly identical to those obtained by the Ebisuzaki method in the main text

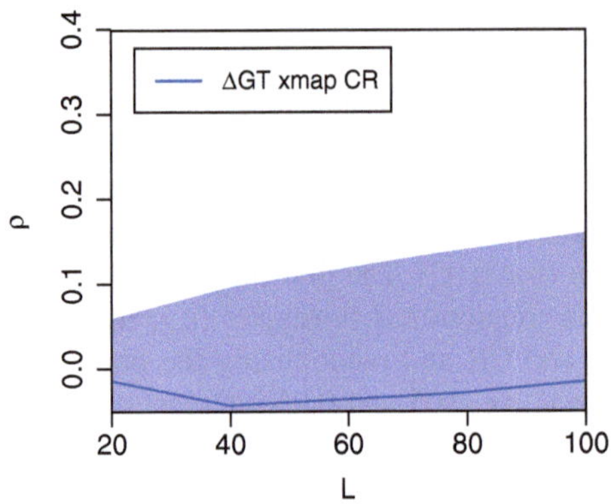

Fig. 6 Results of CCM analysis between the CR time series and the annual variations in global temperature (ΔGT) generated by the CCSM4 NCAR model (an IPCC AR5 model). As expected, because this model does not include any mechanism for cosmic rays to affect temperature, CCM shows there is no identifiable causality—there is no significant cross mapping between the historical cosmic rays time series using ΔGT from the model

significance beyond the surrogates demonstrates that CR has no discernable causal effect on the overall warming pattern for the twentieth century. The analysis shows definitively that the dominant warming signal on this century-long time scale is not a measurable consequence of dynamic forcing by CR.

Some comments on the choice of the aa index as a proxy for cosmic rays: An argument can be made that the aa index may not be the best proxy for cosmic rays. However, ground measurements of cosmic rays flux are significantly correlated with the a index (Perry 2007; Cliver et al. 1998; Stuive and Quay 1980 to mention a few). The reason we don't use the actual ground flux data here is that ground

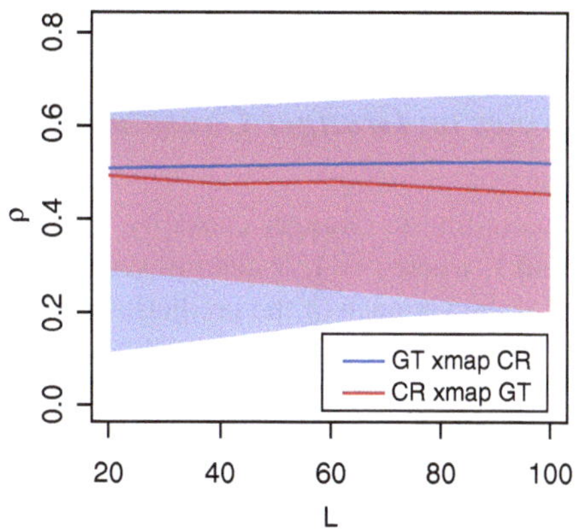

Fig. 7 CCM analysis between CR and raw GT time series with the secular warming trend shows no convergence and no significance with surrogates generated as in Fig. 3. Thus, although CR is statistically correlated with GT ($\rho = 0.38$), it shows no measurable causal effect on the twentieth-century warming trend

measurements of cosmic rays flux did not begin but after 1950. This makes the sample size too small. Indeed, we have repeated CCM between global temperature and actual measurements of cosmic rays flux in Climax Colorado. We find the same evidence for causality but the statistical significance is lower ($0.15 < p < 0.1$). In addition, we have applied CCM to all possible pairs in the aa index-sunspot number (ISN)—Climax cosmic rays flux network. We find significant causality between all pairs, which will indicate synchrony in the network. This is indeed a very interesting but preliminary result, which will be further explored elsewhere.

For completeness, a traditional Granger causality (Sauer et al. 1991) analysis was implemented. Because the S-map test demonstrated the CR time series is from a nonlinear dynamic system and not a purely stochastic one, Granger's test should not apply (Sugihara et al. 2012). Granger causality requires separability (dynamic independence of system parts) and is therefore not defined for interdependent dynamic systems. Thus it is not surprising that the Granger test fails to detect any meaningful association (though the non-sensible case for temperature affecting CR is slightly stronger by Granger's test; see method and results in Appendix 1).

4 Conclusions

In this review paper we present the basic principles behind the convergent cross mapping method to detect causality and we provided an example with the proper steps to establish statistical significance for causality between two time series. For more information the reader is directed to the references. The R package with the codes for the various calculations is available at https://CRAN.R-project.org/package=rEDM.

Appendix 1

Notes on Granger Causality

According to Granger causality, given two simultaneously recorded time series X_t and Y_t where $t = 1, N$ denotes sampling times, we say that Y has causal influence on X if the variance of the prediction error of X given Y is less than the variance of the prediction of X not given Y. This means that if prediction of some output improves with the addition of an input, then the input Granger causes the output. In its original formulation Granger causality is based on linear prediction of stochastic time series.

There are several ways to test for Granger causality. The approach used here uses the autoregressive specification of a bivariate vector autoregression. For a given lag m, we estimate the following unrestricted equation by ordinary least squares:

$$X_t = c + \sum_{i=1}^{m} a_i X_{t-i} + \sum_{i=1}^{m} b_i Y_{t-i} + e_t$$

where *a, b*, and *c* are coefficients and *e* is a residual. The null hypothesis that "*Y does not Granger-cause X*" is then constructed as

H_o: $b_1 = b_2 = \ldots = b_m = 0$

We also estimate the equation

$$X_t = c + \sum_{i=1}^{m} a_i X_{t-i} + w_t$$

and compare the sum of squared residuals

$$\mathrm{RSS}_1 = \sum_{t=1}^{N} \widehat{e}_t^2$$

and

$$\mathrm{RSS}_2 = \sum_{t=1}^{N} \widehat{w}_t^2$$

The statistic $S = \frac{(\mathrm{RSS2}-\mathrm{RSS1})}{\mathrm{RSS1}} \frac{(T-2m-1)}{m}$ is approximately equal to $F_{m,T-2m-1}$, and it is statistically significant at a p level of

$$p = 1 - prob\,(F_{m,T-2m-1})$$

In our case if X is ΔGT and Y is CR the p value assuming an AR-1 (AR-2) process is 0.82 (0.97). If X is CR and Y is ΔGT, the respective p values are 0.81 and

0.64. Thus neither X nor Y Granger-causes the other. In fact, the variance explained in the prediction error is less than 1% regardless which variable is used to predict the other. Considering higher order AR processes does not improve these results.

References

Bond, G., et al. 2001. Persistent solar influence on North Atlantic climate during the Holocene. *Science* 294: 2130–2136. doi:10.1126/science.1065680.

Cliver, E., V. Boriakoff, and J. Feynman. 1998. Solar variability and climate change: geomagnetic AA index and global surface temperature. *Geophysical Research Letters* 25 (7): 1035–1038.

Duplissy, J., et al. 2010. Results from the CERN pilot CLOUD experiment. *Atmospheric Chemistry and Physics* 10: 1635–1647.

Ebisuzaki, W. 1997. A method to estimate the statistical significance of a correlation when the data are serially correlated. *Journal of Climate* 10: 2147–2158.

Enghoff, M.B., J.O.P. Pedersen, U.I. Uggerhoj, S.M. Paling, and H. Svensmark. 2011. Aerosol nucleation induced by high energy particle beam. *Geophysical Research Letters* 38: L09805. doi:10.1029/2011GL047036.

Foukal, P., C. Frohlich, H. Spruit, and T.M.L. Wigley. 2006. Variations in solar luminosity and their effect on the earth's climate. *Nature* 443: 161–166. doi:10.1038/nature05072.

Granger, C.W.J. 1969. Investigating causal relations by econometric models and cross-spectral analysis. *Econometrica* 37: 424–438.

Harrison, R.G., M.H.P. Ambaum, and M. Lockwood. 2011. Cloud bases height and cosmic rays. *Proceeding of the Royal Society of London* 467: 2777–2791. doi:10.1098/rspa.2011.0040.

Kikby, J., et al. 2011. Role of sulphuric acid, ammonia and galactic cosmic rays in atmospheric aerosol nucleation. *Nature* 476: 429–433. doi:10.1038/nature10343.

Marsh, N., and H. Svensmark. 2000. Low cloud properties influenced by cosmic rays. *Physical Review Letters* 85: 5004–5007.

Nayaud, P.N. 1972. The aa indices: A 100-year series characterizing magnetic activity. *Journal of Geophysical Research* 17: 6870–6874.

Neff, U., S.J. Burns, A. Mangini, M. Madelsee, D. Fleitmann, and A. Matter. 2001. Strong coherence between solar variability and the monsoon in Oman between 9 and 6 kyr ago. *Nature* 411: 290–293. doi:10.1038/35077048.

Perry, C.A. 2007. Evidence for a physical linkage between galactic cosmic rays and regional climate time series. *Journal of Advances in Space Research* 40 (3): 353–364. doi:10.1016/j.asr.2007.02.079.

Rawal, A., S.N. Tripathi, M. Michael, A.K. Srivastava, and R.G. Harrison. 2013. Quantifying the importance of galactic cos mic rays in cloud microphysical processes. *Journal of Atmospheric and Solar - Terrestrial Physics* 102: 243–251.

Sauer, T., J. Yorke, and M. Casdagli. 1991. Embedology. *Journal of Statistical Physics* 65: 579–616.

Shaviv, N. 2003. The spiral structure of the Milky Way, cosmic rays, and ice age epochs on Earth. *New Astronomy* 8: 39–77.

Stark, J., D.S. Broomhead, M.E. Davies, and J. Huke. 2003. Delay embeddings for forced systems. II. Stochastic forcing. *Journal of Nonlinear Science* 13: 519–577.

Stuive, M., and P.D. Quay. 1980. Changes in the carbon-14 attributed to a variable Sun. *Science* 207: 11–19.

Sugihara, G. 1994. Nonlinear forecasting for the classification of natural time series. *Philosophical Transactions of the Royal Society A* 348: 477–495.

Sugihara, G., and R.M. May. 1990. Nonlinear forecasting as a way of distinguishing chaos from measurement error in time series. *Nature* 344: 733–741.

Sugihara, G., et al. 2012. Detecting causality in complex ecosystems. *Science* 338: 496–500.

Svensmark, H. 1998. Influence of cosmic rays on Earth's climate. *Physical Review Letters* 81: 5027–5030.

Svensmark, H., J.O.P. Pedersen, N.D. Marsh, M.B. Enghoff, and U.I. Uggerhoj. 2007. Experimental evidence for the role of ions in particle nucleation under atmospheric conditions. *Proceedings of the Royal Society* 463 (2078): 385–396.

Svesmark, H., M.B. Enghoff, and J.O.P. Pedersen. 2013. Response of cloud condensation nuvlei (>50 nm) to changes in ion-nucleation. *Physics Letters A* 377: 2343–2347.

Takens, F. 1981. Detecting strange attractors in turbulence. In *Dynamical systems and turbulence*, Lecture notes in mathematics, ed. D.A. Rand and L.-S. Young, vol. 898, 366–381. New York: Springer-Verlag.

Usoskin, I.G., N. Marsh, G.A. Kovaltsov, K. Mursula, and O.G. Gladysheva. 2004. Latitudinal dependence of low cloud amount on cosmic ray induced ionization. *Geophysical Research Letters* 31: L16109.

Yu, F. 2005. Quasi-unary homogeneous nucleation of H_2SO_4-H_20. *The Journal of Chemical Physics* 122: 074501.

Zhang, R., A. Khalizov, L. Wang, M. Hu, and W. Xu. 2012. Nucleation and growth of nanoparticles in the atmosphere. *Chemical Reviews* 112: 1957–2011.

Randomnicity: Randomness as a Property of the Universe

Anastasios A. Tsonis

Abstract This paper is a concept paper, which discusses the definition of randomness, and the sources of randomness in the mathematical system as well as in the physical system (the Universe). We document that randomness is an inherited property of mathematics and of the physical world, shaping all observed forms and structures, and we discuss its role.

Keywords Determinism • Randomness • Natural processes • Fractals • Chaos • Nonlinear processes

1 Prologue

It is dawn and the battlefield is waiting. It is sometime in the twelfth century B.C. and a critical moment in the Trojan War must be decided. Paris seduced and ran away with Helen, the wife of the king of Sparta, and now Menelaus, the king, and a unified Greek army has invaded the Trojan land and is asking for revenge. The war has been dragging on for years, and Troy is not falling. In fact, it appears that the Trojans, led by Hector, are gaining the upper hand. Somebody from the Greek army has to step in and fight man-to-man with Hector. Who will it be? The decision will be left to chance. Each of the volunteers marks his own lot, then the lots are put in a helmet and are shaken. A lot is drawn from the helmet and identifies the soldier who will fight Hector. It is Ajax.

In Homer's Iliad and in many other early epics such decisions were often left to chance. The concept of randomness appears to have been an integral part of the actions and feelings of early cultures. At the same time they believed that the gods controlled every little detail (determinism) and therefore nothing was left to chance. This, however, is not a paradox. Randomness in early civilizations emerges

A.A. Tsonis (✉)
Department of Mathematical Sciences, Atmospheric Sciences Group, University of Wisconsin - Milwaukee, Milwaukee, WI 53201, USA

Hydrologic Research Center, San Diego, CA, USA
e-mail: aatsonis@uwm.edu

A.A. Tsonis (ed.), *Advances in Nonlinear Geosciences*,
DOI 10.1007/978-3-319-58895-7_28

as part of God. It is controlled only by God thus eliminating human intervention and allows the will of God to apply. Thus, randomness cannot be separated from God (determinism). Randomness and determinism are established early in the human mind as being interconnected and associated with something bigger, like God, who is boundless and everywhere at any time all the time.

2 Introduction

> Others there are who believe that chance is a cause but that it is inscrutable to human intelligence as being a divine thing and full of mystery.
> Aristotle, Physics Book II, 4

Over 2500 years ago Aristotle ponders on what many other philosophers have pondered throughout time. Exactly what is randomness and why is it there? This paper presents a synthesis of facts from the mathematical and physical systems, which clearly establish randomness as a property of nature and that what we see and experience around us emerges from the interplay of rules and randomness. This paper is arranged in four parts. Part 1 deals with randomness in the mathematical system. Part 2 deals with randomness in the physical system (the Universe). In Part 3 the connections between the sources of randomness in these two systems will be discussed. In part 4 we will discuss the role of randomness in the physical world. More details and discussions can be found in my book *Randomnicity: Rules and Randomness in the Realm of the Infinite* (Tsonis 2008).

3 Some Introductory Examples

Consider an equilateral triangle of side size L and the following iteration: Take the middle third of each side and replace it with two $L/3$ length sides forming a smaller equilateral triangle on each original side. We now have a "star" with 12 sides. Repeat the process for each of the 12 sides, and keep on repeating for the new sides ad infinitum. This process will result in a closed boundary, which is called the Koch island or Koch snowflake (Fig. 1a). This boundary is an *exact fractal* (Mandelbrot 1983), but a far cry for real boundaries such as coastlines. We can "improve" on that boundary by introducing randomization of the iteration process, for example, rather than forming the equilateral triangle with the same orientation at each step, we may choose the orientation at random. This leads to a boundary that is an improvement in the right direction (Fig. 1b), but still it's a far cry from the actual coastline. However, we only need to be a little more creative with our randomization technique (Peitgen and Saupe 1988; Peitgen et al. 1992) before we can generate a *random fractal* whose details will be indistinguishable from natural coastlines. One such example is shown in Fig. 1c. Comparison with the coast of England shows striking similarities at all scales (Mandelbrot 1983).

Now let us consider the case of lightning. Lightning is the result of dielectric breakdown of gases, which occurs when some region of the atmosphere attains

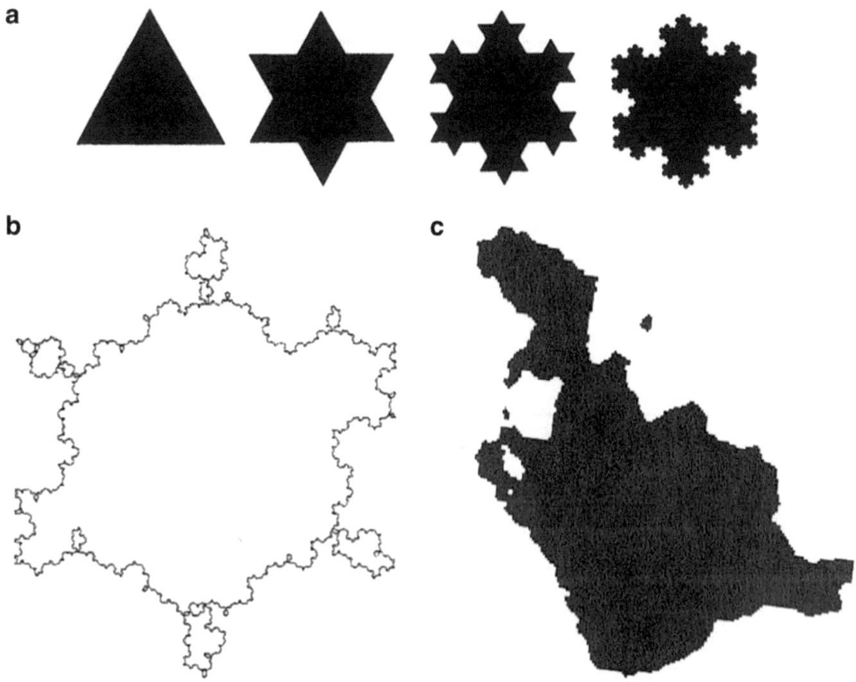

Fig. 1 (**a**) Constructing a Koch island or snowflake. (**b**) A randomized Koch island. Figure courtesy of Professor Heinz-Otto Peitgen. (**c**) A simulation of a random fractal coastline. Figure courtesy of Professor Benoit Mandelbrot

a sufficiently large electric charge. Basically, a strong concentration of negative charge at the cloud base induces through friction a positive charge at the surface. Once this is in place, if the electrical potential between the cloud base and the ground reaches a sufficiently high value, then some negative charge is propelled toward the ground. This cloud-to-ground discharge is called the stepped leader because it appears to move downward in steps. When the stepped leader has lowered a high negative potential near the ground, the electric field at the ground is sufficient to cause an upward-moving discharge, which carries ground potential up the path previously forged by the stepped leader. By doing so, the return discharge illuminates and drains the branches formed by the stepped leader. This luminous part of lightning is called the return stroke. Therefore, both the stepped leader and the return stroke are usually strongly branched downward. The branching character of lightning exhibits a striking presence of structure at many different length scales. Every branch, for example, looks like a lightning itself and so does every branch of a branch. Indeed lightning has been documented to be a random fractal (Tsonis and Elsner 1987).

The dielectric breakdown on the atmosphere can be modeled by considering that the driving force is the electrical potential, which satisfies the Laplace equation. We

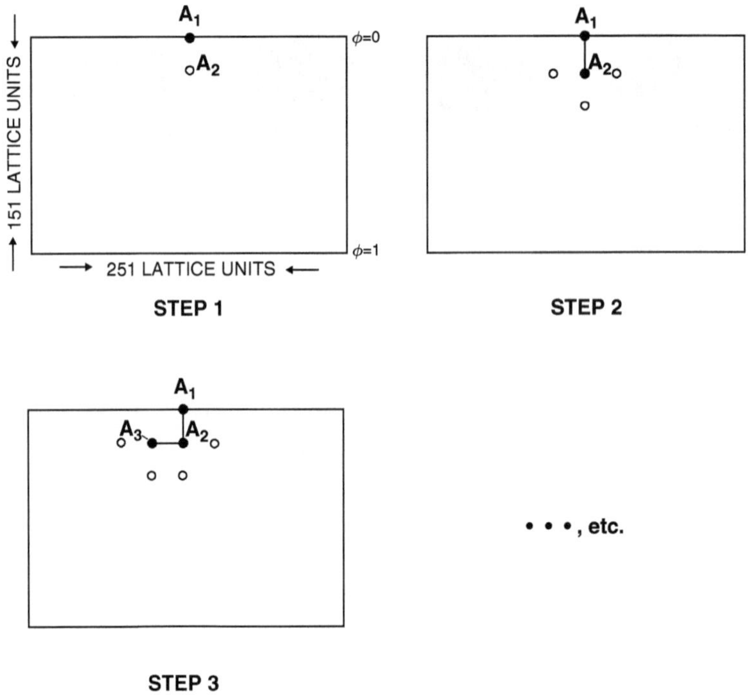

Fig. 2 Illustration of the model used to generate lightning (see text for more details)

start with a two-dimensional lattice in which the potential, ϕ, of the top and bottom row is fixed at the values 0 and 1, respectively. Only the middle point of the top row (A_1) is capable of growth (Fig. 2). Given this arrangement the Laplace equation $\nabla^2\varphi = 0$, where $\nabla^2 = \partial^2/\partial x^2 + \partial^2/\partial y^2$, is solved. In a two-dimensional lattice (i,j) the solution of this equation is obtained by iterating the following equation using successive over-relaxation

$$\varphi_{i,j} = \frac{1}{4}\left(\varphi_{i+1,j} + \varphi_{i-1,j} + \varphi_{i,j+1} + \varphi_{i,j-1}\right).$$

All the immediate non-zero potential neighbors of A_1 are then considered as possible candidates, one of which will be added to the evolving pattern. In Fig. 2, the possible candidates are shown by the open circles and the evolving pattern is shown by the connected black dots. In step 1 there is only one possible candidate. Therefore, point A_2 will be added to the discharge pattern. Since the discharge attempts to neutralize the difference in potential between the top and bottom row (like the discharge in the atmosphere neutralizes the negative and positive charges at the cloud base and ground, respectively), the discharge pattern is assumed to have zero potential. With point A_2 in the picture, the Laplace equation is solved again. This produces new values for the potential at each point. Now there are

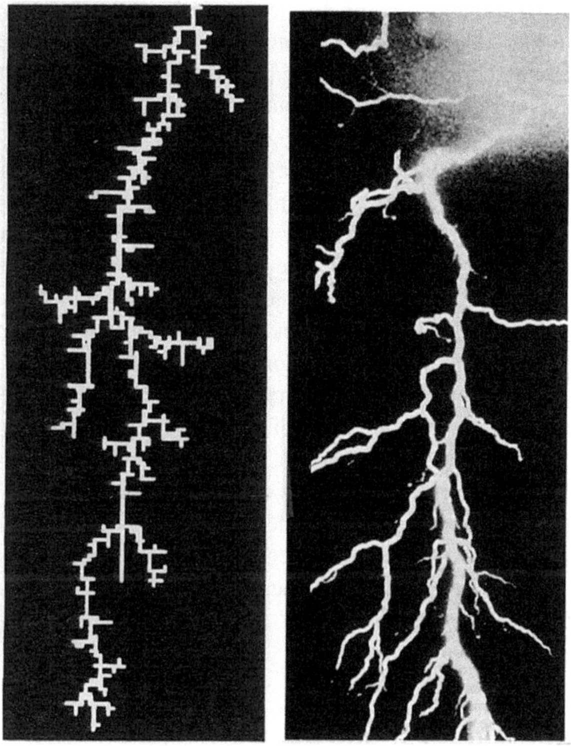

Fig. 3 Simulated (*left*) and actual (*right*) lightning

three possible candidates ($N = 3$). Each one of these candidates is then assigned a probability, which depends on the value of the potential at each point, m. Once each point of the lattice has been assigned a solution each of the $m = 1,N$ possible candidates is given this probability of selection, P_m

$$P_m = \phi_m^2 / \sum_{m=1}^{N} \phi_m^2$$

Given these probabilities a point is chosen randomly and is added to the evolving pattern. The above procedure is then repeated until a point of the bottom row is added to the discharge pattern. The patterns generated from such a procedure not only look similar to real lightning but they also have the same fractal dimension (Fig. 3). The actual (right) and computer generated (left) structural properties of lightning are identical.

In addition, the model is suggesting that no two lightning are alike. Since each step is associated with a probability, if we multiply the probabilities of all chosen points we will get the probability that the resulted structure will occur.

The probability of the above computer-generated lightning is a staggering ten to the power of minus one thousand (Tsonis and Elsner 1987; Tsonis 1987). This incredibly small number suggests that even if there were one million lightning bolts at every location on Earth every 1 s it would take the age of the universe to see the same lightning again.

Exact fractals, like Euclidean structures (straight lines spheres, cubes, etc.), are almost never observed in nature. Randomness eliminates such possibilities. This is also supported by the study of *cellular automata.* Cellular automata (Wolfram 2002) are systems whose evolution is described not by equations but by very simple, computer-program-like rules. They can provide an alternative to more complicated systems described by differential equations. By studying thousands of completely deterministic cellular automata, Wolfram identified five types of evolution: steady states, periodic structures, exact fractals, chaotic evolutions, and evolutions characterized by a mixture of regular and irregular structures (often referred to as the *edge of chaos*). When, however, randomness is introduced to the automata, the exact fractals class does not emerge (see also Tsonis 1996). Clearly then, unless randomness is combined with rules, no realistic forms of natural phenomena will emerge. Numerous other examples can be given from all areas of science to support this statement. Then the obvious question arises: What is randomness and where is it coming from?

4 Randomness in the Mathematical System

In a seminal paper in 1931, Gödel proved that there are mathematical statements that cannot be proved within the current mathematical system. Gödel proved that if all mathematical statements could be proved (which will indicate that the formal mathematical system is complete) then the system will be an inconsistent system. This self-reference about the mathematical system also proves that consistent mathematics is an incomplete system. This means that in a consistent mathematical system there will always be uncertainty about certain statements. This uncertainty introduces a form of randomness into the formal mathematical system. Formally, Gödel's Incompleteness Theorem, is expressed as:

> For every consistent formalization of arithmetic, no matter how complex, there exist arithmetic truths improvable within that formal system,

or in a somewhat simpler form:

> In today's mathematics there are true statements about numbers that cannot be proved.

A nice everyday example of the principle underlying Gödel's theorem is given by Douglas Hofstadter in his monumental book *Gödel, Escher, Bach: an Eternal Golden Braid* (Hofstadter 1979). Consider a phonograph, which is playing a record in a room. The phonograph produces sounds, which are sent out to its surrounding environment. A sound is a vibration. These vibrations, as well as other vibrations from other sources, are reflected by the walls and propagate back to the phonograph. In this way the reflected vibrations may affect the phonograph's operation. Obviously, the stronger the vibrations a record produces the greater the effect they have on the phonograph. As such for any record player there may be records, which cannot be played because they may cause its indirect self-destruction.

The system of mathematics can become less incomplete by adding more rules. There are numerous cases where mathematical statements were proven only after new insights (new rules) were discovered. The *Fermat Conjecture* is the most celebrated such example. Proposed by Pierre Fermat in 1665, it states that the equation $x^n + y^n = z^n$ (a Diophantine equation) does not have a positive integer solution when n is an integer greater than 2. Thousands of mathematicians wrestled with this problem unsuccessfully until Andrew Wiles finally proved it in 1995. Why did it take 330 years before anyone could prove the conjecture? Because there were areas of mathematics, specifically the theory of elliptic curves, which had to be discovered before anyone could prove the conjecture.

However, unless an *infinite* number of rules are added one cannot be certain that the system will not be incomplete. What that means is that there is no *finite* set of rules that can be consistently added to the system to make it complete. A consequence of this is that a procedure, which decides that any mathematical statement is true in a finite number of steps, does not exist. Think of the system as a photograph. Many years ago photographs were black and white and blurry. As such there were "truths" (for example, the color of the sky) that could not be "proven" by them. As technology improved (read: more rules were added), photographs became more realistic (or more complete). Still, however, unless we have an infinite resolution, the details in a scene that is being photographed cannot be known exactly (for example, individual molecules cannot be seen).

Since the time of Euclid, the dream of mathematicians was to reduce mathematics to a set of basic axioms from which, through inference, all theorems could be proven. Gödel's theorem shook the foundations of mathematics by showing that this is not possible. The implications are startling. Apart from philosophical issues such as "can we ever know the truth from reasoning?," it implies that whether a mathematical statement is true or false may not be known. Moreover, since there can be an infinity of such statements, Gödel's theorem implies that the element of uncertainty, and thus randomness, is interweaved with axioms, theorems, and the whole structure of mathematics. This naturally brings up the following question: What exactly is randomness?

4.1 *Randomness of the First Kind*

Imagine that we are given or we observe a pattern-less sequence that has been generated by some rule. If we are not able to extract the rule, what is the difference between such a sequence and a truly random sequence? Many will argue that for all practical purposes there is no difference. Thus, our inability to extract the rules makes predicting future digits impossible and thus constitutes a source of randomness. Here, however, we have to be careful with what we mean by "our inability to extract the rules." Do we mean that in principle we could get to the rules but we do not have the knowledge to get to them, or that there is no possible way to get to the rules? If the former is true then strictly speaking the sequence is not random. It is just waiting to be "debugged." In this case the rules are actually reversible and thus in principle we can go backwards and find the rules from the sequence. If the later is true then the procedure to construct the sequence is irreversible, and thus, while we can go from the rules to the sequence, we cannot ever go from the sequence to the rules. It is this inability to recover the rules that will generate truly random sequences. But how such irreversibility occurs?

Consider again the first mathematical operation, which for simplicity we call it operation O_1: Start with a number. If this number is even, multiply it by 3/2; otherwise add 1 and then multiply by 3/2. This generates the second number. Repeat the above step to produce the third number and so on. Operation O_1 results in a sequence of odd and even numbers, which we will denote by S_1. Once we have S_1 we apply the second operation, which we will call operation O_2, according to which an odd number is replaced with 1 and an even number with 0. This leaves us with a sequence of 0's and 1's, which we will denote as sequence S_2. If the starting number was the number 1, then operation O_1 produces the following sequence S_1 of odd and even numbers.

S_1: 1, 3, 6, 9, 15, 24, 36, 54, 81, 123, 186, 279, 630, 945, 1419, 2130, 3195, 4794...

Subsequently, operation O_2 produces the following sequence S_2

S_2: 110110001101011010 1...

The above two operations define a mathematical system, which is isolated from external influences and in which the initial condition is simple and well defined. Sequence S_2 does not appear to have a coherent pattern (this becomes even more apparent if we continue the process for many steps). However, since it is generated according to a set of rules it is not random. Given the rules, any future value can be calculated or predicted. There is, however, a catch here. If you were given part of S_2 but not the rules that generated it, would you be able to predict the next digit? The answer to this question is no, and Fig. 4 explains why.

Given sequence S_1, a patient person might at some point figure out operation O_1. In this case we can go from operation O_1 to sequence S_1 and vice versa (this reversibility is indicated by the bidirectional arrow between operation O_1 and sequence S_1). Similarly, one might speculate that the zero's and ones represent even and odd numbers, respectively. This will allow us to go from operation O_2 to sequence S_2 and vice versa.

However, once we have figured out operation O_2 there is no possible way to go to sequence S_1 because for each zero there is an infinite even numbers and for each one there is an infinite odd numbers to choose from and the initial number can be anything. This irreversibility is indicated by the unidirectional arrow between sequence S_1 and operation O_2. We thus see that operation O_2 injects an uncertainty that inhibits us from recovering the rules of the construction and makes

Fig. 4 A diagram of an irreversible set of mathematical operations (see text for details)

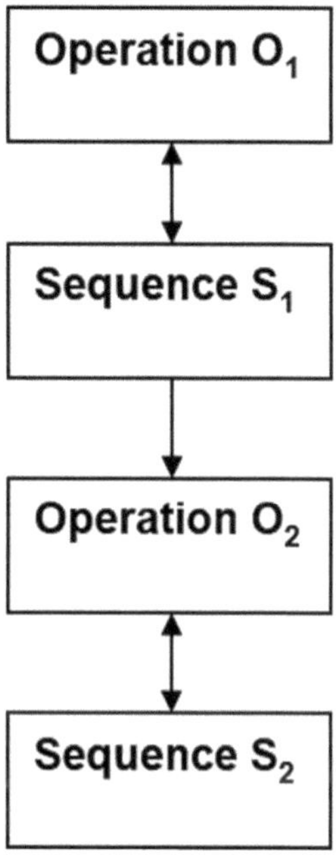

S_2 maximally random. While not all sequence constructions may be irreversible, the above example clearly demonstrates that extracting underlying rules may not always be attainable.

The loss of information, which results from this irreversibility, is not just a well thought mathematical trick. Information initially contained in a system could indeed get lost during its evolution. Imagine two compartments separated by a diaphragm one filled with water at 40 F and the other with water at 90 F. If the diaphragm is removed the two water samples mix and produce a sample, which is at a uniform temperature throughout. In this final sample all the information about the initial temperatures is lost and cannot possibly recovered no matter how knowledgeable we are.

Order and predictability arise from rules. Randomness and unpredictability arise from the absence of rules. This source of randomness is, however, ideal if not trivial. In the mathematical system and in the physical world there is always some kind of an underlying rule(s). Here we see that unpredictability and thus randomness may arise from irreversible programs or procedures, which inhibit us from getting to the rules not just from the absence of rules. We will call this randomness, *randomness of the first kind.*

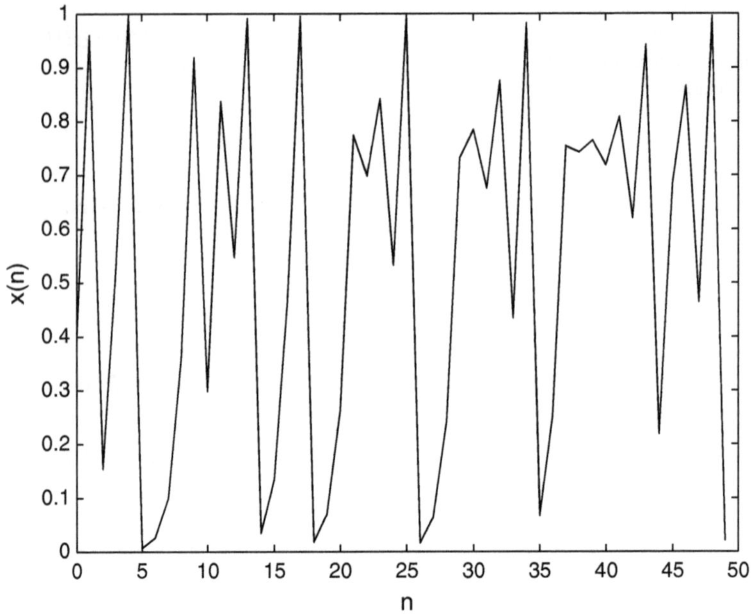

Fig. 5 Solution of the logistic equation from the initial condition 0.4

4.2 Randomness of the Second Kind

Consider an *iterative process* where the same operation is repeated using every time the result of the previous step as the starting point. The most famous such process is given by the logistic equation $x_{n+1} = 4x_n\,(1-x_n)$, which has been used to study population dynamics. The number 4 is the only parameter of this equation. Since this system is described by only one equation it is obviously a very simple system. For $x = 0$ or $x = 1$ all subsequent values become zero and for x greater than 1 or less than -1 the population becomes negative. Thus, for nontrivial dynamics (i.e. requiring that the population does not become extinct or negative) the values of x must range between 0 and 1. Iterating the logistic equation from an initial value of 0.4 results in the evolution shown in Fig. 5

What we observe is that x goes up and down in an apparently irregular way. No apparent pattern is evident. For all practical purposes this signal is random. Nevertheless, it was generated from a well-defined initial condition (0.4) and a very simple rule.

Figure 6 shows x as a function of the time step for initial condition 0.4 as well as for initial condition 0.7. We observe that the two evolutions are different. They are aperiodic and they do not converge to the same result. Somehow the system remembers its initial condition forever. The evolution of this simple system is clearly dependent on the initial condition. It gets even more interesting. Let us assume that

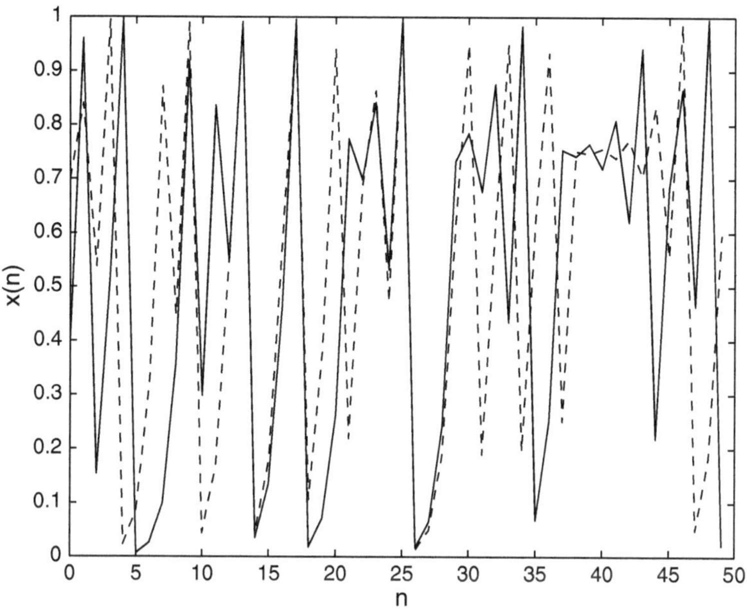

Fig. 6 Solution of the logistic equation from the initial conditions 0.4 and 0.7

the second initial condition was not 0.7 but was 0.405 (about 1% different from the first initial condition). If we compare the two evolutions again (Fig. 7), we see that they start very close to each other, but they soon diverge and follow completely different paths. Not only is the system sensitive to the initial condition, it is sensitive to even the tiniest of fluctuations. And not only does the system not forget a tiny fluctuation, it actually amplifies it and soon the two evolutions diverge significantly.

As we have seen earlier, aperiodicity does not necessarily mean unpredictability. For example, the π digits are aperiodic but we can predict any digit we want. But how about sensitivity to the initial conditions? Could sensitivity to initial conditions create randomness and make a system unpredictable? From the above example one can forcefully argue that the initial condition can be specified exactly. For example, we may specify 0.4 as our initial condition. Then the equation will produce all future values. Thus, one may argue, sensitivity to the initial conditions is not a condition for unpredictability. This is a strong argument but there is a little problem with it.

The five first values of the evolution of the equation $x_{n+1} = 4x_n(1 - x_n)$ from the initial condition $x_0 = 0.4$ are:

$x_0 = 0.4$
$x_1 = 0.96$
$x_2 = 0.1536$
$x_3 = 0.52002816$
$x_4 = 0.9983954912280576$
$x_5 = 0.00640773729417263956570612432896$

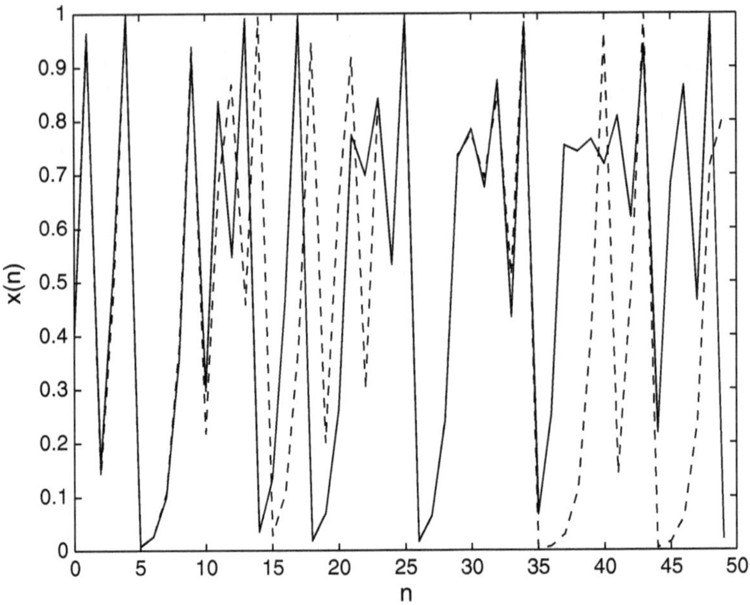

Fig. 7 Solution of the logistic equation from the initial conditions 0.4 and 0.405

What do you observe? The digits after the decimal point increase (in fact double) with every iteration. After seven iterations the result carries 128 digits. After twelve iterations there are 2048 digits! The number of digits is actually given by 2^n, where n is the number of iterations. Calculating exactly out to only 100 steps will require a computer that will carry calculations with 2^{100} decimal points. This number is approximately equal to 10^{30}, which is one trillion times greater than the age of the universe in seconds. Computers do not routinely handle more than a hundred digits. So what does the computer do when the iteration reaches the point where the digits are more than the digits the computer can carry? It simply rounds off the result or it chops off the extra digits. That in effect makes the result an approximation to what the result would have been if the computer had the ability to carry calculations with unlimited number of digits. This approximation will now play the role of a fluctuation, which will be amplified and soon lead to an evolution that will be completely different than the actual one. *Thus, only if we had infinite precision and infinite power we will be able to predict such systems accurately.* Because we do not have that, for systems that are sensitive to the initial conditions, the exact state of the system after a short time cannot be known. The outcome of such systems after that time is simply random as small fluctuations amplify enough to dominate the evolution of the system. Thus, a future state is unpredictable. Note that the logistic equation is what we call a *nonlinear* equation.

We thus see that randomness is created even if we know the rules. In this case the source for the randomness is not rule irreversibility or our inability to find the rules but our inability to have infinite precision and infinite power. This kind of randomness has been termed *chaos* and it is distinctly different from the first kind. Chaos is strictly a property of nonlinear systems.

4.3 Randomness of the Third Kind

Now let us consider the following example. The distance between your home and the shopping mall is fixed. It is always the same, never varies. We all know that if the speed of an object is constant, then the time that it will take to go from A to B is equal to the distance between A and B divided by the speed. It follows that as long as the speed does not change, the time that it takes to travel the distance AB will always be the same. Now assume that you travel with your car (our "system" in this example) from your home (A) to the mall (B). You know that the distance is five miles so you figure that at a constant speed of 50 miles per hour it will take you one tenth of an hour or 6 min to reach the mall. Would you bet money on such a prediction? I hope not, because it will never be exactly 6 min. A slower driver in front of you, a driver that suddenly decides to "cut" in front of you, a yellow traffic light that forces you to decide whether to stop or accelerate, the presence of a police car, the sound of a honk, and many other "external" factors will cause you to depart from the constant speed of 50 miles per hour. Since the number of these factors is not fixed, each time you go to the mall it will take a different length of time. This makes the actual length of the trip very unpredictable, which means that the duration of the trip is a random number. Thus, even though we know the rules of the system (so there is no randomness of the first kind) and we can assume that our system is not chaotic (which excludes randomness due to sensitivity to the initial conditions) we still end up with randomness.

The above example introduces us to a mechanism for randomness, where randomness is explicitly introduced into the underlying rules of the system. It corresponds to saying that there is some kind of external environmental component whose essentially uncountable "agents" continually affect the system with their actions. Such processes are called *stochastic* processes. The word stochastic comes from the Greek word στοχαστης, which refers to the person who learns about future events or hidden things by means not based on reason.

Our "system" here may be thought of as a mathematical system described by a set of simple rules or equations whose evolution can in principle be computed by solving the equations. When the system is exposed to external influences, however, its behavior is modified. But what is this external "noise?" Where is it coming from? In our example, the environment is represented by the other drivers and the traffic lights system, which can also be thought of as simple systems. In this case then, what we have is many simple systems interacting. Each system is very simple but the collective behavior of many interacting systems may be very complicated.

It is then logical to assume that in our example the "system" is a subsystem of a grand system (possibly the universe) where many subsystems operate according to their rules and interact between each other. As subsystems interact they exchange information. Information received by one subsystem from another may interfere with its rules, thus producing an unexpected result. Such interactions, especially in a large number of subsystems, create an extremely complex behavior that can only be studied using probability theory. This "stochasticity" is our third kind of randomness: Randomness generated by the continuing effects of the environment.

It is interesting to note here that under this scenario very complicated behavior and randomness can be generated even if we start with no randomness of any kind. The theory of nonlinear dynamical systems has clearly established, for example, that many systems, which exhibit a very regular (periodic) behavior, become irregular and aperiodic when they are coupled with an external force, which is also very regular. This may also lead to randomness of the first kind or to chaos. Whatever the case might be one thing is certain. Very simple rules can either alone or in combination with other simple rules create randomness and unpredictability. And because we often are dealing with an infinite number of interacting "agents," the only way to study and predict their collective behavior may only be done stochastically.

It is also interesting to make some connections with real life at this point. In life, accidents (randomness) happen (1) when we do not know how things work (we do not know the rules), (2) when we do not pay full attention (we do not compute accurately), and (3) when we interact with people who affect our lives. Doesn't this look like randomness of the first, of the second, and of the third kind, respectively?

Finally, we should keep in mind that behind all mechanisms of randomness lurks the notion of infinity. Whether it is the absence of infinite knowledge or infinite power or the interplay of infinite agents, one cannot avoid infinity. It is the arena where the interplay of rules and randomness takes place. If this arena disappears all evolutions are doomed to repeat. For example, it has been shown that because the computer can only carry a finite number of digits in its calculations, the round-off error will force a non-periodic trajectory to coincide with a point in the past rather than simply coming very close to it. Once this occurs the evolution has no choice but to repeat (Tsonis 1991). Consider again the first five exact values resulted when we start iterating the logistic equation from an initial condition of 0.4.

$x_0 = 0.4$
$x_1 = 0.96$
$x_2 = 0.1536$
$x_3 = 0.52002816$
$x_4 = 0.9983954912280576$
$x_5 = 0.00640773729417263956570612432896$

If we assume that the calculator or computer used to perform these calculations can only carry one decimal point then the first value will be truncated to 0.9. If the value of 0.9 is used rather than the actual value of 0.96 to calculate the second iterate we get a new value of 0.36 which will be truncated to 0.3. Continuing like this we will find that the fifth iterate is truncated to 0.9, which is the value of the first iterate. From this point on we simply start over and the evolution becomes periodic.

5 Randomness in the Universe

5.1 Quantum Mechanics

The fact that light has properties of waves was known since the time of the British physicist Thomas Young who, as early as 1801, performed the two-slit experiment demonstrating that light like waves in the sea creates diffraction and interference patterns. However, almost a hundred years later Einstein observed that when light falls on a metal plate, the plate ejects a shower of electrons. He further observed that the shorter (the more energetic) the wave, the higher the speed with which the electrons are ejected. This is not what would happen if light were a wave. This is more like what happens when two particles collide.

Then, light appears to behave as both particle and wave. This establishes the wave-particle duality of light and proves that electromagnetic waves can behave as particles. *But can particles behave as waves*? This question was posed by the French physicist de Broglie who suggested that electrons, particles of a certain mass, could be treated as systems of superposed waves or as wave packets. Wave packets do not just describe pure waves or just pure particles but a combination of both. With such a description, de Broglie was able to study the quantum-mechanical motion of a particle and to predict the magnitude of the wavelength of an electron. And here lies the beauty of scientific ingenuity and reality. Scientists can always propose a theory, which based on mathematics will make predictions. But it is not until the predictions are verified that a theory becomes accepted. In the case of de Broglie's hypothesis, it was not long before experiments not only verified that electrons have characteristics of waves (they create interference patterns) but also recovered the predicted wavelength of the electron. Subsequent experiment went even further to show that even larger particles, such as molecules, behave as waves and that their wavelengths are exactly those predicted by de Broglie's theory. Soon after that the Viennese physicist Schrödinger developed the equation of motion of a particle whose solutions were the de Broglie waves. Before long, in 1927 the German physicist Werner Heisenberg starting from the hypothesis that an electron is a wave packet obeying a wave function proved his famous uncertainty principle, which ever since has made quantum mechanics as a mystic science as we will ever have.

In classical mechanics what specifies the complete state of a particle is its velocity and position. If we know these two variables we can solve the equations of motion and predict the position of the particle at any time in the future. Such complete determinism is the most fundamental aspect of classical mechanics. The uncertainty principle states that when it comes to subatomic particles, such as photons and electrons, one cannot measure exactly the position and velocity of a particle. More specifically it says that measuring the position with high accuracy results in a great uncertainty in the value of the velocity and vice versa. Consequently, we cannot ascertain the exact position of a particle without losing information about the velocity and vice versa. In quantum mechanics we cannot have well-defined states for the position and velocity; we can only have a *quantum state*, which is a combination of position and velocity. Since the position and velocity can only be known approximately, this state is defined by probabilities of the position and velocity. For example, "the particle's position is most probably somewhere there and its velocity is most probably around that value." Thus, quantum mechanics introduces an element of unpredictability or randomness in the scientific description of subatomic particles. Our universe emerges as intrinsically non-deterministic and unpredictable.

5.2 Chaos

Above, when we discussed randomness of the second kind, we demonstrated the sensitivity to the initial conditions and the definition of chaos using a simple mathematical equation that does not have a direct relation to a physical problem. This property, however, is not just the property of a mathematical system. It is found in natural systems as well. Back in the early seventeenth century, the German astronomer Johannes Kepler published his first law, which stated that the orbit of an object around an attracting body is an ellipse with the attracting body located at one of the foci. The ellipse remains constant in space, the speed, however, of the orbiting body varies. According to Newton's gravitational law, the force of attraction is proportional to the product of the masses of the two objects and inversely proportional to the square of their distance. Since the orbit is an ellipse the distance between the two bodies is not the same at all times. As such the gravitational pull varies; it is greatest at the pericenter and smallest at the apocenter. From Newton's second law it then follows that the speed of the orbiting object varies accordingly. Nevertheless, the position and speed of the orbiting object are determined at any time and they are regular. They repeat exactly after a fixed time interval.

The situation, however, becomes a bit more complicated when there are more than two bodies in the picture. For example, Earth attracts the moon while both are attracted by the Sun. What is the motion of the moon in this case? The problem can be exactly described by a set of nonlinear equations. However, the problem has no analytic solution. In other words we are not yet able to find a solution using standard mathematical approaches. The only way to solve such problems is numerically. If the calculations are done with sufficient numerical accuracy and for short time intervals, we can track the motion of the objects for a long time. This procedure can today be done efficiently with a computer. At the time of Kepler and Newton, however, this was not possible and both of them, while aware of the problem, saw this irregularity as a nuisance. It was not until the early twentieth century when the French multi-scientist Henri Poincare showed that the numerical solution to the three-body problem is very irregular and very sensitive to the initial condition. In fact Poincare discovered chaos but due to the unavailability of computers he could not study this problem in detail. In 1925 a glimpse into the complexity of this problem was provided by a computation carried out by 56 scientists under Elis

Stromgren at the Observatory of Copenhagen, which showed a solution to the so-called restricted three-body problem, which deals with the orbit of a moon under the gravitational influence of two planets. This work, which was published in 1925, took 15 years to complete (due to lack of computers). Today such a computation will take a few hours in a desktop computer. Nevertheless, for the first time it was realized that irregular behavior can be observed in a very simple systems that describes a natural phenomenon and that sensitivity to the initial conditions will make the behavior of the system practically unpredictable. This is the same property we discussed with the logistic equation, which we termed chaos.

The theory of chaos had to wait several decades until the development of fast computers allowed such calculations. Then, in 1963, Edward Lorenz, an atmospheric scientist at MIT, who was trying to explain why weather is unpredictable, reduced the complicated physics of the atmospheric circulation into three simple nonlinear differential equations (a differential equation describes changes of a variable in time), which modeled the behavior of a fluid layer heated from below. This is an approximation of what happens basically every day in the lower atmosphere in our planet. The Sun rises, and the surface of planet absorbs solar radiation and gets warm. Subsequently, the air gets warm by contact with the warmer surface and rises. This rising motion leads to turbulent motion. When Lorenz solved the equations and plotted the results he was surprised to see that this turbulent motion was behaving quite randomly and never repeating exactly. In addition, Lorenz found that this system is sensitive to the initial conditions. As with the logistic equation, evolutions from two slightly different initial conditions soon diverge and follow different evolutions. If we think of these two slightly different initial conditions as the true state of the atmosphere and what we actually measure (measurements always include some error, so we never really measure the true state), then their divergence indicates loss of predictability. This was the first time that somebody provided a scientific reason for why weather cannot be predicted with accuracy after a few days. Lorenz published his results in the highly respectable *Journal of Atmospheric Sciences* (Lorenz 1963). At that time, however, meteorologists were occupied with other problems and did not pay attention to this remarkable paper. It took more than a decade before mathematicians and physicists discovered the paper for the theory of chaos to take of and develop to one of the most important scientific theories of the twentieth century.

Does this mean that chaos is a major property of our universe? The problem with answering this question is that while individual systems might be chaotic, observations are often the result of many systems (some of which are chaotic some of which regular) interacting and affecting each other. For example, when we measure the vertical speed of the air inside a cloud, what system do we probe? Is our system the cloud itself? Or is it the atmosphere or maybe the earth or even the solar system? As we discussed above, in this case the actual chaos may be masked and what we get is stochasticity. Nevertheless, in the last three decades laboratory experiments as well as measurements of natural processes have shown that many phenomena in many areas of science are chaotic. This evidence makes sensitivity to the initial conditions a fundamental property of nature.

Chaos has been called by the late American physicist Joseph Ford "Gödel's child." Just as Gödel's theorem tells us that there will always be questions in any particular logical system that cannot be answered, chaos tells us that there are physical questions that cannot be answered like, what the weather is going to be in New York on November 19, 2021 (or some other date far in the future). For such a prediction, the initial state of each molecule in the atmosphere worldwide should be available to a precision that exceeds the limits imposed by quantum theory.

We should mention here that the unpredictability associated with chaos in natural systems, like weather for example, is more complicated than that of an abstract mathematical system, like the logistic equations, where the initial condition can be specified exactly. In natural systems we have to *measure* the initial condition. For example, to make a weather prediction we measure the temperature, pressure, moisture, and other variables, and then, we set the system (equations) in motion and see what happens. Measurements as we all know are subject to error. Every time I travel to my office I pass places where digital thermometers show the temperature. Somehow they all differ. This may be due to the natural variability of the temperature field, but I have noticed that the two thermometers in my house also never agree. Instruments are simply not exact. This will cause an uncertainty in the measurements used to specify the initial condition of the atmosphere. In this case we will start with an error. That error will couple with the round-off error introduced by the computer and things will go bad even faster. Not to mention that the initial state of the atmosphere is measured only at certain locations, thereby missing a lot of information in between. This results not only in an inexact initial condition but also in an incomplete one.

5.3 The Supreme Law

Imagine a container with two compartments, A and B, separated by a partition. Also imagine that A is full of air while B is empty. If we remove the partition what will happen? Obviously the air will expand to occupy both compartments. The opposite phenomenon where the air in a room suddenly accumulates in one half leaving the other half empty never happens. The impossibility of such events is due to the second law of thermodynamics, often hailed as the supreme law of nature.

Let us consider in more detail the example with the two compartments. Before the partition is removed the particles that make up the air are all in compartment A. This picture actually represents an ordered state, simply because there are restrictions for the particles. When the partition is removed the particles have no restrictions and they can move anywhere they wish. Eventually they uniformly occupy both compartments.

This is a state of equilibrium and from that point on, even though the particles are free to move all over, chances are that the same number of particles will be found in A and in B. This equilibrium state represents a state of lower order or higher disorder. Now would you call this process a reversible process? In other words would you expect that the particles would just by themselves return to the original ordered state? Some will argue that based on probability theory there is always a chance that somehow all the particles will be found in compartment A again, but I would not bet money on this. Even with a limited number of particles this may take the age of the universe before it happens. We can, therefore, safely assume that this process is irreversible. This irreversibility is directly related to the increase of disorder. Physically, this expresses the second law, which says that during an irreversible process the *entropy* of a system increases. Here, just to be on the safe side, we must mention that our system of the two compartments is considered isolated. In other words, it is alone and not interacting with anything else. In this case we cannot argue that an external force can be invoked, which will physically move all particles back to compartment A, thereby decreasing the entropy. By the way, entropy comes from the Greek word εντροπια, which means the "inner behavior."

In nature processes tend to be irreversible. In fact, unless a process is very much controlled by an experiment it is always an irreversible one. A cloud forms, it rains, and then dies out. You never see the opposite. A cup falls and breaks. You never see broken fragments rising and forming a cup. The interpretation of this law for the fate of our universe is fairly straightforward. Assuming that our universe is an isolated system, all the transformations that happen within it result in a steady increase of entropy with time. As such the universe is evolving toward a state of maximum entropy. Therefore, once the maximum entropy has been reached it cannot be increased any further. That simply means that there cannot be any changes anymore. Thus, the maximum entropy corresponds to equilibrium and the second law describes the general tendency of the universe to reach equilibrium.

All this sounds very matter of fact. Unlike quantum mechanics, nothing is weird about our discussion in this section. You may be wondering that in this case there is no place for randomness. Let us consider that we are dealing with four particles all of them being initially in compartment A. In how many different ways can we arrange four particles in A and no particles in B? The answer is obviously in one way. Four particles in A and zero in B. Now we know that if we remove the partition the particles will move around and will occupy both compartments uniformly (i.e. without a preference in either compartment). Thus, we would expect that at any time two particles will be in A and two in B. In this case, in how many different ways can we have two particles in A and two particles in B? The answer to this question is six.

The number of different ways to arrange particles in the two compartments is called the *number of complexions*. Thus, we see that during our irreversible process of particle dispersion, where the entropy is increasing, the number of complexions is also increasing. In the late 1800s the Austrian physicist Ludwig Boltzmann proved that this number of complexions is directly related to the entropy of the system and indeed as the number of complexions increases entropy increases proportionally.

Are the above possibilities the only ones? Could we not have three particles in A and one in B or vice versa? Of course we can. In this case the number of complexions is four.

Thus, altogether (including the possibility that all particles are in B) there are 12 possible particle configurations. Six out of those 12 configurations correspond to maximum disorder, or maximum entropy, two correspond to minimum entropy and four to some intermediate value.

It follows that the most probable state is the state of maximum entropy. It also follows that the irreversibility of natural changes does not result from certainty but from *probability*. There is a higher probability to tend toward the state of maximum entropy than otherwise. We have thus discovered that the essential way in which systems evolve is statistical and that in nature irreversibility is associated with randomness. And since from the small-scale order we go to the large-scale disorder, unlike in quantum mechanics where randomness defines the micro-cosmos, the second law tells us that probability rules the macro cosmos. For completeness, Boltzmann's mathematical relationship between entropy (S) and the number of complexions (P) is

$$S = k \ln P + \text{constant}$$

where

$$P = N!/N_1!N_2!$$

and N_1 is the number of particles in A and N_2 the number of particles in B.

Finally we should mention that since all subatomic particles obey quantum mechanics and all matter is made out of these particles, quantum mechanics is considered the most fundamental theory of nature. The rules of classical mechanics, which are followed by large objects (regular or chaotic), should somehow emerge from quantum mechanics. It is like an impressionist painting, which though fuzzy when viewed close-up, produces a coherent picture when viewed from afar. The connection between quantum mechanics and the macro-scale has not been achieved yet. This may be because quantum theory is not complete. Or it may be that we don't understand it completely. Nevertheless, quantum mechanics has seen many successes and applications in many areas. These include the prediction of the existence and subsequent discovery of the particle positron, the explanation of the formation of a positron and an electron when electromagnetic energy interacts with matter, the operation of transistors, and the development of lasers. Due to that, while it may be that it needs refining, quantum mechanics and its randomness is here to stay. Quantum mechanics, chaos, and the second law are prime examples that our universe in its infinite space does not just obey rules but that it is also inherently random. Here again, as with the mathematical system, we find the concepts of infinity, randomness, and rules, interweaved and working together.

5.4 *Randomness of the Fourth Kind?*

Consider the game of chess. In chess 32 agents interact according to specific rules. For the masters, the game evolves according to a plan but there may be moments where for the next move more than one possibility may exist. Because of the limited time between moves, the player cannot possibly go through all the possible configurations and often has to use an "educated guess" or his free will. Based on the choice, the outcome may be different. This uncertainty in the final result is randomness introduced by free will. A high-speed computer, on the other hand, could run all possible configurations and possibilities in the allotted time. In this case the computation is faster than the evolution of the game and there is perfect predictability: the computer will beat a human opponent all the time. As another example, imagine a soccer player leading an attack toward the opponents' net. During his run, he often has to pass the ball to one of his teammates. The outcome of the attack depends on which teammate will get the ball. Our player can pass the ball to several of his teammates, but as he is advancing toward the opponents' net he has little time to compute all the possibilities open to him. Here again the computation is slower than the evolution of the game and hence the player cannot make an accurate prediction. As is often the case, our player makes a choice, which even though he may be using his best judgment may not be the best choice. He simply uses his free will.

The issue of free will is rather controversial. By definition, free will is the conviction that humans have the capacity to choose their actions. It is a very divisive issue among philosophers, and this author is the last scientist who will argue with philosophers.

However, in the scientific realm (which is of interest here), free will implies that decision making is not completely and necessarily determined by a physical prior cause(s). If we accept the view of a completely deterministic universe then there is simply no free will or randomness whatsoever. Everything has a cause, and the only reason we don't understand or cannot predict is absence of complete knowledge. If on the other hand we reject determinism all together, then everything that happens is independent of what happened before. What modern science and mathematics pointing, however, is that both these two extremes are just extremes. The discussion in this paper presents plenty of evidence that in our universe and in the mathematical system that describes it, determinism and randomness coexist. I personally cannot but accept the fact that free will exists in humans and that free will choices do not necessarily require prior causes. In this case free will actions may through humans introduce randomness in the universe. I will not go into the "weird" topic of whether an electron has free will (I will leave this to philosophers) but it is easy to explain how free will emerges in a deterministic universe.

We can thus see that there could be instances when humans would inject randomness into their environment. The open question is whether or not it is only humans that have free will or that it is also a property of the universe as a whole. Could it be, for example, that the universe is a cellular automaton (as Stephen

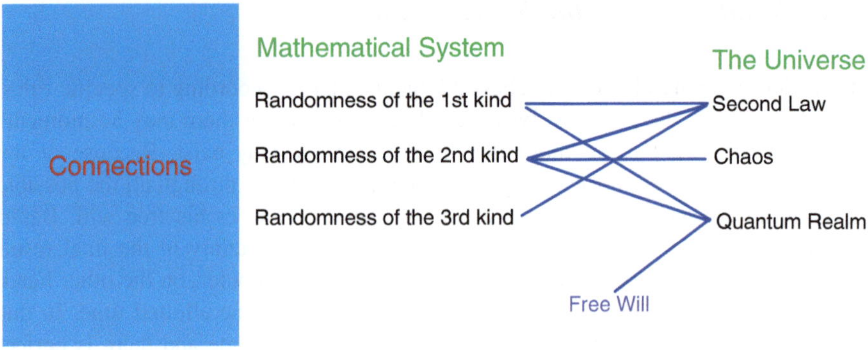

Fig. 8 Possible connections between the mathematical and physical systems

Wolfram contents,) and that it is impossible to devise a simulator that runs faster than it, thereby, causing free will to emerge? I will not attempt to get myself into such an abstract and philosophical issue but I will quote Ray Kurzweil from his "Reflections on Stephen Wolfram's *A New Kind of Science*": "…it should be noted that it is difficult to explore concepts such as free will and consciousness in a strictly scientific context because these are inherently first-person subjective phenomena, whereas science is inherently a third person objective enterprise. There is no such thing as the first person in science, so inevitably concepts such as free will and consciousness end up being meaningless. We can either view these first person concepts as mere illusions, as many scientists do, or we can view them as the appropriate province of philosophy, which seeks to expand beyond the objective framework of science" (Kurzweil 2003).

After all the discussion so far, you may wonder why nature chooses to be irregular and unpredictable. Why is our universe not simple and regular? Well, this is indeed an interesting question, which would be answered soon, but first we have to make connections between the sources of randomness in the mathematical system and those in the physical system.

6 Connections

Figure 8 summarizes the possible connections between the mathematical and physical systems. First, let us take quantum mechanics. The randomness introduced by the uncertainty principle results from assuming that a photon or an electron is a system of superposed states or a wave packet. These wave packets make use of the particle-wave duality. We know that the duality is a fact. But we are still unable to explain *why* the duality exists. In this regard one may argue that we simply cannot recover at this point the rules and as such randomness of the first kind is generated. One might also conjecture that due to the inherited uncertainty in quantum mechanics, it will be subject to chaos. The area of quantum chaos is of great interest in science today. Indeed, there have been many indications that quantum systems display chaotic characteristics. This links quantum mechanics with randomness of the second kind. Now suppose that, as is implied by modern developments in theoretical physics, the universe began as a ten-dimensional bubble of space out of which only four (time and the three spatial) dimensions expanded to form the universe we live in. The other six dimensions simply compacted to form what we call subatomic particles. In a sense then these particles live in a ten-dimensional space, whereas they are observed in a four-dimensional space. Then, what we observe is a projection of an object in a lower dimension. Such a projection may result in observations that cannot be explained. For illustration purposes consider the trajectory on the top Fig. 9. This trajectory is embedded in a three-dimensional space. If we project this trajectory onto a two-dimensional space we will obtain the result shown on the bottom of the Fig. 9. Now the trajectory appears to intersect itself at one point. Every point in the two-dimensional picture corresponds to a point in the three-dimensional picture except for the intersection point, which represents two points in the three-dimensional picture. Thus, in a lower dimension those two separate states appear superposed. If something like this applies to our universe, then we simply do not have the complete picture right. This in turn implies that we either do not know all the rules or we do not have enough information about the initial state. These possibilities create again

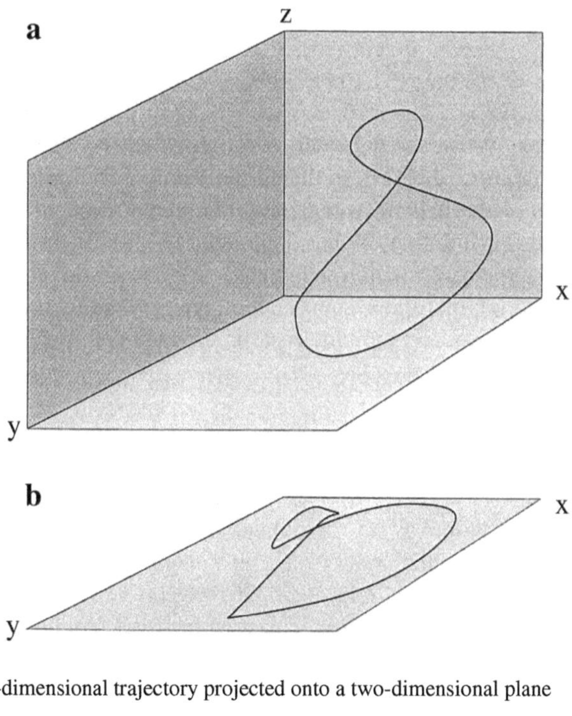

Fig. 9 A three-dimensional trajectory projected onto a two-dimensional plane

randomness of the first and second kind. Finally, let us ponder on the many-words interpretation of quantum mechanics. According to this interpretation the random quantum processes cause the universe to split to as many copies as the possible outcomes. Thus, when our soccer player is ready to pass the ball, quantum effects at his brain lead to a superposition of many possibilities. All possibilities happen, but they happen at different universes. In our universe what the player chooses (free will) appears as a slight randomness. The splitting of the universes can then be seen as a computationally irreducible process producing randomness from free will.

6.1 On Now to the Second Law

Recall that the second law dictates that for irreversible processes the entropy increases. As we discussed earlier, this leads to the inevitable introduction of probability and randomness in the macro-scale. Imagine you have a cup of warm water and a cup of cold water. To start with, you have an amount of information that specifies the difference in temperature between the two cups. You then pour the water from both cups into a pan. What do you get? Simply, you get lukewarm water. This is an irreversible process in which the entropy increases. Does this lukewarm water retain any of the original information? Apparently, not. You cannot say anything about the original temperatures anymore. This demonstrates that irreversible process and entropy increase are associated with loss of information. The same phenomenon occurs in chaotic systems. When we measure an initial condition we may measure it with some uncertainty. Nevertheless, we do have some information about the initial condition. For example, we may measure the outside temperature with some error but we will have a pretty good idea of how warm or cold it is. Whatever information we have in the measurement of the initial condition is, however, lost in the future through the amplification of the uncertainty we have in that measurement. This is the same as saying that the system loses whatever information was supplied to it. It will then appear that the second law may be connected with randomness of the first and of the second kind. Alternatively, assuming that the system undergoing an irreversible transformation is an assembly of many, many individual particles, each one obeying some simple rules and interacting with every other particle, leads to stochasticity, the third kind of randomness.

Compelling arguments can therefore be made that the same ways for generating randomness suggested by pure mathematical systems may apply to real physical systems.

As we mentioned before a key ingredient in all mechanisms of randomness is infinity. In that regard randomness may be thought as an infinite-dimensional (unrestricted) system, whereas rules can be thought as representing relatively low-dimensional (restricted) systems. Thus, while rules confine the dynamics, randomness acts without limits. This interplay creates most of the time "something" in between. This "something" in between is often referred to as *complexity,* and it is the major characteristic of our universe.

7 The Role of Randomness

Imagine a group of small children in a playground. Small children have no fear of having an accident. They have not yet developed this feeling. As such they tend to move around irregularly changing activities and bumping into each other constantly. Such a setup is always vulnerable to accidents. The question is how do we minimize the chance for an accident? One way to eliminate accidents is to have one caretaker per child. In this case, the caretaker is on constant alert guiding each child in all its activities and making sure that no harm will come to it. This scenario amounts to complete determinism with no randomness allowed in the system. This solves the problem, but then all of us would have to become professional babysitters. Simply, this solution is not efficient. It requires too much effort. A more efficient solution would be to limit the area of activity (say, by putting a fence around the playground) and have a few caretakers supervise the children. In this case accidents may happen (and they do) but they will be much less frequent compared to the number of accident when there are no rules or limits (randomness only).

Now imagine a parcel of air near the surface that is a perfect cube. As this cube begins to rise it expands, its relative humidity increases, and eventually becomes saturated. After that, as the parcel continues to rise, a cloud begins to form. But, what happened to our initial parcel during this process? As we all know the shape of a cloud is complex and no two clouds are alike. Given the fact that cubic clouds have never been observed, the initial cube simply gives away to some irregular structure. But why? One could imagine a process whereby each molecule in the original cube moves in such a way as to always form a cube. We could actually devise artificial rules that will have every molecule follow such an evolution. But can you imagine the effort that our atmosphere will have to make in order to achieve this? Nature will have nothing to do with processes like that. Instead, like the example with the children, the rules are set and within these rules the molecules are left to move randomly thus generating irregular cloud shapes.

Let's now consider different examples. Languages were one of the first necessities for humans. We simply had to communicate. But how do you think that languages evolved? Take a little baby that begins to learn how to speak. What are the first words? If there is something universal in our cultures it is that babies all over the world begin muttering simple repetitive syllables like ba-ba, ma-ma, or something similar. They then proceed by becoming more elaborate. There is evidence that human language evolved similarly. In very primitive language words were made of very repetitive basic sounds. In fact, successful decipherment of ancient languages was based on finding repetitive units (Robinson 2002). This is true for music as well. The organization of music normally involves basic material that may repeat exactly or with variations, may alternate with other material, or may proceed continually to present new material. Composers strike a balance between unity and variety, and all pieces contain a certain amount of repetition. As with languages, music may also have evolved from very simple repetitive musical blocks. In fact, early music was very repetitive.

The same has happened to the blueprint of life. Gene evolution is one of the most important aspects of evolutionary and molecular biology. Early in the 1970s the Japanese-American biologist Susumu Ohno advanced his ideas about a possible mechanism of evolution by *gene duplication* (Ohno 1970). In short, Ohno suggested that modern sequences arose from small pieces of genetic material (often called primordial blocks), which found a way to duplicate. Once this was possible, further duplication generated longer and longer sequences that led to the construction of genes. The same mechanism resulted in the generation of novel genes, by gene duplication, and new species by whole genome duplication. However, duplication alone does not produce any novelty. Because of that, another mechanism was at work together with the simple duplication. This mechanism is random *mutation*. As the primordial blocks duplicated they also mutated, meaning that they made slight changes in the repetition pattern. These random mutations are the key ingredient of one of the most powerful theories in the history; Darwin's theory of evolution. In short, according to Darwin, life evolved by natural selection and random mutations. Natural selection is the idea that individual species possess some variation that gives them an advantage over other species when it comes to survival. For example, imagine that in an isolated island populated by different species of birds an environmental fluctuation has caused plants that produce small seeds to die, but plants baring large seeds to survive. Because of some random mutation in their past some of these birds have developed big beaks. Those birds have an advantage in picking up the large seeds compared to those birds that did not have this variation and remained with small beaks. Thus, the birds with the big beaks survive and the birds with small beaks get extinct. Environmental fluctuations and random mutations determined which species lived and which species died. The actual mechanism for this randomness in mutations is still an open question. It is widely believed that mutations arise from pure environmental factors. In this case the randomness may be due to stochasticity. Recent analysis of DNA sequences has also suggested that within this stochasticity some evidence can be identified suggesting that a component of this randomness may be connected to chaotic processes (Tsonis et al. 2002). Whatever the source of randomness, however, the point is that life with its entire splendor is the result of a very simple rule (duplication) plus randomness.

The similarities in the properties of languages, music, and DNA may indicate that all of them have employed a similar construction process in their evolution: repetition and mutations (randomness). And apparently, this process does not apply only to languages, music, and DNA. Given the plethora of self-similar structures in nature (which are created by repeating a certain operation over and over again), it would appear that there is something fundamental in evolving by copying or repeating an operation and modifying it at random, and that this process was adopted by both nature and humans in the early stages of evolution.

And why, you may ask, such a procedure became the favored one? The study of languages, music, and DNA provides an interesting insight into this question. All three of them share a common property. They all transmit information. Furthermore, it is reasonable to assume that they all transmit information effectively and efficiently. Something that is effective and efficient uses the least amount of effort to do what is supposed to do. There is no reason, for example, for nature to adopt a very complex and expensive mechanism to transmit information or to do an operation. A simple and economical procedure would be much more desirable. And what can be simpler than repetition? It may not, thus, be surprising that once the art of repetition or copying was "learned" it become a fundamental mechanism in nature and in human dynamics. But, since pure repetition does not create innovations, randomness was introduced to "spice" things up. I do not mean to imply that this is the only mode operanti in the universe or that other more complicated rules were not introduced later, but clearly simple rules and randomness do not just coexist but they synchronize to produce an efficient and economic universe. It is interesting to note here that many mathematical systems obeying simple rules (such as cellular automata) have been reported, which copy or replicate themselves and which through replication construct more complex structures (Langton 1986). The period doubling observed in the dynamics of the logistic equation is also an example of how duplication leads to complex behavior.

The above can be summed up by what Zipf called the *principle of least effort* or what I call the *principle of minimum energy consumption* (Zipf 1949). Mathematical and physical support for the minimum energy consumption of minimum dissipation principle is provided by the work of Ilya Prigogine. Ilya Prigogine was born in Moscow a few months before the revolution. His family left Russia in 1921 and after spending a few years in Germany settled for good in Belgium. His work in non-equilibrium thermodynamics (Prigogine 1980) won him the Nobel prize in chemistry in 1977. Part of this work is the famous theorem of minimum entropy production, which states that when a system cannot reach equilibrium, but operates near equilibrium, the system settles to a state of minimum dissipation. Natural systems (and for that matter social economic and other systems) can operate at equilibrium, near equilibrium, and far from equilibrium. From all these sates we can argue reasonably that the most preferred state is the near equilibrium state (far from equilibrium represents extreme situations and complete equilibrium means no more ability for changes). Accordingly, while the minimum energy consumption or minimum dissipation principle is not a universal principle it does apply to most phenomena observed in nature.

As Howard L. Resnikoff puts it in his book *The Fusion of Reality* "Fermat's classical variational principle of 'least time' and Maupertuis and Hamilton's principle of 'least action' express the parsimony of nature in a mathematical form: the evolution of a physical system follows that path amongst all conceivable alternatives that extremizes, i.e. maximizes or minimizes, a suitable cost function such as time, action, or energy. Thus, the path of a ray of light through an optically inhomogeneous medium minimizes the time required to pass from the initial position to its emergent point" (Resnikoff 1989). In the same issue he argues that since the final state of an irreversible process is rather unpredictable (due to so many numbers of possible configurations), the final state of maximum entropy is a priori quite unknown. In this case, any measurement of that state yields the maximum information possible about the system (simply because before that there is no available information). In a sense this represents a minimum effort to know something about the system. Therefore, maximization of entropy (and thus the second law) is consistent with the principle of the least effort.

8 Summary

We started our adventure into randomness by looking exclusively at our formal mathematical system and we saw that even in this pure and strictly logical system one cannot do away with randomness. Rules and randomness are blended together and are engulfed in the notion of infinity. Staying within the mathematical system and employing simple mathematical models, we then discussed the three possible sources of randomness: randomness due to inability to find the rules, randomness due to inability to have infinite power (chaos), and randomness due to stochastic processes. Subsequently we expanded from the mathematical system to our physical world and we found out that randomness, through the quantum mechanical character of small scales, through chaos, and because of the second law of thermodynamics, is an intrinsic property of nature as well. We subsequently argued that the randomness in the physical world is consistent with the three sources of randomness suggested from the study of simple mathematical systems. Finally we suggested the principle of least effort or the principle of minimum energy consumption as the underlying principle behind this combination.

We can thus conclude that no matter how randomness comes about, randomness and rules are bound together. They operate together. They synchronize together. They shape our Universe and produce the reality we see and feel everyday in our lives. Randomness emerges as a property of the Universe. This synergy between rules and randomness makes them both equally important in the Universe. One cannot exist without the other. While rules impose boundaries randomness acts between boundaries. They interweave together like facts and fiction in a historical novel. And overlooking this weaving is infinity, the one ingredient behind all mechanisms generating randomness. Possibly, it is what may make them one and the same thing.

I started with a quote by Aristotle, which I found appropriate to introduce the discussion. I would like to end it by another quote by Aristotle, which I find appropriate to the paper's summary.

> Since nothing accidental is prior to the essential neither are accidental causes prior. If, then, luck or spontaneity is a cause of the material universe, reason and nature are causes before it.
>
> Aristotle, Metaphysics, Book XI, 8

References

Hofstadter, D.R. 1979. *Godel, Escher, Bach: an eternal golden braid*. New York: Basic Books.

Kurzweil, R. 2003. *Reflections on Stephen Wolfram's 'a new kind of science'*. www.Kurzweilai.net/articles/art0464.html

Langton, C.G. 1986. Studying artificial life with cellular automata. *Physica* 22D: 120–149.

Lorenz, E.N. 1963. Deterministic nonperiodic flow. *Journal of the Atmospheric Sciences* 20: 130–141.

Mandelbrot, B.B. 1983. *The fractal geometry of nature*. New York: Freeman.

Ohno, S. 1970. *Evolution by gene duplication*. Heidelberg: Springer-Verlag.

Peitgen, H.-O., and D. Saupe. 1988. *The science of fractal images*. New York: Springer-Verlag.

Peitgen, H.-O., H. Jurgens, and D. Saupe. 1992. *Chaos and fractals: new frontiers of science*. New York: Springer-Verlag.

Prigogine, I. 1980. *From being to becoming*. New York: Freeeman.

Resnikoff, H.L. 1989. *The fusion of reality*. New York: Springer-Verlag.

Robinson, A. 2002. *Lost languages*. New York: McCraw Hill.

Tsonis, A.A. 1987. Some probabilistic aspects of fractal growth. *Journal of Physics A: Mathematical and General* 20: 5025–5028.

———. 1991. The effect of truncation and round-off on computer generated chaotic trajectories. *Computers and Mathematics with Applications* 21: 93–94.

———. 1996. Dynamical systems as models of physical processes. *Complexity* 1 (5): 23–33.

———. 2008. *Randomnicity: rules and randomness in the realm of the infinite*. London: Imperial College Press.

Tsonis, A.A., and J.B. Elsner. 1987. Fractal characterization and simulation of lightning. *Contributions to Atmospheric Physics* 60: 187–192.

Tsonis, A.A., F. Heller, and P.A. Tsonis. 2002. Probing the linearity and nonlinearity in DNA sequences. *Physica A* 312: 458–468.

Wolfram, S. 2002. *A new kind of science*. Champain IL: Wolfram Media, Inc..

Zipf, G. 1949. *Human behavior and the principle of least effort*. Cambridge MA: Addison-Wesley.

Insights in Climate Dynamics from Climate Networks

Anastasios A. Tsonis

Abstract This review is a synthesis of work spanning the last 25 years. It is largely based on the use of climate networks to identify climate subsystems/major modes and to subsequently study how their collective behavior explains decadal variability. The central point is that a network of coupled nonlinear subsystems may at times begin to synchronize. If during synchronization the coupling between the subsystems increases the synchronous state may, at some coupling strength threshold, be destroyed shifting climate to a new regime. This climate shift manifests itself as a change in global temperature trend. This mechanism, which is consistent with the theory of synchronized chaos, appears to be a very robust mechanism of the climate system. It is found in the instrumental records, in forced and unforced climate simulations, as well as in proxy records spanning several centuries.

Keywords Networks • Climate subsystems • Synchronization • Climate networks

1 Introduction

The flowchart in Fig. 1 provides the outline of this review. The story starts in the mid-1980s when new and exciting approaches to nonlinearly analyze time series made their appearance in atmospheric sciences. At that time very few in the atmospheric sciences community had heard terminology such as "fractals," "chaos theory," "strange attractors," and the like. Soon, reports of "fractality" and "low dimensionality" in climate records and other geophysical data begun to surface. These climate records represented dynamics over different time scales ranging from very long (thousands of years; Nicolis and Nicolis 1984) to very short (hours; Tsonis and Elsner 1988). Virtually every report suggested underlying attractors of

A.A. Tsonis (✉)
Department of Mathematical Sciences, Atmospheric Sciences Group, University of Wisconsin - Milwaukee, Milwaukee, WI 53201, USA

Hydrologic Research Center, San Diego, CA, USA
e-mail: aatsonis@uwm.edu

A.A. Tsonis (ed.), *Advances in Nonlinear Geosciences*,
DOI 10.1007/978-3-319-58895-7_29

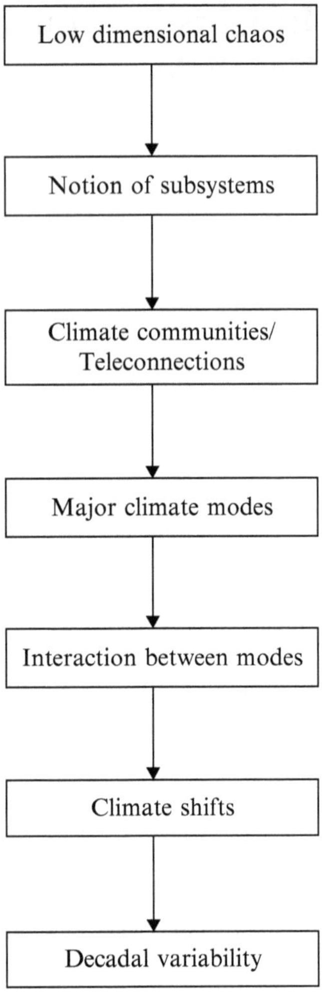

Fig. 1 Flowchart of the outline of this review

dimensions between 3 and 8. These early results suggested that climate variability might indeed be described by relatively a few differential equations. This resulted in both enthusiasm and hope that climate variability may be tamed after all, and in fierce opposition. Fortunately, this "tug of war" did not eliminate interest in this new theory; rather it led to a deeper understanding of the nonlinear character of nature and to new insights about the properties of the climate system. This review is a small part of what we have learned so far and it largely draws from our work over the years.

The initial opposition to those dimension estimates seemed to be that in all these studies the sample size was simply too small. While this issue has been debated extensively (Smith 1988; Nerenberg and Essex 1990; Tsonis 1992; Tsonis et al. 1994), it still remains contentious. In a sense, it is naïve to imagine that our climate system (a spatially extended system of infinite dimensional state space) is described by a grand attractor, let alone a low dimensional attractor. If that were true, then all observables representing different processes should have the same dimension, which is not likely the case based on the myriad of reported dimensions. In Tsonis and Elsner (1989), it was suggested that if low dimensional attractors exist they are associated with subsystems each operating at different space and/or time scales. In his study on dimension estimates, Lorenz (1991) concurs with the suggestion of Tsonis and Elsner (1989). These subsystems may be nonlinear and exhibit a variety of complex behaviors. All subsystems are connected with each other, as in a web, with various degrees of connectivity. Accordingly, any subsystem may transmit "information" to another subsystem thereby perturbing its behavior. This "information" plays the role of an ever-present external noise, which perturbs the subsystem, and, depending on the connectivity of a subsystem to another subsystem, the effect can be dramatic or negligible. Subsystems with weak connectivities will be approximately "independent" and as such they may exhibit low dimensional chaos. It is also possible that the connectivity between subsystems may vary in time and this effect may dictate the variability of the climate system.

Thus, evidence of low dimensional chaos leads to the notion of climate subsystems. Given this, the question arises. If subsystems exist in the climate system what are they and what physics can we infer from them?

2 Searching for Subsystems

Answers on the nature, geographical basis, and physical mechanisms underlying these subsystems are provided by recent developments in graph theory and networks. Networks relate to the underlying topology of complex systems with many interacting parts. They have found many applications in many fields of sciences. In the interest of completeness short introduction to networks is offered next.

A network is a system of interacting agents. In the literature an agent is called a node. The nodes in a network can be anything. For example, in the network of actors, the nodes are actors that are connected to other actors if they have appeared together in a movie. In a network of species the nodes are species that are connected to other species they interact with. In the network of scientists, the nodes are scientists that are connected to other scientists if they have collaborated. In the grand network of humans each node is an individual, which is connected to people he or she knows. There are four basic types of networks.

a. *Regular (ordered) networks*. These networks are networks with a fixed number of nodes, each node having the same number of links connecting it in a specific way to a number of neighboring nodes (Fig. 2, left panel). If each node is linked to all other nodes in the network, then the network is a fully connected network. When the number of links per node is high, regular networks have a high (local) clustering coefficient. In this case loss of a number of links does not break the network into non-communicating parts. In this case the network is stable, which may not be the case for regular networks with small local clustering. Also, unless

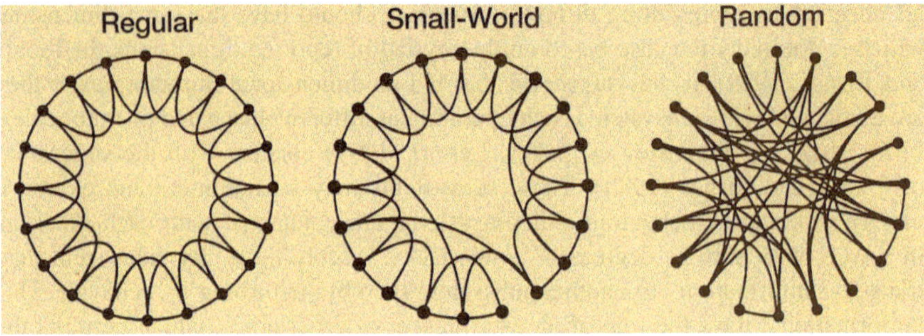

Fig. 2 Illustration of a regular, a small-world, and a random network (after Watts and Strogatz 1998)

networks are fully connected, they have a large diameter. The diameter of a network is defined as the maximum shortest path between any pair of its nodes. It relates to the characteristic path length, which is the average number of links in the shortest path between two nodes. The smaller the diameter, the easier is the communication in the network.

b. *Classical random networks*. In these networks the nodes are connected at random (Fig. 2, right panel). In this case the degree distribution is a Poisson distribution (the degree distribution, p_k, gives the probability that a node in the network is connected to k other nodes). The problem with these networks is that they have very small clustering coefficient and thus are not very stable. Removal of a number of nodes at random may fracture the network to non-communicating parts. On the other hand, they are characterized by a small diameter. Far away nodes can be connected as easily as nearby nodes. In this case information may be transported all over the network much more efficiently than in ordered networks. Thus, random networks exhibit efficient information transfer but they are not stable.
c. *Small-world networks*. In nature we should not expect to find either very regular or completely random networks. Rather we should find networks that are efficient in processing information and at the same time are stable. Work in this direction led to a new type of network, which was proposed twelve years ago by the American mathematicians Duncan Watts and Steven Strogatz and is called *small-world* network (Watts and Strogatz 1998). A "small-world" network is a superposition of regular and classical random graphs. Such networks exhibit a high degree of local clustering but a small number of long-range connections make them as efficient in transferring information as random networks. Those long-range connections do not have to be designed. A few long-range connections added at random will do the trick (Fig. 2, middle panel). The degree distribution of small-world networks is also a Poisson distribution.
d. *Networks with a given degree distribution*. The "small-world" architecture can explain phenomena such as the six-degrees of separation (most people are friends with their immediate neighbors but we all have one or two friends a long way

away), but it really is not a model found often in the real world. In the real world the architecture of a network is neither random nor small-world but it comes in a variety of distributions such as truncated power-law distributions, Gaussian distributions, power-law distributions, and distributions consisting of two power-laws separated by a cutoff value (for a review see Strogatz 2001). The most interesting and common of such networks are the so-called *scale-free* networks. Consider a map showing an airline's routes. This map has a few hubs connecting with many other points (super nodes) and many points connected to only a few other points, a property associated with power-law distributions. Such a map is highly clustered, yet it allows motion from a point to another far away point with just a few connections. As such, this network has the *property* of small-world networks, but this property is not achieved by local clustering and a few random connections. It is achieved by having a few elements with large number of links and many elements having very few links. Thus, even though they share the same property, the architecture of scale-free networks is different than that of "small-world" networks. Such inhomogeneous networks have been found to pervade biological, social, ecological, and economic systems, the internet, and other systems (Albert et al. 1999; Liljeros et al. 2001; Jeong et al. 2001; Pastor-Satorras and Vespignani 2001; Bouchaud and Mezard 2000; Barabasi and Bonabeau 2003). These networks are referred to as scale-free because they show a power-law distribution of the number of links per node. Lately, it was also shown that, in addition to the power-law degree distribution, many real scale-free networks consist of self-repeating patterns on all length scales (Song et al. 2005). These properties are very important because they imply some kind of self-organization within the network. Scale-free networks are not only efficient in transferring information, but due to the high degree of local clustering they are also very stable (Barabasi and Bonabeau 2003). Because there are only a few super nodes, chances are that accidental removal of some nodes will not include the super nodes. In this case the network would not become disconnected. This is not the case with weakly connected regular or random networks (and to a lesser degree with small-world networks), where accidental removal of the same percentage of nodes makes them more prone to failure (Barabasi and Bonabeau 2003).

The topology of the network can reveal important and novel features of the system it represents (Albert and Barabasi 2002; Strogatz 2001; da F. Costa et al. 2007). One such feature is communities (Newman and Girvan 2004). Communities represent groups of densely connected nodes with only a few connections between groups. It has been conjectured that each community represents a subsystem, which operates relatively independent of the other communities (Arenas et al. 2006). Thus, identification of these communities can offer useful insights about dynamics. In addition, communities can be associated with network functions such as in metabolic networks where certain groups of genes have been identified that perform specific functions (Holme et al. 2003; Guimera and Amaral 2005).

Recently, concepts from network theory have been applied to climate data organized as networks with impressive results (Tsonis et al. 2006, 2007, 2008; Tsonis and Swanson 2008; Yamasaki et al. 2008; Gozolchiani et al. 2008; Swanson and Tsonis 2009, Elsner et al. 2009; Tsonis et al. 2011).

Figure 3 is an example of a climate network showing the area-weighted connectivity (number of edges) at each geographic location for the 500 Hpa height field. More accurately it shows the fraction of the total global area that a point is connected to. This is a more appropriate way to show the architecture of the network because the network is a continuous network defined on a sphere [see Tsonis et al. 2006 for details]. These data are derived from the global National Center for Environmental Prediction/National Center for Atmospheric Research (NCEP/NCAR) atmospheric reanalysis data set (Kistler et al. 2001). In Fig. 3 we observe two very interesting features. In the tropics it appears that all nodes possess more or less the same (and high) number of connections, which is a characteristic of fully connected networks. In the extratropics it appears that certain nodes possess more connections than the rest, which is a characteristic of scale-free networks. In the northern hemisphere we clearly see the presence of regions where such super nodes exist in China, North America, and Northeast Pacific Ocean. Similarly several super nodes are visible in the southern hemisphere. These differences between tropics and extratropics have been delineated in the corresponding degree distributions, which suggest that indeed the extratropical network is a scale-free network characterized by a power-law degree distribution (Tsonis et al. 2006). As is the case with all scale-free networks, the extratropical network is also a small-world network (Tsonis et al. 2006).

An interesting observation in Fig. 3 is that super nodes may be associated with major teleconnection patterns. For example, the super nodes in North America and Northeast Pacific Ocean are located where the well-known Pacific North America (PNA) pattern (Wallace and Gutzler 1981) is found. In the southern hemisphere we also see super nodes over the southern tip of South America, Antarctica, and South Indian Ocean that are consistent with some of the features of the Pacific South America (PSA) pattern (Mo and Higgins 1998). Interestingly, no such super nodes are evident where the other major pattern, the North Atlantic Oscillation (NAO) (Thompson and Wallace 1998; Pozo-Vazquez et al. 2001; Huang et al. 1998), is found. This, however, does not indicate that NAO is an insignificant feature of the climate system. Since NAO is not strongly connected to the tropics, the high connectivity of the tropics with other regions is masking NAO out (Tsonis et al. 2008). Indeed if we consider only the extratropics the resulted network is dominated by NAO (Fig. 4).

This is also indicated by the community structure of the 500 HPa network (Fig. 3) shown in Fig. 5 (for details, see Tsonis et al. 2011). The total number of communities is 47. Many of these communities, however, consist of very few points in the boundaries between a small number of dominant communities (think of a country whose population is not only dominated by two races but also includes small groups of other races). Evidently the effective number of communities is, arguably, four (delineated as purple, blue, green, and yellow-red areas). We observe that three of the effective four communities correspond to a latitudinal division 90 S–30 S,

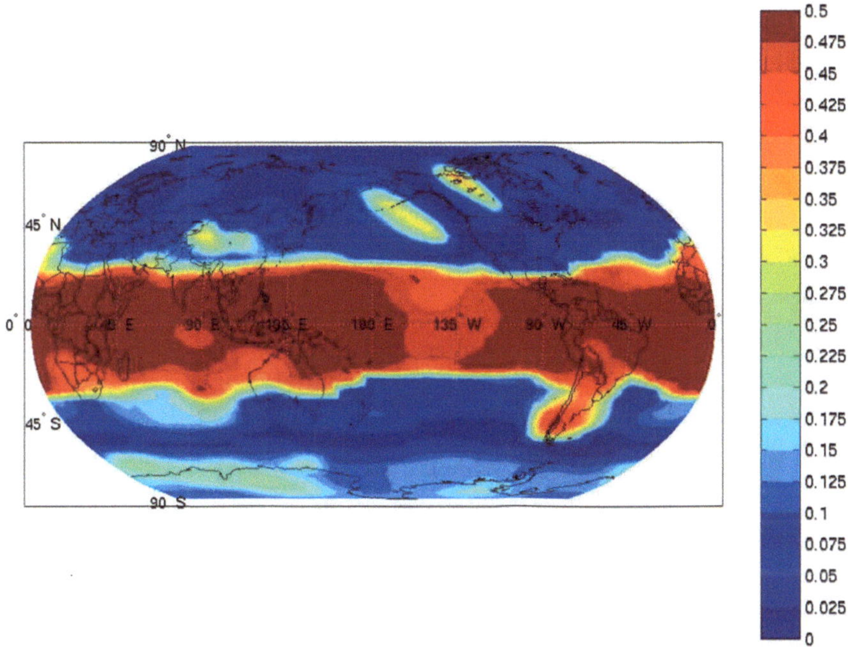

Fig. 3 Total number of links (connections) at each geographic location. More accurately it shows the fraction of the total global area that a point is connected to. This is a more appropriate way to show the architecture of the network because the network is a continuous network defined on a sphere. The uniformity observed in the tropics indicates that each node possesses the same number of connections. This is not the case in the extratropics where certain nodes possess more links than the rest. The definition of a link is based on cross-correlations at lag zero (r) between the time series of any pair of points (nodes). Note that since the values are monthly anomalies there is very little autocorrelation in the time series. A pair is considered as connected if the absolute value of their cross-correlation $|r| \geq 0.5$. This criterion is based on parametric and non-parametric significance tests. According to the t-test, a value of $r = 0.5$ is statistically significant above the 99% level. In addition, randomization experiments where the values of the time series of one node in a pair are scrambled and then are correlated to the unscrambled values of the time series of the other node indicate that a value of $r = 0.5$ will not arise by chance. The choice of $r = 0.5$ while it guarantees statistical significance is somewhat arbitrary. We find that while other values might affect the connectivity structure of the network, the effect of different correlation thresholds (between 0.4 and 0.6) does not affect the conclusions. Obviously, as the threshold $| r | \to 1$ we end up with a random network and as $r \to 0$ we remain with just one fully connected community. The use of the correlation coefficient to define links in networks is not new. Correlation coefficients have been used to successfully derive the topology of gene expression networks (Farkas et al. 2003) and to study financial markets (Mantegna 1999). Other ways to define a link exist. Donges et al. (2009a, b), for example, have used the mutual information instead when they construct the networks. We believe that any way to define a link is adequate if it delineates features of the system. In our case it is consistent with the known features in the climate systems such as ENSO, NAO, and PNA

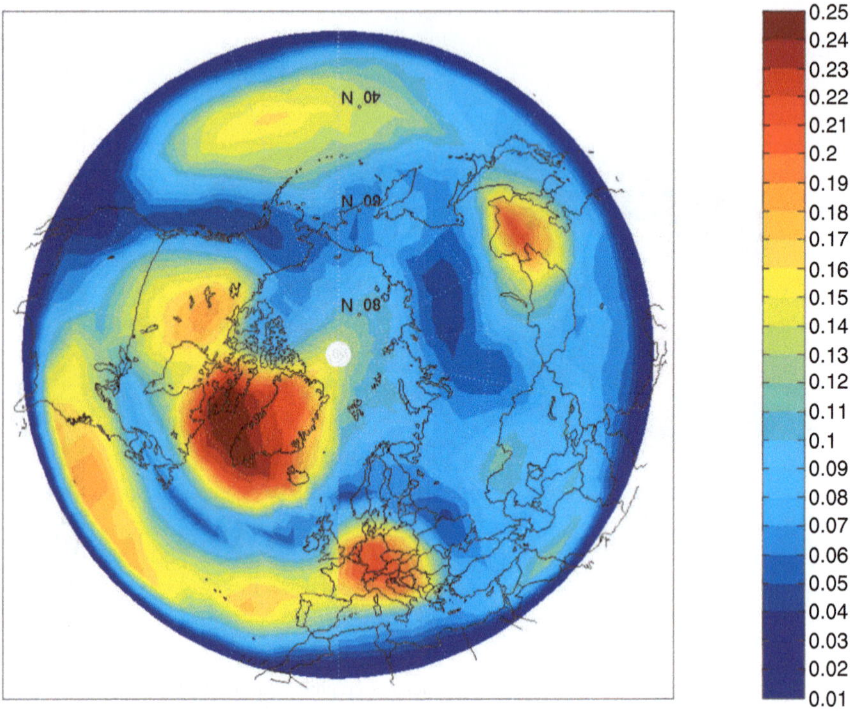

Fig. 4 Same as Fig. 3 but only for the extratropics (north of 30°)

30 S–30 N, and 30 N–90 N. This three-zone separation is not a trivial separation into northern hemisphere winter, southern hemisphere summer, and the rest of the world, because when we repeat the analysis with yearly averages rather than seasonal values we also see evidence of this three-zone separation. This separation is consistent with the transition from a barotropic atmosphere (where pressure depends on density only; appropriate for the tropics–subtropics) to a baroclinic atmosphere (where pressure depends on both density and temperature; appropriate for higher latitudes). Another possibility is that it reflects the well-known three-zone distribution of variance of the surface pressure field. Within the third community (green area) another community (yellow-red) is embedded. This community is consistent with the presence of major atmospheric teleconnection patterns such as the Pacific North America (PNA) pattern and the North Atlantic Oscillation (NAO) (Wallace and Gutzler 1981; Barnston and Livezey 1987). We note here that NAO (which has been lately suggested of being a three-pole pattern rather than a dipole; Tsonis et al. 2008) and AO (Arctic Oscillation; Thompson and Wallace 1998) are often interpreted as manifestations of the same dynamical mode, even though in some cases more physical meaning is given to NAO (Ambaum et al. 2001). In any case, here we do not make a distinction between NAO and AO. We note that similar results are obtained for other observed fields (such as the surface

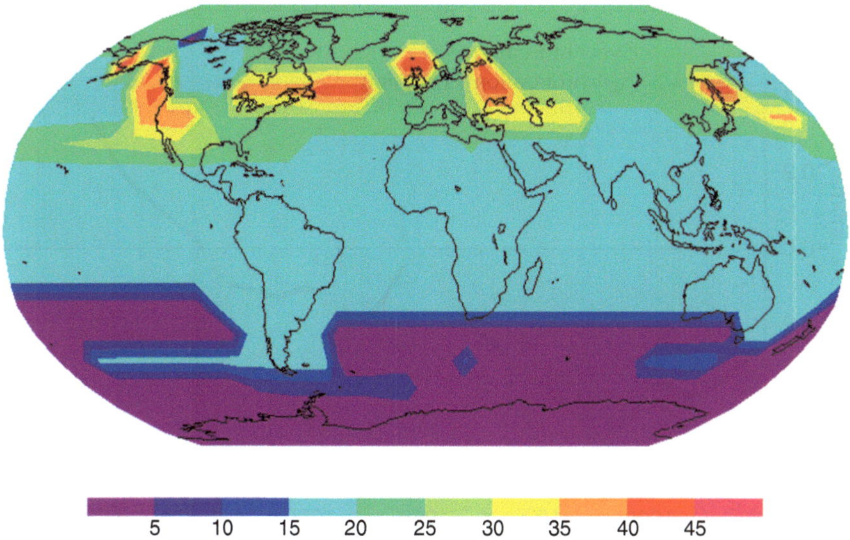

Fig. 5 Community structure of the network in Fig. 3. The number below the shading key indicates the total number of communities (see text for more details)

air temperature and sea level pressure, where influences of ENSO and PDO are present), as well as in model-simulated fields (Tsonis et al. 2011). We note that in spatially extended systems it is possible that spatial correlation may produce spurious 'small-world' networks (Bialonski et al. 2010; Hlinka et al. 2012; Palus et al. 2011). For our climate networks we have shown (Tsonis et al. 2011) that the network structure derived from spatio-temporal surrogate data on a sphere, which are spatially correlated with a de-correlation distance of 3000 Km, is not consistent with the network structure of the observed fields. This provides confidence that our networks and their structures are not an artifact of spatial correlations.

It is interesting to compare Figs. 3 and 5. Apparently there are similarities (the three-zone separation, for example), but the community algorithm identifies NAO clearly whereas in Fig. 3 as we mentioned earlier NAO is masked. Due to barotropic conditions in the tropical areas communication via gravity waves is fast and as result the information flows very efficiently resulting in a fully connected network in the tropics. In the extratropics super nodes are found in locations where major teleconnection patterns are found, which in turn define distinct communities in the network. It may be that in spatially extended systems with spatial correlations extending over a characteristic scale, the connectivity pattern is related to community structure. In any case since the presence of super nodes makes the network stable and efficient in transferring information, it has speculated and shown that indeed teleconnection patterns act as climate stabilizers. Tsonis et al. (2008) have shown that removal of teleconnection patterns from the climate system result in less stable networks, which makes an existing climate

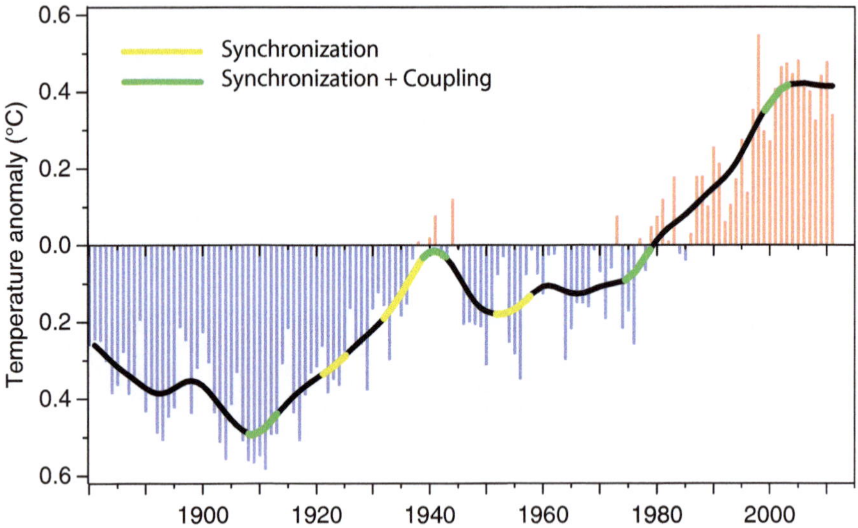

Fig. 6 Summary of synchronization events, coupling between the modes during these events, and climate shifts. See text for details

regime unstable and more likely to shift to a new regime. Indeed they showed that this process may be behind the climate shift of the 1970s, and related to the dynamical mechanism for major climate shifts discussed above. Figure 7 shows 500 HPa anomaly composites for three 5-year periods in the 1970s and early 1980s. In the early 1970s (top panel) the 500 HPa anomaly field is dominated by the presence of a wave-3 pattern with both the PNA and NAO (in its negative phase) being very pronounced. In the mid-1970s (middle panel) this field is very weak and both NAO and PNA have for all practical purposes disappeared. After that (lower panel), the field becomes strong again but a new wave-2 pattern with a very pronounced positive NAO has emerged. This shift is known as the climate shift of the 1970s. According to the Tsonis et al. (2007) mechanism for major climate shifts climate modes may synchronize. Once in place, the synchronized state may become unstable and shift to a new state. The results of Tsonis et al. (2008) and those in Fig. 6 suggest a connection between stability, synchronization, coupling of major climate modes, and climate shifts. This point is the subject of our continuing work in this area and more results will be forthcoming in the future.

In summary, the results outlined in this section suggest that climate networks are characterized by super nodes and a small number of communities, which relate to major teleconnection patterns/climate modes. Having established this, we proceed with our discovery of a mechanism for climate shifts based on the interaction of major climate modes.

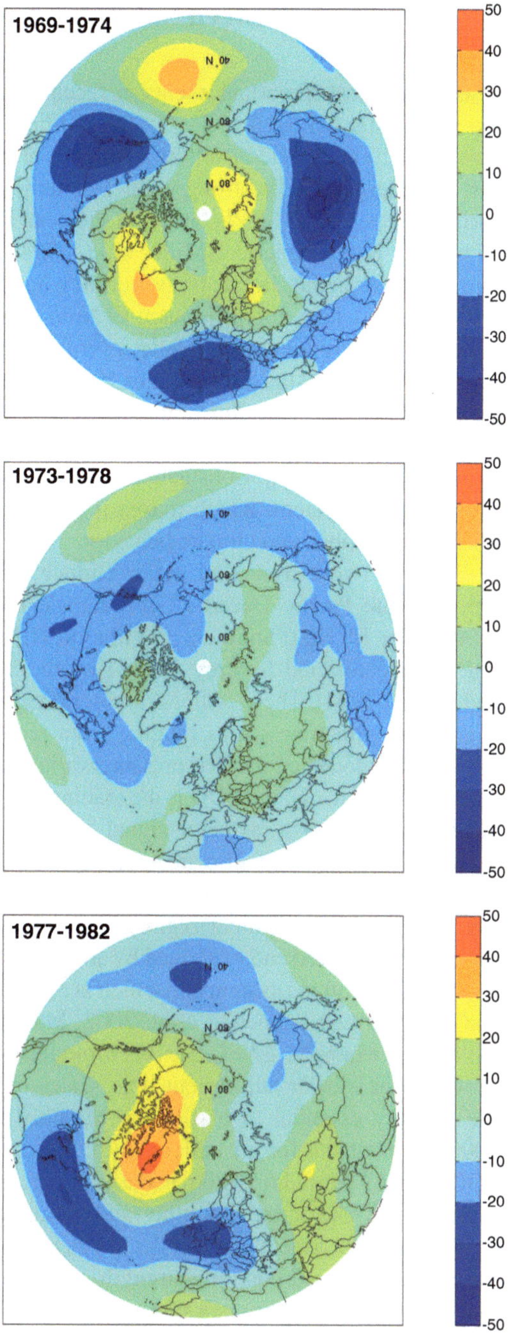

Fig. 7 Five hundred HPa anomaly field composites for the period 1969–1974 (*top*), 1973–1978 (*middle*), and 1977–1982 (*bottom*). On the top a wave-3 pattern is visible with PNA and NAO in its negative phase being present. In the middle, both NAO and PNA have for all practical purposes disappeared. In the bottom the field emerges as a wave-2 pattern with NAO in its positive phase. As we explain in the text this transition (known as the climate shift of the 1970s) is consistent with our conjecture that removal of super nodes makes the (climate) network unstable and more prone to failure (breakdown of a regime and emergence of another regime)

3 Interaction Between Subsystems

One of the most important events in recent climate history is the climate shift in the mid-1970s (Graham 1994). In the northern hemisphere 500-HPa atmospheric flow the shift manifested itself as a collapse of a persistent wave-3 anomaly pattern and the emergence of a strong wave-2 pattern. The shift was accompanied by sea-surface temperature (SST) cooling in the central Pacific and warming off the coast of western North America (Miller et al. 1994). The shift brought sweeping long-range changes in the climate of the northern hemisphere. Incidentally, after "the dust settled," a new long era of frequent El Niño events superimposed on a sharp global temperature increase begun. While several possible triggers for the shift have been suggested and investigated (Graham 1994; Miller et al. 1994; Graham et al. 1994), the actual physical mechanism that led to this shift is not clear. Understanding the dynamics of such phenomena is essential for our ability to make useful prediction of climate change. A major obstacle to this understanding is the extreme complexity of the climate system, which makes it difficult to disentangle causal connections leading to the observed climate behavior.

First a network from four major climate indices was constructed. The indices represent the Pacific Decadal Oscillation (PDO), the North Atlantic Oscillation (NAO), the El Niño/Southern Oscillation (ENSO), and the North Pacific Index (NPI) (Barnston and Livezey 1987; Hurrell 1995; Mantua et al. 1997; Trenberth and Hurrell 1994). These indices represent regional but dominant modes of climate variability, with time scales ranging from months to decades. NAO and NPI are the leading modes of surface pressure variability in northern Atlantic and Pacific Oceans, respectively, the PDO is the leading mode of SST variability in the northern Pacific and ENSO is a major signal in the tropics. Together these four modes capture the essence of climate variability in the northern hemisphere. Each of these modes is assumed to represent a subsystem involving different mechanisms over different geographical regions. Indeed, some of their dynamics have been adequately explored and explained by simplified models, which represent subsets of the complete climate system and which are governed by their own dynamics (Elsner and Tsonis 1993; Schneider et al. 2002; Marshall et al. 2001; Suarez and Schopf 1998). For example, ENSO has been modeled by a simplified delayed oscillator in which the slower adjustment time scales of the ocean supply the system with the memory essential to oscillation. Monthly mean values in the interval 1900–2000 are available for all indices (http://jisao.washington.edu/data sets, for NAO, PDO, and El Nino, http://climatedataguide.ucar.edu/guidance/north-pacific-index-npi-trenberth-and-hurrell-monthly-and-winter, for NPI).

An important aspect in the collective behavior of coupled nonlinear oscillators is synchronization and coupling strength. The theory of synchronized chaos predicts that in many cases when such systems synchronize, an increase in coupling between the oscillators may destroy the synchronous state and alter the system's behavior (Heagy et al. 1995; Pecora et al. 1997). It should be noted that in those studies coupling strength is determined by a parameter which is allowed to increase and the focus is in the perfect synchronization among the modes (i.e., the cross-correlation between outputs of the synchronized coupled systems is one), rather than weaker types of synchronization, such as phase synchronization (Boccaletti et al. 2002; Maraun and Kurths 2005) or clustered synchronization (Zhou and Kurths 2006), which are also important in climate interactions. In view of this theory we investigated whether our climate modes synchronize and when they do how synchronization relates to coupling strength between the modes. It is vital to note that synchronization and coupling are not interchangeable; for example, it is trivial to construct a pair of coupled simple harmonic oscillators whose displacements are in quadrature (and hence perfectly uncorrelated), but whose phases are strongly coupled (Vanassche et al. 2003). In our case, synchronization is defined from the sum of cross-correlations of all pairs in the network over a sliding time window, and coupling is measured by how well the phase between pairs of climate modes is predicted using information about the current phase (Tsonis et al. 2007). Note that according to our definition of coupling strength, if the modes are perfectly synchronized, their states are equivalent and thus coupling strength cannot increase further. Since our network of modes represents signals of a complex physical system where noise is also present, synchronization cannot be perfect but statistically significant (for details, see Tsonis et al. 2007). As such it is possible for the modes to enter into a synchronized state in a period when the coupling strength is decreasing and that de-synchronization may not happen when coupling strength is maximum.

The results from the observations are summarized in Fig. 6. This figure shows the yearly anomaly values of global temperature (blue negative anomalies, red positive anomalies). The black solid line is a smoothed version of this record. It is evident from the smoothed version that on decadal time scales there are times when the global temperature trend is shifting from negative to positive and vice-versa. These "shifts" are superimposed on a low frequency signal known as "global warming." Here we are not interested on the origins of the low frequency signal. Rather we are interested in the departures from this signal over decadal time scales. The part of the black line that is colored yellow indicates that the four climate modes are synchronized during a period when the coupling between the modes is *not* increasing. The part colored green indicates periods when the modes are synchronized and the coupling is increasing. Thus, we see that the network synchronized six times. In the periods 1908–1913, 1921–1925, 1932–1943, 1952–1957, 1975–1979, and 1998–2003. In the periods 1921–1925, 1932–1938, 1952–1957 synchronization is not associated with an increasing coupling strength and no change in the temperature trend is taking place. However, in the periods 1908–1913, 1939–1943, 1975–1979, and 1998–2003, synchronization is associated with an increase in coupling strength. As the modes keep on synchronizing and the coupling strength keeps on increasing, at some coupling threshold the synchronized state is destroyed and climate shifts into a new state characterized by a reversal in global temperature trend. This mechanism appears to be an intrinsic mechanism of the climate system as it is found in both control and forced climate simulation (Tsonis et al. 2007; Wang et al. 2009). It also appears to be a very robust mechanism. In all 13 synchronization events found in the observations and model simulations, once the modes begin to synchronize while the coupling is increasing, de-synchronization and the impeding shift happen at some coupling strength threshold. Due to noise/uncertainties in the data, synchronization cannot be perfect and this threshold is not always the same or always maximum at de-synchronization. Once the modes are de-synchronized the coupling may continue to increase as the modes may fall into phase with each other. This is consistent with the general theory of synchronized chaos where coupling strength may keep on increasing after de-synchronization. No shift ever occurred when during the synchronous state the coupling strength was decreasing. Lately Tsonis and Swanson (2011) extended their analysis to consider proxy data for climate modes going back several centuries. While noise in the proxy data in some cases masks the mechanism, it was found that significant coherence between both synchronization and coupling and global temperature exists. These results provide further support that the discussed here mechanism for climate shifts is a robust feature of the climate system.

The above results refer to the collective behavior of the four major modes used in the network. As such they do not offer insights on the specific details of the mechanism. For example, do small distance values (strong synchronization) result from all modes synchronizing or from a subset of them? When the network is synchronized, does the coupling increase require that all modes must become coupled with each other? To answer these questions Wang et al. (2009) split the network of four modes into its six pair components and investigated the contribution of each pair in each synchronization event and in the overall coupling of the network. It was found that one mode is behind all climate shifts. This mode is the NAO. This north Atlantic mode is without exception the common ingredient in all shifts and when it is not coupled with any of the Pacific modes no shift ensues. In addition, in all cases where a shift occurs NAO is necessarily coupled to north Pacific. In some cases it may also be coupled to the tropical Pacific (ENSO) as well, but in none of the cases NAO is only coupled to ENSO. Thus, results indicate that not only NAO is the instigator of climate shifts but that the likely evolution of a shifts has a path where the north Atlantic couples to north Pacific, which in turn couples to the tropics. Solid dynamical arguments and past work offer a concrete picture of how the physics may play out. NAO with its huge mass re-arrangement in north Atlantic affects the strength of the westerly flow across mid-latitudes. At the same time through its "twin," the arctic Oscillation (AO), it impacts sea level pressure patterns in the northern Pacific. This process is part of the so-called intrinsic mid-latitude northern hemisphere variability (Vimont et al. 2001, 2003). Then this intrinsic variability through the seasonal footprinting mechanism (Vimont et al. 2001, 2003) couples with equatorial wind stress anomalies, thereby acting as a stochastic forcing of ENSO. This view is also consistent with a recent studies showing that PDO modulates ENSO (Gershunov and Barnett 1998; Verdon and Franks 2006). Another possibility of how NAO couples to north Pacific may be through the five-lobe circumglobal waveguide pattern (Branstator 2002). It has been shown that this waveguide pattern projects onto NAO indices and its features contribute to variability at locations throughout northern hemisphere. Finally, north Atlantic variations have been linked to northern hemisphere mean surface temperature multidecadal variability through redistribution of heat within the northern Atlantic with the other oceans left free to adjust to these Atlantic variations (Zhang et al. 2007).[1] Thus, NAO, being the major mode of variability in the northern Atlantic, impacts both ENSO variability and global temperature variability. Recently a study has shown how ENSO with its effects on PNA can, through vertical propagation of Rossby waves, influence the lower stratosphere and how in turn the stratosphere can influence NAO through downward progression of Rossby wave (Ineson and Scaife 2009). These results coupled with our results suggest the following 3-D super-loop: NAO → PDO → ENSO → PNA → stratosphere → NAO, which captures the essence of decadal variability in the northern hemisphere and possibly the globe.

This co-variability of climate modes and its influence on global temperature has recently been confirmed by a different approach. Wyatt et al. (2011) analyzed the lagged covariance structure of a network of climate indices and discovered the so-called stadium wave; a sequence of lagged atmospheric and oceanic teleconnections leading to northern hemisphere temperature reversals every about 30 years. Lately, Wang et al. (2012) investigate whether the collective role of these modes is extended within a regime, i.e., to shorter time scales. They applied nonlinear prediction in order to assess directional influences in the climate system. They showed evidence that input from four major climate modes from the Atlantic and Pacific improves the prediction of global temperature and thus these modes Granger cause global temperature. Moreover, they found that this causality is not a result of a particular mode dominating but a result of the nonlinear collective behavior in the network of the four modes.

4 Conclusions

The above synthesis describes some new approaches that have been applied lately to climate data. The findings presented here and in the references may settle the issue of dimensionality of climate variability over decadal scales, as they support the view that over these scales climate collapses into distinct subsystems whose interplay dictates decadal variability. At the same time these results provide clues as to what these subsystems might be. As such, while 'weather' may be complicated, 'climate' may be complex but not complicated. Moreover, it appears that the interaction between these subsystems may be largely responsible for the observed decadal climate variability. A consequence of these results is that a dynamical reconstruction directly from a small number of climate modes/subsystems may be attempted to extract differential equations which model the network of major modes. Such an

[1]In (Elsner 2007) it is shown that global temperature Granger causes (leads) North Atlantic SST. It may be that the discrepancy between these two studies lies in the bi-directionality between the two variables, which is often the case in Granger causes.

approach may provide an alternative and direct window to study decadal variability in climate. Work in this area is in progress and will be reported in the future elsewhere.

References

Albert, R., and A.-L. Barabasi. 2002. Statistical mechanics of complex networks. *Reviews of Modern Physics* 74: 47–101.

Albert, R., H. Jeong, and A.-L. Barabasi. 1999. Diameter of the World Wide Web. *Nature* 401: 130–131.

Ambaum, M.H.P., B.J. Hoskins, and D.B. Stephenson. 2001. Arctic oscillation or North Atlantic oscillation? *Journal of Climate* 14: 3495–3507.

Arenas, A., A. Diaz-Guilera, and C.J. Perez-Vicente. 2006. Synchronization reveals topological scales in complex networks. *Physical Review Letters* 96: 114102.

Barabasi, A.-L., and E. Bonabeau. 2003. Scale-free networks. *Scientific American* 288: 60–69.

Branstator, G. 2002. Circumglobal teleconnections, the jet stream waveguide, and the North Atlantic Oscillation. *Journal of Climate* 15: 1893–1910.

Barnston, A.G., and R.E. Livezey. 1987. Classification, seasonality, and persistence of low-frequency atmospheric circulation patterns. *Monthly Weather Review* 115: 1083–1126.

Bialonski, S., M.-T. Horstmann, and K. Lehnertz. 2010. From brain to earth and climate systems: Small-worldnetworks or not? *Chaos* 20: 013134.

Boccaletti, S., J. Kurths, G. Osipov, D.J. Valladares, and C.S. Zhou. 2002. The synchronization of chaotic systems. *Physics Reports* 366: 1–101.

Bouchaud, J.-P., and M. Mezard. 2000. Wealth condensation in a simple model of economy. *Physica A* 282: 536–540.

da F. Costa, L., F.A. Rodrigues, G. Travieso, and P.R. Villas Boas. 2007. Characterization of complex networks: a survey of measurements. *Advances in Physics* 56: 167–242.

Donges, J.F., Y. Zou, N. Marwan, and J. Kurths. 2009a. The backbone of the climate network. *EPL* 87: 48007.

———. 2009b. Complex networks in climate dynamics. *European Physical Journal* 174: 157–179.

Elsner, J.B., and A.A. Tsonis. 1993. Nonlinear dynamics established in the ENSO. *Geophysical Research Letters* 20: 213–216.

Elsner, J.B. 2007. Granger causality and Atlantic hurricanes. *Tellus* 59A: 476–485.

Elsner, J.B., T.H. Jagger, and E.A. Fogarty. 2009. Visibility network of United States hurricanes. *Geophysical Research Letters* 36: L16702. doi:10.1029/2009GL039129.

Farkas, I.J., H. Jeong, T. Vicsek, A.-L. Barabási, and Z.N. Oltvai. 2003. The topology of the transcription regulatory network in the yeast Saccharomyces cerevisiae. *Physica A: Statistical Mechanics and its Applications* 318: 601–612.

Gershunov, A., and T.P. Barnett. 1998. Interdecadal modulation of ENSO teleconnections. *Bulletin of the American Meteorological Society* 79: 2715–2725.

Graham, N.E. 1994. Decadal scale variability in the tropical and North Pacific during the 1970s and 1980s: observations and model results. *Climate Dynamics* 10: 135–162.

Graham, N.E., T.P. Barnett, R. Wilde, M. Ponater, and S. Schubert. 1994. On the roles of tropical and mid-latitude SSTs in forcing interannual to interdecadal variability in the winter Northern Hemisphere circulation. *Journal of Climate* 7: 1500–1515.

Gozolchiani, A., K. Yamasaki, O. Gazit, and S. Havlin. 2008. Pattern of climate network blinking links follow El Nino events. *Europhysics Letters* 83: 28005.

Guimera, R., and L.A.N. Amaral. 2005. Functional cartography of complex metabolic network. *Nature* 433: 895–900.

Heagy, J.F., L.M. Pecora, and T.L. Carroll. 1995. Short wavelength bifurcations and size instabilities in coupled oscillator systems. *Physical Review Letters* 74: 4185–4188.

Hlinka, J., D. Hartman, and M. Palus. 2012. Small-world topology of functional connectivity in randomly connected dynamical systems. *Chaos* 22 (3): 033107. http://arxiv.org/abs/1206.3963v1.

Holme, P., M. Huss, and H. Jeong. 2003. Subnetwork hierarchies of biochemical pathways. *Bioinformatics* 19: 532–543.

Huang, J.P., K. Higuchi, and A. Shabbar. 1998. The relationship between the North Atlantic Oscillation and El Nino Southern oscillation. *Geophysical Research Letters* 25: 2707–2710.

Hurrell, J.W. 1995. Decadal trends in the North Atlantic oscillation regional temperature and precipitation. *Science* 269: 676–679.

Jeong, H., S. Mason, A.-L. Barabasi, and Z.N. Oltvai. 2001. Lethability and centrality in protein Networks. *Nature* 411: 41–42.

Ineson, S., and A.A. Scaife. 2009. The role of the stratosphere in the European climate response to El Nino. *Nature Geoscience* 2: 32–36.

Kistler, R., et al. 2001. The NCEP/NCAR 50-year reanalysis: monthly means, CD-ROM and documentation. *Bulletin of the American Meteorological Society* 82: 247–267.

Liljeros, F., C. Edling, L.N. Amaral, H.E. Stanley, and Y. Aberg. 2001. The web of human sexual Contacts. *Nature* 411: 907–908.

Lorenz, E.N. 1991. Dimension of weather and climate attractors. *Nature* 353: 241–244.

Mantegna, R.N. 1999. Hierarchical structure in financial markets. *The European Physical Journal B* 11: 193–197.

Mantua, N.J., S.R. Hare, Y. Zhang, J.M. Wallace, and R.C. Francis. 1997. A Pacific interdecadal climate oscillation with impacts on salmon production. *Bulletin of the American Meteorological Society* 78: 1069–1079.

Maraun, D., and J. Kurths. 2005. Epochs of phase coherence between El Nino/Southern oscillation and Indian monsoon. *Geophysical Research Letters* 32: L15709. doi:10.1029/2005GL023225.

Marshall, J., et al. 2001. North Atlantic climate variability: phenomena, impacts and mechanisms. *International Journal of Climatology* 21: 1863–1898.

Miller, A.J., D.R. Cayan, T.P. Barnett, N.E. Craham, and J.M. Oberhuber. 1994. The 1976–77 climate shift of the Pacific Ocean. *Oceanography* 7: 21–26.

Mo, K.C., and R.W. Higgins. 1998. The Pacific-South America modes and tropical convection during the southern hemisphere winter. *Monthly Weather Review* 126: 1581–1596.

Nerenberg, M.A.H., and C. Essex. 1990. Correlation dimension and systematic geometric Effects. *Physical Review A* 42: 7065–7074.

Newman, M.E.J., and M. Girvan. 2004. Finding and evaluating community structure in networks. *Physical Review E* 69: 026113.

Nicolis, C., and G. Nicolis. 1984. Is there a climatic attractor? *Nature* 311: 529–532.

Palus, M., D. Hartman, J. Hlinka, and M. Vejmelka. 2011. Discerning connectivity from dynamics in climate networks. *Nonlinear Processes in Geophysics* 18 (5): 751–763.

Pastor-Satorras, R., and A. Vespignani. 2001. Epidemic spreading in scale-free networks. *Physical Review Letters* 86: 3200–3203.

Pecora, L.M., T.L. Carroll, G.A. Johnson, and D.J. Mar. 1997. Fundamentals of synchronization in chaotic systems, concepts, and applications. *Chaos* 7: 520–543.

Pozo-Vazquez, D., M.J. Esteban-Parra, F.S. Rodrigo, and Y. Castro-Diez. 2001. The association between ENSO and winter atmospheric circulation and temperature in the North Atlantic region. *Journal of Climate* 14: 3408–3420.

Schneider, N., A.J. Miller, and D.W. Pierce. 2002. Anatomy of North Pacific decadal variability. *Journal of Climate* 15: 586–605.

Smith, L.A. 1988. Intrinsic limits on dimension calculations. *Physics Letters A* 133: 283–288.

Song, C., S. Havlin, and H.A. Makse. 2005. Self-similarity of complex networks. *Nature* 433: 392–395.

Strogatz, S.H. 2001. Exploring complex networks. *Nature* 410: 268–276.

Suarez, M.J., and P.S. Schopf. 1998. A delayed action oscillator for ENSO. *Journal of the Atmospheric Sciences* 45: 549–566.

Swanson, K.L., and A.A. Tsonis. 2009. Has the climate recently shifted? *Geophysical Research Letters* 36: L06711. doi:10.1029/2008GL037022.

Thompson, D.W.J., and J.M. Wallace. 1998. The Arctic oscillation signature in the wintertime geopotential height and temperature fields. *Geophysical Research Letters* 25 (9): 1297–1300.

Trenberth, K.E., and J.W. Hurrell. 1994. Decadal atmospheric-ocean variations in the Pacific. *Climate Dynamics* 9: 303–319.

Tsonis, A.A., and J.B. Elsner. 1988. The weather attractor over very short time scales. *Nature* 33: 545–547.

———. 1989. Chaos, strange attractors and weather. *Bulletin of the American Meteorological Society* 70: 16–23.

Tsonis, A.A. 1992. *Chaos: from theory to applications*. New York: Plenum.

Tsonis, A.A., G.N. Triantafyllou, and J.B. Elsner. 1994. Searching for determinism in observed data: a review of the issues involved. *Nonlinear Processes in Geophysics* 1: 12–25.

Tsonis, A.A., K.L. Swanson, and P.J. Roebber. 2006. What do networks have to do with climate? *Bulletin of the American Meteorological Society* 87 (5): 585–595. doi:10.1175/BAMS-87-5-585.

Tsonis, A.A., K.L. Swanson, and S. Kravtsov. 2007. A new dynamical mechanism for major climate shifts. *Geophysical Research Letters* 34: L13705. doi:10.1029/2007GL030288.

Tsonis, A.A., K.L. Swanson, and G. Wang. 2008. On the role of atmospheric teleconnection in climate. *Journal of Climate* 21: 2990–3001.

Tsonis, A.A., and K.L. Swanson. 2008. Topology and predictability of El Nino and La Nina networks. *Physical Review Letters* 100: 228502.

Tsonis, A.A., G. Wang, K.L. Swanson, F.A. Rodrigues, and L. da F. Costa. 2011. Community structure and dynamics in climate networks. *Climate Dynamics* 37: 933–940. doi:10.1007/s00382-010-0874-3.

Tsonis, A.A., and K.L. Swanson. 2011. Climate mode co-variability and climate shifts. *International Journal of Bifurcation and Chaos* 21: 3549–3556. doi:10.1142/S0218127411030714.

Vanassche, P., G.G.E. Gielen, and W. Sansen. 2003. Behavioral modeling of coupled harmonic oscillators. *IEEE Transactions on Computer-Aided Design of Integrated Circuits and Systems* 22: 1017–1027.

Verdon, D.C., and S.W. Franks. 2006. Long-term behavior of ENSO interactions with the PDO over the past 400 years inferred from paleoclimate records. *Geophysical Research Letters* 33: L06712.

Vimont, D.J., D.S. Battisti, and A.C. Hirst. 2001. Footprinting: a seasonal connection between the tropics and mid-latitudes. *Geophysical Research Letters* 28: 3923–3926.

Vimont, D.J., J.M. Wallace, and D.S. Battisti. 2003. The seasonal footprinting mechanism in the Pacific: Implications for ENSO. *Journal of Climate* 16: 2668–2675.

Wallace, J.M., and D.S. Gutzler. 1981. Teleconnections in the geopotential height field during the northern hemisphere winter. *Monthly Weather Review* 109: 784–812.

Wang, G., K.L. Swanson, and A.A. Tsonis. 2009. The pacemaker of major climate shifts. *Geophysical Research Letters* 36: L07708. doi:10.1029/2008GL036874.

Wang, G., P. Yang, X. Zhou, K.L. Swanson, and A.A. Tsonis. 2012. Directional influences on global temperature prediction. *Geophysical Research letters* 39: L13704. doi:10.1029/2012GL052149

Watts, D.J., and S.H. Strogatz. 1998. Collective dynamics of 'small-world' networks. *Nature* 393: 440–442.

Wyatt, M.G., S. Kravtsov, and A.A. Tsonis. 2011. Atlantic multidecadal oscillation and northern hemisphere's climate variability. *Climate Dynamics* 38 (5-6): 929–949. doi:10.1007/s00382-011-1071-8.

Yamasaki, K., A. Gozolchiani, and S. Havlin. 2008. Climate networks around the globe are significantly affected by El Nino. *Physical Review Letters* 100: 228501.

Zhang, R., T.L. Delworth, and I.M. Held. 2007. Can the Atlantic Ocean drive the observed multidecadal variability in northern hemisphere mean temperature? *Geophysical Research Letters* 34: L02709.

Zhou, C.S., and J. Kurths. 2006. Dynamical weights and enhanced synchronization in adaptive complex networks. *Physical Review Letters* 96 (16): 164102. doi:10.1103/PhysRevLett.96.164102.

On the Range of Frequencies of Intrinsic Climate Oscillations

Anastasios A. Tsonis and Michael D. Madsen

Abstract The purpose of this work is to establish the limits of natural oscillations in the climate system, i.e., not attributed to alleged anthropogenic effects. To this end we considered many proxy climate records representing the state of climate in the past when human activity was not a factor.

Keywords Climate oscillations • Natural variability

1 Introduction and Data

Twenty different reconstructed short-length proxy temperature records, six instrumental temperature records as well as five long-length proxy temperature records (four of which are ice-core reconstructed temperature records and the other reconstructed temperature record being from marine benthic oxygen isotopes) were analyzed in this study. The twenty reconstructed proxy temperature records represent annual means and range in length, location, and type. The six instrumental temperature records are monthly mean records and were all located in central Europe. They range in length from 231 to 247 years. Four of the long proxies are ice cores and one is a global marine benthic oxygen isotope record. Three of them have uneven time interval, while in two of them the values are spaced 500 years apart.

The details of the records used in this paper are as follows: Laguna Aculeo, Chile, summer mean sediment pigments, (856–1997 AD) (Von Gunten et al. 2009); Baffin Island, Canada, summer mean sediment thickness, (752–1992 AD) (Moore et al. 2003); Canadian Rockies, Canada, summer mea tree-ring thickness, (950–1994 AD) (Luckman and Wilson 2006); Firth, Alaska, summer mean tree-ring thickness,

A.A. Tsonis (✉)
Department of Mathematical Sciences, Atmospheric Sciences Group,
University of Wisconsin - Milwaukee, Milwaukee, WI, USA

Hydrologic Research Center, San Diego, CA, USA
e-mail: aatsonis@uwm.edu

M.D. Madsen
Department of Mathematical Sciences, Atmospheric Sciences Group,
University of Wisconsin - Milwaukee, Milwaukee, WI, USA

A.A. Tsonis (ed.), *Advances in Nonlinear Geosciences*,
DOI 10.1007/978-3-319-58895-7_30

(1073–2002 AD) (Anchukaitis et al. 2013); Canadian Rockies, tree-ring thickness (950–1994 AD) (Luckman and Wilson 2006); Iceberg Lake, Alaska, annually varve thickness, (442–1998 AD) (Loso 2008); Gulf of Alaska, summer mean tree-ring thickness, (724–1999 AD) (Wilson et al. 2007); Idaho, USA, annually July mean tree-ring thickness, (1135–1992 AD) (Biondi et al. 2006); North Andes, South America, annual mean tree-ring thickness, 1640–1987 AD), South Andes, South America, annual mean tree-ring thickness, (1640–1993 AD) (Villalba et al. 2006); Beijing, China, summer mean stalagmite thickness, (−665–1985 AD) (Tan et al. 2003); Central Europe, annual mean documentary data, (1005–2001 AD) (Glaser and Riemann 2009); China, annual multi-proxy reconstruction, (1000–1950 AD) (Shi et al. 2012); Cold Air Cave, South Africa, 5-year smoothed annual stalagmite isotope, (1635–1993 AD) (Sundqvist et al. 2013); European Alps, summer mean tree-ring and sediment thickness, (1053–1996 AD) (Trachsel et al. 2012); Lake Silvaplana, Switzerland, summer mean visible reflectance spectroscopy of lake sediment, (1175–1949 AD) (Trachsel et al. 2010); Slovakia, Europe, summer mean tree-ring, (1040–2011 AD) (Büntgen et al. 2013); Sweden, Europe, summer mean tree-ring, (1107–2007 AD) (Gunnarson et al. 2011); Tornetrask, Sweden, annual tree-ring, (500–2004 AD) (Grudd 2008); West Qinling Mts., China, annual tree-ring, (1500–1995 AD) (Yang et al. 2013); Spannagel Cave, Europe, stalagmite thickness, (-9–1935 AD) (Mangini et al. 2005); Paris, France, monthly mean instrumental, (1764–2000 AD) (Météo France 2012); Hohenpeißenberg, Germany, monthly mean instrumental, (1781–2013 AD) (Climate Research Unit CRU 2012); Kremsmunster, Austria, monthly mean instrumental, (1767–2013 AD) (Auer et al. 2007); Munich, Germany, monthly mean instrumental, (1781–2011 AD) (Deutscher Wetterdienst DWD 2012); Prague, Austria, monthly mean instrumental, (1771–2013 AD) (Czech Hydrometeorological Institute CHMI 2012); Vienna, Austria, monthly mean instrumental, (1775–2013 AD) (Climate Research Unit CRU 2012); Dome Fuji, Antarctica, ice core, (−339500–750 AD) (Kawamura et al. 2007); EPICA Dome C, Antarctica, ice core, (−800,000–1900 AD) (Jouzel et al. 2007); GISP2 ice core, central Greenland, ice core, (−48000–1850 AD) (Alley 2004); Global 1Ma Temperature, marine benthic oxygen isotopes, (−1067900–2000 AD) (Bintanja et al. 2005); Vostok, Antarctica, ice core, (−470766–2000 AD) (Petit et al. 1999).

For the analysis here, all six instrumental monthly records were converted to yearly mean records. The uneven records were interpolated to fill in missing values and to create 500-year-interval records. For interpolation we employed the piecewise cubic spline interpolation function in Matlab® (interp1).

2 Method and Results

In this study, we used the simple method of discrete Fourier transform (DFT) as our method for spectral analysis. DFT converts finite, equal spaced time domain samples, temperature records, into a finite combination of complex sinusoids ordered by their frequencies. Note that interpolation can result in enhancing lower frequencies and reducing higher frequency components (Schulz and Mudelsee 2002). To verify that our interpolation has little to no effect on the frequency components, our interpolated temperature records' DFT spectral analyses are compared to the spectral analysis using the Lomb–Scargle periodogram method. We found that both peak frequency and intensity are comparable between the two methods.

Each temperature record used in this study was first detrended using the Matlab® function (detrend). In order to obtain more frequency steps in the DFT spectral analysis, zero padding was applied to both ends of the temperature records to create temperature records of equal length of $N = 10000$ time steps. Then for each temperature record we employed the discrete Fourier transform using the fast Fourier transform function in Matlab® (fft). The output of this function was a combination of complex sinusoids in the form $A + Bi$, where A and B are a pair of harmonic predictors which can be found using:

$$A_k = \frac{2}{N}\sum_{i=1}^{N} y_i \cos\frac{2\pi k i \Delta t}{T}$$

$$B_k = \frac{2}{N}\sum_{i=1}^{N} y_i \sin\frac{2\pi k i \Delta t}{T}$$

$$\text{for} : k = 1, \frac{N}{2} - 1$$

$$A_{\frac{N}{2}} = \frac{1}{N}\sum_{i=1}^{N} y_i \cos\frac{\pi N i \Delta t}{T}$$

$$A_0 = \frac{1}{N}\sum y_i$$

$$B_0 = B_{\frac{N}{2}} = 0$$

where Δt is the time interval, $y_n = y(t_n)\ _{n=1,N}$, and $T = N\Delta t$. To find the variance associated with a given pair of harmonic predictors (C_k):

$$C_k = \frac{{A_k}^2 + {B_k}^2}{2}$$

C_k gives us the power values for the power spectrum. The power values for each spectrum were then normalized by dividing by the area comprised by the whole spectrum. For this to happen it first must be pointed out that by using this method, only half of the spectrum is retrieved as the second half is just a mirror image of the first half. Therefore, in order to be able to normalize each spectrum by dividing by the area of the whole spectrum, we must double the area of the first half of the spectrum. Since the focus of this study is on climate periodicities, each graph has an upper frequency limit of 0.04 year^{-1} or periodicity of 25 years.

In order to obtain significant peaks within the DFT power spectra, we must estimate an appropriate 95% confidence level. For this study the 95% confidence level was established by using 1000 Monte Carlo synthetic runs using fractional Brownian motions (fBms). It has been shown in the past (Koscielny-Bunde et al. 1998), and was verified here for all records, that temperature records do indeed have properties of fractional Brownian motions with an exponent (also referred to as the Hurst exponent) greater than 0.5. This is statistically desired because in this case surrogate data can be generated to assist in the statistical significance of the results. The Hurst Exponent can vary between 0.0 and 1.0. The range between 0.5 and 1.0 corresponds to persistence while the range between 0.0 and 0.5 corresponds to anti-persistence. First, in order to use fBms as surrogates, each temperature record must be examined to verify that it is indeed an fBm. To calculate the Hurst exponent (Feder 1988) of a time series:

$$y_n = y\left(t_n\right)_{n=1,N}$$

First, find the mean of the time series:

$$M = \frac{1}{N}\sum_{i=1}^{N} y_i$$

Then calculate the deviations from the mean:

$$\begin{aligned} x_1 &= y_1 - M \\ x_2 &= y_2 - M \\ &\cdots \\ x_n &= n - M \end{aligned}$$

Next, calculate the cumulative sums:

$$\begin{aligned} Z_1 &= x_1 \\ Z_2 &= x_1 + x_2 \\ &\cdots \\ Z_n &= \textstyle\sum_{i=1}^{n} x_i \end{aligned}$$

Compute the range:

$$R_n = \max\left[Z_n\right] - \min\left[Z_n\right]$$

Compute the standard deviation:

$$S_n = \sqrt{\frac{1}{n}\sum_{i=1}^{n} (y_i - M)^2}$$

The rescaling range $\frac{R_n}{S_n}$ can be used to estimate the Hurst exponent (H).

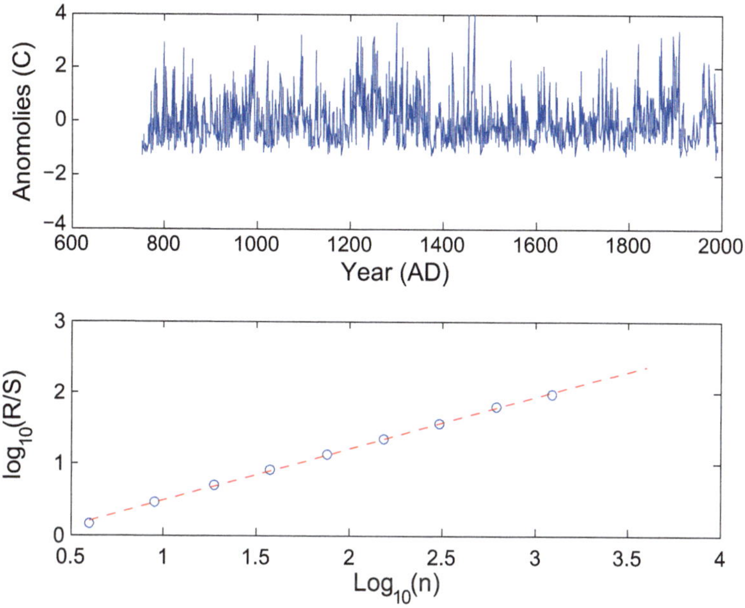

Fig. 1 Proxy record from Baffin Island, Canada (*top*) and its Hurst analysis (*bottom*). The results indicate that this record has properties of a fractional Brownian motion with an exponent of about 0.72

$$\frac{R_n}{S_n} = Cn^H$$

where C is a constant. From here:

$$\log\left(\frac{R_n}{S_n}\right) = \log(C) + H\log(n)$$

Then the slope of the linear regression line between $\log\left(\frac{R_n}{S_n}\right)$ $vs \log(n)$ gives the Hurst exponent H.

Figure 1 shows an example of the data used. It is from Baffin Island, Canada and it is a proxy sediment thickness record (top). The bottom graph shows the results of a Hurst analysis, which indicates that this record is indeed a fractional Brownian motion with an exponent of about 0.72 indicating persistence. We found that all the records used here are fBms with an exponent greater than 0.5. As was mentioned above, this result is consistent with earlier results base on temperature records (Koscielny-Bunde et al. 1998).

Figure 2 shows the statistical procedure used here to produce statistically significant periodicities in the data. First, the spectra of the proxy record were produced (blue line). Then we produced 1000 surrogate Brownian motion with

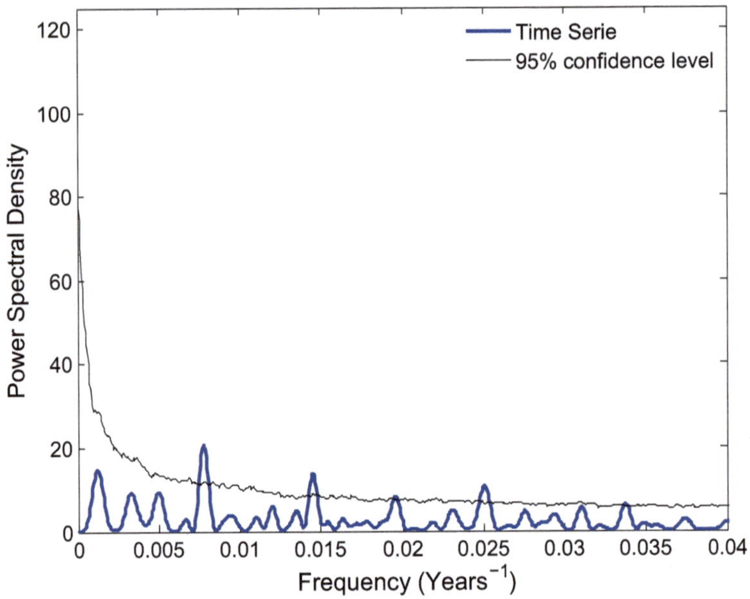

Fig. 2 Spectra of the proxy record in Fig. 2

the same exponent as the proxy data and calculated their spectra. Any peak in the original proxy data above the 95% percentile of these 1000 surrogates (black line) was then considered as a significant oscillation. In this case we have three significant oscillations at about frequencies 0.0075, 0.014, and 0.025 years^{-1} (or periodicities 130, 70, and 40 years). Figure 3 shows all the significant periodicities of all record but the last 5 (long interpolated records) and Fig. 4 shows all of the records (in red the last five very long records indicating astronomical Milankovitch forcing).

The important conclusion from this study is that there seems to exist two types of natural oscillations in the climate system. Those internal to the climate system ranging up to 1000 years and those of much longer period attributed to the Milankovitch cycles. There may still be oscillations in between but the data available here cannot resolve them. Yet the major conclusion is that long time-scale oscillations that cannot be attributed to human activity are present in proxy climate records.

This study is consistent with a much earlier study (Zhuang 1991), which used 13 different isotope records from the SPECMAP project http://gcmd.nasa.gov/records/GCMD_EARTH_LAND_NGDC_PALEOCL_SPECMAP.html (Fig. 5). The similarity between our results and those independent results is striking.

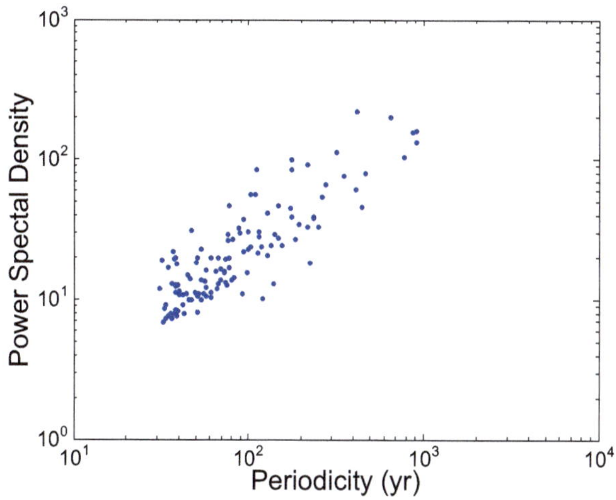

Fig. 3 Significant periodicities. All records but the last five

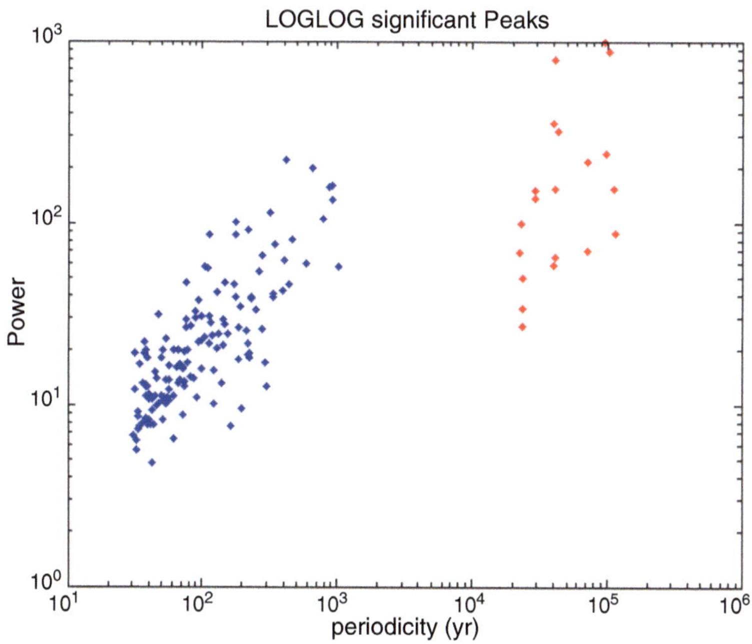

Fig. 4 Same as Fig. 3 but for all records

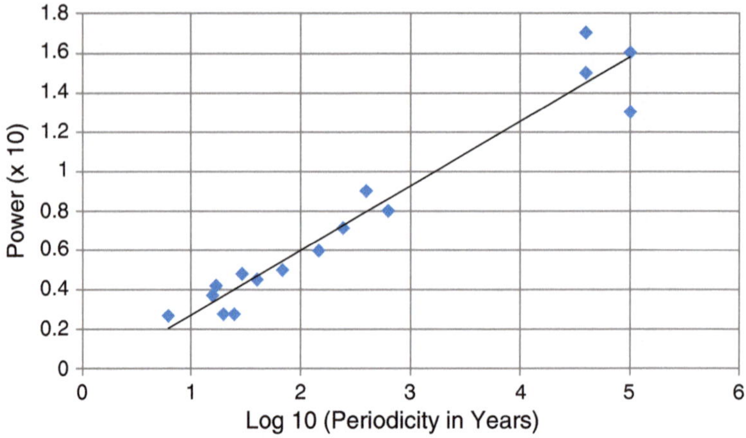

Fig. 5 Same as Fig. 4 but for an independent data set

References

Alley, R.B. 2004. *GISP2 ice core temperature and accumulation data. IGBP PAGES/World data center for paleoclimatology data contribution series #2004-013*. Boulder CO, USA: NOAA/NGDC Paleoclimatology Program.

Anchukaitis, K.J., R.D. D'Arrigo, L. Andreu-Hayles, D. Frank, A. Verstege, A. Curtis, B.M. Buckley, G.C. Jacoby, and E.R. Cook. 2013. Tree-ring-reconstructed summer temperatures from northwestern north america during the last nine centuries. *Journal of Climate* 26 (10): 3001–3012. doi:10.1175/JCLI-D-11-00139.1.

Auer, I., et al. 2007. HISTALP–historical instrumental climatological surface time series of the Greater Alpine Region. *International Journal of Climatology* 27: 17–46.

Bintanja, R., R.S.W. van de Wal, and J. Oerlemans. 2005. Modeled atmospheric temperatures and global sea levels over the past million years. *Nature* 437: 125–128. doi:10.1038/nature03975.

Biondi, F., et al. 2006. *East-central Idaho July Temperature Reconstruction. IGBP PAGES/World Data Center for Paleoclimatology Data Contribution Series # 2006-039*. Boulder CO, USA: NOAA/NCDC Paleoclimatology Program.

Büntgen, U., et al. 2013. Filling the Eastern European gap in millennium-long temperature reconstructions. *Proceedings of the National Academy of Sciences of the United States of America* 110 (5): 1773–1778. doi:10.1073/pnas.1211485110.

Climate Research Unit (CRU). 2012. *University of East Anglia (UK)*. Available at: http://www.metoffice.gov.uk/hadobs/crutem4/data/download.html

Czech Hydrometeorological Institute (CHMI). 2012. *143 06 Praha 4 Czech Republic*. Available at: http://zmeny-klima.ic.cz/klementinum-data/

Deutscher Wetterdienst (DWD). 2012. *Frankfurter Straße 135, 63067 Offenbach (Germany)*. Available at: www.dwd.de

Feder, J. 1988. *Fractals*. New York: Plenum Press.

Glaser, R., and D. Riemann. 2009. A thousand-year record of temperature variations for Germany and Central Europe based on documentary data. *Journal of Quaternary Science* 24: 437–449. ISSN 0267-8179. doi:10.1002/jqs.1302.

Grudd, H. 2008. Tornetrask tree-ring width and density AD 500-2004: a test of climatic sensitivity and a new 1500-year reconstruction of north Fennoscandian summers. *Climate Dynamics* 31: 843–857. doi:10.1007/s00382-007-0358-2.

Gunnarson, B.E., H.W. Linderholm, and A. Moberg. 2011. Improving a tree-ring reconstruction from west-central Scandinavia: 900 years of warm-season temperatures. *Climate Dynamics* 36 (1–2): 97–108. doi:10.1007/s00382-010-0783-5.
Jouzel, J., et al. 2007. *EPICA Dome C Ice Core 800KYr Deuterium Data and Temperature Estimates. IGBP PAGES/World Data Center for Paleoclimatology Data Contribution Series # 2007-091*. Boulder CO, USA: NOAA/NCDC Paleoclimatology Program.
Kawamura, K., et al. 2007. *Dome Fuji Ice Core Preliminary Temperature Reconstruction, 0-340 kyr. IGBP PAGES/World Data Center for Paleoclimatology Data Contribution Series # 2007-074*. Boulder CO, USA: NOAA/NCDC Paleoclimatology Program.
Koscielny-Bunde, E., A. Bunde, S. Havlin, H.E. Roman, Y. Goldreich, and H.-J. Schellnhuber. 1998. Indication of a universal persistence law governing atmospheric variability. *Physical Review Letters* 31 (3): 729–732.
Loso, M.G. 2008. Summer temperatures during the Medieval Warm Period and Little Ice Age inferred from varved proglacial lake sediments in southern Alaska. *Journal of Paleolimnology* 41 (1): 117–128. doi:10.1007/s10933-008-9264-9.
Luckman, H., and R.J.S. Wilson. 2006. *Canadian Rockies Summer Temperature Reconstruction. IGBP PAGES/World Data Center for Paleoclimatology Data Contribution Series # 2006-011*. Boulder CO, USA: NOAA/NCDC Paleoclimatology Program.
Mangini, A., C. Spötl, and P. Verdes. 2005. Reconstruction of temperature in the Central Alps during the past 2000 yr from a d18O stalagmite record. *Earth and Planetary Science Letters* 235 (3–4): 741–751. doi:10.1016/j.epsl.2005.05.010.
Météo France. 2012. Available at: http://france.meteofrance.com
Moore, J.J., et al. 2003. *Baffin Island 1250 Year Summer Temperature Reconstruction, IGBP PAGES/World Data Center for Paleoclimatology Data Contribution Series # 2003-075*. Boulder CO, USA: NOAA/NGDC Paleoclimatology Program.
Petit, J.R., et al. 1999. Climate and Atmospheric History of the Past 420,000 years from the Vostok Ice Core, Antarctica. *Nature* 399: 429–436.
Schulz, M., and M. Mudelsee. 2002. REDFIT: estimating red-noise spectra directly from unevenly spaced paleoclimatic time series. *Computational Geosciences* 28: 421–426.
Shi, F., B. Yang, and L. Von Gunten. 2012. Preliminary multiproxy surface air temperature field reconstruction for China over the past millennium. *Science China Earth Sciences* 55 (12): 2058–2067. doi:10.1007/s11430-012-4374-7.
Sundqvist, H.S., K. Holmgren, J. Fohlmeister, Q. Zhang, M.M. Bar, C. Spitl, and H. Kirnich. 2013. Evidence of a large cooling between 1690 and 1740 AD in southern Africa. *Scientific Reports*. doi:10.1038/srep01767.
Tan, M., et al. 2003. *2650-Year Beijing Stalagmite Layer Thickness and Temperature Reconstruction, IGBP PAGES/World Data Center for Paleoclimatology Data Contribution Series # 2003-050*. Boulder CO, USA: NOAA/NGDC Paleoclimatology Program.
Trachsel, M., et al. 2012. Multi-archive summer temperature reconstruction for the European Alps, AD 1053–1996. *Quaternary Science Reviews* 46: 66–79. doi:10.1016/j.quascirev.2012.04.021.
Trachsel, M., M. Grosjean, D. Schnyder, C. Kamenik, and B. Rein. 2010. Scanning reflectance spectroscopy (380ñ730 nm): a novel method for quantitative high-resolution climate reconstructions from minerogenic lake sediments. *Journal of Paleolimnology* 44 (4): 979–994. doi:10.1007/s10933-010-9468-7.
Villalba, R., et al. 2006. *Southern Andes Temperature Reconstructions. IGBP PAGES/World Data Center for Paleoclimatology Data Contribution Series # 2006-024*. Boulder CO, USA: NOAA/NCDC Paleoclimatology Program.
Von Gunten, L., M. Grosjean, B. Rein, R. Urrutia, and P. Appleby. 2009. A quantitative high-resolution summer temperature reconstruction based on sedimentary pigments from Laguna Aculeo, central Chile, back to AD 850. *The Holocene* 19 (6): 873–881. doi:10.1177/0959683609336573.
Wilson, R., G. Wiles, R. DíArrigo, and C. Zweck. 2007. Cycles and shifts: 1,300 years of multi-decadal temperature variability in the Gulf of Alaska. *Climate Dynamics* 28: 425–440. doi:10.1007/s00382-006-0194-9.

Yang, F., et al. 2013. Multi-proxy temperature reconstruction from the West Qinling Mountains, China, for the past 500 years. *PLoS One* 8 (2): e57638. doi:10.1371/journal.pone.0057638.

Zhuang, J. 1991. A study of the variability of the global climate system. Master of Science thesis, Department of Geosciences, University of Wisconsin-Milwaukee, USA.

The Prediction of Nonstationary Climate Series by Incorporating External Forces

Geli Wang, Peicai Yang, and Anastasios A. Tsonis

Abstract Almost all climate time series have some degree of nonstationarity due to external forces of the observed system. Therefore, these external forces should be taken into account when reconstructing the climate dynamics. This paper presents a novel technique in predicting nonstationary time series. The main difference of this new technique from some previous methods is that it incorporates the driving forces in the prediction model. To appraise its effectiveness, some prediction experiments were carried out using the data generated from some known classical dynamical models and climate data. Experimental results indicate that this technique is able to improve the prediction skill effectively.

Keywords Spatio-temporal series • Nonstationarity • Driving force • Climate prediction

1 Introduction

Most real-world time series have some degree of nonstationarity due to external perturbations of the observed system. Recent studies have pointed out the non-stationarity character of the climate system. For instance, Tsonis (1996) analyzed low-frequency (decadal to multi-decadal) variability of global precipitation over the past century and found that the fluctuations about the global mean have increased significantly, while the mean values have not changed. Their results imply that the second-order moment of the precipitation has changed on those scales, and that the global precipitation process was nonstationary over the past century. In another case,

G. Wang (✉) • P. Yang
Key Laboratory of Middle Atmosphere and Global Environment Observations,
Institute of Atmospheric Physics, Chinese Academy of Sciences, Beijing, 100029, China
e-mail: wgl@mail.iap.ac.cn

A.A. Tsonis
Department of Mathematical Sciences, Atmospheric Sciences Group,
University of Wisconsin - Milwaukee, Milwaukee, WI 53201, USA

Hydrologic Research Center, San Diego, CA, USA

A.A. Tsonis (ed.), *Advances in Nonlinear Geosciences*,
DOI 10.1007/978-3-319-58895-7_31

Trenberth (1990) found that the observed winter Pacific mean sea-level pressure underwent an abrupt change from year 1976 to year 1977.

In recent years, increasing effort has been devoted to devising methods to analyze and predict nonstationary time series, which were often addressed by either identifying some local stationary segments or applying particular techniques to transform the nonstationary signal into a stationary one. For example, Wang and Yang (2005) and Yang et al. (2010) presented two differential techniques, which are called as "compound reconstruction modeling" and "segregation modeling," respectively, to predict nonstationary time series. The first technique was applicable to multi-variable time series consisting of a predicted nonstationary time series and its control time series. This method relied upon extracting some stationary (or approximately stationary) segments from the predicted time series by using state similarity in the reconstructed space of the control time series, and, then, building the prediction model based on these separated segments. The second method was applicable to single-variable time series. This method functioned by decomposing the predicted nonstationary time series into a finite number of mode components by the empirical mode decomposition method and then making and accumulating the predictions of each mode component for one of the original time series.

Though the above techniques used some new procedures to cope with nonstationarity, the basic idea used in all these studies was to remove or reduce the nonstationarity of the predicted system using some mathematical techniques, thereby improving the prediction. In fact, the essential cause of nonstationarity is the time-dependent changes in the external forces (Manuca and Savit 1996). Thus, the most effective way to remove the nonstationarity may be to incorporate all the driving forces in the reconstructed dynamical system considering them as the state variables of that system. Based on this principle, we present an algorithm to incorporate driving forces to predict the nonstationary climate time series.

Following is a brief introduction of the algorithm for establishing the prediction model. To test its effectiveness, we carried out several prediction experiments on the given time series generated from some known classical dynamical systems and climate data, which are discussed next. Finally, a brief discussion is provided.

2 Method

In the field of nonlinear time series analysis, the most important aspects are the state space reconstruction theory (Packard et al. 1980) and the embedding theorem. According to the latter, developed by Takens (1981), for a given single-variable time series, one can use a couple of appropriate values of the embedding dimension and the delay time to convert the series into a phase trajectory in state space. Takens' theorem holds only for an autonomous dynamical system. For the nonstationary case, however, we could still embed the external force components in the same state space (Stark 1999). The dynamics on the reconstructed trajectory is equivalent to that of the original system that generated the time series, based on this trajectory, we can use this time series and its lags to establish a prediction model to predict the future state of the system.

Assuming a nonstationary process composed of two series $\{x_i\}_{i=1,2,\ \ldots\ ,n}$ and $\{\alpha_i\}_{i=1,2,\ \ldots\ ,n}$, the former being the state variable and the latter for an external forcing. With a selected time lag τ, we embed the time series in an $m_1 + m_2$ dimensional phase space and express the reconstructed state trajectory as

$$\vec{y}(i) = \left\{x_i, x_{i-\tau}, \ldots, x_{i-(m_1-1)\tau}; \alpha_i, \alpha_{i-\tau}, \ldots, \alpha_{i-(m_2-1)\tau}\right\}_{i=1,2,\ldots,N}, \quad (1)$$

or simply as

$$\vec{y}(i) = \left\{\vec{x}_i; \vec{\alpha}\right\}_{i=1,2,\ldots,N}. \quad (2)$$

Here m_1 and m_2 are the given embedding dimensions for $\{x_i\}_{i=1,2,\ \ldots\ ,n}$ and $\{\alpha_i\}_{i=1,2,\ \ldots\ ,n}$, respectively, $N = n - (\max(m_1, m_2) - 1)\tau$ is the number of phase points on the trajectory. Based on this trajectory, we built a model to predict the above process. The model is expressed as a map:

$$X(t + P) = f_P(X(t)) \quad (3)$$

where the prediction step p, which was considered as 1 in this study, and f_P is a desired mapping assumed to be a quadratic polynomial; now, the task is to find the cost function $\eta = \sum_{k=1}^{N}\left[f\left(\vec{x}_k, \vec{\alpha}_k\right) - x_{k+1}\right]^2$, which is reached its minimum value. For more details, one can refer to the studies of Farmer and Sidorowich (1987) and Casdagli (1989).

3 Experiments

We applied the approach referred above to perform some prediction experiments using several given nonstationary time series. We begin with time series from ideal nonstationary systems, since the data length and precision can be controlled and guaranteed.

3.1 *Modified Logistic Map*

The first group consisted of three ideal time series from the following logistic map:

$$x_{t+1} = \mu_t x_t (1 - x_t) \quad (4)$$

where μ_t is a parameter that changes with time. As we know for the logistic model, when the value of μ_t varies between 3.57 and 4.0, it should exhibit chaotic behaviors. If we let μ_t change within the following three cases (see Fig. 1):

$$\mu_t^{(1)} = 3.95 - 0.4e^{-2.5t} \quad (5)$$

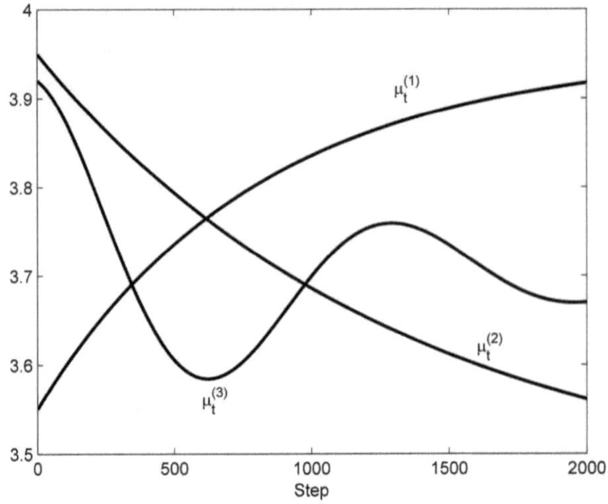

Fig. 1 Three different driving forces

Table 1 RMSE comparison of the prediction experiments

m_2	m_1	$S^{(1)}$	$S^{(2)}$	$S^{(3)}$
0	1	2.12	2.89	0.70
0	2	1.39	1.69	0.41
0	Mean[a]	1.76	2.29	0.56
1	1	0.61	0.63	0.32
1	2	0.63	0.62	0.29
1	Mean	0.62	0.63	0.31

[a]Indicates the averaged values for $m_1 = 1$ and $m_1 = 2$

$$\mu_t^{(2)} = 3.45 + 0.5\mathrm{e}^{-1.5t} \tag{6}$$

$$\mu_t^{(3)} = 3.7 + 0.22\mathrm{e}^{-2t}\cos(3\pi t) \tag{7}$$

then we should obtain three different nonstationary time series with chaotic behaviors written as S_1, S_2, and S_3, respectively.

The following prediction experiments are based on 2000 data points from Eqs. (4)–(7). The first 1900 data points are used to build the prediction model, while the last 100 data points are used to test the predictions by using the root mean square error (RMSE). In all of the experiments, the lag τ equal to one, while the embedding dimensions of the observations $\{x_t\}$, m_1, and of the external force $\{\mu_t\}$, m_2, were set at 1, 2, 0, and 1, respectively. The case of m_2 equal to zero means that the external force was not taken into account in the prediction model, or, in other words, the predictions were based on stationarity. Table 1 shows cases of RMSE resulting from the experiments. From Table 1, we can see that RMSE is improved when the external force is considered. This indicates that introducing external forces into the predictive model can provide an effective way to predict nonstationary processes.

3.2 Modified Lorenz System

The second experiment was performed with data from Lorenz system:

$$\begin{aligned} \tfrac{dx}{dt} &= -\sigma x + \sigma y \\ \tfrac{dy}{dt} &= r(t)x - y - xz \\ \tfrac{dz}{dt} &= xy - bz \end{aligned} \tag{8}$$

σ was taken as 10 and b as 8/3 in this model, while the Rayleigh number $r(t)$ was regarded as a time-varying driving force factor given by the logistic map $r(t+1) = \mu r(t)(1 - r(t))$, where the value of μ was taken as 3.9, which implied chaotic behavior. We multiplied $r(t)$ by 32 to get a time series whose values ranged from 3.2 to 29.3, and assumed this time series to be the time-varying Rayleigh number to force the Lorenz system. Under the present case, the modified Lorenz system could obey the states varying from state points to chaotic regimes (see Fig. 2); therefore, one nonstationary time series was obtained. The data set consisted of 8000 values of the variable x, the preceding 7200 data were applied to establish the predictive model, while the subsequent 800 points were used to test the prediction. We assumed that m_1 took values from 3 to 5 and m_2 either 0 or 1 (which corresponded to the stationary or forcing model). The experimental results for this

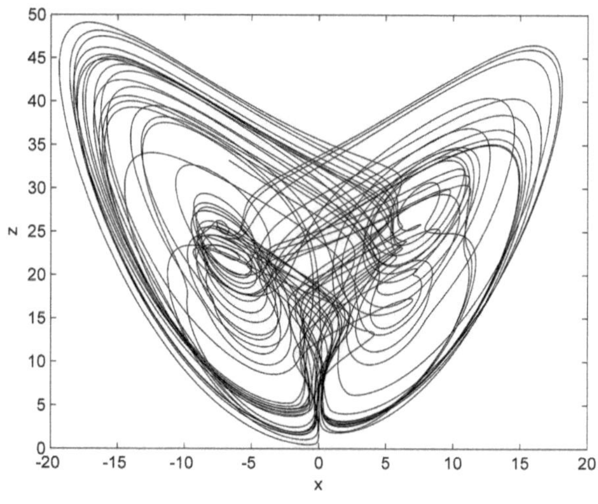

Fig. 2 Projection of the trajectory of Lorenz system in state plane (x, z) for the given time-varying Rayleigh number

Table 2 RMSE comparisons of the prediction experiments

m_2	$\varepsilon_{T=1}$	$\varepsilon_{T=2}$	$\varepsilon_{T=3}$	$\varepsilon_{T=4}$	$\varepsilon_{T=5}$	$\varepsilon_{T=6}$
0	1.21	1.66	5.52	2.47	5.64	12.22
1	0.58	0.88	0.99	2.29	2.44	2.79

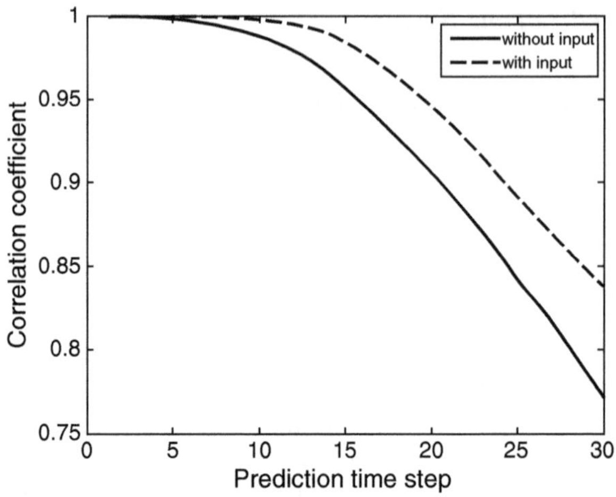

Fig. 3 Predictions of the stationary model and forcing model; all the results are averaged over the three appointed values of m_1

case are listed in Table 2 and Fig. 3. From Table 2, it can be seen that: (1) all RMSE values given by the forcing model were much lower than those by the stationary one, and (2) the growth in error rate with prediction steps for the forcing model was lower than that the stationary one. It can also be concluded that, in comparison with the stationary model, the forcing model had not only higher prediction accuracy, but also better predictability. Figure 3 presents the correlation coefficients between the actual and prediction values. Results show that the forcing model excelled the stationary model, indicating that introducing the driving force into the prediction model could improve the predictive skill effectively.

3.3 *Global Temperature Prediction*

The above approach is successful in improving prediction when inputs are included in "ideal" nonstationary systems. Motivated by this, we examine here whether such approaches are successful when we only have measurements from systems whose formulation is unknown. It has been demonstrated that the collective behavior of the network of four major climate modes (namely the Pacific Decadal Oscillation (PDO), the North Atlantic Oscillation (NAO), the El Niño/Southern Oscillation (ENSO), and the North Pacific Index (NPI)) can account for the decadal climate variability and all climate shifts observed in the instrumental record. More specifically we consider whether the above-mentioned four major climate modes influence global mean temperature in the sense of Granger (1969). Each of these modes involves different mechanisms over different geographical regions. Thus, we treat them as low-order nonlinear sub-systems of the grand climate system exhibiting complex dynamics. Indeed, some of their dynamics have been adequately explored and explained by simplified models, which represent subsets of the complete climate system and which are governed by their own dynamics.

Monthly mean values in the interval 1900–2007 are available for the global mean temperature and all four modes. We first considered the values of the global temperature and embedded it in dimensions 3–5 using $\tau = 1$ month. For each embedding we used the first 103 years (1236 data points) to build the predictive model. The last 4 years (48 data points) were used for predictions and to estimate the correlation coefficient between actual and predicted values as a function of prediction time step. Figure 4 shows, for each embedding dimension, the prediction skill with and without the influence of the inputs of the four climate modes. The results using the non-skill method of persistence are also shown. Note that for five variables and for a range of M possible embeddings for each variable, there exist M 5 combinations. Thus, to keep things simple, the embedding dimensions were set for all variables to either 3, or 4, or 5. Clearly, when the input of the four major

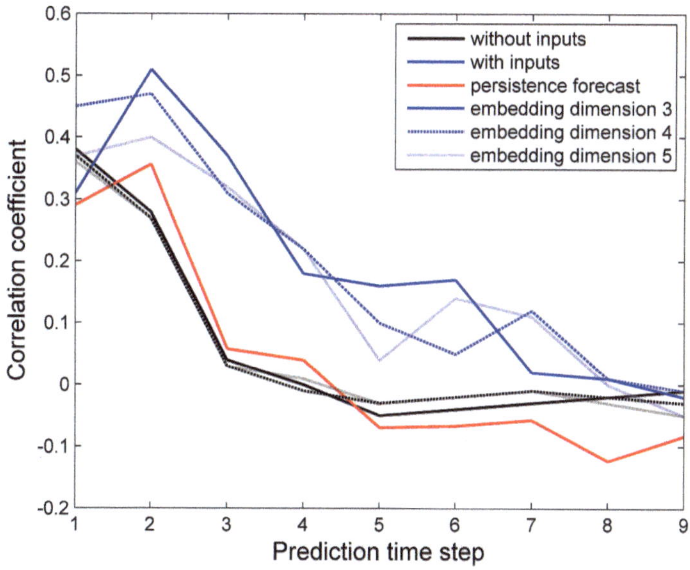

Fig. 4 This figure shows, for each embedding dimension, the correlation between predicted and actual values as a function of the prediction time step (in months) with (*blue lines*) and without (*black lines*) the influence of the inputs. The results using the non-skill method of persistence are also shown (*red line*). The results using all four inputs are superior as the results without inputs are basically the same as persistence

modes is included prediction is dramatically improved. In fact, without the input, the predictive model is only as accurate as persistence. The average correlation over the prediction time step range 1–9 months is improved 125–150% when the inputs are included. The improvement is also observed at embeddings 6 and 7, but due to sample limitations is not as good. In order to address possible effects of nonstationarities in the data we repeated the analysis with detrended data. The conclusions do not change significantly. These results establish for the first time Granger causality between major climate modes and global temperature variability over seasonal time scales.

We then repeated the above analysis but now we used each mode alone as an input. Figure 5 shows the average correlation coefficient as a function of prediction time step over the three embeddings. The solid black line is the average of the black lines in Fig. 4 (not any input considered), the red line represents again persistence, and the broken blue line is the average of the blue lines in Fig. 4 (all four inputs are considered). The green line is the average over the three embeddings and over

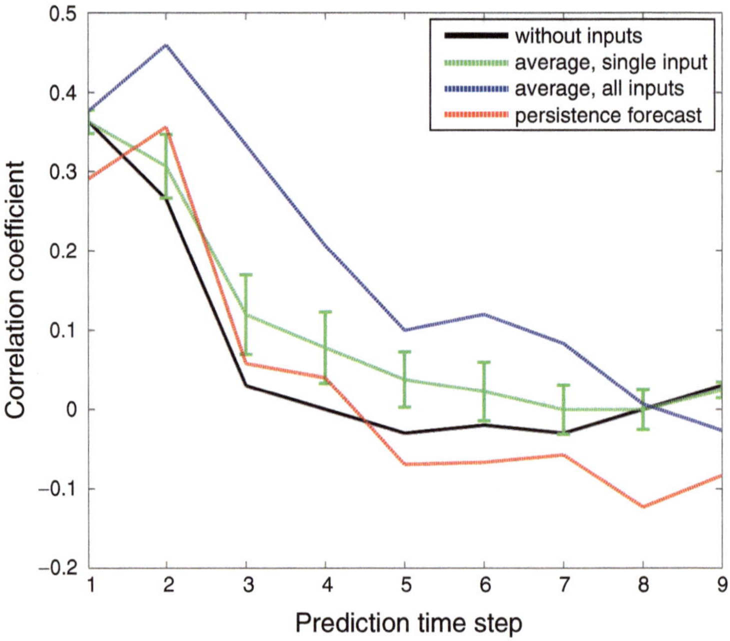

Fig. 5 This figure shows the average correlation coefficient as a function of prediction time step over the three embeddings. The *black line* is the average of the *black lines* in Fig. 1 (not any input considered), the *red line* represents again persistence, and the broken *blue line* is the average of the *blue lines* in Fig. 4 (all four inputs are considered). The *green line* is the average over the three embeddings and over the four modes acting individually. The bars on the *green line* indicate the one standard deviation. Clearly, the *blue line* stands above all other lines indicating that the improvement in predicting global temperature is the result of the collective behavior of the modes in the network and not a result of an individual dominant mode (see text for more details)

the four modes acting individually. The vertical green bars indicate the one standard deviation. While any individual input improves prediction compared to no input or to persistence, the blue line stands above all other lines indicating that the improvement in predicting global temperature is the result of the collective behavior of the modes in the network and not a result of an individual dominant mode.

3.4 *Establishing a Prediction Model of 500 hPa Geopotential High Anomaly with ENSO and Spatio-Temporal Structure*

The above experiments were carried for one single time series, here we extend the approach to the spatio-temporal time series (the global monthly mean geopotential height anomaly at 500 hPa). The 500 hPa geopotential height is one of the most fundamental and widely used meteorological variables for characterizing the general atmospheric circulation. Through short-term climate prediction of 500 hPa geopotential high anomaly has been made some progress in recent decades based on GCMs and statistical techniques, the predictive skill is still not very satisfactory. Uncertainties are still associated with GCMs. For statistical techniques, improvement of prediction skill is still continuing to be hindered partly by the issue of insufficient in observational data.

Observed time series for climate processes are generally too short to satisfy the length requirement, which was referred to as the "data bottleneck" problem in the time series theory. In order to solve this problem, spatio-temporal series have been utilized to reconstruct the dynamical system since late 1980s. This main idea was to consist of the observation data from different locations in physical space, which was applied to estimate the dimensionality of climate attractor and the prediction of the observational field (Essex et al. 1987; Keppenne and Nicolis 1989; Yang et al. 1994; Yang et al. 2000; Wang and Yang 2005). The spatio-temporal series are supposed to be controlled by the same physical law, in other words, the subsequences observed at the different locations are considered to describe an identical dynamical system. That is, by studying this sub-trajectory family, we can obtain the statistical behaviors of the attractor, and thereby predict the dynamics of the spatio-temporal series (Yang et al. 2000; Wang and Yang 2005). Comparing with the method of the single point time series, the spatio-temporal series analysis can efficiently improve the ergodicity of time series.

In this part, we will take advantage of spatio-temporal idea and incorporate El Nino-SOI (ENSO) as the driving force to establish a predictive model for the geopotential height. ENSO was incorporated in the predictive model since it has been widely suggested to be the dominant driving force of inter-annual variability of climate (McPhaden et al. 2006). The effect of ENSO on global climate change has been studied intensively using both models and observational data (e.g., Diaz et al. 2001; Bengtsson et al. 2006; Latif and Keenlyside 2009).

We now proceed with the method based on Eqs. (1)–(3). Suppose that a nonlinear and nonstationary process is composed of two time series

$$x_{ij}\left(t_k\right) = \left\{x_k\left(i,j\right)\right\}\Big|_{i=1,2,\ldots,m;j=1,2,\ldots,n;k=1,2,\ldots l}$$

$$\text{and } \{\alpha_k\}_{k=1,2,\ \ldots\ l}$$

where former is the spatio-temporal series that we are interested in (global monthly mean geopotential height anomaly at 500 hPa in this study), and the latter is for the assumed acting external forcing (SOI index in this study).

Every phase points of a time series is assumed to be embedded into m-dimension space, in such way those trajectories can describe dynamics of the established system. If they are embedded into an $m_1 + m_2$ dimensional phase space with a selected time lagτ, then we can obtain a family of trajectory twining on an identical attractor to shed light on the dynamics of the spatio-temporal system, in which the factor of external force plays a similar role as state variables (Wang et al. 2012). A delay reconstruction with embedding theorem of Takens (1981) is as follows:

$$\vec{E}\left(t\right) = \left\{x_{ij}(t), x_{ij}\left(t-\tau\right), \ldots, x_{ij}\left(t-\left(m_1-1\right)\tau\right); \alpha(t), \alpha\left(t-\tau\right), \ldots, \alpha\left(t-\left(m_2-1\right)\tau\right)\right\}_{t=1,2,\ldots,N} \quad (1)$$

or simply as

$$\vec{E}\left(t\right) = \left\{x_{ij};\alpha\right\}_{t=1,2,\ldots,N} \quad (10)$$

Here, m_1 and m_2 represent the given embedding dimensions for $\{x(t)\}$ and $\{\alpha(t)\}$, respectively, and $N = l - (\max(m_1, m_2) - 1) \times \tau$ is the number of state points on the trajectory.

Considering to introduce those information of the state points which are around the predicted state point, the purpose is trying to use the data from physical space to build a larger state set to ravel "data bottleneck" problem described before. Based on this trajectory for each of the time series at the spatio-temporal state, a predictive model to predict the future state of the system can be established as follows:

$$x_{t+p}^{(i,j)} = \widehat{f}_p\left(\vec{x}_t^{(i,j)}, \vec{x}_t^{(i+1,j)}, \vec{x}_t^{(i-1,j)}, \vec{x}_t^{(i,j+1)}, \vec{x}_t^{(i,j-1)}, \vec{\alpha}_t\right) + \varepsilon_i \quad (11)$$

where p is the prediction time step, $\varepsilon(t)$ is the fitting error, and$\widehat{f}$ is a desired function that was assumed to be a quadratic polynomial here. The following task is to find the cost function $\eta = \sum_{t=1}^{N}\left[\hat{f}\left(x^{(i,j)}(t), \alpha(t)\right) - x^{(i,j)}\left(t+p\right)\right]^2$. The coefficients of the established prediction equation will be obtained by minimizing the cost function.

The global monthly mean geopotential height anomaly at 500 hPa of NCEP/NCAR reanalysis data set was taken as the predicted objects here. The spatial resolution of data is $2.5 \times 2.5°$, its data length and SOI index length are all 720 months (from January of year 1951 to December of year 2010). SOI is one of the traditional ENSO indices, which can address the activity of ENSO and reflect successfully strengthen or weakness of the surface pressure in eastern or western of Pacific. Studies have shown that the sufficient relationship between SOI index and 500 hPa geopotential height by means of wavelet analysis. In the following experiment the SOI index is assumed as the external forcing to establish predicted models for global monthly mean geopotential height anomaly at 500 hPa. In this study we establish three kinds of predicted models. The first is named as the stationary model in which SOI is not included. The second is the forcing model in which SOI is included. The third is as the spatial forcing model in which both SOI index and spatial–temporal information are included. The predictive effectiveness on global 500 hPa monthly anomaly geopotential height is compared between three models.

For each of the 144×73 global points in length of 720 months, data were divided into two parts: the preceding 660 months were applied to construct the predictive model and the following 60 months for testing the prediction accuracy. The 500 hPa monthly mean geopotential height field was predicted for the period from January of 2006 to December of 2010. For establishing the model, parameters used here were assigned as the following values: the time lag τ was taken for 1; the embedding dimensions of 500 hPa monthly mean geopotential height field of m_1 varied from 3 to 7; the embedding dimensions of SOI index of m_2 were set to be 0 as the stationary model, m_2 was set to be in the range from 3 to 5 for both the forcing model and the spatial forcing model. All results were averaged over the embedding dimensions.

Depending on the prediction leading time step, the average correlation coefficient between observed and prediction with the change of prediction steps is shown in Fig. 6, where the dash-dot line represents the result from the stationary model, the broken line is from the forcing model over SOI forcing acting, and the solid line is from the spatial forcing model over the SOI forcing and spatial information which considered act collectively. Clearly, the position of solid line is above the other two lines indicates again that predictability is increased attributed to the ergodicity of time series improved efficiently through the reconstruction of external force by SOI and spatial information into the prediction model.

Since time dependency of the driving forces is the essential cause for nonstationarity and "data bottleneck" problem in time series theory, it is necessary to consider both influences on the prediction of time series. Extension of previous works, the objective of this study is to evaluate the two influence factors that may make a collective role in establishing a predictive model. In essence, the driving forces were regarded as state variables and reconstruct them into the prediction model. Therefore, the reconstructed system becomes to be an autonomous system, and thereby, the prediction is carried out under the frame of the stationary theory.

It is noted that ENSO is as a sole test external forcing and incorporate it into the prediction model in this study, also mechanism and prediction of ENSO itself is on the rise as a challenging topic in atmospheric science due to the special relationship

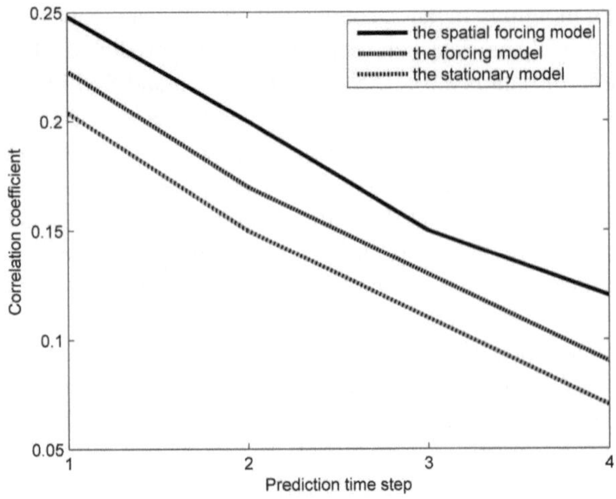

Fig. 6 Dependency of the prediction errors on the leading time step

between El Nino and extreme weather events. Other external forcings should be considered and different external forcings for different partitions in the global need to be taken into account. The experimental results obtained from global 500 hPa monthly geopotential height anomaly and ENSO confirmed the effectiveness of the predictive model, which also shown that the Granger causality (Granger 1969) retains between ENSO and global 500 hPa monthly geopotential height anomaly field.

4 Discussion

Because time dependency of driving forces is the essential cause of nonstationarity, it is necessary to consider its influence on the prediction of time series. However, due to the lack of a complete theory to predict nonstationary process, no general and effective method has yet been developed. One can only choose from some available techniques to remove its nonstationarity for resetting it under the framework of stationary theory. As an attempt to improve the situation, we proposed a new technique and applied it to predict several nonstationary time series with known external forces. The prediction results given by these experiments showed its effectiveness. In essence, the main idea of this technique was to consider all the driving forces as state variables and incorporate them into the prediction model. Therefore, the reconstructed system was changed to be an autonomous system, and thereby, the prediction returned to within the framework of the stationary theory. For the prediction of nonstationary time series with known driving forces, this technique can be used.

Nonlinear prediction is generally successful in identifying chaos and nonlinearity in data because it uses all available points, unlike other methods that exploit only a subset of available points in the attractor (Sugihara and May 1990). Predictions of global mean temperature over a long timescale are very uncertain, both because the climate possesses significant internal variability, and also because the sensitivity of the climate system to natural and anthropogenic effects is difficult to predict. Such an approach as that presented here may provide a compatible and direct window to study external forcings of the climate. Construction of external forcings can be extracted with the technique, for example, convergent cross mapping by Sugihara et al. (2012) to analyze the causality in nonlinear dynamic systems, progress for climate causal relations will be reported.

Acknowledgments This research was supported by the National Natural Science Foundation of China under grants 41275087 and 41575058.

References

Bengtsson, L., K. Hodges, E. Roeckner, and R. Brokopf. 2006. On the natural variability of the pre-industrial European climate. *Climate Dynamics* 27 (7–8): 743–760.

Casdagli, M. 1989. Nonlinear prediction of chaotic time series. *Physica D* 35: 335–356.

Diaz, H., M. Hoerling, and J. Eischeid. 2001. ENSO variability, teleconnections and climate change. *International Journal of Climatology* 21 (15): 1845–1862.

Essex, C., T. Lookman, and M.A.H. Nerenberg. 1987. The climate attractor over short timescales. *Nature* 326: 64–66.

Farmer, J.D., and J. Sidorowich. 1987. Predicting chaotic time series. *Physical Review Letters* 59: 845–848.

Granger, C.W. 1969. Investigating causal relations by econometric models and cross-spectral methods. *Econometrica* 37 (3): 424–438.

Keppenne, C.L., and C. Nicolis. 1989. Global properties and local structure of the weather attractor over Western Europe. *Journal of the Atmospheric Sciences* 46: 2356–2370.

Latif, M., and N. Keenlyside. 2009. El Niño/Southern Oscillation response to global warming. *PNAS* 106 (49): 20578–20583.

Manuca, R., and R. Savit. 1996. Stationarity and nonstationarity in time series analysis. *Physics D* 99: 134–161.

Mcphaden, M.J., S. Zebiak, and M. Glantz. 2006. ENSO as an integrating concept in earth science. *Science* 314 (5806): 1740–1745.

Packard, N.H., J.P. Crutchfield, J.D. Farmer, and R.S. Shaw. 1980. Geometry from a time series. *Physical Review Letters* 45: 712–715.

Stark, J. 1999. Delay embeddings for forced systems: Deterministic forcing. *Journal of Nonlinear Science*. 9: 255–332.

Sugihara, G., and R. May. 1990. Nonlinear forecasting as a way of distinguishing chaos from measurement error in time series. *Nature* 344: 734–741.

Sugihara, G., R. May, H. Ye, C. Hsieh, E. Deyle, M. Fogarty, and S. Munch. 2012. Detecting causality in complex ecosystems. *Science* 338: 496–500.

Takens, F. 1981. Detecting strange attractors in turbulence. In *Dynamical systems and turbulence*, ed. D. Rand and L.S. Young, 366–381. Heidelberg: Springer-Verlag.

Trenberth, K.E. 1990. Recent observed inter-decadal climate changes in the northern hemisphere. *Bulletin of the American Meteorological Society* 7: 988–993.

Tsonis, A.A. 1996. Widespread increases in low-frequency variability of precipitation over the past century. *Nature* 382: 700–702.

Wang, G., and P. Yang. 2005. A compound reconstructed prediction model for nonstationary climate process. *International Journal of Climatology* 25: 1265–1277.

Wang, G., P. Yang, X. Zhou, K. Swanson, and A.A. Tsonis. 2012. Directional influences on global temperature prediction. *Geophysical Research Letters* 39: L13704.

Yang, P., G.P. Brasseur, J.C. Gille, and S. Madronich. 1994. Dimensionalities of ozone attractors and their global distribution. *Physica D* 76: 331–343.

Yang, P., X. Zhou, and J. Bian. 2000. A nonlinear regional prediction experiment on a short-range climatic process of the atmospheric ozone. *Journal of Geophysical Research* 105: 12253–12258.

Yang, P., G. Wang, and J. Bian. 2010. The prediction of non-stationary climate series based on empirical mode decomposition. *Advances in Atmospheric Sciences* 27 (4): 845–854.

The Impact of Nonlinearity on the Targeted Observations for Tropical Cyclone Prediction

Feifan Zhou and He Zhang

Abstract This study examines the impact of nonlinearity on the targeted observations for tropical cyclone prediction. The nonlinearity of the typhoon is determined by comparing the first singular vector (FSV) and the conditional nonlinear optimal perturbation (CNOP), which is the nonlinear extension of FSV. If the similarity between the CNOP and FSV is larger than 0.5, then the typhoon is categorized as weak nonlinearity, otherwise, the typhoon is categorized as strong nonlinearity. First, the impact of nonlinearity on the typhoon targeted observations due to different resolutions is studied. Two typhoons, Meari (2004) and Matsa (2005), with 24 h forecast length are chosen, with 120-, 60-, and 30-km resolutions, respectively. It is found that the nonlinearity of both cases becomes stronger as the resolution increases. However, the sensitive areas identified with lower resolutions are more similar to each other than those identified with finer resolutions. This means that when the motion of typhoon has been described as linear or weakly nonlinear, the sensitive area may be easier to determine. Then, the impact of nonlinearity on the typhoon targeted observations due to different forecast length is investigated. In this part, typhoons Meari (2004) and Matsa (2005) with 60 km resolution are considered with 12-, 24-, and 36-h forecast lengths. We further studied two issues. In the first the initial time is fixed, while in the second the forecast time is fixed. Results show that no matter which issue is considered, typhoon Matsa exhibits stronger nonlinearity than typhoon Meari. Accordingly, Meari is categorized as a linear case, while Matsa as a nonlinear case. In the linear case, the sensitive areas identified for special forecast times (when the initial time is fixed) resemble those identified for other forecast times. Targeted observations deployed to improve a specific time forecast would thus also benefit forecasts at other times. In the nonlinear case, the similarities among the sensitive areas identified for different forecast times were more limited. The deployment of targeted observations in the nonlinear case would therefore need

F. Zhou (✉)
Laboratory of Cloud-Precipitation Physics and Severe Storms (LACS), Institute of Atmospheric Physics, Chinese Academy of Sciences, Beijing, 100029, China
e-mail: zff@mail.iap.ac.cn

H. Zhang
International Center for Climate and Environment Sciences (ICCES), Institute of Atmospheric Physics, Chinese Academy of Sciences, Beijing, 100029, China

A.A. Tsonis (ed.), *Advances in Nonlinear Geosciences*,
DOI 10.1007/978-3-319-58895-7_32

to be adapted to achieve large improvements for different targeted forecasts. For both cases, the closer the forecast time, the higher the similarities of the sensitive areas. When the forecast time is fixed, the sensitive areas in the linear case diverge continuously from the verification area as the forecast period lengthens due to the determination of the subtropical high in the movement of the typhoon, while those in the nonlinear case are always located around the initial cyclone indicating that the main factors affecting the typhoon movements are located within the typhoon. The deployment of targeted observations to improve a special forecast depends strongly on the time of deployment. Generally, it seems that the sensitive areas are easy to be determined in the linear case and more beneficial for the forecast. In the nonlinear case, the identification of sensitive areas is more difficult, which results in harder deployments in targeted observations.

Keywords Nonlinearity • Targeted observation • Tropical cyclone • Prediction

1 Introduction

A targeted observation, which is also called adaptive observation, is aimed to largely improve the forecast skills of the targeted area by placing additional observations in some special areas at some special times. The key point for targeted observation is to determine the observation places which we called sensitive areas. Scientists have proposed many methods to identify the sensitive areas, such as the singular vectors (SV, Palmer et al. 1998; Buizza and Montani 1999), the conditional nonlinear optimal perturbations (CNOP, Mu et al. 2007; Mu et al. 2009), the adjoint sensitivities (ADS, Kim et al. 2004; Wu et al. 2007; Ancell and Mass 2006), the ensemble transform (ET, Bishop and Toth 1999), the ensemble Kalman Filter (EnKF, Hamill and Snyder 2002), the ensemble transform Kalman filter (ETKF, Bishop et al. 2001), and the piece by piece data assimilation method (PBPDA, Huang and Meng 2014). Among these methods, the SV, ADS, ET, and ETKF are in essence linear methods (Rivier et al. 2008), while the CNOP and PBPDA are nonlinear methods. It has been known in several studies that all methods are effective in searching the sensitive areas to improve the targeted forecasts. Some studies also have compared the efficiency of the sensitive areas identified by different methods. Majumdar et al. (2006) found that for strong hurricanes, the sensitive areas identified by ETKF and SV are almost the same, while for weak tropical cyclones, the sensitive areas are different. However, much work about the efficiency in identifying the sensitive areas for weak tropical cyclones remains to be done. Zhou and Mu (2011), Qin and Mu (2011a), Chen and Mu (2012) and Chen et al. (2013) found that CNOP-sensitive areas are more effective than SV-sensitive areas in improving the tropical cyclones' forecasts. The ETKF-sensitive areas and CNOP-sensitive areas have comparable efficiencies in improving tropical cyclones' forecasts (Qin and Mu 2011b). Generally, it is hard to determine which method is the best, as the efficiency varies from case to case, and from time to time.

Since the CNOP method is an extension of first SV (FSV) into the nonlinear region (Mu and Duan 2003; Duan and Mu 2006; Mu and Jiang 2008; Terwisscha van Scheltinga and Dijkstra 2008), and since CNOP-sensitive areas are more effective than SV-sensitive areas, it seems that the impact of nonlinearity is very important in sensitive area identification. In this paper, we will investigate the impact of the nonlinearity on targeted observations by summarizing some previous studies from the viewpoint of nonlinearity. Since the tropical cyclone targeted observations have been widely studied, this paper will focus on the impact of nonlinearity on tropical cyclone targeted observations. The structure of the paper is as follows. Section 2 discusses how to determine the nonlinearity of tropical cyclone. Section 3 describes the tropical cyclone cases and the experimental designs. Section 4 investigates the impact of nonlinearity on tropical cyclone targeted observations with different resolution, while Sect. 5 investigates the impact of nonlinearity on tropical cyclone targeted observations with different forecast time periods. A brief summary and discussion are provided in the final section.

2 Definition of Nonlinearity

In this paper, we define the nonlinearity of the tropical cyclone by comparing the patterns of CNOP and FSV.

2.1 *The CNOP Method*

A thorough description of the CNOP approach to tropical cyclone targeted observation can be found in Zhou and Mu (2012a) and is summarized here.

An initial perturbation $\boldsymbol{\delta X}_0^*$ of vector $\mathbf{X}_0$ is called CNOP if and only if

$$J\left(\boldsymbol{\delta X}_0^*\right) = \max_{\boldsymbol{\delta X}_0^{\mathrm{T}}\mathbf{C}_1\boldsymbol{\delta X}_0 \leq \beta} J\left(\boldsymbol{\delta X}_0\right) \tag{1}$$

where

$$J\left(\boldsymbol{\delta X}_0\right) = \left[\mathbf{P}\boldsymbol{M}\left(\mathrm{X}_0+\boldsymbol{\delta X}_0\right) - \mathbf{P}\boldsymbol{M}\left(\mathbf{X}_0\right)\right]^{\mathrm{T}}\mathbf{C}_2\left[\mathbf{P}\boldsymbol{M}\left(\mathbf{X}_0+\boldsymbol{\delta X}_0\right) - \mathbf{P}\boldsymbol{M}\left(\mathbf{X}_0\right)\right] \tag{2}$$

and $\delta\mathbf{X_0}^{\mathbf{T}}\mathbf{C}_1\delta\mathbf{X_0} \leq \beta$ is a constraint condition of the initial perturbation $\delta\mathbf{X_0}$ with the presumed positive constant β representing the magnitude of the initial uncertainty. $\mathbf{C}_1$ and $\mathbf{C}_2$ are appropriate norms that measure $\delta\mathbf{X_0}$ and its development, respectively. In the discrete form, they can be presented as symmetric positive definite matrices. $\boldsymbol{M}$ is a nonlinear propagator, which propagates initial state $\mathbf{X}_0$ to the state vector at time t $\mathbf{X_t}$. $\mathbf{X_t}$ can be taken as the solution of the nonlinear model $\frac{\partial \mathbf{X}}{\partial t} + F\left(\mathbf{X}\right) = 0$, where F is a nonlinear partial differential operator. $\mathbf{P}$ is a local projection operator and takes the value of 1(0) inside (outside) the verification region. The superscript "T" denotes the transpose of the vectors or matrices. It is clear that the CNOPs depend on the nonlinear model $\boldsymbol{M}$, the initial state vector $\mathbf{X}_0$, and the choice of β, $\mathbf{P}$, $\mathbf{C}_1$, and $\mathbf{C}_2$. Sensitivity studies of the CNOP with respect to β, $\mathbf{C}_1$, $\mathbf{C}_2$, $\mathbf{P}$, and $\boldsymbol{M}(\mathbf{X_0})$ have been investigated in Mu et al. (2009), Zhou and Mu (2011, 2012a, b), and Tan et al. (2010).

2.2 The FSV Method

As above, a detailed description of the FSV approach can be found in the work of Zhou and Mu (2012a) and is summarized here.

Similar to the definition of CNOP, the $\delta\mathbf{X}_0^*$ can be defined as FSV (Ehrendorfer and Errico 1995) from

$$J\left(\delta X_0^*\right) = \max_{\delta X_0^{\mathrm{T}} C_1 \delta X_0 \leq \beta} J\left(\delta X_0\right) \tag{3}$$

where

$$J\left(\delta\mathbf{X}_0\right) = \left[\boldsymbol{PL}\left(\delta\mathbf{X}_0\right)\right]^{\mathrm{T}} \boldsymbol{C}_2 \left[\boldsymbol{PL}\left(\delta\mathbf{X}_0\right)\right] \tag{4}$$

where $\boldsymbol{L}$ is the forward tangent propagator corresponding to $\boldsymbol{M}$. That is, we have

$$\delta\mathbf{X}_t = \boldsymbol{M}\left(\mathbf{X_0}+\boldsymbol{\delta}\mathbf{X_0}\right) - \boldsymbol{M}\left(\mathbf{X_0}\right) \approx \boldsymbol{L}\left(\delta\mathbf{X}_0\right)$$

It can be seen that the FSV is the linear approximation of CNOP, and both CNOP and FSV can be obtained using the same optimization algorithm to facilitate comparison. In this study, the optimization algorithm employed is the spectral projected gradient 2 (SPG2) (Birgin et al. 2001).

2.3 Definition of Sensitive Areas and Nonlinearity

The definition of a sensitive area in this work is similar to that stated in the papers of Buizza et al. (2007) and Zhou and Zhang (2014). A vertically integrated total dry energy function $f(i,j)$ is used:

$$f\left(i,j\right) = \int_0^1 E_{\mathrm{d}}\left(i,j,\sigma\right) \mathrm{d}\sigma \tag{5}$$

where $E_{\mathrm{d}}(i,j,\sigma)$ is the total dry energy (the sum of kinetic energy, available relative potential energy, and available surface potential energy) of the CNOP at grid point (i,j,σ).

The horizontal grid points where the function $f(i,j)$ is higher than a certain value c are defined as the sensitive areas. In this paper, the value c is chosen to be 1/6 percent of the maximum $f(i,j)$.

The similarity between two vectors **X** and **Y** by using the following formula:

$$S_{xy} = \frac{<\mathbf{X},\mathbf{Y}>}{\sqrt{<\mathbf{X},\mathbf{X}>}\sqrt{<\mathbf{Y},\mathbf{Y}>}} \tag{6}$$

The nonlinearity of the tropical cyclones is defined according to the similarity S_{CF} between CNOP and FSV. If the similarity S_{CF} is larger than 0.5, we categorize the tropical cyclone as weak nonlinearity and if the similarity S_{CF} is smaller than 0.5, as strong nonlinearity.

3 Experimental Setup

3.1 The Model and the Cases

The fifth generation Pennsylvania State University–National Center for Atmospheric Research (PSU–NCAR) Mesoscale Model (MM5; Dudhia 1993) is then employed. The initial and boundary conditions are supplied by the National Centers for Environment Predictions (NCEP) FNL (Final) Operational Global Analysis ($1^\circ \times 1^\circ$) interpolated into the MM5 grids. The corresponding adjoint system of MM5 (Zou et al. 1997) is also used, with the following physical parameterizations: dry convective adjustment, grid-resolved large-scale precipitation, the high-resolution PBL scheme, and the Kuo cumulus parameterization scheme.

Two tropical cyclones, Matsa (2005) and Meari (2004) were investigated. For each case, the forecasts were run at 120-, 60-, and 30-km horizontal resolutions with 11 vertical levels. For TC Matsa, the model domain covered 28×28, 55×55, 109×109 (y-direction by x-direction) grids, respectively, for 120- 60-, and 30-km horizontal resolutions. For TC Meari, there were 26×28, 51×55, and 101×109 grids for each horizontal resolution. For each case with the chosen grids, the real physical domain was the same at all resolutions, thus the verification area was chosen to be the same.

To study the time-dependence issues, a set of experiments were designed in which all the parameters were held constant except for the studied time period. In the first issue, the initial time is fixed, and the forecasts have been carried out with 12, 24, and 36 h, respectively. For the Matsa case, the initial time has been set as 1200 UTC 4 Aug 2005 (Fig. 1a), while for Meari case, it has been set as 1200 UTC 25 Sep 2004. In the second issue, the forecast time is fixed as 0000 UTC 6 Aug 2005 for Matsa case, while 0000 UTC 27 Sep 2004 for Meari case. Therefore, the initial times were 1200 UTC 5 Aug 2005, 0000 UTC 5 Aug 2005, and 1200 UTC 4 Aug 2005, respectively, corresponding to 12, 24, and 36 h forecasts for Matsa case (Fig.

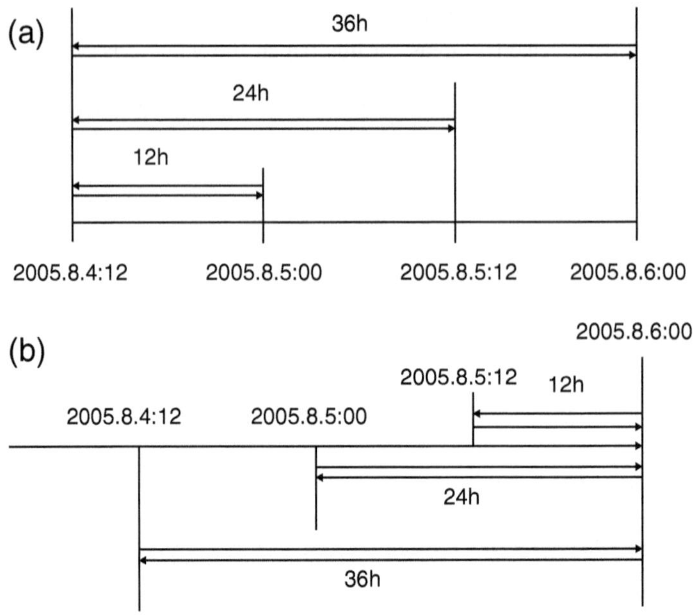

Fig. 1 The design of the optimization time periods for typhoon Matsa (2005) (**a**) for the first approach and (**b**) for the second approach

1b), and 1200 UTC 26 Sep 2004, 0000 UTC 26 Sep 2004, and 1200 UTC 25 Sep 2004, respectively, for Meari.

In this study, the optimization time periods are same as the forecast time periods.

3.2 Initial Constraint and Cost Function

The metric used in the initial constraint condition was the same as that used in the cost function, and both were chosen as the total dry energy. That is, in a continuous expression, we have

$$(\delta \mathbf{X}_0)^{\mathrm{T}} \boldsymbol{C}_1 (\delta \mathbf{X}_0) = \frac{1}{D_1} \int_{D_1} \int_0^1 \left[\boldsymbol{u}_0'^2 + \mathbf{v}_0'^2 + \frac{c_p}{T_r} \mathbf{T}_0'^2 + R_a T_r \left(\frac{\mathbf{p}_{s0}'}{p_r} \right)^2 \right] d\sigma dD_1 \quad (7)$$

and

$$J = \frac{1}{D_2} \int_{D_2} \int_0^1 \left[\mathbf{u}_t'^2 + \mathbf{v}_t'^2 + \frac{c_p}{T_r} \mathbf{T}_t'^2 + R_a T_r \left(\frac{\mathbf{p}_{st}'}{p_r} \right)^2 \right] \mathrm{d}\sigma \mathrm{d}D_2 \quad (8)$$

where the initial perturbation $\delta \mathbf{X}_0$ is composed of $\boldsymbol{u}_0'$, $\boldsymbol{v}_0'$, $\boldsymbol{T}_0'$, and $\boldsymbol{p}_{s0}'$, which are the perturbed zonal and meridional wind components, temperature, and surface pressure at the initial time, respectively. D_1 is the horizontal model domain, and σ represents the vertical coordinate. J is the cost function defined as the total dry energy over the verification area D_2. Here, $\boldsymbol{u}_t'$, $\boldsymbol{v}_t'$, $\boldsymbol{T}_t'$, $\boldsymbol{p}_{st}'$ are components of δX_t, which represents the linear (or nonlinear) development of $\delta \mathbf{X}_0$ at time t. The terms c_p and R_a are the specific heat at constant pressure and the gas constant of dry air, respectively (with numerical values of 1005.7 and 287.04 J kg^{-1} K^{-1}). The reference parameters were the following: $T_r = 270$ K, $p_r = 1000$ hPa.

The initial constraint value β is chosen as 0.03 J/kg for all cases. With this constraint value, the CNOP magnitudes are comparable with the current analysis errors, thus it can be taken as a kind of initial errors. It is noticed that the verification area has been designed the same for one case regardless of the optimization time period to facilitate comparison.

4 The Impact of Nonlinearity on the Typhoon Targeted Observations Due to Different Resolutions

It is easier to understand that an event represents weak nonlinearity when it is described with a low resolution, since in a low resolution many small-scale phenomena have been filtered out. However, if an event is described with a high resolution, small-scale phenomena are present, and their impact on the event becomes important, which would make the event strongly nonlinear. Next, the nonlinearity of the typhoons with different resolutions will be examined.

4.1 Nonlinearity of the Typhoons at Different Resolutions

Figure 2 shows CNOPs and FSVs calculated with different resolutions for Matsa case. We can see that with a low resolution of 120 km, the pattern of CNOP is similar to that of FSV, while with a resolution of 60 or 30 km, the CNOP is much different with FSV. The difference is largest when using 30 km resolution. This means that at a low resolution the motion of typhoon Matsa is basically a linear behavior. However, at finer resolutions the motion of Matsa, is increasingly nonlinear. The similarities are shown in Table 1, where we see that when using 120 km resolution, the similarity between CNOP and FSV is 0.8, which according to the definition in Sect. 2.3 is a weak nonlinear case. However, if we use the 60 or 30 km resolution, the similarities reduce to 0.3 and 0.2, indicating an increasingly nonlinear case.

For Meari case, the result is similar. That is, the lower the resolution, the more similar the pattern of CNOP is to that of FSV. However, in this case with a resolution of 60 km, the CNOP is also much similar to FSV (Fig. 3). This means that the motion

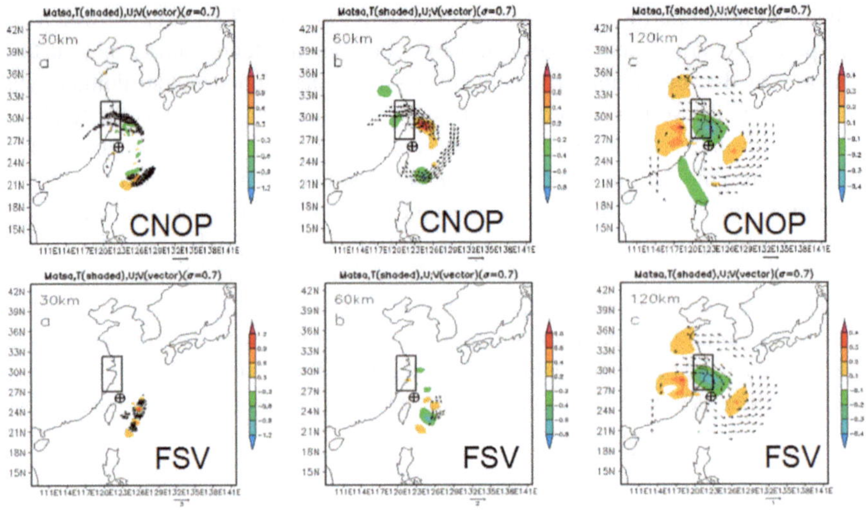

Fig. 2 Typhoon Matsa. The temperature (*shaded*, units: K) and wind (*vector*, units: m s^{-1}) components of CNOP (*the first line*) and FSV (*the second line*) at $\sigma = 0.7$. The *boxes* indicate the verification areas. The *circle* and the *cross signs* indicate the initial position of the cyclone. The first column is at the resolution of 30 km, the second column is at a resolution of 60 km, and the third column is at a resolution of 120 km

Table 1 The similarities between CNOP and FSV

	30 km	60 km	120 km
Matsa	0.2	0.3	0.8
Meari	0.2	0.7	0.8

of typhoon Meari can be described as linear when the resolution is 60 km or lower. The results shown in Table 1 also confirm the results shown in the figures.

In general, comparison among the three resolutions for both cases indicates that the nonlinearity of both cases becomes increasingly stronger as the resolution increases.

4.2 The Sensitive Areas Identified with Different Horizontal Resolutions

Because the CNOP method is a fully nonlinear method, while FSV has adopted linear approximation, and it is demonstrated that reducing initial errors in the CNOP-sensitive areas are more beneficial for improving the typhoon forecast than reducing the initial errors in the FSV-sensitive areas (Zhou and Mu 2011; Qin and Mu 2011a; Chen et al. 2013). In this section, we will focus on the CNOP-sensitive areas, and we will discuss to what extend the CNOP-sensitive areas has been influenced by the nonlinearity of the case.

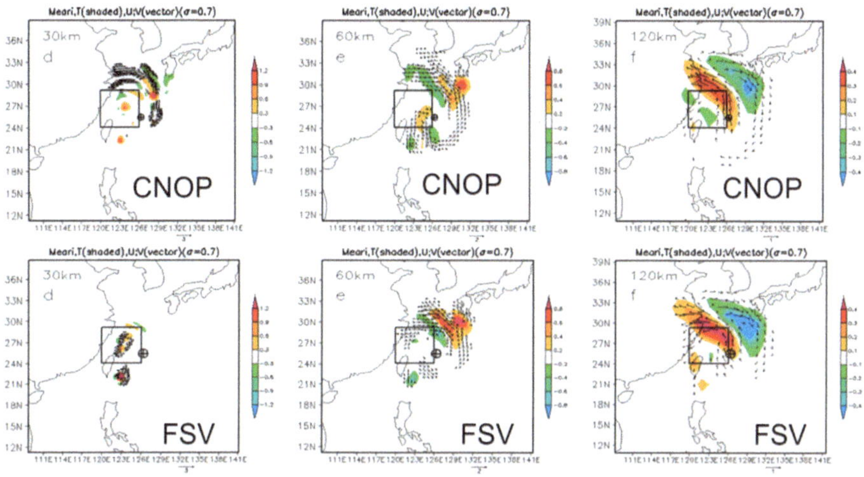

Fig. 3 Same as Fig. 2, but for typhoon Meari

Table 2 The similarities between the sensitive areas obtained at 30-, 60-, and 120-km resolutions for TC Matsa (2005) and TC Meari (2004)

	30 km and 60 km	60 km and 120 km
Matsa	0.70	0.8
Meari	0.55	0.75

For both cases, the CNOP-sensitive areas identified using different resolutions were different from each other (Fig. 4), and the sensitive areas become more localized as the resolution increases; however, common areas occurred at the three resolutions, and the sizes of the common areas were different for different case. In general, the sizes of the common areas are larger between sensitive areas at the lower resolution (Table 2). For both cases, the similarities between the lower resolutions (60 and 120 km) were greater than those between the finer resolutions (30 and 60 km), which simply illustrates that more small-scale activity can be resolved at higher resolutions. It also means that when the motion of typhoon is linear or weakly nonlinear, the sensitive area may be easier to determined, as the sensitive area looks like more stable. From the analysis of the similarities (Table 2), it can be concluded that the sensitive areas identified at lower resolutions are also helpful for improving the forecast at finer resolution. This is a favorable feature of CNOP-sensitive areas as the calculation of CNOP with a high resolution would usually require a long time and a large amount of computer resources. Therefore, when computation conditions are limited, we can use low resolution to calculate CNOP-sensitive areas, which would also be useful for typhoon targeted observations. The above results also illustrate that the linear approximation may be to some extend useful.

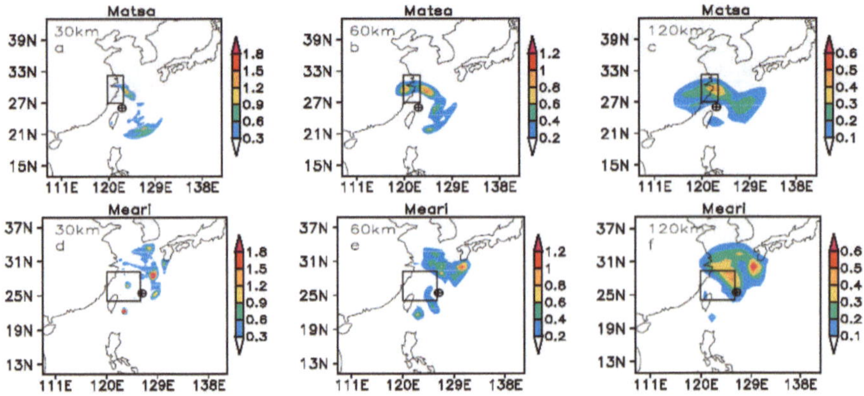

Fig. 4 Sensitive areas identified with different resolutions for Matsa and Meari case for 24 h prediction

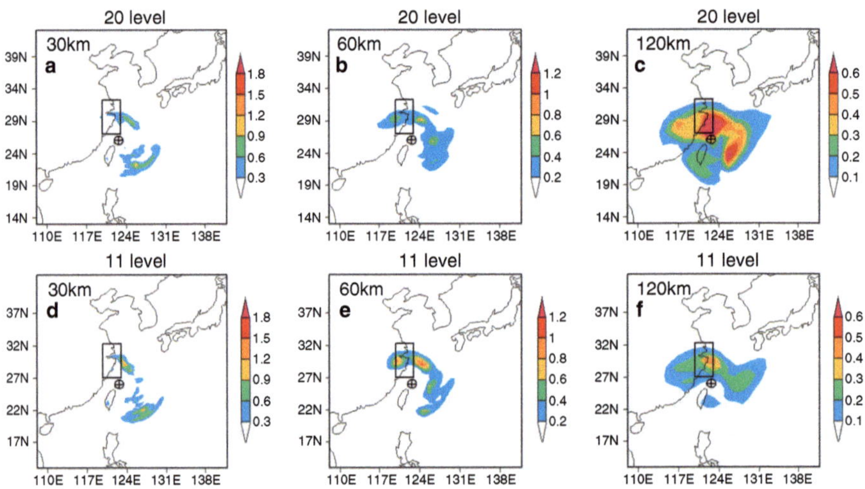

Fig. 5 Sensitive areas identified with different horizontal and vertical resolutions for Matsa case for 24 h prediction

4.3 The Sensitive Areas Identified with Different Vertical Resolutions

It is also interested to study the variations of the sensitive areas with respect to different vertical resolutions. Here, we just focused on the TC Matsa. First, we calculated the CNOPs with 20 vertical levels at 30-, 60-, 120-km resolutions, and then we obtained the CNOP-sensitive areas. Interestingly, we found that the variation of sensitive areas between different horizontal resolutions with 20 levels is similar to those with 11 levels. See Fig. 5. Besides, it is found that at high horizontal

Table 3 The similarities between the sensitive areas obtained with 11 vertical levels and 20 vertical levels, respectively, at 30-, 60-, and 120-km resolutions for TC Matsa (2005)

	30 km 11 l and 30 km 20 l	60 km 11 l and 60 km 20 l	120 km 11 l and 120 km 20 l
Similarity	0.89	0.85	0.80

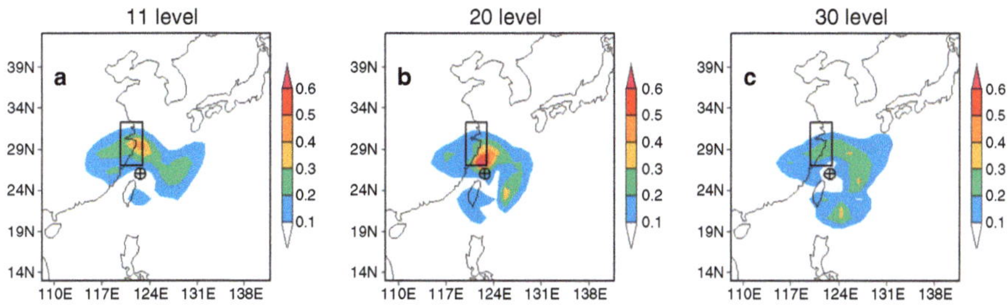

Fig. 6 Sensitive areas identified with different vertical resolutions for Matsa case with 120 km horizontal resolution for 24 h prediction

Table 4 The similarities between the sensitive areas obtained with different horizontal and vertical resolutions, respectively for TC Matsa (2005)

	120 km 11 l and 60 km 11 l	120 km 20 l and 60 km 11 l	120 km 30 l and 60 km 11 l
Similarity	0.782	0.788	0.802

resolutions, the increment of the vertical resolution has little impact on the results, while the increment of the vertical resolution at low horizontal resolutions would cause the result much different. This indicates that with a high horizontal resolution, it is hard to increase the nonlinearity just by increasing the vertical levels, but it is easy to increase the nonlinearity simply by increasing the vertical levels. For example, the sensitive areas identified by the CNOP at 30 km resolution with 20 levels and 11 levels are similar (Fig. 5a, d), and the sensitive areas identified at 120 km resolution with 20 levels and 11 levels present notable difference (Fig. 5c, f). This also can be obtained from Table 3.

Besides, we calculated the CNOPs at 120 km with 11, 20, and 30 vertical levels. Figure 6 shows the results. It is found that when the vertical resolution increases, the pattern of the sensitive areas would become more similar to the sensitive areas identified by the high horizontal resolution (Table 4). This is interesting and sense to us since the increment of vertical resolution cost less than the increment of horizontal resolution as far as the computation cost is considered. Generally, the similarities between different sensitive areas with different horizontal or vertical resolutions are high. This result further confirms that the linear approximation may be to some extend useful.

5 The Impact of Nonlinearity on the Typhoon Targeted Observations Due to Different Forecast Length

In this section, we also studied typhoons Matsa and Meari, but the resolution is fixed at 60 km. As has been showed in Sect. 4.1, when using 60 km resolution and a 24 h forecast length, the motion of Matsa is strongly nonlinear, while the motion of Meari is weakly nonlinear. What is the nonlinearity of the two typhoons when the forecast length is different? Of course, when the forecast time period is short, the linear approximation may be easier to be adopted than the longtime forecast. That is, a weak nonlinearity for a short time forecast while a strong nonlinearity for a longtime forecast. In this section, we will examine to what extend the sensitive areas will be affected by different nonlinearity due to different forecast length in cases with different nonlinearity. Besides, we will study two kinds of issues. The first is the initial time is fixed, while the second the forecast time is fixed (Fig. 1).

5.1 *The Approach with Fixed Initial Time*

First, we check the nonlinearity of the typhoon forecast with different forecast lengths. Figure 7 presents CNOPs and FSVs for the Matsa (2005) case based on the first approach. The CNOPs became progressively more different than the FSVs as the forecast time extended further from the initial time. This result suggests that the nonlinearity becomes strong, especially at longer forecast integrations. Although the difference between the patterns of CNOPs and FSVs for Meari (2004) case (Fig. 8) also becomes larger when the forecast length increases, generally, they are more

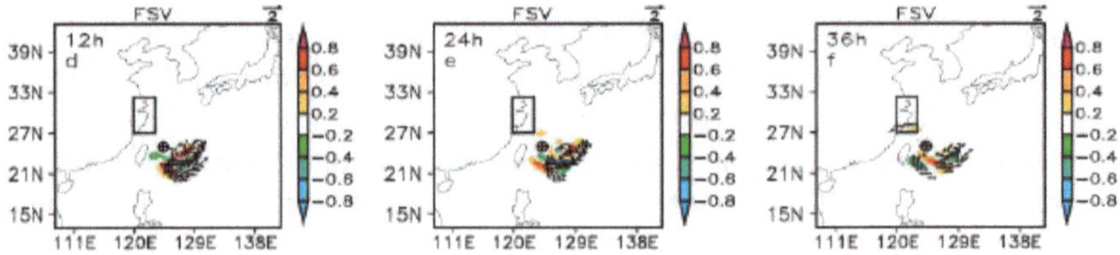

Fig. 7 Typhoon Matsa (2005). The temperature (*shaded*, units: K) and wind (*vector*, units: m s^{-1}) components of CNOP and FSV at $\sigma = 0.7$. The *boxes* indicate the verification areas. The *circled plus* indicates the position of the cyclone at 1200 UTC 4 August 2005

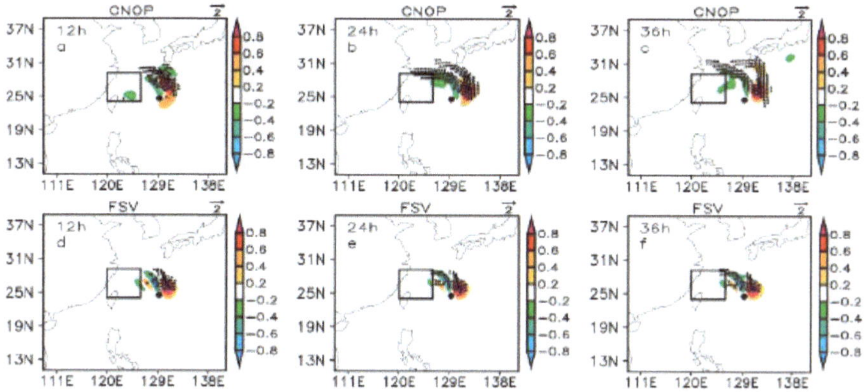

Fig. 8 Same as Fig. 7, but for Meari

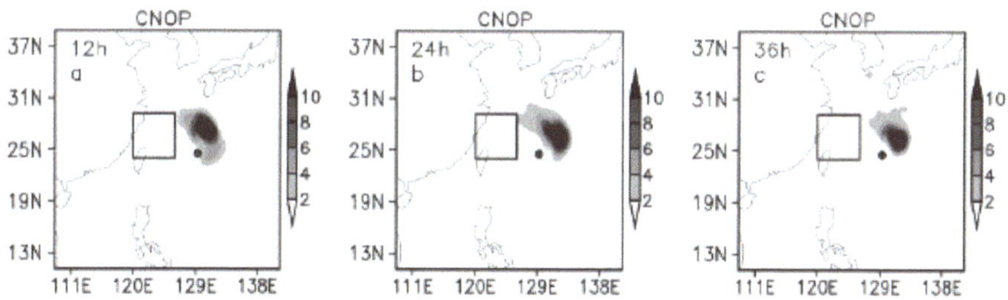

Fig. 9 TC Meari. Same as Fig. 8, but for sensitive areas

similar to each other compared to Matsa case. This suggests that nonlinearity is weak regardless of the optimization time period. In Meari (2004) case, the linear approximation is adoptable (Fig. 8).

Similarly, we only consider the CNOP-sensitive areas next. For Meari (2004) case, the sensitive areas are uniformly located at the northeast boundary of the initial cyclone, regardless of the forecast time (Fig. 9). For Matsa (2005) case, however, the location of the sensitive areas changes significantly as the forecast time extends from 12 to 36 h, with the main part of the sensitive areas shifting from southeast side of the initial cyclone to the northwest side of the initial cyclone (Fig. 10). In addition, for both cases, the closer the forecast times, the higher the similarities of the sensitive areas. Comparison of the sensitive areas identified for Matsa (2005) and Meari (2004) cases revealed several interesting features. In the linear case, the sensitive areas identified for a special forecast time were consistent with those identified for other forecast times when the initial time was fixed. This result means that targeted observations deployed to improve a special time forecast would also favorably affect the forecasts at other times. In the nonlinear case, however, although there were some similarities in the sensitive areas identified for different forecast times, these similarities are limited. This indicates that although the targeted observations

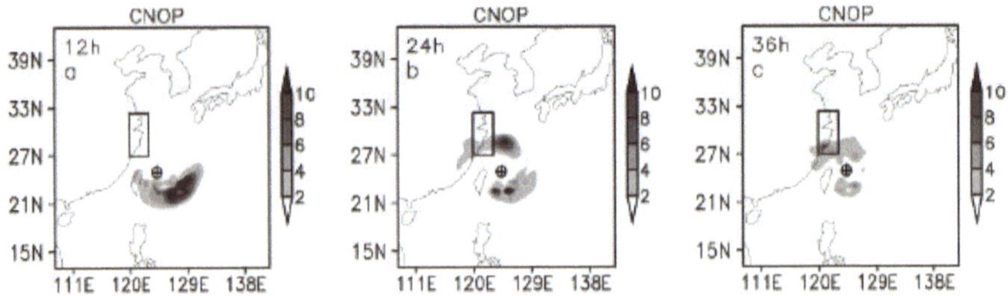

Fig. 10 TC Matsa. Same as Fig. 7, but for sensitive areas

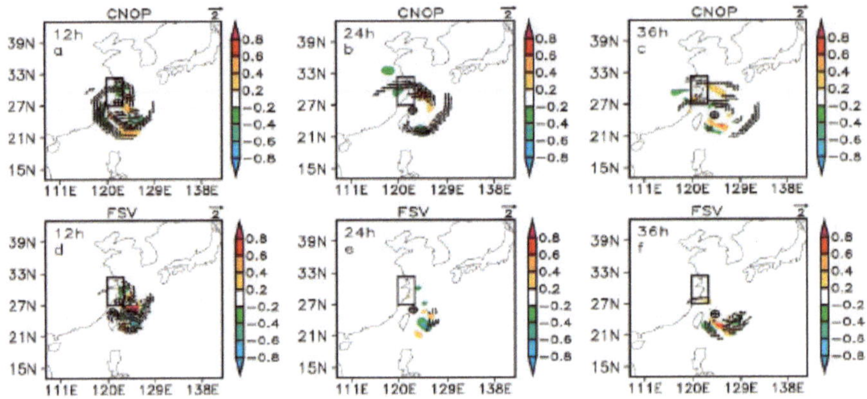

Fig. 11 TC Matsa. Same as Fig. 7, but for the fixed forecast time

deployed for a special time forecast are also beneficial for other times' forecasts, the forecast improvements for other times are limited. In the nonlinear case, therefore, the deployment of targeted observations should be adaptive to obtain the largest improvement for different targeted forecasts, and they should be more widespread in order to achieve the greatest improvement in multiple time forecasts.

5.2 The Approach with Fixed Forecast Time

The investigation of the nonlinearity of the two typhoons with different forecast lengths showed that Matsa (2005) case also presented strong nonlinearity during the studied time period when the second approach was used (Fig. 11). The CNOPs and FSVs differed regardless of when the forecasts were initialized. Same as the first approach, Meari (2004) case maintained its linear features (Fig. 12).

The sensitive areas of Meari case moved to the verification areas as the initial time shifted closer to the forecast time (i.e., as the optimization time period shortened; Fig. 13). The sensitive areas in this case were typically located along the southwestern fringe of the subtropical high at the periphery of the typhoon

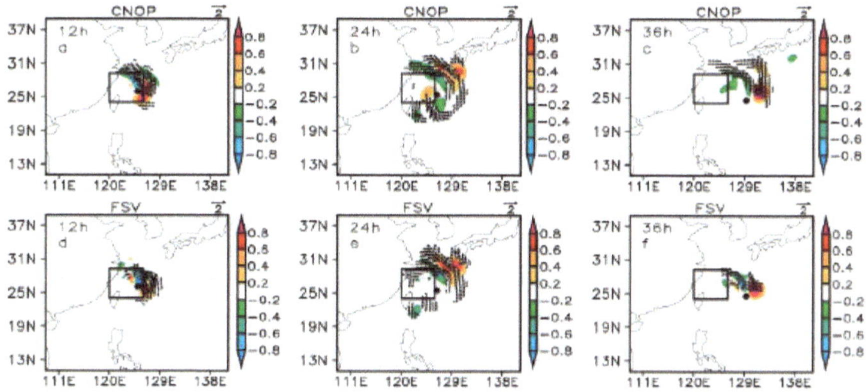

Fig. 12 TC Meari. Same as Fig. 8, but for the fixed forecast time

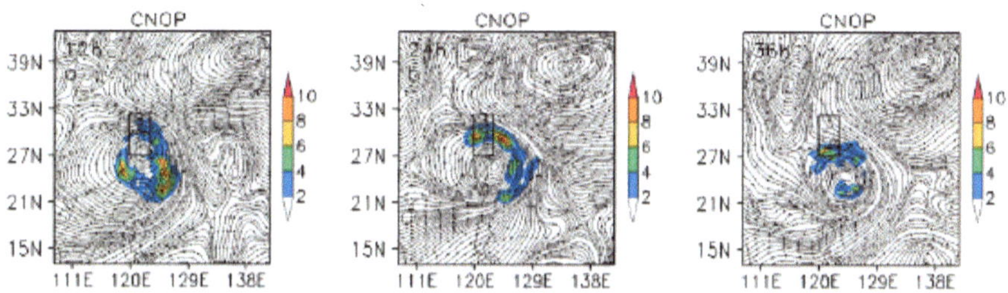

Fig. 13 Sensitive areas of Meari at different initials. The forecast time is fixed

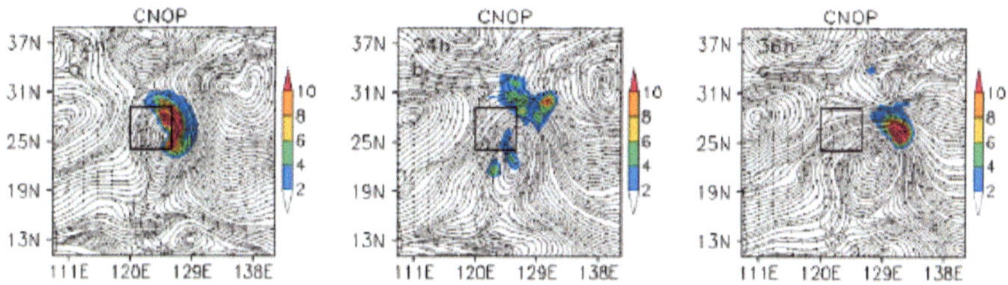

Fig. 14 Same as Fig. 13, but for Matsa case

circulation, especially for the forecasts from initial conditions 24 and 36 h prior to the forecast time (Fig. 13). This proximity suggests that the subtropical high plays an important part in the corresponding targeted forecasts. In the Matsa (2005) case, the sensitive areas fell in disrupted-ring patterns around the initial typhoon centers, and were mainly located inside the typhoon circulation (Fig. 14). This indicates that the targeted forecasts in this case were affected primarily by conditions within the typhoon, while the background fields played a relatively smaller role. The results of these two cases suggest that the deployment of targeted observations intended to improve the forecast at a special time may depend strongly on the time of deployment. The times at which the targeted observations were deployed are thus of crucial importance.

6 Summary and Discussion

This paper investigates the impact of the nonlinearity on tropical cyclone targeted observations by summarizing some previous studies from the viewpoint of nonlinearity. The nonlinearity of the tropical cyclone with different resolutions and different forecast lengths was determined by comparing the first singular vector (FSV) and the conditional nonlinear optimal perturbation (CNOP), which is the nonlinear extension of FSV.

First, the impact of nonlinearity on the typhoon targeted observations due to different resolutions was studied. Two typhoons, Meari (2004) and Matsa (2005), with 24 h forecast length were chosen to be studied with 120-, 60-, and 30-km resolutions, respectively. It was found the nonlinearity of both cases becomes stronger as the resolution increases. However, the sensitive areas identified at lower resolutions were more similar to each other than those identified at finer resolutions. This means that when the motion of typhoon is linear or weaker nonlinear, the sensitive area may be easier to be determined.

Then, the impact of nonlinearity on the typhoon targeted observations due to different forecast length was investigated. In this part, typhoons Meari (2004) and Matsa (2005) at a fixed 60 km resolution were considered with 12-, 24-, and 36-h forecast lengths. We further studied two kinds of issues. The first is the initial time is fixed, while the second is the forecast time is fixed. Results showed that no matter which issue is considered, typhoon Matsa exhibits stronger nonlinearity than typhoon Meari. Thus Meari was assumed to represent a linear case, while the Matsa a nonlinear case.

In the linear case, the sensitive areas identified for a special forecast time were consistent with those identified for other forecast times when the initial time was fixed. This result means that targeted observations deployed to improve a special time forecast would also favorably affect the forecasts at other times. In the nonlinear case, however, although there were some similarities in the sensitive areas identified for different forecast times, these similarities are limited. This indicates that although the targeted observations deployed for a special time forecast are also beneficial for other times' forecasts, the forecast improvements for other times are limited. In the nonlinear case, therefore, the deployment of targeted observations should be adaptive to obtain the largest improvement for different targeted forecasts, and they should be more widespread as to achieve the greatest improvement in multiple time forecasts.

The sensitive areas of Meari case moved to the verification areas as the initial time shifted closer to the forecast time. In the Matsa (2005) case, the sensitive areas fell in disrupted-ring patterns around the initial typhoon centers and were mainly located inside the typhoon circulation.

In general, it appears that the sensitive area is easy to be determined in the linear case and more beneficial for the forecast. In the nonlinear case, the identification of sensitive areas is more difficult, which results in difficult deployments in targeted observations. We conclude that identifying the sensitive areas in strong nonlinear cases is challenging, and more studies are necessary.

Acknowledgements This research was jointly supported by the National Natural Science Foundation of China (Grant no. 41475100), and the Youth Innovation Promotion Association of Chinese Academy of Sciences.

References

Ancell, B.C., and C.F. Mass. 2006. Structure, growth rates, and tangent linear accuracy of adjoint sensitivities with respect to horizontal and vertical resolution. *Monthly Weather Review* 134: 2971–2988.

Birgin, E.G., J.E. Martinez, and R. Marcos. 2001. Algorithm 813: SPG—software for convex-constrained optimization. *ACM Transactions on Mathematical Software* 27: 340–349.

Bishop, C.H., and Z. Toth. 1999. Ensemble transformation and adaptive observations. *Journal of the Atmospheric Sciences* 56: 1748–1765.

Bishop, C.H., B.J. Etherton, and S.J. Majumdar. 2001. Adaptive sampling with the ensemble transform Kalman filter. Part I: Theoretical aspects. *Monthly Weather Review* 129: 420–436.

Buizza, R., and A. Montani. 1999. Targeting observations using singular vectors. *Journal of the Atmospheric Sciences* 56: 2965–2985.

Buizza, R., C. Cardinali, G. Kelly, and J. Thépaut. 2007. The value of targeted observations part II: The value of observations taken in singular vectors-based target areas. *Quarterly Journal of the Royal Meteorological Society* 133: 1817–1832.

Chen, B.-Y., and M. Mu. 2012. The roles of spatial locations and patterns of initial errors in the uncertainties of tropical cyclone forecasts. *Advances in Atmospheric Sciences* 29: 63–78.

Chen, B.-Y., M. Mu, and X.H. Qin. 2013. The impact of assimilating drop-windsonde data deployed at different sites on typhoon track forecasts. *Monthly Weather Review* 141 (8): 2669–2682.

Duan, W.S., and M. Mu. 2006. Investigating decadal variability of El Nino-Southern Oscillation asymmetry by conditional nonlinear optimal perturbation. *Journal of Geophysical Research* 111: Co7015. doi:10.1029/2005JC003458.

Dudhia, J. 1993. A nonhydrostatic version of the Penn State/NCAR Mesoscale Model: Validation tests and simulation of an Atlantic cyclone and cold front. *Monthly Weather Review* 121: 1493–1513.

Ehrendorfer, M., and R.M. Errico. 1995. Mesoscale predictability and the spectrum of optimal perturbations. *Journal of the Atmospheric Sciences* 52: 3475–3500.

Hamill, T.M., and C. Snyder. 2002. Using improved background-error covariance from an ensemble kalman filter for adaptive observations. *Monthly Weather Review* 130: 1552–1572.

Huang, L., and Z. Meng. 2014. Quality of the target area for metrics with different nonlinearities in a mesoscale convective system. *Monthly Weather Review* 142: 2379–2397.

Kim, H.M., M.C. Morgan, and R.E. Morss. 2004. Evolution of analysis error and adjoint-based sensitivities: Implications for adaptive observations. *Journal of the Atmospheric Sciences* 61: 795–812.

Majumdar, S.J., S.D. Aberson, C.H. Bishop, R. Buizza, M.S. Peng, and C.A. Reynolds. 2006. A comparison of adaptive observing guidance for Atlantic tropical cyclones. *Monthly Weather Review* 134: 2354–2372.

Mu, M., and W.S. Duan. 2003. A new approach to studying ENSO predictability: Conditional nonlinear optimal perturbation. *Chinese Science Bulletin* 48: 747–749.

Mu, M., and Z.N. Jiang. 2008. A new method to generate the initial perturbations in ensemble forecast: Conditional nonlinear optimal perturbations. *Chinese Science Bulletin* 53: 2062–2068S.

Mu, M., H.L. Wang, and F.F. Zhou. 2007. A preliminary application of conditional nonlinear optimal perturbation to adaptive observation. *Chinese Journal of the Atmospheric Sciences* 31: 1102–1112. (in Chinese).

Mu, M., F.F. Zhou, and H.L. Wang. 2009. A method to identify the sensitive areas in targeting for tropical cyclone prediction: Conditional nonlinear optimal perturbation. *Monthly Weather Review* 137: 1623–1639.

Palmer, T.N., R. Gelaro, J. Barkmeijer, and R. Buizza. 1998. Singular vectors, metrics, and adaptive observations. *Journal of the Atmospheric Sciences* 55: 633–653.

Qin, X.-H., and M. Mu. 2011a. Influence of conditional nonlinear optimal perturbations sensitivity on typhoon track forecasts. *Quarterly Journal of the Royal Meteorological Society* 138: 185–197.

———. 2011b. A study on the reduction of forecast error variance by three adaptive observation approaches for tropical cyclone prediction. *Monthly Weather Review* 139: 2218–2232.

Rivier, O., G. Lapeyre, and O. Talagrand. 2008. Nonlinear generalization of singular vectors: Behavior in a baroclinic unstable flow. *Journal of the Atmospheric Sciences* 65: 1896–1911.

Tan, X.W., B. Wang, and D.L. Wang. 2010. Impact of different guidances on sensitive areas of targeting observations based on the CNOP method. *Acta Metallurgica Sinica* 24: 17–30.

Terwisscha van Scheltinga, A.D., and H.A. Dijkstra. 2008. Conditional nonlinear optimal perturbations of the double-gyre ocean circulation. *Nonlinear Processes Geophysics* 15: 727–734.

Wu, C.C., J.H. Chen, P.H. Lin, and K.H. Chou. 2007. Targeted observations of tropical cyclone movement based on the adjoint-derived sensitivity steering vector. *Journal of the Atmospheric Sciences* 64: 2611–2626.

Zhou, F.F., and M. Mu. 2011. The impact of verification area design on tropical cyclone targeted observations based on the CNOP method. *Advances in Atmospheric Sciences* 28 (5): 997–1010. doi:10.1007/s00376-011-0120-x.

———. 2012a. The impact of horizontal resolution on the CNOP and on Its identified sensitive areas for tropical cyclone predictions. *Advances in Atmospheric Sciences* 29: 36–46. doi:10.1007/s00376-011-1003-x.

———. 2012b. The time and regime dependences of sensitive areas for tropical cyclone prediction using the CNOP method. *Advances in Atmospheric Sciences* 29: 705–716. doi:10.1007/s00376-012-1174-0.

Zhou, F.F., and H. Zhang. 2014. Study of the schemes based on CNOP method to identify sensitive areas for typhoon targeted observations [J]. *Chinese Journal of Atmospheric Sciences (in Chinese)* 38 (2): 261–272.

Zou, X., F. Vandenberghe, M. Pondeca, and Y.-H. Kuo. 1997. Introduction to adjoint techniques and the MM5 adjoint modeling system. *NCAR Tech. Note*, NCAR/TN-435_STR.

Index

A.A. Tsonis (ed.), *Advances in Nonlinear Geosciences*,
DOI 10.1007/978-3-319-58895-7

O

P

MIX
Papier aus verantwortungsvollen Quellen
Paper from responsible sources
FSC® C105338

If you have any concerns about our products,
you can contact us on
ProductSafety@springernature.com

In case Publisher is established outside the EU,
the EU authorized representative is:
Springer Nature Customer Service Center GmbH
Europaplatz 3, 69115 Heidelberg, Germany

Printed by Libri Plureos GmbH
in Hamburg, Germany